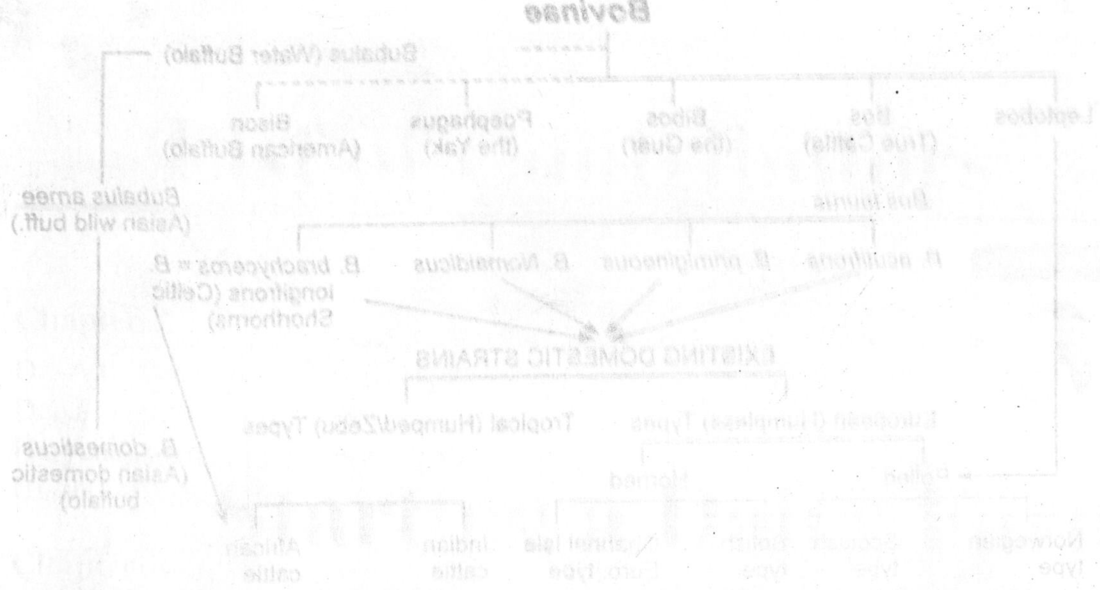

DAIRY BOVINE PRODUCTION

DAIRY BOVINE PRODUCTION

Dr. C.K. THOMAS
BVSc, MSc, PhD, FAvHF
Formerly Professor
Livestock Production Management
Kerala Agricultural University
Chelaparampath
Pakkanamparambu, Vithanassery
NEMMARA PO, Palakkad Dt., Kerala
dr.thomaskuriakose@hotmail.com

Dr. N.S.R. SASTRY
BVSc, MSc, PhD, FAvHF
Formerly Professor
Livestock Production & Management
CCS Haryana Agricultural University
303 Vinayagar Mansion
9-21-1/8 CBM Compound
VISAKHAPATNAM-530003
nsrsastry@gmail.com

Dr. G. RAVIKIRAN
BVSc, MSc, PhD
Senior Scientist
National Insititute of Animal Nutrition & Physiology
BANGALORE
rkgorti@rediffmail.com

KALYANI PUBLISHERS
LUDHIANA • NEW DELHI • NOIDA (U.P.) • HYDERABAD • CHENNAI
KOLKATA • CUTTACK • GUWAHATI • COCHIN • BANGALORE

KALYANI PUBLISHERS

Head Office

B-1/1292, Rajinder Nagar, **Ludhiana**-141 008 • Ph : 0161-2760031
E-mail : kalyanibooks@yahoo.co.in

Administrative Office

4779/23, Ansari Road, **Daryaganj,** New Delhi-110 002
Ph : 011-23271469, 23274393, 23278688
E-mail : kalyani_delhi@yahoo.co.in • kalyani_delhi@kalyanipublishers.in

Works

B-16, Sector-8, **NOIDA** (U.P.)

Branches

1, Mahalakshmi Street, T. Nagar, **Chennai**-600 017 • Ph : 044-24344684
Gopabandhu Lane, Behind Govt. Bus Stand, Badambadi,
Cuttack-753 012 (Odisha) • Ph : 0671-2311391
3-5-1108, Narayanaguda, **Hyderabad**-500 029 • Ph : 040-24750368
10/2B, Ramanath Mazumdar Street, **Kolkata**-700 009 • Ph : 033-22416024
Arunalaya, 1st Floor, Saraswati Road, Pan Bazar, **Guwahati**-781 001 • Ph : 0361-2731274
Koratti Parambil House, Convent Road, **Kochi**-682 035 • Ph : 0484-2367189
No. 24 & 25, 1st Floor, Hameed Shah Complex, Cubbonpet Main Road, **Bengaluru**-560 002

P 31131 05
S 37946 05

First Edition, 1991
Reprinted, 2000, 2005
Second Revised Edition, 2012
Reprinted, 2020, 2023

ISBN : 978-93-272-1828-2

PRINTED IN INDIA

At Kalyani Printings, B-15, Sector-8, NOIDA

Preface

(SECOND EDITION)

Since our first edition came out India's milk production achieved the distinction of being the first in the world. In India milk emerged as the product among all the outputs, including wheat, from agriculture and allied sectors with the highest value contribution. More important, small-scale dairy farming has become even more economically relevant for landless, marginal farmer and small farmer rural households, with even greater economic empowerment of farmwomen.

These developments call for even greater need for strengthening research and development in dairy production, especially, human resource development. This, and the persistent goading from professional peers too, some with even suggestions and ideas that our book be updated, made us attempt revising the same. We took the help of Dr. G. Ravikiran, a young and dynamic serving scientist in this revision work, to bring in current views. We sincerely hope that this edition of the Dairy Bovine Production too will be found useful by students, teachers and field workers.

C.K. THOMAS
N.S.R. SASTRY
G. RAVIKIRAN

Preface

(FIRST EDITION)

India with an annual milk production of 43.9 million tonnes in 1986-87 ranks third in the world after Soviet Union and U.S.A. Still the per capita milk availability in India is only 130 g per day. Similarly, the average milk production of cows is one of the lowest in the world. On the one hand bovine keeping still remains as the most important corollary to crop farming in the farmer's household. On the other, the production systems and techniques used remain underdeveloped.

One of the important causes for the paradoxical situation is the improper dissemination of technical information on modern and scientific dairy farming that is based on our own field and reserach experiences and therefore suited to our farming conditions. Students of dairy farming have to still depend on Western text books. It is difficult to find ready information on indigenous dairy production aspects for field and research workers. Many of the advanced techniques available in the text books currently followed do not have much relevance in Indian conditions.

Though the Indian Council of Agricultural Research, National Dairy Development Board, Gosamvardhana and other organisations have brought out one or a few publications covering some aspects of dairy production, there has been no comprehensive text book that covers all aspects of dairy cattle and buffalo production in India in one volume. The present book is an attempt in this direction.

The authors have tried to analyse and bring out the research findings of Indian workers on economics, genetic improvement, feeding, reproduction, management, health cover, etc., as reported in scientific journals and other publications. It is these analyses which form the basis of the individual chapters of the book. At the same time, the fundamental basics of each main topic have been dealt sufficiently enough to give adequate theoretical background for understanding the applied aspects.

The whole text has been divided into seven broad sections appropriate to the seven important components of dairy production. Within each section, the subject has been divided into convenient chapters based on different topics. At the end of each section are given a list of books and other references for further reading.

The book is primarily aimed at providing comprehensive basic and applied technical information on dairy production to undergraduate students of veterinary, agriculture, dairy science, home science and agricultural cooperation and banking colleges. It would also be useful as a help book for post-graduate students. Besides, farm managers, field workers, dairy administrators, planners and enlightened farmers should find the book very useful in their respective spheres of

Preface

(FIRST EDITION)

India with an annual milk production of 43.9 million tonnes in 1986-87 ranks third in the world after Soviet Union and U.S.A. Still the per capita milk availabaility in India is only 130 g per day. Similarly, the average milk production of cows is one of the lowest in the world. On the one hand bovine keeping still remains as the most important corollary to crop farming in the farmer's household. On the other, the production systems and techniques used remain underdeveloped.

One of the important causes for the paradoxical state of affairs is the improper dissemination of technical information on modern and scientific dairy farming that is based on our own field and reserach experiences and therefore suited to our farming conditions. Students of dairy farming have to still depend on Western text books. It is difficult to find readily information on indegenous dairy production aspects for field and research workers. Many of the facts and techniques available in the text books currently followed do not have much relevance to local conditions.

Though, the Indian Council of Agricultural Research and the Central Council of Gosamvardhana and other organisations have brought out, once in a while, publications covering some aspects of dairy production, there has been no comprehensive text book that covers all aspects of dairy cattle and buffalo production in India in one volume. The present book is an attempt in this direction.

The authors have tried to analyse and bring out the research findings of Indian workers on economics, genetic improvement, feeding, reproduction, management, health cover, etc., as reported in scientific journals and other publications. It is these analyses which form the basis of the individual chapters of the book. At the same time, the fundamental basics of each main topic have been dealt sufficiently enough to give adequate theoretical background for understanding the applied aspects.

The whole text has been divided into seven broad sections appropriate to the seven important components of dairy production. Within each section, the subject has been divided into convenient chapters based on different topics. At the end of each section are given a list of books and other references for further reading.

The book is primarily aimed at providing comprehensive basic and applied technical information on dairy production to undergraduate students of veterinary, agriculture, dairy science, home science and agricultural cooperation and banking colleges. It would also be useful as a help book for post-graduate students. Besides, farm managers, field workers, dairy administrators, planners and enlightened farmers should find the book very useful in their respective spheres of action. It is further hoped that the book would be very useful to such personnel in sister developing countries that are concerned with teaching, research and developement of dairy bovine production.

The authors express their gratitude to several individuals and organisations, especially to the organisations they belong, for providing counsel and permission to use material, data, diagram, etc., for inclusion in the book as well as encouraging them. Dr. A.C. Gangwar of Haryana Agricultural University, Dr. A. Mathews of Indo-Swiss Cattle Breeding Project, Kerala and Dr. S. Saseendranath of Kerala Agricultural University have contributed material to Chapters 2, 8 and 20, respectively. The authors do realize that there can be some shortcomings in the book, which are mostly due to factors beyond their control. They look forward to suggestions from readers towards further improvement of the book.

We dedicate this book to the millions of 'Sacred Cows' and 'Not so Sacred' but more productive buffaloes of India.

1991

C.K. THOMAS
N.S.R. SASTRY

Contents

SECTION V
GENETIC IMPROVEMENT

SECTION VI
HOUSING, HYGIENE AND HEALTH

SECTION VII
HERD MANAGEMENT

List of Contributors

Chapter 2

Dr. A.C. GANGWAR (Former) *Professor & Head*
Department of Agricultural Economics
Haryana Agricultural University
Hisar-125004

Chapter 5 & 6

Dr. D.V. REDDY, *Professor & Head*
Department of Animal Nutrition
Rajiv Gandhi College of Veterinary & Animal Sciences
Pondichery-605009

Chapter 8

Dr. ABRAHAM MATHEW (Former) *General Manager*
Kerala Livestock Development Board
Trivandrum, Kerala

Chapter 20

Dr. M.R. SASEENDRANATH, *Professor & Head*
Department of Veterinary Epidemiology & Preventive Medicine

and

Dr. USHA NARAYANA PILLAI
Associate Professor
Department of Clinical Medicine
College of Veterinary & Animal Science,
Mannuthy-680651, Kerala

Palaeontological evidences from excavations in Turkistan indicate that up to 6000 BC all the domesticated cattle were the long-horned, large sized *Bos taurus* type. But, in course of time a short-horned smaller sized cattle made its appearance. The generally accepted view is that this small, short-horned breed was evolved from the long-horned *Bos nomadicus* type found about 8000 BC. It is also hypothesised that the ox of Turkistan, which was undergoing this change of form around 7000 BC finally reached central Europe in the stunted form of *Bos brachyceros*. There were no cattle living under domestication in Europe until about 5000 BC. *Bos primegenius*, the other near contemporary of *nomadicus* and *leptoboes*, reached Europe in Pleistocene times. But the contribution of it in the making of any of the present day cattle breeds is doubtful. The short-horned *Bos nomadicus* was spread by tribal migrations before 6000 BC to India, Persia, Mesopotamia and Egypt.

The water buffalo (*Bubalus*) extends from the country of its origin, i.e., in Asia, and extends to India, Malaya, Indo-China, the East Indies and China. In Italy and Egypt, the animal is raised for milk. But many other forms like the Carabao of the Philippines are used for draught especially in rice fields. Of the *Bibos* group, the gaur reared and regularly used in the hills of India and Malaya is used as mithun. The mithun (*Bos frontalis*) is reared by the hill tribals of Assam and Northeastern ... as milch animals and also for ... prestige and ... respects. The ... ...

Section I : INTRODUCTION

CHAPTER 1

Dairying in India

1.1. CATTLE THROUGH AGES

For millions of years before known history, cattle roamed on the Earth at will, wherever landmasses were contiguous. Cattle were domesticated for the first time in southern Turkistan sometime around 8000 B.C. The purpose of domestication was probably for draught. But soon man learned to use domesticated cattle for milk and meat. Dairy products followed. It is but natural for such ancient records to be vague and incomplete. Fossils and other remains form the only clues to the early history of cattle.

1.1.1. Place of Cattle in the Animal Kingdom

Dairy cattle and buffaloes belong to family *Bovidae* as shown in the taxonomic classification given below.

Class	:	*Mammalia* (mammals)
Sub-class	:	*Eutheria* (placental animals)
Order	:	*Ungulata* (hoofed animals)
Sub-order	:	*Artiodactiles* (even-toed animals)
		Pecora (true ruminants)
Family	:	*Bovidae* (hollow-horned animals)
Sub-family	:	*Bovinae*

The *Bovidae* are the most important family of hoofed animals and are of more recent origin. All the different types of cattle *(Bovina* or *Bos)* and buffaloes *(Bubalina* or *Bubalus)* come under the Sub-family *Bovinae* but they do not interbreed. The horns of buffaloes are basically angular in section while those of cattle are oval or round (Mason, 1974). Cattle are hoofed ruminants with even number of digits. Sheep, goats and antelopes form their nearest relatives. Their family group

---

Mason, I.L. 1974. *Mohen-jo-daro and the Indus Civilization.* Arthur Probsthian, London.

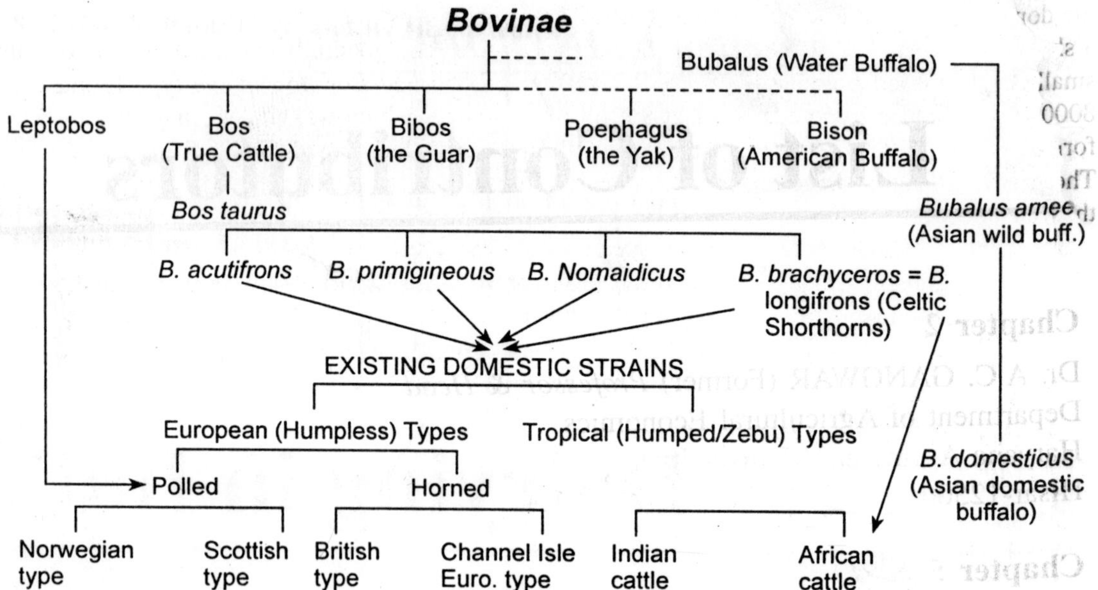

Fig. 1.1. Present day cattle evolution pattern over thousands of years along the trail of migrations.

Bovinae originates from the Latin *Bos bovis*—an ox. Present day cattle got evolved over thousands of years along the trail of migrations as depicted in Fig. 1.1 (Kelly, 1943).

One of the lines in the group, namely *Leptobos* has become extinct. Many feel that the present day polled cattle evolved from *leptobos.* The water buffalo *(Babalus)* forms a sub-generic group. The other members of the family are considered closer to one another. Linnaeus considered all five kinds as belonging to the same genus, *Bos.* Other taxonomists classify them independently as follows :

1. *Bos* : the cattle proper
2. *Bibos* : *gaur (gayal)* and *banting*
3. *Poephagus* : Yak
4. *Bubalus* : water buffalo
5. *Bison* : Buffalo of N. America and of the Caucuses

The sub-families derived from the *Bos* rootstocks form the animals that are now referred to as cattle. Evidences are not that clear to give distinct picture of the component strains of each type of cattle. However, there are indications to show that *leptobos* and *taurus* types were contemporary. The former were high polled and the females were hornless. The latter were large with long horns.

While long horned wild cattle were undergoing domestication in south Turkistan, cattle of *leptobos* type probably were undergoing domestication in Central India. There are also evidences to indicate that tame breeds of the type were introduced to Egypt and later to Europe from India. Evidence to this effect is found in the polled skulls from new stone-age deposits in North Germany and in Switzerland and from Neolithic, Bronze and later deposits in Britain.

Kelly, R.B. 1943. *Zebu cross cattle in North Australia.* Bulletin No. 172. C.S.I.R. Australia. Melbourne.

'alaeontological evidences from excavations in Turkistan indicate that up to 6000 BC all .nesticated cattle were the long-horned, large sized *Bos taurus* type. But. in course of time snort-horned smaller sized cattle made its appearance. The generally accepted view is that this l, short-horned breed was evolved from the long-horned *Bos nomadicus* type found about) BC. It is also hypothesized that the ox of Turkistan, which was undergoing this change of m around 7000 BC finally reached central Europe in the stunted form *of Bos brachyceros.* .ere were no cattle living under domestication in Europe until about 6000 BC. *Bos primegenius,* .ie other near contemporary of *nomadicus* and *leptobos* reached Europe in Pleistocene times. But the contribution of it in the making of any of the present day cattle breeds is negligible. The long-horned *Bos nomadicus* was spread by tribal migrations before 6000 BC to India, Persia, Mesopotamia and Egypt.

The water buffalo *(Bubalus)* extends from the northern parts of Africa through West Asia to India, Malaya, Indo-China, the East Indies and China. In India and Egypt, they are kept mainly for milk. But many other forms like the *Carabao* of South East Asia are used for draught especially in rice fields. Of the *Bibos* group, the *gaur* reared and remained in the hilly forests of India and Malaysia. The *gayal* is a semi-domesticated breed of the nocturnal *gaur* kept by the tribals of Assam. The *banting* of Burma and Indonesia resembles the European ox in many respects. They are used mainly for meat production.

1.1.2. Origin of Cattle and Buffalo Breeds of India

Many zebu breeds are said to have entered India from West Asia between 2200 and 1500 BC through the northern passes and to have fanned out along the route taken by Vedic Aryan invaders. However, Marshall (1931) wrote that the geographical distribution as well a chronological stages of arrival of zebu cattle at different localities show that zebu might have originated in the Indian peninsula. Johnson (1951) supported Marshal's theory, suggesting the diversity of zebu types in the Indian peninsula and their high degree of adaptation to local environment indicate to this. There is also some evidence that suggests that when Vedic Aryans invaded the Indus valley, humpless cattle were present in larger numbers than the humped cattle. It is, however, most likely that zebu originated in the Indus valley and were then taken to West Asia by traders and migrants, as there was considerable commerce between the Indus and Mesopotamia valleys at the zenith of the Mohenjodaro civilization (2500 to 1900 BC) (ICAR 1964).

It is a fact that the cow, the buffalo, the elephant etc. were domesticated in India long before 3000 BC. In the pre-Aryan ages, two varieties of cattle were found in India, the massive long-horned and humped form and a smaller form with short horns, which may have been humpless. The latter variety was found in the upper strata of Mohenjodaro site. But there are abundant remains of humped bulls in every stratum which indicates that at that time, the indus valley must have been very rich in this fine breed of Cattle, which is closely allied, if not identical with the magnificent white and grey breed still common in Sindh, Gujarat and Rajputana, but wholly different from the small humped cattle of central India.

The Asian buffalo (genus *Bubalus)* was widely distributed in Europe and South Asia in the Pleistocene period. When the climate became drier, it was restricted to India, Indo-China and some of the southeast Asian islands. There are now three distinct wild buffalo species in Asia,

Marshall. J. 1931. *Mohen-jo-daro and the Indus Civilization.* Arthur Probsthian. London.
Johnson. C.W. 1951. *Cattleman* 38 : 17
I.C.A.R. 1964. *Agriculture in ancient* India. ICAR. New Delhi.

out of which only the *arni* or Indian wild buffalo is domesticated (Mason. 1974). *Arni* is presently distributed in northern India and possibly in Sri Lanka and Indo-China. The domestic buffalo is known by the specific name *Bubalus arnee Forma Bubalis,* according to Bohlken's terminology. Figures of tame (and so possibly domesticated) buffaloes appear on seals both in the Indus Valley and in Mesopotamia from about the middle of the third millennium B.C. However, exactly when the buffalo was domesticated is not known.

1.1.3. The Cow in India

Much before written history began, cattle became an integral part of the Indian way of life; cow had been eulogised in the past and our ancient books are full in her praise. Even today they are indispensable to our way of life because we get from them the traction power for our agricultural operations, cheap but highly valuable organic manure for preserving soil fertility and milk, the life sustaining fluid for young, and old. India was once a vast pastoral country where domestic animals were synonymous to wealth. Dairying was one of the principal occupations of the people in an almost entirely rural based life. So, it remains to a large extent even now. But with the arrival of the British rule, the process of urbanization was started. Ever since, some of our villages started growing into towns and cities with an exodus of people from the villages. Commercial dairy farms started sprouting in urban areas. A reversal of this trend became evident only recently which was under the influence of concerted government action. But even now a lot of milk is produced in our cities—a thing of the past in all advanced countries.

With the increase in human population, more and more of the grazing land was progressively brought under the plough. Besides, the forest department formed by the British rulers closed the forest and semi-forest grazing lands to men and animals. Though this step must have served a useful purpose in preserving forest resources, it nevertheless contributed to the deterioration of cattle. Rearing cattle and sheep as a livelihood became a distinct occupation in course of time and was done by nomadic communities, who moved with their herds from place to place in search of green pastures. With the extension of agriculture and closure of forest grazing lands, these nomadic tribes began to disappear and raising of cattle was eventually taken up by the crop farmers themselves.

1.2. TYPES AND BREEDS OF CATTLE AND BUFFALOES

1.2.1. What is a Breed?

The different species of animals can further be divided into breeds, which in their turn comprise of families, lines, genetic groups etc. Kravtchenko (1973) described breed as a sufficiently large group of animals developed by the efforts of man, members of which have a common ancestry and possess similar morphological, physiological and economic characters that are inherited more or less firmly.

As the above definition implies, man can further change the nature of a breed in a desired direction or use it in crossbreeding with reasonable confidence about the expected results. This is so because the nature of inheritance or the 'gene fund' *(see* Chapter 15) of a breed can be guessed with some certainty, as members within a breed possess similar heritable characters. The uniformity among individuals within a breed, however, depend on the 'age' of the breed *i.e.* how long man undertook selection in that breed.

Mason, I.L. 1974. *loc. cit.*

Kravtchenko, N.A. 1973. *Breeding Farm Animals* (Russian). Kolos Publisher, Moscow.

The system of genotypes *(see* Chapter 14) in a breed is characterised by the presence of certain specific genes in characteristic frequencies, presence of linkages or absence of them, presence of heterozygotes and homozygotes, occurrence of mutations, as well as existence of certain correlations between selection traits. Homozygosity of certain genes and a sort of balance in the number of heterozygotes for other genes gives relative stability and reliability to inheritance of selection traits. At the same time mutations and gene recombinations will be creating new reserves of genetic variation within a breed. This permits systematic selection in desired direction for further improvement of a breed.

Thus a breed is not just a result of the evolutionary process of animals but a medium of animal improvement. Selection of a breed or breeds and using them in further breeding plans is a prerequisite for improvement of genetic production potential of dairy cattle.

1.2.2. Classification of Breeds

According to Payne (1970) the cattle breeds of the Indian sub-continent and Sri Lanka can be classified as follows :

1. Short-horned zebu (*Bachaur, Bhagnari, Gaolao, Hariana, Krishna Valley, Mewati, Nagori, Ongole* and *Rathi*).
2. Leteral-horned zebu (*Gir, Dangi, Deoni, Nimari, Dhanni, Red Sindhi* and *Sahiwal*).
3. Lyre-horned zebu (*Kankrej, Malvi, Tharparkar* and *Hissar*).
4. Long-horned zebu (*Amritmahal, Hallikar, Kangayam* and *Killari*).
5. Small short-horned or lyre-horned zebu (*Kumauni, Lohani, Ponwar, Shahabadi, Punganoor* and *Sinhala*).

The Indian water buffaloes are of riverine type. Mason (1974) classified buffaloes of India and Pakistan as follows :

1. Murrah group (*Murrah, Nili-Ravi, Kundi*)
2. Gujarat breeds (*Surti, Mehsana, Jaffarabadi*)
3. Uttar Pradesh breeds (*Bhadawari, Tarai*)
4. Central Indian varieties (*Nagpuri, Pandharapuri, Manda, Jernagi, Kalahandi Sambalpur, Parlakhemundi*)
5. South Indian buffaloes (*Toda, South Kanara*)

It is customary in India to classify cattle breeds according to their utility. Milch or dairy breeds (Sahiwal, Red Sindhi. Gir. Deoni) are good yielders of milk (1500-2200 liters per lactation). The draft breeds (such as the Mysore cattle) are good at fast and sustained work but are poor milk yielders. Cows of dual-purpose or general utility breeds (Tharparkar, Hariana, Ongole etc) are moderately fair yielders of milk (1100 to 1700 liters) while bullocks are useful work animals. Keeping the scope of this book in mind only the milch and more important breeds of general utility type are described in the next section.

As regards Indian buffaloes, only the Murrah group and Gujarat breeds are important from dairying point of view. The other breeds of buffaloes are a type of dual-purpose breeds. They are mostly used as work animals in the paddy growing areas of east and southern India.

Pyne, W.J.A 1970. *Cattle production in tropics.* Vol. 1. Longman. London.
Mason, I.L. 1974 *loc. cit.*

1.2.3. General Features of Indian Cattle

The nature of genetic diversity between Indian breeds is quite substantial. This is mainly due to the varied ecological conditions, limited selection by man and genetic drift in the small local populations. The alleles for productivity, particularly for milk, appear to have been lost under extremely poor management conditions and ill-organised choice of sires in the rural areas (Murthy, 1969). Majority of cattle and buffaloes are poor producers of milk (Table 1.1), although their milk is generally richer in fat. The situation is only very slightly better now.

TABLE 1.1. Distribution of Breeding Milch Animals in India According to their Milk Yielding Capacity (Singh and Parnekar, 1966)

Daily milk yield (kg)	*Per cent cows*	*Per cent buffaloes*
Below 1 kg	94.1	19.2
1 to 2 kg	5.5	61.2
2 to 3 kg	0.2	9.6
Above 5 kg	0.2	10.0

However, Indian breeds of cattle are hardy and strong. They are valued throughout the world because they are able to resist diseases, withstand hot climate, thrive on low planes of nutrition and recuperate rapidly. These qualities combine well with the productivity of cattle of European origin. Indian cattle, therefore, are extensively exported to other tropical countries where European cattle cannot maintain good condition without certain proportion of blood of Indian breeds in them.

The most sought after Indian breeds are Ongole. Kankrej and Sahiwal. Broadly speaking, the quality of cattle depends on the environmental conditions, particularly on rainfall, temperature and humidity. Indian cattle are generally found to do well in North West dry region (Figure 1.2). They are small (in body size) and nondescript in areas of heavy rainfall, such as the coastal or the northeast hilly areas of the country.

Cattle of good breeds are found in the semi-arid parts of Punjab. Haryana, Rajasthan, parts of Gujarat and low to medium rainfall areas of Tamil Nadu. Maharashtra. Karnataka and Andhra Pradesh in the Central Plateau region (Figure 1.2).

Within the limits of the same State, different types of cattle may be seen. Thus, in Gujarat one finds the excellent Kankrej cattle in Kathiawar and southeast of Rann of Kutch where the climate is drier, while elsewhere within the same State one finds inferior cattle. In all the States of India uneconomic classes of cattle occur in the humid regions. In hilly areas, where rainfall is heavy, as in the northeastern parts of India, all the cattle are of poor breed.

The total population of well-defined breeds has not been estimated accurately. Roughly 25 per cent of the cattle population is good and conforms to the characteristics of one or the other recognized breed. In each breed, as in any population, there are gradations of quality. A small proportion is always typical and conforms to the required characteristics of the particular breed. Others deviate in various degrees from the breed type.

Murthy, B.R. 1969. Genetic consequences of different systems of crossbreeding in India. In: *Seminar on Animal Breeding*. ICAR, New Delhi.

Singh, H. and Parnekar, M. 1966. *Basic facts about cattle wealth and allied aspects.* Central Council of Gosarmardhana. New Delhi.

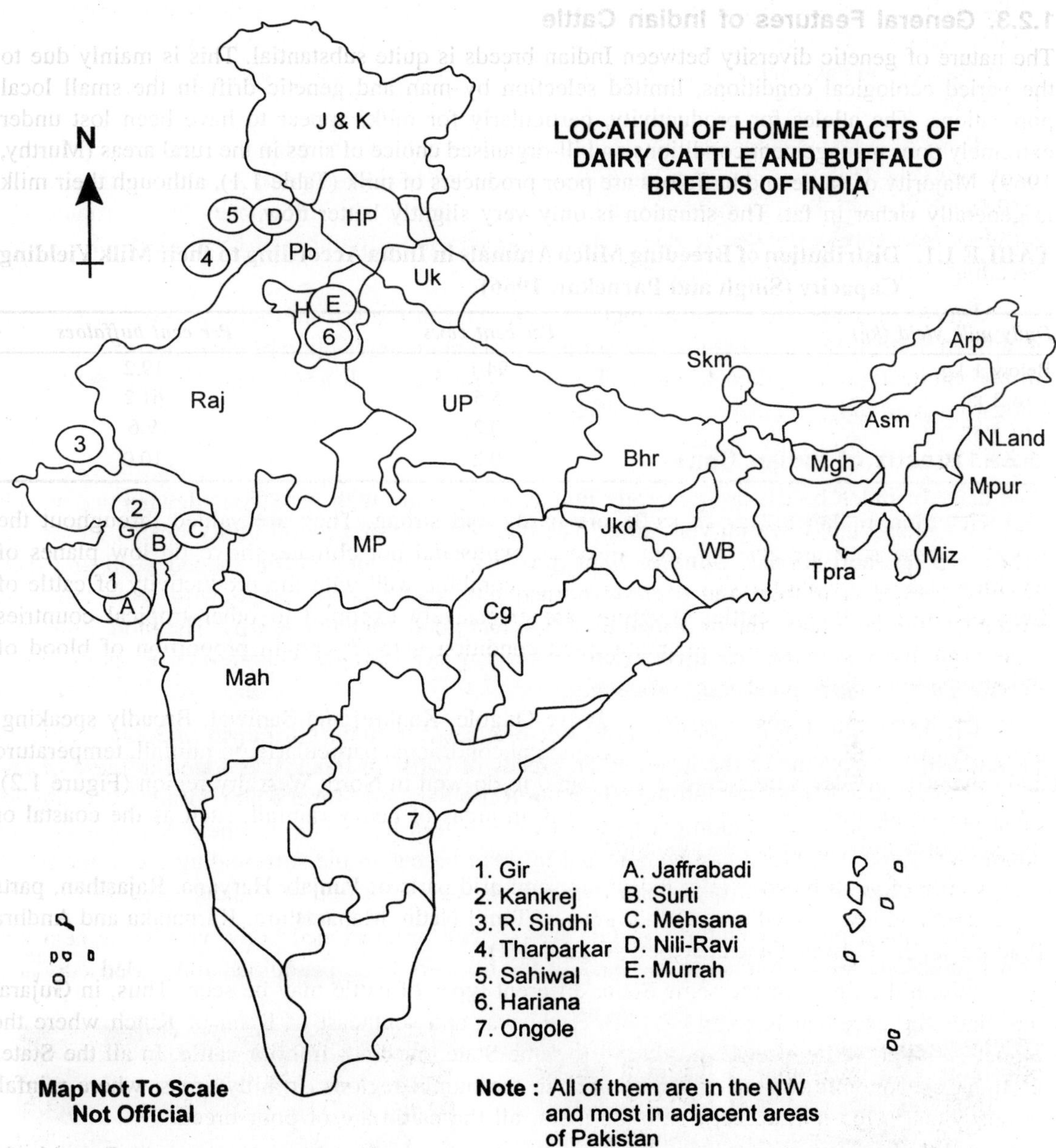

Fig. 1.2. Location of home tracts of main dairy cattle and buffalo breeds of India

Some of the well-defined breeds are not intrinsically different from one another, at least in their production performance. Their distinctiveness is in respect of conformation and external appearance. Besides these breeds, it is possible that there are other types that have not been properly catalogued.

In India, the buffalo is looked upon as the milch animal and is the mainstay of Indian dairy industry. The cow is mainly regarded as the mother of the working bullock. Hence, a majority of the well-defined breeds of cattle in India are of draft type. Superior herds (dairy-wise) of Red Sindhi, Sahiwal, Tharparkar and a few other breeds are maintained in this country, though they are indigenous to Pakistan. These, as also the Gir are essentially dairy breeds, the best ones native to tropics.

Cows of these breeds are economical producers of milk and the males are generally indifferent work bullocks. There are considerable differences in the performance of animals of these breeds at different farms. To a certain extent this may be due to the differences in environment in different places, but may also be due to the differences in the number of generations of selective breeding, which each breed has had.

Breeds of cattle and buffaloes that are important in India from dairying point of view are described in the following sections. For propriety, standard description of characteristics as found in the publications of Indian Council of Agricultural Research and Council for Scientific and Industrial Research have been mostly used.

1.2.4. Breeds of Indian Cattle

The dairy or milch breeds are generally ponderous in build with pendulous dewlap and sheath, and have lateral and often curved horns. The dual purpose or general utility breeds are of two types, *viz.* (*i*) short-horned, white or light gray cattle with long coffin shaped skull and face, slightly convex in profile; and (*ii*) lyre-horned gray cattle deep bodied with wide forehead, prominent orbital arches, flat or dished in profile and good draught capacity. The more important general utility breeds that are used in cross-breeding are Hariana, Ongole and Kankrej. These breeds are also fairly good milk producers.

Gir. Most probably, the Gir breed of cattle originated from the Gir forests of South Kathiawar in Saurashtra and is one of the best-known breeds of cattle in India. It is found in more or less pure form all over Saurashtra and the adjoining Rajasthan and Maharashtra. Graded Gir animals occur over a wide area, including a large part of western Rajasthan, northern Maharashtra and southern Gujarat. Gir blood can be detected in other breeds in the surrounding areas due to the fact that nomadic herdsmen take them over a wide area in search of suitable grazing

The typical physical features of Gir are––a well proportioned body, very prominent and broad forehead; curved horns turning back at the tip; ears long, pendulous and curled outwards

Fig. 1.3. Gir bull

Fig. 1.4. Gir cow

with a typical 'curled-leaf appearance; long and whip-like tail; and prominent hip bones. The body colour of Gir varies from shades of red and white to almost black and white or entirely red or white (Figures 1.3 and 1.4). Adult males weigh up to 544 kg and females up to 390 kg.

Cows of this breed are good milk yielders; maximum yield reported being 3,175 kg in a lactation period of 325 days. Good herds like those at Hosur in Tamil Nadu and Morvi in Gujarat yield on an average 1,675 kg. The herd of the Poona Military Dairy Farm has recorded a yield of 7.5 kg milk per head per day. The bullocks of this breed are heavy, powerful but are medium paced in movement. Gir is also one of the best beef breeds and they are exported to other countries, particularly tropical countries, for crossbreeding purposes.

Red Sindhi. The home tract of this breed is the Sind province in Pakistan. Red Sindhi or simply *Sindhi* is one of the most distinctive cattle breeds of India. It is found in the pure form only in its place of origin and in other areas it is found in an impure state due to infusion of their blood by the grey coloured breeds. However, excellent herds of this breed are found at several places in India. As the name implies, the typical body colour of Red Sindhi breed is red, varying from dark red to fawn yellow. The animals are medium sized, compact and well proportioned, with a rather heavy hump, dewlap and sheath (Figures 1.5 and 1.6). Their horns are thick at the base and emerge laterally and curve upwards. Ears are drooping and moderately sized. Their udder is capacious and well proportioned. Adult males weigh 454 kg and adult females 320 kg.

Fig. 1.5. Red Sindhi bull

Fig. 1.6. Red Sindhi cow

Red Sindhi cows are perhaps the most economical milk producers amongst Indian breeds and in yield they are only second to Sahiwal. Their milk yield can be as high as 2,440 kg in a lactation period of 300 days and the average yield of several herds in this country is 1,725 kg. They are fairly regular breeders. Bullocks of this breed are medium sized, compact animals with good bones and muscles. They are steady workers, useful for slow field operations and carting.

Because of its capacity to resist common diseases (particularly insect-born blood protozoan diseases), adaptability to varying climatic conditions and their small and compact body, Red Sindhi cattle were extensively used in India for grading up local nondescript stock in some parts of the country, particularly in Assam, Orissa, Kerala and parts of Tamil Nadu. They are also extensively exported to several Asian and Latin American countries, as well as to America and Australia. The Australian Milking Zebu cattle have some proportion of blood of Red Sindhi and Sahiwal in them. In India, outstanding herds of Red Sindhi cattle are maintained at the Government

Farm, Hosur, the Military Farms at Bangalore and Hyderabad, and the Agricultural Institute, Allahabad.

Sahiwal. This breed originally belonged to the Montgomery district in Pakistan. However, several pedigree herds are maintained in Punjab, Haryana, Delhi, Uttar Pradesh and Bihar in India, as this breed thrives well in all parts of the country, giving high milk yields.

Sahiwal is a heavy breed with a symmetrical body and loose skin. Animals are long, rather fleshy, having short legs, broad fore-head, stumpy horns, large, heavy and loose dewlap and loose sheath (Figures 1.7 and 1.8). Their ears are medium sized with black hair on the fringe,

Fig. 1.7. Sahiwal bull

Fig. 1.8. Sahiwal cow

and hump in males, straight hip bones high and wide apart, tail long and whip like, and the udder capacious and rather loosely attached to the body. Their predominant body colour is reddish brown with or without white splashes. Bulls weigh 545 kg and cows 410 kg.

Good specimens of this breed yield 2,725 to 3,175 kg of milk in a lactation period of 300 days. There are individual cows that yield as high as 4,355 kg. Bullocks of this breed are lethargic but useful for slow work. Like Red Sindhi, pedigree Sahiwal bulls are in great demand in other tropical countries of the world for breeding purposes.

Tharparkar. This breed is sometimes known as Thari or White Sindhi. Tharparkar hails from the semi-desert tracts of southwest Sindh in Pakistan. Animals of this breed are also found on the less developed farms in Kutch in Gujarat and in Jodhpur and Jaisalmer of Rajasthan. The best herd of Tharparkar cattle is located at the National Dairy Research Institute farm at Karnal, where this breed has been under controlled breeding for several generations and of late are being used for crossbreeding with Brown Swiss.

Tharparkar cattle have a deep, strongly built, medium sized and well-proportioned body (Figures 1.9 and 1.10). Their limbs are short, straight and strong with firm joints. They have a moderately long face with a comparatively broad poll. Their head is of medium size; forehead being broad and flat or is slightly convex above the eyes. Their ears are somewhat long, broad and semi-pendulous. The horns of Tharparkar are medium sized and tail long with a black switch reaching to fetlocks. Their fore and hindquarters are somewhat drooping. The characteristic body colour of Tharparkar cattle is white or gray. Adult Tharparkar bulls and cows weigh about 545 kg and 385 kg respectively.

Fig. 1.9. Tharparkar bull

Fig. 1.10. Tharparkar cow

Tharparkar is undoubtedly one of the best dual purpose breeds of India; tending to be more of a dairy breed. Some farms in India reported as high as 2,000 to 2,500 kg of milk yield per lactation of 300 days. Yields as high as 4,375 kg have also been reported for some individual Tharparkar cows. The average yield of the breed can safely be considered as 1,350 to 1,400 kg. Tharparkar bullocks are quite useful work animals.

Hariana. The Hariana cattle are widely distributed in many parts of the country and reputed as excellent dual-purpose cattle. The Hariana a *Bos indicus* breed used for draft purposes in northern India where they are found. They are well suited to fast road work, being able to pull a one ton load at 2 miles per hour and cover 20 miles a day. While females are kept primarily for breeding of oxen, they are also milked. The breed averages 1400 kg of milk per lactation but high yielding animals will produce over 2300 kg in a single lactation. The breed belongs to the short-horned type of zebu and is gray or white. The average cow weighs 310 kg and the average bull 430 kg.

Hariana breed is most widely used in crossbreeding with exotic breeds in this country. It is bred particularly in Hissar, Karnal, Rohtak and Gurgoan districts of Haryana State and in the Union Territory of Delhi. A variant of this breed, known as Hissar breed, is seen in Hissar district. The Hissar breed is somewhat different in build from Hariana breed. In utility, the former is more of a draft breed. Hariana breed can also be found in pure form in parts of Punjab and Rajasthan (particularly in Alwar and Bharatpur districts) and in western Uttar Pradesh.

The body of Hariana cattle is proportionate, moderately long and compact in appearance. Their head is carried high and their horns are short, curving upwards and inwards. A prominent bony prominence at the centre of the poll is considered a typical characteristic of the breed. Ears are small and only slightly pendulous. Dewlap is small. Hump is well developed in the males. Udder in cows is well developed. Their tail is long and tapering. Like Tharparkar, the Hariana cattle are also white or gray in body colour. Body weights of adult bulls and cows of this breed are 500 and 375 kg, respectively.

There are several farms where good quality Hariana cattle are reared, the best one being at the Haringhata Farm near Calcutta.

Ongole. The history of the Ongole dates back 2,000 years before Christian times. It has a great similarity with the Gaolao breed of Madhya Pradesh and also has a resemblance to the Bhagnari type of cattle in the north of India. This similarity is not surprising in view of the fact that these breeds lie along the path taken by the Rig Vedic Aryans in their march from the north to the south of India. Ongole or Nellore breed is found in Nellore, Guntur and Prakasam districts of Andhra Pradesh. In these districts, cultivators breed and rear them on specially grown fodder and residues of grain crops.

Ongole cattle are large sized animals with loosely knit frames and long body (Figures 1.12 and 1.13). Their limbs are long and muscular. Forehead is broad between the eyes, ears long, horns stumpy and dewlap large and fleshy. Hump in males is well grown and erect. The predominant colour of Ongole cattle is white with dark grey markings on face. The animals are rather heavy in build, males weighing up to 570 kg and females up 435 kg.

Cows of this breed are medium milk producers, producing about 1,360 kg per lactation of 300 days. Individual animals can yield as high as 2,200 kg. Bullocks are powerful and majestic in appearance and highly valued for sustained heavy ploughing work and carting. They are however not suitable for fast work. Because of their hardiness, disease resistance and capacity to thrive on scanty and dry fodders, Ongole cattle were exported to tropical America and other countries in large numbers, mainly for upgrading local stock and evolving suitable beef breeds.

Kankrej. The home tract of Kankrej breed is the area southeast of the Rann of Kutch in Ahmedabad district in Gujarat State and from Deesa in the east to old Radhanpur State in west, particularly along the Banas and Saraswati rivers. Professional nomad cattle breeders breed it. The breeding tract of Kankrej (also known as Wadhiar) is a low-lying land and in some places below mean sea level. The skill of the native breeders as well as the efforts of the former Bombay government resulted in creating a large population of pure Kankrej cattle in their native tract.

Kankrej is the heaviest of the Indian breeds of cattle. Animals have a powerful body with broad chest, straight back and a well developed hump. Their forehead is comparatively broad; slightly dished in the centre. Their face is short with slightly upturned nose. The typical colour of Kankrej cattle is silver grey to iron grey. It can sometimes be steel black. Adult Kankrej males weigh nearly 590 kg, while females weigh 450 kg. Kankerj cows produce on an average 1,360 kg per lactation. Kankrej bullocks are powerful and fast draught cattle. These cattle are resistant to tick fever and they show very little incidence of contagious abortion and tuberculosis. It has also been observed that the red color is recessive.

1.2.5. Buffalo Breeds of India

As mentioned earlier, from dairying point of view, the Murrah group and the Gujarat breeds of buffalo only are important. Buffaloes are semi-aquatic animals, and are most abundant in humid localities. However, the best dairy buffaloes can be seen in arid and semi-arid areas. They are large, massive and clumsy creatures with exceptionally short and thick legs and conspicuous hooves (Cockrill, 1974). Their horns are thick, flat, curved or straight and marked with rings that are indicative of the buffalo's age. Described below are five important Indian buffalo breeds that are well known for their milking qualities.

Cockrill. W.R. 1974. *Husbandry and health of the domestic buffalo.* FAO. Rome.

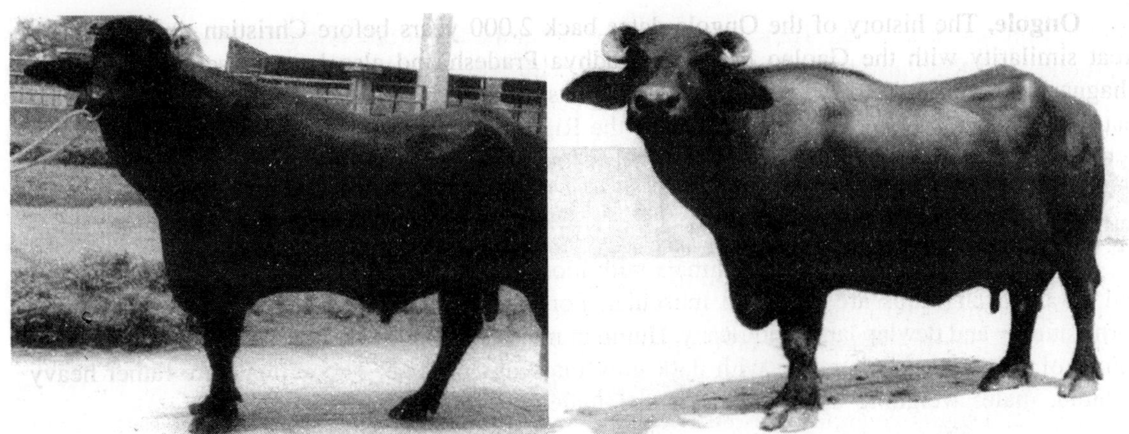

Fig. 1.11. Murrah buffalo bull **Fig. 1.12.** Murrah buffalo cow

Murrah. The tract of the Murrah buffalo is in the southern parts of Haryana *i.e.* Rohtak, Karnal, Hissar, Jind and Gurgaon districts and Delhi. The name Murrah (meaning curled) is due to the typical character of the breed, namely the tightly curled horns. The breed is now distributed practically all over North India, if not the whole of India, and is reared at several places in pure form. It is also widely used in grading up of local buffaloes.

The Murrah buffaloes are heavily built with a comparatively light head (Figures 1.11 and 1.12). Their horns are short and tightly curled. The udder in females is well developed. Wide hips, somewhat drooping fore and hindquarters and a fairly long tail are other characters of Murrah. Murrah buffaloes have a jet-black body colour; a white switch of the tail is considered normal for the breed. Murrah bulls weigh up to 570 kg while buffalo cows weigh about 450 kg. Cows weighing as much as 525 kg are a common occurrence in their home tract, where buffaloes are fed specially grown leguminous and non-leguminous fodders. The Murrah she-buffaloes are very efficient producers of milk and butter fat. Average lactation yields in herds vary from 1,360 to 2,270 kg and several individuals on several farms yielded more than 3,000 kg. Butterfat content in milk of Murrah buffaloes is around 7 per cent.

Nili-Ravi. Animals of this breed are mostly seen in Ferozepur and Amritsar districts of Punjab, that too in small numbers. However, large numbers of Nili-Ravi are found in the adjoining areas of Pakistan. No pedigree records of this breed are maintained anywhere in India. In body size and potential for milk production Nili-Ravi is similar to Murrah but is distinct from it because of its characteristic in wall eyes and white markings on the forehead, face, muzzle and legs (Figure 1.13). Its head is elongated, bulging at the top and depressed between eyes. Horns "are small and coiled lightly. She-buffaloes have a

Fig. 1.13. Nili-Ravi buffalo cow

long, thin and fine neck. Their udder is well developed and tail is pretty long, reaching up to ground. Nili-Ravi also has a black body colour with white markings referred above. Adult males and females weigh about 570 and 460 kg respectively. Males of this breed are used for heavy draught work. Average yield of she-buffaloes is 1,585 kg per lactation. Their lactation period is rather short.

Mehsana. This breed is intermediate to Murrah and Surti breeds in its appearance and body conformation. These animals are mostly found in the Mehsana district of Gujarat and adjoining parts of Maharashtra. It is also found in Banaskantha and Sabarkantha districts of Gujarat, its type varying from place to place. This could be due to the grading up with Murrah practiced in some parts. Mehsana buffaloes are good milch animals, mature early and breed regularly.

Animals of this breed are longer than Murrah, their limbs are lighter, head longer and heavier and horns generally less curved at the end than in Murrah but are longer. The common body colour is black or fawn grey. Many animals have white marking on face and extremities. Buffaloes of this breed weigh nearly as much a Murrah buffaloes under ideal feeding conditions. Breed average for body weight of adult males and females is 560 and 425 kg.

Mehsana buffaloes are good milk yielders, producing up to 1,360 kg per lactation. They are popular for urban milk production and form backbone for some dairies in Gujarat and Maharashtra.

Surti. The home tract of this breed is Kaira, Baroda and Nadia districts of Gujarat. It is considered as an economical producer, is extremely docile and is mainly handled by women in villages. The mainstay of AMUL dairy in Anand is the Surti buffalo. Compact body, sickle-shaped horns and two white bands—one below the neck and other around the brisket, are typical characters of Surti (Figure 1.14). Their bodies are compact, well shaped and the barrel is wedge shaped. The head of Surti buffalo is long and broad and rounded between the eyes. Surti bulls weigh around 500 kg while she-buffaloes weigh up to 410 kg. Average lactation yield of Surti buffaloes can be as much as 1655 kg. The fat content in their milk ranges from 7.5 to 9.0 per cent.

Fig. 1.14. Surti buffalo cow

Jaffarabadi. This is the heaviest breed of Indian buffaloes. Its home tract is in Saurashtra region of Gujarat. Pure Jaffarabadi animals can be seen in villages only as there is no organised farm on which they are maintained in the country. Some people claim this breed to be an efficient converter of roughages into milk and fat, its potentialities being on par with Murrah breed. Jaffarabadi buffaloes are massively built with a rather shorter body. Females look somewhat loosely knit. Their head and neck are massive, forehead very prominent and horns heavy, inclining to droop on each side of the neck. The horns are very lightly curled in comparison with Murrah. The usual body colour of Jaffarabadi buffaloes is black. Adult males weigh on an average 590 kg and females 455 kg. Jaffarabadi buffalo-cows are good milkers, individual animals giving as high a yield as 2,450 kg per lactation, with high fat content. Males of this breed are used for heavy draught work.

1.2.6. *Bos taurus* (Humpless) Breeds Used in India

For improving the milk production potential of indigenous cows some exotic dairy breeds of cattle have been and are being extensively used in India for crossbreeding purpose. In view of the quicker progress that can be achieved by way of crossbreeding, it has become a national policy in this country. Military dairy farms were the first to introduce some well-known European breeds such as Shorthorns, Ayrshires and Holstein Friesians about 50 to 60 years ago. The exotic breeds that are being used in planned crossbreeding projects presently are Holstein Friesian, Jersey, Brown Swiss, Red Dane and German Fleckvieh (Spotted Mountain Cattle).

Holstein Friesian. This breed was developed in the northern parts of Netherlands, especially in the province of Friesland. Perhaps this is the most widely distributed breed of dairy cattle in temperate zone, as well as in some tropical countries. As a matter of fact, now we have several different distinct genetic groups within this breed such as the American Friesians, Israeli Friesians, the Russian Friesians (the Black Pied cattle) etc.

Fig. 1.15. Holstein Friesian Cow

Holstein cattle are ruggedly built and they possess large udders. They are the largest dairy breed and mature cows weigh as much as 700 kg. They have typical markings of black and white that make them easily distinguishable (Figure 1.15). This is by far the best dairy breed. Average production of cow is 6,000 to 7,000 kg per lactation. However, the fat content in their milk is rather low—3.45 per cent. Average production for all Holsteins enrolled in official U.S. production-testing programs in 1987 was 17,408 pounds of milk, 632 pounds of butterfat and 550 pounds of protein per year.

Jersey. The Jersey breed originated on the Island of Jersey, one of the Channel Islands between France and England. The Jersey is one of the oldest dairy breeds, having been reported by authorities as being purebred for nearly six centuries. Jersey breed is also widely distributed in Europe and America. In India this breed has acclimatized well and is widely used in cross-breeding with indigenous cows. They are economical producers. Jersey is the smallest dairy breed and hence is more suited for crossbreeding with zebu cattle. Mature Jersey cows weigh around 450 kg. Heifers grow rapidly and mature early, calving as early as at the age of 24 months sometimes.

Fig. 1.16. Jersey cow

The typical body colour of Jersey cattle is reddish fawn. They have compact and angular body (Figure 1.16). The average milk production of Jersey cows is 3,000 to 4,000 kg with a fat content of 5.3 per cent. The highest recorded milk yield was 11,381 kg.

Brown Swiss. The mountainous region of Switzerland is the place of origin of the Brown Swiss breed. It is rather less refined than other dairy breeds, but is famous in its home tract for its rugged nature and good milk production. In spite of its ruggedness the animal is quite docile and easily manageable. They were originally developed for draft as well as for milk. Brown Swiss heifers are rather slow maturing. However, this breed seems to be more suitable than other *Bos taurus* breeds for introduction in tropics. In India excellent crossbred cattle have been obtained by crossing this breed with recognized Indian breeds of cattle; the Karanswiss cattle of the National Dairy Research Institute, Karnal is one example. Average production per cow is about 4,500 liters with an average fat content of 4 per cent. Maximum milk production recorded so far is 14,024 kg in 365 days, thrice milking.

Red Dane. The Danish Red is of the Baltic Red cattle type and originated on the islands off the coast of Denmark. The breed was developed from North Slesvig Red, with Angeln and Ballum, crossed with the local island cattle. The typical body colour of this Danish breed is red, reddish brown or even dark brown. It is also a heavy breed; mature males weighing up to 950 kg and mature females—600 kg. Red Dane combined well with Hariana cattle (at Haryana Agricultural University) and Sahiwal cattle (at Punjab Agricultural University). Cow No RS 110 (see the photograph elsewhere) is among the best of crossbred cows developed in this country. Lactation yield of Red Dane cattle varies from 3,000 to 4,000 kg with a fat content of 4 per cent and above.

1.3. CATTLE AND BUFFALO POPULATION TRENDS

In general most of the world livestock are found in the developing countries. Nearly 65 per cent of the world cattle and almost all of the world buffaloes are in the developing countries in the tropical zones of Asia, Africa and Latin America (Table 1.2).

TABLE 1.2. Number of Cattle and Buffaloes in Different Regions of the world

Region	*Population 1000's*			
	Cattle	*(% of world's)*	*Buffaloes*	*(% of world's)*
South-east asia	38,824	2.84	14,429	8.63
South Asia including India	2,88,428	21.10	1,24,303	74.38
India	**2,21,900**	**16.24**	**95,100**	**56.90**
Asia- pacific	4,92,074	36.01	1,61,504	96.64
Rest of the world	8,74,589	63.99	5,623	3.36
World	13,66,664	100.00	1,67,126	100.00

Source : FAO RAP publication 2003/10 (Selected Indicators of Food and Agriculture Development in Asia-Pacific Region 1992-2002)

In India itself there are 1/7 of the world cattle and 1/2 of the world buffaloes. Pattern of change in the country's bovine population since 1951 (Table 1.2) shows that increase in buffalo population is much greater than that of cattle. India ranks first in buffalo and second in cattle population in the world according to Government of India data (*Annual report, 2007-08, Dept of Animal Husbandry, Dairying & Fisheries, GOI*).

However, their quality is poor; the yield is as low as 1/5th that of dairy cattle *(Bos taurus)* in the developed countries in the temperate zone and Oceania. In contrast, the buffalo is mainly an animal of Asia and in India it is the mainstay of dairy industry.

For each 1,000 buffaloes, the cattle were 3,529 in 1956, which decreased during the following years to 1892 in 2003. This trend shows increasing liking for buffalo despite advent of crossbreeding between zebu and *Bos taurus* cattle (Table 1.3 & 1.4).

TABLE 1.3. Percentile change in cattle and buffaloes used as work animals between 1997-2003

Work animals in '000s	*1997*	*2003*	*% Increase/decrease*
Cattle	58,917	56,362	–04.3
Buffaloes	6,802	5,834	–14.2

Source : Livestock census, 2003.

TABLE 1.4. Percentile changes in the milch animals in India between 1997-2003

Milch Animals(000)	*1997*	*2003*	*% Increase/decrease*
Crossbred Cattle	8,356	11,231	34.41
Indigenous Cattle	49,874	46,856	– 6.05
Buffaloes	43,732	47,224	10.51

Source : Livestock census, 2003.

Population projections (Table 1.5) beyond 2003 indicate a reduction in cattle and increase in buffaloes by the year 2020 AD and an improvement in their production potential due to development measures being undertaken currently. Further there was definitive shift towards buffaloes and crossbred cattle as preferred animals for milk production (Table 1.4). The use of male cattle and buffaloes for work purpose is also on the decline indicating increasing use of cattle and buffaloes as dairy animals rather than work.

TABLE 1.5. Trends in increase in cattle and buffalo populations (Millions) in India from 1951 Onwards

Year of Census	*Cattle*		*Buffalo*		
	Number	*% increase*	*Number*	*% increase*	*Cow:buffalo*
1951	153.3	—	43.4	—	3.5:1
1956	158.7	(+) 2.2	44.9	(+) 3.6	3.5:1
1961	175.6	(+) 10.6	51.2	(+) 14.0	3.4:1
1966	176.2	(+) 0.3	52.9	(+) 3.3	3.3:1
1972	178.9	(+) 1.5	57.9	(+) 9.5	3.0:1
1977	181.1	(+) 1.7	60.4	(+) 4.3	3.0:1
1982	190.9	(+) 5.4	68.8	(+) 13.9	2.8:1
1987	192.0	(+) 0.6	73.0	(+) 6.1	2.6:1
1992	204.6	(+) 6.6	84.2	(+) 15.3	2.4:1
1997	198.9	(–) 2.8	89.9	(+) 6.8	2.2:1
2003	185.2	(–) 6.9	97.9	(+) 8.9	1.9:1
Projections					
2008	191	(+) 3.13	107	(+) 9.30	1.79:1
2013	187	(–) 2.09	117	(+) 9.36	1.60:1
2017	182	(–) 2.67	128	(+) 9.40	1.42:1
2022	176	(–) 3.30	139	(+) 8.59	1.27:1
2027	169	(–) 3.98	151	(+) 8.83	1.20:1

Source : 1. FAO Production Year Book. Vol. 32, 1978 and Vol. 39. 1985. and Vol. 2004.

Among the Asian countries (Table 1.6) the largest cattle population (45.09%) of the region) is in India followed by China (21.58%) and Bangladesh (4.88%). The largest buffalo population is also in India (58.88% of the region) followed by Pakistan (14.86%) and China (13.78%). Though the number of cattle is more than the number of buffaloes in most countries of the region, buffaloes are more than cattle in Philippines, Vietnam and Pakistan indicating the economic importance of buffaloes to those countries (Table 1.6). However, from milk production point of view the buffaloes of the Indian sub-continent are the most important. It may be noted that though China has the largest area in Asia, it is in India, the third largest country of the continent, that largest population of bovines are found. This reflects the role of these animals in India's economy.

TABLE 1.6. Cattle and Buffalo Population in Asia and the Pacific in 2002 (Thousand)

Country	*Cattle*	*Buffaloes*
Cambodia	2,924.5	625.9
Indonesia	11,200	2,300
Myanmar	11,551	2,552
Philippines	2,547.8	3,122
Thailand	4,640.4	1,800
Viet Nam	4,063	2,814.5
Bangladesh	24,000	830
India	2,21,900	95,100
Nepal	6,978.7	3,700.9
Pakistan	22,857	24,000
Kazakhstan	4,281.7	9
Uzbekistan	5,400	—
China	1,06,175	22,248.9
Iran (Islamic Rep. of)	8,738	523.5
Mongolia	2,053.7	—
Australia	30,500	—
Japan	4,564	—
New Zealand	96,32.5	—
Total in the region	4,92,074.4	1,61,503.8
As % of World population	36.00	96.64

Source : FAO, 2002.

In Asia 30-45% cattle are work animals, the highest percentage of utilization of cattle for work being in South and South-East Asia. In West Asia, about 25% cattle are used for meat production as well. As regards buffaloes, the main dairy bovine in Asia, work is an important production, though less than that of cattle. The buffalo is mainly a work animal in all countries east of the Indian sub-continent. Within India, there are considerable differences in the number of cattle and buffaloes in various regions of the country (Table 1.7).

The largest concentration of cattle is in the medium rainfall western parts of the Peninsula (26.33%), more humid eastern parts of the Peninsula (23.14%) and the Ganga plains of Bihar and UP (19.36%). Buffaloes are in large number in the Ganga plains (31.89%), northwest semi-arid region (30.47%) and the peninsular region, east and west (34.85%). In general, the bovines less densely populate the western and eastern high altitude regions and the coastal areas.

TABLE 1.7. Distribution of Cattle (Crossbred and Indigenous) and Buffaloes in different State and Union Territories of India as per Livestock Census 2003, agro-climatic regionwise (See Fig. 1.17)

State/Union Territory	Crossbred Cattle					Indigenous Cattle					Buffalo				
	Total '000	As % of country's cattles	Adult males as % of total	Adult females as % of total	Young stock as % of total	Total '000	As % of country's cattles	Adult males as % of total	Adult females as % of total	Young stock as % of total	Total '000	As % of country's cattles	Adult males as % of total	Adult females as % of total	Young stock as % of total
N.W. Temperate Himalayan Region															
Himachal Pradesh	677	0.37	0.04	0.19	0.13	1559	0.84	0.36	0.28	0.21	774	0.79	0.01	0.51	0.27
Jammu & Kashmir	1320	0.71	0.08	0.31	0.32	1764	0.95	0.27	0.35	0.33	1039	1.06	0.05	0.60	0.41
Region Total	**1997**	**1.08**	**0.12**	**0.50**	**0.45**	**3323**	**1.79**	**0.62**	**0.64**	**0.53**	**1813**	**1.85**	**0.06**	**1.11**	**0.69**
N.W. Semi-Arid Region															
Chandigarh	5	0.00	0.00	0.00	0.00	1	0.00	0.00	0.00	0.00	23	0.02	0.00	0.02	0.01
Delhi	58	0.03	0.00	0.02	0.01	34	0.02	0.00	0.01	0.01	231	0.24	0.01	0.14	0.09
Gujarat	639	0.35	0.03	0.18	0.13	6785	3.66	1.41	1.27	0.99	7140	7.29	0.14	4.32	2.84
Haryana	573	0.31	0.03	0.15	0.13	967	0.52	0.16	0.19	0.17	6035	6.16	0.17	3.06	2.93
Punjab	1531	**0.83**	0.08	**0.43**	0.32	508	0.27	0.12	0.08	0.07	5995	6.12	0.10	3.35	2.67
Rajasthan	464	0.25	0.01	0.14	0.11	10390	5.61	1.19	2.51	1.91	10414	10.63	0.09	5.78	4.77
Region Total	**3270**	**1.77**	**0.15**	**0.92**	**0.69**	**18685**	**10.09**	**2.88**	**4.06**	**3.15**	**29838**	**30.47**	**0.51**	**16.67**	**13.30**
Gangetic Plains															
Bihar	1274	0.69	0.04	**0.35**	0.30	9455	5.11	1.55	1.69	1.87	5743	5.86	0.28	2.95	2.64
Jharkhand	145	0.08	0.01	0.04	0.03	7513	4.06	1.77	1.09	1.19	1343	1.37	0.59	0.43	0.35
Uttar Pradesh	1634	**0.88**	0.10	**0.39**	0.40	16917	**9.14**	2.93	2.96	3.25	22914	23.40	1.64	11.43	10.32
Uttaranchal	228	0.12	0.01	0.06	0.05	1961	1.06	0.35	0.35	0.35	1228	1.25	0.05	0.75	0.46
Region Total	**3281**	**1.77**	**0.15**	**0.83**	**0.78**	**35846**	**19.36**	**6.60**	**6.09**	**6.67**	**31228**	**31.89**	**2.56**	**15.56**	**13.77**
N.E.Hilly, Heavy Rainfall Region															
Assam	440	0.24	0.02	0.10	0.12	7999	4.32	1.24	1.28	1.79	678	0.69	0.19	0.24	0.26
Arunachal pradesh	13	0.01	0.00	0.00	0.00	445	0.24	0.07	0.08	0.10	11	0.01	0.00	0.01	0.00
Manipur	69	0.04	0.01	0.02	0.02	349	0.19	0.07	0.05	0.06	77	0.08	0.02	0.03	0.03
Meghalaya	23	0.01	0.00	0.01	0.00	744	0.40	0.13	0.16	0.11	18	0.02	0.01	0.01	0.01
Mizoram	9	0.00	0.00	0.00	0.00	27	0.01	0.00	0.00	0.01	6	0.01	0.00	0.00	0.00
Nagaland	243	0.13	0.02	0.07	0.04	208	0.11	0.03	0.04	0.05	34	0.03	0.01	0.01	0.01

Sikkim	80	0.04	0.01	0.02	0.02	79	0.04	0.01	0.02	0.02	2	0.00	0.00	0.00	0.00
Tripura	57	0.03	0.00	0.01	0.02	702	0.38	0.11	0.12	0.15	14	0.01	0.00	0.01	0.01
Region Total	**934**	**0.50**	**0.05**	**0.24**	**0.22**	**10553**	**5.70**	**1.66**	**1.75**	**2.28**	**840**	**0.86**	**0.24**	**0.30**	**0.32**
Western Penninsular Region															
Goa	12	0.01	0.00	0.00	0.00	63	0.03	0.01	0.01	0.01	37	0.04	0.01	0.02	0.01
Karnataka	1602	**0.87**	0.02	**0.51**	0.33	7936	4.29	1.58	1.51	1.19	3991	4.08	0.13	2.41	1.54
Madhya Pradesh	317	0.17	0.01	0.08	0.08	18595	**10.04**	3.53	3.26	3.26	7575	7.74	0.29	4.25	3.20
Maharashtra	2776	**1.50**	0.13	**0.80**	0.58	13527	**7.30**	3.31	2.05	1.94	6145	6.28	0.28	3.77	2.23
Chhatisgarh	253	0.14	0.05	0.04	0.04	8629	4.66	1.63	1.48	1.55	1598	1.63	0.96	0.34	0.33
Region Total	**4960**	**2.68**	**0.21**	**1.43**	**1.03**	**48750**	**26.33**	**10.07**	**8.31**	**7.95**	**19346**	**19.76**	**1.66**	**10.78**	**7.31**
Eastern Penninsular Region															
Andhra Pradesh	1107	0.60	0.07	0.29	0.23	8193	4.42	2.06	1.20	1.16	10630	10.86	0.52	5.97	4.36
Orissa	1063	0.57	0.08	0.24	0.26	12840	**6.93**	2.72	2.01	2.21	1394	1.42	0.58	0.43	0.42
Pondicherry	63	0.03	0.00	0.02	0.02	16	0.01	0.00	0.00	0.00	4	0.00	0.00	0.00	0.00
Tamilnadu	5140	**2.78**	0.13	**1.40**	1.25	4001	2.16	0.57	0.82	0.76	1658	1.69	0.06	0.92	0.71
West Bengal	1119	0.60	0.03	0.30	0.28	17794	**9.61**	2.85	3.20	3.55	1086	1.11	0.62	0.28	0.22
Region Total	**8492**	**4.59**	**0.31**	**2.24**	**2.03**	**42844**	**23.14**	**8.20**	**7.24**	**7.69**	**14772**	**15.09**	**1.78**	**7.60**	**5.71**
West Coast and the Islands															
Andaman & Nicobar	13	0.01	0.00	0.00	0.00	51	0.03	0.01	0.01	0.01	16	0.02	0.01	0.01	0.01
Dadra & Nagar Haveli	1	0.00	0.00	0.00	0.00	49	0.03	0.02	0.01	0.00	4	0.00	0.00	0.00	0.00
Daman & Diu	0	0.00	0.00	0.00	0.00	4	0.00	0.00	0.00	0.00	1	0.00	0.00	0.00	0.00
Kerala	1735	**0.94**	0.01	**0.47**	0.46	387	0.21	0.01	0.09	0.11	65	0.07	0.01	0.02	0.03
Lakshadweep	2	0.00	0.00	0.00	0.00	2	0.00	0.00	0.00	0.00	0	0.00	0.00	0.00	0.00
Region Total	**1751**	**0.95**	**0.01**	**0.47**	**0.46**	**493**	**0.27**	**0.03**	**0.11**	**0.12**	**86**	**0.09**	**0.02**	**0.03**	**0.04**
Total	**24686**	**13.33**	**1.01**	**6.64**	**5.68**	**160495**	**86.67**	**30.07**	**28.19**	**28.41**	**97922**	**100.00**	**6.82**	**52.06**	**41.08**

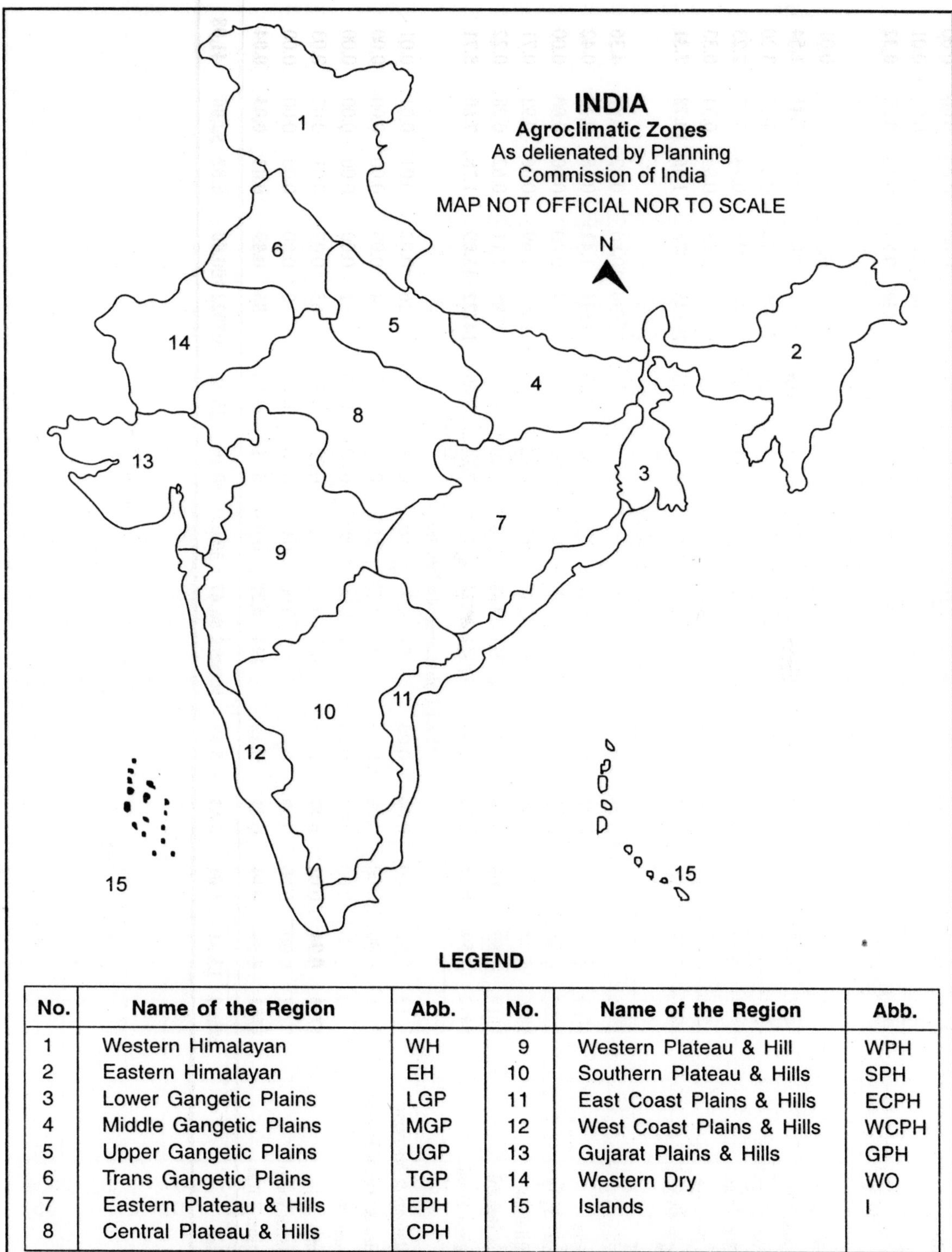

LEGEND

No.	Name of the Region	Abb.	No.	Name of the Region	Abb.
1	Western Himalayan	WH	9	Western Plateau & Hill	WPH
2	Eastern Himalayan	EH	10	Southern Plateau & Hills	SPH
3	Lower Gangetic Plains	LGP	11	East Coast Plains & Hills	ECPH
4	Middle Gangetic Plains	MGP	12	West Coast Plains & Hills	WCPH
5	Upper Gangetic Plains	UGP	13	Gujarat Plains & Hills	GPH
6	Trans Gangetic Plains	TGP	14	Western Dry	WO
7	Eastern Plateau & Hills	EPH	15	Islands	I
8	Central Plateau & Hills	CPH			

Fig. 1.17. Agro-climatic regions of India as delineated by the Planning Commission of India

There are, however, distinct differences in the utilization patterns of cattle and buffaloes in various regions as evidenced by the differences in the proportions of male, female and young stock. The proportion of young stock out of total stock, especially in case of buffaloes, is much higher (41%) in India in comparison to other dairying countries. This could be a reflection of the longer growth period of our stock as well as higher percentage of stock being reared compulsorily in the absence of slaughter. This is despite the well known higher young stock mortality in the country.

Nearly 31 per cent of total cattle population is adult males that are used mostly for work. This trend is reflected in all the regions of the country, though the proportion was lower in high altitude and N E hilly regions (Table 1.7), whereas, 52 per cent buffaloes are females used for milk production. In the semi-arid region comprising of Delhi, Gujarat, Haryana, Punjab and Rajasthan buffaloes are used mainly as dairy animals.

1.4. MILK PRODUCTION TRENDS

Indian Council of Medical Research recommends a daily requirement of 10 oz (280 g) of milk per head. Sukhatme, on the other hand, recommended 7 oz (200 g) milk per head per day taking into account that a judicious mixture of vegetable proteins in Indian diet can bring down the requirement of milk. Against this, the total estimated availability of milk was only enough to average 241 g per head per day in 2005-06. Here we must also understand that a great deal of inequalities exist in the distribution and consumption of milk to make the availability of milk much below the recommended level for a very large section of our population. The situation is not that much due to lack of growth in milk production as due to increase in human population. In fact milk production in the country increased by two and half times since Independence. Barring the years of severe draught (around 1966) the rise in all the milk production parameters has been continuous, thanks to the development efforts of the government in various Five Year Plans (Figure 1.18).

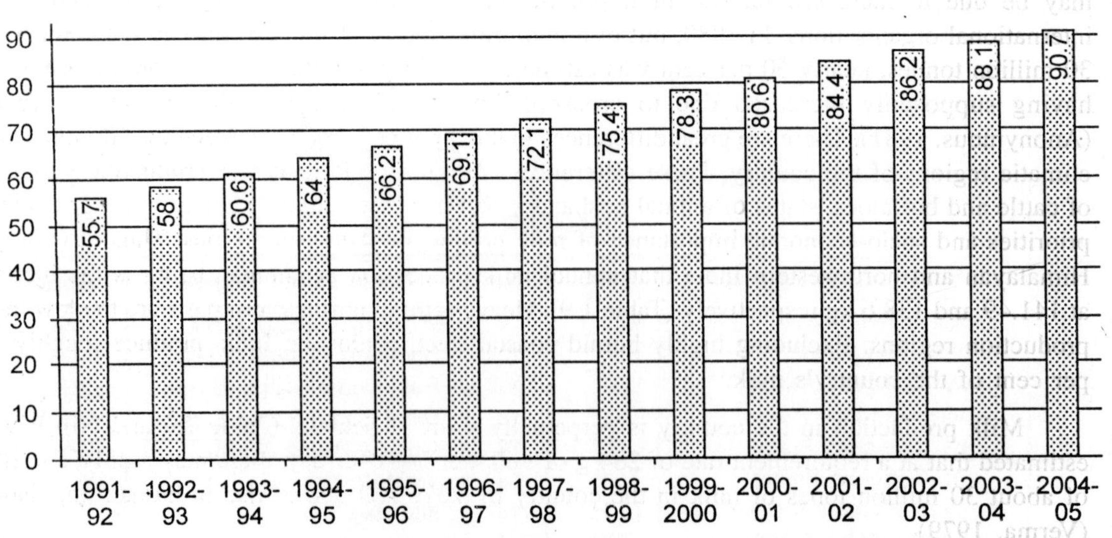

Fig. 1.18. Milk production trends in India over the years.

TABLE 1.8. Milk Production Trends (Million Tonnes) in India and the World from Cattle and Buffaloes

Year	India			World		
	Cow	Buffalo	Cow:Buff	Cow	Buffalo	Cow:Buff
1974-76	10.90	14.30	1:1.32	390.80	23.40	1:0.060
1979-81	13.20	17.30	1:1.31	423.00	27.50	1:0.065
1984	17.10	20.40	1:1.19	453.00	31.40	1:0.069
1985	18.50	21.80	1:1.18	458.10	32.50	1:0.071
1999	32.80	43.00	1:1.31	483.50	65.41	1:0.135
2000	34.00	44.40	1:1.31	491.20	67.40	1:0.137
2001	35.60	46.60	1:1.31	497.60	70.40	1:0.141
2002	36.20	48.00	1:1.33	508.90	72.28	1:0.142
2003	36.50	50.10	1:1.37	518.50	75.54	1:0.146
2004	37.50	50.74	1:1.35	523.40	76.55	1:0.146

Soruce : F.A.O. 2006.

In India milk produced by buffaloes and cows was in the ratio of 1.3:1 in 1975 (Table 1.8). According to estimates, the milk produced by buffaloes was only 26.7 per cent more than that produced by the cows during 1961-65, which increased to almost 32 per cent higher than that from cows in the year 1978. From 1980 proportion of cow milk in total milk production is rising, presumably due to increase in number of crossbred cows. This was reversed in the late 1990's with milk produced by buffaloes gradually increasing. This trend was also seen at world with buffalo milk production increased by hundred percent.

An increase in the proportion of buffalo milk was also observed at the global level. This may be due to increased interest in improving productivity of buffalo by F.A.O. and other international organisations. In 1980, out of a total estimated production of milk in the country of 30 million tonnes, nearly 60 per cent was estimated to be from buffaloes, the share of cow milk having supposedly increased due to crossbreeding in this species with European breeds (Anonymous. 1981). There are great differences in the amount of milk produced in different agro-climatic regions of the country, These differences are due to differences in production potential of cattle and buffaloes, status of animal husbandry practices, soil-climatic conditions, agricultural priorities and socio-economic importance of milk production, between regions. Thus, it is in the Himalayan and northwestern India that annual milk production per human being is the highest at 141.49 and 138.6 kg respectively (Table 1.9). Heavy rainfall northeastern parts are the low milk production regions. Excluding highly humid coastal tract, peninsular India produce roughly 50 per cent of the country's milk.

Milk production in the country is perpetually short of demand (Table 1.9 and 10). It was estimated that at a requirement rate of 284 g of milk per head per day, there may still be shortfall of about 30 million tones of milk in the country by 2020 AD due to rise in human population (Verma, 1979).

Dairying in India-1981. Indian Dairy Association. New Delhi, pp. 21-22.

TABLE 1.9. Bovine Population and Annual Milk Production Trends in Different Agro-Climatic-Livestock Regions in India

Region	*Adult (in Milk) Bovine Population (millions)*			*Annual milk production (million tonnes)*	*Milk Production/ sq. km. area (tonnes)*	*Milk available humanbeing (kg)*
	Crossbred Cows	*Indgenous Cows*	*Buffaloes*			
North-West	2.644	9.145	20.865	56.914	47.93	131.86
Central	0.202	5.594	3.189	8.625	16.49	79.79
Southern	3.322	3.601	5.795	19.789	31.08	88.57
North-East	1.231	8.216	2.272	12.059	26.97	50.41
Himalayan	0.780	1.070	1.199	3.485	10.52	141.49
ALL India	8.177	27.626	33.319	100.869	30.68	98.21

Note. 1. All India population as per Census, 2001.
2. Milk Production 2006-07 as per Dept of Animal Husbandry, Dairying & Fisheries, GOI

North-West: Uttar Pradesh, Chandigarh Rajasthan, Haryana, Delhi, Punjab, Dadra &Nagar haveli, Daman & Diu, Goa, Gujarat and Maharashtra; **Central:** Chhatisgarh, Jharkhand and Madhya Pradesh; **Southern:** Andhra Pra-desh, Karnataka, Kerala, Lakshadweep, Pondicherry and Tamilnadu; **North-East:** Andaman & Nicobar, Arunachal Pradesh, Assam, Bihar, Manipur, Meghalaya, Mizoram, Nagaland, Orissa, Sikkim, Tripura and West Bengal; **Himalayan:** Himachal Pradesh, Jammu & Kashmir and Uttarakhand.

TABLE 1.10. Projections of Human Population Growth and Milk Requirements by 2020 AD in India

Year	*Population (millions @ 1.67% growth rate)*	*Milk requirements @ 284 g per person per day (million tones)*	*Expected milk production @4% yearly growth (million tones)*	*Additional requirement (million tones)*
1990	763.1	77.8	41.80	35.9
1995	801.9	81.8	47.10	34.7
2000	912.0	93.0	63.10	29.9
2005	1094.0	94.0	94.00	0
2010	1188.45	153.1*/106.4**	114.37	–38.73/7.97
2020	1402.45	271.0*/165.8**	169.29	–101.71/3.49

Source : Verma, O.S. 1979. Indian Dairyman 8:725; *TIFAC: Food and Agriculture: Technology Vision 2020
** India Development Report, 2002; 2010 and 2020 are calculated as per table like @ 1.67 and 4 % growth. 1991-2001 annual growth rate in population was 1.93% (Census 2001) another source it is 1.58% as on 2008 estimated figures. 2008 population is 1147.99 (https://www.cia.gov/library/publications/the-world-factbook/print/in.html).

1.5. CATTLE DEVELOPMENT IN INDIA

Nineteen forties, the years which saw the labour for the birth of a new nation, India, also witnessed efforts for resurrection of dairying in this country. In 1943, imported skim milk powder was distributed to Bombay hotels to conserve fresh milk for domestic consumption. Starting of long distance transporting of milk from Anand to Bombay in 1945 followed this. Independence spurred further activity and in 1949 Arey Milk Colony was established near Bombay to transfer city cattle from crowded housing areas in the city, which started distribution of bottled milk on a large scale in 1950. Systematic cattle development work was planned and executed only with the successive five year plans

With the completion of Operation Flood by NDDB, the pace of investment in dairy sector has slowed down. The allocation for dairy development by the central and state governments has also diminished over the last two plans. While de-licensing and the subsequent decision to do away with the concept of milk sheds were expected to boost private sector investment in dairying, it has not happened. Further, no effort was made in investing on technology for the development of value added and innovative milk products. As a result in the first four years of 10th five-year plan, the growth rate of milk has been less than 3% per annum.

Various programmes of cattle development have been taken up like key village scheme, hill cattle development scheme, central cattle breeding farms, frozen semen centres, regional forage demonstration farms, Rinderpest eradication programme, intensive cattle development projects, etc. Out of all the schemes launched, key village scheme and intensive cattle development projects stand out in their totality of development effort and in creating awareness among people.

1.5.1. Key Village Scheme

The All India Key Village Scheme was launched in 1951 as a part of the first 5-year plan. This was the first step initiated for systematic cattle improvement in the country. It comprised of a comprehensive programme incorporating all aspects necessary for improving the productivity of cattle as under :

1. Introduction of superior bulls
2. Castration of undesirable inferior bulls
3. Use of artificial insemination
4. Milk recording
5. Improved fodder production
6. Prevention and treatment of diseases and
7. Distribution of mineral feed supplements.

There were 496 Key Village Blocks in the country. Originally, a Key Village Block covered a contiguous population of 2,500 cows/buffaloes of breedable age, which was later expanded to cover a population of 5,000, and this has been further raised to 10.000. These changes were made to incorporate more area and animals under this scheme. While this programme helped in increasing milk production in certain areas, it did not make a real impact on milk production. This was because the key villages were not supported by organised marketing facilities necessary for giving incentive for increased milk production.

1.5.2. The Intensive Cattle Development Project (ICDP)

While the Key Village programme was in progress, it was realised that for increasing milk production commensurate to the needs, greater efforts are needed. It was found necessary to augment milk production in areas around large dairy plants so as to enable the latter to work to their full capacity and to meet the growing demands for milk. In 1965 the scientists' panel on animal husbandry recommended crossbreeding indigenous cattle with *Bos taurus* breeds on a large scale for rapid increase in the milk production of cattle in India. Embodying these aspirations and recommendations, the Intensive Cattle Development project was launched in 1965.

The Key Village Blocks were continued and extended and improved further so that the general programme of cattle development is continued without break. At the same time, some of the chronically ailing Key Village Blocks were closed down. In addition to the Key Village Scheme, some large ICDPs were also taken up on the same line of package programme for agricultural production. These projects were located both in the breeding tracts of indigenous breeds of cattle and buffaloes, which could be developed in the future and also in the milk shed areas of large dairy projects to enable dairy plants to produce to their installed capacities. In the former areas, the policy of breeding was selection among native breeds. These areas had to be linked up with milk supply schemes. In the latter areas, crossbreeding with suitable exotic breeds and/or grading up with improved indigenous breeds was practiced.

Under each ICDP it was envisaged to cover 100,000 buffaloes/cows of breedable age so that it will have a significant impact on milk production in the area. Concrete efforts were made to cover all the aspects of cattle development, which have a bearing on increasing the productivity of cows, and link it to procurement and marketing. An ICDP comprised of a network of 100 stockmen centers duly supported by regional artificial insemination centers and a central semen bank. It had a budgetary allocation of Rs 13.1 million for 5 years. To start with, there were 31 ICDPs in 1964, which increased to 62 in 1973 in the 5th five-year plan. Later 51 new ICDP projects were initiated in the country.

1.5.3. Small Farmers' Development Agency (SFDA)

Each unit of the SFDA cover at least 38,000 families of small and marginal farmers and agricultural labourers. The agency gives subsidies under **two** heads : (*a*) Cattle purchase subsidy scheme, and (*b*) Calf subsidy scheme. The various aims and objectives of SFDA are as follows :

1. Identification of the eligible beneficiaries, sanction of the subsidy and recommendation for loan.
2. Purchase of good quality milch animals at reasonable price from an area outside the project.
3. Assured arrangement for veterinary services.
4. Assured supply of good quality concentrates.
5. Assured purchase of milk by cooperatives.
6. Training farmers in scientific cattle keeping.

1.5.4. Operation Flood

It is a gigantic dairy development project propounded by Government of India for developing dairy industry in the country. The Indian Dairy Corporation is implementing this project through its subsidiary, the National Dairy Development Board. Operation Flood-I, the first phase, was wholly financed by selling milk reconstituted from free gifts of 1,26,000 MT of skim milk powder and 42,000 MT of butter oil donated out of the bulging surpluses of EEC countries. The Operation Flood-I which was originally meant to be completed in 1975 actually took 9 years from 1970-1979 for its completion at a total cost of Rs. 1,160 millions. The Operation Flood-II, which started in 1978, was scheduled to be completed in 1985 at a cost of Rs. 4830 millions. This was to be partly financed by selling donated skim milk powder and butter oil which was to be imported in much larger quantities than in Operation Flood-I and a substantial amount was to come as a loan from the World Bank. All the 147 cities with a population of over 100,000 in the country would get a milk supply scheme. Operation Flood-III also concluded.

The funds generated by selling the reconstituted milk in the four metropolitan cities are to be used for all round dairy development in villages. Besides it aims at flooding the cities with reconstituted milk making the maintenance of traditional city dairy stables by private operators uneconomical. This is expected to cause removal of the cattle from cities into the countryside where they really belong.

The cooperative approach has been chosen for dairy development in villages. This cooperative set-up is developed on the pattern of the famous Anand Milk Union Limited (AMUL) or the Kaira District Milk Producers' Cooperative Union in Gujarat, which has caused considerable increase in milk production in the villages as well as quality of buffaloes. The cooperative is three-tiered, *viz..* the village cooperative, the district cooperative union and the State level cooperative federations. The milk producer-members, each owning 1 to 3 buffaloes, collectively own, operate and control the whole venture. The cooperative processes and sells the milk procured from members directly (avoiding the middle-men) on the one hand and on the other, provides at reasonable cost at the farmers' door-step all the necessary dairy inputs – cattle feed, fodder seeds, artificial insemination service, health cover etc.

The Anand pattern, the template for the Operation Flood ventures elsewhere in the country, allows utilization of local talents and resources in a most profitable manner at the grassroots level. It has been transplanted with reasonable success in other parts of the country. Based on it, dairy development projects are being implemented in other Asian countries.

1.5.5. Dairy Development Schemes of States

Under these schemes of various nature that are in operation since the 4th Five Year Plan period, provision is made for supply of one milch cow/buffalo to small and marginal farmers of landless labourers at subsidized rates. However, it has been observed that scheme of providing one buffalo cannot give adequate income to a loanee as it is not a viable economic unit. Similar scheme is also financed by the Primary Cooperative Credit and Service Societies (now known as Mini Banks in some States). Since 1976-77 only these societies grant such loans. The milk supply societies would continue to arrange for collection and testing of milk at society level and also distribute cattle feed and other inputs. This primary society can advance loans for milch cattle even to non-agricultural members (labourers etc.). These loans are to be advanced only if satisfactory arrangements for lifting and testing of milk through a milk supply society or by creating a special milk cell in the primary society have been made.

Quantum of Loan. Loans up to ₹ 2,000 for purchase of milch cattle may be advanced without insisting on security or mortgage of land or charge on land, provided arrangements are made through an approved marketing organization for sale of milk arising out of loans so borrowed and recovery of dues out of sales proceeds. Sureties of two persons will be needed. Amount proposed for the purchase of milk buffalo is, however, far inadequate than prevalent prices of buffaloes. The loan may be advanced for a period of three years. *Repaying Capacity.* It is presumed that a member will be able to repay the installment of loan and interest (10.5%) out of his incremental income. This incremental income of a farmer in case of one animal would be ₹ 2,000 though it seems to be higher than what a farmer can obtain from one buffalo. He can utilize 50 per cent of this incremental income towards repayment of installment and interest, *i.e.,* ₹ 1,000 will be available with him for this purpose each year.

Cattle Mortality Fund. In order to cover risk of death of the milch cattle, the Haryana State Cooperative Milk Producers Federation Ltd. has undertaken insurance of milch cattle. The contribution under this scheme should not exceed 5 per cent of the loan taken for purchase of milch cattle. The amount deducted would be remitted to the Milk Federation. In case of death of an animal, the Federation would settle the claim of the concerned member out of this fund. Some other states too offer such facilities.

Mode of Compensation. In all cases of mortality of milch cattle purchased out of loan, the amount of loan outstanding against the beneficiary on the date of mortality or a week prior to mortality along with interest thereon shall be paid by-the federation out of mortality fund without any regard to cause of death. If the cow goes dry during lactation period due to some disease/ injury and the owner has been complying with all terms and conditions of loan prior to it then the value of cow rendered useless would be assessed and difference between the price and the loan outstanding with interest would be paid from mortality fund, provided the animal was treated by a qualified veterinary doctor in time and his certificate is obtained.

Duration of Benefit. The benefit shall not be extended to a beneficiary for a period exceeding 3 years or the duration of period of loan advanced whichever is earlier. The amount of the compensation will be directly paid to the Cooperative Society or Bank on behalf of the beneficary and the Bank would write off the loan against him.

Loan for 2-3 Cow Dairy Units. Now emphasis has been shifted from one milch animal to 2-3 milch animals (2 buffaloes and one crossbred cow). The objective of this programme is to set up viable dairy units, so that landless labourers and small and marginal farmers can make good a living and get self employment. Financial assistance is arranged from different projects (SFDA, DPAP, IARD and DADP etc.) Medium term loans are made available by the Cooperative Bank/Commercial Banks on the advice of above stated agencies. Under the scheme a subsidy of 25 per cent and 33.5 per cent is given to small farmer/marginal farmer and landless labourer beneficiaries respectively. Interest is charged at the rate of 4 per cent per annum. All other interested persons can also borrow from these banks for the establishment of dairy units but interest rate will be 10 per cent and no subsidy will be given. Table 3.16 shows details of the scheme.

Mini Dairy Units Scheme for Medium Sized Herds

Another new scheme entitled 'Mini Dairy Unit' has been initiated by the Government of Haryana since 1978-79 with a total allocation of ₹ 128.90 lakhs for a period of 5 years *i.e.,* 1978-83. The main objectives of the scheme are: (*i*) to provide self-employment to educated unemployed and underemployed rural youth; (*ii*) to raise the socio-economic standard of weaker section of the society; (*iii*) to increase production of milk for matching the requirement of the consumers in general and the milk plant of the state in particular : (*iv*) to meet the requirement of quality dairy products and of balanced diet to all sections of people both urban and rural areas for adequate nutrition.

Eligibility. The following beneficiaries will be eligible for setting up of 'Mini Dairy Unit':

(*i*) Rural educated unemployed youth (at least matriculate), ex-servicemen and members of schedule caste, who will be able to set apart at least one acre of land with assured irrigation facilities for growing latest varieties of green fodder.

(*ii*) Farmers owning up to 5 acres of irrigated land or owning not more than 7 acres of unirrigated land. The beneficiaries shall undergo training in dairy farming for 21 days and be sponsored by the State Department of Animal Husbandry after qualifying such training. The beneficiary shall have to execute a bond of 3 years with state department to sell their surplus milk to the nearest state milk plant. Fifty per cent of sale proceeds of milk will be deducted towards repayment of loan, interest and insurance subsidy. The beneficiary will have to contribute 25 to 15 per cent on his own. The 'Mini Dairy Units' are to be set up in a cluster of 10 units to facilitate marketing of milk.

Disbursement of Loan. The financial assistance will be disbursed directly to the supplier of animals along with the marginal contribution of the beneficiaries towards the cost of animals. The loan for construction of shed shall be released in three stages and shed should be constructed as per guidelines indicated in the scheme.

Purchase of Animals. The animals may be purchased in two lots for keeping the milk production continuous. Further, the animals should be purchased by a sub-committee comprising of the beneficiary, State Department of Animal Husbandry official and Bank official, preferably from cattle fairs being held in the state at various places.

Rate of Interest. Interest at the rate of 10.5 per cent shall be applied. However, since the state government shall subsidize the interest, the differential amount of interest over and above 4 per cent shall be got reimbursed by the bank from the state Department of Animal Husbandry.

Dairy Schemes: Larger Loans from Commercial Banks

There is wide variation in the terms and conditions of these schemes. The margin requirements vary between 5 and 50 per cent, the rate of interest ranges from 10.5 to 15 per cent and the repayment period from 10 months to 3 years. The repayment period for the construction of cattle shed is from 3 to 5 years. Apart from the hypothecation of animals and cattle insurance, some banks have sought mortgage of land. A few banks have also asked for guarantees. Thus these considerations have excluded tenant and landless labourers from availing the benefits of these schemes, though, to begin with, these schemes were taken up for their benefit also. Such security-based restrictive eligibility criteria should be dropped, particularly when the credit schemes are meant for low-income group.

Not all dairy schemes, as they are formulated (or implemented), aim at financing the purchase of exotic high yielding milch animals. Banks, however, cannot be blamed altogether for this feature. In many areas such animals are just not available. Nor can the available support for veterinary services; marketing arrangements for milk, or managerial capabilities of the borrowers to look after the exotic breed can be considered satisfactory in all places. What the schemes, however, could be criticized for is that even when they were formulated for the purchase of exotic breed, not enough attention has been paid to these aspects. Consequently, assumptions made about milk yield often appear unrealistically high. Though most of these schemes are formulated for the cultivators, while working out the details the dairy enterprise has been treated in total isolation and not in the context of the crop enterprise. Similarly, repayment capacity has been worked out without any provision for domestic consumption of milk.

1.6. SCOPE AND FUTURE OF DAIRYING IN INDIA

For developing countries, dairy bovines are important, not just because they give milk—the wholesome food, but also for their draft power, meat, hides, manure and other products: though the cow as a meat producer is a taboo only in India due to religious reasons. The dairy cow is more efficient than other farm animals in converting feedstuffs that are not directly consumed by man (forages and cereal by-products) into edible human food. Efficiency of conversion of feed protein and energy into edible protein and energy respectively, 25 percent and 17 percent by dairy cattle, 4 percent and 3 percent by beef cattle (edible cuts), 26 percent and 18 percent by hens (eggs), 23 percent and 11 percent by broilers and 14 percent and 14 percent by swine (Wedin *et al.*, 1975)

Contribution of cattle and buffaloes to India's economy is enormous (Table 1.11 and Table 1.12). The contribution livestock and fisheries to GDP of the country is 5.26% in 2006-07. The value of output from livestock and fisheries was ₹ 2507610 million in 2006-07 *i.e.*; about 31.7% of the total contribution of Agriculture and allied sectors to GDP in 2006-07*. According to the Central Statistical Organisation (2008), the value output from cattle and buffaloes, through milk was ₹ 1443860 million in 2006-07 at current prices (57.58% of the total livestock contribution). It was ₹ 21.6 million in 1970-71 and ₹ 880920 million in 1999-2000. This was much more than the contribution of Paddy@ ₹ 850320 million, Wheat @ ₹ 667910 million and Sugarcane @ ₹ 284880 million.

TABLE 1.11. Percentage Contribution of Livestock Sector to Agriculture and National GDP (at 1993-94 prices)

Year	*Percent Contribution of Livestock Sector to*	
	Agriculture GDP	*National GDP*
1990-00	24.39	5.59
2000-01	25.92	5.67
2001-02	25.91	5.70
2002-03	29.27	5.70
2003-04	27.26	5.40
2004-05*	24.90*	4.70*

*At 1999-2000 prices

Source : Planning Commission, XI plan document.

TABLE 1.12. Growth rate in Gross Domestic Product (In Percent) (At 1993-94 prices)

Year	*Agriculture and allied sector*	*Livestock Sector*
1994-95	5.08	5.49
1995-96	–1.13	4.33
1996-97	10.10	4.19
1997-98	–2.82	2.01
1998-99	6.87	4.61
1999-00	–0.11	3.42
2000-01	–0.40	5.87
2001-02	6.46	6.40
2002-03	–7.99	3.95
2003-04	10.31	2.73
2004-05*	0.70	4.31

*At 1999-2000 prices

Source : Planning Commission, XI plan document

Wedin, F. Hodgson, H.J. and Jacobson, N.L. 1975. J. Anim. Sci. 41: 667; Gil. S.S. 1979. *Ind Dairyman.* 31:269.

Milk is the largest agricultural commodity in India for the last few years. In addition, the income from hides and dung/fuel were respectively ₹ 25190 and 155960 million in 2006-07 as against ₹ 1,200 and ₹ 7,600 million in 1981-82. If the contribution of bullocks (about 57 million pair to agriculture and transport is also taken into consideration, it will be clear that cattle and buffaloes play a very important role in the nation's economy. In some areas, like parts of Gujarat, Maharashtra and Rajasthan, where dairying has been developed as an important activity, milk accounts for major part of the income of rural households. Indian dairy sector had a growth rate of 4.5 per cent per annum during the period 1973-74 to 1980-81, and 5.48 per cent during 1980-81 to 1990-91 but 4.11 per cent only during 1991 to 2000-01 (Bhasin, N.R. 2008, 11th Five Year Plan Document).

National Sample Survey (61st round—2004-05) estimated employment in animal husbandry to be 11.44 million in principal status and11.01 million in subsidiary status. This constitutes 5.5% of total working population in the country. Out of the 22.45 million engaged in animal husbandry sector 16.84 million are females. About 70-75 per cent of households maintaining cattle and buffaloes belong to the category of small and marginal farmers and landless agricultural labourers. Dairying in India is, thus small man's business and does not yet show much of commercialization in rural areas. The picture is likely to change by a decade or two, thanks to the governmental efforts to develop dairying in such areas of the country where potential for milk production enhancement (milk shed areas) exists.

The special advantage of the cattle is that they convert large quantity of agricultural waste and byproducts into milk, so essential for balanced human nutrition and provide gainful employment to large agricultural labour force in the villages during their spare hours. In the mixed farming pattern, cattle led to increase in crop production. The manure available from cattle maintains fertility of the soil. Cultivation of short-term fodder crops fill the gaps in crop rotations to provide better utilization of the land. The leguminous fodder, which provides excellent feed to livestock, also enriches the soil. The dairy enterprise in addition to providing higher income ensures a constant and regular daily income throughout the year. The other advantage is the lower degree of risk involved in comparison to crop farming which is very vulnerable to fluctuations in weather and price.

Indeed the dairy bovine occupies an economically important role in the lives of rural folk. A Punjab Agricultural University study (Gill, 1979) has shown that greater returns per acre can be obtained from dairy farming (Table 1.13). Dairying has a crucial economic value for landless labourers and small farmers (Table 1.14). Nearly 50 per cent of rural households possessing bovines fall under the category of marginal farmers and agricultural (landless) labourers and 25 per cent that of small farmers.

Experience of Gujarat and Andhra, where an integrated development of dairying took place in some selected regions, has proved the potential scope for dairying in improving the lot of rural poor (Tables 1.13 to 15). In Kaira and Mehsana districts of Gujarat weaker sections of populace are contributing about 68 per cent to the total milk production by owning only 1 or 2 buffaloes per household.

Bhasin, N.R. 2008. Indian Dairyman 60:6; Ann. Rep., Dept of AH, Dairying and Fisheries, GOI 2007-08.
Jain, M.M. 1979. *Indian Dairyman.* 31: 259.
Thomas, C.K. (1998). Role of livestock in Sustainable agriculture in India. Proc. Dominant Farming Systems and Eco-Friendly Technologies for Hills : Implications and Sustainability. Ed. Kapoor, A.C. Publ. Sood Enterprises. Palampur.

Dairying contributes about 65 per cent to the total income of landless labourers and 35 per cent to that of marginal and small farmers (Jain, 1979). The other factors which make the rural poor most suitable base for dairy development is the low but regular (in fact daily) earnings in dairying that can gainfully employ the low wage earning unemployed and underemployed work-force with its traditional know-how of dairying (Prasad, 1978).

TABLE 1.13. Returns for Fixed Factors Under Dairy Farming and Different Crop Rotations (Gill. 1979)

Enterprise	*Returns for Fixed Factors Per Acre (₹)*
Paddy-Wheat	1743.84
Maize-Wheat	1471.04
Groundnut-Wheat	1900.83
Cotton-Wheat	2350.82
Dairying with buffaloes	3278.36
Dairying with crossbred cows	5601.91

Gill. S.S. 1979. *Indian Dairyman.* 31 : 269.

TABLE 1.14. Milch Animals per Holding of Different Sizes in Rural Areas

Size of land holding (acres)	*%of total rural households*	*Average value (Rs.) of all assets/ household*	*%share in total production assets other than land*		*Number of females over 3 years per 100 households*	
			Cattle	*Buffalo*	*Cattle*	*Buffalo*
Landless	9.64	199	22.9	32.60	11.40	18.40
0.01–0.50	27.78	164	45.75	18.65	6.28	23.07
0.50–1.00	7.45	345	51.69	21.28	14.67	43.52
1.00–1.25	4.47	461	46.37	28.02	21.54	48.04
1.25–2.50	13.28	650	52.22	20.42	23.01	56.27
2.50–5.00	15.49	1070	49.26	22.80	38.46	80.91
5.00–7.50	7.89	1571	48.60	20.89	49.92	108.28

*Assets include : (*i*) Cattle, (*ii*) Buffalo, (*iii*) Other Livestock, (*iv*) Non-farm business equipment, (*v*) Agricultural Machinery, and (*vi*) Transport equipment.

Source : 26th round of N.S.S. Report No. 215.

TABLE 1.15. Changes in the in-milk bovine stock per 100 rural households

Category	*In-milk bovine stock per 100 rural households*			
	1971-72 (26th round)	*1981-82 (37th round)*	*1991-92 (48th round)*	*2002-03 (59th round)*
Landless	16	7	6	1
Marginal	33	28	41	40
Small	64	48	69	63
Semi-medium	92	74	80	84
Medium	142	106	102	126
Large	225	153	130	208
ALL	54	37	46	36

Source : NSSO reports (numbers 215, 338, 408 and 493).

Prasad, A.G. 1978. *Indian! AgrL Econ.* 30:252.

In Mehasana district of Gujarat, dairying has changed the village economy to such an extent that more than half of the cash income earned by milk producers in the villages comes from that occupation alone (Saxena *et al,* 1978). Dairying has not only attracted capital from outside the district but also helped a capital formation within the district. Due to increase in the number of milk producers' cooperatives in the area, the share capital evidently increased. Dairying has not only increased the income but it has also increased employment in the district. Besides increase in number of milch animals of better breed, the milk production also increased appreciably. With the result, the per capita consumption of milk in the villages also increased to a greater extent.

Thus dairying can play and is playing a vital role in improving the economic conditions of the village poor and it would bring further improvement in their economic condition in the near future. As identified by the National Commission on Agriculture, there is a good scope for rapid development of milk production through rural poor by increasing availability of necessary inputs. This may serve the twin objectives of our development plans—(*a*) growth with social justice, and (*b*) enhancement of milk production. Estimated demand for milk in 2021-22 is 172 million tones. To achieve this, milk production has to grow @ 4% per annum.

1.7. DAIRY DEVELOPMENT PERSPECTIVE PLAN-2010

The National Dairy Development Board has drawn up a Perspective Plan-2010 for Cooperative Milk Unions, with the objective to raise milk procurement and increase cooperative share in marketing of milk and milk products. 80 Milk Unions have submitted their Perspective Plans to NDDB with an outlay of about Rs. 900 crore*. NDDB has approved Plans of about 60 Milk Unions with an investment outlay of Rs.750 crore*. Formalities of entering into agreements/ hypothecation of assets are being completed. The Perspective Plan has four thrust areas :

1. Strengthening the cooperative businesses
2. Enhancing productivity
3. Managing quality
4. Building a national information network

Strengthening the Cooperative Business. During the year, NDDB continued its efforts to strengthen cooperatives through institution building, increased participation of women, farmers' induction, and enhanced milk procurement and marketing.

NDDB seeks to strengthen the dairy cooperative movement by significantly increasing women dairy farmers' participation in dairy cooperatives. The activities include communication, education, extension and training of members, thrift, and health and income generation. NDDB has initiated pilot women's thrift and credit cooperative projects in Shajapur (MP) and Tirupati (AP). By November 2001, 107 Women Thrift Cooperatives (WTC) with a membership of 8628 had already been formed.

NDDB conducts Farmers' Induction Programme throughout the year to expose farmers from all over the country to the latest developments in dairy farming. They are taken on field visits to village societies; milk processing and cattle feed plants. The classroom session also

Saxena, B.C., Jain. J.P. and Arya. S.R.S. 1978. *Indian Farming*. Aug., 1978. p. 34.
* Crore = 10 million.

includes discussion on clean milk production and good animal management practices. Till December 2001, 2430 farmers including 578 women participated in 26 Farmers' Induction Programmes.

The focus of the project is on the emergence of a well-informed constituency of members, responsive managing committee, members of the village dairy cooperative societies, a pro-active Board of Directors of unions, committed cooperators as well as competent and responsible staff, aimed at building diary cooperatives into self-reliant and thriving business enterprises which will be responsive to economic and social expectations of members. In order to sensitise the Board of Directors regarding business of the Union and evolve long-term goals, NDDB has been conducting Business Orientation Programmes.

Towards the end of the programme the Board drafts value and mission statements and prepares long terms goals, which is discussed later and finalised with professional managers at an interface workshop. Till December 2001, 12 orientation programmes were organised wherein 164 members participated. NDDB also undertook training programmes for the staff union to enable them to facilitate village level institution building programmes.

Enhancing Productivity. In order to ensure maximum economic returns to milk producers and sustained increase in milk production, NDDB has adopted a scientific and integrated approach towards breeding, feeding, health care and husbandry practices.

Animal Breeding. For genetic improvement of the country's cattle and buffaloes, NDDB has developed appropriate selection programmes—the Dairy Herd Improvement Programme Actions (DIPA) and the Open Nucleus Breeding System (ONBS) integrated with artificial insemination. DNA fingerprinting and karyotyping are being undertaken to identify and select superior healthy milch animals. NDDB has supported the Milk Unions to establish semen production, liquid nitrogen delivery and artificial insemination (AI) facilities. During 2001-02, approximately 22,000 DCSs provided AI services and carried out approximately 5.5 million inseminations (Provisional). Approximately, 11 cooperative semen stations produced 7 million frozen semen doses.

Animal Nutrition. Bypass protein and fats have been developed under a collaborative research programme with the Australian Centre for International Agricultural Research (ACIAR). A pilot plant for making bypass protein has been set up in Anand. Another pilot mineral mixture plant has been developed and installed. During the year, 47 cattle feed plants with a total installed capacity of 6255 MT per day produced on an average 4565 MT of cattle feed with an overall capacity utilization of 73 per cent. In addition, 11 Urea Molasses plants produced and sold 2 lakh blocks. Under NDDB's Fodder Seed Production Programmes implemented through dairy cooperatives and NGOs, a total of 45,000 quintals of good quality seeds were sold to milk producers through cooperatives.

Animal Health. NDDB's current animal health research focuses on standardisation of molecular methods for disease diagnosis and development of recombinant vaccines against Brucellosis, Foot and Mouth Disease (FMD), Haemorrhagic Septicaemia (HS), Calf Diarrhoea and Jonhe's Disease. The serological screening of breeding animals for Blue Tongue (BT) and Infectious Bovine Rhinotracheitis (IBR) by NDDB is underway on organised farms. Attempts are being made to develop a recombinant vaccine against ticks infesting dairy cattle. In addition to regular prophylactic vaccination of animals, NDDB also assisted cooperatives in mastitis and brucellosis control.

Managing Quality. Quality initiatives taken up in the cooperative dairy sector under the Perspective Plan-2010 of NDDB include a massive awareness programme on Clean Milk Production (CMP) covering one lakh village DCSs across the country. Through a range of informational and educational inputs, including a book titled "Doodh ki Kahani Gai ki Zabani' published in Hindi and in 10 other state languages, producers are made aware of hygienic milking and milk handling practices. So far, 44 dairy plants and 7 cattle feed plants have obtained ISO/HACCP accreditation. NDDB has designed and developed user-friendly kits for testing raw milk at the village level DCS, the reception dock of a dairy plant and also at the consumers' doorstep.

Building a National Information Network. NDDB continues its endeavor to provide real time information to dairy cooperatives by setting up information systems at various levels using modern technology. The scope of the data encompasses demography, agriculture and land use, livestock population, export and import of dairy commodities, milk production and productivity of animals.

Information support system. Under information support system of EC programme, NDDB is establishing a network with the milk unions; federations and marketing dairies through implementation of Internet based Dairy Information System (IDIS). More than 75% of the Unions are entering data regularly through IDIS. An information system involving Geographical Information System (GIS) software and software for animal enumeration and milk production estimation, society information, veterinary health care and artificial insemination service has been developed and tested and the activities are operational in 33 unions – 12 each in the western region and southern region, 5 in the northern region and 4 in the eastern region. The GIS provides decision support and helps in strategic planning in areas of milk procurement, veterinary health, epidemiology and other operational matters. All Gujarat and Karnataka Unions are already using GIS for procurement management and for other applications.

NDDB has become the sole founder of the cooperative services from the SAARC region from 21st November 2001. The Dot Coop Limited Liability Company (DCLLC), a fully owned subsidiary of National Co-operative Business Alliance, US (NCBA) in partnership with the International Co-operative Alliance (ICA), manages the cooperative services.

Milk and Milk Product Order–1992. The Govt. of India had promulgated the Milk and Milk Product Order (MMPO) 1992 on 9.06.1992 under the provisions of Essential commodity Act, 1955 consequent to de-licencing of the Dairy sector in 1991. As per the provisions of this order, any person/dairy plat handling more than 10000 liters per day of milk or 500 MT of milk solids per annum needs to be registered with the registering authority appointed by the Central Government. The main objective of the order is to maintain and increase in supply of liquid milk of desired quality in the interests of the general public and also for regulating the production, processing and distribution of milk and milk Products.

Recognizing the necessity for suitable amendments in Milk and Milk Product Order 1992 for faster pace of growth in the Dairy sector, Govt. of India has amended Milk and Milk Product Order 1992 from time to time in order to make it more liberal and oriented to facilitate the dairy entrepreneurs. The Govt. of India has notified the amendment proposals in the official gazette on 26/03/2002. Now there is no restriction on setting up of new capacity, while noting that the requirement of registration is for enforcing the prescribed standards of quality and food safety. The salient features of the new amendments made are as follows:

- The provision of assigning milk-shed has been done away with.
- The registrations under MMPO-92 will now cover sanitary, hygienic condition, quality and food safety.
- The provision of inspection of dairy plant has been made flexible.
- The provision to grant registration in 90 days has been reduced to 45 days.
- The power of registration of State registering Authority has been raised from 1.00 LLPD to 2.00 LLPD.
- Altogether the Central and the State Registering Authorities have registered 688 units with combined capacity of 803.74 LLPD in Cooperative, Private and Government Sector as on 31.3.2003 (Table 16).

TABLE 16. Dairy development physical progress

Particulars	*1998-99*	*1999-2000*	*2000-01*	*2001-02**
Societies Organized ('000) (Anand Pattern)*	81.0	83.7	98.0#	99.2#
Farmer members (lakh)*	101.4	105.2	108.3	108.9
Average rural milk procurement (lakh kg/day)	135.8	157.4	165.5	163.56
Liquid milk marketing (lakh litres/day)	121.3	129.0	134.0	135.82

*Cumulative till November, 2001 (provisional); # = includes conventional societies and Taluka Unions
lakh = 100,000

1.8. CENTRAL HERD REGISTRATION SCHEME

The Central Herd Registration Scheme is continuing with four Units located at Rohtak (Haryana), Ajmer (Rajasthan), Ahmedabad (Gujarat) and Ongole (Andhra Pradesh). These four units together cover the States of Haryana, Uttar Pradesh, Delhi, Rajasthan, Gujarat, Maharashtra and Andhra Pradesh. The cattle breeds of Hariana, Gir, Kankrej and Ongole and buffalo breeds of Murrah, Surti, Mehsana and Jaffrabadi are registered under the Scheme depending on the confirmation to breed characteristics and prescribed milk production norm. Owners of registered animals are provided certificate and prizes/incentives to encourage conservation of indigenous breeds and production of high quality cows and buffaloes (Table 1.17). The Scheme has following major objectives.

(*a*) To survey and study production records of cows and buffaloes in their breeding tracts.

(*b*) Cataloguing of superior germplasm of the selected breeds in their breeding tracts as well as breeding farms through milk recording and registration.

(*c*) To disseminate information on registered animals for sourcing of breeding material and inter-farm and inter-breeder exchange of animals.

(*d*) To promote conservation of selected indigenous breeds by providing incentives to breeders.

(*e*) To regulate sale/purchase of elite animals to ensure and enforce quality control for further propagation of breeding.

(*f*) To propagate and awaken consciousness amongst the breeders to improve their economics through livestock improvement by constant publicity, incentive and formation of breeder associations.

TABLE 1.17. The details of the units and the bovine breeds covered are as under Herd Registration Scheme

S. No.	*Name of the unit and year of establishment*	*No. of recording centers State-wise*	*No. of villages covered*	*Breeds covered*
1.	Rohtak (1962)	33—Haryana, UP, Delhi	651	Haryana, Murrah
2.	Ahmedabad (1969)	43—Gujarat, Maharashtra	541	Gir, Kankrej, Surti, Jaffrabadi, Mehsana
3.	Ajmer (1972)	8— Rajasthan	87	Gir, Hariana, Murrah
4.	Ongole (1979)	8— Andhra Pradesh	168	Ongole, Murrah
	Total	92— Centers in 07 States	1,447	Breeds-08

1.9. NATIONAL PROJECT FOR CATTLE & BUFFALO BREEDING (NPCBB)

Genetic improvement in bovines is a long term activity and Government of India has initiated a major programme "National Project for Cattle and Buffalo Breeding" (NPCBB) from October 2000 over a period of ten years, in two phases each of five years, with an allocation of ₹ 402 crore for Phase-I. In order to consolidate gains made during Phase-I, Phase-II has been initiated from Dec 2006 with an allocation of ₹ 775.87 crore. The NPCBB envisages genetic up gradation on priority basis with a focus on development and conservation of important indigenous breeds. The project envisages 100% grant-in-aid to implementing agencies.

The Objectives of the scheme is to : (*a*) Arrange delivery of vastly improved artificial insemination service at the farmers' doorstep. (*b*) Bring all breedable females among cattle and buffalo under organized breeding through artificial insemination or natural service by high quality bulls within a period of 10 years. (*c*) Undertake breed improvement programme for indigenous cattle and buffaloes so as to improve the genetic makeup as well as their availability.

Components. (*a*) Streamlining storage and supply of liquid nitrogen by sourcing supply from industrial gas manufacturers and setting up bulk transport and storage systems for the same; (*b*) Introduction of quality bulls with high genetic merit; (*c*) Promotion of private mobile A.I. service for doorstep delivery of A.I.; (*d*) Conversion of existing stationery government centres into mobiles centers; (*e*) Quality control and certification of bulls and services at sperm stations, semen banks and training institutions; (*f*) Study of breeding systems in areas out of reach of A.I., and (*g*) Institutional restructuring by way of entrusting the job of managing production and supply of genetic inputs as well as liquid nitrogen to a specialized autonomous and professional State Implementing Agency.

Progress of the Scheme. At present 28 States and one Union Territory are participating under the project. Financial assistance to the tune of ₹ 398.36 crore has been released to these States upto 2007-08. During 2009 allocation of ₹ 89.70 crore has been made under the scheme.

Constitution of State Implementing Agencies (SIAs). Since inception of the project in October 2000, 26 SIAs have been constituted under the project. These agencies are implementing the project with professional approach. In case of small States who are unable to constitute viable SIAs, funds have been released to the State Governments for implementation of the project.

Increase in coverage of breedable animals. Semen production in the country has increased from 22 million straws (1999-2000) to 44 million straws (2007-2008) and the number of inseminations has increased from 20 million to 41 million. As per the impact analysis report submitted by NABARD, overall conception rate has increased from 20% to 35%.

Development of minimum standard protocol for semen production (MSP). In order to produce frozen semen of the uniform quality, a Minimum Standard Protocol (MSP) for semen production has been developed in consultation with experts from BAIF, National Dairy Development Board (NDDB), NDRI (Karnal) and Central Frozen Semen Production and Training Institute (CFSP&TI) and the same has been made effective from 20th May, 2004.

ISO Certification of semen station locations

- Ooty (TCMPF)
- ABC Saloon (NDDB)
- Jagudan Mehsana Dairy)
- Hissar, Gurgoan, Jagadhari (HLDB)
- Shyampur (Uttranchal)
- Bhattain (Punjab Milk Fed)
- SSCC Hessarghata
- Dharwad (Karnataka)
- Bidaj (NDDB)
- AR&DA, Amul Dairy
- Urulikanchan Pune (BAIF)
- Haringhata, Salboni, Beldanga (West Bengal)
- Nabha, Ropar (Punjab)
- Nandani (KMF)
- SLBTC Hessarghata
- CFSP&TI Hessarghata (GOI Karnataka)
- Mattupatty, Dhoni, Kulathupuzha (Kerala)
- Chitale Bhilwadi (Private)
- Bhadbhada (Madhya Pradesh)
- Bassi (Rajasthan)
- Cuttack (Orissa)
- Vizag, Nandyal, Karimnagar, Banwasi (Andhra Pradesh)

Semen Station at Mattupatty, Dhoni, Kulathupuzha (Kerala), Haringhata (West Bengal, Salboni, Beldanga (West Bengal) and Bhadbhada (Madhya Pradesh) are also HACCP certified semen stations.

Strengthening Infrastructure for Quality and Clean Milk Production. Centrally sponsored scheme initiated during 10th five year plan with an outlay of ₹ 30 crores with the objectives of :

- Creation of necessary infrastructure for the production of quality milk and milk products from the farmer level to the point of consumption.
- Improvement of milking procedure at the farmer level
- Training and strengthening of infrastructure (by providing seamless buckets, milking pails, disinfectants etc) to create mass awareness about importance of clean milk production.
- Basis of funding under this scheme is 100 percent grant-in-aid for all the components except bulk coolers for which financial assistance is in the ratio of 75:25 between the GOI and the respective dairy co-operative society/Milk Union/State Dairy Federation. During 2006-07 an amount ₹ 35 crores has been provided under this scheme.

1.10. DAIRY/POULTRY VENTURE CAPITAL FUND

A venture capital fund for dairy and poultry sectors has been created by the government for providing financial assistance for improving the quality of milk handled by the non-organized sector/small scale milk producers in rural areas scattered all over the country. It is being implemented through NABARD. Under this scheme fianancial assistance will be provided for the following :

- To promote setting up of small dairy farms for milk production
- To bring structural changes in the unorganized sector; processing and marketing of pasteurized milk at village level
- To bring about upgradation of quality and traditional technology to handle milk on a commercial scale
- To boost the unorganized poultry sector in states where development is in primitive state and also to give incentive and create infrastructure facilities for export of poultry products by organized sector from advanced states
- Establishment of poultry and breeding farms with low input technology and also for other poultry species
- Setting up of poultry feed plant/ laboratory
- Setting up of egg grading, packing and storing facility for promoting export
- Marketing of poultry products (specialized transport vehicles, cold storage etc.)

Scheme funding/Pattern of Investment : Entrepreneur's Contribution at 10%, Revolving Fund's (GOI)share : 50% (Interest free), Bank's share : 40% of total outlay-interest as applicable to agriculture loans;

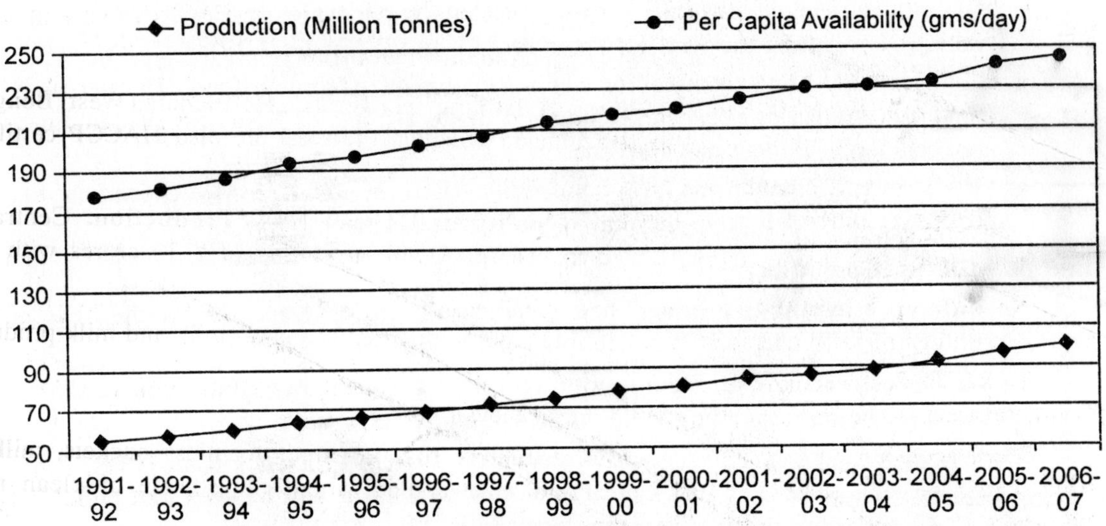

Fig. 1.19. Dynamics of annual milk production and per capita availability of milk in India over the years

Data Source : Department of Animal Husbandry, Dairying & Fisheries, Ministry of Agriculture, GoI.

1.11. INTEGRATED DAIRY DEVELOPMENT PROJECT IN NON-OPERATION FLOOD, HILLY & BACKWARD AREAS

This is a plan scheme launched in 1993-94 with 100% grant-in-aid with the objectives of

- Development of milch cattle/buffaloes
- Increasing milk production
- Procurement, processing and marketing of milk in a cost effective manner
- Ensuring remunerative prices to the milk producers
- Generating additional employment opportunities
- Improving social, nutritional and economic status of residents of comoparatively more disadvantaged areas

1.12. LIVESTOCK INSURANCE SCHEME

A centrally sponsored scheme Livestock Insurance was implemented on pilot basis during the years 2005-06 and 2006-07 of the 10th Five Year Plan and 2007-08 of the Eleventh Plan in 100 selected districts across the country. Under the Scheme, crossbred and high yielding cattle and buffaloes were insured at maximum of their current market value. The premium of the insurance was subsidized to the tune of 50% by the Central Government. Rest 50% of the premium was borne by the beneficiaries.

1.13. INDIAN DAIRYING: SOME CONTRADICTORY LOOKING FEATURES

- India is the world's top milk producer but its per capita animal productivity is a meager 1.5 liters per day.
- The national average animal productivity is 1.5 liters per day as against 4.2 and 6.4 liters per day in buffalo and crossbred cow respectively in Operation Flood areas.
- Milk production doubled in the last two decades, yet per capita availability of milk at 240 grams is lower than the world average of 285 grams (Figure 1.19).
- High per capita milk availability, yet many remain undernourished. The actual milk intake per day varies between 20 grams in many districts of Eastern region to almost a liter per day in Punjab and Haryana.
- Milk is India's number one farm commodity based on producer price exceeding that of paddy, yet did not receive the due support from planners

Buffalo Milk Vs. Cow Milk

No difference in nutritive value. There is practically no difference in the nutritive value and digestibility of milk and milk products obtained from cow and buffalo.

Lower cholesterol content. Significantly, cholesterol content of buffalo milk is 0.65 mg/g as compared to the corresponding value of 3.14 mg/g for cow milk.

More proteins. Animal bioassays have shown the Protein Efficiency Ratio (PER) value of buffalo milk proteins to be 2.74 and that of cow milk as 2.49. It will be seen that buffalo milk has about 11.42 per cent higher protein than cow milk.

More important minerals. Buffalo milk is also superior to cow milk in terms of important minerals, namely calcium, iron and phosphorus which are higher by 92 per cent, 37.7 per cent and 118 per cent.

More vitamin A. Buffalo metabolizes all the carotene into vitamin A, which is passed on to milk as such.

More viable, commercially. Buffalo milk is commercially more viable than cow milk for the manufacture of fat-based and SNF-based milk products, such as butter, ghee and milk powder because of its lower water content and higher fat content. Most significantly, the lower cholesterol value should make it more popular in the health conscious market. By the virtue of greater opacity of casein micelles, coupled with higher levels of colloidal proteins, calcium and phosphorus, buffalo milk is more densely white and has superior whitening properties as compared to cow milk. Therefore, unlike the cow milk (which is pale-creamy yellow in color) and cow milk fat (which is golden yellow in color), buffalo milk and milk fat are distinctively whiter. UHT-processed buffalo milk and cream are intrinsically whiter and more viscous than their cow milk counterparts, because of conversion of greater levels of calcium and phosphorus into the colloidal form. Buffalo milk is, therefore, more aptly suitable for the production of tea and coffee whiteners than cow milk. Higher innate levels of proteins and fat render buffalo milk a more economical alternative to cow milk for the production of casein, caseinates, whey protein concentrates and a wide range of the fat-rich dairy products.

Better whey proteins. Proteins of buffalo milk, particularly the whey proteins, are more resistant to heat denaturation as compared to the cow milk proteins. Dried milk products prepared from buffalo milk exhibit higher levels of un-denatured proteins when processed under similar conditions. In general, the reconstitution behavior of dried milk products made from buffalo milk is indistinguishable from those made from cow milk. However, dried buffalo milk may be preferred over dried cow milk for those technological applications where higher levels of un-denatured whey proteins would be more desirable.

Better cheese. Cheese made from buffalo milk displays typical body and textural characteristics. More specifically, where chewing and stringing properties are specially desired as in the case of Mozzarella cheese, buffalo milk is technologically preferable over cow milk. In Italy, recently legislation has been introduced to restrict use of term "Mozzarella" only to those products exclusively made from buffalo milk (without admixture with cow milk). Certain traditional cheese varieties, such as *paneer* in India or pickled cheeses from the Middle-East countries, are best made from buffalo milk.

Better health foods. The presence of higher levels of various bioprotective factors, such as immunoglobulins, lactoferrin, lysozyme, lactoperoxidase as well as bifidogenic factors, render buffalo milk more suitable than cow milk for the preparation of a wide range of special dietary and health foods.

Thus buffalo is not just the main stay of milk production in India. Buffalo milk has great nutritional, technical and economic significance for Indian dairy Industry.

CHAPTER

2

Dairy Farm Management

2.1. ECONOMIC PRINCIPLES APPLICABLE TO DAIRY BUSINESS

If dairy farming has to be adopted on the commercial basis, it must be economically sound and viable. Therefore, one who wants to adopt dairy farming as business must acquaint himself with the economic principles. Economic principles form the basis for making decisions and choices on matters relating to all aspects of dairy farming. These principles are not peculiar to dairy or crop farming or to any other aspect of farm business. They are universally applicable to all aspects of management. Only the application varies. In practice, the use of economic principles requires reasonably complete and accurate records and common sense. In the following section the important economic principles, which are helpful in making correct choices with respect to livestock enterprises, are discussed.

2.1.1. The Principle of Diminishing Returns

The principle of diminishing returns is of prime importance in economic analysis. It is an important guiding factor in dairy farming to decide the level to which the farmer can push up milk yield per cow or buffalo or marketing weight per animal. It comprises of the basis for the entire framework of marginal analysis, the most comprehensive and powerful tool in economics. The principle of diminishing returns refers to the amount of additional output obtained as additional inputs of variable factors of production added to a fixed amount of other factors. In general, total output or yield increases most rapidly with the first units of input, then continues to increase but at a decreasing rate until it reaches the point of maximum total output and finally additional input results in a decline in total output. Should a farmer stop using additional dose of input after the point of diminishing returns has been reached? The principle of diminishing return refers to diminishing physical returns. The inputs and yields (output) are expressed in physical units such as liters of milk, kilograms of concentrates or fodder. The question can be properly answered, if one understands the concept of the factor product relationships amongst total, average and marginal yield of products. Figure 2.1 shows the general concept of the law.

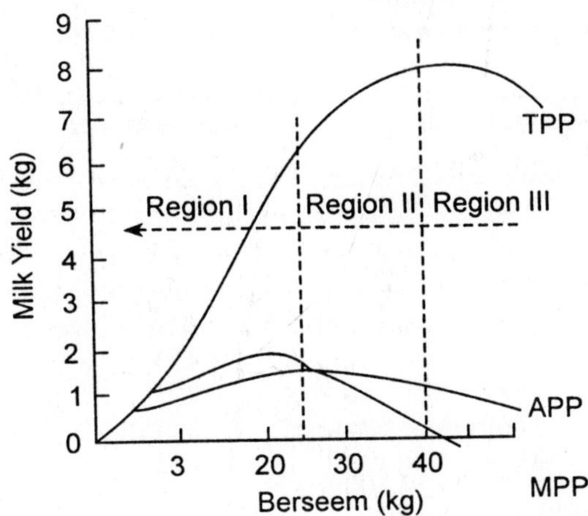

Fig. 2.1. Input-output relationship in dairy farming

Total product (TP) is the amount of output, which results from different quantities of variable inputs. Assume that the output Y is milk yield and that input X denotes quantity of feed being fed to a buffalo or cow. If increasing amount of feed is fed to the animal, the output increases quite rapidly in the beginning, then less rapidly and finally reaches a maximum, after which declines.

Average product (AP) is simply the total product divided by the quantity of input used (Y/X). The shape of the total product curve determines the shape of the average product curve.

Marginal Product (MP) refers to the increase in output with the added quantity of input. It is actually the ratio of change in the total product as related to change in input ($\Delta Y/\Delta X$). The word *marginal* is used here as synonymous with additional. Marginal product curve increases quite rapidly in the beginning, reaches a maximum and then declines. It always intersects the average product curve at the point where the latter curve is at its maximum. The marginal product curve touches the horizontal axis X or becomes zero when total product curve is at its maximum.

The relationships that are described above are physical in nature. Furthermore, these physical relationships isolate the relevant and irrelevant ranges. It will never be profitable to use more inputs than necessary to get maximum total output. Similarly, it is also not advisable to stop production when average production is at peak. This will result in under-utilization of fixed inputs. Thus so long as physical returns per unit of input are increasing, it will be profitable to increase output at least to the point where output per unit is maximum (12.71 litre in the example given in Table 2.1). But it is not always necessary that maximum physical output will also be the most profitable level of output. To find out the most profitable level of output or most economical level of input, it is necessary to consider the prices of inputs and outputs. Data given in Table 2.1 shows the relationship between the level of concentrate feeding and milk yield obtained by the cultivators in Haryana in the year 1975-76. This Table is based on data derived from the equation :

$$Y = 5.5039 + 1.8957X - 0.1243X^2$$

where Y stands for milk production per day in kilogram and X for the quantity of concentrates fed to milch buffalo per day in kilogram.

TABLE 2.1. Estimated Total Average and Marginal Physical Products a Different Levels of Concentrates

Concentrate (kg)	*Total Milk Yield (Litres)*	*Average milk yield (Litres)*	*Marginal Milk Yield (Litres)*
0	5.50		1.78
1	7.28	7.28	1.52
2	8.80	4.40	1.27
3	10.07	3.36	1.03
4	11.10	2.78	0.77
5	11.87	2.37	0.53
6	12.40	2.07	0.26
7	12.66	1.81	0.05
8	12.71	1.59	– 0.21
9	12.50	1.39	– 0.47
10	12.03	1.20	

In general it is not always possible to demonstrate all the three stages of production, *i.e.*, total production increasing at increasing rate, increasing at diminishing rate and final decrease, in cultivators' data. The main reason seems to be that the zero level assumed is not in real sense a zero, because all the animals consume some ration for the maintenance in one or the other form.

The principles of diminishing economic return differ from that of diminishing physical returns in that an initial increase in added returns occurs more rapidly in terms of economic returns than that of physical returns. Similarly diminishing economic returns develop more quickly than diminishing physical returns. In other words, the point where it is not economic to-continue to feed concentrate is reached before the point of zero response in yield is reached.

Let us now discuss about the principle of most profitable level of milk output or most economical level of input (concentrate in this case). In determining most profitable level of concentrate the basic economic principle involved is *"added cost should be equal to the added returns."* Average returns do not reveal the correct picture; therefore, always added or marginal returns are considered. In this example the value of the additional milk must be equal to added cost of concentrate. In Table 2.2 are given calculated values of added physical returns as well as economic returns at various prices of milk.

Varying prices of milk are used to demonstrate the effect of price on profitable level of concentrate consumption. When price of milk is taken as Rs. 1.50 per litre, the highest profitable level of concentrate, which could be fed to a buffalo, is 4 kg. If we have considered physical returns (concentrate in kg and milk in litres), it would seem optimal to give 8 kg of concentrate to a buffalo. When price of milk is Rs. 2.00 per litre, it would be profitable to feed up to 5 kg of concentrate. Thus, if the price of milk is raised or price or concentrate is reduced, it is advisable to feed more concentrate and achieve higher milk yield. Similarly if the price of milk is reduced or price of concentrate is increased, it will be advisable to use less concentrate as it will give less milk output.

Sharma, V.N. and Thomas, C.K. (1981). Economic level of concentrate supplementation in the ration of buffalo heifers. Indian J. Anim. Sci. 51 : 157-163.

TABLE 2.2. Returns from Varying Levels of Concentrate for Buffalo Milk Production in Haryana

Concentrate	*Total milk yield (litres)*	*Added or marginal milk yield (litres)*	*Added cost of the concentrates @ Rs. 1.50 kg*	*Added value of milk at different prices @ Rs. 1.50/litre*	*@ Rs. 2.00/litre*
0	5.50		0		
1	7.28	1.78	1.50	2.67	3.56
2	8.80	1.52	1.50	2.28	3.04
3	10.07	1.27	1.50	1.90	2.54
4	11.10	1.03	1.50	1.54	2.06
5	11.87	0 77	1.50	1.16	1.54
6	12.40	0.53	1.50	0.79	1.06
7	12.68	0.28	1.50	0.42	0.56
8	12.71	0.03	1.50	0.06	0.06
9	12.50	0.21	1 50	–0.31	–0.42
10	12.03	0.47	1.50	–0.70	–0.94

In addition to change in prices, which may affect the maximum profit point in the consumption of concentrate, other factors influencing the most profitable level of input use are the quality or breed of animals and quality of inputs. The same amount of feed can perhaps bring higher output of milk from high yielding breeds of animals.

2.1.2. Resource Substitution

The principle of diminishing returns indicates how production will vary as additional resources are applied per animal. Another economic principle that has application to many problems in animal husbandry is the principle of cost minimization, which is also known as the principle of substitution. This principle provides for determination of least-cost method of production.

The principle of substitution says that, if the output of a product remains constant and the product is of equal quality, it is economical to substitute one resource for another resource when the cost of the substitute is less than the cost of the resource it replaced. Farmers may face delicate situations when they are considering the ways and means of minimization of costs. It may be possible that two inputs or resources substitute each other-at constant rate without affecting the level of output. In all such cases it is profitable to use all of the one input or all of the other input whichever is cheaper. In agriculture or animal husbandry many possible combinations of two inputs or resources or practices substitute one for another at diminishing rate. For example, farmers can feed cows or buffaloes with a large amount of green fodder and a small amount of concentrates, a medium amount of both green fodder and concentrate or any other combination of the two. A minimum amount of green fodder is, however, necessary in any feeding programme. Above this minimum level, various combinations of green fodder and concentrates are possible for producing a given level of milk output. In such cases, the farmer is interested in minimizing his feeding cost by substituting the cheaper feed for the more costly one.

Table 2.3 gives different combinations of concentrate (X_1) and fodder (X_2) nutrition-wise required (hypothetical) to produce 10 kg of buffalo milk. It should be noted in advance that all combinations available from the calculations are not physically possible in actual practice. The actual scope of substitution is narrow. In this example the fodder at a constant rate substitutes concentrate.

TABLE 2.3. Feed Combinations to Produce 10 kg Buffalo Milk

Combination of concentrate and breseem		*Total cost of the ration in Rupees when price of concentrate is taken ₹ 2.00 per kg and of berseem ₹ 0.25 per kg = Value of X_1 + X_2*
Concentrate X_1	*Besreem (kg)* X_2	
10.00	0	20.00 + 0.00 = 20.00
6.00	10	12.00 + 2.50 = 14.50
4.00	20	8.00 + 5.00 = 13.00
3.00	30	6.00 + 7.50 = 13.50
2.50	40	5.00 + 10.00 = 15.00
1.25	50	2.50 + 12.50 = 15.00

Note : The animal is fed straw *ad libitum.*

It is possible to find out the least cost ration simply by finding the cost of each combination of concentrate fodder and selecting the one with the lowest total cost.

For example if concentrate costs ₹ 2.00/kg and berseem costs ₹ 0.25/kg, the total cost of first combination is ₹ 20.00. In this manner the total cost for all the combinations given in Table 2.3 can be found out. It is evident from Table 2.3 that out of all the combinations, combination of 4 kg concentrate and 20 kg berseem is the cheapest. Hence, the farmer can select this combination.

The other method of determining the least cost ration combination is through marginal analysis which has some special advantages : (1) it is quicker to determine least cost by this method, (2) data are sometimes presented in this form, and (3) it represents a scientific way of decision making.

TABLE 2.4. Marginal Rate of Substitution and Least Cost Combination of Inputs for Producing 10 kg Buffalo Milk

Combination of concentrate and berseem		*Change in concentrate*	*Change in barseem*	*MRs of X_2 for X_1*	*Price ratio (PX_2/PX_1)*
Concentrate (kg) X_1	*Berseem (kg)* X_2	ΔX_1	ΔX_2	($\Delta X_1/\Delta X_2$)	
10.00	0				0.25/2.00 = 0.125
6.00	10	4	10	0.40	
4.00	20	2	10	0.20	
3.00	30	1	10	0.10	
2.50	40	0.5	10	0.05	
2.25	50	0.25	10	0.025	

PX_1 is price of concentrate, which is assumed equal to ₹ 2.00 per kg.
PX_2 is price of berseem, which is assumed equal to ₹ 0.25 per kg.

The procedure to find the least cost ration combination is illustrated below :

(1) Estimate the marginal rate of substitution (MRS) by dividing the number of units of the replaced resources (concentrate) by the number of units of the added resources (berseem). That is MRS of berseem (X_2) for concentrate (X_1) is :

$$\frac{\text{Number of units of replaced resources (concentrate)}}{\text{Number of units of added resource (berseem)}} = \frac{X_1}{X_2}$$

MRS of various combinations is given in Table 2.4, column 5.

(2) Compute Price ratio : The Price ratio is equal to

$$\frac{\text{Cost per unit of added resource}}{\text{Cost per unit of replaced resource}} = \frac{PX_1}{PX_2}$$

Assuming the price of concentrate (PX_1) is ₹ 2.00 kg and Price of fodder (PX_2) is ₹ 0.25 per kg. Then price ratio is equal to :

$$0.25/2.00 = 0.125$$

(3) Find out the point where above ratios are equal when

$$\frac{\text{Number of units of replaced resources } (X_1)}{\text{Number of units of added resource } (X_2)} = \frac{\text{Cost per unit of added resource } (PX_2)}{\text{Cost per unit of replaced resource } (PX_1)}$$

The above relationship indicates that the cost of replaced concentrate is equivalent to the cost of the added resource *i.e.* berseem. Using the MRS calculated in Table 2.4 for various levels of concentrates and berseem, it is possible to calculate least cost combination of these inputs, which would maximize profit when producing 10 kilogram of milk per buffalo per day. In the example given in Table 2.4, price ratio 0.125 falls between the MRS values 0.20 and 0.10. The application of least cost principle indicates that this quantity of milk (10 kg) can be produced at a minimum cost if the buffalo is fed with a ration combination somewhere between 3 to 4 kg of concentrates and 20 to 30 kg of berseem.

Always be certain that value of the resource replaced is greater than the value of the resource added. If we follow the conditions outlined above for substitution and price ratio, we will automatically select low cost method of production.

2.1.3. Opportunity Costs

The principle of diminishing returns helps in finding out the most profitable level of resource when resources are unlimited. Most farmers have limited resources. Therefore, their main task is to divide or allocate their limited resources among different enterprises in such a way as to get maximum returns. If the farmer has excess of the resources at his disposal, he will not have any difficulty in deciding which enterprise to raise, he would not even compare one enterprise with another. Instead, he would produce all crops and livestock products, which physically suit his farm conditions, weather, climate etc. The rule is simple. Select all products, which are adopted and expand output as long as the principle of added returns and added cost (diminishing economic returns) permit.

Most farmers cannot consider the number and size of the enterprise in this manner because their resources are limited. Hence, enterprises must be considered together, expansion of one enterprise generally requires an equivalent contraction in another and the question is one of which enterprise will give the greatest income, The principle of equimarginal returns states that the income will be greatest, if each unit of the scarce resource is used in producing the products where it adds the most returns. Putting in another way, the principle states that the best combination of the enterprises from one limiting resource is attained when they are organized in such a way that we cannot change the use of single unit of resource without reducing the income. The principle does not say that resources should always be used where they bring in the greatest average returns but where they bring in the greatest added or marginal returns. The best combination of enterprise is not attained when we select the most profitable enterprise.

Suppose a farmer has ₹ 4,000 and he wants to adopt three different enterprises, *i.e.* buffalo, cow and poultry, which can physically be adopted on his farm. What amount of money should he invest in each enterprise so as to get maximum returns ? The possibilities are given in Table 2.5.

TABLE 2.5. Added or Marginal Returns From Using Different Amounts of Capital for the Three Enterprises

Amount of Capital used	*Addition to net income from last Rs 1,000*		
(Rs)	*Buffalo*	*Cow*	*Poultry*
1.000	1.300	1,190	1,200
2.000	1,000	950	1.100
3.000	900	900	800
4,000	800	850	500
Total Returns from Rs 4.000	4,000	3,890	3.600
Average Return per rupee invested	1.00	0.96	0.90

On the basis of the average returns, our advice to the farmer would be to have only buffalo because average returns per rupee here invested is ₹ 1.00 as compared to ₹ 0.96 from cow and Rs 0.90 from poultry. But this advice gives wrong information about combinations of the enterprise. Let us now apply the principle of added or marginal returns for each addition of resource.

According to this principle the first ₹ 1,000 of his ₹ 4,000 should be invested in Buffalo because added net returns are high (₹ 1,300) from this enterprise, the second ₹ 1,000 should be invested in poultry because it adds ₹ 1,200 to net returns and third ₹ 1,000 of ₹ 4,000 on cow and last ₹ 1,000 of ₹ 4,000 again in poultry. With the help of equimarginal principle, we add ₹ 1,300 + 1,200 + 1,190 + 1,100 to the net returns. Thus total returns with the use of this principle are ₹ 4,790. Any other way of investment will result in lesser total income from these three enterprises.'

The principle is also known as opportunity cost principle because it considers the value of one enterprise sacrificed as a cost in selecting another enterprise. In our example ₹ 1,000 from buffalo will be opportunity cost, if the farmer decided to invest more on cow enterprise. Because if he decides to invest more on cow enterprise, he has to sacrifice an income of ₹ 1,000 from buffalo.

2.1.4. Enterprise Combination

By tradition, the Indian farmer produces a number of products on the farm. It includes various types of crops, few milch cattle and fruit trees. The availability of more than one enterprise on the farm poses an important management question to the farmer—what combination of different enterprises to be followed for maximum profit? The problem is to determine a suitable combination of products for a given outlay of resources. In other words, how the resources can be best utilized by having a suitable combination. The combination of enterprises is influenced by the relationships that exist between and among enterprises. Following relationships are generally found.*

Competitive enterprises are those, which compete mutually for the use of resources, *i.e.,* land, labour, capital etc. The increase in one enterprise results in a decrease in other enterprise. An example is buffalo and cow farming, because both enterprises require similar types of resources.

Supplementary enterprises. Some enterprises supplement each other, yet do not add directly to the production of each other. A small number (two-three) milch animals or a small poultry unit is supplementary to crop production. Similarly a few goats or sheep are also supplementary enterprises to crops. As no competition is involved, supplementary enterprises can be selected on the basis of net marginal returns of each individual enterprise. The inclusion of supplementary enterprises does not necessitate an increase or decrease of another enterprise, but will facilitate the utilization of resources left idle. A supplementary enterprise will become competitive if it is expanded beyond a limit. For example, if the number of milch animals keeps on increasing, the dairying will compete for the resources.

Complementary enterprises. Some enterprises complement each other, *i.e.,* one adds to the production of other. Examples are leguminous and cereal crops. The former increase the yield of the latter by adding nitrogen and organic matter to the soil even as they are harvested from the field. Livestock dung and urine can be used to increase the yield of-crops. Just like supplementary enterprise, they also become competitive if these enterprises extended beyond a limit. The selection of complementary enterprises depends on their total net returns, taken together, *i.e.,* crop rotation rather than individual crops.

In deciding enterprise combination, the benefit of complementary and supplementary relationship should be fully exploited. Therefore, the problem is only in selecting competitive enterprises.

Procedure of Combining Two Enterprises

If the farmer has adequate resources at his disposal, he can adopt as many enterprises as he desires. He will grow all the combinations of crops and livestock, which suit his physical resources. In such circumstances he will be guided by the simple economic principle, added cost = added returns (MC – MR). But this rarely happens in reality because majority of the farmers have limited resources at their disposal and number of opportunities open to them are always many. In such circumstances, they have to make selection of the enterprises rationally. Similar to cost minimization problem, here also farmer is guided by the following three considerations :

(*i*) The marginal rate of substitution of one enterprise for another,

(*ii*) The prices of the products, and

(*iii*) The cost of production of the competing products. If the cost of production of the two enterprises is same, the first two steps will guide the farmer.

Enterprises, which are competitive for use of resources, may substitute for each other either at increasing or at constant rate. The marginal rate of substitution of one enterprise for another refers to the amount of one enterprise displaced as another is added. Enterprises substitute at increasing rate when an increase in one requires larger and larger reduction in the other. They

*For details *see* Heody E.O.; *Economics of agricultural production and resource use.* Prietice Hall Inc. New York 1961 pp. 201:234.

substitute at constant rates when uniform increase in one requires a uniform reduction in the other. Increasing rate of substitution is always found when diminishing marginal productivity takes place in producing either or both of the two competing products. Constant rates of substitution are always found when constant marginal productivity prevails in the production of the product. In such cases either of the enterprise will give maximum returns to the farmer. In case of increasing rate of substitution one has to find out optimum combination of the enterprise. Such a relationship is shown in Table 2.6.

It is evident from Table 2.6 that the farmer has in all, facilities to maintain just 10 milch animals. The possibilities open to him range from to have no buffalo and ten crossbred cows, and to have ten buffaloes and no crossbred cow or any combination of the both but having ten milch animals in all. Milk yield, which the farmer will get from the various combinations, is given in the second and the third columns of the Table. To know the combination of the buffaloes and crossbred cows, which will give maximum return, following estimates will be required:

(*i*) Determination of marginal rate of product substitution (MRPS) for another, *i.e.*, buffaloes for crossbred cows (Y_1 for Y_2) which is equal to, $\Delta Y_2/\Delta Y_1$.

(*ii*) Estimation of price ratio PY_1/PY_2,

(*iii*) Determination of equilibrium between two :

$$\frac{\Delta Y_2}{\Delta Y_1} = \frac{PY_1}{PY_2}$$

A simpler way to know the maximum profitable combination of the enterprises is to compute total returns for each possible output and then select the combination, which is giving maximum returns. When price of the buffalo milk is ₹ 2.00 per litre and cow milk is ₹ 1.00 per litre it is profitable to have the third combination, *i.e.*, 2 buffaloes and 8 crossbred cows. From this combination farmer will get maximum return. Next combination also gives same returns. So he may or may not go for the next combination. Here also MRPS is equal to the price ratio $\Delta Y_2/\Delta Y_1 = PY_1/PY_2$. At this point MRPS of Y_1 for Y_2 is equal to (1200/600) = 2 and price ratio is also (2/1) = 2. Thus third combination will give maximum returns. But as the prices of the product changes, the most profitable (optimum) combination of enterprises also changes. When prices of the cow milk are raised from ₹ 1.00 to ₹ 1.25 per litre and prices of the buffalo milk remain the same (₹ 2.00 per litre) it will be profitable to have 9 cross-bred cows and one buffalo. Similarly if the prices of the cow milk are held constant and buffalo milk are raised the price ratio will increase and it will be profitable to the farmer to have more buffaloes than crossbred cows. Thus prices of the products will affect the combination of the enterprises, if the cost of raising is held constant. If the cost of production is not constant better to consider net returns than gross returns.

2.1.5. Cost Principle

Nothing is more important in farming than the relationship of cost and income and most farmers keep in mind the cost of producing additional unit of output. At given level of prices, a farmer can increase his income from his dairy farm in two ways, *i.e.*, (*i*) by increasing production and/ or (*ii*) by reducing cost of production. In a competitive market as is generally found in case of farm products, prices are not under the control of the individual producer because there are large number of producers. Therefore, second way of increasing income is most common in case of dairying. The cost here refers to the outlay of funds used in the production. Cost in dairy farming

TABLE 2.6. Possibilities of Combining buffalo and Crossbred Cows on a Farm (Hypothetical Data)

Production possibilities of combining 10 milch animals on a firm		*Milk Yield in litres*		*Marginal changes in production*		*Marginal rate of products substitution of Y_1 for Y_2*	*Gross returns (₹) at different prices*					
							Buffalo milk @ ₹ 2.00 per litre	*Cow milk @ ₹ 1.00 per litre*	*Total gross returns*	*Buffalo milk @ ₹ 2.00 per litre*	*Cow mlk @ ₹ 1.25 per litre*	*Total gross returns*
Buffalo	*Crossbred cow*	Y_1	Y_2	ΔY_1	ΔY_2	$\Delta Y_2/\Delta Y_1$						
0	10	0	12,000				0+	12,000	12,000	0+	15,000	15,000
1	9	800	10,800	800	1,200	1.50	1,600+	10,800	12,400	1,600+	13,500	15,100
2	8	1,500	9,600	700	1,200	1.71	3,000+	9,600	12,600	3,000+	12,000	15,000
3	7	2,100	8,400	600	1,200	2.00	4,200+	8,400	12,600	4,200+	10,500	14,700
4	6	2,600	7,200	500	1,200	2.40	5,200+	7,200	12,400	5,200+	9,000	14,200
5	5	3,000	6,000	400	1,200	3.00	6,000+	6,000	12,000	6,000+	7,500	13,500
6	4	3,300	4,800	300	1,200	4.00	6,600+	4,800	11,400	6,600+	6,000	12,600
7	3	3,500	3,600	200	1,200	6.00	7,000+	3,600	10,600	7,000+	4,500	15,100
8	2	3,600	2,400	100	1,200	12.00	7,200+	2,400	9,600	7,200+	3,000	10,200
9	1	3,650	1,200	50	1,200	24.00	7,300+	1,200	8,500	7,300+	1,500	8,800
10	0	3,700	0	50	1,200	24.00	7,400+	0	7,400	7,400+	0	7,400

can be divided into two main categories, *i.e.*, fixed and variable costs. Fixed costs as the term indicates are fixed, they are constant and do not change with output. They are a result of past commitments, of cost already 'sunk', or overhead costs. They are expenses that must be paid even if nothing is produced; they do not increase under higher yields and high production. It includes depreciation on buildings, equipments, and livestock (that part only which does not vary with their use but results from the passage of time) insurance charges, wages of the permanent labour, interest on fixed investment (buildings, equipment and livestock) and taxes. *Variable costs,* as the term indicates, vary with the level of production and with what is produced. In contrast to fixed cost, they have not been 'sunk', hence producer is free to determine whether or not expenses should be incurred. This cost does not occur if producer produces nothing. Variable costs includes such inputs as feeds and fodder, veterinary expenses, casual labour wages, repair and maintenance of buildings and equipments, interest on working capital etc.

In a short period, it will be profitable to produce whenever gross returns (production price) are greater than variable costs. Since fixed costs do not change, even if there is no production, farmer will go on producing as long as returns are sufficient to cover variable costs of the livestock enterprises. In the long period, however, returns from livestock enterprises must be equal to variable plus fixed costs, i.e. total costs. Once these costs are covered, one will be little interested in the average or per unit cost of production. Once a farmer has determined that production is profitable, the next question is how much in the way of resources it will pay to use. The principle of equi-marginal cost is the proper economic principle to answer this question. Net returns or income will be highest when marginal cost (MC) is equal to marginal returns (MR).

In terms of diminishing returns principle, the question to be answered normally in dairy farming is "How far should we produce?" We seldom try to produce with the lowest per unit costs. The point of lowest average cost is not necessarily the point of maximum profit. We are willing to increase per unit cost if added (marginal) return is greater than the added cost (MC). This is apparent from Table 2.7, according to which 3 kg of concentrate is the point of lowest average cost (AC), *i.e.*, ₹ 1.05. But it is profitable to feed 5 kg. of concentrates because added returns (₹ 1.54) is more than the added cost ₹ 1.50. The net returns are also highest (₹ 9.96) at this point. Beyond this level of maximum net revenue the added cost will be more than the added returns. The maximum profit from an enterprise does not necessarily occur when milk yield is highest. Milk yield reaches a maximum of 12.71 kg at 8 kg concentrate feeding. The net returns at this point are ₹ 7.14. Thus maximum yield is not always consistent with income maximization.

A farmer generally faces two types of situations: (*i*) when selling prices covers total costs, and (*ii*) when selling prices are less than total cost but more than average variable costs. In a short period, since variable costs are important in decision-making, variable cost must be covered by the selling price if milk production is to be profitable in short period. Even if the selling price or MR does not cover total average costs of production, it is economical as long as MR is more than MC. In the long period, however, all costs become variable costs; therefore, unless all fixed costs are covered over long period, the farmer will either switch over to other enterprises or suffer a continuous loss in income. Change in input prices cause changes in production in the opposite direction. A decrease in input prices will result in higher output. Similarly farmer will increase production, if input prices remain constant and output prices are raised. Thus farmer must have production incentive, which is possible either by increasing output prices or decreasing input prices.

TABLE 2.7. Production and Cost Relationship in Milk Production Business

Input (Concen-trate kg)	Total Milk Produc-tion (Lit.) (Y)	Additional Milk Produc-tion (lit)	Total Variable Cost ₹ 1.50 per kg (TVC)	Total Fixed Cost (TFC)	Total Cost (TC) = (TFC + TVC)	Average Variable Cost (TVC/Y) = (AVC)	Average Fixed Cost (TFC/Y) = (ATC)	Average Total Cost (TC/Y)	Change Variable Cost (VC)	Marginal Cost (DVC/sY) (MC)	Total Revenue (Y.PY) PY = ₹ 2.00 per lit. t(TR)	Net Revenue (TR–TC) + (NR)	Marginal of added Returns (V.PY) = (MR)
1	2	3	4	5	6	7	8	9	10	11	12	13	14
0	5.50		0	6.28	6.28	—	1.14	1.14			11.14	4.86	
1	7.28	1.78	1.50	6.28	7.78	0.20	0.86	1.07	1.50	0.84	14.56	6.78	3.56
2	8.80	1.52	3.00	6.28	9.28	0.34	0.71	1.05	1.50	0.99	17.60	8.32	3.04
3	10.07	1.27	4.50	6.28	10.78	0.44	0.62	1.07	1.50	1.18	20.14	9.36	2.54
4	11.10	1.03	6.00	6.28	12.28	0.54	0.57	1.11	1.50	1.46	22.20	9.40	2.06
5	11.87	0.77	7.50	6.28	13.78	0.63	0.53	1.16	1.50	1.95	23.74	9.96	1.54
6	12.40	0.53	9.00	6.28	15.28	0.73	0.51	1.23	1.50	2.03	24.80	9.52	1.06
7	12.68	0.28	10.50	6.28	16.78	0.83	0.49	1.32	1.50	5.35	25.36	8.56	0.56
8	12.71	0.03	12.00	6.28	18.28	0.94	0.49	1.44	1.50	5.00	25.42	7.14	0.06
9	12.50	0.21	13.50	6.28	19.78	1.08	0.50	1.58	1.50		25.00	5.22	–0.42
10	12.03	–0.47	15.00	6.28	21.28	1.25	0.52	1.76	1.50		24.06	2.78	–0.94

Relationship between Total Fixed Costs, Total Variable Costs and Total Costs

For studying relationship between these costs and principle of diminishing returns the data presented in Table 2.7 will be considered.

Total fixed costs are shown in 5th column of Table 2.7. These costs are the same for all the levels of milk production.

Total variable costs (TVC) are computed by multiplying the amount of variable input used by the per unit price of input. TVC is zero when milk output is zero. It is given in column 4 of Table 2.7. TVC increases as milk output increases.

Total costs (TC) are the sum of total variable costs and total fixed costs (TVC + TFC) and are given in column 6 of Table 2.7.

Relationship between Average Fixed Costs, Average Variable Cost and Average Total Cost

The average of various costs are presented in column 7, 8 and 9 of Table 2.7.

Average fixed cost (AFC) is estimated by dividing total fixed cost by quantity of milk produced (TFC/Y). It is the fixed cost per unit of output. AFC will vary with the output. As output increases AFC decreases, since milk output changes and fixed costs remains the same.

AFC decreases per unit increase of milk output. When output is zero, AFC cannot be estimated because division by zero is not possible.

Average variable cost (AVC) is worked out by dividing TVC by the milk output. AVC varies with the level of milk production. AVC is inversely related to average product (AP) *i.e.* when AP decreases AVC increases. AVC declines initially due to economies of large-scale production and then rises due to diseconomies of the scale of production.

Average total cost (ATC) can be computed in two ways : (*i*) by dividing total cost by total milk yield, and (*ii*) by adding AVC + AFC. ATC decreases as milk output increases; attains a minimum and increases thereafter. The spreading of TFC among an increasing quantity of milk causes initial decrease.

Marginal cost (MC) is the change in total cost in response to a unit increase in milk output. It is change in cost associated with an increase or decrease of an additional quantity of milk *i.e.*, $\Delta TC/\Delta Y$. It is given in column 11 of Table 2.7. As milk output increases MC first decreases due to efficient use of variable input and then increases due to inefficient use of variable input.

Procedure for Estimation of Cost Parameters

The method of estimating the different parameters of cost used in the economic studies to estimate the cost of milk production is as follows:

1. ***Maintenance cost of the animal*** includes cost of feed, hired and family labour, depreciation on animal, depreciation on capital investment, veterinary aid, interest on capital invested and working capital, miscellaneous recurring expenditure less income from dung.

2. ***Feeding cost*** includes purchased feed as well as home grown feed. Homegrown feeds are valued at the market price, though they can actually be cheaper.

3. ***Cost of labour*** is actual time spent in the care of animals and apportioned for each animal. The cost of the hired labour is based on the actual wage paid to the labour. Family labour cost is estimated on the basis of prevalent wage rate for the permanent labour.

4. ***Depreciation on animal.*** There are various methods to estimate the depreciation. Straight-line method is most simple and commonly used in the economic studies. It is assumed that the value of an animal appreciates up to the age of the 5 years in the proportion of 1:3:5:7:10 and depreciation is charged for the remaining life of the animal.

5. ***Depreciation of cattle shed*** is charged on the value of the cattle shed @ 2 per cent for the *pucca* (permanent) shed and @ 5 per cent for the *kuchha* (temporary) shed per year. The cost is apportioned among different categories of animals on the basis of the value of the animal.

Depreciation on equipment and machineries is charged @ 10 % per annum

6. ***Cost of veterinary aid*** Actual expenses incurred over the animal are charged.

7. ***Interest on capital*** is charged on the fixed capital assets and the initial value of the animal at prevalent bank rate. In case of working capital the rate of interest is also charged at prevalent bank rate but it is charged for half of the year only. In case of animals in milk, no interest is charged on the working capital because the farmer derives regular income from the sale of milk.

8. ***Miscellaneous expenses*** include recurring expenditure on rental value of land on which cattle shed is built, cost of repairs, service charges, chains, ropes, etc.

9. ***Quantity of dung produced by*** the animal must be valued on the basis of market rates. The income from dung is apportioned at the ratio 1 : 2 : 3 : 4 for 3 to 6 month old calf, 6-12 month old calf, heifers and adult animals, respectively.

The estimates for the gross cost are obtained by adding the cost of all components, *viz.* feed labour, depreciation on animal and cattle shed, interest on fixed investment. The net cost is estimated by deducting income from dung from the gross cost.

2.1.6. Principle of Comparative Advantages

This principle is actually an extension and application of the principle of specialization and diversification. The physical and economic conditions influencing production vary from country to country, region to region and from farm to farm. Each farmer utilizes his resources in the production of those items which have the maximum advantages, the fewest disadvantages and which can be produced at the lowest relative cost. Physical factors such as soil type, topography, climate, and distance from market; biological factors such as diseases, insects, economic factors including yield per animal, output per man, and demand for the product and social factors play a role in determining comparative advantage. When a combination of favourable factors extends over a substantial area, we begin to have regional specialization with the producers concentrating largely on the production of simple or relatively few commodities. Other needed items are purchased from other regions, which enjoy cost advantages for those items. There are two types of advantages in raising farm products :

(1) Absolute advantages and (2) comparative or relative advantages. Absolute advantage is studied within the region and comparative advantage is studied between the regions. In spite of the comparative advantages, farmers in our country prefer to diversify the use of resources to minimize the risk and meet most of the family needs from the products grown on the farm.

2.1.7. Time Principle, Cost Compounding and Discounting

This principle deals with decisions over time. The time problem arises because certain resources last longer than one production season or period and portion of their cost become fixed. For example, a calf requires 3-4 years before it starts giving returns. Such problems can be solved either by compounding of the cost or discounting the income.

A cash investment grows over time due to compounding of interest charges. If ₹ 100.00 is put in a savings bank account with interest at 5 per cent per annum, it will amount to ₹ 127.63 by the end of five years, if the interest is not withdrawn or compound interest is charged. A cash investment made on a farm grows over a period in a similar manner.

In discounting procedure the present value of the future income is determined. The process is opposite of compounding. Using the following equation can derive this:

$$PV = \frac{1}{(1+r)^n}$$

where PV refers to the present value of the future amount, I is the future amount, r is the interest rate and n refers to the number of years into the future with which we are concerned. This equation is used when future income is discounted annually. If the income is discounted m times in a year, r should be divided by m and n multiplied by m. Application of the equation can be illustrated by determining the present value of ₹ 127.63 to be received at the end of 5 years assuming the discounting rate as 5 per cent per annum. Substituting these values in the above equation :

$$PV = \frac{(127.63)}{(1+0.05)^5} = \frac{127.63}{1.2763} = 100$$

Farm investments, machinery and equipment, breeding stock (milk cows or buffaloes etc.) are examples of costs. Since cost increases as a result of opportunity cost accumulation, the following equation may be used for compounding interest :

$$S = s(1 + i)^n$$

where, S represents the sum at the end of n period, s the amount that is invested for n period, i is the interest rate. In case compounding is done semi-annually, 2 will divide the interest rate, and 2 *multiply n*. This equation is valid for a lump sum investment.

To find out the real cost of rearing dairy heifers data from the National Dairy Research Institute, Karnal is used with some modification (Table 2.8). The data are adjusted for a period of 3 years, cost of the remaining period is left uncompounded. The rate of interest used for the compounding is assumed to be 8 per cent per annum. The value of the figure $(1 + i)^n$ can be obtained from any standard book on finance.*

Using the above equation, the real cost of rearing Tharparkar heifer is estimated as follows:

$$S = 796\ (1 + 0.08)^3 + 591\ (1+ 0.08)^2 + 816(1 + 0.08) + 491.$$

By substituting the value of $(1 + i)^n$ from the tabulated values (*see* any book on finance) the equation will be as follows:

$$S = 796(1.2597) + 591(1.1664) + 816(1.0800) + 491.$$

$$S = 1002.72 + 689.34 + 881.20 + 491 = 3059.34$$

The real cost of rearing a Tharparkar calf will be Rs 3059.34, and not Rs 2694 as shown in the column of total cost. Similarly, real rearing costs of heifers of other breeds are also

Nelson, A.G. and W.G. Murray: *Agricultural Finance,* Iowa State University Press, Ames, Iowa (U.S.A.), fifth edition. Appendix Table 1, pp. 512-523.

estimated. These values will help the farmer in deciding whether he should rear a heifer himself or purchase a milch animal near the time of calving.

TABLE 2.8. Cost (Rs.) of Rearing Dairy Heifers to First Calving for Different Breeds of Cattle and Buffalo

Breed	*Cost in different years*				*Total Cost*	*Estimated real cost (using compound method)*
	I	*II*	*III*	*Remaining period*		
Tharparkar	796	591	816	491	2,694	3059.34
Shaiwal	797	589	804	453	2,643	3012.31
Red Sindhi	796	590	756	356	2,274	2863.38
Cross-bred cow	910	668	696	—	2,274	2677.17
Buffalo	717	583	876	480	2,656	2,879.69

Source : *Studies in Dairy Economics and Statistics* (1962-72), Vol. I. Division of Dairy Economics and Statistics, National Dairy Research Institute, Karnal, pp 17-25.

Capital can be available with farmers either in a limited quantity or in unlimited quantity. The relevant rate for discounting is the market interest rate in case of farmers having unlimited capital. It represents his opportunity cost in using his funds. The farmer has exploited all the available opportunities of the investment and the only alternative left to him is to deposit the money in the bank or lend it to other farmers. The same rate of discounting could not be used for farmers having limited capital. The appropriate discounting rate will depend on the opportunity returns of using funds, in different enterprises.*

Suppose a farmer plans to start dairying as main enterprise. He can keep the same number of buffaloes and produce the same amount of milk with same feed and labour if he builds a cattle shed costing ₹ 6,000 which will last 50 years (permanent), or if he builds one costing ₹ 5,000 which will last only 25 years (semi-permanent). If he replaces the semi-permanent one at the end of 25 years, his total cost for the cattle shed will be ₹ 10,000 as compared to ₹ 6,000 for the permanent one, which represents the cheapest method of housing his buffaloes and producing milk.

If he has adequate capital and if discounting is done at market interest rate of 5 per cent, his comparison is as follows: he can (1) invest ₹ 6,000 in permanent cattle shed and need not have to replace it or (2) invest ₹ 5,000 in the cattle shed lasting 25 years only and keep aside ₹ 1,475 at 5 per cent interest for 25 years. A sum of ₹ 1,475 kept aside at 5 per cent interest rate will provide him exactly ₹ 5,000 to replace his semi-permanent cattle shed at the end of 25 years. Therefore, the comparison is between ₹ 6,000 for the permanent cattle shed and ₹ 5,000 + ₹ 1,475 or a total of ₹ 6,475 for the semi-permanent shed. The permanent cattle shed is cheaper for a farmer having abundance of capital at his disposal. We arrived at the ₹ 1,475 by using the discounting procedure stated earlier. To find out the present value which will discount to ₹ 5,000 in 25 years with an interest rate of 5 per cent [*i.e.*, $PV = 5000/(1 + 0.05)^{25}$] = ₹ 1,475, the amount which if kept aside at compound interest, will provide the amount for replacing short duration cattle shed in 25 years.

*Heady, E.O. and H.R. Jonson. *Farm Management Economics*, Printice Hall of India (Pvt) Ltd., New Delhi, 1954, pp. 628-629.

Now suppose the farmer has limited amount which can earn 25 per cent returns elsewhere in his business. Then the farmer must compare ₹ 6,000 of the permanent cattle shed with ₹ 5,000 for the semi-permanent cattle shed plus the amount invested at possible current earning rate of 25 per cent in his business which will replace the semi-permanent cattle shed in 25 years. The equation will now be $PV = 5000/(1+0.25)^{25}$ = ₹ 20.00

This is the amount, which if the farmer invests in the business and earns 25 per cent returns will provide ₹ 5,000 for the new cattle shed in 25 years. ₹ 5,000 for one semi permanent cattle shed now plus ₹ 20,000 to replace the second one in 25 years, is less than ₹ 6,000 for the 50 years permanent cattle shed. Therefore, he has to opt for the semi permanent cattle shed and spend balance in the business.

2.2. DAIRY ENTERPRISE PLANNING AND BUDGETING

Successful management of dairy farming is planning and putting into operation a combination of resources such as land, labour, capital and management to get maximum income from the enterprise. Dairy farm management includes two main divisions : *organization* and *operation.*

Organization is planning the kind of animals that are to be maintained on the farm. Organization is a foundation upon which the day-to-day and year-to-year operations are built Organization and operation both are highly important in dairy industry. Excellent operations of a poor plan of organization will result in failure of the business. Similarly an excellent orgainzation plan, if implemented poorly, will also result in the failure of the business.

Any scheme of action prepared in advance is a plan. A dairy enterprise plan, thus, is a scheme for the operation and organization of a dairy business. Dairy plan, thus, may relate to any contemplated change in the method or practice followed on the farm or it may relate to complete reorganization of the business. Dairy budgets may be prepared merely to compare the cost and returns resulting from the adoption of new methods or the use of new equipment, but they are also prepared to determine the prospective expenses and income resulting from complete change in the choice and combination of livestock enterprises and a complete reallocation of the resources on the dairy farm.

2.2.1. Why a Dairy Plan ?

Dairy planning is a process of observations, appraisal and analysis of judging the merits of new and old practices or new and old plans and then deciding which should be followed in the near future. This process could be learned much better if the results of the thinking and reasoning are noted down on a piece of paper in a systematic and organized way. Thus, a written dairy plan must show the animals to be raised on the farm, the practices to be followed, the plan for the use of labour, the investment to be made in equipment, building and similar details about the use of other resources. It is better to commit mistakes on the paper rather than in practice.

Planning is of great importance to the farmers as well as to the national planners. Here we are mainly concerned with the planning of the individual farm. Dairy farm planning helps the farmers in the following ways :

(*i*) It helps to find out a better combination of livestock and crops.

(*ii*) It helps in selecting the most profitable practices, which are suited to their local conditions.

(*iii*) It helps in getting the jobs to be completed in time so that they do not conflict with one another.

(*iv*) It helps in avoiding wastage of resources. There is wastage of resources if they are not properly and efficiently used.

(*v*) It gives a clear idea to the farmers regarding yield, production and income, which he can reasonably expect from the livestock enterprises.

(*vi*) It helps in estimating expected requirements of various inputs such as fodder, feeds etc.

(*vii*) It helps in estimating the capital requirements of the business and ultimately in assessing the credit needs. This also provides the guidelines when he should borrow and when he should return the borrowed funds as he will have a clear idea regarding expenses and income of each month related to his dairy business.

(*viii*) It furnishes a guide to go by and standards to compare the progress made from the past and the farmer can look with confidence towards future.

2.2.2. Steps in Planning

Step I. Survey of Local Conditions

One should gather all the relevant information from secondary sources such as survey reports, livestock census, meteorological records etc. If there is any experimental station, information regarding various types of livestock and other inputs should be studied carefully to acquaint oneself with the practices of the region.

Step II. Inventory of the Resources

The main resources, which affect returns from the dairy enterprise, are (*a*) land, (*b*) water, (*c*) labour, (*d*) working capital, (*e*) machinery and equipment, (*f*) farm livestock and (*g*) managerial ability.

(*a*) Land. Land is required to construct cattle shed-and grow feeds and forage crops on the farm. For the construction of the cattle shed, land must be free from water logging, salinity or alkalinity. Otherwise building constructed on it will depreciate earlier. Information regarding soil type, irrigation and drainage facilities should also be acquired. All this information will be required in deciding crops to be grown to meet the demand of feeds and fodder for the dairy animals.

(*b*) Water. Availability of good quality water is more important on the dairy farm than those on an arable farm. Requirements of the water must be assessed for the livestock and farm crops separately. Information is needed regarding the source of water, quality of water, availability of the water in summer months etc. Information regarding area commanded is very much important. This restriction should be carefully examined in the alternate plans. If new plan requires more water than what is available, it will become an unrealistic plan.

(*c*) Labour. Labour resources are of two kinds: Fixed and Variable. Fixed labour is labour, which is used on farm throughout the year; it includes both permanent workers and family labour. Variable labour is labour used on seasonal basis, *i.e.*, casual labour that can be hired when needed.

(*d*) Working capital or operating capital. This refers to the capital needed for the purchase of inputs such as feeds and fodders, seed, manure and fertilizers, chemicals needed for the fodder crops and medicines for livestock. The information is required regarding how much money is available with the farmer for this investment and how much additional amount he can borrow from outside, if his own funds are inadequate.

(*e*) **Machinery and equipments.** Information is needed regarding types of machinery and equipment available along with their capacity. Sometimes machinery and equipment available with the farmers are just suited to his present plan. Any change in dairy plan may not suit him because he does not have additional capacity of the machinery and equipment.

(*f*) **Farm animals.** Information will be needed regarding kinds of animals maintained on the farm, breed, age, value, veterinary facilities available etc.

(*g*) **Management ability.** Obtaining information on management is probably the most difficult step in dairy enterprise planning because there is no precise method of measuring a person's management ability, as the plan is to be made to fit the man and not to the paper. A simple plan of organization that can be put into operation with little mental efforts may succeed better financially for some farmers than a complicated plan, which requires many wise operational decisions. However, the farmer, who has the will and ability to do the latter, can get more profit. So a good rule is : "make the plan to fit the man".

Step III. Selection of Expected Prices

The planning process involves estimating the future. Hence the prices which are used in planning must be based on expectations of future. Expectations should be established for all products, which appear to merit considerations in planning. One guide to future price is the past trend. Prices generally move upward, downward or remain stable and hence current year often tells something about next year. However, the next year can also be far different from the current year. But it is not necessary that our expectations should be entirely correct but error must be minimised. It is always safe to use conservative prices rather than inflated prices with respect to the price of output.

Step IV. Determining the Most Profitable Farming (Animal + Crop) System

In India farmers generally select herd size, which is best suited to their cropping plan. This is because of the reason that they maintain dairy animals to meet their family requirements. However, with the improvement in milk yield, dairy farming in India in near future will be adopted on commercial basis. In such a situation planning will be one of the important tools for extending it.

Step V. Selecting Most Profitable Cropping System

It is important to know the cost and returns of the important crops grown in the region for food, feeds and fodder. Maximum profit can be attained when, after meeting the feed and fodder requirements of the livestock, farmer devotes major part of the area to high profitable crops which could be grown on the farm.

Step VI. Estimating New Capital Requirements

So far the discussion was about crop and livestock system, which the farmer should follow. Capital is one of the important constraints in India. Even with the existing technology, the income of the farmer is less than it can be achieved by providing additional capital to meet the requirements of good quality of inputs. Capital becomes further more restrained when new technology is adopted.

Step VII. Compare the New Plan with the Old Plan

Net income, labour income and management returns of both old and new livestock plans should be compared. This comparison is necessary, so as to decide whether one should follow new plan or continue with his old plan.

So far we have discussed various steps involved in livestock planning. The most commonly used tools for livestock planning are budgeting and linear programming. More detailed discussion of these techniques is beyond the scope of this book. However, we will discuss budgeting in brief.

2.2.3. Budgeting What is Budgeting ?

The two words *farm planning* and *budgeting* are closely interlinked. A budget is used to evaluate and a plan is the basis for a budget. Thus it is possible to have a plan without a budget, but not a budget without a plan. Because of this relationship the term budgeting has come to be referred to the combined process of planning and budgeting. A dairy enterprise budget is a plan for future use of resources; it includes the plan for the system of farming for the coming year or period of years. It shows the type of dairy animals to be kept, feeds and fodder crops to be grown and estimated production receipt and expenses from various resources. More specifically budgets are financial statements of estimated receipt and expenses from the production of one or more products for a selected period of time. The primary purpose of a budget is to anticipate and control expenses and to allocate resources so as to maximize income.

A budget may be partial or total. A partial budget enables the farmer to evaluate the desirability of a change in one enterprise or a single practice. Partial budgeting is thus a comparative study of cost and returns analysis from a change or part of dairy business. This change can be made through careful selection of alternate methods of production or practices or relative profitability and does not affect total organization vitally. It helps in decision-making process whenever small changes in the existing organization are contemplated.

A total budget provides a basis for reorganization of the entire system in order to increase returns from the resources employed in the production and, ultimately income of the farmer. Therefore, all aspects of the business organization are considered at the same time. The long run complete budget provides the general framework within which short run or year-to-year plan is adjusted.

Example. Would it be more economical to have crossbred cows or buffaloes on a farm? One can answer this question easily by preparing partial budget for both the enterprise (Table 2.9).

TABLE 2.9. Enterprise Budget for a Crossbred Cow and Buffalo

Items	*Buffalo*	*Crossbred cow*
1. Feed (₹)	2,506.31	1,893.88
2. Depreciation or replacement value (₹)	375.00	450.00
3. Labour (₹)	833.87	793.22
4. Veterinary charges (₹)	3.80	6.70
Gross Cost	3,718.98	3,143.80
Income from dung	36.50	20.00
Net cost	3,682.48	2,123.80
Milk yiled in a year (lit)	2,187.77	2,786.90
Value of the milk product @ ₹ 2.00 per littre for buffalo milk and @ ₹ 1.50 per litre for cow milk *i.e.*, gross income	4,375.55	4,244.23
Net income (₹)	693.07	2,120.43

The net income from crossbreds in this case is ₹ 1,427.36 more than that from buffaloes. Recent studies in Farrukhabad district of Uttar Pradesh have also shown that the cost of maintaining the cross bred cow was less compared to buffalo (Atul Chandra and Agarwal, 2000). Therefore, it is profitable for the farmer to keep crossbred cow on the farm rather than buffalo. By partial budgeting technique similar comparisons in other aspects, like economics of farm rearing versus purchasing herd replacement etc. can be made.

It is not necessary that one should make detailed analysis for partial budget. It can easily be estimated by using the following information :

Added costs (₹)	....
Reduced income (₹)	....
Sub Total	....
Reduced costs	....
Added income	....
Sub Total	....
Change in net income	....

It is not also necessary to consider all costs. One can proceed directly if he knows items of costs and returns, which do change, as those, which remain unchanged, will not affect the net change in income. Thus partial budgeting may be used for major farm adjustment, as well as current operational problems. But as the number of items that are involved in an analysis increases the partial budget will approach a complete budget.

Total or Complete Budgeting

Complete budgeting is used when a complete reorganization of the farm is contemplated. Such reorganization generally occurs infrequently and the decisions have such long run effect that it is important to use best possible tools. The main purpose of this type of budgeting is to compare profitability of different kinds of organizations. The present organization generally constitute a base for the comparison. If adequate information has been kept, it may not be necessary to prepare a new budget for the existing organization. In case a planner is starting from scratch he will not have an existing organization from which to work up. The best thing under such a situation is to prepare a tentative plan on long-term basis and divide the plan into annual plans. The emphasis in the annual plans is on operational improvements, while the long term plan takes care of structural improvements also, which are of a permanent type. Total or complete budgeting brings about progressive change in income. It takes an entire view of the dairy farm as a whole. It considers complementary, supplementary, and competitive relationships between enterprises. This method allows substitution between resources and avoids any vital part of organization being left out of consideration. Limitations are that it requires more time and efforts and more basic data in accurate form. To illustrate complete budgeting an example is worked out. The assumption regarding costs of inputs, parameters etc. in this plan are based on published literature. One can substitute this with latest data from his own locality for working out the total budgeting of his dairy enterprise. Assumptions in the present case are as follows :

Atul Chandra and Agarwal. S.B 2000. Cost and returns from milk production in Farrukhabad district of Uttar Pradesh. Indian-Journal-of-Dairy-Science. 53(4): 310-316.

1. Lactation yield in 305 days	2,500 litres in 1st year and 3,000 in subsequent years
2. Intercalving-period	13 to 15 months
3. Age at first calving	2½ years
4. Cost of cultivation of green fodder	@ ₹ 6,000/hectare
5. Dry fodder	To be purchased @ ₹ 1500/tonne and to be fed @ 2 kg/adult unit
6. Labour charges	@ ₹ 6 per day
7. Concentrate cost	@ ₹ 8,000/tone to be fed @ 2 kg/day
8. Price of milk	@ ₹ 10/litre for cow milk and ₹/litre for buffalo milk
9. Appreciation value of female	@ ₹ 500/ each
Appreciation value of male calf	@ ₹ 100/ each
10. Value of dung	@ 12 tonnes/adult unit and value @ ₹ 50/tone
11. Feeding green fodder	Produced on farm and fed @ 40 kg/adult unit
12. Insurance	Crossbred cow ₹ 800/year and ₹ 600/year for buffalo
13. Veterinary care	₹ 200 animal/year
14. Depreciation	@ 20% year for animals; 10% per year for equipment; 20% per year for building
15. Cost of Construction	@ ₹ 70 per sq. ft; Cow shed and Store cost ₹ 5,000 per animal

2.2.4. Budget Estimates for Starting Dairy Farm

Task. Budget to be worked for a commercial dairy farm with 100 crossbred cows and followers to be established nearby Gurgaon city. For this purpose 17 hectares of land with irrigation facilities will be acquired. There is good market for the milk.

Purchase of high yielding animals. Since it is desired to start a dairy farm with 100 crossbred cows along with followers, it is essential that good quality cows are purchased in order to get profit from such an enterprise. To begin with, freshly calved cows at an average cost of Rs 1,500 each are purchased.

Herd strength. Probable herd strength for five year period along with adult units in each year is given in Table 2.10. As the crossbred cows are good breeders, it is possible to keep them in milk for 10 months and dry for 2 months in an year on an average. Thus 75 per cent will be in milk at any time of the year. Male calves are reared up to the age of one year and sold and female calves are reared for replacement purpose. Artificial insemination is practiced in herd by getting good quality semen with best germplasm to reduce the cost of maintaining good quality bulls.

During subsequent years of plan, few fresh cows are added by fresh purchase and also replaced from raised heifers thereby bringing higher productivity in the herd. The herd strength of animals in milk is not increased up to five years but can be increased afterwards to get still more profit. Surplus heifers and old cows are sold.

Feeding. One of the most important principles for economic feeding of dairy cattle is that they should be fed with sufficient quantity of nutritious and palatable fodder. Adult cows are

TABLE 2.10. Strength of Herd 1st year to 5th year

Year		Crossbred cows		Suckling		1 to 2 years age		2 to 3 years age		Bullocks	Total	Remarks
		In milk	Dry	♂	♀	♂	♀	♂	♀			
1st year	Strength	100	—	48	48	—	—	—	—	4	200	(1) 5 per cent mortality in Calves
	Adult Units	100	—	16	16	—	—	—	—	4	136	(2) Cow = 1 Adult, Heifers 2 years and above = 2/3 adult, Calf 1 to 2 years = 1/2 Adult, Below 1 year = 1/3 adult
2nd year	Strength	75	25	38	37	—	48	—	—	4	227	
	Adult Units	75	25	13	12	—	24	—	—	4	153	(3) 75 per cent in milk, 25 per cent usually dry, 48 male calves will be sold at the age of one year.
3rd year	Strength	75	25	38	37	—	36	—	48	4	157	
	Adult Units	75	25	13	12	—	18	—	32	4	179	(4) 20 cows will be purchased. 20 cows will be sold. 32 calves will be sold.
4th year	Strength	75	25	28	37	—	36	—	36	4	251	(5) 20 poor cows will be sold 20 heifers will be used for replacement 38 male calves will be sold. 28 heifers will be sold.
	Adult Units	75	25	13	12	—	18	—	24	4	171	
5th year	Strength	75	25	33	37	—	36	—	36	4	251	(6) 20 poor cows will be sold. 20 heifers will be used for replacement 38 male calves will be sold. 16 heifers will be sold.
	Adult Units	75	25	13	12	—	18	—	24	4	171	(7) After 5th year milch cows will be increased by replacement and herd milk average will be increased due to good care and management of farm-raised better heifers.

provided with 40 kg good quality green fodder produced on farm, along with 2 kg dry fodder/ day (purchased) to make up dry matter requirement. Among dry fodders fed are wheat straw *(Bhusa)* and Jowar/Bajra straw *(Karbi)*. They are given concentrates @ 1 kg/2.5 kg milk production.

Concentrates include grains, oil cakes, brans, meals, *churies* etc. Suitable mixtures depending upon the cost and availability will be prepared and fed as per requirement of various classes of animals. The cost of feeding concentrates is given in Table 2.11.

Farm. The enterprise begins with 17 hectares of purchased land under irrigation. A tractor of 25 HP will be needed to have intensive cropping pattern, speedy sowing, harvesting, handling etc. Two pairs of bullocks will also be kept for field operations. A tube-well with 10 H.P. motor will facilitate irrigation in the whole area. Out of 17 hectares, 15 hectares will be kept under fodder cultivation and 2 hectares under building, roads, channels, drains, residence and for future expansion. After 3rd year of plan, 4 more hectares of land will be added and put under cultivation. The land will be purchased @ ₹ 8,000/hectare. Crop rotations* to get maximum output/hectare and supply of green fodder round the year are given in Table 2.12. On an average, it is estimated to get fodder @ 132 tonnes/hectare and cost of cultivation is estimated @ ₹ 3,000/hectare (which will include manures, fertilizers, insecticides, pesticides, seeds, oil and lubricants etc). Yearly expenditure on cost of fodder cultivation is given in Table 2.13.

Housing. Expensive house is not necessary in the beginning. In order to protect animals from heat or cold, putting asbestos sheets supported by angle irons with facilities of feeding and watering in the shed with sufficient loafing space for each class of animals separately provides covered space. The cost per adult unit is considered @ ₹ 200 each. Besides this, implement shed cum storeroom for hay and labour quarters are also provided. Costs of building are given in Table 2.13.

Labour requirements. Even though the size of herd and area of land is quite big, it is necessary to have minimum number of labourers to carry out farm work as well as cattle unit chores very efficiently. A distinct division is to be made in allotment of both types of work. For this a unit of 15 labourers along with tractor driver is enough. After 3rd year of plan, when additional 2 hectare of land will be included, 5 more casual laboured are employed for a period of six months to carry out both the types of work. Out of these two labourers are provided with quarters on the farm so that they can look after emergency work. The duty of tractor driver is assigned for operational work of field as well as running of motor on tube well. If tractor is free, it can be utilized for custom work through which extra income will be available which is not taken into consideration in this plan. Two pairs of bullocks are also needed for day-to-day minor operations on the field. The details of labour requirement and cost of labour is given in Table 2.14.

Fixed capital. Fixed capital is required for the purchase of land, construction of buildings, and purchase of livestock and implements. The details are given in Table 2.13. A tractor of 25 H.P. along with driller, disc harrow and trolley is necessary. Besides this, a reaper, soil inversion plough, two *deshi* (indigenous) ploughs, wooden planks and a pneumatic tyre cart is a must as two pairs of bullocks are kept on the farm. Manually operated chaff cutter, few milking

---

* These rotations are mostly for the NW semi-arid parts of India. Farmers in other parts can formulate crop rotations suitable to their locality in consultation with local agricultural experts.

TABLE 2.11. Concentrate Requirement and Cost

Class of Animals	*Rate of feeding kg/day*	*Quantity per year in tones*	*I year*		*II year*		*III year*		*IV year*		*V year*	
			No. of animals	*Feed required (tones)*	*No. of animals*	*Feed required (tones)*	*No. of animals*	*Feed required (tones)*	*No. of animals*	*Feed required (tones)*	*No. of animals*	*Feed required (tones)*
I. Milch animals : Production	1 kg/2.5 kg milk for 305 days	0.91	100	91	75	68.25	75	68.25	75	68.25	75	68.25
II. 1. Pregnancy : Dry cows to be given fodder only except during the last 70 days of pregnancy period where 1 kg/day concentrate is given	1 kg/day for 70 days	0.07	25	1.75	25	1.75	25	1.75	25	1.75	25	1.75
2. Heifers	1 kg/day for 60 days	0.06	—	—	48	2.88	84	5.04	72	4.32	72	4.32
3. Bullocks	1 kg/day	0.365	4	0.75	4	0.73	4	0.73	4	0.73	4	0.73
4. Suckling calf starter	1.5 kg/day for	0.27	96	25.5	75	20.25	75	20.25	75	20.75	75	20.25
	Total			119.00		93.86		96.02		95.80		95.30
Cost of (₹) feeding @ ₹ 1000/tonnes				119000.00		93860.00		96020.00		95800.00		95300.00

TABLE 2.12. Plan for Fodder Cultivation

Sl. No.	*Crop Rotation*	*Hectareage under crop*	*Green fodder yield in tonnes*	*Remarks*
I.	1. Maize + Cowpea	1	600	1. The rotation and timings of sowing will be such that green fodder will available as per requirement of animals throughout the year.
	2. Maize + Cowpea + Teosinte	1	700	2. After 3rd year of plan, 2 hectares more land is added, so two more rotations will be added to give additional yield of green fodder to meet the increased strength of the herd.
	3. Berseem + Mustard	1	1,500	3. 300 to 400% intensity of crop rotation will be adopted so that three to four crops in a year could be taken from the same piece of land.
			2,800	
II.	1. Maize + Cowpea	1	600	
	2. Jowar + Guar	1	900	
	3. Oats + Pea	1	1,000	
			2,500	
III.	1. Sweet Sudan + Cowpea, 3 cuts	0.5	1,000	
	2. Berseem + Oat, 6 cuts	0.5	1,000	
			2,000	
IV.	1. Sweet Sudan + Cowpea, 3 cuts	1	1,000	
	2. Oats + Pea, 2 cuts	0.5	1,250	
	3. Mustard	0.5	500	
			3,750	
V.	1. Teosinte and Bajra + Cowpea	1	2,000	
	2. Berseem + Oats	1	2,000	
			4,000	
VI.	1. Maize + Cowpea	1	600	
	2. Maize + Cowpea + Teosinte	1	700	
	3. Oats + Mustard	1	500	
			1,800	
VII.	1. Hybrid Napier (Perennial)	0.5	1,500	
	2. Lucerne (Perennial)	0.5	1,500	
			3,000	
	Total	**15 hectares**	**19,850**	
	Yield/hectare		132 tonnes	

pails, buckets, feeding troughs, hose pipes, ropes, chains, shovels and some medicines need be purchased.

Working capital. Requirements of working capital to meet the expenses of salary, cost of fodder production, concentrates, dry fodder, replacement, medicine and detergents, depreciation and interest is shown in Table 2.14. Working capital is needed to meet the day-to-day expenses of the business. It is related with the production.

2.2.5. Economics of Dairy Farming

The economics of dairy farming has been worked out on the basis of data given in Tables 2.13 and 2.14. Fixed cost includes depreciation on buildings, machinery and equipments, animals and interest on fixed cost.

TABLE 2.13. Fixed Capital Particulars Needed on the Proposed Farm

Item	I year	II year	III year	IV year	V year	Remarks
A. Land	1,26,000.00	1,26,000.00	1,58,000,000	1,58,000.00	1,58,000.00	A. 17 hectares land purchased @ ₹ 8,000 per hectare, 15 hectares under fodder cultivation and 2 hectares under buildings, sheds, roads, further expansion.
B. Buildings						
Cattle sheds	14,540,00	14,540.00	14,540.00	14,540.00	14,540.00	
Implement shed-cum-store room 10 × 5 metres	2,500.00	2,500.00	2,500.00	2,500.00	2,500.00	
Store-room (fodder) 8 × 5 metres	2,000.00	2,000.00	2,000.00	2,000.00	2,000.00	B. 4 more hectares of land will be purchased in 3rd year of the farm due to increase insize of herd.
Labour quarters two (pucca)	3,000.00	3,000.00	3,000.00	3,000.00	3,000.00	
	22,040.00	**22,040.00**	**22,040.00**	**22,040.00**	**22,040.00**	
C. Livestock						C. Cost of buildings calculated @ ₹ 200 per adult unit; covered portion is abestos sheets supported with angle iron horizontally and hollow iron pipes vertically. Sheds for cows, heifers, calves, bullocks kept separate with loafing area, which is fenced with feeding and watering.
Cows 100	2,50,000.00	2,50,000.00	2,50,000.00	2,50,000.00	2,50,000.00	
Bullocks 4	4,000.00	4,000.00	4,000.00	4,000.00	4,000.00	
	2,50,400.00	**2,54,000.00**	**2,54,000.00**	**2,54,000.00**	**2,54,000.00**	
D. Equipment						
Tractor 25 HP with seed drill, cultivator and trolley	40,000.00	40,000.00	40,000.00	40,000.00	40,000.00	
Bullock cart	1,500.00	1,500.00	1,500.00	1,500.00	1,500.00	
Hand operated implements	1,000.00	1,000.00	1,000.00	1,000.00	1,000.00	
Tube-well with 10 HP motor	10,000.00	10,000.00	10,000.00	10,000.00	10,000.00	D. 100 crossbed milch cows will be purchased @ 2,500 each and 4 bullock @ ₹ 1,000 each.
Cattle yard equipment and daily appliances	1,000.00	1,000.00	1,000.00	1,000.00	1,000.00	
	53,500.00	**53,500.00**	**53,500.00**	**53,500.00**	**53,500.00**	
Total Fixed Capital	**4,55,540.00**	**4,55,540.00**	**4,55,540.00**	**4,55,540.00**	**4,55,540.00**	
E. Depreciation						
Depreciation on buildings 5% and upkeep 2%	1,542.80	1,542.80	1,542.80	1,542.80	1,542.80	
Depreciation on tube well 5%	500.00	500.00	500.00	500.00	500.00	
Depreciation on Tractor 10%	2,000.00	2,000.00	2,000.00	2,000.00	2,000.00	
Depreciation on crossbred cows and bullocks 16.67%	42,431.80	42,431.80	42,431.80	42,431.80	42,431.80	
Interest on fixed cost @ 15% per annum	68,331.00	68,331.00	73,131.60	73,131.60	73,131.60	
	1,24,715.60	**1,14,715.60**	**1,19,515.60**	**1,19,515.60**	**1,19,515.60**	
Total Fixed Cost	**5,70,255.60**	**5,70,255.60**	**6,07,055.60**	**6,07,055.60**	**6,07,055.60**	

Note. Cost of land and buildings varies greatly with the location (village/town/city).

TABLE 2.14. Working Capital Recurring Expenditure (Rs.) on the Proposed Farm

S.No.	Item	I year	II year	III year	IV year	V year	Remarks
1.	Cost of fodder	45,000.00	45,000.00	54,000.00	54,000.00	54,000.00	1. 40 kg/adult unit per day 40 × 136 × 365 = 1,980 tonnes. 2. Production 132 tonnes per hectare. 3. 15 hectares will be required. 4. Cost of production of fodder @ ₹ 3,000 per hectare. 5. Four hectares of land will be purchased during 3rd year as size of animals is increased. Extra cost will be ₹ 9,000. 6. All dry fodder requirements will be purchased and fed @ 2 kg per day per adult animal. Cost of dry fodder is ₹ 120 per tonne. 7. (*i*) One tractor driver @ ₹ 200 per month, (*ii*) 15 labourers for farm and cattle yard @ ₹ 6/day. (*iii*) 5 casual labour @ ₹ 6/day for 6 months from third year as strength is increased and more land is also purchased.
2.	Cost of concentrate	1,18,990.00	93,860.00	96,020.00	95,300.00	95,300.00	
3.	Cost of dry fodder	11,913.00	13,402.80	15,680.40	14,779.60	14,979.60	
4.	Labour charges	35,250.00	35,250.00	40,650.00	40,650.00	40,650.00	
5.	Cost of medicines and detergents	500.00	500.00	500.00	500.00	500.00	
6.	Cost of replacement and repair for farm cattle yard	500.00	500.00	500.00	500.00	500.00	
	Total	**2,12,143.00**	**1,88512.80**	**2,07,350.00**	**2,05,929.60**	**2,05,929.60**	
7.	Interest on the working capital @ 14% per annum (for 6 months)	14,850.00	13,195.04	14,514.90	14,415.03	14,415.03	
	Total working capital	**2,26,993.00**	**2,01,707.84**	**2,21,864.90**	**2,20,344.63**	**2,20,344.63**	

Total value of building, livestock and equipments is not included in the fixed cost as income generated by them is spread over a longer period. Working cost includes all the expenses given in Table 2.14. Returns include income from sale of milk, animals and manure (Table 2.15). Summary of the cost and returns is given in Table 2.16. It is evident from this Table that a unit of 17-hectare dairy farm can give per year a net income ranging from ₹ 40,000 to ₹ 54,000. In a period of five years the net returns from the above plan will be about ₹ 2,30,000. Thus the plan will yield a fairly good return.

TABLE 2.15. Receipts from Sale of Milk, Animals and Manure on the Proposed Farm

A. Receipt from sale of Milk

Number of animals in milk	*Rate of production per year (Litres)*	*Total production (Litres)*	*Total receipts @ 1.50/Litre (₹)*
100	2500 litres/year	2,50,000	3,75,000.00
75	3000 litres/year	2,25,000	3,47,250.00
75	-do-	2,25,000	3,47,250.00
75	-do-	2,25,000	3,47,250.00
75	-do-	2,25,000	3,47,250.00

B. Receipt from sale of animals

Number & Class of animals for sale			*Price per animal (₹)*			*Total Receipt (₹)*	*Remarks*
Male calf	*Cows*	*Heifers*	*Male calf*	*Cows*	*Heifers*		
—	—	—	100	1000	500		
48	—	—				4,800	
38	20	—				23,800	
38	20	28				37,800	
38	20	16				31,800	

C. Receipt from sale of manure

Adult Units	*Total manure production @ 12 tonnes per adult unit (tonnes)*	*Total receipts (₹)*	*Remarks*
136	1,632	8,160	Manure sold @ ₹ 5/- tonne
153	1,846	9,230	
179	2,148	10,740	
171	2,052	10,260	
171	2,052	10,260	

For calculating receipts in a particular locality one should take into consideration prices in that locality.

2.3. ANALYSIS OF PROFITABILITY OF DAIRY ENTERPRISES

How the profitability of a dairy farm as a whole depends on the relationship between fixed costs and the total gross margin is shown in the previous sections. It has also been shown how the total gross margin depends on the efficiency with which the farm is organized and managed. Implied in the above discussions is the fact that better farmers are able to produce much higher gross margins than others from a given level of fixed costs.

TABLE 2.16. Summary of Costs and Returns on the Proposed 100 Cow Dairy Farm

Item	*1st Year*	*2nd Year*	*3rd year*	*4th year*	*5th year*	*Total Cost and Returns*
A. Cost incurred in Milk Production						
(*i*) Fixed Cost (₹)	114,615.60	114,715.60	119,515.60	119,515.60	119,515.60	587,878.00
(*ii*) Working Cost (₹)	226,993.00	201,770.84	221,864.90	220,344.63	22,344.63	1091,318.00
Total Cost (₹)	341,608.60	316,486.44	341,380.50	339,860.23	339,860.23	1679,196.00
B. Returns						
(*i*) Milk (₹)	375,000.00	347,250.00	347,250.00	347,250.00	347,250.00	1764,000.00
(*ii*) Sale of Animals (₹)		4,800.00	23,800.00	32,800.00	31,800.00	93,200.00
(*iii*) Manure (₹)	8,160.00	9,230.00	10,740.00	10,260.00	10,260.00	49,650.00
Total returns (₹)	383,160.00	361,280.00	381,790.00	390,310.00	389,310.00	1905,850.00
C. Net Returns (₹)						
(B – A))	14,551.40	44,793.56	40,409.50	50,449.77	49,449.77	22,6654.00

Note. Data in this and preceding tables are for illustration only and are on the conservative side.

It is important, further, to consider the factors determining the gross margin produced and the ultimate profitability of a dairy herd, and methods that can be used for analysis of the enterprise's profitability or otherwise. Some common methods of analysis of profitability of dairy enterprises are discussed in this section. Gross margins per cow and per hectare are by far the most common and easily understandable measures of efficiency of a dairy enterprise.

2.3.1. Business Records—Essential Guidelines

Farming—a very complicated business—involves a variety of operations to be carried out periodically (daily, weekly, seasonally, so on) throughout the year. Money, materials and labour are spent on them. Some operations give returns immediately (milking), some do so later (insemination) and yet others may be non-return giving "necessary evils" (*e.g.* dehorning).

In order to get a complete picture of the dairy farm business, a farmer must keep different types of records in a systematic and classified way. Major farm business record books and instructions for filling them, as recommended ,by the economists at the National Dairy Research Institute, are reproduced below (NDRI, 1977).* These are simple, concise and even high school students can easily fill them.

Instructions for Filling the Books

1. Farm inventory. Inventory is a statement of farm assets with values assigned to it at a particular date. It contains values of land, buildings (residential, cattle shed and godowns etc.), wells and tube wells, position of draught and productive stock, farm produce in stock and farm supplies. Inventory is opened in the beginning of the year and closed at the end of the year. Thus it gives the changes in values of net worth during the accounting year.

2. Financial transactions. Farmers should keep a proper record of all financial transactions made during the year date-wise. Accounts of loans taken, loans given, amounts and the parties

*NDRI (1977). *Know your farm business.* Publication No. 89, National Dairy Research Institute, Karnal, 1977.

involved etc. should all be maintained. The records of the outstanding loans, if any, and the interest should also be kept.

3. Farm supplies. A record of all farm supplies used in production should be kept. Values of stocks of home produced and purchased fodder and feeds, with price and the source of purchase should also be recorded date-wise.

4. Expenditure on fodder and feeds. On a dairy farm, fodder and feeds are the major items of expenditure. All expenditure on maintenance of cattle with details of different types of fodder and feed, quantity and value for each category of stock should be recorded separately.

5. Labour. A record should be maintained of all labour engaged and expenditure involved for each operation along with a description of family labour, permanent, hired and temporary labour. All values of requisites for the permanent labour should also be accounted for.

6. Miscellaneous expenditure on livestock. Items of miscellaneous expenditure on livestock such as servicing, veterinary, electricity, and water charges, should be recorded in separate columns with descriptions of different diseases, types of treatments taken and the expenditure etc.

7. Expenditure on farm production. Every farmer should maintain proper records of the expenditure on home produced fodder and feeds so that he may know the cost of home-produced fodder and compare it with the purchased fodder and feeds. It should include all the inputs like seed, fertilizer, manure, human labour, irrigation, bullocks or tractor power etc. as well as the yield.

8. Production from livestock. Besides recording expenditure on production, the farmer should also have a systematic account of farm production and income from the produce. He should date-wise record the total production, total sales, with prices of both main and by-products.

9. Farm sales. The farmer should keep a separate book for the sale of the farm produce which should include the commodity sold, the market where sold, prices received and marketing cost involved.

10. Business Analysis. At the end of the year the farmer should total all the expenditures on production from different accounts maintained by him. Similarly he should also total the income from all the sources during the year and finally he should compare the total cost, total returns and margin of profit or loss in the business. The most common yardstick of measuring the economic efficiency are the Net Farm Income/year, Net Farm Income per hectare and per milch cattle, and the Return to Capital per ₹ 100 of Investment. He should also analyse with the help of different records, what were the factors associated with his profit or loss in the business so that he may improve upon the weak points in the subsequent planning.

BOOK NO. 1

Farm Inventory

Sr. No.	*Particular*	*Number*	*Date of Purchase*	*Value on 1st Jan.*	*Value on 31st Dec.*	*New purchases/ Sale in the year*
1.	Land owned (in acres) :					
2.	Trees :					
3.	Buildings :					
	(*a*) Residential:					
	(*b*) Cattle-shed:					
	(*c*) Godowns :					

4. Irrigation :
   (*a*) Well :
   (*b*) Tube well :
5. Livestock :
   (*a*) Draught cattle :
      (*i*) Bullocks (cattle) :
      (*ii*) Bullocks (Buff) :
   (*b*) Milch cattle :
      (*i*) Cows :
      (*ii*) Buffaloes :
      (*iii*) Young stock :
      (*iv*) Bulls :
6. Implements and Equipments :
   (*a*) Farm :
      Tractor :
      Plough :
      Sickles :
      Hand-hoe *(Khurpa)*:
      Spade *(Kassi)*:
   (*b*) Cattle :
      Feeding Troughs :
      Ropes & Chains :
      Pails & Buckets :
      Fencing :
      Pegs (Iron):
7. Farm produce (in stock):
8. Farm supplies :

BOOK NO. 2
Financial Transactions

Year ....

Month	*Date*	*Loans*						*Remarks*
		Taken		*Given*		*Outstanding*		
		Source of Agency	*Amount* ₹	*Particulars*	*Amount* ₹	*Amount* ₹	*Interest* ₹	

BOOK NO. 3
Farm Supplies

Year ....

Month	*Date*	*Home Produce*			*Purchased*			*Remarks*
		Particulars	*Qty. kg.*	*Value* ₹	*Particulars*	*Qty. kg.*	*Value* ₹	

BOOK NO. 4
Farm Supplies

Year ....

Month	*Date*	*Fodders and Feeds*									*Remarks*
		Green		*Dry*		*Concentrate*		*Silage*		*Total*	
		Qty. kg	*Value ₹*	*Qty. kg*	*Value ₹*	*Qty. kg*	*Value ₹*	*Qty. kg*	*Value ₹*	*Value ₹*	

BOOK NO. 5
Expenditure on Labour

Year ....

Wage Rate/Day : M (Male) : F (Female) :

Month	*Date*	*Family Labour*	*Hired Labour (Hr)*								*Total wages*
			Permanent		*Casual*		*Contract*		*Total*		
		Man Hrs.	*M*	*F*	*M*	*F*	*M*	*F*	*M*	*F*	

BOOK NO. 6
Miscellaneous Expenses on Livestock

Year ....

Month	*Date*	*Servicing (₹)*	*Vety. (₹)*	*Electricity (₹)*	*Water (₹)*	*Consumables (₹)*	*Total (₹)*	*Remarks*

BOOK NO. 7
Expenditure on Fodder Production

Year ....

Month	*Date*	*Value of seed*	*Value of Fertiliser & FYM*	*Human Labour (₹)*		*Bullocks/ Tractor (₹)*	*Irrigation (₹)*	*Misc. (₹)*	*Total*
				Family	*Hired*				

BOOK NO. 8
Production from Livestock

Year ....

Month	*Date*	*Total milk*		*Value of dung cakes sold (₹)*	*Value of F.Y.M. sold (₹)*	*Remarks*
		Qty. (kg)	*Value (₹)*			

BOOK NO. 9
Farm Sales

Year ....

Month	Date	Particulars of the produce sold (Qty. in kg)	Particulars of the party	Value		Remarks
				Cash	Credit	

BOOK NO. 10
Business Analysis

Year ....

Fixed cost (₹)		Variable cost (₹)			Total cost (₹)	Returns			Total Returns
Depreciation on livestock	Interest on fixed	Fodder & feeds	Labour	Misc.		Milk (₹)	Dung (₹)	Misc. (₹)	
1	2	3	4	5	6 (1 to 5)	7	8	9	10 (7+8+9)

2.3.2. Analysis of the Business

If a farmer has limited capital (mostly the case) the maximum profit may be forthcoming even if he stops short of the point where added costs become greater than added income. With limited capital, the decision is not entirely on whether added investment will give the greatest profit. Such a situation is more or less common to peasantry in the developing countries. Every alternative investment opportunity should be investigated before further investments are made. The very soundness of the enterprise need also be analysed. Such an analysis is important to the farmer as well as organisations lending money to farmers. Some of the measures of farm efficiency are as follows :

1. Net farm income. Among the various income terms, *net farm income* is the most commonly used one. This figure is obtained by subtracting operating and fixed expenses (feed and livestock costs, value of unpaid labour—family or farmer's own—plus decrease in the value of inventory, if any) from the sum of the value of milk sold, sale of animals, sale of manure and sale of feeds and fodder, if any. This figure shows net earnings from the use of all the farmer's own resources, including his capital, land, labour and his management. This is the income, which is available to the farmer to make further investment on the farm or elsewhere.

Net farm income = Total returns − Total costs.

Using the figures given in Table 2.16 (5th year) the net farm income can be estimated as follows :

Total Returns	₹ 3,89,310.00
Total cost	₹ 3,39,860.00
Net Farm income	₹ 49,450.00

2. Net cash income. It is obtained by subtracting the sum of cash purchases from the sum of cash sales. This is the income which farmer earns over his cash expenses. This measure is good for short periods, whereas in the long run, the farmer is interested to get maximum net farm income. From data given in Table 2.13 to 2.15 the net cash income is calculated thus :

Cash expenses :	***Amount (₹)***
1. Salary and wages of the labour	40,650
2. Cost of fodder	54,000
3. Cost of repairs	500
4. Cost of medicines	500
5. Cost of concentrates	95,300
6. Cost of dry fodder	14,990
	2,05,930

Cash Receipts :	
1. Sale of milk	3,47,250
2. Sale of animals	31,800
3. Sale of manure	10,260
	3,89,310

Net cash income = ₹ 3,89,310 – 2,05,930 = ₹ 1,83,380

3. Net income per hectare. Return above all costs divided by the number of hectares of the farm. This measure could be used to make comparison between crops and livestock enterprises. From data given in Table 2.15.

$$\frac{\text{Net farm income}}{\text{Number of hectares of farm}} = \frac{49{,}450}{19} = ₹\ 2{,}603$$

4. Net farm income per adult unit

$$\frac{\text{Net farm income}}{\text{Number of adult units on farm *}} = \frac{49{,}450}{136} = ₹\ 364$$

This measure of efficiency could be used to compare the merits of different dairy animals or efficiency of the various dairy farms of the region having similar type of milch animals. This measure is not quite adequate to make comparison between two regions if the climatic conditions are not similar.

5. Net return per Rs 100 worth feed fed to livestock. There are many ratios, which have been used to measure the role played by dairy cows in the farm organization and the efficiency with which they are managed. Net return per ₹ 100 worth of feed fed to cows gives an idea of overall success.

**See* Table 12.13.

This measure reflects three important management aspects :

(i) the extent, to which the most profitable combination of enterprises has been achieved,

(ii) the extent by which the livestock and their product have been marketed most advantageously, and

(iii) the efficiency with which production has been obtained from the dairy animals (or efficiency of cows).

This measure reflects only the influence of feed cost and not of any other cost. To find out this ratio it is necessary to find out the value of feed including concentrates and roughages. Total value of the feed is divided by the total value of production and multiplied by 100. Using data from Tables 2.13 to 2.15

$$\frac{\text{Estimated value of concentrates and roughage}}{\text{Estimated receipts from sale}} \times 100 = \frac{1,64,280}{3,89,310} \times 100 = 42 \text{ per cent}$$

6. Capital per ₹ 100 gross return. This ratio is computed to measure the total amount of capital used per ₹ 100 gross return (Value of products sold). The intention is to know how much capital is required to yield Rs. 100 of gross income.

Capital	*Amount* (₹)
1. Land	1,26,000
2. Buildings	22,040
3. Livestock	2,54,000
4. Equipment	53,500
5. Expenses (Fixed + working)	3,39,860
Total	7, 95,400
Gross Income	3, 89,310

$$\frac{\text{Gross Income}}{\text{Capital invested on farm}} \times 100 = \frac{3,89,310}{7,95,400} \times 100 = 48.95\%$$

Gross income ratio does not reflect the influence of annual expenses, which must be deducted to arrive at net farm income. Further, the gross income relates to a particular year, whereas capital invested (Items 1 to 4) will continue to yield income for over a period of years.

7. Rate of turnover. Net income is the most significant efficiency measure because it measures the farmer's success in attaining his economic goals. It would be better to estimate the net income ratio with respect to total cost (fixed + operating). The capital investment will remain intact because depreciation and interest are charged on the investment.

$$\frac{\text{Net farm income}}{\text{Total expenses}} \times 100$$

$$\frac{49,450}{3,39,860} \times 100 = 14.55 \text{ per cent}$$

Thus it is evident from the above ratio that the net turnover is 14.55 per cent per annum.

There are methods eminently suited to the loan giving organisations for evaluation of the viability of dairy projects loaned by them, such as—Net Present Value, Internal Rate of Return, Benefit cost Ratio, Pay-back Period, Repayment Capacity etc. Since it is out of the scope of this book to discuss them, the reader may refer to standard books on *Farm Economics and Management* for details. A list of such books is given at the end of this chapter.

2.4. FEATURES OF DAIRY FARMING IN INDIA

Milk production in India is in the hands of millions of small-scale operators distributed over vast areas and their efforts are poorly coordinated. Perhaps it is a misnomer to call such activities as dairy farming; milk production being a secondary, albeit important, to the overall farming operations of the farmers. Described in this section are the systems, sizes and modes of operations of milk producing units in India as well as their economic aspects.

2.4.1. Systems of Farming

Running of a farm is different under different conditions. Depending on the mode of ownership and organisation, farms can be classified as—(*i*) individually owned, (*ii*) cooperative. (*iii*) government or (*iv*) collective. The type of farming can be (*i*) specialised farming. (*ii*) diversified farming, or (*iii*) mixed farming, depending on how income is derived from a farming enterprise.

Dairy farming can be run under a mixed farming set up, or it can be a specialised enterprise. Under **mixed farming**, crop-production is combined with dairy (or any other animal) farming. Livestock enterprises are complementary to crop production and vice versa, depending on the relative emphasis given to one or the other. Agricultural economists consider that for a farm to be called a mixed farm, 10-15 per cent of its gross income must be contributed by livestock components.

A **specialised farm** is one on which 50 per cent or more of the receipts are obtained from a single source. A specialised dairy farm is one on which the main source of income is milk, while there can also be income from other subsidiary sources like sale of manure, sale of extra fodders, grains etc. Under a given set of agro-climatic conditions the selection of farming system depends on—(*i*) demand and supply of products and marketing facilities, (*ii*) distribution of income over the year, (*iii*) opportunities of family labour employment; (*iv*) fixed and variable costs, and (*v*) net profit per unit area per unit time.

Studies conducted at various places have shown that dairy farming is equally or even more profitable than crop farming. In order to produce more milk at a lower cost, initially we should develop a system of fodder production in a mixed farming set-up in rotation with cultivation of high yielding varieties of grain and cash crops, with a view to obtain the highest income per hectare of land, per animal and per unit of time.

Studies conducted at different centres of the country have clearly shown that generally, economy-wise, mixed farming and specialised dairy farming stand at par, but above **arable farming** (cash and grain crops) alone. The fuller employment of family labour is not possible under arable farming conditions. A combination of crop and dairy farming promises employment of labour round the year. Secondly, a constant and regular flow of income from milk production is an added' advantage for increasing farm efficiency. Another advantage of mixed farming is a lower-degree of risk and uncertainty in production and income as compared to crop farming alone

(Patel, 1976). N.D.R.I. studies (Sundaresan, 1975) showed that the net return per hectare in case of milk production through high yielding cows was of the order of ₹ 5,000 as against ₹ 3,000 to 3.500 through paddy and wheat cultivation with modern technology.

Dabas *et al.* (1971) made an attempt to compare a 25-acre farm under arable farming with another farm of the same size with dairy farming under similar agro-climatic and management conditions. In arable farming under the following conditions net profit per acre was ₹ 749.

(*i*) Cropping intensity—200 per cent
(*ii*) Crops grown—paddy, maize, *chari,* sugarcane, peas, wheat, berseem and potato.
(*iii*) One pair of bullocks.

On the other hand, in dairy farming under the following conditions the net return per acre was ₹ 1,854.

(*i*) home-grown fodders.
(*ii*) crossbred cows with 75 : 25 :: Milch : dry.
(*iii*) average daily, milk yield 7.5 litres.
(*iv*) annual replacement of 15 per cent of the stock.
(*v*) culling animals after 6th lactation.
(*vi*) feeding concentrates to cows giving above 4 litres per day only.

Moreover, the income from dairy farming is further likely to go up, in the opinion of these authors, after 5-6 years of operation, when the purchased herd would be completely replaced by home-breds and weaning practiced in the 2nd generation cows.

Singh and Chibbar (1971) too, based on a six-year study (1962-68) on a 3-hectare farm, reported dairy farming to be more economical. In their study the income per hectare from (*i*) specialised farming with 8 Tharparkar cows, (*ii*) specialised farming with 8 Murrah buffaloes, (*iii*) mixed farming with 5 Tharparkar cows, and (*iv*) arable farming with one Tharparkar cow was respectively, ₹ 1,808, ₹ 1,301, ₹ 1,250 and ₹ 1,107. The authors opined that if crossbred cows can be procured, income from dairy farming can be as high as ₹ 3,600 per hectare.

The relative economic efficiency of different types of milch animals like purebred cows, crossbreds and buffaloes will also be different. Gill and Gill (1974) studied the economics of mixed farming with dairy component having crossbreds and buffaloes in the Ludhiana district of Punjab. Two types of farms were selected, *viz.* bullock operated farms having less than 3 hectare as operational holding and mechanized farms, having 3 or more hectares as operational holding. Two-crossbred-cow dairy units and five-buffalo dairy units were studied on bullock-operated holdings. Similarly, four-crossbred-cow dairy units and one-buffalo dairy unit were selected on mechanized farms. For each category a minimum of 5 high producing milch animals were introduced. Thus a total of 12 farms having mixed farming with dairying were studied.

Patel, R.K. 1976. Economics of dairy firming, mixed farming, arable farming, Proc. Summer Inst. *Forage Quality Evaluation and Processing Techniques,* College of Animal Sciences, Haryana Agricultural University, Hissar, p. 205.
Sundaresan. D. 1975. *Indian Fmg.,* 25: 12.
Dabas, I.S., Singh, B. and Rastogi, B.K. 1971. *Indian Dairyman,* 23: 165.
Singh, G.D. and Chibbar, S.S. 1971. *Indian Dairyman,* 24: 49.
Gill. S.S. and Gill, G.S. 1974. *Indian J. Dairy Sci.,* 27: 206.

The results (Table 2.17) showed that dairying was more paying than general cropping on bullock operated as well as mechanized farms and that milk production was comparatively much paying with crossbred cows than with buffaloes. The cost of milk production for crossbred cows was ₹ 0.84 per litre as against ₹ 0.99 per litre for buffaloes. Studies conducted in the last two decades showed that the cost of milk production at the farmer household was lower in crossbred cows as compared to buffaloes and local cows at the farmers households in different areas of India (Table 2.18).

In an All India Co-coordinated Research Project (sponsored by the Indian Council of Agricultural Research) on Economics of milk production, three systems of farming. *viz..* dairy

TABLE 2.17. Comparative Economics of Dairy Farming and Arable Farming (Gill and Gill, 1974)

Farming systems	*Total Expenditure ₹/ha*	*Gross income ₹/ha*	*Net profit ₹/ha*
A. With Murrah buffaloes			
1. Bullock operated farms :			
(*a*) Dairying	18,128.98	20,107.24	1,978.26
(*b*) Cropping	4,790.65	4,795.22	4.57
2. Mechanized farms :			
(*a*) Dairying	28,243.62	28,165.99	– 77.63
(*b*) Cropping	4,579.99	4,400.75	– 179.24
B. With crossbred cows :			
1. Bullock operated farms :			
(*a*) Dairying	21,453.14	32,315.44	10,862.30
(*b*) Cropping	4,185.13	4,717.73	532.60
2. Mechanized farms :			
(*a*) Dairying	17,344.04	20,751.18	3,407.14
(*b*) Cropping	4,595.46	4,236.87	– 358.59

TABLE 2.18. Cost of Milk Production at farmer households in different areas of India

Author	*Location*	*Cost of Milk Production (₹/litre)*		
		Local	*Crossbred cow*	*Buffaloe cow*
Gupta and Agarwal, (1996)	Himachal Pradesh	5.20	4.68	5.85
Sanjay Kumar *et al.* (1999)	Kumoan hills	5.36	—	2.38
Atul Chandra and Agarwal, (2000)	Farrukhabad district of Uttar Pradesh	—	6.83	7.58
Bhowmik *et al.* (2006)	Tripura	19.26	9.40	—
Singh R.N. *et al.* (2006)	Tribal areas of Udaipur, Rajashtan	14.26	—	12.59
Rishikanta Singh and S.B. Agarwal (2007)	Imphal West District of Manipur	19.01	11.02	—
Smita Sirohi *et al.* (2007)	Karnal district of Haryana	—	7.90	10.68

TABLE 2.19. Economics of Different Systems of Farming Under All India Coordinated Project on The Economics of Milk Production
(Av. of two years data, 1973-74 and 1974-75) (Anon., 1975)

Sr. No.	Centres	Unit	Total Exp. (₹/ha)	Gross income (₹/ha)	Net profit/loss (₹/ha)
A. With crossbred cows					
1.	Allahabad	SDF	7,775.75	7,918.25	142.50
		MF	6,932.00	7,123.25	191.00
		AF	4,289.50	4,789.00	499.50
2.	Hyderabad	SDF	5,359.25	7,990.25	2,631.00
		MF	5,131.00	7,845.25	2,712.25
		AF	3,718.00	5,717.75	1,999.75
3.	Jhansi	SDF	6,468.00	8,049.50	1,531.50
		MF	4,583.25	6,436.25	1,853.00
		AF	5,193.00	3,650.25	457.25
4.	Midnapore	SDF	8,630.25	8,944.25	314.00
		MF	5,375.50	4,958.25	– 427.25
		AF	2,902.75	3,657.25	754.50
5.	Poona	SDF	12,678.00	16,977.75	4,299.75
		MF	9,415.75	17,480.50	8,064.75
		AF	7,089.50	13,192.75	6,103.25
	Average	SDF	8,182.25	9,976.00	1,793.75
		MF	6,287.40	8,766.30	2,478.90
		AF	4,238.55	6,201.40	1,962.85
B. With buffaloes					
1.	Hissar	SDF	6,149.00	8,969.00	2,820.00
		MF	4,312.00	6,183.00	1,871.00
		AF	3,147.50	4,318.25	1,170.75
2.	Jabalpur*	SDF	10,037.50	7,737.25	– 2,300.25
		MF	8,455.75	7,892.75	– 563.00
		AF	5,582.50	6,178.75	596.25
3.	Udaipur	SDF	9,577.00	10,915.00	1,338.00
		MF	8,231.00	13,186.00	4,955.00
		AF	6,232.75	10,255.50	4,022.75
	Average	SDF	8,587.83	9,207.08	619.25
		MF	6,999.58	9,087.25	2,087.67
		AF	4,987.58	6,917.50	1,929.92

SDF = Specialized Dairy Farming; MF = Mixed Farming; AF = Arable Farming

* Animal performance poor due to reproductive disorder.

farming, mixed farming and arable farming were compared at various centres (Table 2.19) using crossbred cows or buffaloes as dairy animals. Two-hectare area was put under each system of farming. Five to eight 3-4 and 1 crossbred cows/buffaloes were kept as dairy animals under specialized dairy farming (SDF), mixed farming (MF) and arable farming (AF), respectively.

TABLE 2.20. Level of Milk Production and Cost of Milk Production Using Crossbreeds and Buffaloes at different Centres (Anon., 1975)

Particulars	*Years*	*Milk Yield and cost of milk production average (cost-bred cows)*						*Milk yield and cost of milk production (buffaloes)*			
		Allahadad	*Hyderabad*	*Jhansi*	*Poona*	*Midnapore*	*Average*	*Hissar*	*Jabalpur*	*Udaipur*	*Average*
Total milk production per animal per annum (litres)	72-73	1,711	1,803	1,309	2,385	1,722	1,786	1,037	2,218	1,196	1,484
	73-74	1,794	1,889	1,408	2,125	1,708	1,785	1,385	904	1,477	1,255
	74-75	1,544	1,549	1,356	1,952	—	1,600	403	1,440	1,221	1,021
Average		1,683	1,747	1,358	2,154	1,715	1,721	942	1,521	1,298	1,253
Cost of milk production (₹/Litre)	72-73	1.45	0.86	1.77	1.60	1.66	1.47	1.34	1.68	2.26	1.76
	73-74	1.58	1.15	1.75	1.48	1.24	1.44	1.40	2.31	2.11	1.94
	74-75	1.65	0.65	1.62	1.84	—	1.44	1.82	2.71	1.99	2.18
Average		1.56	0.89	1.71	1.64	1.45	1.45	1.52	2.25	2.12	1.96

Recommended crop rotations for respective zones were followed at different centres (Anon. 1975).

In general, better lactation performance of the dairy animal, irrespective of species, resulted in lower cost of production of milk (Table 2.20). There are centre-to-centre differences, however, (Faroda *et al.*, 1978).

It may be appropriate now to remember the actual farming practices obtaining in rural India. A vast majority of farms in our country can be considered as mixed farms with an established complementary relationship between crops and livestock. Crop by-products like straws, cereals and gram husks, bran, oil cakes etc. are used for feeding milch animals while their dung is used for manuring the fields. While cash and grain crops give income at the end of the harvesting season, milch animals give income throughout the year. Mixed farming also gives balanced labour load for the farmer and his family throughout the year. Ultimately it results in higher returns from a unit of land holding. The need of the hour is to encourage farmers—small and big—to combine a more realistic livestock (dairy) component in their farming systems. Needless to say, good dairy animals have to be made available to the farmers. Top agricultural scientists of the country like Dr. Swaminathan and Dr. Sundaresan saw mixed farming as the only answer to our problems of rural poverty, underemployment and unemployment. Net returns were higher in mixed farming and at the same time mixed farming also resulted in generation of more employment (Sheokand *et al.*, 2000) (Table 2.21). It is also more sustainable.

TABLE 2.21. Comparative economics of different farming systems under paddy-wheat farming sequence in rice-belt of Haryana (Sheokand *et al.*, 2000)

Particulars	*Arable farming*	*Buffalo rearing*	*Mixed farming*
Net returns/ha (₹)	6,326	39.4	10,966
Net returns/₹ 1000 invested	223	241	250
% net returns over gross expenditure	22.29	24.10	24.99
% returns over other variable costs	39.69	33.87	39.99
% labour expenses over total expenditure	12.90	18.99	25.36

2.4.2. Scale of Operations

There are different means of describing the scale of dairy enterprises, *viz.*, depending on how much of milk is produced and marketed by the unit daily; number of dairy animals maintained or herd size; or area of the holding over which the dairy enterprise is based. A dairy farmer venturing into the business should have at least a broad idea about the likely financial requirements and receipts from dairy farming at different scales of operation.

The average herd size in most of the developing countries is small because : (*i*) the economic status of the average peasant does not permit him to keep more cows; (*ii*) many of the farmers

Anonymous. 1975. Summary Annual Report of the coordinator of "All India Coordinated Research Project to determine the economics of milk production under intensive dairy farming conditions in relation to high yielding varieties of cereals and cash crops" (1974-75). 3rd Workshop. IGFRI, Jhansi.

Faroda. A.S. 1978. Comparative economics of different systems of farming with crossbred dairy animal. Proc National Level Seminar on Management of Exotic and Crossbred cattle. 6-12. Mar. 1978. *College of Anim Sci.* HAU. Hissar, p. 104.

consider dairying only as an inconsequential secondary to crop farming; (*iii*) the milk drinking tradition is not strong; and (*iv*) milk marketing facilities are generally poor.

The total expenditure on the farm naturally increases with increase in number of animals, but not all components of cost increase directly proportionately. For example, labour costs, expenditure on buildings etc., per animal can actually come down with increase in herd size, though there is no accurate information on this aspect, based on actual data from farmer's holdings.

TABLE 2.22. Livestock Husbandry Status and Milk Production Costs on Buffalo Dairy Units of Different Sizes in Haryana

S.No.	Parameter	Herd Size		
		1-2	3-5	5-7
1.	Per cent households	85.8	11.9	2.3
2.	Per cent buffaloes in milk (as a whole for the class)	82.6	73.4	68.8
3.	Total crop area : Area under fodder crops (ha)	6.3 : 1.3	8.7 : 2.2	10.0 : 2.5
4.	Green fodder fed/day/head kg	12.4	11.2	9.8
5.	Dry fodder fed/day/head (kg)	7.6	7.0	6.0
6.	Concentrate fed/day/head (kg)	1.6	1.4	1.1
7.	Daily milking average (kg)	5.67	5.95	5.58
8.	Overall average milk yield (kg)	4.95	4.45	3.82
9.	Net cost of maintenance/buffalo/day (₹)	8.31	7.92	7.49
10.	Per litre cost of milk production (₹)	1.75	1.87	1.87

Source : *Economics of milk production in Haryana State.* Res. Bull. No. 4. Dept. Agri. Econ., HAU. Hissar. 1979.

Surveys conducted in the high milk potential area of Krishna district in Andhra Pradesh and in Uttar Pradesh have shown that, on mixed farms, an average herd size of 5 bovines seems to give optimal financial returns. Too large a size of herd (above 10) does not seem to be ideal for the Indian peasant, whose financial and physical resources are generally meagre. Further, farmers with small herds (1-3 animals) are those who seem to maintain the animals best (Table 2.22). Net returns per litre of buffalo milk were found to be highest in the small herd size milk producers followed by medium and large herd size milk producers in West Godavari district of Andhra Pradesh (Hymajyothi *et al,* 2003) Slightly different trend was observed as regards net returns per buffalo per year in suburban and rural areas around Madras city (Table 2.23). While 5 animals

TABLE 2.23. Net Return Per Buffalo Per Year in Urban and Rural Areas Around Madras at Various Sizes of Animal Holding at 1972 Prices (Natarajan, 1978)

Size of animal holding	Net return/buffalo/year (in rupees)	
	In Urban areas	In rural areas
1	811.00	446.00
2	799.00	422.00
3	—	421.00
4	710.00	423.00
5	927.00	374.00

Source : Proc. Indo-Soviet Symposium on Buff. Breeding, 16-18. Nov. 1978, NDDB. Anand. p. 216.

holdings were most profitable in sub-urban areas, the returns slightly decreased with increase in animals per holding (1 to 5) in rural areas. This was attributed to the interplay between demand and supply and price of milk in the two areas.

From a study of three, sub-urban medium sized private dairy units, Singh and Krishna (1975) reported that all input costs for running the dairy enterprise were lowest and net returns per animal and per litre of milk were highest on a 25 animal unit : the efficiency of the organisation falling as the herd size decreased from 25 to 12 animals (Table 2.24).

Thus, while bigger herds can be viable when located nearer to towns and cities, under village conditions smaller (3 to 5 animals) dairy units seem to give best returns per rupee invested. In fact, many studies in India have revealed that dairy farming is ideally suited for landless agriculture labour, marginal and small farmers as means of improving their overall income (Sundaresan, 1981).

TABLE 2.24. Costs and Returns of Dairy Enterprises on Three Farms in Uttar Pradesh (Singh and Krishna, 1975)

S.No.	Particulars	Number of cows on farm		
		12	20	25
1.	Cost of maintenance of milch animals including fodder production (₹)	3,600.00	6,000.00	7,500.00
2.	Cost of labour, rent, water, electricity, veterinary services etc. (₹)	405.00	763.00	1,155.00
3.	Quantity of milk produced and sold locally (lit.)	2,860.00	5,550.00	7,290.00
4.	Total returns from milk and milk products (₹)	6,585.00	12,487.00	16,502.00
5.	Net returns per litre of milk produced (₹)	0.90	1.03	1.08

Sundaresan, D 1981. *Strategy for milk production enhancement in the decade of eighties.* N.D.R.I., Karnal. p. 21.

In Kerala, along the south west coast of India, the landless tea estate labourers earned more returns per cow than settler farmers; the poorest earners being the land owning farmers in plains, for whom dairying is of secondary importance (Patel *et al.,* 1978). Further, net returns per cow were higher with Brown Swiss × Non-descript crosses than with non-descript cows and more so in areas where animal husbandry extension service was operating (Table 2.25). In such areas, purebred Brown Swiss cows yielded highest returns.

The dairy enterprise can be more remunerative if such small-scale operators organise themselves into cooperatives. Kumar *et al.,* (1975) have recorded that in Uttar Pradesh, dairy farmers that have organised themselves into a cooperative derived more returns than individual private dairy farmers with similar type of dairy animals (Table 2.26).

A study (Thakur, 1975) from Gujarat too confirmed that landless rural folk earned as much as 65 to 75 per cent and small farmers 25 to 30 per cent of their total income from dairying. The proportion of income earned from dairying goes down with increase in size of holding. Small farmers in such villages where dairy cooperatives were operating produced double the quantity per animal and marketed surplus milk to that of their counterparts in villages where there were no cooperatives.

Thakur, D.S. 1975. Impact of dairy development through milk cooperatives—a case study of Gujarat. *Indian J.Agri. Econ.,* 30 : 83.

Thus large number of small herds of good dairy animals organised into cooperatives seems to be the ideal solution for milk production enhancement as well as for giving better deal to the prospective target farmers—landless labourers, small and marginal farmers. Further labour utilization was better when farmers organized themselves into cooperatives (Rao and Singh 1999). Reasons for the greater success of co-operating farmers are not difficult to seek.

TABLE 2.25. Annual Net Returns Per Milch Cow (In Rupees)* on Holdings of Different Categories of Farmers in the Coastal State of Kerala (Patel *et al.*, 1978)

Category of farmer	*In areas where extension programme is in existence*			*Control (no extension programme)*	
	Non-descript	*Brown Swiss*	*Graded cows*	*Non-descript*	*Graded cows*
Farmers in plains (Established, main occupation crop farming)	–391.62	577.25	145.04	– 26.35	– 60.33
Settler farmers (recently settled on hills, taken up dairying)	222.87	875.67	579.44	257.22	83.22
Tea labourers (milk production important side business)	298.10	620.68	422.63	234.51	411.06

$*I = Y (P_y - C)$,

where, I = Annual net returns to fixed farm resources per cow.

Y = Annual milk production per milch cow.

P_y = Price per litre of milk.

C = Estimated net cost per litre of milk, excluding family labour and fixed costs, as these are not real costs to the farmer.

TABLE 2.26. Costs and Returns From Dairy Holdings Forming into a Cooperative and Dairy Units Not Forming into Cooperative in Rural Areas of Kanpur District. Uttar Pradesh (Kumar *et al.*, 1975).

Items	*Cooperative holdings* ₹	*Non-cooperative holdings* ₹
1. Fixed costs/buffalo	400.00	600.00
2. Maintenance cost/buffalo	1510.00	1530.00
3. Gross returns/buffalo/year	835.00	189.00
4. Animal Husbandry status on the holding	Very good	Ordinary

Cooperative dairy farmers need not depend on the middlemen and professional milk traders to market their milk. The milk producers sell their surplus milk twice directly through their milk cooperative located in their own village and get paid for it every day or as decided by themselves. Besides, the milk cooperative is in a position to employ technical personnel for their common benefit and provide to milk producers in each village various technical inputs like balanced cattle feed, fodder seeds, artificial insemination service, veterinary care etc. as part of their milk production enhancement and milch animal improvement programmes. Technical inputs provided by the Anand Milk Union Limited in Kaira district of Gujarat, to their member producers have paid rich

Patel, R.K. Kumar. P., Voegele, K., Sukumaran Nair, and Sreekumar Nair, G. 1978. Economics of crossbred cattle. *Indian Dairyman,* 30:31.

dividends by way of increased milk production per animal and procurement per year (Table 2.27). Needless to say, the economic status of the farmers in the Cooperative Union area improved appreciably over the years. Efforts are now being made under the Operation Flood Programme of the National Dairy Development Board to create similar cooperative ventures in other milk potential areas of the country with variable success.

TABLE 2.27. Particulars Showing Improvement in Productivity of Surti Buffaloes in Kaira District and Milk Procurement by Anand Milk Union in Gujarat in Relation to Technical Inputs Provided by the Union to Producers During 1961-74 (After. Chotani. 1978)

Year	*Balanced cattle feed sale (MT)*	*Lucerne fodder seed distributed (MT)*	*Artificial inseminations performed with semen of butter bulls*	*Infertility cases treated*	*No. of cases treated by Vets by routine and special visits*	*Average lact. yield (lit.)*	*Buff, milk procured (Tonne)*
1961	—	—	1.3	101	30.5	837	35.58
1964	3,945	6.1	31.6	243	76.4	*1,343	60.64
1965	17,226	11.0	41.8	85	81.9	1,415	65.91
1966	28,587	86.6	87.4	2,534	116.9	1,379	71.60
1967	29,895	167.1	194.3	2,524	132.3	1,455	80.64
1968	41,145	73.0	137.8	3,868	166.7	1,514	113.15
1969	57,558	173.7	152.0	3,618	215.4	1,541	113.88
1970	39,721	174.5	157.5	1,0024	234.2	1,536	118.23
1971	41,974	175.6	156.8	5,426	226.1	1,554	132.55
1972	51,313	246.2	157.3	7,140	240.8	1,588	141.52
1973	50,730	281.3	150.4	8,195	242.9	1,451	105.62
1974	59,868	195.5	192.6	17,163	256.0	1,510	121.16

Chotani, A.A. 1978, "Buffalo productivity in relation to milk requirements in the country". *Proc. Indo-Soviet Symp. Buffalo Br.*, NDDB, Anand, 16-18.

2.4.3. Types of Dairy Bovines

From the discussions in the preceding sections one might realize that the type of dairy animal (high or low yielder; zebu, crossbred or buffalo cow), has a considerable influence on the ultimate success of the enterprise. An N.D.R.I. estimate puts 2,000 litres/lactation in cows and 1,500 litres/ lactation in buffaloes as the break-even point. Animals yielding less give only loss. Whatever may be the breed, only high yielders who are also regular breeders should be selected. Crossbred cows and members of some zebu and buffalo breeds as well, meet these requirements. However, their relative efficiencies are different.

It was estimated (Patel, 1973) that the cost of production per litre of milk falls gradually as level of production increases both in case of indigenous and crossbred cows (Table 2.28). Recent studies by Smita Sirohi *et al.*, 2007 at the farmer household in Karnal district also reinforces earlier findings (Table 2.29) Experience at NDRI farm (Table 2.30) and other farms

Patel. R.K. 1973. "Importance of cattle and buffaloes in our rural economy and the challenges ahead." Paper presented at the *1st Annual Conference on Animal Production*, 21-23 April, 1973, Calcutta.

have clearly established the fact that the net cost of production per litre of milk was lowest in the high yielding crossbreds. A closer scrutiny of the production cost patterns (Table 2.29) reveals that, though the feed costs for crossbred and exotic cows are only marginally higher (55 per cent of total cost in comparison to 47 per cent in zebu and 52 per cent in buffaloes), the replacement costs in their case (4-9 per cent) are considerably lower than that in case of indigenous animals (about 17.5 per cent). Replacement costs go up when cows calve at longer intervals and calves grow and mature slowly or experience high morbidity and mortality. Also, this would mean that an efficient dairy cow is one, which remains productive for longer periods in her life.

If good crossbred cows are not available readily, which is mostly the case, one may start with reasonably good indigenous cows. These can in turn be bred by artificial insemination using semen of crossbred or exotic bulls available at government artificial insemination centres. One can then raise one's own crop of crossbreds, which can be expected to be in production from 4th year of operation onwards. Further, in India, buffaloes of excellent quality can be purchased from villages in the home tract of the famous buffalo breeds like Murrah (Haryana), Nili-Ravi (Ferozepur), Mehsana and Surti (Gujarat).

Start with really high yielding (8-10 kg/day) cows or buffaloes. Feed costs per litre of milk comes down as the yield of animals increases. Good milk yield (above 10 litre/day), sufficiently long lactation length (300 days) and reasonably short intercalving period (13-15 months) are the desirable criteria. Based on a study conducted by the Institute for Techno Economic studies in and around Madras city, Natarajan (1978) indicated the-desirable level of production of Murrah buffaloes for various rates of return to be as follows: 10 per cent rate of return—653 to 971 kg/year; 50 per cent rate of return—1,523 to 2,226 kg/year in urban and sub-urban areas. Similar desirable levels of production for rural areas were 800-1286 kg/year for 10 per cent rate of return and 1,631-3,190 kg/year for 50 per cent rate of return.

TABLE 2.28. Average Cost Per Litre of Milk for Different Levels of Milk Production during 1969-70 (Patel, 1973) at NDRI, Karnal

Production level	*Breeds*				
(litres)	*Tharparkar*	*Sahiwal*	*Redsindhi*	*Crossbred*	*Overall*
>800	2.93	2.47	2.35	2.90	2.60
801-1200	1.76	1.77	1.62	1.40	1.70
1201-1600	1.44	1.41	1.30	1.10	1.38
1601-2000	1.27	1.18	1.11	0.95	1.19
2001-2400	1.06	1.06	0.92	0.81	1.01
2401-2800	0.96	0.96	0.80	0.74	0.87
2801-3200	1.07	0.97	0.87	0.71	0.88
3201-3600	1.01	0.92	—	0.66	0.77
3601-4000	0.95	—	—	0.64	0.71
>4000	—	0.96	—	0.63	0.73
Overall average	1.24	1.11	1.20	0.72	0.99

Kumar, R.; Chauhan, Y.S. and Singh, I. 1975. A comparative study of private and cooperative units in Kanpur district of U.P. Indian J. Agri. Econ., 30 : 143.

Nirarajan, B. 1978. Proc. Indo-Soviet Symposium on Buffalo Breeding, 16-18, Nov., 1978, NDDB, Anand, p. 216.

TABLE 2.29. Cost of milk production per litre for different production levels from crossbred cow and buffalo in Karnal district of Haryana (₹/day) (Smita Sirohi *et al.*, 2007)

Particulars	*Crossbred cow*				*Buffalo*		
	Average Productivity range (litres/day)				*Average Productivity range (litres/day)*		
	≥20	**15-20**	**10-15**	**5-10**	**≥10**	**5-10**	**Up to 5**
Average Milk yield (Lit)	24.19	16.97	12.03	7.33	11.54	7.29	3.66
Net cost of milk production/litre	3.93	4.72	5.46	8.37	5.30	6.79	11.66

TABLE 2.30. Cost of Milk Production for Different Breeds of Cattle And Buffaloes Maintained at N.D.R.I., Karnal During 1978-79. (Sundaresan, 1981)*

Cost components (Paise/litre)	*Purebred Zebu like Sahiwal and Tharparkar (No. = 73)*	*Purebred Exotic like Brown Swiss (No. = 20)*	*Crossbred cows (No. = 344)*	*Murrah buffaloes (No. = 148)*
Feed cost	93.35 (47.9%)	88.62 (55.3%)	70.17 (55.5%)	131.46 (52.8%)
Labour cost	44.02 (22.6%)	32.34 (20.2%)	29.50 (23.4%)	50.17 (20.2%)
Supervisor cost	3.25 (1.7%)	2.43 (1.5%)	2.09 (1.7%)	4.55 (1.8%)
Replacement cost	34.69 (17.8%)	6.20 (3.9%)	11.73 (9.3%)	42.51 (17.0%)
Veterinary service and medicines cost.	6.52 (3.3%)	21.03 (13.1%)	4.59 (3.6%)	2.30 (0.9%)
Miscellaneous costs	12.91 (6.6%)	9.61 (6.0%)	8.27 (6.5%)	18.00 (7.2%)
Gross cost	194.74 (100%)	160.23 (100%)	126.35 (100%)	248.99 (100%)
Net cost	190.68	156.33	123.44	241.34
Average yield per day (kg)	5.55	7.42	9.54	3.96

Note : Approximate average market price per. litre of buffalo milk was ₹ 2.50 and cow milk was ₹ 1.65 during this period. This would give an idea about the margin of returns per litre of milk from different types of animals.

Based on an extensive economic analysis of milk production under mixed farming system in villages in Aligarh district of Uttar Pradesh, Pandey (1980) estimated the net cost per litre of milk in rural areas to be ₹ 1.98, ₹ 1.61, ₹ 1.57 and ₹ 1.18 respectively in case of low, medium, high yielding buffaloes and crossbred cows. At the then existing market price of milk even profits from high yielding buffaloes ought to be considerable. For detailed analysis of costs and returns and the. conditions mentioned in the above study, one may consult Table 2.31. Similar trend in cost of milk production was found in the villages around Karnal district of Haryana, by Smita Sirohi *et al.* (2007) (Table 2.29).

Thus higher yields and regular reproduction irrespective of the breed of the cow are the factors that determine returns from dairying. Persistent lactations is another key factor (Sastry *et al.*, 1979). It was observed by these workers (Table 2.32) that the cost of production per litre of milk varied directly proportional to the average number of days the buffaloes were in milk, other factors being constant. During 1974-75 during which period on this farm the buffaloes were in milk on an average just for about 122 days, the cost of production per litre of milk was the highest—₹ 2.13. It was lowest (₹ 1.23) during 1975-76 when buffaloes were in milk on an average for about 270 days. This was so in spite of the fact that during 1975-76 per day milk yield of buffaloes was lowest (3.9 litres) as compared to 5.2 litres in 1974-75. This shows the relatively greater economic importance of persistency of milk yield in dairy buffaloes.

TABLE 2.31. Category-wise Cost of Milk Production and returns per animal during an Inter-calving Period in Aligarh District of Uttar Pradesh at 1973-79 Prices (Pandey, 1980)

Particulars and cost/returns (₹)	*Low yielding buffaloes (in rupees)*	*Medium yielding buffaloes (in rupees)*	*High yielding buffaloes (in rupees)*	*Crossbred cows (in rupees)*
Current value of animal	1,184.00	1,589.00	1,919.00	2,246.00
1. Interest on investment on animal	190.70 (9.6)	251.78 (10.8)	301.58 (10.7)	320.44 (11.2)
2. Replacement cost/depreciation on animal	204.32 (10.3)	269.76 (11.6)	323.13 (11.5)	343.32 (12.0)
3. Interest and depreciation on fixed costs	84.39 (4.3)	72.64 (3.1)	91.66 (3.3)	81.75 (23)
4. Imputed value of labour @ ₹ 6 per man per day.	410.85 (20.7)	495.44 (21.2)	550.42 (19.5)	646.29 (22.5)
5. Feed and fodder costs	1083.97 (54.6)	1230.83 (52.7)	1536.44 (54.5)	1450.28 (50.6)
6. Other costs	10.50 (0.5)	14.82 (0.6)	15.91 (0.5)	24.37 (0.8)
7. Total costs	1984.73	2335.27	2819.04	2866.45
8. Total costs excluding labour costs	1537.88	1839.83	2268.62	2220.16
9. Average milk yield (litres)	898	1299	1652	2086
10. Returns from milk (₹)	1526.60	2208.30	2774.40	2481.60
11. Value of calf and manure (₹)	210.60	238.90	257.50	415.30
12. Gross returns (10 + 11)	1737.20	2447.20	3031.90	2896.90
13. Net returns excluding labour cost	163.22	607.37	763.28	676.24
14. Net returns including labour cost	–247.53	111.93	212.86	30.45
15. Net cost (7-11)	1774.13	2096.36	2561.54	2451.15
16. Net cost per litre of milk	1.98	1.61	1.57	1.18
17. Net return per litre of milk	–1.28	0.09	0.13	0.20

Note : (*a*) Figures in brackets indicate costs as percent of total cost.

(*b*) Labour costs under village conditions is never an out-of-pocket cost as members of the farmer's family generally take care of milch animals as a routine chore (refers to item 4, 8, 13 and 14 above).

(*c*) The price of buffalo milk and cow milk was ₹ 1.70 and ₹ 1.20 per litre respectively (refers to item 17 above).

TABLE 2.32. Average Number of Days the Buffaloes were in Milk and Average Annual Milk Production per Buffalo during the 5 years under Study alongwith other Relevant Economic Characters on a Small Dairy Farm at Hissar (Sastry *et al.*, 1979)

Particulars	*1971-72*	*1972-73*	*1973-74*	*1974-75*	*1975-76*	*Average*
Number of buffaloes	5	8	8	7	8	7.4
Average annual milk production per buffalo (kg)	1005.1	988.5	1175.5	652.0	1064.2	979.9
Average number of days in milk/year	206.6	225.4	268.0	121.5	270.1	225.6
Wet average (kg)	4.8	4.4	4.4	5.2	3.9	4.3
Dry average (kg)	2.8	2.7	3.2	1.7	2.9	2.7
Total annual milk production in the unit (kg)	5025.3	7908.3	9404.1	4424.3	8513.2	7055.4
Cost of production of 1 litre of milk (₹)	1.82	1.33	1.30	2.13	1.23	1.39

$$\text{Wet average} = \frac{\text{Annual milk production}}{\text{Number of days in milk during the year}} \qquad \text{Dry average} = \frac{\text{Annual milk production}}{365}$$

Hence, not only one should start with dairy animals of the type discussed above, but also strive to keep them in such a production state. Further, one should start with young animals so that the animals remain in herd longer. Either choose pregnant heifers (performance of their dams and other relatives should be good) or select first calvers by recording their actual milk yield by three consecutive milking.

2.4.4. Cost of Raising Replacements

Only very limited information is available on this from developing countries. Sharma and Thomas (1981) observed that raising Murrah buffalo replacement can fetch a profit varying from ₹ 465 to ₹ 1,066 per animal if they are sold immediately after first calving. The profitability depended on the level of concentrate feeding (see Table 2.33).

TABLE 2.33. Approximate Net Returns from Buffalo Calf Raising up to First Calving under Different Levels of Concentrate Feeding

Level of concentrate feeding (percentage) of body weight	*Cost of feeding per month (₹)*	*Compounded cost of feeding from birth to first calving (₹) @ 12%*	*Compounded cost of raising buffalo up to first calving (₹) @ 12%*	*Compounded returns from buffalo calf raising up to first calving (₹) @ 9%*	*Net returns from buffalo calf raising up to first calving (₹)*	*Net returns per rupee spent (₹)*
0.0	25.43	1,695.08	2.173.18	3.000.00	826.82	0.38
0.2	35.36	1,651.49	2,117.29	3,183.57	1,066.28	0.50
0.3	38.12	1,765.45	2,263.40	3,193.92	930.52	0.41
0.4	45.24	1,822.07	2,335.99	3,257.60	921.61	0.39
0.5	54.42	2,175.00	2,788.46	3,253.03	464.57	0.16

TABLE 2.34. **Cost and returns from milk production in Farrukhabad district of Uttar Pradesh. (Atul Chandra and Agarwal, S.B. 2000. Indian-Journal of Dairy Science. 53(4): 310-316)**

Particulars	*Landless*		*Small Farmer*		*Marginal Farmer*		*Larger Farmer*		*Overall*	
	Crossbred cow	*Buffalo*	*Crossbred cow*	*Buffalo*	*Crossbred cow*	*Buffalo*	*Crossbred cow*	*Buffalo*	*Crossbred cow*	*Buffalo*
A. Variable Cost										
Green Fodder	1.53	1.06	1.08	1.10	1.05	1.17	1.18	1.14	1.14	1.11
Dry Fodder	2.76	2.32	2.02	2.95	1.71	2.04	1.66	2.00	1.96	2.41
Concentrate	2.17	1.75	1.93	1.88	1.84	2.11	2.02	1.96	1.95	1.91
Feed Cost	**6.46**	**3.74**	**5.03**	**5.93**	**4.60**	**5.32**	**4.86**	**5.10**	**5.05**	**5.43**
Human Labour	2.35	1.74	1.60	1.89	1.31	1.52	1.58	1.73	1.61	1.78
Miscellaneous	0.02	0.05	0.09	0.06	0.05	0.06	0.07	0.07	0.07	0.06
Total Variable Cost	**8.83**	**7.12**	**6.71**	**7.88**	**5.96**	**6.90**	**6.51**	**6.90**	**6.73**	**7.27**
B. Fixed Cost										
Interest on Fixed Capital	0.32	0.44	0.21	0.51	0.15	0.45	0.14	0.41	0.19	0.46
Depreciation on Fixed assets	0.44	0.27	0.34	0.30	0.32	0.28	0.40	0.23	0.36	0.28
Fixed Cost	**0.76**	**0.71**	**0.55**	**0.81**	**0.47**	**0.73**	**0.54**	**0.64**	**0.55**	**0.74**
Gross Cost (A + B)	**9.59**	**7.83**	**7.26**	**8.69**	**6.43**	**7.63**	**7.05**	**7.54**	**7.29**	**8.01**
Net Cost (Gross Cost-value of dung)	**8.99**	**7.42**	**6.86**	**8.21**	**5.85**	**7.23**	**6.65**	**7.16**	**6.83**	**7.58**

TABLE 2.35. Economics of cow milk production—A microlevel study in Andaman & Nicobar Islands. (B. Ganesh Kumar, S. Jeyakumar, R.B. Rai, Raj Vir Singh, Zachariah George and Kuntola Roy, (2006; Indian Journal of Dairy Science 59(6) : 395-400)

Investment on Fixed Assets in Dairy Farming (₹/Cow)

Investment	Local Cows				Crossbred Cows			
Farm size :	*Small*	*Medium*	*Large*	*All Farms*	*Small*	*Medium*	*Large*	*All Farms*
Milch Animal	5,300.00 (57.43)	6,000.00 (88.11)	9,666.67 (87.37)	6,988.89 (77.36)	14,142.86 (62.37)	17,778.00 (81.03)	18,862.00 (84.59)	16,927.62 (75.89)
Dairy Buildings	3,667.67 (39.73)	625.00 (9.18)	1,316.00 (11.89)	1,869.22 (20.69)	8,071.43 (35.60)	3,666.00 (16.71)	3,312.00 (14.85)	5,016.48 (22.49)
Dairy Equipments and Machinery	262.33 (2.84)	185.00 (2.72)	81.67 (0.74)	176.33 (1.95)	460.70 (2.03)	494.77 (2.26)	124.62 (0.28)	360.03 (1.61)
Total	9,229.00	6,810.00	11,064.34	9,034.45	22,674.99	21,938.77	22,298.62	22,304.13

Figures within parenthesis indicate percentage to total cost.

TABLE 2.36. Economics of milk production in Imphal West district of Manipur. (Kh Rishikanta Singh and S.B. Agarwal, 2007; Indian Journal of Dairy Science 60(6):441-446)

Fixed investment (₹) in dairying per household

Item	Herd Size Category			
	Small (one milch cow)	*Medium (2.3 milch cows)*	*Large (4 and above milch cows)*	*Overall*
Cattle shed/stores	5,992.85 (36.55)	14,300.00 (22.15)	30,214.28 (21.18)	16,835.71 (22.58)
Dairy equipment	993.21 (6.06)	2,909.38 (4.51)	5,148.00 (3.61)	3,016.86 (4.05)
Milch cow	9,411.90 (57.39)	47,364.52 (73.34)	1,07,314.30 (75.21)	54,696.91 (73.37)
Total investment	16,397.96 (100.00)	64,573.90 (100.00)	1,42,676.58 (100.00)	74,549.48 (100.00)

Figures within parenthesis indicate percentage to total cost.

2.4.5. Some Economic Observations Under Village Conditions

Most of the above economic patterns of dairy farms are based on performance of animals on institutional farms. While these give important insight into economics of dairying, it is necessary to study economic performance under actual farmers' conditions. Hence, results of economics of dairying on farmer's holding in different parts of the country are reproduced below for understanding the field conditions.

TABLE 2.37. Economics of buffalo milk production in Hisar district of Haryana state. (A. Bharadwaj, V.B. Dixit and R.K. Sethi, 2006; Indian Journal of Dairy Science: 59(5) 322-327).

Maintenance cost of milk buffaloes per day in different herd size groups

Item	*Herd Size*		
	Small (1-2 animals)	*Medium (3-4 animals)*	*Large (5 & above)*
Average number of milch buffaloes	1.38 ± 0.05	3.18 ± 0.06	5.36 ± 0.19
Percent households	67.00	24.50	8.50
Average size of land holding (ha)	3.97 ± 0.50	12.13 ± 2.05	15.07 ± 2.82
Variable costs (₹)			
Green fodder	14.05 ± 0.36(23.00)	14.30 ± 0.51 (21.03)	14.17 ± 0.72 (20.71)
Dry fodder	4.20 ± 0.27 (6.87)	4.45 ± 0.51 (6.54)	3.78 ± 0.48 (5.52)
Concentrate	25.29 ± 1.15 (41.38)	33.45 ± 1.45 (49.19)	34.14 ± 2.56 (49.90)
Total Feed cost (₹)	**43.54 ± 1.32 (71.25)**	**52.20 ± 1.66 (76.76)**	**52.09 ± 2.86 (76.13)**
Labour cost	10.88 ± 0.28 (17.80)	8.60 ± 0.28 (12.65)	8.31 ± 0.39 (12.15)
Veterinary and service charges	0.56 ± 0.02 (0.92)	0.59 ± 0.03 (0.87)	0.50 ± 0.05 (0.73)
Interest on fixed capital	4.24 ± 0.15 (6.94)	4.60 ± 0.20 (6.76)	5.16 ± 0.30 (7.54)
Depreciation on fixed capital	1.89 ± 0.08 (3.09)	2.01 ± 0.08 (2.96)	2.36 ± 0.04 (3.45)
Total cost (₹)	**61.11 ± 1.50 (100.00)**	**68.00 ± 1.70 (100.00)**	**68.42 ± 2.90 (100.00)**
Income from dung	2.25	2.25	2.25
Income from young stock	1.86 ± 0.07	1.95 ± 0.10	1.87 ± 0.12
Net cost (₹)	**57.00**	**63.80**	**64.30**
Milk yield(lt)	5.88	6.07	6.20
Sale price (₹)	11.65	11.70	11.80
Gross return (₹)	68.50	71.02	73.16
Net profit (₹)	11.50	7.22	8.85
Net profit (₹/lt)	**1.96**	**1.19**	**1.43**
Breakeven output (lt)	5.00	5.56	5.56
Percent breakeven output to total output	85.03	91.60	89.68

TABLE 2.38. Economics of buffalo milk production in West Godavari District of Andhra Pradesh—a case study. (Hymajyothi, S., Reddy, S.U., Raju, V.T. 2003; Indian-Journal of Dairy Science. 56(4): 258-260).

Item	*Herd Size*		
	Small (1-2 animals)	*Medium (3-4 animals)*	*Large (5 & above)*
Gross cost (₹/day)	31.71	42.60	53.25
Net cost (Gross cost—value of dung) ₹/day	28.01	38.60	48.00
Milk Yield in litres	3.52	4.87	6.10
Cost of Milk (₹/lt)	7.96	7.93	7.87

TABLE 2.39. Economics of milk production in Imphal West district of Manipur (Kh Rishikanta Singh and S.B. Agarwal, 2007; Indian Journal of Dairy Science 60(6): 441-446)

Gross maintenance cost per day per milch animal with cost of milk production per litre in local cow and crossbred cow in Imphal West district of Manipur

Items	*Local Cow*				*Crossbred cow*			
	Herd size category				*Herd size category*			
	Small (one)	*Medium (2-3)*	*Large (4 & above)*	*Overall*	*Small (one)*	*Medium (2-3)*	*Large (4 & above)*	*Overall*
Total feed cost (₹)	9.95	13.51	8.16	10.54	41.56	47.70	46.61	45.29
Total variable cost (₹)	18.57	22.62	16.21	19.13	57.48	62.29	60.17	59.98
Total fixed cost (₹)	3.29	4.17	3.83	3.76	13.70	12.68	12.55	12.97
Gross cost (₹)	21.86	26.79	20.04	22.89	71.18	74.97	72.72	72.95
Net cost (₹)	21.09	25.65	18.88	21.87	69.93	73.34	71.12	71.40
Milk yiled/milch cow (lt.)	0.85	1.16	1.45	1.15	5.50	6.56	7.30	6.48
Cost/liter of milk (₹)	24.81	22.11	13.02	19.01	12.71	11.17	9.74	11.02
Gross return (₹)	15.38	20.88	26.10	20.73	77.71	85.28	94.90	86.24
Net return (₹)	– 5.70	– 4.77	7.22	– 1.14	7.78	11.94	23.78	14.84

TABLE 2.40. Economic analysis of milk production in tribal area of Udaipur, Rajasthan. (Singh, R.N.; Chauhan, A.K.; Sharma, S.P., 2006; Indian Journal of Dairy Science. 59(5): 328-336)

Item	*Local Cow*	*Buffalo*
Gross cost of maintenance (₹/day/milch animal)	21.89	30.70
Imputed value of dung (₹/day/milch animal)	0.65	0.74
Milk Yield (lt.)	1.49	2.38
Net cost of milk production (lt.)	14.26	12.59
Net returns (₹/Litre)	– 3.05	2.84

TABLE 2.41. Economics of cow milk production—A microlevel study in Andaman & Nicobar Islands. (B. Ganesh Kumar, S. Jeyakumar, R.B. Rai, Raj Vir Singh, Zachariah Ceorge and Kuntola Roy, 2006; Indian Journal of Dairy Science 59(6): 395-400).

Cost of Milk Production and net returns in local and crossbred cows maintained by different categories of farmers

Items	*Local Cows*				*Crossbred cows*			
Farm size :	*Small*	*Medium*	*Large*	*Overall*	*Small*	*Medium*	*Large*	*Overall*
Total cost of milk production (₹)	24.31	15.89	18.37	21.45	39.71	39.70	39.83	39.75
Milk Yield (Lt.)	3.27	3.38	4.00	3.55	4.93	4.91	5.50	5.11
Cost of Milk Production per litre (₹)	7.17	4.35	4.44	5.79	7.90	7.94	7.13	7.65
Net returns (₹)	15.95	22.81	21.90	18.47	17.69	20.21	23.83	21.10

TABLE 2.42. Economics of milk production: variations across productivity levels. (Smita Sirohi; Joshi, B.K.; Yogendra Kumar, 2007; Indian Journal of Dairy Science; 124-128)

Cost of returns for milk production from crossbred cow and buffalo in Karnal district of Haryana (Rs./day)

S. No.	*Particulars*	*Crossbred cow*						*Buffalo*				
		Average productivity range (litres/day)				*Lactating*	*Milch*	*Average productivity range (litres/day)*			*Lactating*	*Milch*
		≥ 20	*15-20*	*10-15*	*5-10*			*≥ 10*	*5-10*	*Upto 5*		
A.	**Total Fixed Cost**	**12.45**	**13.19**	**12.44**	**13.43**	**12.88**	**12.70**	**10.99**	**9.96**	**9.29**	**9.84**	**9.37**
	(1) Depreciation on fixed assets	7.42	7.81	7.71	8.80	8.01	8.03	6.96	5.92	6.06	6.21	5.91
	(2) Interest on fixed capital	5.03	5.38	4.74	4.63	4.88	4.68	4.03	3.54	3.23	3.63	3.46
B.	**Total Variable cost**	**84.19**	**68.26**	**54.44**	**49.05**	**58.00**	**53.85**	**51.41**	**40.51**	**34.23**	**41.77**	**38.26**
	(1) Feed cost	79.08	62.87	48.44	42.72	52.03	47.77	45.94	34.80	29.16	36.15	32.69
	(*i*) Dry fodder	11.09	9.38	10.24	9.44	9.86	9.52	10.02	11.00	11.55	10.87	10.51
	(*ii*) Green fodder	21.75	16.75	12.22	10.90	13.68	13.13	12.67	8.48	6.10	8.91	8.42
	(*iii*) Concentrate	46.24	36.75	25.98	22.39	28.49	25.13	23.15	15.32	11.51	16.37	13.77
	(2) Labour	3.13	3.66	4.43	5.04	4.42	4.50	4.73	5.00	4.53	4.92	4.93
	(*i*) Hired	2.14	2.31	2.94	3.07	2.81	2.82	2.25	2.22	2.19	2.22	2.19
	(*ii*) Family	0.98	1.35	1.49	1.96	1.61	1.68	2.48	2.75	2.34	2.70	2.74
	(3) Vet. & Misc. expenses	1.99	1.72	1.57	1.31	1.55	1.57	0.74	0.71	0.55	0.70	0.63

C. Gross cost (A + B)	**96.64**	**81.45**	**66.88**	**62.48**	**70.88**	**66.55**	**62.40**	**50.47**	**43.52**	**51.61**	**47.63**
D. Average Milk yield (Lit)	24.19	16.97	12.03	7.33	12.41	8.27	11.54	7.29	3.66	7.53	4.37
E. Value of Milk	211.30	143.86	104.91	63.90	107.60	71.71	116.27	73.24	36.14	75.66	45.37
F. Value of dung	1.59	1.37	1.17	1.10	1.28	1.22	1.20	0.97	0.88	1.00	0.96
G. Net cost of milk production per litre (C – F)/D	3.93	4.72	5.46	8.37	5.61	7.90	5.30	6.79	11.66	6.72	10.68
H. Net profit margin (5) (E + F – C)/E	**55.02**	**44.34**	**37.36**	**3.95**	**35.32**	**8.91**	**47.36**	**32.42**	**–17.97**	**33.11**	**–2.83**
I. Income per Rupee of variable expense (E + F)/B	2.53	2.13	1.95	1.33	1.88	1.35	2.28	1.83	1.08	1.84	1.21

REFERENCES

Atul Chandra and Agarwal.S.B 2000. Cost and returns from milk production in Farrukhabad district of Uttar Pradesh. Indian Journal of Dairy, Science. 53(4): 310-316.

Gupta, J.N., Agarwal. S.B. 1996. Economics of milk production in Himachal Pradesh. Indian Journal of Dairy Science. **49**(9): 556-561.

Hymajyothi, S., Reddy, S.U., Raju, V.T., 2003. Economics of buffalo milk production in West Godavari District of Andhra Pradesh—a case study. Indian Journal of Dairy Science. **56**(4): 258-260

Kh Rishikanta Singh and S.B. Agarwal 2007. Economics of milk production in Imphal West district of Manipur. Indian Journal of Dairy Science **60**(6):441-446

Pandey, R.N. 1980. *Economic significance of buffaloes under the mixed farming system.* Proc Summer Inst. Buffalo Management Systems, HAU, Hissar, p. 42.

Pranajit Bhowmik; Smita Sirohi; Dhaka, J.P. 2006. Gains from crossbreeding of dairy cattle in the North East: micro evidence from Tripura. Indian Journal of Agricultural Economics. **61**(3): 305-313.

Rao. B.D and Singh. C.B. 1995. Impact of the Operation Flood Programme on income and employment levels of rural households in Guntur district of Andhra Pradesh. Indian Journal of Dairy Science. **48**(2): 122-128.

Sanjay-Kumar; Mahesh Chander; Harbola, P.C. 1999. An economic analysis of production and disposal pattern of milk and milk products in Kumaon Hills. Indian Journal of Animal Health. **38**(2): 147-150.

Sastry, N.S.R., Yadav, R.S. and Pal, R.N. 1979. An appraisal of the economics of a Small dairy farm with buffaloes, *H.A. J. Res.* **9**: 266.

Sharma, V.B. and Thomas, C.K. 1981. Economic level of concentrate supplementation in the ration of buffalo heifers. *Indian J. Anim. Sci* **51**: 157.

Sheokand.R.S., Narinder Singh and Sheoran, O.P. 2000. Comparative economics of different farming systems under paddy-wheat farming sequence in rice-belt of Haryana. Haryana Agricultural University Journal of Research. 2000; **30**(1/2): 49-53.

Singh .R.N., Chauhan, A.K., Sharma, S.P 2006. Economic analysis of milk production in tribal area of Udaipur (Rajasthan). Indian Journal of Dairy Science; **59**(5): 328-336.

Singh. R.B. and Krishna, P.V. 1975. "Economics of dairy farms—a comparative study of three private dairy farms in U.P." Indian J. Agril Eco., **30**: 154.

Smita Sirohi, Joshi, B.K., Yogendra Kumar (2007). Economics of milk production: variations across productivity levels. Indian Journal of Dairy Science. **60**(2): 124-128.

CHAPTER 3

Starting a Dairy Farm

Starting a dairy farm, like starting any other enterprise, should be based on a sound knowledge about the business, a reconnaissance and an assessment of the demand and marketing facilities for the produce—milk; a study about the ease and cost of availability of inputs including finance; and a genuine love for the job.

3.1. OBTAINING TRAINING AND EXPERIENCE

No amount of training can make a dairyman out of a person unless he possesses a genuine love for cows. Through ages, it is man's love for animals that guided him to domesticate and rear cows, laying foundations for the future civilization in the process. Wherever dairying prospered, there have been men and women who loved cows. In the Indian villages care of dairy animals is lovingly done by the kind-hearted women folk.

It is this love for cows, which helps the dairy farmers stoically undertake the harshest chores on a dairy farm and bear long hours of tiring work and lack of a "holiday". It is this love that makes the dairyman not to bother about the stench of cow dung, slurry and filth. It is this love, which brings out emotional attachment between him and his animals and makes him feel "delicate" towards a new born calf, a cow ill or having difficult labour and makes him worry over impending shortages of feeds or an epidemic. In short, a person having no liking for animals can never be a good dairy farmer. He can neither realise the problems himself nor can appreciate the same when presented to him by others.

Further, to be a successful dairy farmer, one must have a good knowledge about all aspects of scientific management of dairy stock as well as management of the farm business. To acquire such skills, there is no substitute for practical experience. This practical ability to do the job on the farm with one's own hands is a must if one wants to prosper in dairying.

Watching how the other dairy farmers are performing he can learn this art. One can work as a sort of apprentice on a couple of already established farms. Another way is to watch how the neighbours are approaching and doing the job. One may visit demonstration units put up periodically by the extension wings of Agricultural Universities, Animal Husbandry Departments, Dairy Development Corporations and other government organisations. The above organisations also periodically organise cattle or dairy fairs and undertake short and long term training programmes on dairy farming. Such organisations exist in each State and may be contacted by the prospective dairyman.

The minimum that the enthusiast can do is to contact local veterinarian or the extension official of the locality (generally located in the Panchayat Samiti or Community Development Block) or ask for and listen to information broadcast from the local radio station.

3.2. LOCATION OF THE FARM

Discussions in the previous chapter might have given the reader an idea about the relative importance of the dairy animals, the land and the farmer to the overall efficiency of the dairy enterprise. Having understood these fineries, if one decides to go into dairying, the next logical decision to be taken is where to start the farm.

The greatest demand for milk is where there is a large concentration of people who themselves are not producers of milk, *i.e.,* in towns and cities. However, in urban areas the conditions are not exactly ideal nor economical for production of milk. Land may not be available (or available at high rentals) for animal stables and fodder production. The producer has to purchase perhaps all the feed items (greens, dry fodder and concentrates) at higher prices. Yet many city dairy stable owners in our cities mint money as they can sell milk (at times adulterated) at high prices. They can do so because of the great gap in demand and supply of milk.

In rural areas costs of key inputs are lower but so are the demand for milk and marketing facilities. Villages located close to cities on highways leading into cities are ideally suited for dairy business. Indeed, bulk of the commercial milk producers in the country comprise of this category, which market milk readily in cities either by themselves or, more often, by middlemen. As many State governments are spreading milk collection networks into rural hinterlands, it should now be possible to operate dairies even in interior villages. Wherever such facilities exist, the dairy farmers must make best use of them. Alternatively, all producers of a village or locality may form their own cooperative marketing facilities.

3.3. LAND REQUIREMENTS

While the size of the land available puts a limit to the number of cows that can be maintained, there is considerable scope to increase the carrying capacity (number of cows per unit area) of land by following suitable agronomic practices. More number of cows (especially high yielder) per unit area means more income per unit area. Milk production is dependent on fodder production and milk yielding potential of the herd. A hectare of land producing 6 tons fodder per year can support only 4 cows, whereas a hectare producing 30 tons per year can maintain 20 cows. On an average, an annual fodder yield of 15 tons is by no means a difficult proposition when irrigation is assured. One has to take into consideration the normal yields/ha of various fodders in one's own locality while planning one's dairy farm.

Level of milk production of cows may vary from 5 to 10 liters or more per day. Thus a hectare of land with a carrying capacity of 10 cows yielding on an average 5 liters per day will produce 13,687.5 liters per year (Note that only about 75 per cent of cows—7 or 8 in this case—are expected to be in milk at any given time). The milk production can be doubled if either the carrying capacity is raised to 20 cows by producing more fodder per hectare of by keeping cows with a daily milk yield of 10 liters. Better management and breeding can make such cows available. The size of the farm, the efficiency of fodder production (and carrying capacity of land) and the level of production of cows are key factors for a successful dairy enterprise. Estimated milk yield "per hectare" at different levels of carrying capacity and with different types of cows are given in Table 3.1.

The land-fodder-cow-milk production axis can be easily realised from this Table. The farmer can increase daily milk production (and so cash returns) either by improving carrying capacity of land (by improving agronomic practices) or by improving milk production of cows (by improving husbandry); best results can be achieved by improving both. Further, with increasing carrying capacity per hectare there is more dung production, which can be put back in the fields as manure, thus permitting limited use of chemical fertilizers. If the reproductive efficiency of cows is poor, there may be less than 75 per cent of cows in milk. This means less milk producing cows per hectare and less income. For this reason is the need for maintaining high yielding regularly calving cows.

Apart from land required for production of fodder, land is also required for animal houses, stores, roads etc. Space requirements for different classes of cattle and buffaloes are discussed in Chapter 19. One may need to set apart roughly 10 per cent area for roads, space between buildings etc. Estimated values of land required for dairy farms producing 30 and 50 liters of milk per day with different dairy animals are given in Table 3.2. These calculations are based on actual production figures of animals and crops under field conditions as reported in literature. One can draw up one's own estimates by using similar figures for the locality.

Grazing is looked upon as an important and viable feeding programme for milch cows. In India, though grazing is the main (though meagre) source of green fodder for bovines all over the country, it cannot be considered on economic terms vis-à-vis dairying as, except on a few organised farms, it is not part of commercial milk production. Nor are pastures specially cultivated and kept in prime condition. We have only over-grazed natural grasslands, which are really green only after rains. Rest of the year they serve mostly as loafing areas.

3.4. MOBILISATION OF FINANCES

(Material on dairy loans in this section is based on an article by Dr. A.C. Gangwar; Proc. *Summer Inst. Buff. Management Systems,* H.A.U, Hissar, 1980, p. 242).

After getting convinced of the technological and economic feasibility of dairying as business, the would-be dairy farmer has to look for money. This could be and generally is the most difficult-to-obtain input. Personal and parental investment, joint or co-operative ventures and institutional financing are the common means. The central and many state governments, through various development schemes, extend either directly or through commercial banks loans for dairying to landless labourers, small and marginal farmers, members of weaker sections or even to unemployed graduates.

Government dairy development schemes of some type or other are in operation in most of the developing countries. So are schemes of some wings of international organisations like the Food and Agricultural Organisation. Under such schemes dairy farmers are provided with cash loans, subsidised inputs and technical advice.

TABLE 3.1. Estimates of total milk production per hectare at various levels of fodder yield/hectare and yield/Cow (After Relwani, L.L., 1972, *Lush green fodder yields*. NDRI Publication No. 96, NDRI, Karnal)

Green yeild hecate/ year (q)	*Carrying capacity (cow/ha)*	*Cows actually in milk*	*When milk yield/cow/day is 5 kg (with no extra concentrate)*		*When milk yield/cow/day is is 7.5 kg (with 1 kg concentrate extra)*		*When milk yield/cow/day 10 kg (with 2 kg concentrate extra)*		*Production of dung tones/year @ 16 kg/cow/day*
			Milk yield per day (kg)	*Milk yield per year (kg)*	*Milk yield per day (kg)*	*Milk yield per year (kg)*	*Milk yield per day (kg)*	*Milk yield per year (kg)*	
600	4	3.0	15.0	5,475.0	22.5	8,212.5	30.0	10950.0	23.36
900	6	4.5	22.5	8,212.5	33.8	12,337.0	45.0	16425.0	35.04
1,200	8	6.0	30.0	10,950.0	45.0	16,425.0	60.0	21900.0	46.72
1,500	10	7.5	37.5	13,687.5	56.3	20,549.5	75.0	27375.0	58.40
1,800	12	9.0	45.0	16,425.0	67.5	24,637.5	90.0	32850.0	70.08
2,100	14	10.5	52.5	19,162.5	78.8	28,762.0	105.0	38325.0	81.76
2,400	18	12.0	60.0	21,900.0	90.0	32,850.0	120.0	43800.0	93.44
2,700	18	13.5	67.5	24,637.5	101.3	36,974.5	135.0	49275.0	105.12
3,000	20	15.0	75.0	27,375.0	112.3	40,989.5	150.0	54750.0	116.80

TABLE 3.2. Estimates of Land Required and Fodder Production and Animal Housing for Two Dairy Farms Planning to Produce 30 and 50 litres of milk dialy with different Diary Bovines.

Particulars	*30 littres/day*			*50 litres/day*		
	Buffalo	*Zebu cow*	*Crossbed*	*Buffalo*	*Zebu cow*	*Crossbred*
1. Average body weight of cow (kg)	480	250	375	480	250	375
2. Average daily milk yield (kg)	6	4	8	6	4	8
3. No. of milch cows required to produce desired quantum of milk daily	5	8	4	8	13	6
4. No. of total cows to be maintained to take care of dry cows (wet : dry cows being 65 : 35 in buffaloes, 70 : 30 in Zebu cows and 75 : 25 in crossbreds)	8	12	6	13	16	8
5. Likely numbers of followers to be maintained (female calves of various ages, males sold, mortality too considered)	4	6	3	6	7	5
6. Digestible crude protein requirements of herd per day as per Sen and Ray standards (kg)						
(*a*) DCP for maintenance/cow/day	0.290	0.168	0.240	0.290	0.168	0.240
(*b*) DCP for milk production/cow/day	0.378	0.204	0.360	0.378	0.204	0.360
(*c*) DCP per calf (150 kg)/day	0.350	0.350	0.350	0.350	0.350	0.350
Total daily DCP need for the herd	6.744	6.564	4.650	10.784	8.402	6.550
7. Daily green fodder (1.2 per cent DCP level) requirement to meet 80 percent of the DCP requirement + 20 percent for wastage, shortfalls etc. (q)	5.75	5.50	4.00	8.60	7.00	5.50
8. Annual green fodder requirement (tons)	210	200	150	320	270	200
9. Land required to produce requisite quantum of green fodder annually at an average yield of 75 tons/ha/year (ha)	2.800	2.660	2.000	4.260	3.600	2.660
10. Land required for buildings for animals, spaces between buildings, roads, etc. (ha)	0.015	0.020	0.010	0.023	0.025	0.014
11. Total land requirement (ha)	2.815	2.680	2.020	4.283	3.625	2.674
Rounded up value (ha)	3.00	2.75	2.2	4.50	3.75	2.75

Dairying is a capital-intensive farming enterprise. The individual who is establishing himself in the dairy business needs a relatively large amount of money on a long term basis for : (*i*) purchase of land and dairy animals, (*ii*) construction of animal houses; and (*iii*) to obtain equipment and supplies. In addition, one also requires short-term loans as running capital for purchase of feed, fertilizers, seeds etc. Interested farmers may take advantage of the various government dairy development schemes (see section 1.5) for loans etc.

Bankers Rating of Borrowers Characteristics (Table 3.3)

In addition to the information contained in the financial statement, the lender considers carefully other characteristics of the borrower in deciding whether or not to grant a loan.

TABLE 3.3. Banker's Rating of 15 Borrower Characteristics

Characteristics	*Upper third*	*Middle third*	*Lower third*
Group I			
1. Managemental practices	77	20	3
2. Financial progress	70	28	2
3. Loan history with bank	67	28	5
4. Ambition	58	37	5
5. Farming experience	44	49	7
Group II			
6. Financial backing	38	32	30
7. Formal credit rating	33	32	35
8. Scale of farm operation	31	48	21
9. Family and martial status	16	39	45
10. Buying habits	11	54	35
Group III			
11. Flexibility of farm business	11	49	40
12. Level of family living	5	33	62
13. Family background	4	25	71
14. Mechanical skills	2	18	80
15. Status in community	2	6	92

The most important criterion for deciding on credit application should be the performance of the borrower in managing his business and repayment of loans on schedule. The borrower may be asked to prepare and provide a complete and accurate financial statement based upon a good set of farm records. This must be a realistic summary of all assets and all liabilities. Credit is based on faith and mutual trust. Any one who is withholding information about him, which is essential to the banker for an accurate appraisal of his financial status, is not worthy of a loan. He should also give a detailed plan as to how the money is to be used and how the loan is to be repaid. One cardinal principle for banks in giving loan is "loan only as much money as the earning ability of the borrower's business will allow him to repay on a reasonable schedule."

Money may be loaned for use in profitable productive units only, such as high producing cows and land. Lending money for buildings that are unnecessarily elaborate and extensive or for equipment and machinery that cannot be fully employed should be avoided. Banks may not encourage too much of short-term loans since these debts with tight repayment schedules may limit borrowing power for long-term loans.

It has been observed that a large amount of sum advanced by banks is overdue against the borrowers. The major reasons for the poor recovery of dairy loans as reported by the commercial banks are as follows : (*i*) natural calamities such as floods or drought and death of animals due to epidemics (*ii*) delay in the disbursement of loans due to factors beyond the control of bank (mainly delays in the completion of documentation and other formalities) and consequent mis-utilization of loans; (*iii*) defective loan appraisal particularly in terms of assumptions made with respect to yields, input costs and output prices; (*iv*) unsatisfactory marketing arrangements resulting in borrowers getting lower than expected price for milk; (*v*) unforeseen expenses of the borrowers; (*vi*) borrower's tendency to repay loans from other agencies which carry higher rates of interest; (*vii*) lack of earnestness on the part of bank staff to recover loans; and (*viii*) wilful default.

3.5. FEASIBILITY TESTS FOR DAIRY CREDIT

In order to judge whether borrowing is necessary and feasible, certain tests have been prescribed. Such feasibility tests involved in credit analysis fall into three groups pertaining to returns, repayment capacity and risk-bearing ability—the three R's. Borrowers and financing agencies should examine the following three questions carefully:

(*i*) Will investment produce sufficient returns to cover additional costs?

(*ii*) Will the borrower have sufficient repayment capacity to pay back the loan when it is due; and

(*iii*) Does the borrower have the risk-bearing capacity to meet the risk and uncertainties involved in using borrowed funds?

If the answer to these questions is 'yes' from the borrower and financer, there is a sound case for advancement of loan. If the answer is negative for question one (above) it indicates that the use of borrowed funds is not profitable. Hence there is no point in using owned capital or borrowed funds. The negative answer to question two and three indicates that the borrower may fail to fulfill his obligations and hence loan should never be advanced or borrowed by the borrower under such conditions. These three economic principles of credit management are discussed below.

Return. The first 'R' of credit, *viz.*, Return, has great significance to the borrower as well as the financier. Both should be satisfied that the credit produces sufficient returns to cover the principal and interest. The procedure, which is commonly used to determine capital and credit needs, constitutes an important part of planning and budgeting techniques. Livestock planning is to guide the borrower in selection of right type of livestock or their combinations, as to increase the income. Budgeting is determining the capital requirements and income from livestock.

Repayment Capacity. The second 'R' of credit comprises of the test of economic feasibility, which determines the amount, which the borrower would spare for repayment of loan from his total earning. Self-liquidating loans are those funds, which are used for short-term purposes such as operating expenses, which are used up in one production season. They become component of the cost; hence generally do not pose any problem of recovery. Non-liquidating funds are partially held up in a single production process but continue for several years, *e.g.*, livestock and cattle shed.

Both the proposals have been critically examined. Although they will generate sufficient income for the repayment of loan, adequate amount will not be left for the maintenance of the family if loan is repayable in 4 years as proposed in the schemes. In both the cases one lactation is treated equal to one year, which is not correct. To have 4 times income in loan period, about 1,920 days are needed which is equal to 5 years and 4 months. Hence recovery of loan in 4 years could not be effectively enforced. It is suggested that period may be raised to 5-6 years instead of 4 years if not more. Further, if subsidy on interest and insurance is not allowed, loan advanced for establishment of Mini Dairy Unit will not generate adequate income for repayment of the loan, although cost-benefit analysis show about 90 paise return over investment. Loan installment is quite heavy, hence period of repayment need be increased.

In the first proposal, no provision is made for cattle shed. Where small and marginal farmers and landless labourers intend to take up dairying, there is need for the provision of an additional sum of Rs 5,000 for the construction of cattle shed. Such loan proposals generate adequate income to meet loan instalment of cattle shed too.

Risk-bearing ability. It is another economic feasibility test of a loan. Risk bearing ability means the ability to withstand unexpected low income and unpredictable expenses and losses. Risk bearing ability provides the "last line of defense" in the use of credit. Variation in income occurs as a rule rather than exception in dairying. Such variability in income has to be accounted for, before arriving at fairly stable and reliable estimates of repaying capacity. Insuring animals against death can eliminate part of risk but yield variability is to be overcome with varying success by scientific husbandry.

In conclusion, dairy loans should be considered beneficial both to the borrowers and banks. The study conducted by the Centre for Management of Agriculture of the Indian Institute of Management, Ahmedabad (George and Srivastava, 1975) on the impact of the State Bank of India initiated scheme for purchase of dairy animals in Baroda district in 1972, has shown that the said loaning scheme was viable and feasible from the point of view of its impact on the direct beneficiaries (farmers and banks) and the indirect beneficiaries (co-operative societies and the dairy). The scheme assured farmers of a continuous source of income through self-perpetuation of cattle loan. The bank recovered all the expenses incurred in running the scheme. The feasibility criteria of the scheme ensured that the investment in cattle development schemes was justified on economic considerations and, at the same time, provided built-in tool for demonstrating that it was self-liquidating.

3.6. MANAGER AND MANAGEMENT

Management means skillful use of means of production. The agency that carries out this activity is the manager. On dairy farms, management comprises of carrying out by and/or through the manager a number of operations aimed at usage of such means of production as land, livestock and capital for efficient and economic production of milk.

3.6.1. Role of the Dairy Farm Manager

The farm manager or the owner-husbandman is the central figure around whom the whole farm activity revolves. His quality, skill and honest role is, thus, a key factor for the success of the dairy enterprise.

Qualities of a good manager. The farm manager or the farmer has to be a leader. A leader is a person who can get the right things done in the way he wants without causing any bad feeling

---

George. P.S. and Srivastava. U.K. 1975. Institutional finance for dairy development. *Indian J. Ag. Econ.*, 30:90.

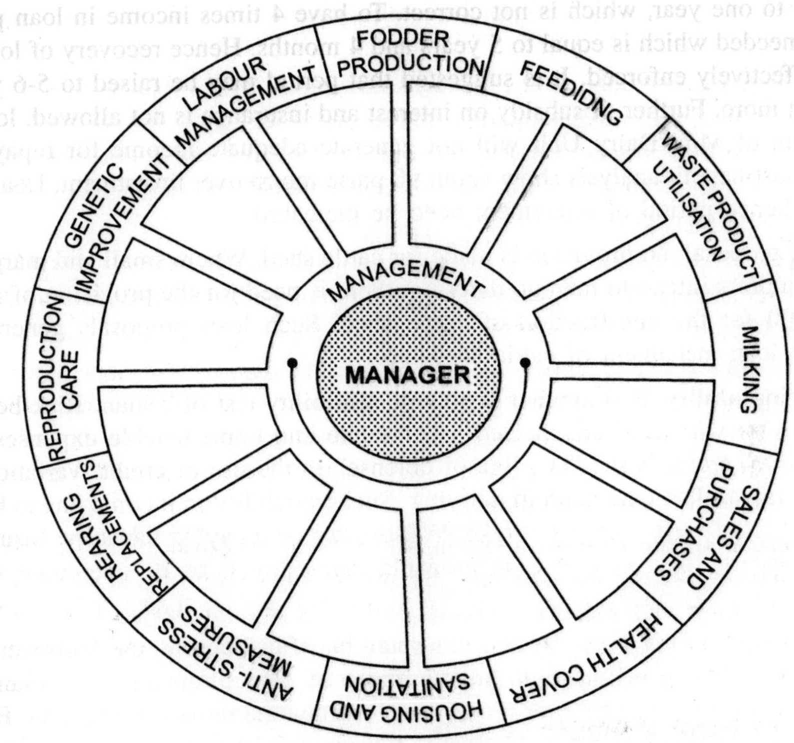

Fig. 3.1 The wheel of dairy enterprise management: man-management—management component relationships.

among the people who do it for him. A good manager should, like a general, lead his staff and not drive from behind or be so indecisive as to be led by subordinates. Subordinates should have confidence in his managerial judgment and abilities.

A manager can possibly be of one of the three types, *viz.*, authoritarian, democratic or inspiring (catalytic). The first two types are equally not favoured as the ruthless application of power or yielding to everybody's opinion is nothing but ruinous. A good manager gets things done by tact, persuasion or sometimes by controlled use of harsh words, depending on the situation. At the same time he should avoid involvement in non-managerial routine or physical work. His ability to do a thing in a better way by demonstration, however, should never be in doubt.

Animal science is a vast subject and as manager in a big establishment one can come from any of the disciplines—animal husbandry, technology, nutrition, veterinary medicine etc. Consequently, there is every temptation that his priorities maybe biased towards the discipline, from which he comes; but undoubtedly this will ruin the business. Special care has to be taken to devote needed attention to all the departments, so that lopsided development is avoided.

A manager's duty, apart from management and co-ordination for harmonious running of the farm, is to lay down policies and assure their execution or implementation. An appraisal of the situation, where you stand in the enterprise is a good starting point. This, one can know if he maintains proper records of the farm, suitably modifying the policies after studying the records so that the income is increased. Originality and ingenuity has to be shown. A static farm devoid of original ideas is a lifeless entity.

Sound management always ensures continuity and ordinarily the second in command is groomed in a fashion that he can shoulder the responsibility at a short notice. If the manager is out of commission by accident or death the work on the farm should not come to an abrupt stop. Apart from this, there should be formalised routes within the organisation for the permanent replacement of its manager.

The manager should never let it get into his head that he is indispensable. The *Iron Man of India* once remarked, "All the persons who are now resting in graves were once indispensable but the world did not fall apart on their leaving".

Added to the above qualities are sound health, sound judgement, and sound and firm decisions. Nothing pays more than to carry the employees along with the manager. If any one has a good suggestion, the manager must consider and may adopt it. He should not brush it aside as it has come from a low paid person. Their suggestions are valuable as they are practical workers and know the basic difficulties.

The manager should be down to earth with a sound practical mind. His main consideration in making decisions should be economy. By proper supervision he must try to economise the working of the farm and thereby increase profit. He should devise time and money saving functions of men and machines so that more can be obtained with the available resources. This can be achieved by organising time motion studies.

3.6.2. Role of Farm Manager

The manager's activities include planning, organising, directing, coordinating and controlling the enterprise.

Planning. An advance plan regarding the four pillars of successful animal enterprise, *viz.*, breeding, feeding, management and disease control and marketing needs to be drawn. The purchase, selection (minimum level of production of animals to be retained in the herd), breeding (future of the enterprise depends on this) and disposal of animals (culling), cultivation of crops to supply green fodder throughout the year, preservation of surplus forage, procurement and storage of concentrates and other feeds, are some of the details. Off season purchases always cost much more than purchases at harvest. Such minor considerations need advance planning. Otherwise one may have to pay through one's nose.

A thorough study of records should be made and new plans drawn after remedying the shortcomings if any. Each day's work should be planned at least one day earlier. The various farm operations for a day (*see* Chapter 22) should be put down in a register. Depending on the inclination and capacity of each labourer duties should be allotted to each of them and not like a round peg in a square hole. The quantum of work assigned to each should be such that he is kept busy throughout without being overworked.

Organising. The guiding motto is proper men at the proper place and at the proper time. It is the manager's concern to know the norms and the quantum of unit of work that a labourer is expected to finish in certain unit of time. On this basis work has to be assigned. A practical man will know these aspects more precisely. The manager should probe the psyche of his sub-ordinates deeply to find out their temperaments and aptitudes so that he can allot to each person the type of work he likes.

Direction. Nothing confuses more than inconsistent orders given and renewed at short intervals. Orders should be specific and simple. A good officer does not do most of 'bossing'. He delegates some of his powers to his subordinates, supervises and inspires them. If some one has done a good job, he has to be told so. If someone has failed, the manager should find out how he has failed before firing him. May be there is managerial deficiency.

Co-ordination and control. Many problems can be solved easily, if there is good communication between each other, *i.e.,* the manager and the staff working with him. Any bad work or non-cooperative link in the organisation can put the entire operations out of gear. Co-ordination has to be achieved between persons working in different sections like feeding and feed supply, milking and milk handling, detection of animals in heat and breeding operations, maintenance of equipment and utensils, calf care and disease control. Strict respect for the time schedule is not debatable.

A manager should watch the trends in functioning. If he sees that production is going down, he should take immediate remedial measures. The schedules where immediate changes are required, he should, if necessary, go and check himself and do the needful before the problem is aggravated.

Good management means doing a lot of small things correctly all the while. Small things, but important things. Think, judge and decide on your own, based on previous experience. Self-evaluation and evaluating the work of other staff help in improving the conditions. He should also try to clarify the relations and maintain harmony among employees.

Owner-manager. Like an owner-driven taxi, an owner-managed farm will be more efficient and profitable, provided the owner possesses the necessary skill and technical knowledge. It is better that he should consider the following points before appointing himself as the manager.

1. Does he possess the necessary technical knowledge and training?
2. Has he sufficient knowledge of accountancy and book-keeping?
3. Is he having sufficient experience with fuel operated machines, *i.e.,,* tractors etc?

If the answers for the above questions are in the affirmative, the owner can appoint himself as the farm manager. However, this calls for great endurance, patience and perseverance on his part. On the other hand, if he lacks technical know-how it is advisable to get himself equipped by undergoing a short duration course at institutions like Agricultural Universities, Veterinary Colleges, National Institutes, etc. He can also get the help of Animal Husbandry Extension Officers in the initial stages.

The manager should be able to manage the system (milk production), as well as the enterprise (business aspects) well. While a good system management and a good enterprise management yield the highest profit margin, a poor enterprise management of a good system can be more catastrophic than good enterprise management of a poor production system (Slater and Throup, 1981). It is further relevant to remember that the enterprise or system never fails on its own accord. It is the Man in Manager who can fail them.

3.6.3. What is Management?

Management is the art of contriving to succeed (in an enterprise). If the whole dairy enterprise can be compared to a wheel the manager is at the hub of the wheel (*see* Figure 3.1). The dairy business itself comprises of performing several practices, which can be grouped (based on

Slater, K. and Throup, G. 1981. *Dairy Farm Business Managements.* Farming Press Ltd. Ipswich, p. 52.

commonness) into different components. The various components are along the rim of the wheel. The spokes of the wheel are the management efforts of the manager to succeed in each of the dairy farm management components.

Considering that only when all the components are carried out well the rim remains round and smooth; any slackness in any component will cause an inward dent or bend in the rim. This makes the progress of the wheel difficult, *i.e.*, the enterprise limps and cannot give good profit. One can imagine how difficult will be the progress if there are more bends in the rim.

Another factor that can impede the smooth progress of the wheel (dairy enterprise) is the path along which it moves (the locality where the farm is); the path (locality) should be even (ideal for dairying).

3.6.3. Cost of Establishment of a Dairy Farm

It is important to have some idea about the costs involved in starting a dairy farm and the likely profit-loss scenario of the farm in the next few years. Only then one will get the confidence to go into dairy farming.

Given below are such cost estimates computed by experts of the Department of Animal Husbandry, Haryana and National Bank for Agriculture Development (NABARD) and other financing organisations for guidance. These are actually the costs of a Unit of diary farm for which such organisations extend loans. This information will help in formulations of projects for submission to financial institutions.

3.6.3.1. *Estimates of (Dept of AH and Dairying, Govt of Haryana)*

TABLE 3.4. Cost estimate for establishing a model dairy of 20 milch animals

A. CAPITAL COST

		₹
1.	Cost of 20 Cross Breed Cows/Murrah Buffaloes @ ₹ 35,000/- per animal including transport charges	7,00,000.00
2.	Cost of animals shed and store cum office	2,36,000.00
3.	Milking equipment with site	50,000.00
4.	Cost of animal cooling unit	14,000.00
	Total Fixed Expenditure	**10,00,000.00**
	Less : Margin @ 15% *i.e.* borrowers contribution	1,50,000.00
	Bank's Finance (Term Loan)	8,50,000.00
	B. RECURRING EXPENDITURE	
1.	Cost of Green Fodder 30 kg/day/animal @100 per quintal on average basis	1,53,300.00
2.	Cost of dry fodder 10 kg/day/animal @ ₹ 250 per quintal	1,82,500.00
3.	Cost of feed/concentrates 5 kg/day/animal @ ₹ 900 per quintal	1,34,685.00
4.	Labour charges @ ₹ 3,000 p.m. for 2 labourers	72,000.00
5.	Miscellaneous expenses @ ₹ 1,500 per animal p.a. (as veterinary expenses, electricity and water supply expenses)	30,000.00
	Total	**5,72,485.00**
	Cost of Feed for Calves	
1.	Cost of Green Fodder 1.5 kg/day/calf/ @ ₹ 100 per quintal	76,650.00
2.	Cost of Dry Fodder 1 kg/day/calf @ ₹ 250 per quintal	18,250.00
	Total	94,900.00

Total Recurring Expenditure	6,67,385.00
Less Margin @ 25% *i.e.,* borrowers contribution	1,66,846.00
Working Capital (as cash credit limit) Maximum ₹ 2,50,000 (working capital provided for 6 months recurring expenditures)	2,50,270.00
INCOME	
Sale of Milk (Average milk in litres per annum × No. of animals × Approximate price of milk/l) *i.e.,* 3500 × 20 × 13	9,10,000.00
Value of Manure	10,000.00
Total Income	9,20,000.00
Income/Expenditure Analysis	
Total Income	9,20,000.00
Total Expenditure	6,67,385.00
Income after Expenditure p.a.	2,52,615.00
Monthly Income	21,051.00
Less Loan Installment	15,000.00
Net Monthly Income	6,051.00

Note : 1. This is a model Scheme keeping in view the prevailing market rates. However, Branch Manager is authorized to sanction variations ± to the extent of 15%.

2. Working Capital is calculated assuming that the farmer will keep stock of feed etc. for 6 months. However, the branches can consider the stock period for a minimum period of one month to maximum period for 6 months and accordingly working capital can be calculated and limit fixed in consultation with the farmer.

3.6.3.2. *Estimates and Report of NABARD*

1. Why do Dairy Farming ?

Dairying is an important source of subsidiary income to small/marginal farmers and agricultural labourers. The manure from animals provides a good source of organic matter for improving soil fertility and crop yields. The bio gas from the dung is used as fuel for domestic purposes as also for running engines for drawing water from well. The surplus fodder and agricultural by-products are gainfully utilised for feeding the animals. Almost all draught power for farm operations and transportation is supplied by bullocks. Since agriculture is mostly seasonal, there is a possibility of finding employment throughout the year for many persons through dairy farming. Thus, dairy also provides employment throughout the year. The main beneficiaries of dairy programmes are small/marginal farmers and landless labourers. A farmer can earn a gross surplus of about ₹ 12,000 per year from a unit consisting of 2 milking buffaloes. The capital investment required for purchase of 2 buffaloes is ₹ 18,223. Even after paying a sum of ₹ 4294 per annum towards repayment of the loan and interest the farmer can earn a net surplus of ₹ 6000-9000 approximately per year. (For details see model scheme enclosed). Even more profits can be earned depending upon the breed of animal, managerial skills and marketing potential.

According to World Bank estimates about 75 per cent of India's 940 million people are in 5.87 million villages, cultivating over 145 million hectares of cropland. Average farm size is about 1.66 hectares. Among 70 million rural households, 42 per cent operate upto 2 hectares and 37 per cent are landless households. These landless and small farmers have in their possession

53 per cent of the animals and produce 51 per cent of the milk. Thus, small/marginal farmers and land less agricultural labourers play a very important role in milk production of the country. Dairy farming can also be taken up as a main occupation around big urban centres where the demand for milk is high.

2. Scope for Dairy Farming and its National Importance

The total milk production in the country for the year 2001-02 was estimated at 84.6 million metric tonnes. At this production, the per capita availability was 226 grams per day against the minimum requirement of 250 grams per day as recommended by ICMR. Thus, there is a tremendous scope/potential for increasing the milk production. The population of breeding cows and buffaloes in milk over 3 years of age was 62.6 million and 42.4 million, respectively (1992 census).

Central and State Governments are giving considerable financial assistance for creating infrastructure facilities for milk production. The ninth plan outlay on Animal Husbandry and Dairying was ₹ 2345 crores.

3. Financial Assistance Available from Banks/NABARD for Dairy Farming

NABARD is an apex institution for all matters relating to policy, planning and operation in the field of agricultural credit. It serves as an apex refinancing agency for the institutions providing investment and production credit. It promotes development through formulation and appraisal of projects through a well organised Technical Services Department at the Head Office and Technical Cells at each of the Regional Offices.

Loan from banks with refinance facility from NABARD is available for starting dairy farming. For obtaining bank loan, the farmers should apply to the nearest branch of a commercial or co-operative Bank in their area in the prescribed application form which is available in the branches of financing banks. The Technical Officer attached to or the Manager of the bank can help/give guidance to the farmers in preparing the project report to obtain bank loan.

For dairy schemes with very large outlays, detailed reports will have to be prepared. The items of finance would include capital asset items such as purchase of milch animals, construction of sheds, purchase of equipments etc. The feeding cost during the initial period of one/two months is capitalised and given as term loan. Facilities such as cost of land development, fencing, digging of well, commissioning of diesel engine/pumpset, electricity connections, essential servants' quarters, godown, transport vehicle, milk processing facilities etc. can be considered for loan. Cost of land is not considered for loan. However, if land is purchased for setting up a dairy farm, its cost can be treated as party's margin upto 10% of the total cost of project.

4. Scheme Formulation for Bank Loan

A Scheme can be prepared by a beneficiary after consulting local technical persons of State animal husbandry department, DRDA, SLPP etc., dairy co-operative society/union/federation/ commercial dairy farmers. If possible, the beneficiaries should also visit progressive dairy farmers and government/military/agricultural university dairy farm in the vicinity and discuss the profitability of dairy farming. A good practical training and experience in dairy farming will be highly desirable. The dairy co-operative societies established in the villages as a result of efforts by the Dairy Development Department of State Governments and National Dairy Development Board would provide all supporting facilities particularly marketing of fluid milk. Nearness of dairy farm to such a society, veterinary aid centre, artificial insemination centre should be ensured. There is good demand for milk, if the dairy farm is located near an urban centre.

The scheme should include information on land, livestock markets, availability of water, feeds, fodders, veterinary aid, breeding facilities, marketing aspects, training facilities, experience of the farmer and the type of assistance available from State Government, dairy society/union/ federation.

The scheme should also include information on the number and types of animals to be purchased, their breeds, production performance, cost and other relevant input and output details with their description. Based on this, the total cost of the project, margin money to be provided by the beneficiary, requirement of bank loan, estimated annual expenditure, income, profit and loss statement, repayment period, etc. can be worked out and shown in the project report. A format developed for formulation of dairy development schemes is given as Annexure I.

5. Scrutiny of Schemes by banks

The scheme so formulated should be submitted to the nearest branch of bank. The bank's officers can assist in preparation of the scheme for filling in the prescribed application form. The bank will then examine the scheme for its technical feasibility and economic viability.

(A) Technical Feasibility—this would briefly include—

1. Nearness of the selected area to veterinary, breeding and milk collection centres and the financing bank's branch.
2. Availability of good quality animals in nearby livestock market. The distribution of important breeds of cattle and buffaloes are given in Annexure II. The reproductive and productive performance of cattle and buffalo breeds is given in AnnexureIII.
3. Availability of training facilities.
4. Availability of good grazing ground/lands.
5. Green/dry fodder, concentrate feed, medicines etc.
6. Availability of veterinary aid/breeding centres and milk marketing facilities near the scheme area.

(B) Economic Viability—this would briefly include—

1. Unit Cost—The average unit cost of dairy animals for some of the States is given in Annexure IV.
2. Input cost for feeds and fodders, veterinary aid, breeding of animals, insurance, labour and other overheads.
3. Output price, *i.e.* sale price of milk, manure, gunny bags, culled cows, male/female calves, other miscellaneous items etc.
4. Income-expenditure statement and annual gross surplus.
5. Cash flow analysis.
6. Repayment schedule (*i.e.* repayment of principal loan amount and interest).

Other documents such as loan application forms, security aspects, margin money requirements etc. are also examined. A field visit to the scheme area is undertaken for conducting a techno-economic feasibility study for appraisal of the scheme. Model economics for a two animal unit and mini dairy unit with ten buffaloes are given in Annexure V and VI.

6. Sanction of Bank Loan and its Disbursement

After ensuring technical feasibility and economic viability, the scheme is sanctioned by the bank. The loan is disbursed in kind in 2 to 3 stages against creation of specific assets such as construction of sheds, purchase of equipments and machinery, purchase of animals and recurring cost on purchase of feeds/fodders for the initial period of one/two months. The end use of the fund is verified and constant follow-up is done by the bank.

7. Lending terms—General

1. Unit Cost. Each Regional Office (RO) of NABARD has constituted a State Level Unit Cost Committee under the Chairmanship of ROs-in-charge and with the members from developmental agencies, commercial banks and cooperative banks to review the unit cost of various investments once in six months. The same is circulated among the banks for their guidance. These costs are only indicative in nature and banks are free to finance any amount depending upon the availability of assets.

2. Margin Money. NABARD had defined farmers into three different categories and where subsidy is not available the minimum down payment as shown below is collected from the beneficiaries.

Sr. No.	*Category of Farmer*	*Level of predevelopment return to resources*	*Beneficiary's Contribution*
(*a*)	Small Farmers	Upto ₹ 11,000	5%
(*b*)	Medium Farmers	₹ 11,001–₹ 19,250	10%
(*c*)	Large Farmers	Above ₹ 19,251	15%

3. Interest Rate. As per the RBI guidelines the present rate of interest to the ultimate beneficiary financed by various agencies are as under.

No.	*Loan Amount*	*CB's and RRB's*	*SLDB/SCB*
(*a*)	Upto and inclusive of ₹ 25000	12%	As determined by SCB/SLDB subject to minimum 12%
(*b*)	Over ₹ 25000 and upto ₹ 2 lakhs	13.5%	-do-
(*c*)	Over ₹ 2.0 lakhs	As determined by the banks	-do-

4. Security. Security will be as per NABARD/RBI guidelines issued from time to time.

5. Repayment Period of Loan. Repayment period depends upon the gross surplus in the scheme. The loans will be repaid in suitable monthly/quarterly instalments usually within a period of about 5 years. In case of commercial schemes it may be extended upto 6-7 years depending on cash flow analysis.

6. Insurance. The animals may be insured annually or on long term master policy, where ever it is applicable. The present rate of insurance premium for scheme and non scheme animals are 2.25% and 4.0% respectively.

3.6.3.3. *Format for Submission of Schemes*

A. General

1. Name of the sponsoring bank
2. Address of the controlling office sponsoring the scheme

3. Nature and objectives of the proposed scheme
4. Details of proposed investments

S.No	*Investment*	*No. of units*
(*a*)		
(*b*)		
(*c*)		

5. Specification of the scheme area (Name of District & Block/s)

S.No	*District*	*Block*

6. Names of the financing bank's branches :

S.No	*Name of the Branch/District*
(*a*)	
(*b*)	
(*c*)	

7. Status of beneficiary/ies: (indidivual/Partnership/Company/Corporation/Co-operative Society/ Others)
8. In case of area based schemes, coverage of borrowers in weaker sections (landless labourers, small, medium & large farmers as per NABARD's norms, SC/ST, etc.)
9. Details of borrowers profile (Not applicable to area based schemes
   (*a*) Capability
   (*b*) Experience
   (*c*) Financial Soundness
   (*d*) Technical/Other Special Qualificaitons
   (*e*) Technical/Managerial Staff and adequacy thereof

B. Technical Aspects

1. **Location, Land and Land Development.**
   (*a*) Location details of the project
   (*b*) Total Area of land and its cost
   (*c*) Site map
   (*d*) Particulars of land development, fencing, gates, etc.
2. **Civil Structures.** Detailed cost estimates along with measurements of vaious civil structure
   — Sheds
   — Store room
   — Milk room
   — Quarters, etc.

3. **Equipment/Plant and Machinery**
   (*a*) Chaff cutter
   (*b*) Silo pit
   (*c*) Milking machine
   (*d*) Feed grinder and mixer
   (*e*) Milking pails/milk cans
   (*f*) Biogas plant
   (*g*) Bulk coolers
   (*h*) Equipment for manufacture of products
   (*i*) Truck/van (price quotations for the above equipments)
4. **Housing**
   (*a*) Type of housing
   (*b*) Area requirement
   —Adults
   —Heifers (1-3 years)
   —Calves (less than 1 year)
5. **Animals**
   (*a*) Proposed species
   (*b*) Proposed breed
   (*c*) Source of purchase
   (*d*) Place of purchase
   (*e*) Distance (kms.)
   (*f*) Cost of animal (₹)
6. **Production parameters**
   (*a*) Order of lactation
   (*b*) Milk yield (ltrs. per day)
   (*c*) Lactation days
   (*d*) Dry days
   (*e*) Conception rate
   (*f*) Mortality (%)
   Adults
   Young stock
7. **Herd projection (with all assumptions)**
8. **Feeding**
   (*a*) Source of fodder and feed
   Green fodder
   Dry fodder
   Concentrates
   (*b*) Fodder crop rotations
   —Kharif
   —Rabi
   —Summer

(*c*) Fodder cultivation expenses

(*d*) Requirement and costs :

	Quantity required (kg/day)		
Cost (₹/kg)	*Lactation*	*Dry Period*	*Young Stock*
Green Fodder			
Dry Fodder			
Concentrates			

9. **Breeding Facilities**
   (*a*) Source :
   (*b*) Location :
   (*c*) Distance (km.) :
   (*d*) Availability of semen :
   (*e*) Availability of staff :
   (*f*) Expenditure per animal/year
10. **Veterinary Aid :**
   (*a*) Source
   (*b*) Location
   (*c*) Distance (km.)
   (*d*) Availability of staff
   (*e*) Types of facilities available
   (*f*) If own arrangements are made—
      (*i*) Employed a veterinary doctor/stockman/consultant
      (*ii*) Periodicity of visit
      (*iii*) Amount paid/visit (₹)
   (*g*) Expenditure per animal per year (₹)
11. **Electricity**
   (*a*) Source
   (*b*) Approval from SEB
   (*c*) Connected load
   (*d*) Problems of power failure
   (*e*) Arrangements for generator
12. **Water**
   (*a*) Source
   (*b*) Quality of water
   (*c*) Abvailability of sufficient quantity for drinking, cleaning and fodder production
   (*d*) If investment has to be made, type of strucutre, design and cost
13. **Marketing of milk**
   (*a*) Source of sales
   (*b*) Place of disposal

(*c*) Distance (km.)
(*d*) Price realised—(₹ per liter of milk)
(*e*) Basis of payment
(*f*) Periodicity of payment

14. **Marketing of other products**
(*a*) Animal—age
—place of sale
—price expected

(*b*) Manure—Qty./animal
Price/unit (₹)

(*c*) Empty gunny bags
—Number
—Cost/bag (₹)

15. **Beneficiary's experience**
16. **Comments on technical feasibility**
17. **Government restrictions, if any**

C. Financial Aspects :

1. **Unit Cost**

Sr. No.	*Name of the Investment*	*Physical units and specification*	*Unit cost with component wise break-up (₹)*	*Whether approved by state level unit cost committee*
Total				

2. Down payment/margin/subsidy (Indicate source and extent of subsidy) :
3. Year-wise physical and financial programme :

Year	*Investment*	*Physical Units*	*Unit Cost (₹)*	*Total Outlay (₹)*	*Margin/ Subsidy (₹)*	*Bank loan (₹)*	*Refinance Assistance (₹)*
1	**2**	**3**	**4**	**5**	**6**	**7**	**8**
Total							

4. Financial viability (comment on the cash flow projection on a farm model/unit and enclose the same.) Particulars :
(*a*) Internal Rate of Return (IRR) :
(*b*) Benefit Cost Ratio (BCR) :
(*c*) Net Present Worth (NPW) :

5. Financial position of the borrowers (to be furnished in case of corporate bodies/partnership firms)
   (*a*) Profitability Ratio :
      (*i*) GP Ratio
      (*ii*) NP Ratio
   (*b*) Debt Equity Ratio :
   (*c*) Whether Income Tax and other tax obligations are paid up to date :
   (*d*) Whether audit is upto date (enclose copies of audited financial statements for the last three years)
6. Lending Terms :
   (*a*) Rate of Interest
   (*b*) Grace Period
   (*c*) Repayment Period
   (*d*) Nature of Security
   (*e*) Availability of Government guarantee wherever necessary

D. Infrastructural Facilities

1. Availability of technical staff with bank/implementing authority for monitoring
2. Details of—
   (*a*) technical guidance
   (*b*) training facilities
   (*c*) Govt support/extention support
3. Tie-up arrangements with marketing agencies for loan recovery
4. Insurance—
   —Type of policy
   —Periodicity
   —Rate of premium
5. Whether any subsidy is available, if so amount per unit.
6. Arrangements for supply of green fodder and cattle feed.

3.6.3.4. *Practical profitable dairy unit to begin with (3 Cows and 2 Buffaloes)*

Assumptions

1. **Cows :**
   - Production level: 3000 lits. per lactation
   - Cost : ₹ 15,000 each
   - Period between successive calving : 14 months
   - Pure milk sold @ ₹ 10 per lit
2. **Buffaloes**
   - Production Level: 2400 lit.
   - Cost : ₹ 20,000 each
   - Period between successive calving : 16 months
   - Pure milk sold at ₹ 14 per lit

3. **Feed Cost**: Dry fodder @ ₹ 1.50 per kg, Green fodder @ 0.70 per kg, Concentrate @ ₹ 8.00 per kg.
4. **Feeding**: Dry fodder @ 5 kg/day; Green Fodder @ 20 kg/day; Concentrate @ 2 kg/day for maintenance ration, needs to be adjusted with production, general thumb rule for Cow : Milk production/3; Buffalo: Milk production/2.5); Mineral Mix @ 30 gms per day (₹ 50 per kg.)
5. **Insurance**: Crossbred cow @ ₹ 800 per animal per year; Buffaloes @ ₹ 600 per animal per year.
6. **Depreciation**: @ 20% per year for animals; 10% per year for Equipments; 20% per year for building
7. **Veterinary Care:** @ ₹ 200 per animal per year.
8. **Labor (Mostly family) :** @ ₹ 5000/ year for 5 animals
9. **Interest:** @ 15%
10. **Cost of Construction:** @ ₹ 70 per sq. ft., Cow shed and Store cost ₹ 5000 per animal.

A. Capital Investment

Cross	*Bred Cows 3 numbers*	*Buffaloes 2 numbers*
Cost of Animals	₹ 45,000	₹ 40,000
Buildings	₹ 15,000	₹ 10,000
Equipment	₹ 3,000	₹ 2,000
TOTAL	₹ 63,000	₹ 52,000

Capital Investment Grand Total : ₹ 1, 15,000.

B. Fixed Expenses

Cross	*Bred Cows 3 numbers*	*Buffaloes 2 numbers*
Interest on Capital @ 15%	₹ 9,450	₹ 7,800
Depreciation of Livestock @ 20% per year	₹ 9,000	₹ 8,000
Depreciation on Building @ 10% per year	₹ 1,500	₹ 1,000
Depreciation on equipment @ 10% per year	₹ 300	₹ 200
Insurance	₹ 2,400	₹ 1,200
TOTAL	₹ 22,650	₹ 17,300

Total Fixed Grand Total : ₹ 39,950

C. Recurring Expenses

Cross	*Bred Cows 3 numbers*	*Buffaloes 2 numbers*
Green fodder @ 20 kg/day Cost 0.70/kg	₹ 15,330.00	₹ 10,220.00
Dry fodder @ 5 kg/day Cost 1.50/kg	₹ 8,212.50	₹ 5,475.00
Cost of Concentrate (Maintenance feeding) Cost ₹ 8 per kg	₹ 17,520.00	₹ 11,680.00
Labor	₹ 3,000.00	₹ 2,000.00
Veterinary care @ ₹ 200 per animal/year	₹ 600.00	₹ 400.00
Misc. Expenses ₹ 500/animal	₹ 1,500.00	₹ 1,000.00
TOTAL RECURRING	₹ 46,162.50	₹ 30,775

TOTAL RECURRING GRAND TOTAL = ₹ 76,937.50

Total Expenses (B + C) : ₹ 1,16,887.50

D. Income

1. Sale of Milk

	Cross Bred cows	Buffaloes
Total Milk Produced	9,000 l	4,800 l
Fed to calves	300 l	265 l
Total Milk Sold	8,700 l	4,535 l
Milk Sold as Pure Milk	@ 60% (5,220 l)	@30% (1,360 l)
	₹ 52,200	₹ 19,040
Milk Sold as Mixed Milk	@ ₹ 12/l @ 40% (3,480 l)	@ 70% (3,175 l)
Total	6,655 l	Amount : ₹ 79,860
Total Income from Milk	₹ (52,200 + 19,040 + 79,860) = ₹ **1,51,100**	

2. Sale of Calves

	Cross Bred cows	Buffaloes
Calves available for sale	2 Female + 1 male	1 Male + 1 Female
Amount	₹ 900/ (@ ₹ 300/calf)	₹ 600/ (@ ₹ 200 for male ₹ 400 for female)
Total Income from Sale of Calves	**₹ 1,300**	

3. Sale of Manure @ ₹ 500/ animal/year

Cross Bred cows	Buffaloes
₹ 1,500	₹ 1,000

Total income from sale of manure : ₹ 2,500

Total income = ₹ (1,51,100 + 1,300 + 2,500) = ₹ 1,54,900

PROFIT = NET INCOME – TOTAL EXPENSES (Fixed + Recurring)

= ₹ 38,012.50

Profit per animal/year ₹ 7,602.40

Profit per animal/month ₹ 633.50

CHAPTER 4

Problems of Dairy Production in Hot Regions

There are very few places in the world and none in India where the natural climate continuously remains optimum for animals. Many of the components of the surroundings impose serious limitations on the animals' productivity unless man designs structures and environmental control systems to modify them. The effect of the climatic environment on the input-output ratio and economic optimisation of shelters and other environmental control systems are areas of vital importance to dairy production industry in the tropics.

4.1. THE PHYSICAL ENVIRONMENT

Environment has been conveniently sub-divided into physical and social. The physical environment includes factors like thermal conditions, space, pressure, light, sound and equipment. The social factors are the number of animals per pen, housing system, behaviour and butting order. Among the physical components of the environment, thermal factors are so important to animal comfort that they have been separately classified.* Major components of thermal environment are : *air temperature, humidity, air movement and solar radiation.* Precipitation is another important meteorological factor which has profound direct and indirect effect on thermal conditions and animal productivity. The instantaneous values of these meteorological factors or the group of these meteorological conditions during a short interval (like one day) is termed as 'weather'. 'Climate' is defined as the long-term averages of these meteorological variables. In other words, it comprises of the average values of weather conditions of a region. It is represented by the average values of various meteorological observations over several years. *Microclimate* refers to the physical environment in the immediate vicinity of the animal.

---

*Bond, T.E. and Kelly, C.F. (1960). *Environment of animals. Year-book Agric.,* U.S. Department of Agric., Washington, D.C.

While planning dairy animal management it will be necessary to consider both micro and macro climate. Considering only the direct effects of microclimate will be an over simplification of the problems because the indirect effects of the macroclimate operating through feed supplies, feed quality, parasitism, and other factors are equally or more important. These factors confound the direct effects of climatic elements and under many situations far overshadow their effects. Thus, evaluation of the direct and indirect effects of climate independent of each other is necessary.

4.1.1. Major Climatic Components

Various components of climate like air temperature, humidity, solar radiation, wind velocity and precipitation produce a combined effect on the 'thermal comfort' of animals and their performance. But, to understand fully, we have to study each factor separately.

4.1.1.1. *Air Temperature*

The most important single climatic element that is concerned in thermal comfort of dairy cattle and buffaloes is, undoubtedly, the air temperature. It is but natural that the environment that immediately surrounds an animal would influence the heat exchanged by it and the environment. The laws of heat and mass transfer, in general, control this exchange of energy. In fact, the thermal properties of the ambient (surrounding) air influence directly or indirectly all the four modes of heat transfer, *viz.*, conduction, convection, radiation and evaporation.

The skin temperature of cattle is about 33°C. Heat flows as per Newton's Law of cooling to the cooler atmosphere. Newton's Law states that *the rate of cooling of a body is directly proportional to the temperature difference* $t_1 - t_2$ *between body and environment.* Thus, as air temperature rises above the comfort range, the temperature gradient between animal body and atmosphere and thereby scope of cooling, diminishes. If air temperature exceeds skin temperature, heat will flow in a reverse direction as sometimes happens in hot-arid regions.

Apart from the heat from the air, the heat of hot ground and floors is a major factor. The dry uncovered soil and concrete floors heat up rapidly under intense solar radiation and add to the heat load on the animal when the animal lies down. However, after sunset the ground cools down rapidly giving the cows an avenue to get rid of their body heat through conduction to the cooled ground.

Measurement. There are two major systems of measurement : *the Fahrenheit scale and the Centigrade or Celsius scale.* The conventional measures are the maximum temperature, the minimum temperature and the average of these two—mean daily temperature.

Mean daily temperature alone makes very little sense as it does not give any indication of the diurnal variation. In hot-humid regions the mean daily temperature is generally low, but the difference between the daily maximum and minimum temperature may be only 5-8°C with the minimum temperature above 25°C. Dairy cattle and buffaloes are likely to be under some degree of thermal stress throughout the day and most part of the night under such circumstances. In hot dry climates, in spite of a much higher daily mean temperature, at night the mercury falls sufficiently to within comfortable range of dairy cattle. A diurnal variation of 20-22°C in air temperature allows animals to restore thermal balance during the latter half of night before encountering the hazards of high environmental temperature the next day.

Faulty enclosed housing in hot arid regions instead of doing good, imparts continuous thermal stress on dairy cattle in the same way as hot humid tropical climate. Even though the direct solar radiation is cut off and the average air temperature reduced, the increased minimum temperature keeping it above the comfort zone throughout, the increased humidity, radiation from the walls and reduced air movement make such a housing extremely uncomfortable for dairy animals.

A continuous record of air temperature with a thermograph and calculating the percentage of time in a day exceeding the comfort range of the animals (30°C) will be a better indicator of thermal comfort for animals.

4.1.1.2. *Atmospheric humidity*

The water vapour content of the air surrounding animals has significant effect on their capacity for heat loss. The rate of cooling by evaporation from the respiratory tract and the skin depends on the water vapour content of the air (humidity). As and when atmospheric temperature increases, the importance of the evaporative channel of thermolysis progressively increases, so much so, when the atmospheric temperature equals or exceeds the animal's body temperature, evaporation is the only available means of heat-loss. If this environmental air is also saturated with moisture, physical means of heat-loss is not possible. *From 21°C for European cattle and 27°C for zebu cattle, humidity becomes a problem affecting their production when the vapour pressure is above 20 mm Hg or the relative humidity exceeds 60 per cent.*

The indirect effects of atmospheric humidity are also serious. High humidity poses many problems to dairy animals by increasing chances of disease incidence and parasitism and by lowering nutritive value of forages.

Measurement. Humidity is generally expressed as relative humidity (R.H.), vapour pressure or dew point. Relative humidity is the ratio of moisture actually present in the air to the maximum amount it would hold at saturation at that temperature. It is generally expressed as a percentage. The amount of water vapour present in unit volume of air or the partial pressure of water vapour as the fraction of the total barometric pressure expressed as millimeters of mercury (absolute humidity or vapour pressure) is a more realistic and meaningful measure of humidity. In animals, since evaporation from lungs and skin takes place at near constant temperature to a *boundary layer* of more or less stationary air at the same temperature, absolute humidity gives a better indication of the ability of the atmosphere to accept water vapour from skin and the lungs.

4.1.1.3. *Solar radiation*

Radiant energy is transmitted by the sun in the form of electromagnetic waves. This falls on the animal directly or indirectly and has the following three components :

1. Direct solar radiation consisting of ultraviolet, visible and short infra-red radiation.
2. Solar radiation reflected from clouds and other particles in the sky which also consists of the same components but having more of short infra-red.
3. Radiation emitted by hot ground, walls, roofs and other constructions and the horizon. These objects absorb the solar energy and get heated up and emit long wave infra-red radiation during the day as well as the night.

Out of the total radiation, approximately 50 per cent comes from the first two sources and the remaining from the third.

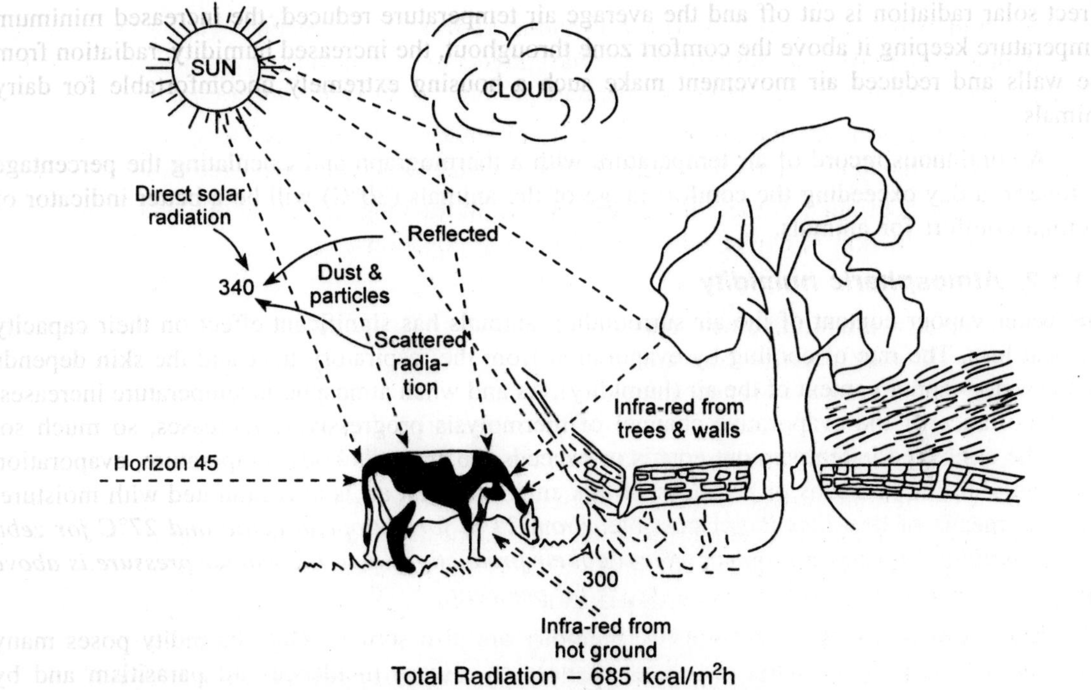

Fig. 4.1. The radiation energy flow to a dairy cow standing out in the sun on a summer day (k cal/m²h)

The ultra-violet portion of the spectrum with a wavelength of 100 to 4000 Angstroms (Å) (10^{-10} metres) contribute only one per cent of the radiant heat load. The visible range (4000 to 7000 Å) contribute 40 to 45 per cent and the infra-red range (10^4 to 10^7 Å) 50 to 60 per cent.

A white coat of hair can reflect most of the incidence of visible light radiation and therefore was considered superior by some. But it also increases the penetrance of the sun rays and therefore the beneficial effect due to high reflectivity is somewhat nullified. In the final analysis it is not the colour of the coat, but its texture and density that matter more.

Figure 4.1 represents the approximate heat load on animals from various sources. In a cloudless tropical environment the part of animal body exposed directly to the sun gets. a total radiation load of 600-700 kcal/m^{2}h.

Measurement. For a person interested in knowing approximately the total heat load on animals (excluding direct solar radiation) a Vernon globe (black globe) thermometer is useful. For more precise measurement of the intensity of radiation different types of radiometers are made use of. Radiometers generally can be sorted into 2 groups : (*a*) those mainly dealing with the direct solar radiation and some reflected rays and (*b*) those dealing with all the components.

An example for the former is the pyranometer and for the latter, net radiometer. For more details on instrumentation the readers may see Hafez (1968), Esmay (1969) and National Academy of Science (1971).

4.1.1.4. Wind

Air movement over the body surfaces of animals facilitates heat dissipation, through evaporative as well as non-evaporative channels. The latter involving heat transfer from the animal's body surface (solid) into air (fluid) is a typical example of thermal convection. The transfer of heat through non-evaporative route also includes some conduction. Air movement aids in heat dissipation through this pathway as long as air temperature is lower than skin temperature.

At high temperatures the reverse may be true. Under such circumstances, only evaporative channel is available to the animals for heat dissipation which is highly facilitated by air movement.

Thermolysis through evaporation takes place from the lung as well as the skin. Evaporative heat loss from the skin is explained by the following equation :

$$q^e = K^e A^e V^n (P_s - P_a)$$

where, q^e = evaporative heat loss
K^e = evaporative constant
A^e = effective wetted area
V^n = Velocity component to some exponential power
P_s and P_a = vapour pressure of skin surface and the air.

The exponential power n of air velocity has only been quantitatively determined experimentally. This equation demonstrates the importance of wind velocity in thermolysis through evaporative channels. It also shows the importance of surface area per unit body mass as well as the capacity of the animal to supply moisture to the skin to keep it wet. If moisture supply to skin is slow, the effect of air movement on heat loss at high environment temperature is limited.

Drier regions of the earth in general have windy and gusty climate. In hot dry regions, this may cause dehydration and heat gain through reverse convection. In such regions, the ideal wind velocity during day time is 8 kmph or less. But generally such regions have much higher wind speed. At night a wind velocity of 8-16 kmph would be beneficial to facilitate heat loss and attain thermal balance.

Under hot-humid conditions which are notorious for low air movements any means of enhancing air movement is beneficial. Every effort should be taken to avoid interfering with air movements below 5 kmph and even irregular eddies.

Measurement. Wind velocity is generally measured by two types of instruments. One type known as directional (propeller; swinging vane) and semi-directional (cup) anemometers convert the kinetic energy of moving air mass into mechanical energy and record with the help of a rotary impeller or show as static pressure against a counter weighted vane in the velometer. The other type known as 'thermoanemometers' is non-directional and depends on the cooling power of the moving air on a heated body. Heated globe anemometer, Wilson thermoanemometer, Anemotherm and Alnor thermoanemometer, are some instruments of this type. For more details consult Hafez (1968), Esmay, (1969) and National Academy of Science (1971).

4.1.1.5. Rain

Rain has more indirect effects than direct effects on bovines. Under hot conditions, rain falling on the animals or retained by the hair evaporate, thereby cooling the animal body. But rain also contributes to the humidity of the atmosphere, thereby adding to the heat stress. Animals seem to react more strongly than man to the physical impact of rain. It is a common observation to see animals leaving grazing and seeking protection from rain. Milk yield is often depressed following heavy rainfall.

The indirect effects of rain is far more important in dairy production. The average annual rainfall, the distribution of rain from month to month and the year to year fluctuations are all important factors influencing feed supplies and forage quality. Rainfall is the most important single factor that affects rate of growth and nutritive value of natural forages.

Thus in rainfed grasslands and cultivated areas, total rainfall and its distribution are critical factors influencing dairy productivity. Apart from this, precipitation also affects dairy animals indirectly through its influence on diseases and parasitism.

Measurement. Precipitation is measured most frequently by the instrument, *rain gauge.* Routine recordings of rainfall is available from all the meteorological stations.

4.1.2. Classification of Climate

There are many systems of classification of climate depending on the purpose for which it has been done. Obviously a classification for application in aviation cannot be used in animal husbandry.

The climate of a region is determined by its latitude, altitude, proximity to sea, soil conditions, mountain ranges and other topographical features.

In relation to dairy production, we are more interested in the relative importance of climatic elements in terms of energy transfer between the animal and the environment. The principal climatic elements in terms of energy transfer between them are : air temperature, humidity, radiation and air movement.

Some of the important climatic classifications are those proposed by Koppen, Thornthwaite, Williamson and Payne, Webster and Wilson and Lee. For more details readers are directed to Lee (1953) and McDowell (1972). All the classifications demarcated broadly climatic types based on the air temperature and the moisture content. For example, Lee designated regions according to the mean temperature and humidity prevailing during the warmest and coolest months as follows :

A. (Mean temperature of months)
   Over 30°C — hot
   20-29°C — warm
   10-19°C — temperate
   under 100°C — cool

B. Humidity (only during warm and hot periods) in mean monthly dew point
   over 20°C — wet
   16 – 19°C — humid
   under 15°C — dry

The above system is reasonably suited for interpreting animal comfort. But it is not always possible to keep both temperature and vapour pressure, relative humidity or dew point in mind while interpreting significance of these for their effect on animal comfort. Therefore there had been many attempts to combine more than one environmental variables into a single expression of comfort. One of the most widely used ones is the climograph (Fig. 4.2A) in which the average monthly ambient temperature and average monthly vapour pressure are used to plot the climographic diagram. Another important concept which combines these two climatic factors is the temperature humidity index (THI). U.S. Weather Bureau has suggested the following formula for computing THI :

$$THI = 0.72\ (C_{db} + C_{wb}) + 40.6$$

where, C_{db} = dry bulb temperature (°C)
and C_{wb} = wet bulb temperature (°C)

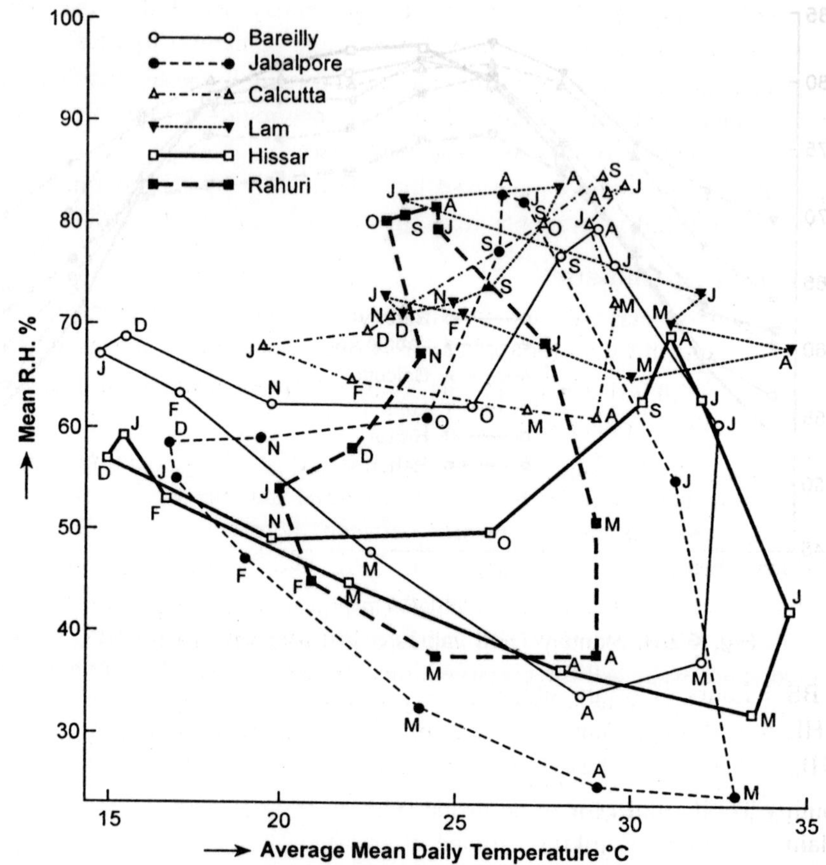

Fig. 4.2-A. Climograph of some stations.

On the other hand Bianca suggested the following equation :

$$\text{THI} = 0.35\ C_{db} + 0.65\ C_{wb}$$

These two are better indicators of the physical environment than the ambient temperature or humidity alone. But they do not take into account the diurnal variation in ambient temperature and humidity nor the effect of solar radiation.

Effective temperature (ET) is an index incorporating the effect of solar radiation in addition to ambient temperature and humidity.

$$\text{ET} = K_a\ (db)\ K_b\ (rh) + K_c\ (sr)$$

where K_a, K_b and K_c are constants depending on the units of measurement, *db* = dry bulb temperature, *rh* = relative humidity and *sr* = solar radiation. But the measurement of solar radiation poses problems and requires sophisticated instruments. Taking this into account a modification of the THI concept known as 'Temperature humidity sunshine index' (Thomas and Acharya, 1981) has been developed which is as follows :

$$\text{THSI} = \frac{(0.75\,\text{BS})\,\text{THI}_e + (24 - 0.75\,\text{BS})\,\text{THI}_m}{24}$$

Thomas, C.K. and Acharya, R.M. (1981). Note on the effect of Physical environment on milk production in *Bos indicus* × *Bos. taurus* crosses. Indian J. Anim. Sci. 51: 351-356.

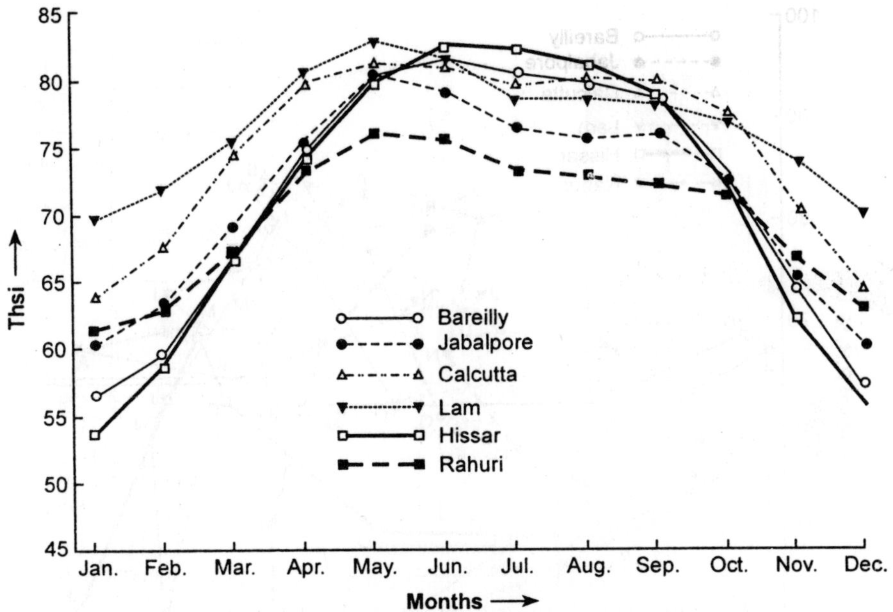

Fig. 4.2-B. Monthly THSI values at different stations in India

where, BS = hours of bright sunshine.

THI_e = THI values based on evening (3.00 pm) dry and wet bulb readings

THI_m = THI values based on morning (9.00 am) dry and wet bulb readings.

The monthly THSI values for some of the stations are presented in figure 4.2 B. THSI was found to explain variations in milking average of crossbred cows to a greater degree than either maximum temperature or relative humidity alone or simple THI values.

4.2. REACTION OF DAIRY CATTLE AND BUFFALOES TO HOT ENVIRONMENTS

4.2.1. Homeothermy

Homeotherms are a group of animals which maintain their deep body temperature essentially constant. In the case of an inert body, basic thermodynamic forces establishes an equilibrium in which the body attains the same temperature of the surrounding environment. The same laws are applicable and cannot be circumvented in the case of homeothermic animals. Therefore, these animals will have to make compensating physiological and behavioural adjustments to keep a constant deep-body temperature.

Animals utilize the chemical energy which is solar energy trapped by plants through the process of photosynthesis. While converting this energy in their feed to maintenance, production, growth and work, a portion of the feed energy is wasted as heat. This metabolic heat plus the heat it gains from the environment through various processes of conduction, convection and radiation results in considerable thermal gain to the animal. Heat, in excess of its needs for maintaining its level of body temperature has to be dissipated, if heat balance is to be maintained.

Thomas, C.K. and Pearson, R.A. (1986). Effect of ambient temperature and head cooling on energy expenditure, feed intake and heat tolerance of Brahman and Brahman × Friesian cattle working on thread-mills. Animal Poduction. 43 : 83-90.

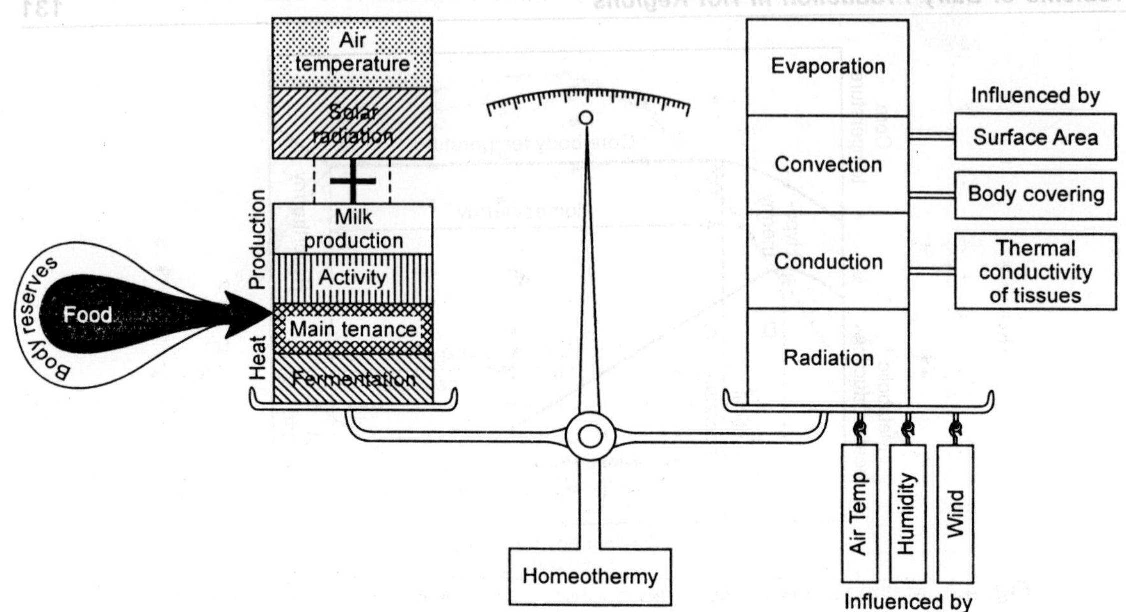

Fig. 4.3. Factors involved in homeothermy of animals.

For this, the animal depends on its physiological and behavioural faculties. This balancing act by which a constant normal core temperature is maintained is known as homeothemy. Figure 4.3 represents the various factors involved in the homeothermy of dairy animals.

Homeotherms resort to various physiological and behavioural adjustments to compensate for environmental and internal heat gain or losses. These are orchestrated by the central nervous system by way of its thermal sensors, afferent paths, integrative elements and efferent neuro-humoral pathways. Thermogenesis (heat production) is regulated by increasing or decreasing hormonal secretions, changes in muscle tonus, shivering and by voluntary muscular activity. Thermolysis (heat loss) is controlled by changes in cutaneous circulation and evaporative losses. The animals will drink more water, reduce feed intake as well as production. In the long term, the type of hair coat changes making it lighter with less insulation. There are also postural changes and behavioural adjustments. They will try to increase surface area, seek shade and reduce physical activity.

4.2.2. Environmental Temperature and Thermal Balance

Effective environmental temperature (which represents the collective thermal stress on the animal from ambient temperature, solar radiation, humidity and air movement) has a very key influence on the homeothermy of dairy animals. Figure 4.4 represents the behaviour of core body temperature and heat production under the influence of effective environmental temperature.

4.2.2.1. *Zone of Thermoneutrality*

In this range of environmental temperature (B-B) the animal can maintain homeothermy with physical compensations alone. When the effective temperature goes up beyond point B (the upper critical temperature) the body temperature increases and, as the regulatory mechanisms break-down, an increase in heat production more or less according to the Van't Hoff effect takes place.

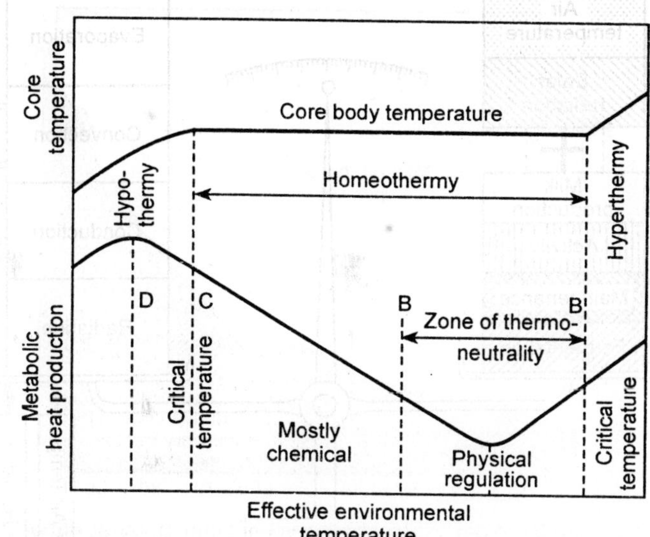

Fig. 4.4. Matabolic heat production and core body temperature as influenced by environment temperature (Adapted from Kleiber, 1961)

The Van't Hoff effect represents the two-fold or three-fold increase in the speed of thermodynamic chemical reactions when the temperature increases by 10°C. The actual critical temperature vary from species to species and breed to breed and also within breeds between individuals. The upper critical temperature is reached very quickly on crossing the upper limit of zone of thermoneutrality.

When the effective temperature goes down below B, the animal increases heat production through chemical means to compensate for the heat loss. Soon a point (C) will be reached when the increasing heat production can no longer keep the core body temperature in the normal range and it begins to decline. Maximum heat production is attained at point D. Any further decline in environmental temperature will result in breakdown of homeothermic mechanisms resulting in slowing down of thermogenic reactions resulting in reduced heat production. Body temperature declines further.

4.2.3. Physical and Physiological Factors in Homeothermy

4.2.3.1. *Regulating Centres*

Scientific opinion generally agrees on the fact that the centres of heat regulation are situated in the basal part of diencephalon (hypothalamus). Receptors located in the periphery and deep in the body collect and through afferent pathways transmit this information to the CNS where it is integrated. Thermoregulatory centres in the hypothalamus react quickly by activating various thermoregulatory mechanisms.

To maintain homeothermy, there has to be sensitive and efficient mechanisms which balance any change in heat gain by an equivalent change in heat loss. There is continuous production of heat in the animals' body due to metabolism. This indicates that the net exchange of heat with the surroundings must be in the direction of heat loss.

The regulation by the hypothalamus of the rate of heat loss involves two major pathways—the quick-acting neural and the delayed but more permanent humoral. Most of the neural thermo-

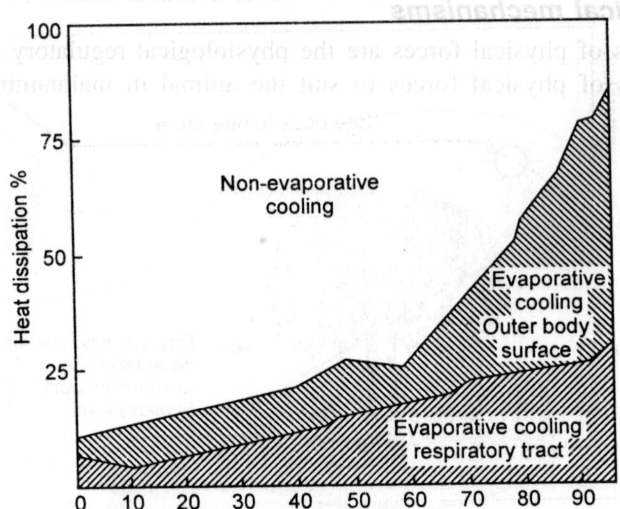

Fig. 4.5. Evaporative cooling from the respiratory tract, outer body surface and total non-evaporative cooling in cows. (Adapted from Kleiber and Brody, 1950).

regulatory mechanisms act through the sympathetic nervous system. These mechanisms involve (1) shift of circulating blood from core to periphery and *vice versa* through regulating the size of capillaries in the skin and sub-cutis, (2) activating sweat glands, changing general posture of the animal, and (4) increase in respiratory activity. The humoral command chain also emanates from the hypothalamus. Chemical agents known as releasing hormones activate production of various hormones in the anterior and posterior pituitary. These by controlling hormone production and release from other endocrine glands like thyroid, adrenal, gonads and pancreas directly regulate thermoregulatory mechanisms. These involve more stable changes in basal metabolic rate (influencing heat production), water balance, electrolyte balance, fat deposition, hair growth, feed intake and milk production.

4.2.3.2. *Physical Means of Thermal Exchange*

Energy is exchanged by the animal body to the environment as described below :

1. When there is a temperature gradient, energy is transferred from particle to particle. This is known as *conduction.*
2. When there is non-uniformity in temperature in a liquid or gas, heat transfer will take place by circulation of the particles. This process is called *convection.*
3. *Radiation* which is defined as the exchange of energy through space without heating the intervening medium, is another means.
4. *Evaporation.* When water from the surface vapourises (change from the liquid state to gaseous state) it removes the latent heat of vapourisation from the surface (skin or respiratory tract) thereby cooling it.

At higher effective environmental temperatures, homeothermy of bovines depends largely on evaporative cooling. As the environmental temperature rises the animal has to depend more and more on evaporative heat loss (Fig. 4.5). At temperature of 40°C, dairy cows can hardly dissipate 100 per cent of the total heat produced.

4.2.3.3. Physiological mechanisms

Overriding the effects of physical forces are the physiological regulatory processes which tend to change the effects of physical forces to suit the animal in maintaining homeothermy. For

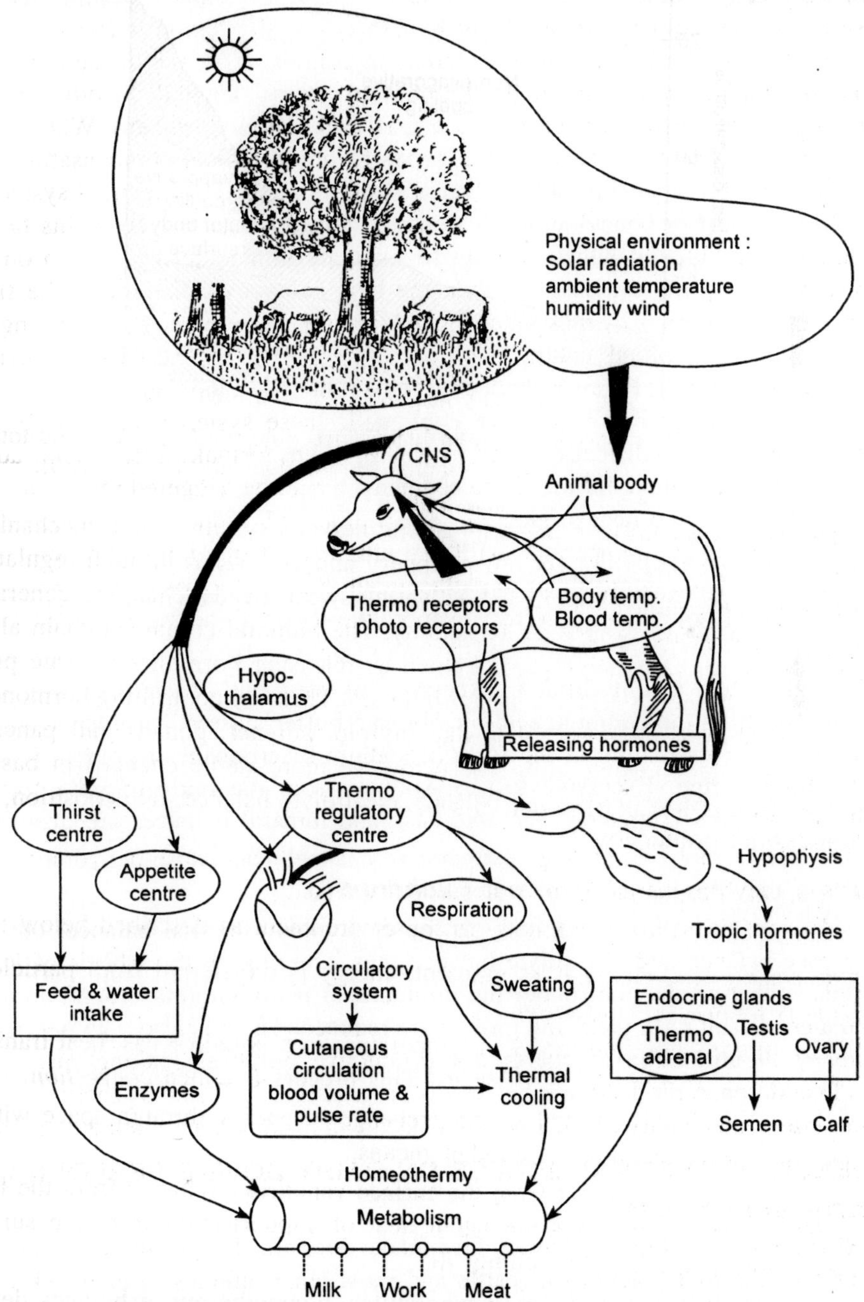

Fig. 4.6. Reaction of dairy cattle to physical environment.

Thomas, C.K., Sorensen, B. and Nauheimer, H. (1984). Reaction of Holstein Friesian cows to high ambient temperature. Proc. International Conf. on Milk Production in Developing Countries. Edinburg, U.K.

convenience of discussion this can be partitioned into two, *i.e.*, (*i*) those influencing heat production and (*ii*) those regulating heat loss.

Animal Response. When the animal is outside the 'comfort zone' it is subjected to various degrees of stress—either cold stress or heat stress. In N—S 30° region heat stress is the most important factor. This results in the animal experiencing 'strain' which is the sum-total of all the compensations or displacements caused by the thermal 'stress' applied. In other words, strain represents the totality of discomfort felt by the animal in a hot environment. When subjected to thermal stress, animal system reacts through a series of physiological compensations which act in the form of a chain reaction (Fig. 4.6). When a stress is applied, the animal system activates some of the physiological or behavioural reactions to neutralize that effect. If this fails, another group of reactions are initiated which attempt to maintain homeothermy and so on until heat balance is restored. If all the systems fail then the heat balance is disturbed. The first line of defence against thermal stress is by making changes in the circulation, *i.e.*, by increasing cutaneous blood flow and increasing blood volume with an aim to bring the heat to the periphery and facilitate evaporative heat loss. This is followed by activation of sweating, and increased respiratory activity. Both aim at increasing evaporative cooling. If these systems fail to bring about the desired compensation, they will feel discomfort forcing them to make behavioural adjustments. If the thermal stress is sustained, hormonal mechanisms would be triggered into action soon with changes in metabolic rate and heat production. The animal will reduce feed intake and increase water consumption. The body temperature will rise often stabilising at a higher level. There will be shifts in body water from intracellular to extracellular and from extracellular to vascular space which will be ultimately mobilised to effect evaporative thermal cooling from the skin and the respiratory tract.

Increased respiratory activity leads to excessive 'blowing out' of carbondioxide through the expired air thus upsetting the carbonic acid-bicarbonate buffer system. In order to maintain acid base balance, bicarbonates mainly in the form of sodium bicarbonate or potassium bicarbonate are excreted through urine. Excessive loss of sodium from the body triggers cortico-steroid hormones like aldosteron. Ultimately, the acid-base and mineral balances are disturbed.

Excessive loss of moisture through evaporative channels can cause dehydration under hot-arid conditions if they are subjected to water scarcity also.

Sustained exposure to hot conditions can upset not only thermal balance, but also water balance, acid base balance and electrolyte balance thus jamming all the wheels of *homeostasis*. In a bid to reduce heat production (under the influence of neurohumoral orders) animals reduce feed intake and as a result milk yield and growth are reduced. There is also a 35 per cent decrease in the efficiency of utilization of energy for productive purposes making many enterprises uneconomic. In addition, reproductive efficiency is progressively affected by thermal stress. Under prolonged heat stress, the vitality of the animal breaks down making it susceptible to many diseases.

4.3. INFLUENCE OF THERMAL STRESS ON DAIRY BOVINE PERFORMANCE

4.3.1. *Climate and Feeding*

In the thermoneutral temperatures, voluntary feed intake does not change with ambient temperature conditions. One of the first reactions of cattle and most other animals to elevated temperatures

---

Thomas, C.K. and Razdan, M.N. (1973). Adaptability of ½ Sahiwal–½ Brown Swiss Cattle to Sub-tropical conditions.

1. Feeding behaviour. Indian J. Anim. Sci. 43 : 5.
2. Physiological reactions. Indian J. Anim. Sci. 43 : 358.

beyond the thermoneutral zone is a decrease in feed intake. The extent of this depression is proportional to the thermal stress. A fall in the volatile fatty acids (VFA), especially acetic acid in the rumen fluid has been found when cattle were exposed to hot environments. This has been attributed to a reduced feed intake. Probably, high body temperature acts as a satiety signal to cattle (Simkins *et al.*, 1965). Recent work has confirmed that the voluntary feed intake of cattle is reduced by thermal stress. Digestible energy consumed by lactating Holstein cows declined by 14 per cent from the level at 21°C after one week exposure to 32°C and subsequently by 4 per cent in the second week. If the animal under heat stress is forcefed through a rumen fistula, it puts on weight and increases milk production. This indicates that it is the satiety centre that is mainly affected by the environmental heat.

Under field conditions, it is not the absolute increase in ambient temperature but the sustained nature of the heat stress that affects animal performance to a greater degree. Thus, under hot-dry conditions, even though, daily maximum temperature during summer exceeds far above that in hot-humid regions, the minimum temperature during the night dip to a comfortable range thereby allowing better feed intake and performance in the former case than in the latter.

Critical temperatures for feed intake depression in different breeds and genetic groups are also important considerations in the feeding behaviour and the type of housing and management a set of animals require. During summer in hot semi-arid Karnal, a thick brick-walled barrack-type shelter was found to increase feed intake and growth rate in Sahiwals, because this shelter helped in keeping the ambient temperature below the critical level for a longer time. On the other hand the same shelter was found to be more stressful than the unsheltered to the Sahiwal-Brown Swiss crossbreds because the temperature inside the shed remained above their critical temperature (which is much lower than in Sahiwals) throughout (Thomas *et al.*, 1969). Those crossbreds kept in the open during summer consumed more of TDN, digested crude fibre and ether extract better and showed a higher nitrogen balance in spite of the exposure to solar radiation and higher ambient temperature during the day. This was partially because the temperature in the open shed fell below the critical level for the crossbreds during the early morning hours.

Another behavioural adjustment by the dairy animals during summer is to reduce feed intake during the day time and increase it during the night. By so shifting the feeding regime, the animal can rearrange the major part of the thermogenic fermentative activity to take place during the cooler parts of the day thus facilitating thermolysis to the cooler surroundings. This mechanism was found to be more in play when the heat stress just crossed the critical limit for a short duration in the day time (Thomas and Razadan, 1973). By this the animal will be able to compensate the reduction in feed intake during the day time by enhancing the intake during the night thereby keeping up the total intake. But when the thermal stress is of longer duration during the day, of more intensity and sustained for many days, a reduction in absolute feed intake takes place and there is not much day to night shift observed. This is because day to night shift in feed intake is of transient nature, most probably nurally controlled which reflects the thermal stress during the day by an immediate response by reducing feed intake. As soon as the thermal stress withdraws itself during the night, there was retrieval in feed consumption. On the other hand, in sustained severe heat stress humoral mechanisms involving the huypothalamus-pituitary-thyroid axis resulting in reduced thryoid activity were triggered which resulted in reduced feed-

Thiagarajan, M. and Thomas, C.K. (1991). Housing effects on cross-bred cows in hot-humid climate. Indian J. Anim. Sci. 61 : 1222-1225.

Thomas, C.K. and Razdan, M.N. (1973). Adaptability of ½ Sahiwal–½ Brown Swiss cattle to sub-tropical conditions. 1. Feeding behaviour. Indian J. Anim. Sci. 43 : 5.

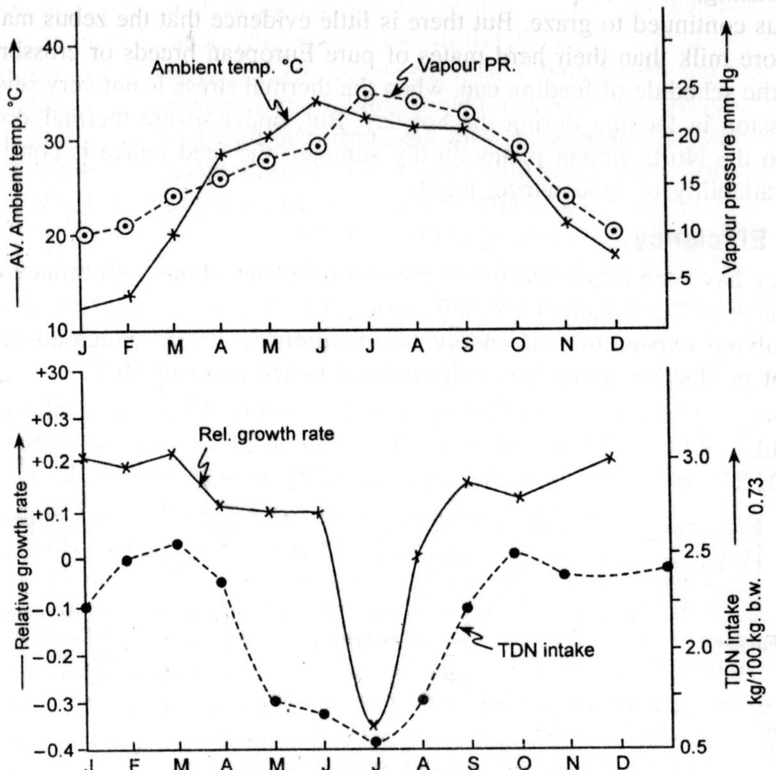

Fig. 4.7. Effect of ambient temperature and humidity on TDN intake and relative growth rate of crossbreed calves.

intake. This superimposed humoral depression, once it got triggered, was not affected by the relatively cooler hours during the nights thus depressing feed intake during nights also.

Figure 4.7 depicts the effect of ambient temperature and humidity (vapour pressure) on TDN intake. As air temperature and vapour pressure increased, TDN intake of crossbred calves decreased. The maximum depression in feed intake was in June, July and August during which increased humidity was prevalent along with high air temperature.

In a hot environment, a ration high in protein content and low in fibre content is beneficial for keeping down body temperature and improving production.

Considerable difference has-been observed between European and Zebu breeds of cattle with respect to temperature at which they reduced feed intake. Feed intake of lactating Jersey, Brown-Swiss and Holstein cows was depressed at 27°C while that of Zebu (Brahman) cows did not decline until 38°C. But it has been observed that even under thermal stress, the European breeds ate equal to or more than Zebu breeds and produced more than them. But it must be borne in mind that these are observations from short-term studies in climatic chambers. Under prolonged exposure, the purebred European animals have been found to lose condition and become susceptible to many ailments.

V.H. Nauhsimer Thoneick, C.K. Thomas and J.H. Weniger (1988). Investigations on energy metabolism of lactating cows under heat stress. III. Effect of long-term high environment temperature on parameters of thermal regulations, food intake and milk yield. Zuchtungskunde. 60 : 376-387. IV. Effect of long-term high environmental temperature on heat production and energy utilization. Zuchtungskunde. 60 : 388-397.

While grazing, the European breed cattle have been found to seek shade under hot sun while the zebus continued to graze. But there is little evidence that the zebus make greater gains or produce more milk than their herd mates of pure European breeds or crossbreds. The day to night shift in the schedule of feeding can, when the thermal stress is not very severe, compensate for the depression in feeding during the hot day. But, under severe thermal stress, as has been experienced in the North Indian plains during summer, the feed intake is considerably reduced in spite of availability of food during nights.

4.3.2. Feed Efficiency

Feed efficiency has been depressed under hot environments along with reduction in feed intake and heat production. The reduction in efficiency may be due to increased physical and physiological activities involving expenditure of energy for thermolysis. These increased activities result in increased heat production above an environmental temperature of 30°C.

McDowell and his co-workers (1969) observed that the efficiency of utilization of digestible energy for milk yield in Holstein cows was 60 per cent at 21°C, but only 40 per cent after 7 days exposure at 32°C and 31 per cent after 14 days exposure (see table 4.1).

TABLE 4.1. Efficiency of Energy Utilization for Milk Yield, and Digestibility for Lactating Holstein Cows at 21°C and after 7 and 14 days Exposure to 32°C. (adapted from McDowell *et al.*, 1969)

Factor	*Average at 21°C*	*Average per cent deviation from 21°C*	
		7 days at 32°C	*14 days at 32°C*
Efficiency of utilization of digestible energy (DE) for milk(%)	60	–35	– 51
DE consumed (Mcal/ day)	38	–14	–18
DE (Meal/day) for production	22	28	34
Maintenance requirements (Mcal/day)	15	–1	–2
Solids corrected milk (Meal/day)	12.6	–28.6	–39.7
Digestibility of D.M.	60	3	5
Calories	59	1	1
Fibre	35	1	12

Experimental evidence points to the fact that cattle may have to make higher energy expenditure for maintenance in a hot environment because of the demands for energy to activate thermolytic systems like increased blood circulation rate, increased respiratory activity, increased action of sweat glands and increased metabolic rate resulting from elevated body temperature. During summer, lactating cows were 10-12 per cent less efficient compared to cooler seasons.

Thus, under circumstances of environmental thermal stress, not only do bovines reduce feed intake, but also return less per unit intake due to a reduction in feed efficiency.

4.3.3. Milk Production

Reports from field and climatic chamber studies generally indicate that thermal stress causes a decline in milk production. The reduction in milk yield is mainly due to the reduced feed intake.

Extensive field studies by Johnson (1958) indicated that the fall in milk production during summer was due mainly to a decline in the quality and quantity of the available forage. In India

McDowall, R.E. 1972. *Improvement of livestock production in worm climates.* Watt. Freeman & Co. San Francisco.

milk production of crossbred cows was lower under heavy rainfall hot-humid regions as compared to hot-dry areas in spite of higher mean ambient temperature. Humidity, and not ambient temperature, was negatively correlated to a significant order with average milk production (Thomas and Acharya, 1981). The reduction in the quality of forages in wet and humid tropical regions might be the reason. (The cows were of comparable genetic make up and were managed and fed according to same standards).

Hot climatic environment affects an animal's productivity both directly and indirectly. Under most circumstances occurring naturally, the latter effects may be more important than the former.

Milk biosynthesis depends on the mammary gland receiving a continuous supply of various nutrients needed, the availability of hormones concerned with lactation and the maintenance of the proper environment.

In the lower elevations (<1000 m) of the N-S 30° latitudes, the environmental heat presents a stress which brings into play a complex of nervous, endocrine and motor functions. All these combine to restore a constant body temperature and to adjust body fluid and electrolyte balance, energy metabolism and behaviour to the needs concomitant with survival in the new environment. The well-being and productivity of the cow under such circumstances will depend on the power of such homeostatic mechanism to operate effectively resulting in the maintenance of normal conditions.

If the high thermal stress conditions were continuous, as they are in most of the N-S 30° latitudes, the indirect effect on the animal through inadequate or poor quality feed, parasitism and disease would become more important. Summer forages in the hot humid areas tend to be low in digestible nutrients but high in moisture and fibre which is less digestible. The coarse nature of the fibre increase the proportion of food energy lost as heat in the 'work of digestion' thus adding further to the heat load on the animal. High moisture content in feed will often result in less dry matter and digestible nutrient consumption by the dairy cows due to the 'filling' caused by added bulk of fodders from it.

The optimum environmental temperature for lactation depends on the breed and the heat tolerance potential of the individual cows. For temperate breeds of cattle 10-18°C has been identified as the optimum range for milk production. Among *Bos taurus* breeds, milk yield starts declining at 21°C in Holstein Friesians and in Jersey and Brown Swiss at 24° to 27°C. On the other hand, milk yield of zebu cattle decline only above 32-34°C. Observations in India on crossbred cows has shown that the productivity of the crossbred of Holstein-Friesien, Brown Swiss, and Jersey with various Indian breeds were not seriously affected by the hot-dry climate in north western parts where ambient temperature exceeded 40°C in summer. On the other hand, in hot-humid coastal and eastern regions of the country with heavy rainfall, the productivity was much lower especially so in the larger breed crosses (Holstein-Friesien and Brown Swiss).

One reason for the poor performance of the larger breed crosses under hot-humid conditions may be that the total energy supply is less than required due to inadequate fodder supply, coarseness of fodder, high moisture, fibre and lignification and less digestibility of nutrients. Also the smaller breeds have more surface area per unit mass for heat dissipation.

McDowell (1972) concluded that if 7000 Mcal of energy could be supplied per cow per year, the improved *Bos taurus* breeds with their better genetic potential for milk production would be more satisfactory in hot climates. He went on to say that the condemnation of European breeds because of low milk yield under hot conditions is not warranted. But this line of argument does not hold good considering the fact that feed intake is depressed due to low thyroid activity under hot conditions and that the heat generated due to specific dynamic action adds to the heat load of the animal compelling it to reduce feed intake and milk production. Perhaps a high level of productivity can be kept up in these animals with a predominantly concentrate feeding as has been achieved in Israel. But most of the poor developing countries which fall within the N-S 30° latitude do not have the resources or feed material to resort to heavy concentrate feeding of cattle. Milk production in these countries have to be roughage oriented utlizing mostly the by-products of arable farming. Under such farming conditions where energy supply is likely to be restricted to 3000 Mcal per year, a small to medium sized breed or crossbreds may have higher net efficiency.

4.3.4. Milk Composition

High ambient temperature causes certain important changes in milk composition of cows. Most important are depressions in yield of milk fat and solids-not-fat (SNF). The characteristics of milk fat also changed under heat stress in that the content of lower chain fatty acids decreased, whereas, that of palmitic and stearic acids increased. At temperatures of 4, 15, 27 and 35°C, same cows produced milk with 8.26, 8.06, 7.88 and 7.58 per cent of SNF respectively. There was a parallel decline in casein content also (Regan and Richardson, 1938).

Lactose percentage also seems to be depressed where cows are exposed to high ambient temperature. The level of pantothenic acid is lowered, Freezing point of milk lowers, but, the pH and levels of ascorbic acid and riboflavin appears to increase under' heat stress.

4.3.5. Growth

A hot environment reduces the growth rate of cattle, probably through a reduction in feed intake and changes in hormone secretion.

Fig. 4.7 illustrates the parallelism between reduction in nutrient intake and growth rate, both being suppressed by thermal stress. However, reduction in growth rate is not a simple consequence of a reduced feed intake. The animal also expends more energy for combating the effects of heat stress. Nitrogen retention is reduced under hot conditions mainly due to reduced digestibility of crude protein (Thomas, 1966) and excessive protein catabolism for energy purposes to compensate for the reduced TDN intake (Vercoe, 1969). The high N content of sweat in a hot environment is also believed to be quantitatively significant.

Hancock and Payne (1955) split 8 pairs of identical twins between temperate New Zealand and tropical Fiji and provided similar conditions of management and feeding. At calving, the Fiji animals were about 10 per cent lighter than the New Zealand ones. But when the initial setback in growth rate following transportation was accounted, the differences between the environments were small. But one factor often ignored is that considerable hydration takes place especially in the extracellular space as a means for combating heat stress (Thomas & Razdan, 1974). Thus a reduced body weight in a hot environment still underestimates the retardation in growth because there is an increase in body water which partly hides the real loss of body solids (Macfarlane and others, 1969).

Different breeds of cattle vary in their growth response under a hot environment. But the results are not conclusive. When the differences in initial weights and inherent differences in growth rate among breeds are adjusted, many conclusions formerly drawn on the basis of absolute gain in body weights would be altered. For example, based on adjusted values, Shorthorns did as well as Brahmans and Santa Gertrud is at 27°C and better at 10°C. Holstein-Friesiens grew equally well at both the temperatures. The Jerseys did better than the Brown Swiss at 27°C although on absolute weight gain basis, Brown Swiss alone was found to fare well among milch breeds at 27°C.

TABLE 4.2. Effect of Thermal Stress on the Weight gains of Sahiwal & Sahiwal-Brown Swiss

	Sahiwal	*Sahiwal-Brown Swiss Crossbred*
Initial weight on 10-1-1968	234.9	286.8
Average weight on 10-4-1968	281.4	326.2
Average gain in weight	46.5	39.3
Percentage of change from initial	16.5	13.7
Initial weight on 10-4-1968	281.4	326.2
Average weight on 10-7-1968	303.5	356.5
Average gain in body weight	22.1	30.3
Percentage of change	7.85	9.29
Initial weight on 10-7-1968	303.5	356.5
Average weight on 10-10-1968	291.0	339.4
Weight change	– 12.5	– 17.1
Percentage of change	– 4.12	– 4.88

Smaller breeds of cattle are better equipped to thrive in hot regions because of larger surface area per unit body mass which enables better heat dissipation. One of the reasons for superior heat tolerance of zebus is their smaller size. But, this does not imply that individuals of larger breeds stunted due to heat stress or other reasons will be more adaptable. *Bos taurus* cattle raised at high temperature conditions would maintain low body temperature than temperate raised ones when exposed to hot conditions thus giving lower response in heat tolerance tests (Johnson, H.D., 1967). But, their feed intake and growth will be as much impaired as the temperate reared ones because of the reduced fitness of high-temperature reared calves.

FURTHER READING

Chapter 1

IDA. 1985. *Dairy Science—A review of research during the last decade.* Indian Dairy Association, New-Delhi.

IDA. 1985. *Dairying in India*—1984. Indian Dairy Association, New Delhi.

Anon. 1985. *Dairy India 1987.* Dairy India Year Book. Publisher : P.R. Gupta, Priyadarshni Vihar, New Delhi.

Whyte, R.O. and Mathur, M.L. 1969. *The Planning for milk production in India.* Orient Longman, New Delhi.

Mishra, S.N. 1978. *Livestock Planning in India.* Vikas Publishing House, New Delhi.

Tandon, R.K. and Dhondyal, S.P. 1967. *Principles and methods of farm management.* Achal Prakashan Mandir, Kanpur.

CSIR. 1970. *Wealth of India—Supplement Livestock.* Council of Scientific and Industrial Research, New Delhi.

Bath, D.L., Dickenson, F.N., Tucker, H.A. and Appleman, R.B. 1978. *Dairy Cattle: Principles, Problems and Profits.* Lea and Febiger, Philadelphia.

Sundaresan, D. 1981. *Strategy for milk production enhancement in the decade of eighties.* NDRI, Karnal.

Slater, K. and Throup, G. 1981. *Dairy farm business management.* Farming Press Ltd., Ipswitch.

Chapter 2

Bebermeyer, P.H., *et al.*, **1965.** *Farm Business Planning Guide for Organization.* FN 6500-October, 1965, University of Missouri, Columbia (U.S.A.)

Crothy, R. 1980. *Cattle, Economics and Development.* Commonwealth Argi. Bureau. Slough, Sl. 2 3 BN, England.

Dabas, I.S., Bhopal Singh and B.K. Rastogi : An Economic Appraisal of Dairy Farming and Arable Farming. *Indian Dairyman*, Vol. XXIII, No. 8, 1971.

Dhondyal, S.P. 1979. *Farm Management Economics—An Introduction to Economic Analysis.* Achal Prakashan Mandir, Parmat, Kanpur, 1979, Chapter, 3.

Gill, S.S. and G.S. Gill. 1975. Financial Management of Dairying with crossbred cows and buffaloes. *Indian Dairyman.* Vol. XXVII, No. 6, 1975.

Heady, E.O. and H.R. Jensen. 1964. *Farm Management Economics.* Prentice-Hall of India (Pvt.) Ltd. New Delhi, 1964, Chapter 4.

Hopkins, J.A. and Earl H. Heady. 1967. *Farm Records and Accounting.* (Fifth edition). Iowa State University Press, Ames, Iowa (U.S.A). 1967, Chapter 3, 11, 13 and 14.

Johl, S.S. and T.R. Kapoor. 1978. *Fundamentals of Farm Management.* Kalyani Publishers, Ludhiana. 1973, Chapter 8.

Mishra, S.N. 1978. *Livestock Planning in India.* Vikas Publishing House Pvt. Ltd., New Delhi.

Chapter 3

Relwani, L.L., P.G. Nair, B.K. Rastogi and M.S. Sidhu. 1979. *How to Make Large Profits on a Dairy Farm* ? National Dairy Research Institute, Karnal, Publication No. 128.

Sastry, N.S.R. and R.N. Pal. 1982. Dairy Farming. Haryana Agricultural University, Hissar, Publication No. 3, June, 1974 (2nd ed. 1982).

Singh. I.J. 1977. *Elements of Farm Management Economics.* Affiliated East-West Press Pvt. Ltd., New Delhi, Chapter 4.

Slater K. and Throup, G. 1981. *Dairy Farm Business Management.* Farming Press Ltd., Ipswitch, England.

Yang, W.Y. 1968. *Methods of Farm Management Investigations.* F.A.O. Agriculture Development Paper No. 80, Food and Agriculture Organization of the United Nations, Rome, Italy, Chapter 3.

Smith, A.J. (Ed.). 1984. *Milk Production in Developing Countries.* C.T.V.M., Easter Bush, Edinburgh, Scotland.

Chapter 4

Bianca, W. 1963. Rectal temperature and respiratory rate as indicators of heat tolerance in cattle. *J. Agric. Sci.* **60** : 113.

Bianca, W. 1965. Reviews of the progress of dairy science. Set. A. Physiology. Cattle in a hot enviormnent. *J. Dairy Res.* **32** : 291.

Bond. T.E. 1967. Microclimate and livestock performance in hot climates. In: *Ground level climatology.* Ed. R.H. Shaw., Amer. Assoc. Adv. Sci., Publ. 86. Washington, D.C.

Hafez E.S.E. 1968. *Adaptation of Domestic Animals.* Lea & Febiger. Philadelphia.

Herz, A. and Steinhauf, D. 1978. The reaction of domestic animals to heat stress. *Animal Research and Development.* 7 : 7.

Johnson, H.D. 1967. Climatic effects on physiology and productivity of cattle. In : *Ground Level Climatology.* Ed. R.H. Shaw. Amer. Assoc. Adv. Sci., Publ. 86, Washington, D.C.

Lee. D.H.K. 1953. Manual of field studies on the heat tolerance of domestic animals. FAO Development Paper No 38. FAO, Rome.

McDowell, R.E. 1972. *Improvement of Livestock Production in Warm Climates.* W.H. Freeman and Co., San Francisco.

Thompson. G.E. 1973. Review of progress of Dairy Science : Climatic physiology of cattle *J. Dairy Res.* **40**: 441.

Section II : FEEDS AND FEEDING

CHAPTER

5

Feeds and Forages for Dairy Bovines

A combination of different feeds and fodders are fed to dairy bovines to furnish the required nutrients. These feedstuffs can be classified into different groups based on their fibre and total digestible nutrients (TDN) content as given in Figure 5.1 below.

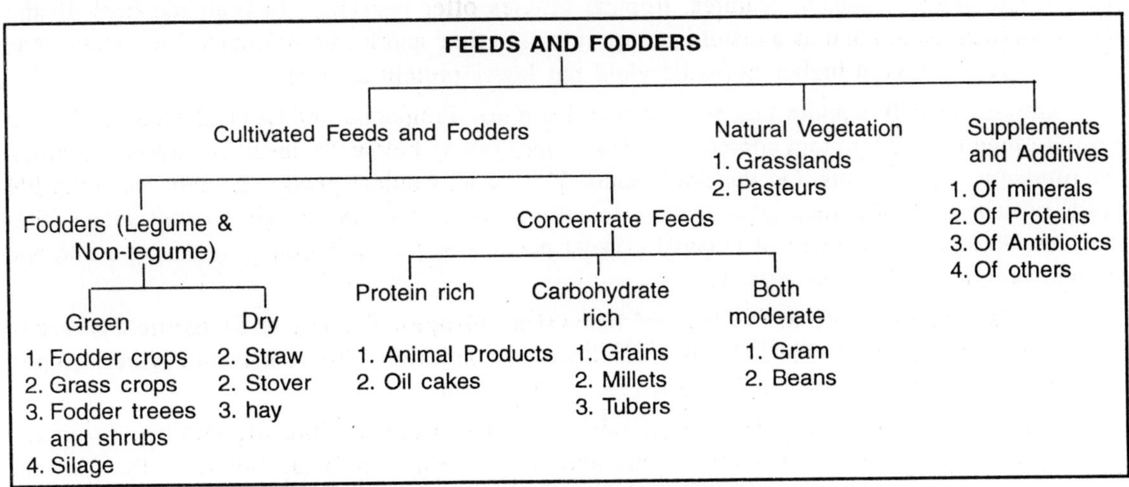

Fig. 5.1. Classification of livestock feeds and fodders

5.1. HERBAGES, FORAGES AND ROUGHAGES

Plant material grown for feeding herbivorous animals has been called herbage. Forage is a term, which is used more or less interchangeably with herbage. Some restrict forage only to that part of the plant that can be eaten by the animals. Roughage would include forages or herbages, but would not be restricted specifically to plants grown to feed animals. They include both leguminous and non-leguminous crops and can be in the form of pasturage, green-chopped feeds (soilage), silage, haylage, hay and straw. In contrast to concentrates, they contain a relatively high proportion of fibre and less of energy. Dairy bovines are well adapted to a high proportion of forages in their ration. They require a substantial amount of fibre for good health and optimum performance. Most of the fibre in the bovine rations is supplied through forages. Forages are also good sources of energy, protein, minerals and vitamins.

Under majority of conditions forages form the cheapest source of nutrients, especially energy, for bovines. Because of this, a maximal forage ration will usually be the most economical. But there are exceptions and the cost relationships vary a lot from situation to situation.

5.1.1. Variation in Feeding Value

Large number of species of plants is used as forages. It is but natural to have wide variation in the composition, nutritive value, physical properties and palatability of these due to, (1) species and strains of plants, (2) stage of maturity, (3) climate and soil, (4) stem-leaf proportion, (5) fertilizer application, and (6) methods of preservation, if done.

1. Species and strains of plant. Morphological and physiological differences between species such as growth habit, leaf-stem ratio, perenniality, and flowering behavior etc. influence forage quality. There are variations between species of plants in basic leaf anatomy, which are related to physiological differences, and in turn to potential nutritive value.

Most of the tropical grasses have a special leaf anatomy known as Kranz anatomy characterised by a radial arrangement of chlorenchyma cells around the vascular bundles. Most of them have a C_4 pathway of photosynthesis. Temperate grasses and both temperate and tropical legumes have C_3 pathways of photosynthesis. They also lack the specialised anatomical features. As a result of these special features, tropical grasses offer resistance to both mechanical and microbial degradation and as a result increase retention time and lower voluntary feed intake. The tropical grasses have a higher potential yield but lower protein content.

Grasses usually contain less protein than legumes. Temperate and tropical legumes have a similar range in crude protein content with few values falling below 9% level, considered minimal for ruminant requirements. On the other hand, 55% of all tropical grasses contain less than 9% crude protein. Only less than 20% of the tropical grasses have crude protein content above 15%. Thus the low protein content of tropical grasses poses a major limitation to intensive cattle and buffalo production in those regions.

2. Stage of maturity at the time of harvesting forages. The key to forage quality lies in the stage of harvesting fodder. Under a given set of conditions, this is a factor, which a dairy manager can easily control, but is often neglected by him.

Lignin in cell wall is a major determinant of the extent of fibre digestibility. Structural carbohydrates in the walls of sclerenchyma and some vascular cells are bound to the phenolic compound lignin. By bonding to fibre, lignin prevents swelling, thereby restricting entry of microbial digestive enzymes. This causes depression in fibre digestibility. Tropical warm climate seems to stimulate more lignin production than temperate climate. Lignin is higher in stem than in leaves. As the plant matures, lignin content increases faster than cell wall content. As a result cell walls become progressively more lignified and less digestible.

The protein content of grasses is usually high in the vegetative stage of growth. It is only when the plant approaches maturity that the protein level drops and lowers the forage quality. With advancing maturity, the decline in protein content is slower in leaf than in stem. Compared to grasses, protein levels in leguminous plants decline slowly with age. With maturity, palatability decreases and percentage of discarded parts increases. However, too early a cutting will lower yields of fodder and increase cost per tonne. Thus in practical feeding, a balance has to be struck between higher quality in early stage and higher quantity in later stage (Table 5.1).

TABLE 5.1. Effect of Maturity of Green-Chopped Lucerne-Brome Forage on Digestibility, DM Intake and Milk Production of Cows

Stage of cutting per cent	*Digestibility of DM kg/day*	*Intake of DM kg/day*	*Milk production needed kg/day*	*Additional grain**
Pre-bud	67	15.5	20.0	1.8
Bud	65	15.0	18.0	2.7
Early bloom	63	14.5	15.5	3.6
Mid bloom	61	14.0	14.0	5.0
Full bloom	59	13.2	12.2	6.4
Late bloom	58	12.7	10.5	7.3
Mature	56	11.9	9.1	8.2

*Data adapted from Hibbs and Conrad (1975). Cows fed free choice 1-2.5 kg grain per day. Grain needed to maintain initial milk production.

3. Climate and soil. Environmental temperature has a very important influence on the nutritive value of forages. High temperature adversely affects the nutritional value of both tropical and temperate grasses. Legumes are affected to a lesser extent by it. High growth temperatures accelerate stem development and the maturation process in plants. This leads to increase in tissue cell wall content and lignification resulting in 5-10 per cent lower DM digestibility.

High levels of solar radiation and moderate soil water deficits are mostly beneficial to the nutritive quality of grasses. Generally, the nutritive quality of legumes, tropical or temperate, is less influenced by environmental factors than that of grasses.

4. Fertilizer application. Fertilizer effects on DM digestibility and associated cell wall and lignin characteristics are small and inconsistent. Nitrogen application may have a positive, nil or negative effect on these. It increases DM digestibility in young tissue and early re-growth, but has the opposite effect on mature tissue. Potassium and phosphorus fertilization rarely seem to have much direct effect on the DM digestibility of forage crops. However, in extremely deficient soils, they may improve the nutritive quality of forage indirectly by stimulating rapid new growth.

5. Leaf-stem proportion. Generally, leaves of a forage crop have a higher content of crude protein, digestible energy, essential minerals, and vitamins but less fibre than stems. Thus the overall nutritional value is usually more in leaves than in stems. Leafy forages have been found to have better palatability also. There is a decrease in leaf: stem ratio with maturity. This is one of the reasons for lowering of nutritional value when the plants mature. The stem and leaf sheath decline in digestibility with maturity at a much faster rate than leaves. These changes take place in leguminous plants at a less rapid rate because there is less lignification of the stems in them and the leaves even less. However, legumes lose more leaves so that the leaf-stem ratio changes more with maturity.

6. Irrigation. Coupled with fertilizer application improves DM yield and nutritive value of forages. Irrigated N-fertilized *Digitaria decumbens* pastures have produced about 2,000 kg annual weight gains a hectare as against 1,300 kg without irrigation.

7. Methods of preservation. Forages are usually preserved as hay or silage. It is impossible to cut, move and store forages without losses occurring in the process. The quality and quantity of preserved forages depend on factors like stage of maturity at the time of cutting, methods of handling and preservation, moisture content and weather conditions while harvesting. It has been reported that DM losses during field drying range from 11 to 60 per cent. Leaching by rain of partially dried hay is very damaging. Leaf shattering is a major cause of nutrient loss in legume hay. While converting into silage, in addition to losses in the field, there can be reduction in nutritive value due to plant respiration, bacterial fermentation and spoilage on the periphery. In well-preserved silage, fermentation losses should not exceed 0.5 to 5.0 per cent of the energy.

5.1.2. Pasture Management

Pasture is the natural feed of bovines. Raising them on pastures is the cheapest because no cost of growing, cutting, transporting or preserving is normally involved. Pastures are widespread and

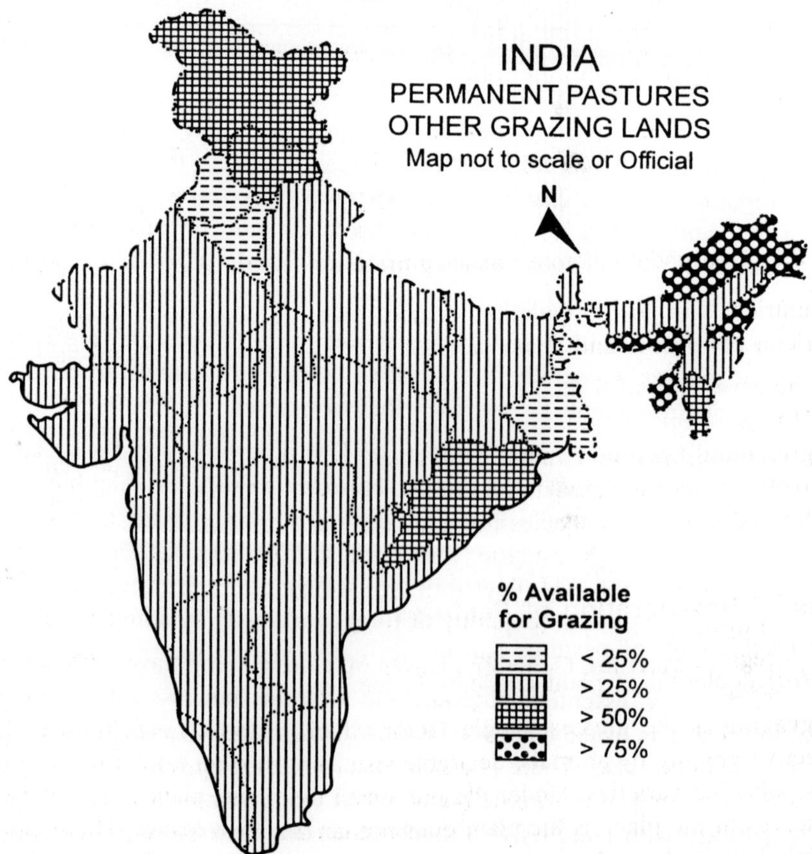

Fig. 5.2. Extent of availability of grazing in different regions of India

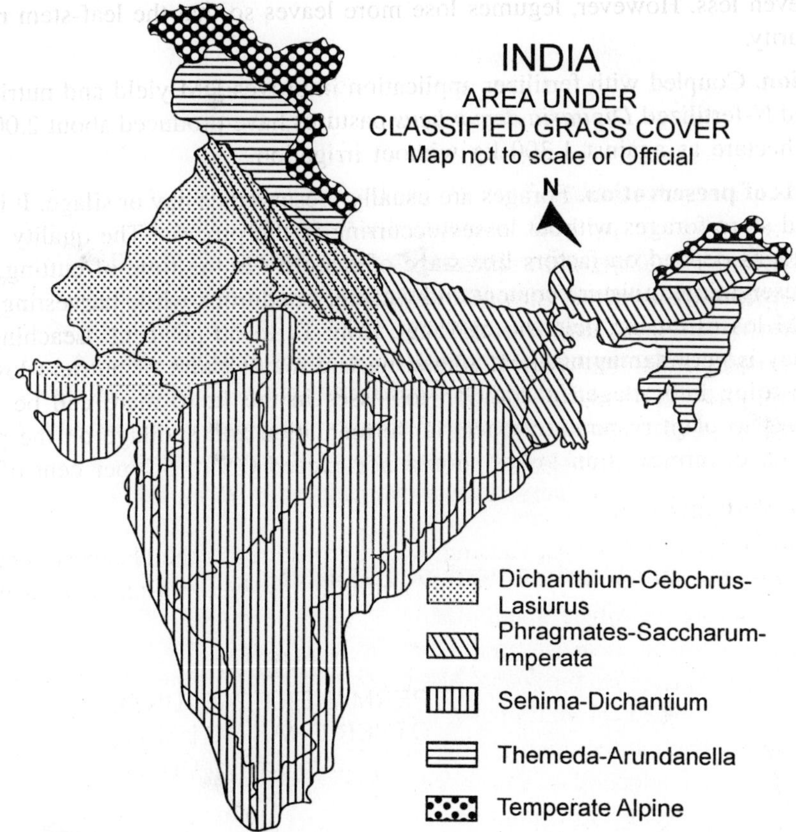

Fig. 5.3. Area under different classified grass covers in different regions of India

popular in countries with low population density (both human and animal). Australia, New Zealand, American continents and some parts of Europe depend widely on pastures.

In India, however, out of a total cultivable area of 131.11 million ha, less than 2 per cent is utilized for forage crops. Population pressure on the land and cheapness of labour, which makes other agricultural operations more profitable are the major factors preventing earmarking of land exclusively for pastures and fodder crop cultivation. Besides in a tropical climate there are only very limited number of species that are suited for use as grazing legumes (See Figures 5.2 and 5.3).

5.1.3. Causes of Deterioration of Pastures

Decline in the productivity of pastures can be attributed mainly to— (*i*) overgrazing; (*ii*) bad drainage; and (*iii*) depletion of plant nutrients in the soil.

(*i*) **Overgrazing** is the greatest single factor which causes deterioration of pastures. On account of selective grazing by animals, desirable species of herbages tend to get depleted much faster than less palatable varieties. Under the burden of overgrazing, there is a continuous drain on the food reserves of the plant as the plant does not get an opportunity to replenish its reserves by photosynthetic activity because the new shoots sprouted would have been immediately eaten. As a result, the plant gets weaker and weaker and finally dies.

(*ii*) **Bad drainage.** While a good supply of soil moisture is advantageous for fodder production, in the humid regions, excess may be detrimental. Waterlogging changes the soil structure and reduces the supply of available oxygen in the soil causing stunting and yellowing of certain crops. Variation exists in the ability of different species and varieties of plants to withstand excess moisture. For example, lucerne and berseem require well-drained soil. Para grass, on the other hand, is well known for its capacity to grow in waterlogged conditions.

(*iii*) **Depletion of soil fertility.** Lack of soil fertility is one of the major reasons for inferior pasture production. Most of the soils available for fodder production are not naturally fertile. Continued use and over-stocking often deplete the soil nutrients. High rainfall and leaching, non-availability of leguminous plants among the crops grown etc., are certain other reasons for poor soil fertility. Many of the good pastures were not naturally fertile. Man through fertilization and cultural practices have made them highly productive. Many pastures are low in productivity not because they do not carry the desired highly productive grass and legume species, but because the fertility demanding types cannot compete efficiently with species whose fertility requirements are lower. The improvement of the fertility status immediately lifts the competitive strength of the highly productive species and in very short time completely alters the botanical balance of the sward in their flavour.

5.1.4. Measures for the Improvement of Pastures

Management of pastures involves manipulation of plant resources and the animals grazing on them to sustain a stable system of production. In a properly managed grassland, no special measures are necessary because the plants and animals grazing in them are in equilibrium. When this equilibrium is disturbed through over-grazing and neglect, denudation and soil erosion take place. The measures that can be employed to prevent this and improve pastures are : (*i*) soil and water conservation, (*ii*) selection of improved fodder species and varieties, (*iii*) fertilizer input, (*iv*) control of grazing through stocking rate and system of grazing, (*v*) weed control, (*vi*) burning, (*vii*) temporary closure and reseeding and fertilizer application.

5.1.4.1. *Soil and Water Conservation on Pastures*

In the denuded overgrazed and depleted grassland, the first step in pasture improvement should be towards conservation of soil and water. Where gullies have been formed due to soil erosion, dykes and bunds will have to be constructed across the gully channels. In sloppy areas terracing and contour bunds will have to be made. These are helpful in high rainfall regions to divert water to the channels with a minimum of soil loss from the grasslands. In semi-arid and arid regions, harrowing, contour furrowing, pitting etc., may be done to conserve maximum extent of rainwater.

In countries like India where manual labour is still relatively cheap: cutting, digging out, burning etc., maybe used for weed control. In limited areas weedicides can be used even though their high cost prevents extensive use of these agents. In many areas, the stubble remaining after harvesting or grazing are set fire to during the summer to encourage new growth during the next season. It is helpful in weed control and encourages better growth during the next season. But it is antagonistic to the process of primary and secondary natural succession. In dry regions especially, burning causes damage to pastures.

5.1.4.2. *Selection of Improved Fodder Species*

Some improved fodder species or varieties are known to have better yield or better nutrient composition. Some change the pattern of fodder production to reduce deficiencies during critical periods of animal requirement. Some improve the efficiency of the operation and make it more economical. Where there are defined warm and cold seasons, fluctuations in seasonal pasture growth may be reduced by the integrated use of species adapted to cold and warm conditions. In some draught prone areas, introduction of draught-resistant species of forages can substantially increase yield. Differences in nutritive values are important criteria for selecting a plant variety. Legumes improve pasture and animal production remarkably. There are differences in the nutritive value between leguminous species. It has been observed in many instances that introduction of legumes into native pasture has substantially increased animal production without any fertilizer input.

5.1.4.3. *Control of Grazing*

The objective of good grazing management should be to maintain the grassland in the peak of productivity and at the same time to satisfy the nutrient requirements of the grazing bovines and maintain soil fertility. Grasses in general are more palatable and nutritious in early stage, but grazing at this stage will harm the plants more because they get less opportunity to build up reserves in the underground parts of the plant. Allowing the plants to mature reduces palatability and nutritive value putting animals grazing on it to disadvantage. Good pasture management involves balancing these divergent needs.This may be achieved through adopting various grazing systems and controlling the stocking rate.

5.1.4.3.1. Systems of Grazing

(*i*) Controlled continuous grazing. The number of livestock allowed for grazing is limited depending on the carrying capacity. Grazing is continuous, but altogether stopped when certain preferred species of grasses reach a minimum. This is better than the uncontrolled grazing going on in many developing countries.

(*ii*) Deferred grazing. This system is applicable where perennial grasses are predominant. It involves allowing seed formation and seed maturing in part of the pasture by delaying grazing there. This gives chance for the perennial grasses to make good growth and accumulate sufficient nutrients in reserve to maintain their vigour and produce new shoots next year. The grassland is divided into 3 compartments. One of them is earmarked for seeding and the other two are subjected to grazing alternately. Grazing is allowed in the seeded compartment at an appropriate stage when the seeds are mature and the plants are not too mature and unpalatable. Grazing at this stage helps in broadcasting of seeds also. Grazing in the other two compartments ensures tender forage, while alternating the plots gives chance to the plants to recoup. One plot out of the three is left for seeding yearly in rotation so that in a period of 3 years each segment gets a period of rest and recuperation.

(*iii*) Rotational grazing. The whole pasture is fenced into convenient -number of plots. The cows are let into one compartment at a time one after the other in rotation at suitable intervals. This brings about uniform grazing. The quick shifting from plot to plot gives the cows the advantage of a steady supply of nutritive young herbage. At the same time the vacant plots get rest for recuperation. The main disadvantage is that this does not provide for seeds to form and mature.

(*iv*) Rotational deferred grazing. This system is beneficial when seeding is desired as in annual species or when perennials do not have sufficient density. The whole pasture is divided into 3 equal compartments and fenced. Each compartment is grazed continuously for 1/3 of the total grazing season and protected through the remaining 2/3 of the season for seed setting and maturing. By the time animals are allowed to graze the 3rd compartment, the seeds would have been shed. When they go back to first and second compartments seeds in those would have also ripened and got shed. In the second year rotation will start from the second compartment and in the third year from the third compartment.

(*v*) Hohenheim System. This is another type of rotational grazing system developed in Germany during the First World War. The pasture is compartmentalized and grasses are raised, with heavy nitrogenous fertilization. The dairy cattle are divided into 3 categories, *viz.,* high yielder, average yielder and low yielders. These three groups of cows are successively let into the compartments in that order and also rotated among plots in such a way that the opportunity for grazing more nutritious immature grasses goes to the high yielder followed by medium producers and lastly to low yielders. The advantages of rotational grazing over continuous set stocking in terms of animal productivity are rarely evident except at high grazing pressures. Under intensive grazing the maximum feeding value of pastures in young leafy stage is best exploited by rotational systems of grazing. Walshe (1975) reported a higher growth rate in calves and 13 to 20 per cent increase in milk production due to rotational grazing.

However, except in particular circumstances, the evidence from temperate pastures shows no advantage from rotational grazing in increasing production per head. Willoughby (1970) concluded that under year round grazing there was no advantage from rotational systems of grazing except in the case of lucerne pastures.

A review of grazing systems in South America and Southern Africa (Paladines *et al.,* 1974; Paladines and Leal, 1979) support the general view that continuous grazing is more productive at moderate stock intensity. In arid and semi-arid grasslands perennial grasses are the main source of feed supply. Drought coupled with overgrazing can result in replacement of these perennial grasses by annual species of lower quality. Under these conditions a flexible grazing system is used to conserve the better quality perennial grasses.

5.1.4.3.2. Stocking Rate

The long term stability of the pasture depends on the stocking rate. In grass-legume pastures, the persistence of the legume component is influenced by the frequency and intensity of defoliation by grazing. It affects seed setting and the competitive relationships between the legumes and grasses in the pasture.

High intensity of stocking will also result in grazing when the plants are too immature with the result that the plants do not get enough opportunity to build up reserves in the underground parts for future regeneration. Persistence of tropical legumes under high stocking rates may be enhanced by deferred grazing over the period of seed production to allow regeneration from the seeds.

5.1.4.3.3. Weed Control

In countries like India where manual labour is still relatively cheap, cutting, digging out, burning etc. may be used for weeding. In limited areas weedicides can be used although their high cost is a serious handicap.

5.1.4.3.4. Burning of grasslands

In many areas, the stubbles remaining after harvesting or grazing are set fire to during the summer to encourage new growth after the next monsoon rains. It is helpful in weed control and encouraging better growth during the next season. But, it is antagonistic to the process of primary and secondary natural succession. Especially in dry regions burning cause damage to pastures.

5.1.4.3.5. Temporary Closure

Protecting the pasture from grazing for sufficient time to allow seed setting and regeneration induces a progressive rejuvenation. Closure for one or two years is necessary in mildly eroded pastures. But in highly degraded ones, closure up to 5 years may be necessary.

5.1.4.3.6. Reseeding and Fertilization

Addition of manures and fertilizers and reseeding with palatable varieties of pasture forages coupled with proper control on grazing can result in quicker regeneration. After sowing a light harrowing may be useful.

TABLE 5.2. Chemical Composition (in-Per Cent DM) of Different Fodder Crops Depending on the Stage of Harvest

Fodder crop	*Stage*	*DM*	*Crude protein*	*NDF*	*Cell*	*ADF*	*Cellulose*	*HC*	*Lignin*	*Ash*	*Ca*	*P*
Berseem	I	1.19	22.6	42.5	57.5	35.2	24.3	7.3	10.2	12.3	—	—
	II	16.2	17.7	49.6	50.4	38.6	25.7	10.9	11.8	11.24	2.2	0.1
	III	32.2	15.5	53.3	46.7	41.7	26.9	11.6	13.1	9.6	—	—
Cowpea	I	11.7	24.2	43.2	56.8	34.5	23.8	8.8	9.9	13.0	—	—
	II	14.3	20.2	49.8	50.2	37.3	25.3	12.5	11.1	11.6	2.4	0.2
	III	17.0	17.1	54.0	46.0	40.8	27.8	13.2	14.2	11.2	—	—
Lucerne	1	20.5	23.4	37.3	62.7	30.6	21.4	7.1	8.2	11.9	—	—
	II	25.4	20.0	43.6	56.4	35.8	24.6	7.8	9.5	10.6	3.1	0.1
	III	29.9	15.9	50.7	49.3	39.8	26.1	10.9	11.1	10.3	—	—
Maize	I	19.5	11.1	55.1	44.9	33.9	26.7	21.2	5.7	10.2	—	—
	II	25.8	9.0	62.3	37.7	38.3	29.5	24.0	6.8	8.9	0.7	0.1
	III	35.4	7.6	66.5	33.5	41.6	31.0	24.8	8.1	6.9	—	—
Jowar	I	18.0	12.5	55.9	44.1	37.9	30.2	18.1	5.8	9.5	—	—
	II	29.9	9.0	65.7	34.3	41.9	32.6	23.8	6.7	8.4	0.4	0.1
	III	38.9	7.2	70.4	29.7	46.0	34.3	24.3	7.9	7.6	—	—
Teosinte	I	14.3	7.0	61.6	38.4	39.3	29.5	22.2	8.1	10.6	—	—
	II	25.9	6.2	67.2	32.8	41.7	31.1	25.5	8.5	8.9	0.7	0.1
	III	34.3	4.8	71.7	28.3	45.1	33.2	25.6	9.3	7.1	—	—
Oats	I	13.1	11.7	47.2	52.8	30.0	22.3	17.3	5.5	9.1	—	—
	II	17.6	10.2	52.2	47.7	32.2	23.2	20.1	7.1	7.5	1.4	0.2
	III	29.1	8.0	69.0	30.4	37.6	26.9	32.1	8.0	9.1	—	—
Hybrid napier	—	25.5	7.1	61.6	39.0	31.7	23.3	29.3	4.2	—	—	—
Guinea grass	—	27.3	10.5	57.1	42.9	28.4	30.3	28.7	4.8	—	—	—
Para grass	—	32.1	6.8	61.8	38.2	28.9	22.2	32.9	2.8	—	—	—
Napier grass	—	34.0	7.4	71.0	29.0	31.8	26.5	39.2	3.7	—	—	—
Wheat straw	—	—	2.6	76.2	23.8	51.8	36.3	24.4	9.7	—	—	—
Rice straw	—	—	2.1	74.7	25.3	53.6	37.0	21.1	8.2	—	—	—

DM—Dry matter, NDF—Neutral ditergent fibre, ADF—Acid ditergent fibre, HC—Hemi-cellulose. Ca—calcium. P—Phoshorus. Data based on Gupta, P C. and Pradhan, K (Personal communication)

5.2. FODEER CROPS

Apart from pastures, certain leguminous and non-leguminous plants are cultivated for the purpose of feeding animals. Usually they are harvested to be fed fresh as green chops or preserved as hay or silage to be fed later. The chemical compositions of some of these crops are given in Table 5.2 and their pictures shown in Figures 5.4 to 5.9.

5.2.1. Leguminous Fodders

Leguminous fodders consist of the stem and leaves of a group of plants belonging to the group Leguminosae These plants by virtue of harbouring bacteria of the Rhizobium group in their root nodules, 'fix' nitrogen from the atmosphere and make it available to themselves, other plants, to animals and man. Generally these fodders are rich in proteins, minerals and vitamins.

Contrary to popular belief, legumes are not consistently superior nutritionally in every respect to all grasses. For example some crops like maize and jowar have equal or more energy value than most of the legumes. Some leguminous crops like Calapogonium are poor in palatability and nutritive value.

Fig. 5.4. Berseem

Fig. 5.5. Lucerne

Fig. 5.6. Cowpeas + Sorghum – 1

Fig. 5.7. Cowpeas + Sorghum - 2

Fig. 5.8. Fodder Maize

Fig. 5.9. Mustard/Rape Fodder

5.2.1.1. *Lucerne* (Medicago sativa)

Also known as *Alfa alfa* (an Arabic word meaning 'the Best'), it is a highly nutritious perennial leguminous crop rich in protein, minerals and vitamins. Originating from South-West Asia, it has world-wide distribution now. In fact no other fodder legume is so adaptable to various soil and climatic conditions. It is persistent, productive, heat and drought tolerant and provides better seasonal distribution. Lucerne persists in a wide range of climatic conditions from extreme heat as prevails in Rajasthan to cold temperatures in Ladakh. It prefers high light intensity and cool temperatures. It thrives best on a well-drained loamy soil rich in lime and pH between 5.5 and 7.5. Drainage is very necessary as the crop is intolerant to water logging and heavy rainfall.

Cultural Practices. The best time for sowing is between early October and late November, in India. A deep fine and firm seed-bed may be prepared by one mould-board ploughing followed by two cultivations, disc harrowings or ploughing with a country plough. Perfect leveling is a must for proper drainage and uniform irrigation.

The seeds have a hard covering. The germination may be improved by pre-treatment to soften their coverings. The seeds may be soaked in water for six to eight hours after which they are rubbed against rough stone to achieve this. Alternatively, they are dipped in concentrated sulphuric acid for 10 to 15 seconds and quickly rinsed four or five times to remove the acid. The seeds are then inoculated with *Rhizobium meliloti,* which is specific for lucerne. Sowing may be done by one of the following methods depending on the circumstances:

(i) **Broadcasting.** This is a method suitable for arid and semi-arid regions with sandy loams. The Land is properly leveled and seed is broadcast at a rate of 25 to 30 kg per hectare and mixed into the soil by cultivation.

(ii) Drilling. The seeds are drilled in rows 30-35 cm apart.

(iii) **Ridge Sowing.** Ridges are made 50-60 cm apart and seeds are dribbled on either side of the ridge. This method is more suitable for areas with high rainfall as the excess water is drained through the furrows. The seeds can also be dropped either by hand or through a funnel fixed to the plough into the furrows opened by a country plough.

Manuring. Fertilization is necessary as Lucerne is a heavy user of phosphorus, potassium, calcium and sulphur. In the first year, farmyard manure at a rate of 20-25 tonnes per hectare may be applied before sowing and well mixed with the soil. At sowing, a fertilizer mixture consisting of 18 kg nitrogen, 70-75 kg phosphate, 40 kg of potassium and 150 g of sodium molybdate maybe applied. In subsequent years, farmyard manure at the rate of 10 tonnes per hectare may be applied. Besides, phosphate at the rate of 70 kg/ha and potassium 40 kg/ha may be added once a year.

Irrigation. Lucerne needs moist soil conditions. Two light irrigations may be given every week after sowing for quick germination. After germination, during summer, irrigation may be given 10 to 12 days apart. During winter, irrigation at intervals of 15-20 days will suffice. Lucerne is generally badly infested with summer weeds. Proper weed control measures like inter-culture may be taken to keep the weeds checked.

Harvesting, nutritive value and yield. The most difficult part in managing the harvest of Lucerne is cutting schedules or intervals between harvests. One established recommendation is to make the first cutting at late bud to one-tenth bloom and succeeding harvests at 55 to 40 days intervals during the growing season. If the first harvest is made when the Alfaalfa plant is in the bud stage, it will increase the nutritive value. Experimental evidences show that for each day cutting is delayed past the bud stage, feed value drops about one per cent daily.

Average yields vary from 85-90 tonnes/ha/year. Alfalfa in the early bloom stage contains about 3.5 per cent DCP (15-20 per cent protein on air-dry basis). This goes down to 2.5 per cent (12 to 14 per cent protein on air-dry basis) at the full bloom stage. Besides it is a very rich source of most of the minerals, carotene and other vitamins. On a ration having a major proportion of Lucerne, the high calcium content of the forage can lead to phosphorus deficiency. When fed *ad libitum* with a phosphorus supplement, Lucerne can not only meet maintenance requirements, but also sustain milk production at 6-7 litres per day or growth of 0.5 kg/day of young calves.

5.2.1.2. *Berseem, Egyptian clover* (Trifolium alexandrinum)

Most likely Berseem originated in Asia Minor but moved to Egypt, slowly. Presently it is grown in many countries like India, Pakistan, Syria, Israel, Italy, South Africa, some South American countries, U.S.A. and Australia. In India this is mostly grown in Punjab, Haryana, Rajasthan, U.P. and other parts of Northern and Western India as an annual Rabi (winter) crop under irrigation.

The crop requires cool to moderately cold conditions, but temperature 4 to 5°C below freezing kills the plants. On the other hand, warm and humid climate is also not suited to grow Berseem. Well drained loam to clay-loam soils rich in calcium and phosphorus and with a neutral to slightly alkaline pH are the most suitable ones.

Cultural practices. Berseem can be sown from middle of September to the end of October in North Indian plains and from the middle of August to first week of September in the hills. The land is prepared by one ploughing followed by 2-3 harrowings. It should be well levelled before sowing. The easiest method of sowing is to let about 5 cm water to stand in the levelled field and broadcasting soaked Berseem seeds uniformly over it. For *Meskawi* variety, which is commonly grown in India, a seed rate of 24-26 kg/ha is sufficient. The seed is soaked in water for 10-12 hours to soften the hard coat of the seed. In new fields, the seed may be smeared with *Rhizobium trifolii* in a 10 per cent jaggery solution. An alternative is to collect 3 to 4 quintals of soil at a depth of 5 to 15 cm from a field where Berseem has been cultivated in the immediate past and to incorporate it in the soil of the new field uniformly. It is not necessary to broadcast the seeds in standing water always. A seedbed can be prepared and the soaked seeds smeared with bacterial culture can be drilled to a depth of 2-3 cm.

Manuring. Before sowing, about 25 tonnes of FYM per hectare may be applied and well mixed with the soil. At the time of sowing fertilizers containing 70-75 kg phosphorus, 25 kg nitrogen and about 28 kg of potash per hectare may be added to the soil.

Irrigation. First two irrigations should be very light and given at weekly intervals to aid germination and establishment of the plants. Afterwards irrigation may be given fortnightly. Chicory is the major weed found in the Berseem crop in India. Immediately after the last cutting, one deep ploughing with soil inversion plough burying deeply the Chicory seeds preventing their germination may be given. At the time of sowing, immersing the seeds in a 20% salt solution makes the Chicory seeds and the immature Berseem seeds to float, which can be manually removed.

Harvesting, yield and nutritive value. First cutting is obtained in two months time, second after 40 days from the first cut, third and fourth at 30 days and fifth and sixth at 20 days intervals. The first-cut fodder has 90% moisture and only 10% DM. It. increases gradually by about 1.5 per cent in every subsequent cut. The yield per hectare average from 75 to 100 tonnes per year.

It is rich in most of the nutrients except phosphorus. It contains 2.5 to 2.8 percent DCP (crude protein—17.5 per cent on DM basis). Being surplus in calcium and deficient in phosphorus, phosphorus rich items like wheat-bran may be fed along with it. Rich in protein with hollow stems incorporating air, it is more bloat producing than alfalfa. If fed mixed with hay or straw, this problem can be minimised. Straw with 10-15 kg Berseem can be a maintenance ration for dairy bovines. If fed *ad libitum* along with some straw, and phosphorus supplement, it can sustain maintenance plus milk production of 5-6 kg. In the first one or two cuttings, the crop may have less than 10 per cent DM. Under such circumstances, addition of dry forages is a must to meet the DM and nutrient requirements of the animal.

5.2.1.3. *Cowpea or Lobia* (Vigna sinensis)

It is indigenous to Central Africa and India. It is a quick growing legume adapted to all types of soils. It grows well in tropical and sub-tropical climates. It can withstand both moderate drought and heavy rainfall. It is susceptible to waterlogging, cold and frost. In the tropical regions, it can be grown during any period provided irrigation is available.

Cultural Practices. One ploughing with soil inversion plough followed by 2 ploughings with a country plough and planking is done to prepare the land. Five to ten tonnes of farmyard manure and 40 kg of phosphate per ha may be applied before the last ploughing. At the time of sowing 12 kg N in the form of urea and 18 kg potash may be applied as a fertilizer mix. The seed may be broadcast or drilled behind the plough at 22-45 cm interval. A seed-rate of 20-30 kg/*ha* is recommended for seed production. For fodder the seed rate may be doubled or even tripled. The crop may be given 3-4 irrigations. Mixed crops of cowpea with maize, jowar, teosinte or bajra are common.

Yield and nutritive value. Cowpea can be harvested 60-75 days after sowing. An yield of 10-15 tonnes per hectare is usually obtained although an yield of 25-30 tonnes is quite within reach. The feeding value of cowpea is high and quite comparable to lucerne. When fed to cows *ad libitum* it can support animals yielding 5-6 kg per day without concentrates.

5.2.1.4. *Stylo* (Stylosanthes gracilis)

This leguminous crop is a native of South America, especially Brazil. It is a successful summer forage crop in Bihar and Kerala. It combines well with perennial grasses like Congo signal, Thin Napier and Guinea. The average yield of the crop is 4 tonnes in the first cut and 30-35 tonnes in the next 12 months. On dry matter basis it has about 10 per cent crude protein.

5.2.1.5. *Centro* (Centrosema pubescens)

It is one of the most promising legumes for pasturelands. Its potentiality for producing profuse viable seeds and its quick regeneration capacity are notable. It is essentially a crop of warm-humid tropics and flourishes best in areas with an annual rainfall of 125 cm or more. It is used in grass-legume mixtures for pasture or ley for hay production. On an average it yields 30-35 tonnes of fodder per year. If cut at proper stage it has 17 per cent crude protein on dry matter basis.

5.2.2. Non-leguminous Forages

Non-leguminous forages are generally rich in energy and low in protein compared to legumes. But most of the crops are highly palatable and highly nutritious and, with some supplementation of protein through concentrates, are excellent balanced bulk feeds for bovines.

5.2.2.1. *Maize* (Zea mays)

An important grain and fodder crop, maize (Fig. 5.3), originated from Mexico. At present it is widely cultivated throughout the world. It is an ideal forage crop as it is highly palatable, nutritious as well as quick growing and high yielding. Alluvial loamy fertile and well-drained soils are ideally suited for maize. It thrives well in warm climates as well as during warm months in temperate regions. Dry-hot winds, heavy rains, cold spells, frost etc. are detrimental and retard growth.

Cultural Practices. Maize is grown mostly as a *kharif* (warm-monsoon season) crop in Indo-Gangetic Plain and North-Western and Central India. Sown with the early rains in mid June, it can be harvested in September-October. In South India it is grown between September and February. The seed-beds are prepared with 2-3 ploughings during which manures and fertilizers are added. Maize responds well to liberal fertilization. Recommended doses are 45-57 kg nitrogen and 22-45 kg of phosphorus in the form of urea and super phosphate. It may either be sown by broadcasting or dribbled in rows behind the plough at a depth of 5 cm. The recommended seed rate is 30-38 kg/ha as a fodder crop and 14-18 kg/ha for grains.

Yield and nutritive value. Maize should be harvested for fodder when the cobs are just being formed when it is 60-70 days old. Yield varies from 15-50 tonnes per hectare depending on soil fertility and cultural practices. Maize is the ideal crop for silage as it is rich in carbohydrates. It contains 8-10 per cent protein and therefore when fed alone is sufficient for maintenance. If a legume is mixed or a protein-rich concentrate is supplemented it is one of the best forms of feeding bovines.

5.2.2.2. *Jowar* (Andropogon sorghum)

It is one of the best drought and heat resistant fodder crops of the world. It can grow over a wide range of climatic and soil conditions. The home of this crop is Africa, India and China. This crop grows best in heavy soils that retain moisture although any soil except very sandy soil is suitable for it.

Cultural Practices. It is generally sown in two seasons, December-January and March to July. As a fodder crop, staggered sowing in different months for feeding as green cuts may be done. Being shallow rooted the crop does not require deep cultivation. One ploughing followed by one or two harrowings are sufficient to prepare the seedbed.

In low rainfall regions Jowar is given little manuring. However where there is 50 cm rainfall or more, application of 5 to 8 tonnes of FYM per ha may be given at the time of land preparation and ploughed into the soil. In areas with adequate rainfall or where irrigation is available, application of 20-25 tonnes of FYM at the time of land preparation and fertilizers to supply 30-35 kg nitrogen and 15-20 kg phosphorus may also be applied, part of it during land preparation and the other part as a top dressing when the crop is about 60 cm tall. As a fodder crop, broadcasting 40-55 kg seed/ha may be used.

Yield and nutritive value. The fodder yield varies from 25 to 50 tonnes per ha; the crop being ready for harvest in 75-90 days. Sorghum can make excellent silage. This crop is low in protein and therefore addition of small quantities of urea while ensiling will improve the quality and nutritive value of the silage. When fed as green fodder, mixing with a leguminous crop will be necessary to, meet maintenance requirements because of the low protein content (4-5 percent on dry basis).

5.2.2.3. *M.P. Chari*

It is an improved variety of Jowar grown in many parts of North India. Apart from higher yields (70-80 tonnes/ha) it has also higher protein content (6-8 per cent protein on dry basis).

5.2.2.4. *Teosinte* (Euchlaena mexicana)

Teosinte is a close relative of maize and a native of Central America. This crop is suited for hot-humid regions with more than 100 cm rainfall although it can be cultivated in less rainfall regions with irrigation. Rich, well drained loamy soils are ideal for it. Generally, it is sown in the months of March to August in North India. In the South, the crop is sown usually in June-July and September-October. Seeds are drilled behind the plough or broadcast, the seed-rate being 30-45 kg/ha. Manuring with 15-25 tonnes of FYM/hectare and fertilizer application with 20 kg N, 30 kg P_2O_5 and 25 kg K_2O per hectare may be given.

Yield and nutritive value. It can be harvested in 90-100 days and yields 35-50 tonnes per ha. The fodder has about 9 per cent crude protein on dry basis and can meet the maintenance requirements of animals when fed alone.

5.2.2.5. *Oats* (Avena sativa)

This is one of the most popular fodder crops of the world. In India it is grown mostly in the northern parts as a *rabi* (cold season) crop. It is ideal for feeding as green chop, silage and hay. It performs well in cool, temperate climate. It grows well on all types of fertile soils.

Cultural Practices. Sown from late September to late November, oats requires well-prepared and firm seed-beds. Two ploughings followed by two harrowings are sufficient. Fertilization and manuring are essential. About 25 tonnes of FYM and nitrogenous fertilizer to yield about 50 kg nitrogen per ha may be ploughed into the soil at the time of preparation of the land. Seed-rate varies from 70 to 100 kg per ha. The crop requires 2-3 irrigations and is harvested immediately after flowering.

Yield and nutritive value. It yields from 25 to 35 tonnes per ha. It has 7-9 per cent protein on dry basis and can maintain animals. With a leguminous crop, it can sustain milk yields up to 5-6 kg per day.

5.2.2.6. *Grasses*

5.2.2.6.1. Hybrid Napier

This is an interspecific hybrid (Fig. 5.4) between Napier grass *(Pennisetum purpureum)* and Bajra *(P. typhoides)*. It grows well in a warm moist climate on clay loam soils. It can be planted in any month except winter months depending on moisture. In North India, late February to early March is the ideal period, since late planting may give only one cut till the end of November, after which it remains dormant.

Cultural Practices. The land is prepared by 2-3 ploughings with the addition of 30-40 tonnes of FYM per ha. Furrows approximately 15 cm deep are made 60 cm apart. Fertilizers containing 45 kg P_2O_5 and 35 kg K_2O per ha may be drilled into the furrows. Rooted slips or stem cuttings having at least 2 nodes on them are planted in the furrows 60 cm between plants. First irrigation is given soon after planting. Thereafter, it may be irrigated at 8-12 days intervals depending on situation. The crop responds to N application and therefore, should be liberally fertilized with ammonium sulphate or urea to supply 200 kg nitrogen per hectare, preferably in split doses, 10-15 days after each cutting.

Yield and Nutritive Value. The grass can be cut 90 days after planting. Thereafter, it is cut every 45-60 days. If allowed to grow, it can attain up to 4 meters height. But with progressive maturity, protein level and digestibility of crude fiber decrease. In that case hybrid Napier will not constitute even a maintenance ration and thus becomes nutritionally inferior. At full maturity it can yield up to 280 tonnes of fodder per ha. But allowing this much maturity for the sake of bulk is not ideal. It should be harvested ideally at 45 days intervals to yield about 150 tonnes of green fodder per hectare.

5.2.2.6.2. Guinea Grass (*Panicum maximum*)

This perennial grass is a native of tropical Africa. It was introduced to India in 1793. It is a grass well adapted to warm humid tropical regions and as such well suited to the southern regions of India. It is extensively grown in Uganda, Kenya, Malaysia, Phillipines, Sri Lanka, West Indies, Brazil, Australia and Trinidad. Guinea grass tolerates shade and comes up well in coconut and mango orchards. It is fairly tolerant to drought and wet conditions, but it does not thrive in water-logged and cold conditions. Rich loamy soil is the best for the grass.

Cultural Practices. Being perennial, the grass remains at the same spot for three to four years. Therefore the land is prepared more thoroughly to make a fine bed and eradicate the weeds. One soil inversion ploughing may be followed by three ploughings with a country plough or disc harrowing. Thirty to forty tonnes of FYM may be incorporated into the soil at the time of ploughing and harrowing. Ridges and furrows may be made or narrow gently sloping flat beds may be prepared. The slips, cuttings or seedlings may be planted 100-120 cm apart on the ridges or flat beds. Rooted slips, stem cuttings and seeds can propagate the grass. The seeds may also be sown in a nursery and the seedlings transplanted in the field with the onset of the monsoon. Fertilizers incorporating 45 kg $P_2 O_5$ and 35 kg $K_2 O$ per hectare may be added at the time of planting and every subsequent year. Nitrogenous fertilizer to the tune of 40 kg N/ha may be added as a side dressing after each cutting.

Harvesting, Yield and Nutritive value. The grass can be harvested three months after planting and thereafter every six to seven weeks. A stubble height of eight to nine cm may be left at the time of cutting to encourage re-growth. The average production varies from 80 to 140 tonnes/hectare per year. Although it has less yielding potential compared to Hybrid Napier, it is more palatable and free from oxalates. It is fairly nutritious with 7-8 per cent crude protein on DM basis.

5.2.2.6.3. Napier Grass or Elephant Grass (*Pennisetum purpureum*)

It is another perennial grass suited for heavy rainfall hot-humid regions and soil with good drainage. With the wide acceptance of Hybrid Napier, Pure Napier lost its significance as a fodder grass in India. It has good palatability and is fairly nutritious. Planting and cultural practices are similar to that of the previous grasses. It yields 100-150 tonnes per ha/year with 6-8 per cent crude protein on dry basis.

5.2.2.6.4. Thin Napier (*Pennisetum polystachyori*)

Thin Napier is a very drought resistant grass adapted to warm and humid regions of India. It may be propagated through seeds or vegetative means. It needs very little attention being very hardy. An yield of 75 to 100 tones per ha per year is obtained. It is fairly nutritious with 7 to 8 per cent crude protein and the palatability is good.

5.2.2.6.5. Dinabandhu Grass (*Pennisetum pedicellatum*) (Other names : Kyasuwa grass, Dheenanath grass, Dina grass)

This is a very hardy and drought resistant grass growing wildly in many parts of Peninsular India and Bihar. Its fodder value was first recognised in Africa and Australia. This establishes quickly under available rain and remains leafy under moderate drought conditions. It is palatable and of moderate nutritive value (7 per cent crude protein on dry basis at optimum stage).

5.2.2.6.6. Para Grass (*Brachiaria mutica*) (Other names : Buffalo grass, Water grass, Angola grass, Mauritius Signal Grass)

It is a native of Brazil but well adapted to all hot-humid regions. It is one of the few species of fodder grasses that can be cultivated in waterlogged soils. Propagation is by planting slips about 60 cm apart. It spreads rapidly by surface runners which root profusely at the nodes and bear erect leafy clumps growing up to 1.5 meters in height. It does not thrive well in acidic soils. Addition of lime will be useful under such conditions. A basal dose of 40-50 tonnes per ha of FYM followed by a top dressing with nitrogenous fertilizer containing 35 kg N/ha after every 2-3 cuttings will be useful. First cut is taken after 90 days and subsequently at the interval of 35-45 days. It can give an yield of 90 to 200 tonnes per ha per annum. It has about 7 per cent CP on dry basis.

5.2.2.6.7. Anjan Grass or Kolukattai Grass (*Cenchrus ciliaris*)

It is a grass indigenous to India and now cultivated in semi-arid regions of Rajasthan, Punjab, Tamil Nadu and other parts of India. Anjan is a tufted perennial grass. It is generally propagated by seed. A well-established rain-fed stand of Anjan gives 9-10 tonnes of fodder/ha/year under very arid conditions. In regions receiving 38-76 cm rainfall, the yield would be 20-30 tonnes per annum. It is highly nutritious with about 11 per cent CP in early stages with a well balanced ratio of calcium to phosphorus. Palatability is good.

5.2.2.6.8. Rhodes Grass (*Chloris gayana*)

A perennial grass from South Africa, it is drought resistant and palatable grass fit for feeding as green chop, hay and pasturage. In India it grows best on light loams. It yields about 12-35 tonnes per ha/year. But if irrigated and manured, it can yield up to 175 tonnes per ha. The grass is leafy and palatable with about 5 per cent protein on dry basis.

5.2.2.6.9. Dhub Grass or Hariali Grass or Bahama Grass (*Cynodon dactylon*)

It is a very hardy ubiquitous creeping perennial grass distributed in all tropical and sub-tropical countries. It is a very good grass for pasture. It can be raised from seed or by planting slips or rooted cuttings. The yield is relatively low, about 20-35 tonnes per ha. But it is one of the most nutritious grasses with 10-12 per cent protein on dry basis. Some varieties may contain as much as 20 per cent protein. It is an excellent pasture grass and can form a productive ration of cows.

5.2.3. Grass Legume Mixtures

Often a number of species of grasses and legumes are mixed to produce leys. They have the advantage of (1) Providing balanced feed (2) Year-round growth and (3) Sustenance of soil fertility. One such mixture includes Congo signal, *Centrosema* and *Stylosanthes*. Another example is a mixture of White clover, Rye grass and Cocksfoot.

5.3. CONSERVATION OF FORAGES

Green fodder will be in excess of needs during certain seasons and scarce in certain others. This is unavoidable however much carefully the cropping scheme has been prepared. In order to feed during the lean months (usually May, June and October-November), green fodder is preserved as silage, haylage or hay.

5.3.1. Silage

An important method of preserving forages is through ensiling. Silage is the product resulting from controlled fermentation of forage crops of high moisture content. Bacteria in the forage utilise the sugars and other readily available carbohydrates for their metabolism producing certain organic acids as end products. This brings down the pH from 6 to the range of 3.5-4.5. This acidity eventually inhibits the bacteria and kills them, thus preserving the silage in a very palatable form so long as air is excluded from the silo. It may however be noted that there is no critical pH value for silage as it depends not only on pH but also on moisture content and temperature.

Advantages

1. It is less risky under doubtful weather conditions;
2. The ensiling process is the only means by which the entire forage plant can be preserved in a succulent form;
3. The crop can be harvested and preserved when it has the maximum nutritive value;
4. Retains higher proportion of nutrients than hay because losses due to shattering and bleaching are minimised;
5. Silage crops have more yield compared to crops used for hay. Earlier cutting at higher levels of digestibility is possible and re-growths are quicker. Thus, more feed nutrients can be grown on unit area used for silage than the same area for most other purposes;
6. Crops can be preserved as silage more cheaply, more quickly and with less labour;
7. Fear of fire is avoided;
8. Practically any forage crop is fit for ensiling. Crops with thick stems and weedy crops can be ensiled. Similarly, crops with less palatability if ensiled with other crops will be eaten well.
9. It requires less storage space than hay;
10. Many by-products can be economically used;
11. Converting crops into silage clears the land earlier for other crops;
12. It is palatable and slightly laxative;
13. It preserves protein and carotene better than hay;
14. Ensiling ensures better storage for a long time;
15. Mechanization from field cutting to feeding is easier with silage;
16. There is a wider choice of feeding methods for silage.

Disadvantages

1. Requires silo and special equipment;
2. Less amount of Vitamin D is present in silage compared to hay;
3. Due to moisture content, tonnage and transporting charges are increased;
4. Additional expenditure are involved for preservatives like molasses;
5. Wet silage can present problems of disposal of effluents;
6. Smell from poorly fermented silages can be foul;
7. Wastage percentage can be high when only small amounts are made at one time.

5.3.1.1. *How to make good silage ?*

Good silage should have a mild, pleasant aroma, an acid taste and a slightly greenish colour. Many factors, which affect the bacterial fermentation, influence the quality of the silage. The important ones among them are discussed below.

1. The crop should contain an adequate level of sugars and starch for acid fermentation. Most of the crops used .for silage production like maize, sorghum and hybrid napier have naturally high sugar content. Avoid inclusion of a high proportion of legumes, as they favour growth of Clostridia. Do not cut immature crops for silage as they have high protein and low sugar content. Crops for silage may be cut late in the day as sugar content increases during the hours of sunlight. When a crop is known to be deficient in sugars add molasses as uniformly as possible before ensiling.
2. The crop should not have high moisture content. Best results are obtained when the dry matter content is between 30 per cent and 36 per cent. Avoid cutting immature crops with dry matter content below 30 per cent as this will-result in excessive seepage losses. Excessive moisture will also lead to low acid concentration and consequent undesirable fermentation. Harvesting of crops when soaked in rain or dew also should be avoided. If necessary wilt the crop in the field before transporting to the silo.
3. ***Exclude air from silage.*** When the crop is filled inside the silo, air is trapped inside the meshes of the forage. Oxidation of sugars takes place with generation of heat. This causes the trapped air to rise and set up currents which draw in fresh air. To minimise this, the silo should be filled rapidly, packed well and sealed properly. By this, the air-free condition needed for Lactobacilli is quickly established while the maximum amount of sugars remain to be fermented.
4. ***Seal the silo as soon as filling is finished.*** This seal will have several functions. Mainly, this prevents exposure to air and other elements, thus keeping the fermented silage stable. In the 1st and 2nd phases, it prevents the warm air from escaping and setting in convective currents bringing in fresh air. In ordinary pit silos without roofs the sealing will prevent rain water entering and diluting the acid which will cause rotting of the silage.
5. Wilting is a controversial aspect in silage making. Excellent silage can be made from crops with higher moisture content provided enough sugars are present for producing enough acidity. But immature crops with high moisture content are invariably always low in sugars. Therefore, in such crops, removal of some moisture by wilting will help in preventing undue dilution of acids produced by the fermentation of limited sugars available. Generally wilting is done to a level of 30-35 per cent dry matter. If dried beyond 35 per cent dry matter, packing is more difficult and air may get into silage mass, resulting in greater spoilage and mold growth. If an additive like molasses is added, less wilting is necessary. An advantage claimed for wilting is that it will increase the dry matter consumption of animals especially under all-silage roughage feeding programme.
6. ***Additives.*** If the crop is having the required concentration of sugars and dry matter, additives are not necessary. Otherwise, many ways of chemically preserving silage or correcting fermentative process by adding certain substances called additives are available. The additives belong to the broad classes given below.

(*a*) Those, which supply additional readily fermentable carbohydrates to produce sufficient organic acids, *e.g.*, molasses, sugar, jaggery.

(*b*) Mineral acids which give a silage with very low pH, *e.g.*, Mixture of sulphuric and hydrochloric acids.

(*c*) *Organic acids.* These help in achieving stable pH more rapidly. Alternatively, a culture of Lactobacilli can be used.

(*vii*) Expose the minimum surface when silage is being fed out. In an opened silo, the aerobic conditions favour the growth of molds on the open face. This is negligible if silage is removed fast enough (say 15 cm per day). But wastage can occur if the face remains exposed for a long time.

5.3.1.2. *Chemical and Bacterial Changes during Ensilage*

For the sake of understanding the ensilage process can be divided into 5 phases, but one should bear in mind that it is a continuous process.

Phase-1. Starts immediately after packing the fodder inside the silo compactly and sealing it off. The plants continue to respire and use up all the oxygen present in the trapped air, giving off carbon dioxide. This checks mold growth. The process also generates heat. The carbon dioxide produced helps in creating anaerobic conditions needed for the growth of bacteria, which produce organic acids.

Phase-2. Acetic acid is produced to a greater proportion during this phase by the fermentation of sugars and starch. The acids produced dissolve in the plant juices.

Phase-3. As the concentration of acetic acid increases, this phase starts with a marked proliferation of *Lactobacilli* resulting in a shift to lactic acid production. Simultaneously the increase in acidity inhibits the bacteria producing acetic acid. The forage settles down during the first few days and there will be also considerable seepage if the forage is high in moisture.

Phase-4. Lasts for a period of 15-20 days during which lactic acid is produced in large amounts. When the concentration of acids reaches a desired level, further bacterial action is checked.

Phase-5. This is the long phase during which the fodder is preserved well because of the acidity formed. This takes place only if the conditions are ideal. The acids produced dissolve in the plant juices. It is the concentration of the acid solution and not the quantity of acid produced that is important—for, the wetter the crop, the greater is the quantity of acid needed for stable preservation. The crops contain bacteria other than Lactobacilli, like the Clostridium species. Clostridum can utilise sugars and lactic acid and produce butyric acid. If there is enough acetic and lactic acid present in the silage to prevent further bacterial growth, no further changes will take place and the silage will be properly preserved. But if there is not enough concentration of acids due to excessive moisture, bacteria like Clostridium will multiply instead of Lactobacilli. This will result in production of butyric acid. They can also break down proteins to ammonia and amines. This process of rotting and decomposition produces a vile-smelling silage of very low palatability.

Chemical Changes. The chemical changes that take place in the crops due to ensiling may be divided into two categories, *viz.*, *(i)* the changes taking place immediately after cutting due to the activity of plant enzymes and *(ii)* those brought about by the action of micro-organisms present on the original herbage or which gain entry at a later stage.

Plant enzymes. Using the oxygen present in the silo, aerobic respiration takes place until the plant sugars or the oxygen supply inside the closed silo are exhausted. Sugars are oxidised to carbon dioxide and water with generation of heat. If the crops inside the silo are not well packed excluding as much air as possible, convective currents are set in and air may permeate into the mass and the temperature will continue to rise. If the rise in temperature is not checked, an inferior brownish black coloured silage is produced. Such a silage will be poor in nutritive value because a large proportion of the soluble carbohydrates have been oxidised in the process and proteins rendered less digestible due to the heat. Proteins are also attacked during this initial stage. About 16 per cent proteins are degraded within the first 24 hours to protein break-down products, mainly amino acids.

Microbial Activity. Aerobic micro-organisms (fungi and bacteria) are the dominant micro floras on fresh herbage, but as anaerobic conditions develop, bacteria able to grow anaerobically replace them. Microbial changes continue even after aerobic respiration has been halted. The micro-organisms present in the herbage mass multiply using the cell-contents of the plants as a medium. As a result of this activity many substances are broken down to simpler products. When conditions become favourable for the lactic acid bacteria, they multiply producing lactic acid. Acidity of the mass increases and at about pH 4.0-4.2 organisms other than the aciduric lactic acid bacteria are inhibited. The lactic acid producing organisms continue to multiply producing more acidity until the pH falls down to 3.7. At this pH, even the aciduric lactic acid bacteria are inhibited and the silage is preserved in a stable condition as long as anaerobic condition prevails.

The lactic acid organisms responsible for acid fermentation in silage can be classified into two groups, *viz.,* homofermentative and heterofermentative. Both types produce lactic acid but the former are more efficient in converting hexoses into lactic acid. Mainly, sugars and fructans present in the crops are fermented to give lactic acid. At a later stage of the ensilage process, hemicelluloses are also degraded. Usually small quantity of acetic acid (0.5 to 4 per cent) is present in any silage. On the other hand, butyric acid will be absent in well made good silage. In badly made silage where acidity has not developed enough to bring down the pH to the required level, butyric acid is invariably present. The chief causes of such inferior silage are either lack of enough soluble carbohydrates in the crop or presence of high moisture content in the plant material.

Along with carbohydrates, about 60 per cent of the proteins are also degraded to simpler compounds. A rapid lactic acid type fermentation with production of satisfactory degree of acidity results in a type of protein degradation where the end products are mainly amino acids. This will not affect the nutritive value of the silage. But in poorly preserved silage, the protein decomposition extends further or takes other pathways to produce various amines such as histamine, tryptamine and phenylethylamine. Many of these can be toxic if absorbed into the circulatory system. The protein degradation products are further attacked by bacteria in badly preserved silage to result in ammonia which is lost from the silo in the gaseous form.

Some changes take place in the mineral content too. Salts of lactic, acetic and other volatile acids with potassium, calcium, magnesium and sodium may be formed. These changes, however, do not affect the availability of these minerals to the animals feeding on them.

In well-preserved silage where the temperature does not rise beyond normal levels, the vitamin A precursor, carotene is not affected. But in badly preserved silage, overheating can result in loss of a major proportion of the vitamin precursor.

During the ensilage, a certain amount of the nutrients are lost in the gaseous form mainly as carbon dioxide. The dry matter loss due to this varies from 2-30 per cent. The soluble and the readily digestible materials are easily attacked by micro-organisms and broken down to gaseous compounds. Therefore, higher the gaseous loss, the lower will be the feeding value of the product.

5.3.1.3. *Types of Silos*

Large number of different types of silos is in existence. They can be classified into two main types, namely, upright and horizontal *(See* Chapter 19).

Upright or tower silos are cylindrical structures made with concrete staves, wooden staves, poured concrete, bricks, metal etc. They have a diameter of 5-10 m and are 10-25 m in height. Fresh forage is blown in at the top and when ready for use, the silage is removed through a series of openings at one side. Losses due to spoilage are minimum in this type. Also it is highly adaptable to mechanical unloading and feeding. However, they are" very expensive. The advantages of reduced spoilage and easier handling should be weighed against the higher cost involved.

The horizontal silos may be in the form of a pit, a trench, a stack, or a bunker. Some may be partly below and partly above ground. *Pit silos* as the name indicates are pits for silage making. In large operations, it involves more labour. *Trench silos* are shallow elongated trenches. *Bunker silos* are usually made of concrete floors and sides. Sides should be well braced because the silage bulk exerts great lateral pressure. Bunker silos are more expensive than pit or trench silos. But silage in pits and trenches are subject to spoilage in areas where the water table is high.

Another horizontal type is the **stack silo.** But this type of silo is more prone to spoilage losses because of problems of packing. Horizontal silos are relatively less expensive and are easily filled and packed with tractors. Self-feeding methods can be applied in these types of silos. Sealing horizontal silos is more important due to large exposed surface area. This may be done with a plastic sheet held down by dirt or with a mixture of dung and soil.

Clamp silos can be used where soil is suitable and the water table is high. A shallow trench is dug and the soil is piled on either side to increase the height of the walls. Silage crops are stacked as high and as compactly as possible. A good seal is applied with plastic sheet.

5.3.1.4. *Crops for silage*

Silage can be made from many crops. Crops like maize, sorghum, oats etc, grasses which are planned mainly for cutting, some leguminous crops mixed with grasses or arable crops and certain arable by-products can be used.

Maize silage. Maize (corn) is increasingly used for silage making. In well adapted areas, corn harvested as silage results in an energy yield per unit area, unsurpassed by any other crop. Maize is equally good as a grain crop, silage crop or green chop material. This is why maize is called the king of crops. In many parts of the world, maize silage is the main forage component of dairy cattle feeding programmes. High digestibility and palatability, and the ease with which quality silage can be prepared without additives make corn the most ideal crop for silage. The recommended dry matter content of maize harvested for silage is between 30 and 35 per cent. If it is harvested when the kernel is in early dent stage, the dry matter content will be proper.

Maize has only 7-8 per cent crude protein on dry matter basis. Similarly it is low in calcium and other minerals. So when maize silage forms the sole roughage, special attention must be paid to supplying these nutrients from other sources.

Sorghum is another crop, which can be readily ensiled. It is similar to maize but slightly lower in feeding value. If the sorghum grains are allowed to mature, much of it will pass out in the dung undigested, thereby lowering the energy value of the total silage. Therefore precaution should be taken that the crop is cut for silage when it has flowered and before grains mature to secure maximum nutrients from the plants.

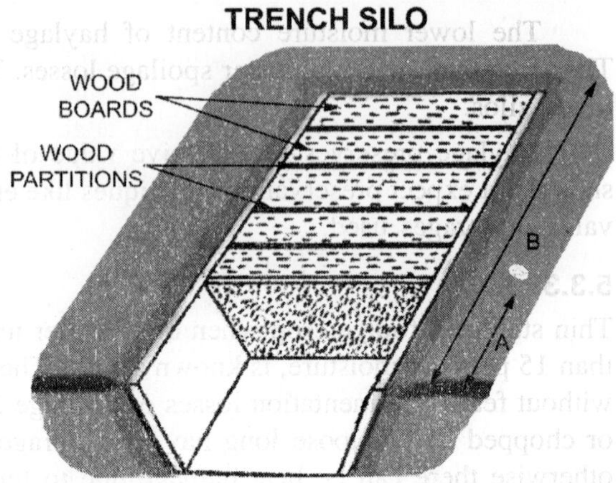

Fig. 5.10. Trench silo of adjustable length

Oats and barley also make good silage. This silage along with legumes forms a good roughage ration. However, alone, silage from these crops is low in protein and calcium.

Grass silage. Grasses like hybrid napier, thin napier, guinea etc. are ensiled either alone or mixed with crops like maize and sorghum or legumes like lucerne. Mostly these crops are deficient in readily fermentable carbohydrates. When leguminous crops are included ammonia produced by bacterial fermentation of protein may neutralize some acidity thereby failing to achieve the required acid concentration.

The legume-grass mixtures can also be high in moisture content. When berseem is used, the stem is hollow and there by result in incorporating more air which prevents anaerobiasis to set in quickly. To overcome these difficulties the following steps are recommended.

(*a*) Lowering the moisture content by wilting.

(*b*) Adding mineral acids.

(*c*) Adding readily fermentable carbohydrates like molasses from which more acid is produced.

(*d*) Making mixed silages with crops like sorghum or maize.

(*e*) Lacerating crops like berseem.

The palatability and utilization of certain less palatable crops like kalapagonium and centrosema and some byproducts like straw can be increased if ensiled along with crops like maize and sorghum. The leguminous crops will increase the protein content.

Another way of increasing the crude protein content of silage produced from crops like maize, sorghum and grasses is to incorporate urea at 0.3 to 0.5 per cent level (by weight). The resultant silage will have 11-12 per cent crude protein.

5.3.2. Haylage

Haylage is low-moisture silage. Grasses and legumes are wilted to contain 50 per cent dry matter before ensiling. This system of preserving fodder is advantageous in areas where rains are frequent in harvesting season. The fodder can be removed off the field and protected inside silos faster than in the case of making hay. It has also the advantage of increasing the dry matter consumption.

The lower moisture content of haylage makes it difficult to pack well to remove air. Therefore there will be greater spoilage losses. To avoid this, haylage is usually made in air-tight tower silos.

Haylage preserves the nutritive value of forages as good as hay or silage. However, one should not expect preservation techniques like ensiling or haylage making to improve the nutritive value of forages used.

5.3.3. Hay

Thin stemmed forage crop when dried either under natural sunlight or artificially to contain less than 15 per cent moisture, is known as hay. The reduction in moisture helps in preserving forages without fear of fermentation losses or spoilage due to mold growth. Storage is done as baled hay or chopped hay or loose long hay. For storage, moisture content should be below 20 per cent, otherwise there can be heat damage due to fermentation or even spontaneous combustion.

5.3.3.1. *Stage of harvesting the crop*

The stage of maturity of the crop at the time of cutting is the most important factor as far as the nutritive value of the hay is concerned (Fig. 5.6). To prepare good hay, the crops should be harvested in the pre-flowering or early flowering stage so that maximum percentage of nutrients is preserved. An early cut means more nutritive value but less yield. Late cuttings, on the other hand, will result in more bulk, but less nutritive value and palatability. A compromise, therefore, has to be struck. An appropriate stage is when the gain due to increased bulk no longer compensates the loss in nutritive value. In most of the cases this is the early flowering stage.

5.3.3.2. *Properties of Good Hay*

1. Good hay is the one made out of forages cut at the proper stage. Delay in harvesting will make the stem woody and lignified with resultant reduction in digestibility and palatability. Besides, a part of the nutrients would have been used up in seed formation.
2. Good hay has a major proportion of the leaves preserved. Leaves are rich sources of proteins, vitamins and minerals. Shattering of leaves during the process of hay making will reduce the quality of hay.
3. Green colour of hay is an indication of its quality. It means that bleaching, leaching, fermentation etc. have been minimised and also that greater proportion of leaves have been preserved,
4. It will be made out of thin-stemmed crops so that it is soft and pliable.
5. The crop used for hay should be free from weeds and poisonous plants.
6. Hay should not have more than 20 per cent moisture when stored. If there is more moisture, fermentation will take place with resultant deterioration in quality and also result in fire hazard.
7. Good hay will be free from dust and mould.
8. It will have an appetising pleasant aroma

5.3.3.3. *Forage crops used for hay*

Both legumes and non-legumes are used for hay making. Fodder crops with thin, soft and pliable stems are more suitable for haymaking.

Lucerne is an ideal leguminous crop for haymaking. It has high dry matter yield, high protein, mineral and carotene content and good palatability. Berseem, cowpea, clovers etc. are other leguminous crops fit for haymaking.

The non-leguminous hay contains less protein, minerals and vitamins. Certain thin stemmed arable crops like oats and barley and grasses like Congo signal, Spear grass and Brome grass are also used for hay making. Hay made from non-leguminous arable crops is usually rich in carbohydrates. Mixed hay is usually made from a mixed crop of leguminous and non-leguminous plants. Such a crop is generally cut early as the various crops do not come into flowering simultaneously. A mixture of Congo signal, Styloxanthes and Centrosema is popular in some heavy rainfall regions.

5.3.3.4. *Hay making*

Production of hay involves—

1. Selection of suitable plant varieties and their cultivation on good soil with proper care.
2. The seed rate should be high to ensure dense stand of the crop which will result in finer stems, heavier yield and less weeds.
3. Cutting the fodder at the proper stage.
4. Curing the hay. In curing the objectives are to preserve maximum nutrients and to remove moisture sufficiently to avoid mouldiness and spoilage. Curing may be done by—(*a*) field curing, (*b*) mow curing (barn drying) or (*c*) dehydration or artificial drying.

5.3.3.5. *Forms of Storing Hay*

Long hay. The field-cured hay is stacked in or stored in a barn. This is very bulky.

Chopped hay. Dry hay is chopped and stored. The bulkiness is not reduced. Besides it is dusty and highly prone to spontaneous combustion. More nutrients are preserved in this form than in long hay. It can be fed mixed with other ingredients. Wastages also are less in this form.

Pelleting. The hay is made into pellets with a pelleting machine. This requires specialized machinery and may not be suitable for developing countries.

Wafering. It is another form of processing hay, which makes handling and automatic feeding easy. Hay is chopped into 3 cm length before wafering or cubing. Pelleting and wafering increase the density of the hay and increase dry matter intake. Besides, these are highly suited to mechanized feeding in intensive systems.

5.4. TROPICAL FORAGES—MEASURES TO OVERCOME THEIR NUTRITIONAL LIMITATIONS

In tropics, animal performance levels often fall short of expectations of predicted performance based on nutritional recommendations taken from temperate climates. The tropical forages are characterized by the following factors.

1. Low digestibility—The process of lignification is very fast and a delay in harvesting the forages by a few days can render it over-mature and of very low digestibility,
2. The carbohydrate component consists of elements of very low solubility (cell wall components) with low rates of degradation by ruminal microbes. The cell content consists of highly soluble sugars which are fast degraded in the rumen,

3. These forages are very low in starch, which is an important component of most rations fed to cattle in temperate region.
4. The levels of total nitrogen and true protein are extremely low.

So for tropical forages the digestion process in the animal must be principally by the rumen fermentation route. Such a situation would be the most natural for the ruminant as it corresponds to the environmental niche for which they are evolved. But, actually, when the diet of a ruminant is such that it is obliged to pass through the rumen fermentation and the only items available for digestion in the small intestine are the body cells of rumen microbes, the voluntary intake falls abnormally. Because of this the level of their intake rarely exceeds that of maintenance. When voluntary intake is lowered, turnover rate of rumen contents also goes down (Fig. 5.11). This results in the reduction of rate and efficiency of microbial growth and therefore, of microbial protein synthesis (Bergen, 1979). The low voluntary intake and poor feed conversion efficiency ultimately leads to poor performance.

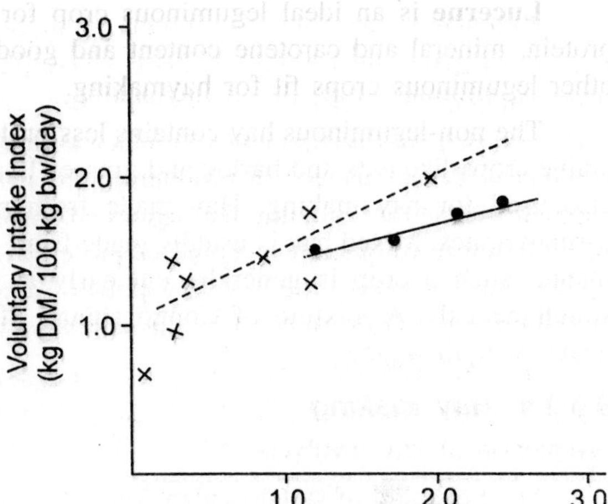

Fig. 5.11. Relationship between rumen liquid turnover rate and voluntary intake. (From Ferreiro *et al.*, 1977, Row, *et al.*, 1979)

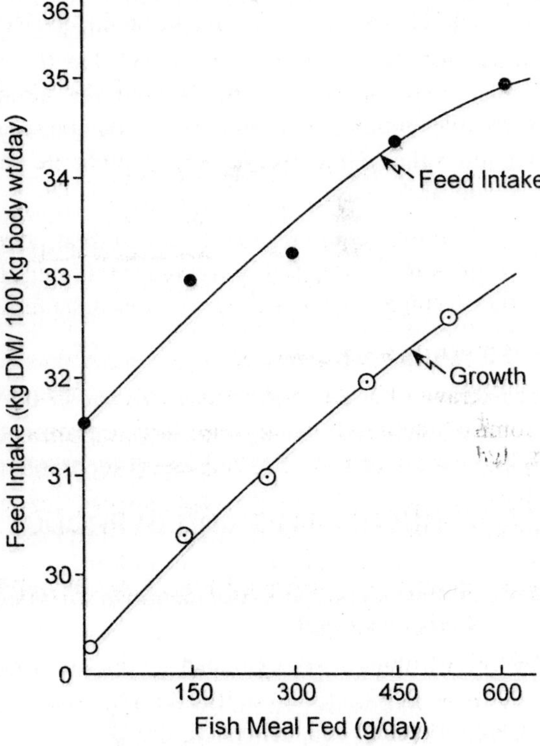

Fig. 5.12. Effect of fish meal (by-pass protein) level on feed intake and growth of bovine. (Adapted from Preston and Ling. 1978 and Gaya *et al.*, 1979)

In the presence of 1 kg per day of rice polishing (providing rumen bypass nutrients) supplemented with urea raised live weight gain from zero to 700 g/day (Alverez and Preston, 1976). Adding urea alone to chopped whole sugarcane had no effect on voluntary intake, and the animals lost weight on all treatments.

The value of 'bypass' protein lies not only in its role as a direct source of amino acids for milk and tissue synthesis, but also on the effect that it has on overall voluntary intake and rumen function. It has been found that supplying a higher level of bypass protein proportionately increased voluntary feed consumption and growth rate (Fig. 5.12). A result of increasing bypass protein in the diet of ruminants receiving a molasses urea diet was to increase rumen turnover rate and efficiency of rumen microbial protein synthesis (Elliot *et al.* 1978a).

"By pass" energy also assumes some significance in the ration of a ruminant in tropics. The amount of dietary starch that escapes rumen fermentation may be a major factor in productivity of ruminants. The tropical grasses and feeds like urea-molasses and sugarcane are very low in starch. Starch as a component of dietary supplement has been shown to increase growth rate in ruminants. When rice polish (containing 10 per cent protein and 60 per cent starch) is used to supplement the diet of cattle on sugarcane based diets, it produces a larger response than to fish meal which has higher percentage of protein. Duodenal canula studies showed that 80 per cent of the starch ingested through rice polish escaped rumen digestion (Elliot *et al.,* 1978b). Rice polish supplied both bypass protein and bypass energy.

The third factor in efficiency of utilization of tropical diets by ruminants is the roughage characteristics of the diet. The incorporation of good quality roughages helps providing an improved 'ecosystem' in the rumen which results in better intake and digestibility. They also stimulate rumen motility and rumen contractions.

5.5. STRAWS

5.5.1. Wheat Straw and Paddy Straw

Byproducts of the grain industry, these cereal straws form the bulk of roughages fed to livestock in India. Due to non-availability of much croplands for fodder production as well as the total destruction of communal grasslands due to overgrazing, these straws are the only staple feed available to the bovines throughout the country. They contain about 40 per cent TDN and negligible amount of protein. Low palatability, high lignin content, dustiness etc. reduce their feeding value. High oxalic acid content in paddy straw has also been found to interfere with calcium absorption.

Certain techniques like (*i*) alkali treatment of straws, *(ii)* soaking in water, *(iii)* impregnation with urea molasses, *(iv)* ammonia treatment, and *(v)* ensiling with other crops, have been tried with varying success to improve the palatability and nutritive value of cereal straws.

5.5.2. Pulse Straws

The straws of some of the pulse crops after the seeds have been removed are also fed to bovines. Some of them like black gram, green gram and *moth* were found to be palatable and nutritious. While the energy value of these is comparable to cereal straws, they have relatively more protein.

5.6. CONCENTRATES AND BYPRODUCTS

Roughages alone cannot sustain milk yield beyond a certain level. Similarly, young bovines are not equipped to utilize large quantities of roughages. In both the situations concentrates of some sort or other will have to be used. They usually include cereals and their byproducts, pulses, oil cakes, byproducts of sugar cane industry, meat industry and brewers industry.

Cereal grains like maize, barley, oats, sorghum, bajra, and wheat (when damaged and unfit for human consumption) are used extensively in concentrate mixtures as a source of energy. They are rich in starch and low in fibre. Their TDN ranges from 75-80 per cent and DCP, 7-10 per cent.

Tapioca chips. In some parts of the country, tapioca chips have replaced grains as an energy source and are as good as grains in energy content.

Pulses. There is only very limited scope for using pulses in animal rations as they are extensively used for human consumption. However, certain pulses like horse gram, Bengal gram and certain byproducts from pulses are used for animal feeding. They are rich in protein, energy and minerals.

Oilseed cakes (meals). Many types of oilseed cakes—the residue after removing the oil from the seeds—is used for animal feeding. They are all rich in protein and reasonably of high-energy content. The protein content vary from 25-50 per cent. There are two types of oil meals available—(*i*) The expeller variety, and (*ii*) the solvent extracted type. While the former has relatively more of energy (due to unremoved oil), the latter is richer in protein. The most popular oilseed cakes are : (*i*) groundnut cake, (*ii*) cottonseed cake, (*iii*) linseed cake (*iv*) coconut cake (*v*) mustard cake and (*vi*) gingely cake. Besides rubber seed cakes are also being used marginally.

Rice bran and wheat bran. Rice bran and wheat bran are extensively used in bovine feeding. They contain 12-14 per cent protein and about 60-65 per cent TDN. This is one of the ingredients in concentrate mixes, which is comparatively cheap. They are also rich in phosphorus and B vitamins.

Sugar industry wastes like molasses and **bagasse** are available in plenty. Molasses is used as an energy source and as an agent that increases palatability. **Bagasse** is the residue of sugarcane left after extraction of the juice. This highly fibrous material can be used for feeding bovines with nutritive value much similar to straws. Impregnating bagasse with urea-molasses has the benefit of increasing palatability and nutritive value.

5.6.1. Concentrates of Animal Origin and Special Supplements

These are mainly protein, vitamin and mineral supplements of high nutritive value. They contain some of the essential amino acids deficient in proteins of plant origin and therefore are necessary in the diet of young bovines to balance the ration. In adult bovines, incorporation of animal proteins are not essential because of rumen synthesis of bacterial protein which is also a good source of these limiting amino acids. The main feeds available from animal sources are meat meal, fish meal, tankage, hoof horn and liver meals, blood meal etc. There are also certain byproducts of the breweries like *distillary wastes,* dried brewer's yeast, which can be used as cattle feeds.

5.6.2. Other Unconventional Feeds of Limited Use

In a frantic search for usable byproducts which could be incorporated in bovine feeds so that the cost can be reduced, a number of products were studied and some were identified to be useful.

Tomato pomace, citrus peels, pineapple wastes, cashew apple wastes, frog-leg trimmings, guar meal, processed chicken excreta, rubber seed cake, silk cotton seed, tapioca waste, Virginia tobacco seed cake, maize gluten feed, sun hemp seeds, tamarind seed, rain tree pods, mahua cake, mango seed kernel etc. are some of the feed items that have been successfully used. For more details about them, the reader is directed to Ranjhan (1980).

Ranjhan, S.K. (1980). *Animal nutrition and feeding practices in India.* Vikas Publ. House Pvt. Ltd., New Delhi.

CHAPTER

6

Nutrition and Feeding

6.1. FEED NUTRIENTS

Substances which when eaten and absorbed by the body, produce energy, promote growth and repair of tissues or regulate these processes or result in the making of certain products, are foods. In a more general sense we use the term 'feed' to describe substances that are eaten. For example, grass and oil cakes are feeds, but not all their components are utilized, part being excreted. Those components of foods that can be utilized by the body to perform various body functions are termed nutrients. More than 100 chemical elements are recognized. Out of these, about 20 singly or in various combinations constitute the feed nutrients.

Sun is the fountain of energy. An infinitesimal fraction of that energy is trapped by plants through photosynthesis, which store part of it as chemical energy. Similarly, they use simple elements in nature like carbon dioxide in air, water and inorganic elements in soil to synthesize complex materials. Animals feed on these materials and use the energy trapped by plants through photosynthesis, for various life processes. Feed of dairy cattle and buffaloes are mostly of plant origin although in the calf ration some materials of animal origin like milk powder, meat meal, bone meal etc. are added.

6.1.1. Functional Classification of Nutrients

From a functional point of view, we can recognize 5 major categories of nutrients-water, energy, protein, minerals and vitamins (Fig. 6.1). In this classification, carbohydrates, lipids, and part of proteins have been clubbed as energy. Proteins, minerals and water contribute towards growth and tissue repair. Proteins, vitamins, minerals and water are mainly concerned about control of body processes. All categories of nutrients are involved in synthesis of products like milk.

6.1.2. Function of Feed Nutrients

These are as follows (See Fig. 6.2)—

1. For growth and repair of body tissues. Certain nutrients, like cement mortar, bricks, wood and steel in building a house, serve as a structural material for building and upkeep of the

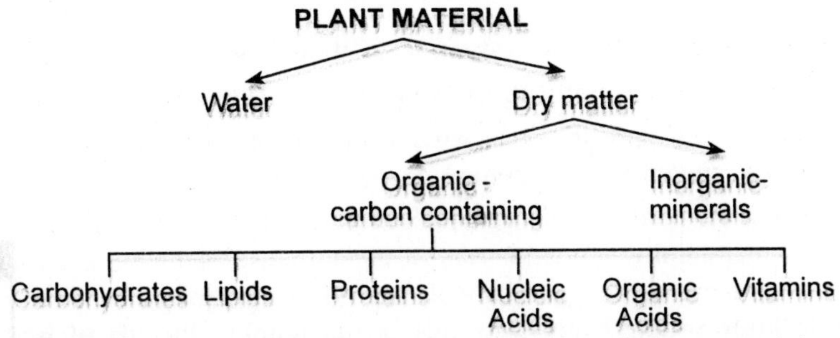

Fig. 6.1. Major livestock nutrient constituents of plant material.

body structure. Proteins, minerals, water and fat form the building material involved in body growth. Due to various vital activities, continuous wear and tear can be expected in the body tissues and therefore there is also a requirement for constant maintenance. Essentially, the same nutrients involved in growth and tissue building are involved in maintenance also.

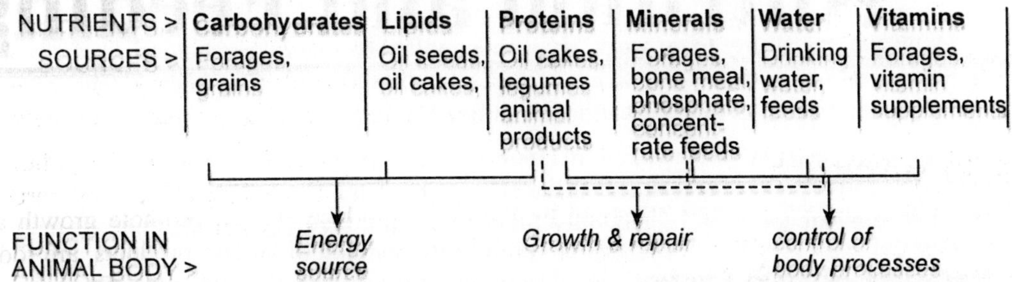

Fig. 6.2. The nutrients : functions and representative feeds in which they are found.

2. As a source of energy. The body needs energy for maintaining vital functions. Growth itself needs energy in addition to the tissue forming nutrients. A pregnant cow needs energy for the growth of the foetus. A lactating cow needs energy-rich nutrients to serve as raw material for the manufacture of various milk components like lactose and milk-fat. Nutrient energy must be supplied to the animals for keeping the body warm and for the capacity to do work. Un-utilizable energy lost as heat often serves the purpose of warming the body. Excess energy is deposited as fat under the skin and around internal organs as a reserve supply of stored energy.

3. For regulation of body processes. Various biochemical processes in the body are integrated by neuro-humoral and enzymatic processes. Proteins in the form of enzymes and hormones, vitamins, minerals, certain amino acids and certain fatty acids as components of enzyme systems, buffer systems etc. regulate the functioning of various systems of the body.

4. Milk production. A certain amount of nutrients, which the animal has consumed and digested is transformed and transferred in mammary gland to form milk. Since milk contains almost all of the nutrients, all the various nutrients are concerned with milk production.

6.2. NUTRIENT REQUIREMENTS

Water. Water is the most essential nutrient for sustaining life. The animal body can lose all the fat and 50 per cent of the protein and still live. Loss of as much as 10 percent water on the other hand will lead to death of the animal. Animal bodies have 70-90 per cent water. Water has several important functions in the animal body.

(*a*) Animals require water for the formation of their body tissues.

(*b*) It gives elasticity and rigidity to the supportive tissues of the animal.

(*c*) It acts as a medium and universal solvent in almost all vital processes like digestion, absorption, intermediary metabolism, excretion and even reproduction.

(*d*) Water through its qualities of high points of latent heat of freezing and evaporation, serves a major role in maintaining homeothermy of the animals. Water can absorb large quantities of heat with limited temperature change. Animals get rid of a major portion of excess body heat through evaporating body water. Water also aids in maintaining the pH of body fluids within normal range. (For water supply to dairy farms refer to Chapter 19).

6.2.1. Energy Requirements

The largest part of the feed components is used in the animal body for the supply of energy (70-80 percent of DM). Consequently a major part of the total expenditure of feeding cattle is on energy.

Energy means 'the ability to do work'. It is not supplied by one nutrient alone. Carbohydrates including fibers, fats and proteins (in excess of tissue synthesis) contribute energy. Fats, although a more concentrated source of energy, cannot be utilized in large amounts by cattle and buffaloes.

Every function, activity, process and metabolic reaction in the body needs energy. A lactating dairy cow may require 15-30 megacalories (Mcal) of energy daily. Due to the above considerations, the practical feeding of the cow is generally planned around meeting energy requirements.

6.2.2. Measures of Feed Energy

In the course of development of the science of nutrition, several measures and systems of evaluating the energy content of feeds have been developed (National Research Council, 1966). The energy requirements of dairy cattle, the energy content in various feedstuffs and practical feeding recommendations with respect to energy have been given in terms of these measures.

Total Digestible Nutrients (TDN). The TDN system as a measure of energy has been in wide use in India for a long time. Even though in other countries they are fast adopting other systems, in this country TDN still remains the most popular one. Its calculation involves chemical analysis coupled with a biological evaluation of the digestibility of each nutrient component in the feed. TDN in a feed is the sum total of the digestible crude fibre, digestible nitrogen free extract, and digestible protein and 2.25 times digestible ether extract. TDN requirements are given as kilograms (kg) whereas TDN content of feeds is expressed as percentages either as fed (fresh) or on dry matter basis. There is a fairly good correlation between TDN and digestible energy (DE). However, TDN partially takes into account urinary energy losses by attributing equal energy value to protein as carbohydrates. Although protein contains more DE than carbohydrates, most of this additional energy is lost through urine.

However TDN does not consider the quality of the end products of rumen digestion (proportions of various VFA produced, their energy value and function) or the large amounts of energy wasted in the digestion of fibrous feeds in the form of gases and heat increment. Ether extract from forages contains largely materials other than true fat and has a lower energy value.

National Research Council (NRC) (1966). *Biological energy interrelationships and glossary of energy terms.* Natl. Acad. Sci.,Washington, D.C.

Multiplying such ether extract with the factor 2.25 and also because of not taking into account losses of energy in crude fibre digestion in the form of gases and heat increment, TDN over-values the energy value of roughages relative to concentrates. The third limitation of TDN is that with higher levels of feeding, the digestibility and consequently TDN content of the same ration declines (Figure 6.3).

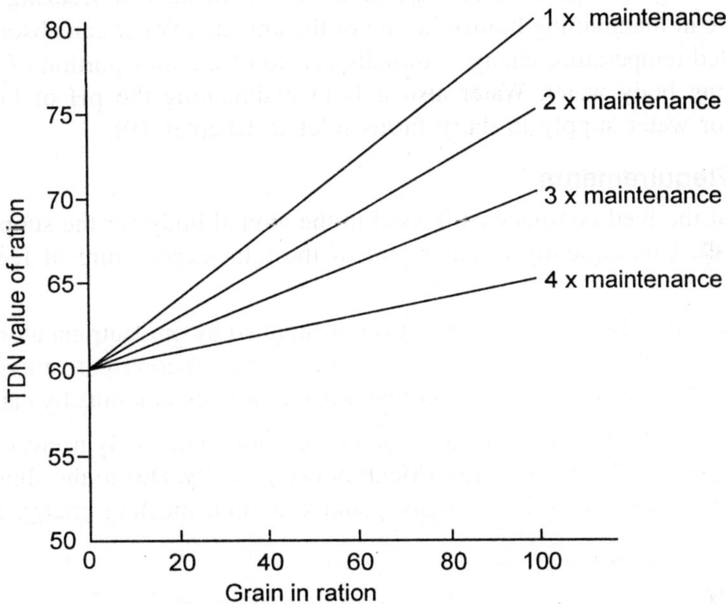

Fig. 6.3. The digestibility and TDN content of rations decline as level of feed intake is increased (Adapted from Tyrrell, H.F. and Moe, P.W., 1975, J. Dairy Sci. 58: 1151-1163).

Other Measures of Energy. Digestible energy (DE), metabolizable energy (ME) and net energy (NE) are based on the caloric value of feeds and therefore usually expressed in kilo calories, kcal (1000 calories) or megacalories, Mcal (1,000,000 calories) (NRC, 1966). The magacalorie is sometimes called the Therm (T).

Gross energy. When feedstuffs are burnt in a bomb calorimeter the heat so produced can be measured precisely. All the gross energy of a feed is not, however, utilized by the animal body. If the energy of the voided dung can be measured the *digestible energy* (DE) for the feed can be known by subtracting it from gross energy. As the excreted dung may contain some absorbed material, which is re-excreted and sloughed intestinal walls and residues of digestive secretions, DE content of a feed represents the apparent digestible energy. DE is similar in concept as TDN, but not the same. Whereas TDN is based on analysis of proximate principles in feed and dung, DE is determined through bomb calorimeter estimations of gross energy and faecal energy.

DE = Caloric content of the feed (gross energy) — caloric content of faeces (faecal energy). Both DE and TDN grossly represent the digested portion of energy giving feed nutrients.

Metabolizable Energy (ME). ME = DE — caloric content of urine and gases.

As the equation shows, ME is determined from DE by subtracting the energy losses through urine and gaseous products like methane formed during digestion (Fig. 6.4). Generally speaking, the energy losses through gases will be more from forages than from concentrates. ME values

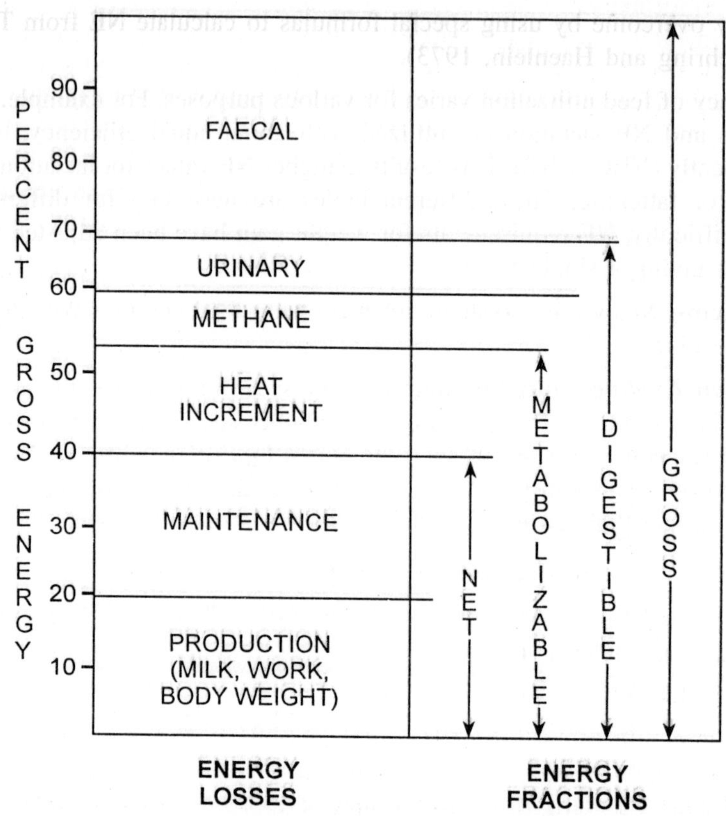

Fig. 6.4. Approximate proportions of gross energy of feed lost by the animal through various channels like faeces, urine, gases of digestion, heat increment and what is finally available to the animal for maintenance and production.

have not been in much vogue in the practical dairy cattle feeding except in the United Kingdom where it was used with a special modification (ARC, 1965).

This diagram also explains the concepts of DE (digestible energy), ME (metabolizable energy) and NE (net energy). (Adapted from NRC 1966).

Net energy (NE). NE = ME – heat increment (calories lost as a result of microbial fermentation and metabolism of ingested feeds).

Net energy is that portion of gross energy available to the animal for maintenance (NEm), for gain (NEg), for reproduction and for milk production (NE lactation or NEl)

Among the energy measures, net energy system is the most precise one for expressing energy requirements of animals and energy of feeds. At the same time it is also the most difficult to determine experimentally. At present, in developed countries NE is the most widely accepted system and other less precise systems have been discarded. However, in India and other developing countries NE is yet to gain full acceptability. This is mainly because the amount of research data on NE content of feeds available in these parts of the world is extremely scarce. This difficulty

Agricultural Research Council (ARC), (1965). *"The Nutritive requirements of Farm Livestock"* No, 2.

has been partially overcome by using special formulas to calculate NE from TDN (More *et al,* 1953) or DE (Nehring and Haenlein, 1973).

The efficiency of feed utilization varies for various purposes. For example, NE maintenance of lactating cows and NE lactation are utilized with about equal efficiency, but NE growth is utilized less efficiently (NRC, 1978). This results in higher NE values for lactation and maintenance than for growth or fattening. Thus, different tables are necessary for different functions. To circumvent this difficulty, NE requirements for weight gain have been adjusted for the difference in efficiency thus giving a single table.

Other Systems. Many other systems for measuring feed energy have been widely used in some parts of the globe.

Scandinavian Feeding Units for example expresses more or less NE, but the information for calculating this is obtained from feeding experiments and not from calorimetry. Yet another energy unit popular in some parts of Europe is the *Starch Equivalent* which is also an approximation of NE. Developed by Kellner in Germany, it represents the quantity of starch that will yield as much energy to the body above maintenance needs as would 100 lbs of the feed in question.

Inter-Relationships between Measures of Energy

The DE requirements can be computed from TDN values as given below (N.R.C. 1978)

1 lb TDN = 2.00 MCal of DE

1 kg TDN = 4.409 MCal of DE

The NE requirement values for lactating cows generally have been directly computed from experimental data. But ME requirements for growth have been computed from NE values for maintenance and gain. Similarly direct ME values of various feeds are not available. They have been computed from DE values.

The ME requirement values for lactating cows generally have been directly computed from experimental data. But ME requirements for growth have been computed form NE values for maintenance and gain. Similarly direct ME values of various feeds are not available. They have been computed from DE values using the equation (N.R.C. 1978):

$$\text{ME (MCal/kg DM)} = -0.45 + 1.01\ \text{DE (MCal/kg DM) or}$$

$$\text{ME (MCal/lb DM)} = -0.20 + 1.01\ \text{DE (MCal/lb DM)}$$

Approximate ME values can also be computed from TDN with the assumption that

1.0 kg TDN = 3.615 MCal of ME;

1.0 lb TDN = 1.64 MCal of ME.

6.3. ENERGY NEEDS

Maintenance Requirements. An animal, even when it is not doing any work or yielding any product, will still be expending some energy in the essential physiological processes of life. Added to this, there is the energy required for repairing wear and tear of tissues as well as that

ARC, London. Moore, L.A., Irvin, H.M. and Shaw, J.C. (1953). *J. Dairy Sci.* 36 : 93-97.

Nehring, K. and Heenlein, G.F.W. (1973). *J. Amin. Sci* 36 : 949-964.

National Researcn Council (NRC) (1978). *Nutrient Requirements of Dairy Cattle.* 5th rev. ed. Natl. Acad. Sci. Washington DC.

needed for incidental muscular movements. The portion of an animal's ration, which would cater to the above needs is termed as the maintenance ration. It has been defined as the feed which would provide the energy and also other nutrients for resting, non-producing, non-growing animals to keep them in normal health without gaining or losing body weight and change of body composition, over a course of time.

Maintenance is a major component of feed cost. In practice cows will be lactating and/or pregnant; young ones growing and possibly pregnant. Thus, partitioning energy requirement for maintenance is a theoretical division rather than a practically applied one.

The maintenance requirement increases with the size of the animal, but not in direct proportion. Experiments have indicated that the energy requirement for maintenance is proportional to 0.75 power of live weight, which has been termed the metabolic body size (Blaxter, 1967). In NRC (1978) feeding standards, maintenance requirement of cattle has been calculated as follows.

Net Energy

(*i*) maintenance requirement of growing stock = $0.077 \times W^{0.75}$ MCal.

(*ii*) maintenance requirement of lactating cows = $0.08 \times W^{0.75}$ MCal

Digestible Energy

(*i*) maintenance requirement = $0.155 \times W^{0.75}$MCal

TDN

(*i*) maintenance requirement = $0.0352 \times W^{0.75}$Mcal; (where W is live weight of the animal)

There are factors other than metabolic body size, like activity, which affect maintenance requirement. Individual variations up to 10% have also been noticed.

Energy Requirements for Growth. Metabolizing, transporting and synthesizing new tissues require energy. Composition of gain is the major factor that causes variation in the energy needs for growth. In the early calfhood, weight gain is mainly through addition of protein, water and minerals and relatively less of fat (Blaxter, 1967). From puberty onwards, as age advances, there is more and more percentage of fat with proportionately less protein, water and minerals. Consequently, as the animal becomes older and larger, unit gain of body weight will require more energy. This is so because the energy needs of fat deposition is far higher than for other constituents. Besides, the maintenance need of a larger animal is more than that of a smaller animal. For example, a 250 kg heifer growing at 500 g per day will require 3 times the energy than a 75 kg calf growing at the same rate.

In dairy cattle, the adult production is influenced by the growth attained by the animal. Although the maximum adult size and development are determined by heredity, nutrition influences attainability. Feeding standards give recommendations for optimum growth at a particular body weight. Under TDN system, these figures are inclusive of maintenance.

Energy For Milk Production. Some of the milk constituents like fat, protein and lactose are energy sources. Their secretion therefore requires energy. Similarly their mobilisation and assembly in milk involves energy expenditure. The energy content of the milk secreted is the major factor influencing energy needs for lactation. The energy content of milk in turn is mostly

Blaxter, K.L (1967). *The Energy Metabolism of Ruminants.* Thomas Springfield. Illinois.

dependent on its fat percentage. To remove the effect of fat percentage, milk is often converted to fat corrected milk or FCM. Frequently milk is corrected to 4 per cent fat using the formula

$$4 \text{ per cent FCM} = 0.4 \text{ milk (kg)} + 15 \text{ fat (kg)}$$

The feeding standards (See appendix tables) for milk production give the energy requirements in terms of fat percentage of milk. Approximately 1 kg of 4 per cent fat-corrected milk requires 0.75 MCal of net energy or 0.33 kg of TDN.

Energy Needs of Reproduction. From the 6th month of pregnancy onwards, dairy cows need extra ration for the growth of foetus, foetal membranes, fluids and uterus. At full term, the total bulk of the foetus and membranes weigh approximately 50 kg in smaller breeds but can be more than 100 kg in larger breeds. Out of this 64 per cent develop in the last two months and 38 per cent after 250th day. Therefore, the energy requirements of a larger cow for reproduction will be more than that for a smaller cow. NRC (1978) estimated this amount to be 3.0-6.0 MCal of additional net energy or 0.58-1.15 kg of additional TDN depending on the size of the cow.

If the cow is in a run-down condition, additional energy should be provided during the last few weeks of the gestation period to recondition her.

Energy Requirements For Work. Work can be done only at the expense of energy. The general belief that work does not need extra protein over the maintenance except when work done, is wrong. As bullocks are not used for draft in developed countries and since very little experimental data is available from developing countries only certain tentative values are available. The standards for feeding working bullocks under normal work and heavy work are given later in the appendix tables.

A concentrate mixture of 12 per cent DCP and 75 per cent TDN when fed @ 0.5 kg per 100 kg body weight along with straw and grass to fill will be sufficient for normal work. Under heavy work the bullocks may be fed 0.75 kg/100 kg body weight per day.

6.3.1. Energy Utilization Efficiency

Dairy cattle are among the best converters of feed energy (Table 6.1). Dairy cattle convert about 55 per cent (Range 53-57 per cent) of DE (above maintenance) and 62 per cent ME (Range 61-64 per cent) to milk energy.

TABLE 6.1. Efficiency of Dairy Cattle in Converting Feed nutrients to Edible Products Compared to other Species of Farm Animals

Species	*Conversion efficiency per cent**	
	Energy	*Protein*
Dairy cattle	17	25
Swine	14	14
Broilers	11	23
Turkeys	9	22
Beef cattle	3	4
Lambs	—	4

*Based on lifetime production and feed conversion (Adapted from Harshbarger, 1975).

In dairy cattle, feed energy is utilized more efficiently for milk production and maintenance as compared to growth and fattening (NRC, 1978) and consequently feed has a higher net energy value for maintenance and lactation compared to growth and fattening. However, in NRC (1978) feeding standards, the feed energy for different body functions are presented in the same units after making suitable compensations for the differences.

The level of feeding and the composition of the ration are two major factors that affect feed efficiency. At higher levels of feed intake, the digestibility goes down. But, this is partially compensated for by reduced urinary and gaseous losses. Generally, a ration high on concentrates and low in roughages results in a type of rumen fermentation in which acetic acid production is reduced and propionate increased.

6.3.2. Energy Balance and Built in Mechanisms for Controlling it

Life is a constant effort on the part of the organism to assimilate into itself certain substances in the environment at lower concentration, but needed by the animal body at a higher concentration or *vice versa.* This effort on the part of the organism to maintain a condition of relatively uniform environment within has been termed as homeostasis. Homeostatic control of every nutrient is essential for survival. But the mechanisms vary.

Energy balance is kept in a way totally different from other nutrients. The major mechanism employed by dairy cattle and buffaloes for this purpose are a change in feed consumption. Energy is absorbed and utilized for body functions under varying levels of intake, with very little change in efficiency and therefore will not serve as a control mechanism. It is apparent that changes in feed intake, since it is used as a mechanism for controlling energy balance, cannot be used as a primary homeostatic control for any other nutrient. It is energy that controls feed intake.

There are two limiting factors in the voluntary feed intake of cattle and buffaloes. With poor quality roughages it is the physical capacity of the alimentary tract. Once filled, it can be re-used only when it becomes empty. Dry matter intake increases with the energy needs of the cow until physical capacity restricts feed intake and with low quality feeds the dairy cow does not consume as much energy as needed thus affecting production. Dry matter intake increases with digestibility to about 63-68 per cent and subsequently declines. Above this digestibility level, physiological and/or chemical factors restrict the amount of usable energy consumed (Conrad, 1964) (Fig. 6.5).

With digestibility below 65-68 per cent, physical capacity limits DM intake, the DM increasing with increase in digestibility. Above 68 per cent digestibility, physiological factors determine maximum consumption thus controlling the energy consumed.

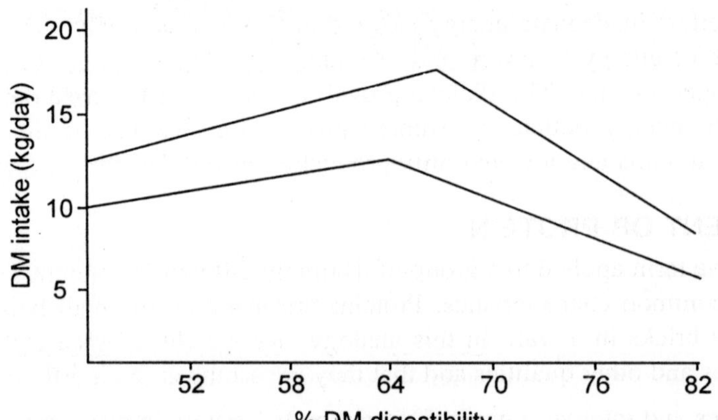

Fig. 6.4. Dry matter intake increases with its digestibility to about 63-68 per cent and thereafter declines due to physiological factors restricting energy intake.

Consequences of Excess and Deficiency of Energy. The feeding standards aim at optimum rather than maximum growth and production. Therefore, feeding above these recommended rates will result in slight enhancement of both. But this will not be usually economical and may result in health problems. Energy in excess of requirements will be deposited as fat. Over-fed heifers will have excessive fat deposition in the mammary gland and around internal organs. This may permanently reduce the secretory tissue in the udder and impair reproduction, thus decreasing their lifetime milk producing ability.

Low energy intake on the other hand causes lowered milk production in lactating cows and if continued results in loss of body condition. When cows are underfed during early lactation, milk production will be lowered throughout the lactation even if fed adequately afterwards (Fig. 6.6).

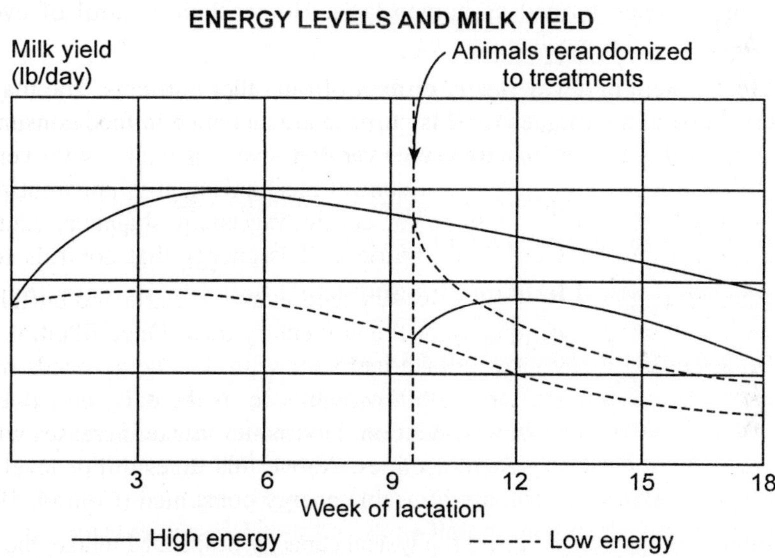

Fig. 6.6. High feeding in early lactation helps in attaining a higher peak which will be maintained (Adapted from Broster *et al* 1969).

In growing heifers inadequate energy slows down growth and delays the onset of puberty. Unless the shortage of energy is severe and extended over long duration, there will be little or no permanent damage to health. The skeletal growth is not affected in mild deficiency of energy. In extreme cases of energy deficiency, stunted growth, delayed age at first calving, delayed reproductive ability in bulls and lowered milk production in first lactation etc will be manifested.

6.4. REQUIREMENT OF PROTEIN

Protein is a collective term applied to a group of distinctly different compounds of large molecular size having some common characteristics. Proteins are made up of small building units known as amino acids like bricks in a wall. In this analogy, one should imagine that the bricks are of various shapes, sizes and other qualities and that they are arranged in all different planes possible.

The soft tissues and internal organs of animal body have protein as a major constituent. The transformation of food protein into body protein is a vital nutritional process which sustains growth and repair. A continuous liberal supply of proteins throughout life is necessary for this.

Young animals require considerably more proteins for growth. Similarly, milch cows secrete 300-600 g of protein per day in milk depending on production.

The animals build their body proteins primarily from the amino acids absorbed from the digested dietary proteins. Absorption will not take place unless food proteins are broken down into amino-acids. There are about 25 amino acids that are classified either as essential or non-essential. The essential ones must be supplied through diet and absorbed in the alimentary tract into the body for tissue synthesis. But the non-essential amino acids can be either supplied through diet or can be synthesized in the body tissues. Some or all of the amino acids may be combined in varying proportions and sequences to produce innumerable number of proteins.

In order to synthesize a particular body protein in the animal system, all the amino acids required for it must be simultaneously present in sufficient quantities. Even if one is missing, all the rest will be wasted for tissue building and de-aminated for liberating energy. The body has the capacity to convert surplus of amino acids into certain other simpler (non-essential) amino acids. On the other hand, the large animal tissues do not have the capacity to synthesize more complicated essential amino acids. About 40 per cent of the proteins in animal tissues consists of non-essential and 60 per cent essential amino acids.

The usage of the terms 'essential' and 'non-essential' does not imply in any way that one group is more necessary and important than the other in metabolism. The distinction only means that the 'essential' amino acids have to be supplied from external sources as such because they cannot be synthesized in the body.

6.4.1. Rumen Function in Protein Utilisation

About 50-60 per cent of the proteins consumed by the ruminants are attacked by rumen microorganisms, which hydrolyze them to peptides and amino acids. They are further fermented to short-chain fatty acids, carbon dioxide and ammonia. While this degradatory process is in progress, synthetic activities in which the rumen micro-organisms absorb various protein degradatory products like ammonia, amino-acids, nitrates, nitrites etc. to form microbial body proteins also goes on side by side. The remaining portion of the proteins which are unaffected by rumen digestion and the microbial proteins pass on to the lower alimentary tract for enzymatic digestion and absorption as in non-ruminants.

The proportion of dietary protein hydrolyzed in the rumen depends on many factors like the solubility of protein, quantity of protein and availability of readily soluble carbohydrates etc. If the rate of hydrolysis is moderate and other needed substances like carbohydrates are available, the microorganisms themselves in the synthesis of microbial tissue proteins will utilize most of the ammonia released.

A major quantity of the ammonia liberated in the rumen is absorbed by its wall and reaches the liver through the portal system. Liver converts ammonia into urea for excretion through urine or recycling via saliva or blood for reuse in the rumen. This recycling of urea back to the rumen provides for conservation of nitrogen for periods of low nitrogen intake.

When poor quality proteins are fed or when nitrogen intake is inadequate, synthesis of microbial proteins in the rumen can be of enormous help. The microbial tissue averages about 65 per cent protein on DM basis. This protein contains all the essential amino-acids. Thus, after digesting and using most of the feed eaten by the cow, the microbes in their turn are digested and eaten by the cow. The cow gains important benefits in this process because the poor quality proteins and NPN generally present in dairy bovine rations are converted to high quality microbial proteins with all the essential amino acids.

If high quality readily soluble proteins are fed to adult ruminants, there may be more loss of *N* through urea or ammonia and the amino acid mixture available to the animal will be less in quantity and poor in quality than what was in the diet. At present certain techniques have been evolved by which high quality proteins can be protected by chemical/physical means to remain insulated and unavailable to microbial activity in the acidic medium of rumen but amenable to enzymatic digestion in the abomasum and lower alimentary tract.

Due to the intervention of rumen microorganisms, the protein requirements of ruminants are relatively simple. Even if the proteins in the diets of these animals are of low biological value, lacking in many essential amino acids, they will be converted into microbial protein of high biological value rich in all essential amino acids.

6.4.2. Utilization of Non-Protein Nitrogenous Substances (NPN)

Protein is a costly item. Therefore, in most parts of the world, protein is one of the most limiting nutrients. Dairy cattle and buffaloes are bestowed with the capacity to utilize NPN for conversion to protein. Substituting protein with NPN for dairy bovine feeding often lowers costs as well as spares natural proteins for non-ruminants and humans.

Protein Synthesis in the Rumen from NPN. In ruminants ammonia is the common denominator in nitrogen metabolism. NPN either found naturally or added as special supplements are attacked by rumen microbes to produce ammonia. The same happens to most of the proteins in their diet also. Rumen microbes for synthesizing their own body protein utilize the ammonia thus formed. The metabolic pathway of utilization of urea, one of the important NPN compound in ruminant feeding, is given below.

NPN substance	*Acting agent & Action*	*End product*
1. Urea	*Microbial Urease*	ammonia (NH_3) + Carbondioxide (CO_2)
2. Carbohydrates	*Microbial Enzymes*	volatile fatty acids (VFA) + keto acids (carbon chains)
3. Ammonia + keto acid (carbon chains)	*Enzymes from Microbes*	amino acids
4. Amino acids	*Enzymes from Microbes*	microbial protein
5. Microbial proteins	*Enzymes of abomasums & small intestine*	free amino acids
6. Free amino acids absorbed from the small intestine		body protein of the ruminant

Toxicity of NPN Substances. Properly fed, urea and other NPN substances are safe and useful nutrients for bovines. But if used improperly, they can be highly poisonous. Excessive levels of ammonia in the blood stream are invariably the cause of such toxicity. A blood ammonia level equivalent to 1.0 mg. nitrogen per 100-ml blood will be toxic (N.R.C., 1976). Consumption of high levels of NPN at short intervals or deficiency of other nutrients required for rapid growth of rumen microbes can cause build up of ammonia levels leading to toxicity.

6.4.2.1. *Guidelines in feeding NPN*

Practices for increasing the efficiency of utilization of NPN reduce toxicity. NPN compounds are useful to the bovines only after microbial conversion to their body protein. Thus, conditions favouring an optimum growth and multiplication of rumen microorganisms favour efficient utilization of NPN also. One of the most limiting factor is the availability of readily fermentable carbohydrates to supply energy and 'carbon chains' for the synthesis of amino acids. A certain amount of starch in the ration will be very useful. Conrad and Hibbs (1968) suggested that for every kg of urea, 10 kg readily fermentable carbohydrates are needed with two-thirds as starch. In addition, other nutrients like essential minerals are also required for proper microbial growth. NPN supplies only nitrogen. On the other hand natural proteins provide substantial amounts of other nutrients like energy and minerals. Thus when incorporating urea like compounds in dairy cattle ration, care has to be taken to meet requirements of other essential nutrients. A thorough mixing of urea and other feed ingredients is necessary to avoid toxicity.

Thumb rules for urea feeding. 1. The amount of supplemental urea should not exceed 1 per cent of the total ration. 2. Urea should not be more than 3 per cent of a concentrate mixture. 3. It should not exceed one-third of the total crude-protein equivalent.

6.4.3. The Protein Needs of Infant Calves

In the baby calf, the rumen is not developed and therefore its digestion is similar to that in the non-ruminant. There is only insignificant level of synthesis of microbial proteins. Therefore, they require high quality proteins with sufficient quantity of all the essential amino acids in their ration. The amino acids shown to be essential for ruminants are (1) histidine, (2) isoleucine, (3) leucine, (4) lysine, (5) methionine, (6) phenylalanine, (7) threonine, (8) tyrosine, and (9) valine.

The infant calves also require a high percentage of protein in their ration. In comparison with its size the young calf grows rapidly and most of the tissues added have a high content of protein in them. Baby calves require a minimum of 22 per cent protein (on DM basis) in their ration. The calf is ill equipped to utilize NPN.

The reticulo-rumen of the calf develops gradually and become functional over a course of 3 to 4 months age. Parallel to the development of rumen, the dependency of the calf on dietary source of essential amino acids declines. The percentage of protein needed in the total diet also declines quite rapidly. At weaning age of 3-4 months they require only 16 per cent protein.

6.5. MINERALS

The animal body contains about 40 mineral elements. Some of them are present purely adventitiously because they happened to be present in the feed consumed, water drunk or air inhaled. No nutritional roles for them have been established. On the other hand, about 19 minerals have been established to have essential metabolic roles in the animal body. Another two also may have essential roles.

6.5.1. Essential Minerals

The following mineral elements have been established as essential for dairy bovines :

(*i*) Calcium (Ca)
(*ii*) Phosphorus (P)
(*iii*) Magnesium (Mg)
(*iv*) Potassium (K)
(*v*) Sodium (Na)
(*vi*) Chlorine (Cl)

(*vii*) Sulphur (S)
(*viii*) Iodine (I)
(*ix*) Iron (Fe)
(*x*) Copper (Cu)
(*xi*) Cobalt (Co)
(*xii*) Manganese (Mn)
(*xiii*) Zinc (Zn)
(*xiv*) Selenium (Se)
(*xv*) Molybdenum (Mo)
(*xvi*) Fluorine (F)
(*xvii*) Chromium (Cr)
(*xviii*) Silicon (Si)
(*xix*) Vanadium (V)

There are many evidences to indicate that tin (Sn) and nickel (Ni) may be also essential. Probably other elements may be added to the required list.

6.5.2. General Functions of Minerals

Minerals play many a vital role in animal system. As constituents of skeletal tissue, minerals impart rigidity and strength to it. Minerals also form part of the soft tissues and body fluids. In dairy bovines, large amounts of minerals are secreted in milk.

Nucleo-proteins, which form the continuity from generation to generation, have phosphorus as a constituent. This element also plays a crucial role in metabolic reactions involving energy transfer acting as a 'ready currency' of energy. Protein synthesis, a vital metabolic activity, involves elements like P, Fe, Mn, Ni, Zn, and Cr. Most of the mineral elements function as part of enzyme systems.

Minerals play essential roles in maintaining acid-base balance, osmotic pressure, optimum pH, membrane permeability and irritability of tissues. The important minerals involved in these functions are calcium, phosphorus, magnesium, sodium, potassium and chlorine. Some of these are so vital that they can make the difference between life and death. For example, regular functioning of the heart is dependent on the changes in irritability of cardiac muscles which in turn is controlled by the proportion of calcium, sodium and potassium ions in the fluid that surrounds it.

Apart from the general roles many minerals have specific metabolic functions to play. The red blood corpuscles are able to transport oxygen from the lungs to the tissues and carbon dioxide from the tissues to the lungs due to the presence of haemoglobin, which contains iron. The only known essential role of cobalt is as part of vitamin-B_{12}. Likewise, perhaps the only essential function of iodine is as a part of the hormone thyroxine, which regulates the metabolic rate in animals.

Minerals are generally classified into *major elements* and *trace elements* depending on the vast differences involved in the quantity required in diet.

6.5.3. Major Elements

The seven essential major elements are calcium, phosphorus, sodium, chlorine, potassium, magnesium and sulphur. All of them are required in the diet substantially more than 100 ppm (parts per million) (0.01 per cent). The average tissue concentrations of these elements also exceed 100 ppm. Concentration of major minerals is generally expressed as a percentage. These elements form part of structural tissues apart from their roles in metabolic processes.

Calcium (Ca). Calcium has been long recognized as a critical nutrient in the practical feeding of dairy bovines. More than 98 per cent of the body calcium is in the bones and teeth.

But the remaining 2 per cent or less in blood and soft tissues performs numerous key functions like blood clotting and tissue excitability. These roles of calcium are so critical that the Ca level of blood plasma is one of the most closely regulated of blood constituents.

Large amounts of Ca are secreted in the milk and naturally a lactating cow will require the element in large quantities. Similarly growth and gestation (growth of foetus), which involve synthesis of calcium containing bone tissue will require additional Ca in the diet. However, precise Ca requirements for all phases of growth, reproduction and production of dairy animals are not available. The situation is not so critical as it might be, however, as body Ca reserves are quite large. Even though the blood level of Ca is very critical, mobilizing from skeletal tissue finely regulates it. Along with this, the ability of the dairy cow to absorb higher percentage of dietary Ca when the diet is deficient in Ca, enables her to perform normally with long periods of inadequate dietary calcium.

However, in growing calves and heifers, a deficiency of Ca in the diet prevents normal bone growth, low bone content of Ca and P, and depressed general growth. In mature cows and buffaloes, prolonged periods of Ca deficient diets cause depletion of bone stores of Ca and P which result in fragile easily broken bones. Eventually this may lead to reduced milk production. But the percentage of Ca in milk is not affected.

All the major feeding recommendations include Ca requirements also. Lactating cows may be given 15-25 g Ca per cow per day for maintenance, plus 2.5 to 3.0 g of the element for every kg milk produced. This requirement can be met if the total ration contains 0.5 to 0.6 percent Ca.

Phosphorus (P). More than 85 per cent of the total P in the body is present in the skeletal tissue and teeth. In addition, P is an extremely important element forming part of organic compounds like nucleo-proteins, phosphoproteins, phospholipids, hexose phosphates etc. Phosphorus is a component of various enzyme systems and being a component of AMP, ADP and ATP it plays a very important role in metabolic reactions involving energy transfers, involving formation and breaking of high energy P bonds. P is also a major mineral component of milk. A major part of the P requirement of humans is supplied through P in milk and milk products.

The metabolism of P is intimately related with Ca. But there are also some major differences in their nutrition and metabolism. For example, dietary deficiency in P reduces level of inorganic P in blood plasma much more than low Ca intake affects plasma Ca.

Symptoms of P deficiency are likely to become obvious during growth and heavy lactation. In growing animals P deficiency is manifested by the symptoms of rickets like swollen, stiff and aching joints, arched back, beaded and deformed ribs and retarded growth., In adult animals the symptoms of P deficiency may not be that pronounced as the animals will draw on their reserves for some weeks before they show any outward signs. Once the reserves are depleted they will exhibit depraved appetite, craving for bones, wood, bark, hair and other foreign material; irregular heats, anaestrus and lowered conception rates. The bones may become porous and fragile making the cows prone to fractures.

Daily dietary requirements of P for dairy cows include 10-20 g for maintenance plus 1.7 to 2.4 g per kg. milk, produced. Thus the requirement of P is slightly less than that of Ca which works out to 0.3 to 0.4 percent of the total ration dry matter.

Calcium : Phosphorus ratio. Not only should dairy animals be fed sufficient Ca and P, but a proper ratio be maintained between the two minerals. The feeding stuffs vary markedly in their

Ca and P contents. Besides, the availability of these elements from the feedstuffs is not necessarily indicated by their gross composition in a feed. The absorbability of these elements in the animal system is dependent upon several factors such as the chemical combination in which they occur, the physical association with other feed constituents, percentage level in the feed, age of the animal and its physiological state, the proportion of Ca and P in the ration, availability of vitamin *D* and presence of deleterious substances like oxalates, fluorine etc. A calcium : phosphorus ratio of 1 : 1 to 2 : 1 seems to be satisfactory to dairy animals. Less vitamin *D* is required when the Ca : P ratio is within this range. If vitamin *D* is adequate a wider ratio may be fed without harm.

Magnesium. Magnesium has a role similar to Ca and P as far as it forms a part of the skeletal tissue. Its nutritional role and metabolism are closely linked to these two elements. The bovine body contains roughly 0.05 per cent of Mg of which about 60 per cent is present in the bones. However, it is not easily mobilised from the skeletal tissue as in the case of the other two elements. Therefore, dietary supply should correspond to requirements. Abrupt changes in giving rations deficient in Mg can produce deficiency symptoms within 2 to 15 days in adult dairy cattle.

In adult ruminants Mg deficiency has been reported in Western countries where during spring season the animals are turned out on young succulent pastures which are deficient in Mg. This condition is known as 'Magnesium tetany', 'Grass Staggers', 'Grass tetany' etc. Other factors like hormonal disturbances and faulty inter-relationship of Ca, P and Mg also contribute towards Grass staggers.

In India calves reared on milk diet without any other supplement for a prolonged period were also reported to have Mg deficiency.

Calves require a minimum of 1 g of Mg per day per 100 kg b.w. Cows require 2.0 to 2.5 g per day for maintenance plus 0.12 g per kg milk produced. Inclusion of 0.97—0.96 per cent Mg in calf diets and 0.20 per cent in cow diets is adequate to meet the requirements.

Sodium and Chlorine (Common salt). It is a universal practice to give supplemental salt to bovines. Without it most dairy animals will not receive sufficient sodium (Na). Although chlorine has many essential roles, a deficiency has not been observed.

The earliest symptom of Na deficiency is pica (an intense craving) for salt. Other signs may not develop for many months due to the efficient conservation of the mineral by the animals by reducing excretion. When Na deficiency finally develops after long periods of depletion, it is marked by a decrease or loss of appetite, unthrifty haggard appearance with rough coat, weight loss, lowered growth rate and lowered milk production. Terminal symptoms include shivering, general weakness incoordination, cardiac arrhythmia and finally death (NRC, 1978). Lactating and growing animals require more salt. Similarly animals in hot regions need more salt because of added excretion in sweat. The Na content of forage varies widely and is influenced to a great extent by the presence of the element in soil. In low-rainfall areas the soil is rich in salt and therefore a deficiency rarely occurs. On the other hand wet-humid-hot tropical soils and grasses and therefore rations of animals raised on them are markedly deficient in salt. They need salt supplementation.

Potassium (K). Until recently, it was believed that there was very little chance of K deficiency in dairy bovines. Present information indicates that dairy cattle require more K than any other mineral element. The major functions of K are to maintain osmotic pressure and acid-base balance in intra-cellular fluids, in electrolyte balance and in the control of nerve and muscle excitability.

Severe deficiencies rarely occur. Forages normally contain high levels of K and therefore, cows receiving sufficient proportion of forages will not normally suffer from any K deficiency. However, cattle fed on a predominantly concentrate ration can suffer marginal deficiencies of K if their diet is not supplemented with K. Deficiency symptoms include general muscular weakness, a marked decrease in feed intake, reduced weight gains, decreased milk yield, pica, loss of hair/ glossiness, decreased pliability of skin and loss of muscle tone. Milk and blood plasma levels of K are also reduced. If the dairy cattle ration contains a minimum of 0.7 per cent K of the total ration dry matter for calves and 1 per cent for lactating cows, deficiencies are not likely to occur.

Sulphur (S). The total body tissues of cattle contain 0.15 percent S and milk 0.03 per cent. The most critical S requirement in dairy bovines is for optimal rumen microbial growth, besides S is a component of essential amino acids cystine and methionine. It also forms part of vitamins thiamin and biotin. It is a constituent of heparin, the anticoagulant of the body and glutathione, which functions as a coenzyme. Dairy bovines require S @ 0.2 per cent of the dry ration. In rations with natural proteins the S requirement is more or less met. But, in rations containing high proportion of NPN, sulphur should be supplemented at a rate of 1 part of S for every 10 parts of N.

6.5.4. Trace Elements (Micro-elements)

These elements are required in very small amounts. Iron, zinc, copper, cobalt, iodine, selenium, manganese, fluorine, molybdenum, chromium, silicon, nickel, vanadium and tin are required in the diet and present in the tissues in levels much lower than 100 ppm. These 14 elements are therefore expressed in ppm or ppb (parts per billion).

The normal feed of bovines generally contains sufficient quantities of these elements to satisfy nutritional needs. However, as soils get depleted due to intensive cultivation, as milk production levels increase and composition of feeds change, more deficiencies are precipitated. A trace mineral mixture may be judiciously added to the total ration or a trace mineral salt-lick may be provided on a free-choice basis to help avoid such deficiencies. A brief summary of the deficiency symptoms and requirements of some of the trace minerals are given in Table 6.2.

6.5.5. Inter-relationships among Minerals

The requirements, absorption, metabolism, excretion, availability at cellular level and maximum safe dietary level of various minerals are inter-dependent. Often these interactions take the form of antagonism between 2 or more elements. An example is Cu and Mo. The mechanism of action of these antagonisms is different in different cases. Most often high level of one affects the absorption of another. In certain other cases high level of one prompts the rapid excretion of another. A third type is through effects at the tissue or functional level, resulting in competition for the active sites.

One of the best known inter-relationships is between calcium, phosphorus and magnesium. Similarly high levels of calcium can increase the requirements of some of the trace elements including Zn and Mn. Most of the mineral interactions have not been studied sufficiently in dairy cattle.

TABLE 6.2. Trace Minerals : Deficiency Symptoms and Requirements of Cattle

Element	*Deficiency symptoms*	*Daily Requirements of cattle*
Iron (Fe)	Anaemia, particularly in calves maintained on milk; seldom seen in adult bovines.	300 mg/kg for growth, 100 ppm in adult rations (2 g).
Copper (Cu)	Anaemia, depressed growth, bone disorders depigmentation of hair, demyelination of spinal cord and lesions in brain with muscular incoordination and gastro-intestinal disturbances with diarrhoea, still births, sudden death with heart degeneration.	5 ppm of ration dry matter, Increased quantity needed with high molybdenum intake.
Cobalt (Co)	Deficiency of Co leads to insufficient production of vitamin B_{12} in the rumen—Therefore cobalt deficiency symptoms are actually vitamin B_{12} deficiency symptoms. They include anaemia, loss of appetite, emaciation, low appetite for grain. Calves unthrifty with poor appetite are the first to exhibit symptoms of deficiency because of low content of vitamin B_{12}.	0.1 ppm in dry ration (2 mg.)
Manganese (Mn)	Poor growth, leg disorders, ataxia (uncoordinated movements), poor fertility and frequent abortions. In new borns, deformed bones, enlarged joints, stiffness, twisted legs, shorter humerus (foreleg), general physical weakness. Deficiency could occur in cows fed high grain-low roughage rations.	*Calves* : 7-10 ppm of ration DM. *Cows* : 20 PPM of ration DM.
Zinc (Zn)	Reduced milk yield, lowered fertility, loss of hair, rough scaly skin, itch, stiff gait, swelling of hocks and knees, soft swelling above rear feet, undersize testicles in bull calves.	*Calves* : 8-10 ppm of ration DM. *Lactating cows:* 40 ppm of ration DM.
Iodine (I)	Thyroxine production is reduced. Goitre in calves which are frequently born hairless and dead; reproduction is affected. Some feeds like cabbage, kale and those of the Brassica family contain goiterogens feeding which produce iodine deficiency symptoms even if the ration has adequate iodine.	0.1 ppm of ration DM. *Non-lactating:* 2 mg/head/day; *Pregnancy* : 0.8 ppm or 8-12 mg, more may be required if feeds containing goiterogens are fed heavily.
Fluorine (F)	Deformed teeth and bones.	Small amounts appear to prevent dental cavities. Toxic above 10 ppm
Selenium (Se)	Nutritional muscular dystrophy, high calf mortality, incidence of retained placenta cases increased.	0.05 ppm to 0.10 ppm. of ration DM. Toxic above 3 ppm
Molybdenum (Mo)	Deficiencies are unknown. But it causes toxicity in excess. Mo toxicity interferes with the availability of Cu and causes copper deficiency symptoms.	Extremely low levels.
Chromium (Cr)	Deficiencies are unknown in bovines.	Extremely low levels.
Silicon (Si); Vanadium (V); Tin (Sn); Nickel (Ni)	They are considered essential because their biochemical roles have been established in laboratory animals and generally all animals require similar trace elements for biochemical functions.	

6.6. VITAMINS

Bovines require the same vitamins and almost to the same extent as in the case of other farm animals like swine and poultry. However, because most of the vitamins are synthesised either in the rumen or tissues they are not dietary essentials for bovines which have a functionally active rumen.

The usual needs of Vitamin *B* and Vitamin *K* of adult ruminants are met by synthesis by rumen micro-organisms. Only in young calves these are dietary essentials because the rumen is not functional. Vitamin *C* is synthesized in the body tissues of dairy bovines and, therefore, not needed in the diet. Under tropical and subtropical conditions where cattle are exposed to ample sunlight, Vitamin *D* is synthesized in the skin under the influence of ultraviolet irradiation from the sun. But, in temperate regions especially during winter when chances for exposure to sunlight are less, Vitamin *D* should be supplied in the diet. Vitamin *A* and *E* should be present in the diet of dairy bovines of all ages and all regions.

Under normal types of feeding, vitamins are not likely to be deficient in the rations of dairy animals and no supplementary vitamins are needed for adults and calves. Dairy cattle also have the capacity to store in the liver sufficient reserves of **fat-soluble vitamins *(A, D, E* and *K)*** so that they are able to tide over short to moderately long periods of deficiency without any apparent ill effects. For example, if cattle have access to plenty of green grass for 6 months in an year, they can store enough of Vitamin *A* to meet the needs for the remaining 6 months of the year even if their ration is of hay and straw (deficient in Vitamin *A)*.

6.6.1. Vitamin A

Vitamin *A* is essential for the health and normal functioning of epithelial tissue in different parts of the body (NRC, 1978). All animals require a dietary source of Vitamin A.

Vitamin *A* plays an important role in vision. *Rhodopsin* (visual purple) in the retina of the eye is a combination of a protein and Vitamin *A*. Rhodopsin when exposed to light is converted to retinene and opsin (visual process). In the darkness rhodopsin is resynthesized. A deficiency of Vitamin *A* will impair this resynthesis and will reduce the ability of the animal to adapt to dim-light (night blindness). This is one of the earliest symptoms of Vitamin *A* deficiency. In a severe Vitamin *A* deficiency, there will be excessive watering, keratitis, softening and cloudiness of cornea, and xerophthalmia characterized by drying of the conjuctiva. Blindness is the ultimate result.

In other tissues, Vitamin *A* deficiency is characterized by stratified keratinization, degeneration and drying out of the epithelial or mucosal tissue of alimentary system, urethra, kidneys, vagina, gonads and respiratory tract (NRC, 1978). Calves suffering from Vitamin *A* deficiency become susceptible to diarrhoea, cold, pneumonia and other infections. Calves are often weak at birth. In cows reproductive problems, shortened gestation and high incidence of retained placenta are seen.

Major source of Vitamin *A* in dairy animals is carotene, a precursor of Vitamin *A* in green plants. Animals can change the yellow coloured carotene to Vitamin *A* in the intestinal wall and liver. Carotene is abundant in the green, yellow or orange coloured parts of plants. For cattle 1.0 mg of Beta carotene is equivalent to 400 IU of Vitamin *A*.

Calves require a minimum of 10 mg carotene or 4,000-5,000 IU of Vitamin *A* per 100 kg body weight per day. Cows require 10-12 mg carotene or 4,000-5,000 IU of Vitamin *A* per 100 kg body weight per day for maintenance. An equal amount would be necessary for lactation cum pregnancy. Thus a cow would require 80-100 mg carotene (32,000-40,000 IU Vitamin *A)* per day.

Vitamin A in Preserved Feeds. The amounts of carotene are relatively high in fresh green forages. Heat, moisture and light are potent agents in oxidizing and inactivating carotene. Thus, hay preserves only a small proportion of the carotene content of fresh grass. Even so, well-preserved hay can fairly meet the Vitamin *A* requirement of dairy animals. Green colour in well-preserved hay is a rough indicator of its carotene content. Carotene content of hay decreases continuously on storage. On the other hand, the acid medium of silage preserves carotene reasonably well. In compounded feeds, the rate at which Vitamin *A* is oxidized and destroyed depends on the nature of the feed, storage conditions and the degree to which the vitamin was stabilized.

6.6.2. Vitamin D

Vitamin *D* is essential for the metabolism of calcium, phosphorus, magnesium and certain other minerals. It plays a key role in synthesizing a protein that binds calcium during its absorption. It is also necessary for mineral deposition in bone formation.

In calves, deficiency of Vitamin *D* results in rickets. This is characterized by retarded mineral deposition in bones with accumulation of osteoid tissue. Joints are swollen and enlarged, and bones fracture easily. Arched backs, straight pasterns and beading of ribs are other symptoms. In acute cases, a still gait with dragging of hind limbs, accumulation of fluids in joints, fast and laboured breathing, high irritability, tetany, anorexia, weakness and stunted growth are observed.

In adult bovines a prolonged vitamin *D* deficiency manifests itself as osteomoalacia. This is characterized by re-absorption of the mineral from the already formed bone, making the bones, soft, weak and prone to fractures very easily. Deficiency of the vitamin also reduces milk production as well as inhibits estrus.

Among the different forms of vitamin D_1 D_2 and D_3 are the most important in dairy cattle rations, and are nutritionally equally useful. The precursor 7-dehydrocholesterol in the animal body is converted to cholecalciferol (Vitamin D_3) by the ultraviolet rays of sunlight. Thus, if the dairy cattle obtain sufficient sunlight with ultraviolet component (280-297 nm), all the required quantity of vitamin *D* will be synthesized in its skin. Vitamin D_2 (calciferol) is the plant form of the vitamin and is present in abundance in sun-cured hay and straw and dead leaves of freshly cut forages. Mature cows require about 1000 IU of the vitamin per 100 kg b.w. daily. This much requirement is normally met from sunlight conversion and from sun-dried fodder under normal conditions of management and feeding in tropical and subtropical conditions. Excess of vitamin *D* has been found to be toxic and therefore supplementing the vitamin in feeds may be done judiciously.

6.6.3. Vitamin E

Many of the tocopherols have vitamin *E* activity with á tocopherol having the maximum activity. This fat soluble vitamin is functionally closely related to selenium; both having anti-oxidant properties. In spite of extensive research, it has not been possible to assign a concrete metabolic role for tocopherols. Many scientists were striving to find a possible common physiological link between diverse pathological lesions and symptoms observed in vitamin *E* deficiency. The present belief is that tocopherols act as antioxidants, preventing oxidation of other vitamins and essential constituents both in the intestine and tissues. Thus, the lack of Vitamin *E* can add to the deficiencies of other nutrients with their characteristic symptoms.

In dairy cattle other than young calves deficiency of vitamin *E* is uncommon. Symptoms of nutritional muscular dystrophy (white muscle disease) and heart muscle lesions are produced in calves in acute vitamin *E* deficiency. Calves may walk with a typical crossing of hind legs; relaxation of the pastern and splaying of toes are also observed.

Vitamin *E* is known to play a role in the successful reproduction in some species of animals. But attempts to establish a practical role for the vitamin in both male and female reproduction in bovines have been unsuccessful. Perhaps this may be because the common feeds of bovines contain an ample supply of the vitamin. Calves require 30-40 mg of the vitamin per day. Feeding 1000 to 2000 I.U. of Vitamin *E* is beneficial in reducing the incidence of oxidized flavour in milk having high copper content.

6.6.4. Vitamin K

The major role of Vitamin *K* is in prothrombin synthesis in liver and consequently it is essential for normal blood clotting. In dairy cattle, ample amount of the vitamin is synthesized in the rumen by microorganisms. Similarly, forages are a good source of the vitamin. Therefore, Vitamin *K* deficiency is not common under normal feeding. A conditioned Vitamin *K* deficiency may develop when bovines are fed mouldy sweet clover caused by dicumarols in it, which reduce the prothrombin. Generalized hemorrhaging and slow clotting from even small wounds characterize this condition known as sweet clover disease. Supplementary feeding of Vitamin *K* will be beneficial in sweet clover disease.

6.6.5. Water Soluble Vitamins

Members of the B-Complex and Vitamin *C* belong to this category.

6.6.5.1. *B-Vitamins*

Formerly all the Vitamins of this group were considered as a single factor—Vitamin *B*. There seem to be about 12 of these vitamins and there may be still unidentified ones. The important ones are—Thiamine, Riboflavin, Biotin, Vitamin B_6 (Pyridoxine, Pyridoxamine and Pyridoxal), Niacin (Nicotinamide), Pantothenic acid, Vitamin B_{12} (Cyanocobalamine), Choline, Folic acid, Lipoic acid, Inositol, and P-aminobenzoic acid.

These vitamins are highly essential metabolically. Dairy bovine tissues must have most of the above-mentioned B-Complex vitamins as cofactors with various enzyme systems for most of the biochemical reactions in intermediary metabolism. However, a dairy cow or buffalo will seldom have a deficiency in B-Vitamins nutritionally because of the following reasons.

(*i*) Ruminal micro-organisms synthesize most of the *B*-Vitamins and make them available to the ruminant.

(*ii*) Green forage, well cured legume hay and silages are good sources of these vitamins.

Even in young calves under normal feeding conditions, B-Vitamin deficiency seldom occurs under natural feeding conditions because milk is a rich source of most of the *B*-Vitamins. Besides, there is some amount of ruminal synthesis of these vitamins even when rumen is not well developed. However, when milk replacers consisting mostly of plant proteins are fed, deficiency of some of these vitamins can arise.

6.6.5.2. *Vitamin C (Ascorbic Acid)*

In species like man, monkey and guinea pigs deficiency of Vitamin *C* leads to a disease known as *scurvy.* But cattle and buffaloes synthesize Vitamin *C* in their tissues. They, under normal feeding conditions do not require a dietary source of Vitamin C. But, on a Vitamin A deficient diet, Vitamin C is needed in the diet of dairy bovines. It is easily oxidized. So even if it is fed it becomes useless because it will be oxidized in the rumen.

6.7. FEEDING PROCESS

Bovines are a family belonging to the sub-order Pecora or true ruminants. They possess evolutionary modifications in their alimentary system to suit their feeding and self-preservatory behavioural needs. Unlike other animals, which have a simple single-compartmental stomach, the ruminant has four compartments to its stomach. With the help of symbiotic microorganisms harboured inside the reticulo-rumen, these animals are equipped to utilize substances such as cellulose, a major plant constituent and urea, a non-protein-nitrogenous (NPN) compound. Animals possessing simple stomach can make only very limited use of these substances. The bovines convert forages and byproducts into milk and meat—two valuable food items for man. But for ruminants, many of the plant materials and residues would have gone waste as a food source. The conversion of these materials into milk and meat is the most compelling justification in favour of keeping ruminants.

6.7.1. Anatomy and Physiology of the Alimentary System

The process of feeding starts with the animal carrying the food material to the mouth (*prehension*), chewing (*mastication*), swallowing (*deglutition*) chewing the cud (*regurgitation*), *digestion, absorption* and *defecation.* Food brought to the mouth by tongue and lips passes through oesophagus to the rumino-reticulum (Fig. 6.7). Inside that part of the stomach the feed gets mixed-up with ruminal contents and is subjected to microbial fermentation. At intervals, parts of the reticulo-ruminal contents are brought back to the mouth, chewed and re-swallowed. The process is termed regurgitation or rumination.

Prehension is the act of bringing food inside the mouth. In bovines this is achieved by the joint effort of lips, tongue and teeth. Out of this, the lips of the cow are not very mobile and thereby not very useful in prehension. The tongue, on the other hand, is used extensively to encircle

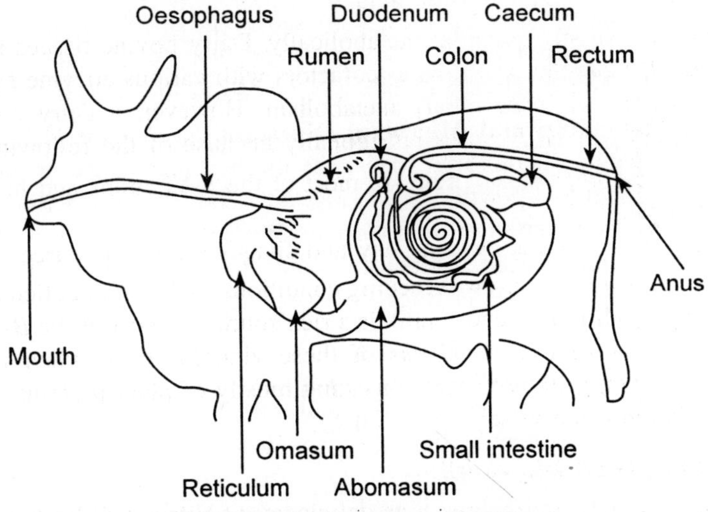

Fig. 6.7. Digestive tract of cattle

the forages and other feeds and bring them inside the mouth. To suit this, the tongue is very strong, rough surfaced, long and mobile.

The forages brought inside the mouth are held between the dental pad above and the incisor teeth below and snapped with a jerky movement of the head. The dental pad is an evolutionary adaptation in bovines to facilitate prehension. In place of upper incisor teeth, these animals have the dental pad, which is a thickened rough epithelial organ which helps in holding the grass tightly between it and the lower incisors.

Mastication. Bovines are in a hurry to swallow feed initially and they chew only slightly, enough to mix the feed with saliva and form a bolus for swallowing. They have no canines. They have six premolars and six molars on each side. Thus the permanent dental formula of cattle is

$$2(I\frac{0}{4}, C\frac{0}{0}, PM\frac{3}{3}, M\frac{3}{3}) = 32$$

The tables of premolars and molars form the surface for finely chewing and decimating the feed. But this complete mastication of feed occurs only after regurgitation.

Swallowing or Diglutition. By this process, chewed or partially chewed food material is transported from the mouth through the oesophagus to the rumen. The oesophagus is a muscular tube slightly more than one metre in length connecting throat and the rumen.

Regurgitation or Rumination. Apart from the initial swallowing ruminal contents are regurgitated back to the mouth through the oesophagus. The food material would have become softer by now due to soaking in the ruminal fluid. It is chewed thoroughly and mixed well with saliva and re-swallowed. Materials that are not initially decimated by microbial action is drawn back into the oesophagus through action of the inter-costal muscles, diaphragm, reticulum and rumen. Negative pressure created draws it back to the mouth. The liquid portion is re-swallowed immediately. The remaining bolus, called a cud, is re-masticated thoroughly and re-swallowed. This process of bringing back the feed from the reticulo-rumen to the mouth, chewing thoroughly and swallowing back is termed as regurgitation or rumination. Salivation plays a significant role in the case of ruminants. In addition to providing lubrication for swallowing, the bicarbonates and phosphates of sodium and potassium present in saliva provide buffering action in the rumen. Large quantity of saliva is secreted by the three pairs of salivary glands; the parotid, the mandibular and sublingual glands. These salivary glands are located at the rear of the mouth.

Rumen and Reticulum and Microbial Digestion. Esophagus leads food to the rumen which is a large receptacle divided into four areas by muscular bands or pillars. The four areas are the dorsal, ventral and two posterior sacs (Fig. 6.8). The feed inside the rumen is agitated in a rotary fashion by the contraction of the muscular bands known as pillars. This helps in mixing the feed well with the rumen fluid. Unlike the stomach in the non-ruminant, rumen is a major area for absorption of nutrients. The finger-like papillae lining the rumen greatly increase the surface area for absorption.

The reticulum is a smaller sac partially separated from rumen. This organ has a thick wall resembling honeycomb. The contents of rumen and reticulum mix freely. Reticulum is connected to omasum, the third compartment of stomach through an opening on the right side of reticulum. Rumino-reticulum is non-functional at the time of birth, but rumen fermentation begins within a few weeks. As the rumen develops, a mixed population of bacteria and protozoa becomes established. These organisms are derived from other animals by cross inoculation.

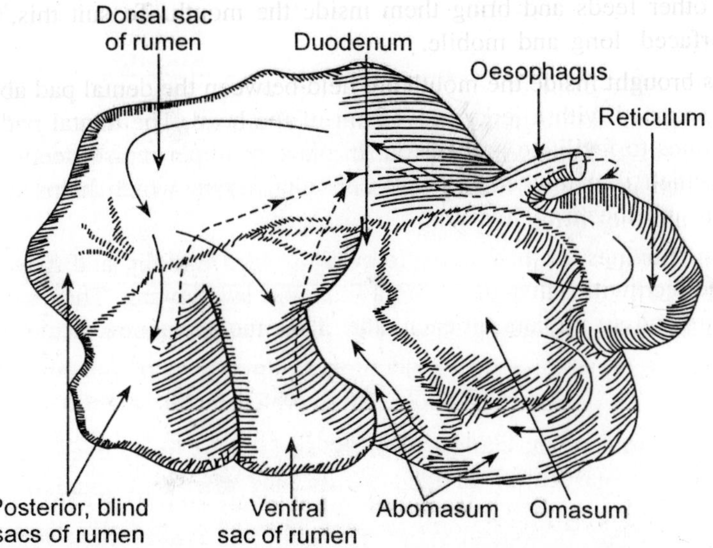

Fig. 6.8. Organs that comprise the stomach of the ruminants: Rumen, reticulam, omasum and abomasum (arrows indicate path of ingesta).

A wide diversity of organisms is found, the proportion of each type depending on the diet of the animal. It has been estimated that the rumen contents have about 10^{10} bacteria and 10^{6} protozoa per gram. There are a large number of species of bacteria. More than 100 species of protozoa have been found in the rumen.

The microorganisms of rumen exist in a state of symbiosis with the bovine animal as well as among themselves. The microbes are benefited by maintenance of constancy of conditions, steady supply of feed materials, constant waste removal and physical protection. The large animal is benefited by the bacterial digestion and transformation of food nutrients.

Microorganisms break-down complex carbohydrates like cellulose to volatile fatty acids. In fact no enzyme secreted by the large animals has the power to digest cellulose. But certain microorganisms possess the enzyme *cellulase, which* splits the cellulose molecule and makes it amenable to further digestion to short-chain **volatile fatty acids (VFA).** The VFA are then absorbed into the blood stream from the reticulo-rumen. The VFA forms the source of energy in ruminants. The digestion of dietary cellulose is the major function of rumen. Mammals do not produce *Cellulase* enzyme in their digestive juices. Ruminants are able to digest cellulose only because of the cellulolytic activity of the microorganisms harboured in the reticulo-rumen. This enables the ruminants to live on coarse fibrous foods, which would not otherwise support life.

In ruminants, most of the dietary carbohydrates are degraded to volatile fatty acids like *acetic acid, propionic acid, butyric acid* etc. The proportion of these vary depending on the diet. Under different types of feed the molecular proportion of various VFAs were estimated as given below (Table 6.3) by Balch and Rowland (1957) in cows.

Acetic acid predominates under most conditions, but substantial amounts of propionic and butyric acids are always formed. Diets rich in readily fermentable carbohydrates like sugars and starch lower the level of acetic acid and increase that of propionate. A probable explanation is the lowered rumen pH, which encourages the proliferation of propionate producing organisms. Lowering of proportion of acetic acid is known to affect milk fat percentage.

TABLE 6.3. The molecular proportion of various VFAs estimated under different types of feeding

Diet	*Rumen VFA (molecular proportion)*			
	Acetic	*Propionic*	*Butyric*	*Higher acids*
Hay (16 lb) + concentrates (24 lbs)	57.1	23.7	12.0	7.2
Free grazing	67.5	18.2	11.1	3.2
Silage	73.7	16.8	6.6	2.9

With protein, there is a synthetic as well as degradatory activity going on simultaneously in the rumen. On one hand protein in feed is broken down to peptides, amino acids, ammonia and amines. Simultaneously microorganisms use these protein-break-down components for building their own body protein (microbial protein). Eventually this microbial protein is available to the bovine when it is passed down the intestinal tract and digested. This is helpful to the ruminant in that poor quality protein or even non-protein-nitrogenous substances like urea are converted into bacterial protein of high biological value. Therefore, no matter what source of protein you feed to the ruminant, it generally will not suffer from lack of essential amino acids, provided the total protein requirement is met.

Protein digestion in the rumen has another aspect. If good quality proteins like milk proteins are fed fermentation by ruminal micro-organisms cause much wastage when fermentation progresses beyond the amino acids stage and ammonia is lost. The quality of such proteins of high biological value is lowered.

In the nursing calf, milk bypasses the reticulo-rumen through the *oesophageal groove* which is obviously, nature's provision against wasteful ruminal fermentation of milk. *Oesophageal groove* extends between oesophagus and omasum. When the calf is nursing, the sides of the groove erect upward by reflex action and form a conduit through which milk passes directly from oesophagus to omasum.

Lipids are needed only in small quantities in the diet of ruminants. It has been estimated that 2 per cent of fat in the total ration is sufficient for bovines. Almost all rations contain this much. Excess fat is known to cause ruminal disorders, reduce appetite and milk fat content. Much of the fat consumed by the ruminant is hydrogenated by bacterial action. Some portion is absorbed and metabolised by the microorganisms and finally becomes microbial lipid which is available to the large animal for digestion in the intestines. Some portion is hydrolysed to fatty acids and glycerol in the rumen. Glycerol is acted upon by microorganisms to become propionic acid, which is absorbed from the rumen. But the long chain fatty acids pass on to lower alimentary tract to undergo further digestion and absorption.

Ruminal microorganisms are also known to synthesize all B-complex vitamins and Vitamin *K*. Vitamin *B*-12 synthesis requires the presence of cobalt in rumen. The tissues of bovines synthesize Vitamin C. In the presence of ultra-violet irradiation from the sun, the skin of bovines synthesizes Vitamin *D* from sterols. Thus, all vitamins, except Vitamin *A* are not dietary essential for adult bovines in which rumen is functional.

Omasum. This organ has many layers of muscular leaves giving it the resemblance to a book. After fermentation, the ingesta pass through the reticulo-omasal orifice into the omasum.

We have not clearly understood the functions of omasum. With the help of the muscular leaves having horny papillae the ingesta is squeezed out and the water and fatty acids so squeezed out are absorbed.

Abomasum is the true stomach in the ruminant. From this organ onwards, the digestion-taking place in ruminants is similar to that in simple stomached non-ruminants. Table 6.3 summarizes the enzymatic digestive activities taking place in abomasum and intestines.

TABLE 6.3. The Digestive Enzymes and their Action

Orgam	*Enzyme*	*Nutrients digested*	*End products of digestion*
Abomasum	Rennin, Pepsin (Activated by hydrochloric acid secreted) Lipase	Milk proteins, Proteins, Fats, Oils and other lipids	Coagulates milk proteins Polypeptides Monoglycerides, glycerol and fatty acids.
Duodenum, Pancreas and bile from liver	Trypsinogen, Chymotrypsinogen, Carboxypeptidase	Protein and protein breakdown products	Peptides and aminoacids.
	Lipase	Fats, oils and other lipids	Monoglycerides, glycerol and fatty acids
	Amylase	Starch, glycogen, dextrin	Maltose, glucose
Small intestinal mucosa	Amino-peptidase	Protein break-down products	Peptides, amino acids
	Dipeptidase	Nucleoprotein	Nucleic acids. Nucleosides
	Nucleosidase	Nucleosides	Purines, phosphoric acid
	Maltase	Maltose	Glucose
	Lactase	Lactose	Glucose + Galactose
	Sucrase	Sucrose	Glucose + Fructose

Gastric juices containing hydrochloric acid, proteolytic enzymes pepsin and rennin and a lipolytic enzyme act upon the fermented food material that enters abomasum. In the presence of hydrochloric acid, pepsin breaks the peptide linkages of proteins to short chains of amino acids known as peptides.

In the young calf, rennin coagulates milk proteins rendering them amenable to further enzymatic digestion. The acidity produced by hydrochloric acid controls the opening of pylorus, the opening from abomasum to duodenum. Until a particular level of acidity is reached, the pylorus remains closed, keeping the abomasal content subject to the action of gastric juice. The lipase acts upon neutral fats and other lipids producing monoglycerides, fatty acids and glycerol.

Duodenum, the first loop of small intestine is an important area as far as digestion is concerned because the pancreatic duct emptying the pancreatic juice and bile-duct emptying bile join here. Besides, the intestinal mucosa also produces many digestive enzymes. Pancreatic juice contains the proteolytic enzymes trypsinogen, and chymotrypsinogen and carboxy peptidase. These are activated in the duodenum to split proteins to peptides and amino acids.

Pancreatic juice also contains amylase, which hydrolyses starch, dextrins and glycogen to maltose and glucose. It also has a lipase, which accelerates the hydrolysis of fats to fatty acids and glycerol.

Bile from the liver and that stored in the gall-bladder plays many essential roles in digestion like activation of pancreatic lipase, emulsification of fats, helping absorption of fatty acids and acting as a buffering agent and a reservoir of alkali for maintaining the proper alkaline pH for the functioning of intestinal enzymes.

Small intestinal mucosa secretes two proteolytic enzymes, *viz.*, *aminopeptidase* and *dipeptidase*. They act upon protein breakdown products ultimately converting them to amino acids. Another set of enzymes produced by small intestinal mucosa, *nucleotidase* and *nucleosidase* break down nucleoproteins to the components; purines, phosphoric acid and the sugar. Intestinal enzymes like *maltase*, *sucrase* and *lactase* act upon the respective sugars as indicated in Table 6.3.

Absorption. The movement of the end products of digestion from the digestive tract into the blood and/or lymph system is known as absorption. The process of osmosis through the semipermeable membranes lining the alimentary canal accomplishes it.

In the ruminants, the end products of carbohydrate digestion are volatile fatty acids. These fatty acids are absorbed from the rumen, reticulum omasum and large intestine. Maximum absorption of VFA takes place in the rumen itself.

Glucose from carbohydrate digestion in the intestines and amino acids and peptides from protein digestion are absorbed directly into the blood stream by the villi of small intestinal wall.

The absorption of lipids is more complicated. Lower fatty acids and choline, which are soluble in water, are absorbed by the intestinal villi. Insoluble fatty acids and amino glycerides are made into an emulsion with the help of bile salts. These combinations form micelles, which penetrate the intestinal mucous membrane and enter the lymphatic system. While doing so the epithelial cells of intestine re-synthesize triglycerides from monoglycerides and fatty acids having more than 14 carbon atoms. These triglycerides join with small amounts of cholesterol and phospholipids to form colloidal complexes known as *chylomicrons*. These mix with lymph to form chyle and enter the systemic circulation through the thoracic duct. Most of the absorptive activity takes place in the reticulo-rumen and in the lower part of small intestine.

Large intestine absorbs water making the dung drier. Digestive enzymes continue to act in the large intestine. Some bacterial enzymes, especially putrefactive type act upon the ingesta giving it the characteristic foul smell.

Defecation. Many excretory products of intermediary metabolism are voided into the large intestine through its wall. These along with undigested feed material, sloughed cells and undigested residue of microorganisms are voided out through rectum and anus by the process known as defecation.

6.8. FEEDING STANDARDS

Conventionally, feeding standard is a table showing what is believed to be the daily requirement of a specifically defined animal for various nutrients. Information with respect to the nutrient requirements of various classes of livestock available from numerous feeding experiments is collected, classified, tabulated and revised from time to time. Such information has come to be known as feeding standards. This undesirable feature of feeding standards, that they are formulated by assembling scattered data from studies conducted at different experimental stations under different conditions, to a large extent, is unavoidable.

With the development of the 'hay values' originated in Germany as early as 1810, the concept of feeding standards came into existence. Greuven in 1859 developed a feeding system based on the proximate analysis of feedstuffs. Another German scientist, Wolff and later Lehmann from the same country developed Wolff's standards and later Wolff-Lehmann Standards based on digestible nutrients.

This line was picked up and further developed in the United States by scientists like Atwater, Henry and Morrison. These have finally found expression as Morrison Standards and have been revised from time to time.

Atwater proposed feeding standards based on the 'Available fuel values' calculated by applying Rubner's factors to digested nutrients. Feeding standards based on net energy values were first proposed by Kellner in Germany, the NE being expressed as starch equivalents. On the same lines, Armsby from USA proposed a feeding standard where net energy values were stated in therms. Somehow these were not widely accepted.

Nutrient requirements of animals are still an active field and therefore, there is always the need for revising the standards periodically. Recognising this need the Committee of Animal Nutritionists of the U.S. National Research Council started a long-term project of preparing and keeping up-to-date by periodic revision a set of modern feeding standards. These are popularly known as the *NRC Standards.* It is believed that the NRC reports representing the pooled judgment of experts should be considered as the most authoritative statement of nutritional needs of farm animals. The British Agricultural Research Council has prepared the BARC standards on similar lines in U.K. Similar standards are available in Scandinavian and other European countries.

Sen and Ray have come out with feeding standards for bovines in India, which has been published by the Indian Council of Agriculture Research as a bulletin (No. 25). These are based on limited experimental data available in the country and average values of lower and higher recommendations of Morrisson Standards where such information is not forthcoming.

6.9. FEEDING THE MILKING COWS

Feed costs constitute about 60 per cent of the total cost of producing milk. Therefore, in today's economy, it becomes imperative to get the maximum out of the feed fed. The old adage,—'you cannot starve a profit out of a good cow and you cannot feed a profit out of a poor cow'—is still a very sound principle. It explains the relationship between feeding, level of production and profit. We can choose different combinations of feeds and fodders, depending on the circumstances, to meet the nutrient requirement of animals. Only thing is that, there should be a proper balance of nutrients and they should be wholesome and palatable.

6.9.1. Calculating Nutrients Requirements

The first step in computing balanced daily rations for milking cows is to determine the amount of each critical nutrient required by the animal, by seeing the respective tables in the appendix. Adding to that for maintenance, an additional set of requirements for milk produced and its fat percentage arrives at the total requirement. Furthermore, first and second lactation cows will be still growing and need additional nutrients for the same. Cows also need extra nutrients during the last quarter of gestation for the growth of the foetus and recouping body reserves for the next lactation.

6.9.2. Formulating a Ration

Next step is to compute a ration, which supplies the desired nutrients and be eaten in adequate amounts to meet all the above nutrient requirements. Forages, being the cheapest source of dairy cattle feeds, should form the basis of their feeding. In general, the dry matter from roughages should not exceed 2 per cent of the cow's live weight nor it should be less than 1 per cent.

Feeding standards give the energy requirements of dairy cows either as TDN or in terms of NE lact. (net energy for lactation). TDN or NE lactation. values of most of the common dairy feeds are also available. If a calculator is available and precise calculation required, certain mathematical formulas can be applied. Otherwise, certain thumb rules along with trial and error approach may suffice.

However well computed, the nutrients supplied through a ration can only be an approximation of the requirements. How far a ration can vary from the requirement ? It is better that the ration supplies at least the minimum of the requirement. However, if it is below that mark by 2-3 per cent it can be ignored. Nutrients in excess of requirement above a certain limit can also be harmful. For example, if energy supply exceeds more than 5 per cent it is undesirable as the energy utilization capacity of an animal is limited. With respect to protein, excess of 5 per cent is a good insurance against feed low in protein. Excess of carotene (Provitamin A) is often unavoidable. Since it is harmless, rations thus high in carotene are accepted as a balanced diet.

Many common feeds have calcium or phosphorus in excess of requirement and therefore excesses of these minerals are sometimes impracticable to avoid. In such cases mineral supplements should be curtailed and effort should be taken to keep the Ca: P ratio between 1 : 1 and 2 : 1.

Example. For the sake of illustration, the example of a 400 kg cow in first lactation producing 10 kg milk daily with 4 per cent fat and 7 months pregnant is worked out below :

Requirements

Requirement	*DP (kg)*	*NE lact (MCal)*	*Ca (g)*	*P (g)*	*Carotene (mg)*
Maintenance: 400 kg cow + pregnancy	0.355	9.7	23	18	76
Growth (20 per cent of above)	0.071	1.9	5	4	15
Milk production	0.510	7.4	27	20	—
Total	0.935	19.0	55	42	91

I. A 400 kg cow should be fed somewhere between 4 and 8 kg of dry-matter through roughage daily. In this instance, let us assume that the roughage consisted of maize silage and lucerne hay as indicated :

	DM (kg)	*DP (kg)*	*NE lac. (M. cal)*	*Ca (g)*	*P (g)*	*Carotene (mg)*
Maize silage 15 kg	5.0	0.17	7.05	12	9	66
Lucerne hay 2 kg	1.8	0.14	2.28	26	4	130
Total	6.8	0.31	9.33	38	13	196
Balance needed from concentrates		0.625	9.67	17	29	—

II. The next step is to determine what type of concentrate and how much to be fed. A concentrate mixture having approximately 1.7 $NE_{lactation}$ may be used. Feeding about 5.5 kg of the concentrate would meet the energy needs. By dividing the DP to be supplied through concentrates by the quantity to be fed (0.625/5.5) it can be seen that the concentrate mixture should have about 11.5 per cent DP.

III. A tentative concentrate mixture as follows may be made to contain about 1.7 NE lact. using the Pearson square method.

Feed	*NE in Feed*	*Required NE*	*Proportion of feeds*	*Proportion of feeds (%)*
Ricebran	1.4		0.2	40
Barley, Maize, GNC in equal parts)	1.9		0.3	60
		1.7		(20 : 20 : 20)
Total			0.5	100

GNC = groundnut oil cake.

The above concentrate ration will give—

Feed	*Proportion*	*DP*	*NE lact.*	*Ca*	*P*
Rice bran	40	1.68	0.56	28	636
Barley	20	1.74	0.38	14	80
Maize	20	1.30	0.42	6	54
GNC	20	8.92	0.36	40	130
Total	100	13.64	1.72	88	900

IV. It can be seen that the above ration is high in protein and low in calcium. The excess in protein can be corrected by replacing high protein GNC with low protein containing coconut cake (CNC). The percentage of CNC to replace GNC can be arrived at as below :

The percentage of DP in the tentative concentrate mixture	13.64
The percent of DP needed	11.50
Excess	2.14 %

Decrease of DP brought about by replacing 1 kg GNC with 1 kg CNC

(0.446 – 0.166) = 0.28

Replacement of GNC with CNC needed to bring down the DP per cent to 11.5

2.14/0.28 = 7.7

Considering the need to provide for some extra protein, this is rounded to 7 percent. Thus the final concentrate mixture will be :

Feed	*Proportion*	*DP (kg)*	*NE lact. (Mcal)*	*Ca (g)*	*P (g)*
Ricebran	40	1.68	0.56	28	636
Maize	20	1.30	0.42	06	54
Barley	20	1.74	0.38	14	80
GNC	13	5.30	0.23	26	85
CNC	7	1.16	0.13	14	40
Total	100	11.68	1.72	88	895

V. The ultimate ration for the cow would be :

Item	*Qty. (kg)*	*DM (kg)*	*DP (kg)*	*NE lact. (Mcal)*	*Ca (g)*	*P (g)*	*Carotene (mg)*
Maize Fodder	15	5.0	0.17	7.05	12	9	66
Lucerne hay	2	1.8	0.14	2.28	26	4	130
Concentrate	5.5	5.0	0.64	9.35	5	49	—
Total		11.8	0.95	18.68	43	62	196

The above ration is still slightly low in calcium and excess in phosphorus. This can be balanced by addition of a calcium supplement in the concentrate mixture. Under practical feeding conditions it is not possible to select much among the roughages or to vary the ingredients in a concentrate mix. A manager can follow a carefully laid out cropping plan to ensure year round supply of a mixture of leguminous and non leguminous forages. For concentrates, generally one has to depend on commercial feed-mixture and therefore it is difficult to control the composition. Depending on the quantity and quality of roughages the concentrate mixture may be fed to the cows on the basis of certain **thumb rules**. One feeding schedule based on **thumb rules** is given below :

FEEDING SCHEDULE

Method I

	Quantity of green grass to be given (kg) for animals weighing			*Concentrate*
	250 kg	*300 kg*	*350 kg*	*(kg)*
I. When grass is available				
Dry cows	25	30	35	Nil*
Milch cows	25	30	35	1.0 kg for every 2.5 kg of milk of average 4% fat
II. When silage is available				
Dry cows	15	20	25	1.0
Milch cows	15	20	25	1.0 kg for every 2.5 kg of milk of average 4% fat

Note. 1. Pregnant cows should be fed an additional quantity of 1.5 kg of concentrates from 7th month of gestation.

2. Growth allowance of 1 kg concentrates may be given to young cows in 1st and 2nd lactation.
3. High yielding cows (more than 8 kg of milk) may be given extra allowance of concentrates at the rate of 0.5 to 1 kg till the peak yield is attained.
4. If additional fodder is available for feeding. 500 g of concentrate can be reduced for every additional 10 kg green grass fed.

* In the case of dry cows, allowance of upto 1 kg concentrates may be given if the condition of the cow is poor/ the fodder is of inferior quality.

The above ration schedule is prescribed on the following assumptions :

1. Green grass contains 25 per cent DM and silage 30 per cent.
2. Green grass has an average of 0.7 kg of DCP and 12.0 kg of TDN per 100 kg.
3. Silage possesses 0.8 kg of DCP and 13.0 kg of TDN per 100 kg.

4. The concentrate mixture contains a minimum of 20 per cent CP, and 65 per cent TDN.
5. Wastage of Fodder up to 10 per cent is anticipated.
6. DM intake is 2.5-3.5 per cent of body weight for various categories.

Method II

In this method, allowance is given to milk fat and prevailing prices of milk and concentrates. The first step is to calculate the total milk fat production during the previous month. This may be done as below :

$$\text{Total fat produced} = \frac{\text{Monthly average milk fat percentage} \times \text{total milk produced during the month}}{100}$$

Monthly milk fat production may be divided by 4 to arrive at the kilograms of concentrates to be fed to the cow daily. For example, if the cow produces 300 kg of 4 per cent milk in a month, the concentrate to be fed daily will be

$$\text{Total milk fat produced during the month} = \frac{300 \times 4}{100} = 12 \text{ kg}$$

$$\text{Daily concentrate allowance} = \frac{12}{4} = 3 \text{ kg}$$

When milk prices are high or when concentrate prices are low in comparison to fodder it is better to use 3 as the denominator. The daily concentrate allowance in such circumstances would be 12/3 = 4 kg. When opposite conditions prevail 5 may be used as the divisor. Then the daily concentrate allowance would be 12/5 = 2.4 kg.

Method III

Many experiments have proved that milk production of 5-7 litres per day can be sustained by feeding high quality roughages alone. Therefore, by employing the previous method we will be over feeding low producers and underfeeding high producers to some extent. To overcome this, a new thumb rule formula as given below may be applied.

Substract 6 from the monthly milk fat production and divide that figure by 2. In the previous example (300 kg milk @ 4 percent fat/month)

$$\frac{12 \text{ kg fat} - 6}{2} = 3 \text{ kg concentrate.}$$

On the other hand for a cow producing 7.5 kg milk per day with 4 per cent fat.

$$\text{Monthly milk fat} = \frac{7.5 \times 30 \times 4}{100} = 9 \text{ kg}$$

$$\text{Concentrate allowance} = \frac{9 - 6}{2} = 1.5 \text{ kg}$$

Method IV

Economic considerations indicate that a farmer can profitably feed more concentrates to his cows when milk prices are high. Higher levels of concentrates are also warranted when the concentrate prices are low. Thus, a thumb rule for economic feeding of concentrates to milking cows should be based on the following considerations :

1. Quantity of milk produced by the cow
2. Fat percentage of milk
3. Market price of milk
4. Market price of concentrates
5. Quality of roughage being fed and
6. Decrease in roughage intake when large quantity of concentrate is fed.

Production of total milk-fat by the cow during the previous month is taken as a basis of feeding as it will account for (*i*) milk production, (*ii*) milk fat percentage, and (*iii*) fluctuations in day to day milk production due to various reasons. Total milk fat is calculated as below :

$$\text{Total milk fat} = \frac{\text{Total milk produced during the month} \times \text{Average milk fat percentage}}{100}$$

Davies and his co-workers from Maryland University have evolved a concentrate-feeding schedule, which takes into account, all the factors mentioned earlier. The same method has been modified to suit the conditions in India. The use of this modified/dairy cow concentrate feeding thumb rule is described below with the help of an example. With total milk production of the cow during the previous month being 300 kg and average fat percentage 4.0, total fat produced was estimated to be 12.0 kg. Current market price of milk (4 per cent fat) is ₹ 18/kg and that of concentrate mixture is ₹ 9.50/kg.

TABLE 6.4. Price ranges of milk and concentrates for deciding the optimum level of concentrate-feeding

Price of concentrate (rupees)	*Price of milk (rupees)*																
	12	**12.5**	**13**	**13.5**	**14**	**14.5**	**15**	**15.5**	**16**	**16.5**	**17**	**17.5**	**18**	**18.5**	**19**	**19.5**	**20**
4.00	A	A	B	B	C	C	D	D	E	E	E	E	E	E	E	E	E
4.50		A	B	B	C	C	C	D	D	E	E	E	E	E	E	E	E
5.00		A	A	B	B	C	C	C	D	D	D	E	E	E	E	E	E
5.50			A	A	B	B	C	C	C	C	D	D	D	E	E	E	E
6.00				A	A	B	B	C	C	C	C	D	D	D	E	E	E
6.50					A	B	B	B	C	C	C	C	D	D	D	E	E
7.00					A	A	A	B	B	C	C	C	C	C	D	D	D
7.50						A	A	B	B	B	C	C	C	C	C	D	D
8.00						A	A	A	B	B	B	C	C	C	C	C	C
8.50						A	A	A	A	B	B	B	C	C	C	C	C
9.00							A	A	A	B	B	B	B	C	C	C	C
9.50								A	A	A	B	B	B	B	C	C	C
10.00								A	A	A	A	B	B	B	B	C	C
10.50									A	A	A	A	B	B	B	B	C
11.00										A	A	A	A	B	B	B	B
11.50										A	A	A	A	A	B	B	B
12.00											A	A	A	A	B	B	B
12.50											A	A	A	A	A	B	B
13.00												A	A	A	A	A	B

Check Table 6.4 to find out the column to use in Table 6.5. Find the feed price of Rs 9.50 in the left hand column in Table 6.4 and move across the chart to the right until the column labeled Rs 18 or nearest to it is reached. The letter in the box where these two intersect is the one to be used in Table 6.5, it being B in this example.

TABLE 6.5. Optimum Level of Concentrate Feeding

A	*B*	*C*	*D*	*E*	*I*	*II*	*III*
	(Total monthly milk fat)				*(Straw to fill & little grass)*	*(Green grass) 25-30 kg & straw to fill*	*(Excellent green fodder to capacity)*
						concentrate (kg)	
6.0	5.5	5.0	4.5	4.0	0.50	0.00	0
6.5	6.0	5.5	5.0	4.5	1.00	0.00	0
7.5	7.0	6.5	6.0	5.5	1.50	0.00	0
8.5	8.0	7.0	6.5	6.0	1.75	0.00	0
9.0	8.5	8.0	7.5	7.0	2.00	0.00	0
10.0	9.0	8.5	8.0	7.5	2.30	0.50	0
11.0	10.0	9.0	8.5	8.0	2.75	1.00	0
12.0	11.0	10.5	9.5	9.0	3.00	1.25	0
13.5	12.0	11.5	10.5	9.5	3.25	2.00	0
15.0	13.0	12.5	11.5	10.5	3.75	2.50	0.50
16.0	15.0	13.0	12.5	11.5	4.00	2.75	1.00
17.0	15.5	14.0	13.0	12.0	4.05	3.00	1.25
19.0	17.0	15.5	14.5	13.0	5.00	3.50	1.80
20.0	18.0	17.0	15.5	14.5	5.50	4.00	2.25
21.0	19.0	18.0	16.5	15.5	6.00	4.50	2.75
22.0	20.0	18.5	17.5	16.0	6.25	5.00	3.00
23.0	21.5	19.5	18.5	17.0	6.75	5.50	3.75
25.0	22.5	20.5	19.0	17.5	7.25	6.00	4.00
26.5	24.0	21.5	20.0	18.5	7.75	6.50	4.50
27.0	25.0	22.5	21.0	19.5	8.25	7.00	5.00
28.5	26.0	24.0	22.0	20.5	8.50	7.25	5.50
29.5	27.0	25.0	22.5	21.5	9.00	7.75	6.00
31.0	28.0	26.0	24.0	22.0	9.05	8.25	6.50
32.0	29.0	27.0	25.0	22.5	10.50	8.50	7.00
33.5	30.5	28.0	26.0	23.0	11.00	9.00	7.25
35.0	31.0	29.0	27.0	25.0	11.50	10.00	7.75
36.0	32.0	30.0	28.0	26.0	11.75	10.50	8.25
34.0	31.0	28.5	27.0	12.25	11.50	9.00	
35.0	32.0	29.5	27.5	12.75	11.75	9.50	
36.0	33.0	30.5	28.0	13.50	12.75	10.00	
	34.0	31.5	29.0	14.00	13.50	10.75	
	35.0	32.0	30.0	14.50	14.00	11.25	
		33.5	31.0	15.00	14.50	11.75	
		34.5	32.0	15.75	15.50	12.25	
		35.5	32.5	16.25	15.75	12.75	
			34.0	16.75	16.00	13.50	
			35.0	18.00	17.50	14.50	

Locate the prevailing feed price in the left hand column and move across to the right until the column labeled the current market price of milk. The letter in the box where these two intersect is the one to be used in Table 6.4.

To find out the daily concentrate ration for the cow producing 12.00 kg milk fat in a month, move down the column B to the figure 12.00 or the nearest. The amount of concentrate mixture to be fed to the cow is given across this in column I. II. or III, depending on the quality of roughages the cow is receiving.

When major portion of the roughage ration comes from straw with little or no green grass, column I is used. When moderately good amount of green grass (25-30 kg) and straw to fill are fed, column II may be used. With excellent green fodder or hay inclusive of 30 per cent or more of leguminous content available to the capacity, column III may be used.

In the case of the cow in question, in the first case 3 kg concentrate is fed, in the second case 1.25 kg and in the third none. This system provides a simple but economically rational system for deciding the daily concentrate ration for dairy cows.

6.10. COW FEEDING CYCLE

6.10.1 Early Lactation

The recently calved high producing cow is unable to eat enough feed to support her milk production. This means that she should have enough reserves of stored nutrients to be drawn to tide over the period of heavy demand in early lactation, during which period the cow loses weight.

6.10.2. Challenge Feeding

This starts two weeks before the expected date of calving. Cows with high milk production potential are fed increasing quantity of concentrates to 'challenge' them to produce at the maximum levels. This "challenge feeding" will condition her digestive system for the increased amount of concentrates of early lactation and provide enough nutrients to initiate lactation on a higher plane. This effect has been found to persist through the lactation curve (see figure 6.9).

For challenge feeding, two weeks before the expected date of calving start feeding 500 g of concentrate mixture. Daily increase this amount by 300-400 g until the cow is consuming 500-1000 g concentrate for every 100 kg body weight (Table 6.6).

After calving, start increasing the concentrate allowance by 500 g a day in the first 2 weeks of lactation until the cow is eating concentrates on an almost free choice level. This is continued until the cow achieves peak yield somewhere in the second month of lactation. After this the milk yield is tested and the concentrate allowance is fixed accordingly.

6.10.3. Feeding the Cow in Mid and Late Lactation

The nutrient deficit period of early lactation is followed by a relatively stable period during which the cow can consume enough feed to meet the various demands for nutrients and the body weight of the cow remains more or less stable. During this period the cow may be fed a well-balanced ration of good quality fodder and concentrates according to milk yield and fat percentage of milk.

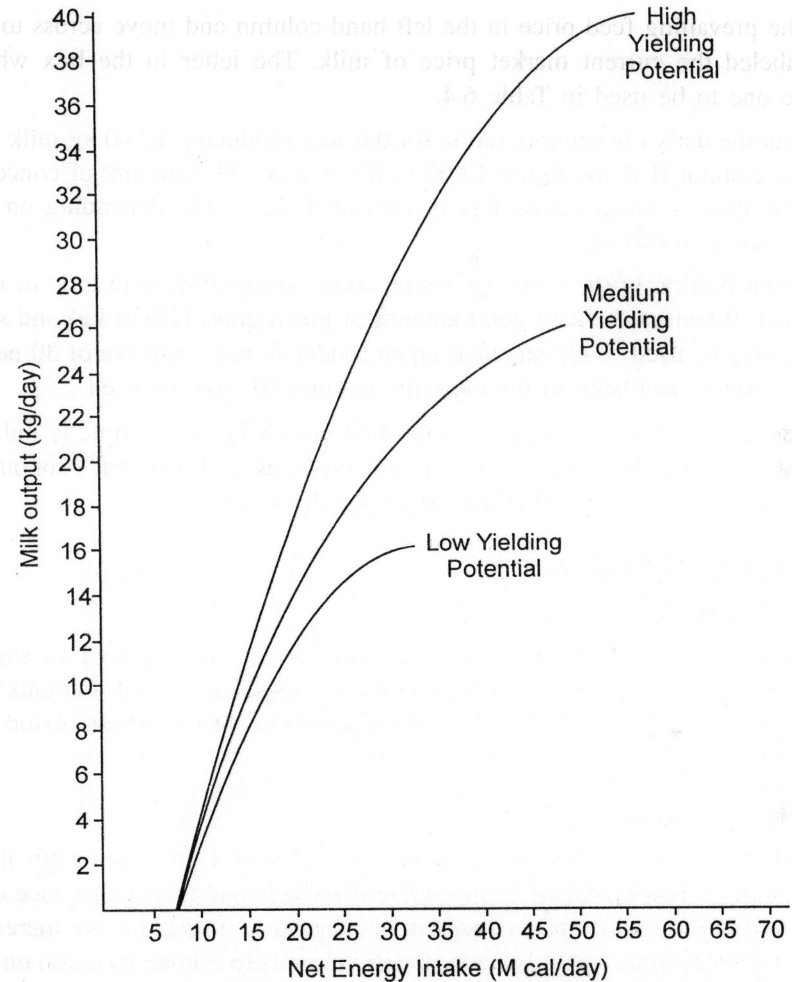

Fig. 6.9. Estimated milk production of 540 kg cows with high medium and low potential at varying estimated net energy intake. (*Courtesy.* D.W. Dean).

TABLE 6.6. Challenge Feeding Schedule

Period	*Concentrate allowance*
1. Last 2 weeks before calving	Starting from 500 g, increase 300-400 g daily until the cow is eating 500-1000 g per 100 kg body weight
2. First 2 weeks, of lactation	Increase 500 g per day to free choice level
3. Second week to peak yield (test day)	Free choice
4. From test day onwards	According to production as per thumb rules, *e.g.*, 1 kg for every 2½ kg milk produced
5. Remaining lactation	Concentrate adjusted to monthly test of milk production
6. All periods	Green fodder and dry fodder given adequately

During late lactation, intake ability of the cow exceeds nutrient needs. This is the time when the cow starts needing extra allowances for the growing foetus. This is also the period when the cow can readily replenish the already depleted body reserves and gain weight very fast. Moreover, it is well established that the feed efficiency for weight gain is maximum during late lactation.

From 7½ months to 10 months of lactation, cows may be fed 1-2 kg concentrate feed in addition to their nutrient requirement for maintenance and milk production to replenish the condition they lost in early lactation.

6.10.4. Feeding the Dry Cow

More than half of the calf's foetal growth takes place during the last quarter of pregnancy—the dry period. Feeding of the cow during this period is often neglected because the cow is not returning anything in terms of milk. Due to this reason, good fodder and concentrates are generally withheld from the dry cow. This will have adverse consequences on her future production, health as well as the development of the calf. A dry cow should be fed adequate quantity of a well-balanced ration for the following purposes :

1. Maintenance of the cow.
2. Forming sufficient reserves of nutrients in the body of the cow for the ensuing lactation.
3. Growth of the foetus and
4. Production of colostrum with high nutritive value.

The condition of the cow can be controlled either by feeding extra nutrients during the latter part of lactation or during the dry period. Concentrate feeding during the dry period is partially dependent on the body condition of the cow at the time of drying off. Cows in poor condition at drying off should be fed additional concentrates during the dry period so that they pick up body condition by the time they calve. A gain in body weight of 20-25 kg for smaller cows and 30-40 kg for larger cows is recommended.

The nutrient needs for foetal development has been recognised and additional nutrients are incorporated in the ration during the last quarter of pregnancy. But a cow produces only 10-15 kg of dry matter in the foetus, foetal membranes and fluids. This is equivalent to 50-100 kg of milk production. Thus it is the requirement for the replenishment of body reserves that constitute the major additional requirement for the dry cow, not foetal development. The foetal growth does produce a steady drain on the mother's milk production, and therefore adds to the nutrient requirements of the cow. But, this requirement is not as great in magnitude as the one needed for building up body reserves.

The cow can be made to regain condition either during late lactation or during the dry period. It has been found that the former is better because of the following reasons.

1. The efficiency of feed utilization for weight gain is better during late lactation than dry period.
2. Many of the problems resulting in ketosis, displaced abomasums, fatty livers, retained placentas, prolapses etc. can be minimised by resorting to an all roughage feeding during the dry period. The roughage should be tough enough to stimulate and restore rumen muscle tone.

Only if the cow is in poor condition at drying off, she should be given extra concentrates to regain condition.

6.11. FEEDING HIGH PRODUCING DAIRY COWS

Physical fill. High producing dairy cows should eat a large volume of nutrients daily to sustain the milk production at that level. This is simply not possible with bulky forages alone due to physical limitation of volume intake. The space in the rumen is limited and once filled; it cannot be reused unless the previous material passes into the lower digestive tract. The obvious answer to this problem is to increase the proportion of concentrates in the ration.

Problems when high levels of concentrates are fed. High levels of concentrates change the microbial and chemical atmosphere in the rumen, often causing dysfunction. In feeding high levels of concentrates to supply inputs for heavy milk output, all care should be taken to maintain an optimum fermentation in the rumen. Rumen fermentation can be influenced in .the feed manger by controlling (*a*) composition on the ration, (*b*) ratio of ingredients in the ration, (*c*) quantity of feed supplied, (*d*) frequency of feeding, and (*e*) physical form of the feed.

Composition and ratio of ingredients in the ration. For a higher producer, the ration should be so computed that while satisfying the nutrient requirement within the capacity to eat dry matter, it should incorporate the maximum possible proportion of forages. This implies that the forage fed should be of high quality. In fact, the desirability of feeding only high quality roughage to high producers cannot be over-emphasized. Increasing concentrates cannot rectify the deficiencies of low quality roughage. In fact, the scope for increasing concentrates is very limited in the case of high yielders.

Role of Crude Fibre in the Ruminant Ration. Crude fibre is very important in the ration of milking cows because it is well known that rumen fermentation leading to acetic acid production is dependent on the percentage of cellulose in the ration. Cows need acetic acid for maintaining normal milk fat percentage as well as total milk production. Ruminant ration should contain a minimum of 20-25 per cent crude fibre. Higher percentages of grain in the ration reduce cellulose digestibility and cause disturbances like depressed milk fat, depressed milk production and in extreme cases damage to the rumen wall, development of acidosis and death. The quality of cellulose in the ration also has its influence in this situation.

Importance of Rumen Protozoa. Rumen protozoa has an important role in moderating the ill-effects of high levels of grain feeding. They take away a large quantity of starch out of circulation from rumen fluid. This prevents an explosive growth of starch splitting bacteria which would cause sudden release of large quantities of organic acids resulting in dangerously low levels of pH. Later on, these starch granules are available to the ruminants at a slower rate as and when the protozoa die and get digested in the lower alimentary canal. Protozoa are highly sensitive to low pH. A large quantity of grains fed in a single dose, will overwhelm the protozoa and the consequent low pH will completely annihilate them. Further lowering of pH can have bad consequences.

6.11.1. Frequency of Feeding

Ruminal Fermentation Curve. Fermentative growth of micro-organisms has five distinct phases, *i.e.*, (*i*) lag phase, (*ii*) acceleration phase, (*iii*) exponential phase, (*iv*) retardation phase, and (*v*) inactive phase. The rumen fermentation resulting from a single feeding is similar to the one in a fermentation vat for alcohol production by batch method.

If ratio of concentrates to fibre is wide in a heavy ration, the rumen pH will drop to very low levels during the acceleration phase adversely affecting the milk fat percentage and total milk production. Acceleration phase is followed by the exponential phase, the duration of which is dependent upon the quantity of grains in the ration. In low producers fed small quantities of concentrates, this phase is short-lived and therefore of little consequence. But when high producers are fed a heavy concentrate ration in one single dose or two, serious consequences can result from a prolonged exponential phase.

Dividing the daily ration into 3 or 4 parts and feeding them in so many installments has been found to be useful in overcoming this problem. By feeding twice in a day, the acceleration and retardation phases are reduced, with 4 times a day they are almost eliminated. Under such a feeding schedule the rumen pH is not unduly lowered. This results in greater digestibility and better utilization of protein apart from preventing explosive release of acids.

Role of Saliva. The sodium and potassium salts of saliva play an important role in buffering ruminal acidity. A high concentrate ration induces less amount of saliva flow compared to roughages. But when feed is given in 4 or 5 installments the proportion of Na and K salts in rumen returns to normal.

Ration Density. Feeding high producers are an act of tight rope walking. One should pack all the nutrients needed for heavy production in limited volume available in the cow's rumen. This implies that the ration should be compact enough. On the other hand, economic considerations as well as need for minimum fibre level in the feed necessitates including roughages to the maximum limit without sacrificing nutritional requirements. These two opposing needs are to be compromised at some point. At what point? One attempt in this line is the concept of ration density × digestibility measure. Experiments have shown that maintaining a density × digestibility value of 35 for the whole ration is desirable for proper rumen fermentation (Table 6.7). In practice one may include feedstuffs having a value above 35 with those having below 35 to strike at an optimum ration.

TABLE 6.7. Showing Density × Digestibility Values of Some Cattle Feeds

Feed item	*Density (wt/ml, g) (D)*	*Digestibility (dg)*	*D × dg*
Cotton seed hulls	0.33	38	12
Maize silage	0.41	65	27
Sorghum silage	0.43	62	27
Concentrates			
Sorghum	0.54	80	43
Maize grain	0.61	73	44
Linseed cake	0.64	70	45
Wheat middlings	0.63	77	48
Oats	0.74	68	50
Barley	0.70	77	54
Soyabean (Whole)	0.74	87	64
Wheat	0.87	77	67

How to arrive at the value for D × dg of 35 ? An example is worked out below :

Sorghum silage	66 per cent	66 × 27/100 = 17.82
Barley	21 percent	21 × 54/100 = 11.34
Linseed cake	13 per cent	13 × 45/100 = 5.85

Overall D × dg value of the ration = 35.01

Mixing Concentrates and Roughages. Traditionally, concentrates are fed at the time of milking. Roughages are offered either before or after milking. In high producers, when concentrates are fed in heavy doses at milking time, the appetite of the cows will be reduced temporarily and they may not eat roughages for some time. Consequently, there will be four different fermentations, two primarily due to concentrates and two primarily of roughages.

The feeding of concentrates separately from roughages during a four-time feeding schedule reduces acetic acid production and increases propionic acid. It has been observed that feeding grain on top of silage increased the fat percentage and milk production. Feeding concentrates either on top of forages or mixed with forages has been found to favour optimum rumen fermentation. This has led to the concept of complete feeds, which incorporates both roughages and concentrates.

6.1.1.2. *Complete Feeding*

In order to simplify feeding of dairy cows complete-diet systems have been introduced. Complete diet is an intimate mixture of concentrates and roughages in a desired proportion, processed in such a way as to preclude selective eating. It forms the sole source of food for the cow. It reduces labour requirements and keeps a tighter control on the cow's nutrition. It also facilitates the application of least-cost methods in ration formulation.

Feeding of complete diets *ad libitum* to dairy cows has been found to be advantageous in that it increases feed intake, preserves milk quality and results in better utilization of nitrogen. These are in addition to the most obvious advantage of preventing acidosis from over-eating of concentrates by high producers.

The complete diet feeding system is radically different from conventional feeding methods in that there is no individual approach in feeding cows. Group feeding is practised. There are also less changes in diet formulation according to milk yield. This has come as a result of the experimental and practical feeding observations that yield and efficiency are not improved by individual rationing compared to flat rate feeding of cows grouped according to milk yield or stage of lactation.

A doubt persisted whether the same concentrate roughage ration can be applied to formulate complete-diets for feeding in different stages of the lactation curve. Some experimental finding by Everson (1974) and Everson *et al.* (1976) indicate that over 2 years of feeding, no advantage in milk yield or fat percentage was observed due to varying concentrate roughage ratio at different stages of lactation over a constant ratio of 60 : 40. Thus a complete diet system based on a single concentrate forage ratio for the whole lactation can be justified.

Some studies have also shown that the use of complete diets has resulted in more efficient use of protein and NPN and higher milk yield. This might be due to the effect of better utilization of rumen ammonia due to increased frequency of feeding in an *ad libitum* system as well as the favourable rumen fermentation that preserves useful rumen micro flora when concentrates and roughages are fed to the rumen simultaneously.

Feed Processing. Attempts have been made to increase feed intake by increasing the rate of passage of ingesta from rumen through processing of feed materials by grinding, chaffing,

---

Everson, R.A. (1974). Diss. Abstr. Int. 13 35 : 7

Everson, R.A., Jorgenson, N.A., Crowley; J.W., (Jensen, E.L. and Berrington, G.P. (1976). J. Dairy Sci. **59** : 1776

pelleting etc. But when particle size of the feed is reduced, there is less mixing of feed and less flow of saliva which indirectly contribute to a type of rumen fermentation leading to a lower pH. The bad effect of such fermentation has been discussed earlier. Such reduction in particle size also results in lowered digestibility. When good quality roughages are subjected to these processing, the consumption does not improve, but digestibility is reduced. In the case of coarse fodder, reduction in particle size increases intake, but, a price has to be paid in terms of reduced digestibility. Thus, it is evident that the ruminant has full capacity to deal with roughages by chewing and rumination and that attempts to augment intake by reducing particle size will have the adverse effect of reduced digestibility. Therefore, chaffing may be done in the case of coarse forages only, to reduce wastages. The reduction in particle size should not be adopted as a means of increasing rate of intake and rate of passage of feed through the alimentary tract in order to meet the increased nutritive requirements of heavy yielding bovines.

6.11.3. Guidelines to Feed High Yielders

1. Ration of a dairy cow should contain a minimum of 20-25 per cent dry matter from forages.
2. In the case of high yielders the forages given should be of superior quality. It should be cut at the proper stage of maturity. Thirty to fifty percent of the roughages should be of leguminous crops.
3. Processing and reducing the forage to smaller bits may be avoided except in the case of very coarse fodder.
4. For high yielders when large quantities of concentrates are to be fed, it should be either fed immediately after feeding roughages or preferably, it may be fed mixed with the roughages.
5. To ensure proper nutrient intake and roughage concentrate ratio, a density × digestibility value of 35 may be maintained.
6. The feeding schedule should be such that it maintains a continuous fermentation in the rumen. The cow should be fed a minimum of 4 times a day, each feeding comprising of both concentrate and roughage.

6.12. LEAST COST OR MAXIMUM PROFIT RATIONS

Computer technology has found application in ration formulation, balancing rations using a combination of ingredients with the lowest possible total cost. Without a computer, it would have been very laborious to make such an attempt. Computer-formulated least-cost rations are neither good nor bad nutritionally. It is the cost, which will be the lowest under the price line of ingredients existing at that time.

Linear programming is the technique employed in the computer analysis for evolving least-cost rations. The computer will give the percentage of selected feed ingredients that are to be included in the feed mixture. This feed mixture formula is valid only under the set of feed prices and ration constraints fed to the computer. The number of feed ingredients selected is equal to or less than the number of constraints included in the programme.

A slight variation from the least-cost formula is the maximum profit ration in which the rations formulated maximise income above feed cost.

Computer formulated least cost rations have their own limitations. There had been instances of animals performing poorly under computer formulated least cost rations. These were either due to improper nutrient specifications or ingredient limitations imposed on the computer.

Apart from formulations of concentrate mixes, computer programming has its application in formulating complete feeds also. Constraints with respect to energy, protein, crude fibre, calcium and phosphorus are generally included. The minimum level of roughage in the feed also will be indicated. One example of constraints for a least cost complete feed is as follows :

Net energy	120-155 MCal/100 kg
Crude protein	12-14 percent
Crude fibre	15-20 percent
Calcium	0.5-0.7 per cent
Phosphorus	0.4-0.6 per cent
Roughage	1.5-2.0 per cent of body weight

The response of dairy cows to increase of nutrients fed depends on the genetic potential for milk production. There are programmes for complete rations incorporating information on such constraints. The information of one such study is depicted in Fig. 6.8. Use of a programme incorporating such information allows the computer to select the optimum level of milk production and the composition and quantity of complete feed to be fed for achieving it under the conditions with respect to price existing at that time. This, in other words, maximizes income over feed cost. Additional feeding will result in more milk production, but the cost of the feed necessary to support production of the extra milk may be more than what it returns in terms of milk.

FURTHER READING

Agricultural Research Council. 1980. *The Nutrient Requirement of Ruminant Livestock:* Commonwealth Agric. Bureau, Farnham Royal.

Broster, W.H., Bines, J.A. and Sutton, J.D. 1978. Forage: concentrate ratios for high yielding dairy cows. In: *Recent Advances in Animal Nutrition*—1978 (ed. W. Haresign and D. Lewis; pp. 99-126. Butterworth., London.

Broster, W.H. and Swan, H. 1979. *Feeding strategy for the high yielding Dairying Cow.* Crossby Lock wood Staples, Granada Publishing. London.

Cullison, A.E. 1978. *Feeds and Feeding.* Prentice Hall of India Private Ltd., New Delhi.

Watson, S.J. and Nash, M.J. 1960. *The Conservation of Grass Forage Crops.* Oliver and Boyd, London.

Whyte, R.O. 1957. *The Grassland and Fodder Resources of India.* I.C.A.R., New Delhi.

Etgen, W.M. and Reaves, P.M. 1978. *Dairy Cattle Feeding and Management.* John Wiley and Sons. New York.

Lloyd. L.E., McDonald, B.E. and Crompton, E.W. 1978. *Fundamentals of Nutrition.* 2nd ed. W.H. Freeman & Co. San Francisco.

Maynard, L.A.. Loosli, J.K., Hintz, H.F., Warner, R.G. 1979. *Animal Nutrition.* Tata Mc Graw Hill Publ. Co. Ltd.. New Delhi.

McDonald, P., Edwards, R.A. and Greenhalgm, J.ssszz F.D. 1975. 2nd Ed. *Animal Nutrition.* Longman. New York.

Morrison, F.B. 1956. *Feeds and Feeding,* 22nd Ed. The Morrison Publishing Co., New York.

Ranjhan, S.K. 1980. *Animal Nutrition and Feeding Practices in India.* 2nd, Ed. Vikas Publ. House. New Delhi.

Ranjhan, S.K. 1980. *Animal Nutrition in Tropics.* Vikas Publ. House, New Delhi.

Dabadghao, P.M. and Shankara Narayanan, K.A. 1973. *The Grass Cover of India.* I.C.A.R., New Delhi.

George Thomas, C. 2008. Forage Crop Production in the Tropics. 2nd ed. Kalyani Publ. Ludhiana.

Section III : REPRODUCTION AND LACTATION

CHAPTER

7

Reproduction

7.1. REPRODUCTIVE ORGANS OF THE BULL

The reproductive organs of the bull (Fig. 7.1) consists of :

1. A pair of testes, the gonads or primary sex organs.
2. The epididymis where sperms are nurtured to maturity providing motility and capacity to fertilize and stored.
3. The accessory sexual glands and ducts including ampullae, seminal vesicles, the prostate and Cowper's glands, and
4. The penis, the copulatory or external genital organ.

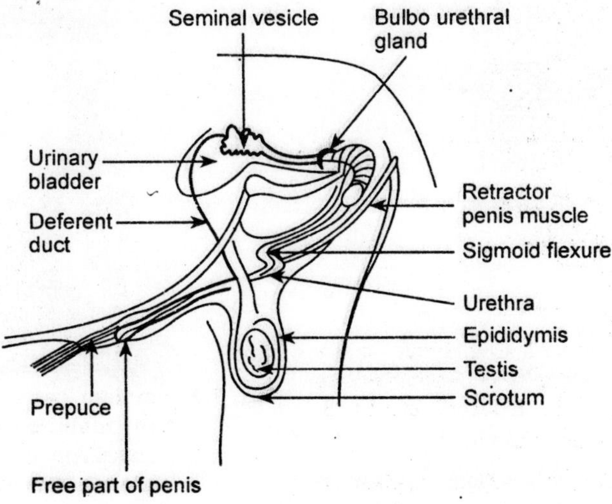

Fig. 7.1. Male reproductive organs of the bovines.

The gametes (spermatozoa or sperms) are produced in the testes which are nurtured and stored in the epididymis (Fig. 7.2). During the course of this maturation, the spermatozoa acquire motility and the capacity to fertilize. At the climax of copulation, the sperms are squeezed along the *ductus deferens* into the urethra, mixed with the secretions of accessory sex glands thus forming the ejaculate or semen which is pumped out into the female genital tract.

7.1.1. The Testes and Scrotum

The pair of testes are the primary sex glands which produce the gamates. It also serves as an endocrine gland secreting the male sex hormone, testosterone.

They are elongated ovoid glands suspended in the scrotum (Fig. 7.3) with their long axis vertical weighing 250-500 g depending on breed and age.

Testes originate near the kidneys but descend into the scrotal sac in late foetal life.

The scrotum is a pendulous pouch containing the testes. It serves to keep the testes 4 to 7°C cooler than body temperature more or less acting as a thermostat. The scrotal skin is thinner and has less hair coverage but a more abundant supply of sweat glands and temperature receptor nerve endings. Due to this, the testicular venous blood gets cooled when it passes near the scrotal skin. This coolness is transferred to the incoming arterial blood for supplying testes through a counter-current heat exchange mechanism. The coiled nature of the artery in the spermatic cord which is surrounded by a plexus of small veins carrying cooled outgoing blood greatly facilitate this heat exchange.

The thermostat-like action of the scrotum is controlled by temperature receptors which evoke physiological responses like sweating and contraction or extension of the scrotum. Scrotum has a double muscle system which under a cold stimulus draws the testes closer to the abdominal

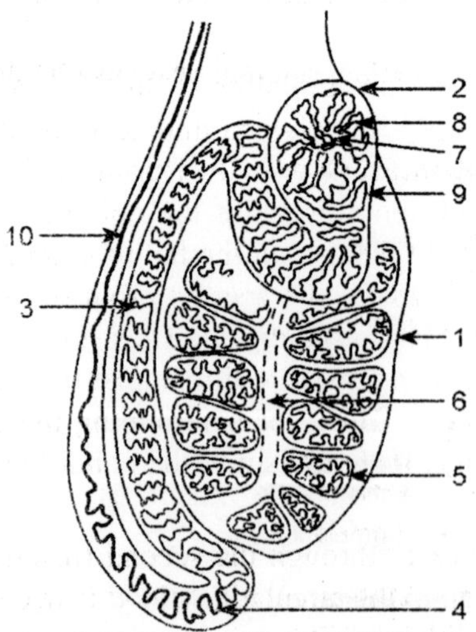

1. Testes. 2. Epididymis—head 3. Epididymis—body 4. Epididymis—tail 5. Lobule with Seminiferous tubules 6-7. Rete testis 8. Efferent ducts 9. Epididymal duct 10. Deferent duct.

Fig. 7.2. Diagram showing the tubular system of testes and epididymis.

Fig. 7.3. Scrotum carrying testicles-By bringing the testicles nearer or away by contraction or relaxation, the scrotum controls the temperature inside the scrotal sac.

wall for warmth, shorten the scrotum, cause the scrotal skin to pucker and contract thus reducing surface area for evaporation. Under hot conditions the reverse of these happen with the scrotum becoming pendulous thereby keeping the testicles as far away from the body as possible. The surface area of the scrotal sac also increases to facilitate evaporative cooling.

The thermoregulatory mechanism of the scrotum is vital for normal sperm production in the bulls. If the temperature of the scrotum attains body temperature or slightly above for only a few hours it can cause temporary infertility after a delay of about 2 months. In very hot weather, the thermal stress may become beyond the capacity of the animal to cope with, which may cause degenerative changes in the germinal epithelium. The spermatozoa which have already left the testes and stored in the epididymis are not much affected by elevated temperature so also the interstitial cells of testes which produce testosteron are only slightly affected and animals retain almost normal libido under such conditions.

The spermatozoa are produced in the *seminiferous tubules* which are highly coiled and compactly arranged in the testes. The *seminiferous tubules* if uncoiled and arranged in a straight line will measure 4-5 kilometers in the bull. The germ cells of the germinal epithelium that line the seminiferous tubules undergo 2 meiotic cell divisions to become the sperm. The mature bull produces on an average 10 billion sperms per day. Older bulls and larger bulls generally produce more sperms. The *seminiferous tubules* converge on to larger tubules known as *rete testes* which is continued from the head of the testis as epididymis. The sperms formed in the *seminiferous tubules* are transported through it to rete testes and from there to the epididymis.

Between the *seminiferous tubules* there are connective tissue, blood vessels, nerves and the *interstetial* or *leydig cells*. These interstetial cells secrete the male sex hormone *testosteron* which is responsible for typical male action and attitude. Because of the production of testosteron in the interstetial cells, the interstetial fluid and lymph in the testes have a higher content of testosteron than blood that supplies them.

Epididymis is a single highly coiled tube connected to the seminiferous tubules through rete testes. In the bull it is more than 30 metres long. Arbitrarily it is divided into head, body and tail although it is a single physiological entity and should be treated as such. During the course of their passage from the anterior end to the posterior end and storage in the epididymis the spermatozoa undergo development and maturation. By the time they reach the tail of epididymis, they are mature and capable of fertilization. It takes 4-8 weeks for the sperms produced in the seminiferous tubules to undergo maturation and reach the tail of epididymis. Thus, factors like high environmental temperature which adversely affect spermatogenesis will manifest their effect on the fertility of the bull only with a lag phase of 4-8 weeks. Thus, for determining the cause of infertility involving spermatogenesis in a bull, an investigator should study the conditions and factors prevailing 1-2 months earlier.

The sperm acquires the capacity to fertilize when it passes through the second (distal) half of the middle pieces of epididymis. Normal level of androgen in the circulating blood is necessary for the survival and maturation of spermatozoa in the epididymis. The capacity for progressive motility is also acquired by the sperms in this organ. A striking morphological change that occurs in the spermatozoa during its travel through the epididymis is the movement of the cytoplasmic droplet along the middle piece and its eventual loss.

Vas deferens is the tube that connects epididymis to the urethra. On palpation it gives a cord-like feeling due to its thick muscular wall. It is convoluted in the beginning but inside the spermatic chord becomes straight. The spermatic chord consists of vas deferens, nerves and blood and lymph vessels encased in a fibrous sheath. It enters the abdominal cavity from the scrotum through the inguinal canal.

7.1.2. The Accessory Sex Glands

Ampullae. The vas deferens from either side dilates towards the urethral end to form the ampullae. Ampullae have secretory glands on their walls. They open into the urethra through slit-like openings. During precoital courtship, sperms stored in the tail of epididymis are transported into the ampullae. Ampullae produce and secrete small amounts of fructose and citric acid to the seminal fluid.

Seminal vesicle is a lobulated gland situated lateral to the ampullae. In the bull they are comparatively large and contribute about 50 per cent of the seminal ejaculate. The secretion of the seminal vesicles is a yellow coloured, slightly opalescent viscous fluid. This secretion is rich in proteins, fructose, citric acid and potassium and has small amounts of inositol and traces of ergothioneine. The duct that transports the secretion of the seminal vesicle unites with the vas deferens at the point where it joins the urethra.

Prostate and Cowper's Glands. In the bull the prostate is inconspicuous and surrounds the urethra where it is joined by the bladder. The gland opens into the urethra through many small ducts. The two Cowper's glands are located above the urethra where it emerges out of the pelvic cavity. The secretions of these two glands are clear and watery. The clear watery fluid that dribbles from the prepuce of the bull before mounting is the secretion from these glands This fluid is known for its antibacterial properties and its function most likely, is flushing of urethra before ejaculation takes place thus rendering semen free from contamination.

7.1.3. Penis

Penis is the copulatory organ of the male. It is enclosed by the prepuce in the non-erect retracted state. In the bull, the penis is fibro-elastic type which is rigid even in the non-erected state. The penis in the bull is S-shaped (sigmoid flexure). On erection, the penis becomes rigid and the sigmoid flexure is straightened out thereby protruding the penis outside.

The *urethra* passes through the length of the penis and acts as a canal for urine and semen. Semen is ejaculated only when the penis is erected by sexual excitement or other stimuli. The engorgement of cavernous tissue on the neck of urinary bladder during erection closes it during ejaculation. This prevents spillage of semen into the bladder.

7.2. THE FEMALE REPRODUCTIVE SYSTEM

The reproductive system in the cow (Fig. 7.4) consists of primary and secondary organs. The primary reproductive organs are the ovaries whereas the secondary organs comprise of the oviducts (Fallopian tubes), the uterus; the cervix, the vagina and external genitalia (vulva). The ovaries produce the female sex cells—ova and some of the important female sex-hormones. The secondary reproductive organs of the cow receive and conduct both the female and male gamates so that they meet and fertilization takes place. They also provide the environment in which the embryo is conceived. implanted, nourished and protected during pregnancy and delivered at the end of the gestation period.

7.2.1. The Ovaries

There are two ovaries on either side suspended from the broad ligament near the kidneys. The size of the ovary depends largely on the age and reproductive state of the cow. They are ovoid in shape and weigh 10-20 g each.

The outer surface of the ovary is covered by a layer of flattened cells known as *germinal epithelium*. Follicles develop from the germinal epithelium by invagination.

The ovarian tissue below the germinal epithelium has two main parts, namely, the *cortex* or parenchymatous layer containing most of the follicles and the *medulla* which is free of follicles (Fig. 7.5). The follicles formed from the germinal epithelium undergo the changes leading to maturation of the *gamete*, ejection of the *ovum* (ovulation) and formation of the *corpus luteum* in the cortical layer. Thus the cortex remains the site of egg production as well as hormone production. The medulla consists of connective tissue and an extensive system of blood and nerve supply.

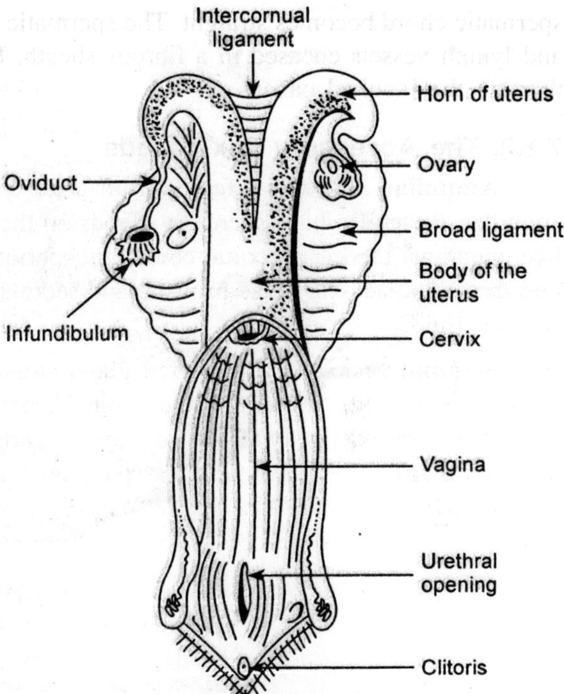

Fig. 7.4. Female reproductive tract of the bovine (Diagrammatic)

7.2.2. The Duct System

The duct system consists of the infundibulum. the oviducts, the uterus, the cervix, the vagina and the vulva (Fig. 7.4). The *infundibulum* is the funnel shaped end of the oviduct which lies in close proximity with the ovary and receives the ova when they are shed from the ovary.

The *oviduct* is a tortuous tube connecting the infundibulum to the uterine horn. It is lined with a thick layer of mucosa consisting of epithelial cells the height of which changes with different phases of estrous cycle. The lumen of the oviduct is also lined with motile cilia beating towards the uterus. Besides, the oviduct has also a muscular and outer serous layers.

7.2.3. The Uterus

The bovine uterus has a *body* and two *horns* (bicornuate) as in most other large animals. In the young non-pregnant cow the uterus lies almost entirely in the pelvic cavity. Only the tips of the horns extend over the pelvic ridge into the abdominal cavity. On the other hand in a cow that has conceived and calved many times the uterus may be larger and lie, to a greater extent, in the abdominal cavity. The inner mucosal surface of the uterus has 80-120 button-like projecting circular areas known as *caruncles*. During pregnancy, corresponding areas on the foetal chorionic membrane known as *cotyledons* interlock with the caruncles providing for the attachment of foetal membranes. The rest of the mucosal membrane is provided with uterine glands.

The cervix may be considered as the entrance door to the uterus. It is a thick-walled muscular organ at the junction between uterus and vagina. It effectively closes the external

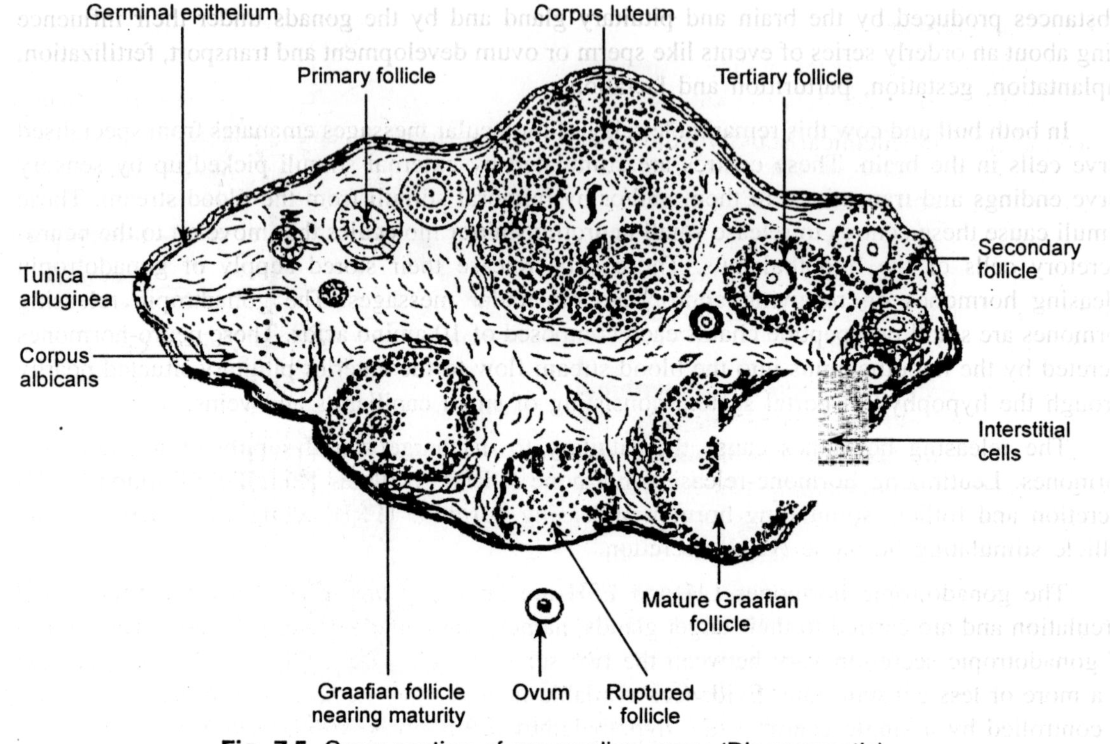

Fig. 7.5. Cross-section of mammalian ovary (Diagrammatic)

opening of the uterus during pregnancy. Thick musculature, mucosal folds and a cervical mucous plug aid in tight closure. At the time of calving the cervix dilates allowing delivery of the calf.

7.2.4. The Vagina

The vagina is the copulatory organ of the female. It is thin walled but tough and highly elastic. The epithelial liming of the mucous membrane of vagina undergoes changes during various phases of estrous cycle. The *vulva* is the external genitalia which consists of the vestibule and the clitoris. The urethra opens into the vestibule. The clitoris is the rudimentary male organ.

7.3. NEUROENDOCRINE CONTROL OF REPRODUCTION

Reproduction is a symphony orchestrating physiological and behavioural events and anatomical changes which proceed in a perfect order and synchrony. The centre of the stage in reproduction is occupied by the gonads : the testes in the male and ovaries in the female.

The *gonads* have two functions. They produce gametes (spermatozoa in the bull and ova in the cows). They also produce the sex hormones. The role of the bull is to provide spermatozoa and ends with mating followed by fertilization. The cow on the other hand goes on to harbour, protect and nourish the developing embryo until parturition and provide milk and protection till weaning to the young calf.

For a multicellular organism to perfectly organise the sequential changes in reproduction in different parts of the body, it is necessary that the cells in widely separated regions of the body transfer information between them. This co-ordination is achieved through the nervous and endocrine systems which themselves are functionally interlocked in many ways. Chemical substances produced by the brain and pituitary gland and by the gonads under their influence bring about an orderly series of events like sperm or ovum development and transport, fertilization, implantation, gestation, parturition and lactation.

In both bull and cow this remarkable relay of molecular messages emanates from specialised nerve cells in the brain. These centres are stimulated by external stimuli picked up by sensory nerve endings and transmitted to them and/or by humoral stimuli from the blood stream. These stimuli cause these centres to release small neurotransmitter molecules that move on to the neuro-secretory cells of the hypothalamus. These cells release their stored supply of gonadotropin releasing hormones on receiving appropriate molecular messages. The gonadtropin releasing hormones are small polypeptide chains each composed of 10 amino acids. These neuro-hormones secreted by the hypothalamus into the blood stream flow to the anterior pituitary situated nearby through the hypophyseal portal system consisting of small capillaries and veins.

The releasing hormones cause the pituitary to discharge stored supply of gonadotropic hormones. Leutinizing hormone-releasing hormone (LRH) regulates leutinizing hormone (LH) secretion and follicle stimulating hormone-releasing hormoue (FRH) along with LRH regulate follicle stimulating hormone (FSH) secretion.

The gonadotropic hormones LH and FSH are large glycoproteins. They enter the blood circulation and are carried to their target glands, namely the gonads (testes or ovary). The pattern of gonadotropic secretion vary between the two sexes. In adult bulls FSH and LH are secreted at a more or less constant rate. Evidences available indicate that the secretion of these hormones is controlled by a single centre in the hypothalamus in or around the arcuate nucleus.

In females also there is a steady rate of gonadotropin secretion. But, this is superimposed by a cyclic increase or surge of LH and FSH secretion resulting in ovulation. In the cow, the medio-basal hypothalamus controls the steady secretion of FSH and LH. The ovulatory surge of LH secretion in the cow is triggered by increased levels of circulating estrogens (above 100 pg/ml) over a period of more than 36 hours. The negative feed-back effect of estrogens on pituitary gonadotropins appears to be mediated through its action on the regulatory centre situated in ventral hypothalamus. Thus, in the female there are two hupothalamic regulatory centres one for controlling the tonic secretion of FSH and LH and the other for triggering a burst of LH secretion and increase in FSH level to result in ovulation.

In the cow the two gonadotropins unitedly stimulate ovarian function. but each has a specialized role. FSH, as the name indicates. stimulate the development and maturation of the ovarian follicles which contain the immature ova surrounded by a multilayered encirclement of granulosa cells. While doing so. the endocrine cells of the follicle are stimulated to produce increasing amounts of estrogens. The immature ovum in the follicle also develops and is brought to the state of maturation needed for ovulation. Under the influence of sustained increasing levels of estrogens. the neuro-endocrine system is triggered to produce the pre-ovulatory surge of LH culminating in ovulation. The mature egg is liberated from the follicle. The empty follicle is transformed into the corpus luteum, a structure which is in effect a temporary endocrine gland. Under the influence of LH. the corpus luteum starts secreting large amounts of progesterone. another major steroid sex hormone of the cow. A third gonadotropic hormone, prolactin is responsible for maintaining the progesterone production by corpus luteum for its usual 17 day life span.

The gonadotropic hormones and their releasing hormones secreted by the hypothalamus are amino-acid and peptide derivatives. They do not enter the target cells because of their size or molecular structure. Instead they attach themselves to specialized receptor sites on the surface of the cell and influence the cell's biochemical activity from outside. The hormone receptors—large protein molecules embedded in the cell membrane—are highly selective in their ability to bind the hormone for which the cells are targets. As soon as the hormone attaches itself to the receptor site on the target cell, increased production of an enzyme *adenylate cyclase* takes place. This enzyme converts adinosine-tri-phosphate (ATP), the universal currency of chemical energy in the cell, into cyclic adinosine-mono-phosphate (cyclic AMP). Cyclic AMP acts as second messenger within the cell to relay and amplify the message of the hormone to the cell's biochemical organisation. Exactly how cyclic AMP acts is not fully known. One hypothesis is that cyclic AMP generated as a result of attachment of a hormone to the cell membrane receptor activate a protein kinase, which on entering the nucleus phosphorelates histone or some other nuclear regulatory protein at a specific site and causes its displacement from DNA of inactive genes. When histones are intimately associated with DNA, they appear to inhibit the expression of DNA's information content by blocking access to the enzymes that effect transcription of DNA to RNA. On phosphorelation the histones move and DNA strands are exposed which become available for transcription into messenger RNA and ultimately translation into proteins. These proteins are other hormones or enzymes which can produce a particular effect.

The gonadal hormones in a 'feed-back' mechanism act back on the brain to induce the appropriate sexual behaviour necessary for successful reproduction.

Because of the gonadal hormone's major and profound effect on sex behaviour and secondary sexual characteristics, they are commonly referred to as sex steroids. But this does not mean that their effect is confined to reproductive system only. They also affect other organs, including bone, muscle, blood and the liver and many metabolic processes like that of minerals and water.

Steroid hormones, unlike the proteinaceous hormones are easily soluble in fat and hence can pass through the fatty cell membranes of the target organs and directly influence the processes there. The cells that are targets for sex steroid hormones have unique large molecules in their cytoplasm. They bind specifically to a particular steroid hormone. These receptors are highly specific in that they do not bind with other steroids as well as that they are found only in the cells which are acted upon by the hormone. After binding, the receptor-hormone complex moves into the nucleus or more likely the binding liberates a catalytic sub-unit of the molecule which move into the nucleus. This liberated protein kinase as in the case of protein-hormones, expose quiescent DNA thereby stimulating the particular messenger RNA production and protein synthesis. In this manner the non-stimulated target cell is converted to the stimulated state.

7.4. CYCLES OF REPRODUCTION

There are 3 cycles involved in the reproductive process of all individuals, *viz.,* (1) the life cycle, (2) annual breeding cycle, and (3) the estrous cycle in the female and spermatogenic cycle in the male.

The life cycle has 4 distinct stages, namely, (1) foetal, (2) prepubertal, (3) reproductive, and (4) seniscent.

7.4.1. Puberty

Puberty is a major milestone in the life cycle of an animal. This is the stage at which animals become sexually mature and the secondary sex characteristics become conspicuous. The term sexual maturity means that the animal is capable of reproduction. Puberty is the age at which the first estrus occurs in the heifer and the bull-calf starts giving semen with viable sperms. The reproductory organs undergo marked increase in size at the time of puberty.

Puberty occurs before mature body size is attained. Under good feeding, a calf attains puberty approximately at 66 per cent of adult body size. Breed, genotype, climate and level of feeding influence the age at puberty. The time of onset of puberty appears to be a function more of body size than of age. In general, smaller breeds like Jersey attain puberty earlier. Restrictions in nutrition delay puberty considerably.

The sudden increase in the size and weight of reproductive tract associated with puberty involves hormonal action. In the heifers progesteron plays a key role in the changes leading to the phasic LH release which initiates the estrous cycles.

7.4.2. The Estrous Cycle

Every 21 ± 3 days from the time of puberty, the cow prepares for a pregnancy. A mature ovum is liberated from the ovary, the cervix becomes receptive to the spermatozoa, the female exhibits behavioural adjustment and attraction to receive the male in copulation, the uterus and Fallopian tubes produce the secretions which are conducive to the transport of ovum and sperms and the endometrial lining of the uterus prepares to receive and nourish a fertilised ovum. The events listed above in sum total is known as estrous cycle which is repeated cyclically throughout the reproductive phase of life, each cycle lasting 21 ± 3 days.

The estrous cycle of the cow has two major phases : (1) one associated with the growth of the follicle known as the *period of the follicle* (the estrogenic phase) and (2) the one associated with the growth of the corpus luteum (luteal or *progestational phase*).

7.4.2.1. *Period of the Follicle*

Under the influence of circulating FSH, a few primary follicles start to develop. After a few days, only one follicle continues to flourish and become fully mature. The follicles, when they mature, produce progressively increasing amounts of estrogen. This phase during which the follicles mature under the influence of FSH and produce estrogen is referred to as the period of the follicle (oestrogenic phase). This phase includes *proestrus* and *estrus* and lasts about 4 days of the cycle.

Proestrus. The proestrus which lasts for about 2 or 3 days in the cow is characterized by follicular growth and estradiol production. Estradiol causes hyperaemia of the tubular genitalia. Engorged vagina and cervix produce a thin serous secretion which appears as a thin clear vaginal discharge.

Estrus. At the end of proestrus, estrus (the period in which the animal exhibits Sex desire) appears. This sexual behaviour is caused by a 'feed-back' mechanism through which the estradiol act on the CNS. During the 12 to 18 hours of estrus the cow manifests various signs of heat which are described below :

Signs of Estrus

1. The cow becomes restless, does not eat and frequently bellows and seldom ruminates.

2. Milk production suddenly goes down.
3. She may search for the male.
4. Traits of homosexuality is shown in which the cow will attempt to mount other cows while other females not in estrus tend to mount the estrus cow which she permits.
5. During estrus the cow is receptive to the act of mating and will stand when the bull mounts her.

   The behaviour of standing quietly while being mounted by the bull or other cows is referred to as the 'standing heat' which is the surest sign of estrus.
6. Mucous discharge may be found sticking to the tail.
7. Hair on the tail head may be pushed forward or denuded from being mounted.
8. There may be dirt on the sides of the abdomen left by the forefeet of mounting animals.

The cow is peculiar in that she has a very short period of sexual receptivity-hardly 12-18 hours. Besides, ovulation takes place 12 to 16 hours after the end of estrus.

Thus, a steadily rising level of LH and a superimposed phasic wave of the same hormone under the influence of estradiol will have the joint effect of a surge in the LH level resulting in ovulation.

7.4.2.2. *Period of the Corpus Luteum*

This includes *metestrus* and *diestrus*. The period immediately following the stoppage of visible heat manifestation is *metestrus* which lasts for 2 or 3 days in the cow. Ovulation takes place during this period.

The process of ovulation is a gradual one. Early in the period of the follicle the ovary secretes in ever-increasing amount, the estrogenic steroids, principally estradiol. The final surge of estrogen production heralds (and also induces by means of hypothalamic recognition of blood estrogen levels) a peak in LH level causing ovulation within 24 hours. How the ovulatory surge of LH brings about ovulation is not yet fully explored. Many theories have been propounded which were rejected one after the other as newer and newer evidences poured in. A theory gaining strength at present is that LH induces the synthesis of a proteolytic enzyme which lyses the walls of the follicle, releasing the ovum. The ovum comes out from the rupture point in a cascade of follicular cells and fluid. The released egg is swept in by the fimbriated end of the oviduct.

Meanwhile the follicular cavity from which the ovum has been released. gets filled with blood and a new type of cells—the luteal cells—proliferate which organise into a temporary endrocrine gland—the *corpus luteum.* Corpus luteum begins to secrete progesterone. under the influence of which the congestion of the tubular genitalia diminishes. Secretions of the genital tract also become viscid and less voluminous.

During metestrus the ovum picked up by the oviduct is transported towards the tubal-uterine junction. Fertilization takes place in the upper portion of the oviduct. If the ovum is not fertilized it undergoes auto-degeneration.

Meterstrus is followed by *diestrus* which is dominated by the effect of corpus luteum. This is the most lengthy phase of estrous cycle lasting about 15 days. Even if there is no pregnancy the corpus luteum develops into a functional organ secreting progesterone and some estrogen. Progesterone causes considerable hypertrophy of myometrium and uterine glands which secrete a viscid fluid for nourishing the zygote. If fertilization takes place and the zygote reaches uterus, the cropus luteum will persist throughout pregnancy. If otherwise, the corpus luteum will start disintegration from day 17, allowing the onset of the next estrous cycle.

The period of the follicle (proestrus and estrus) lasts 3 to 4 days whereas the period of corpus luteum (metestrus and diestrus) about 17 days.

7.4.3. Reproduction at Cellular Level

Cattle like most higher vertebrates procreate by sexual reproduction. In them the body cells are of 2 types. *somatoplasm* and *germplasm*. It is the latter which is directly concerned with reproduction. These cells are present in the testes in the males and ovary in the females. The germinal cells give rise to spermatozoa or ova (the gametes). See chapter 14 for a detailed discussion on the various aspects of reproduction at cellular level.

7.4.4. Spermatogenesis

Spermatozoa are produced in the seminiferous tubules of the testes. The germinal epithelial cells lining the seminiferous tubules multiply by a series of mitotic divisions followed by the two divisions which constitute meiosis and produce the haploid cells. This is followed by a period of metamorphosis and maturation in which both the nucleus and cytoplasm of the future sperm-cell undergoes morphologic changes to become mature spermatozoan with the characteristic head, mid-piece and tail. The cell divisions and the metamorphosis together is known as spermatogenesis. It begins in the seminiferous tubules and is completed in the epididymis.

7.4.5. Ovogenesis

The process of producing the ovum starts early in embryonic life with the growth and invagination of a small area in the germinal epithelium. These germ cells are cut-off from the original invagination and each amoeba-like germ cell migrates into the ovary to become the primary oocyte. These are sorrounded by a layer of follicular cells. The oocytes together with these follicular cells are known as primary follicles. The newborn heifer has approximately 1,50,000 primary follices in the ovary which decline to 75,000 by spontaneous degeneration in a 3-month-old heifer. In an adult cow the number is further reduced to about 60,000. This number continues to decline and in aged cows of 18-20 years of age, it is as low as 1000. It is only an occasional egg that actually ovulates and avails itself of the opportunity to get fertilized. perhaps fewer than 160 in a cow's reproductive life.

The cells lining the follicle multiply during estrous cycle so that eventually a fluid-filled cavity. or *antrum* is formed around the egg. A layer of cells immediately surround the ovum which is connected by a bridge-like cellular growth known as the *cumulus oopherus* to the follicular cell layers.

The first meotic division takes place a few hours before ovulation in which the spindle rotates and throws out half the chromatin and a small amount of cytoplasm as the first polar body. The number of chromosomes thus gets reduced from 60 to 30. The remaining 30 chromosomes replicate once more. But, the resultant second polar body is extruded only after the sperm penetrates the pellucid zone to 'activate' the oocyte.

7.5. FERTILIZATION AND PLACENTATION

Fertilization is an intricate biological phenomenon in itself. This union of the sperm and the ovum generally takes place in the upper half of the oviduct. Sperms must traverse both male and female reproductive tracts to reach upper oviduct to which ovum would have descended by then. During mating or artificial insemination, oxytocin is released by reflex action in response to genital stimulation. This causes estrogen-stimulated uterus and oviduct to produce strong muscular contractions. These contractions propel the sperms towards the upper oviduct within a matter of minutes. Sperm motility does not contribute significantly to the transport of sperms form cervix to the upper oviduct. But, once they reach the upper oviduct. further finer manoeuvres to locate and dock-up with the ovum is dependent mostly on the sperms' own motility

Ovulation takes place about 12 hours after the end of estrus. It takes another six hours for it to travel half-way down the oviduct. The sperms, even though reach the oviduct within minutes after insemination, must be exposed to the female reproductive tract for about 6 hours to attain the capacity to fertilize. This process of preparing the sperms for fertilization is known as *capacitation*. Sperms are viable for 24 hours in the female reproductive tract whereas the ovum remains fertile for only about 10 hours after ovulation. *This implies that mating or insemination between mid-estrus to about 6 hours after the end of estrus will be most fertile.* Sperms entering the female genitalia earlier than that will have to spend more then 24 hours in it before encountering the ovum and will die off. Sperms deposited after this period do not get enough time for capacitation and by the time capacitation is over ovum will be more than 10 hours old in which case degenerative changes would have set in.

Only a few hundred sperms out of millions deposited in the female genitalia reach the upper oviduct. By enzymatic action, the sperm slices through the *cumulus oopherus* and *zona pellucida* and makes contact with the surface of the egg. As soon as one sperm makes contact with the egg membrane certain important structural and functional changes are initiated. A key event is the immediate blockade against the entry of other sperms by which all other sperms are excluded from the ovum. The barriers to fertilization by more then one sperm include changes in the *zona pellucida* making it impenetrable to sperms and alterations in the outer membrane of the ovum which preclude the attachment of more sperms.

The single sperm cell penetrates the outer membrane of the ovum into the cytoplasm in several stages. In the process it loses its tail. The penetration of ovum by sperm activates the egg to complete its second reduction division. The plasma membranes of the head of the sperm dissolve and the nuclear membranes of sperm and ovum fuse to form a zygote. This union of male and female gametes is in essence *fertilization*. It restores the chromosome number to *diploid* state.

All the ova carry *x*-chromosomes whereas only half of the sperms carry *x*-chromosome while the other half *y*-chromosome. Whether a *y*-chromosome-carrying-sperm or a *x*-chromosome-carrying-sperm fertilizes the egg is purely a matter of chance. In the former case (*y*-chromosome sperm) the offspring will be a male and in the latter (*x*-chromosome sperm) female.

After about 36 hours the zygote starts mitotic cell division to become 2 cells. Two days later it would have undergone 2 more cell divisions to become an eight celled mass and at that stage completes its transit through the Fallopian tube to enter the uterus.

The cluster of cells remains unattached for one or 2 days while undergoing division to assume the form of a signet ring : an inner mass of cells encircled by a single row of cells. The developing ball of embryonic cells are called the *blastocyst.* They slowly form a fluid-filled cavity and differentiate into 3 basic embryonic tissues, namely, *endoderm*, *mesoderm* and *ectoderm*. These basic tissues undergo further differentiation to become various organs of the developing calf and the placental tissues. The heart of the calf begins function within 22 days after fertilization and, by 2 months, most of the organs are formed.

The part played by hormones in nidation and placentation cannot be over-emphasised. During estrus, endometrium undergoes proliferation under the influence of estrogenic steroids. After ovulation, the ovarian steroid production is switched over from a predominance of estrogen to a predominance of progesterone. In response, the endometrial cells proliferate and become larger, accumulating secretions. The whole endometrial surface turns into a highly vascular. velvety nest ready to accept, protect and nurture the developing blastocyst if one should arrive from the oviduct. The uterine environment is predominantly under the influence of progesterone produced by the corpus luteum (CL).

For the pregnancy to survive, there must be a source of pregesterone to condition the endometrium throughout. Otherwise, the endometrial cell proliferations are dismantled and the next estrous cycle sets in. Avoiding this crisis requires uninterrupted supply of progesterone. The initial source is the CL. The CL receives a signal from the newly formed blastocyst to continue production of the steroid hormone. If pregnancy had not taken place the CL would have disintegrated for want of this signal. The blastocyst produces a gonadotropic hormone which will induce the CL to continue function. A second critical point lies ahead, however. The progesterone secretion by the CL becomes insufficient in later pregnancy. At this stage (about 200 days of pregnancy), the placenta itself begins to produce sufficient quantity of progesterone to maintain pregnancy. In other words, to pass the first crisis, the embryo produces a gonadotropic substance that ensures continued CL functioning; to meet the second crisis, the developing embryo itself assumes the required endocrine function.

The secretions and fluids present in the uterus nourish the embryo during the first month of its life when it has not yet become attached to the uterine caruncles. The placental membranes grow rapidly and completely enclose the embryo. The amnion completely encloses the embryo which in turn is surrounded by the allanto-chorion. The cavity inside the embryonic membranes around the embryo gets filled and distended with fluid thereby making the membranes press against the endometrium. Placenta develops cotyledons opposite the maternal caruncles and the two unite firmly so that the transfer of nutrient material from the mother's system to the embryo and excretory products from the embryo to the mother become possible. This process is completed in 4 to 5 weeks time.

The umbilical artery and vein emerging from the embryo pass through the amnion and end in capillaries on foetal cotyledons on the chorion, which is in close apposition to the maternal caruncles through finger-like growths known as villai. The arteries of the cow also end in a capillary network in the maternal caruncles. Thus, the maternal and fetal circulation come in close apposition to one another through two closely lying capillary networks. In cattle and buffaloes five layers of cells separate the blood of foetus from that of its mother and because of this, some molecules can not pass through that barrier. For example the placental barrier is impenetrable by vitamin A and globulins (antibodies). Instead, nature provide these vital substances soon after birth through the first milk (*colostrum*).

7.6. PREGNANCY

Pregnancy or gestation is one of the sophistications of evolution found in viviparous mammals. It provides a protected environment to the offsprings in the early stages of their development inside the body of the mother supplying nutrients and a favourable environment and removing waste products. The physiological mechanisms involved are varied and complex. They are co-ordinated by an intricate neuro-humoral mechanism.

7.6.1. Tubal Passage

Following fertilization, the zygote generally takes 3 to 4 days to descend through the Fallopian tube to the uterus. In the cow, usually only one ovum is shed. Upon arrival in the uterus the ovum gets attached to the distal end of the horn on the same side as the ovulating ovary. When 2 eggs are released from a single ovary, one migrates and attaches itself in the uterine horn on the other side.

7.6.1.1. *Period of the Ovum*

The fertilized ovum on reaching the uterus spends a relatively short period of 'free-life' in which it is not attached to the uterine wall. During this 'free-swimming' state prior to attachment, the developing embryo is nourished first by its own yolk and tubal fluids and then by the 'uterine milk'—the secretions of the uterine glands. The eggs of cattle contain very little yolk. The developing conceptus, therefore, depends on *histotroph* (a milky fluid commonly referred to as 'uterine milk')—as a source of nutrients in the early stages of intra-uterine life.

The attachment of the blastocyst (stage of developing embryo) takes 2 to 5 weeks in the cow. In spite of its relatively 'free-living' state, the blastocyst should not be considered as a 'passive object'. The conceptus produces an agent which is responsible for maternal 'recognition' of pregnancy, maintenance of corpus luteum of pregnancy, and anoestrum.

7.6.1.2. *The Period of the Embryo*

During this period the blastocyst undergoes further development with the formation of organs. The embryo more or less reviews some of the major evolutionary milestones the bovine species has come through, for example, at a certain stage, the embryo resembles fish with organs like gills.

Unlike in rodents, the trophoblastic layer of the blastocyst is non-invasive in cattle. Growth of placental tissue occurs mainly in relation to specialized area of the endometrium known as caruncles or cotyledons. There is a pronounced increase in the vascularity of the maternal caruncles. Inter-cotyledonary area shows a marked proliferation of secretory uterine glands. The multiplex or cotyledenory placenta of the cow is formed by the fusion of rounded convex projections of the maternal endometrium (maternal cotyledons or caruncles) with corresponding localized tufts of chorionic villi (foetal cotyledons). This functional units of maternal and foetal cotyledenory fusion is termed 'plascentomes'. Elongated slender chorionic villi develop and become attached to small pits in the endometrial stroma known as 'crypts'. In cows, development of the plascentomes takes 3 to 4 months to be complete. During this period, the plascentomes increase in size and number.

7.6.1.3. *The Period of the Foetus*

It consists of the period when the developed embryo and placenta undergo rapid growth. This period is terminated by parturition.

Absolute weight of the foetus increases little during the first half of intra-utrine life. The foetus makes greatest gain in weight during the last 100 days of pregnancy.

However, foetal weight is far more variable and provides a less reliable index of foetal growth than foetal length. Over the last 200 days of pregnancy the crown-rump (C-R) length of foetus grows in an almost linear relation to foetal age. Even if it follows an elongated S-shape typical of growth curves, the curvings at either end is so minimal that for all practical purposes it can be treated as linear.

7.6.2. Placenta and its Function

Placenta has a meternal element, the endometrium of the uterus and a foetal element, the chorion. Placentation means the apposition or fusion of the endometrium and chorion for physiological exchange.

The foetal part of the placenta, foetal membranes, are developed from an invagination of the developing conceptus, but form no part of the embryo itself. They are shed at birth. There are 4 separate membranes, the *amnion, chorion, yolk sac* and *allantois*. In cattle the placenta is of the allantochorial type, that is, it is formed by the fusion of allantois and chorion. The main functions of placenta are :

(*i*) The most important function of placenta is to transfer nutrients and oxygen from the maternal circulation to the foetal circulation and excretory products in the opposite direction.

(*ii*) The amnion, the inner membrane forms a fluid-filled sac that surrounds the embryo, which forms a protective cushion against external shocks; the foetus floats in the amniotic fluid.

(*iii*) The amniotic fluid prevents foetal adhesions.

(*iv*) At parturition, the amnion, commonly called 'the water bag' opens the cervix by forming a wedge; the amniotic fluid also provides lubrication for delivery.

(*v*) The allantois is an outgrowth of the hind gut through which foetal circulation enters the chorion.

(*vi*) While providing nutrients to the embryo, placenta prevents most of the harmful agents like poisons and some pathogens from entering the foetal circulation.

7.6.2.1. *Classification*

Placentas in different species of domestic animals have been classified on the basis of distribution of chorionic villi into zonary, cotyledonary and diffuse. Cattle and buffaloes belong to the cotyledonary type. Grosser on the other hand had classified placenta on the basis of number of tissue layers intervening between the foetal and maternal blood. He recognised five major types of which cattle was classed under syndesmochorial. In this all the layers of tissue except the maternal epithelium is present. However, Grossers' findings based purely on histological analysis has been superceded by the results of more sophisticated ultrastructural studies of recent times. One such finding by Bjorkman (1973)* showed that the maternal epithelium persists in ruminant placenta. Consequently the placenta of cattle has been reclassified as *epitheliochorial.*

7.7. CALVING

Many factors like (*i*) age and parity of the cow (*ii*) number of the calves carried and (*iii*) species/ breed of the animal (Table 7.1) influence gestation length.

---

*Bjorkman, N. (1973) Acta Anat. 86 (Suppl. 1) : 1.

TABLE 7.1. Gestation length in Different Breeds of Bovines

Breed	*Gestation length (days)*
Cattles	
Bos indicus	
Red Sindbi	283
Sahiwal	286
Gir	286
Tharparker	287
Karkrej	289
Haryana	290
Bos taurus	
Ayrshire	278
Jersey	279
Holstein Friesien	279
Guernsey	284
Brown Swiss	290
Buffaloes	
Murrah	309
Egyptian buffalo	318

7.7.1. Parturition

Parturition is a highly complicated process, regulated by many factors. The regulation of normal delivery is possible only if a well balanced equilibrium between morphological, mechanical, physiological, endocrinological and behavioural parameters of the cow and the calf has been established.

In recent years, considerable evidence has accumulated which suggest that the foetus plays a primary role in the initiation of its own delivery. This hypothesis is nothing original as Hippocrates as early as 460 BC suggested that "at the appointed hour, the foetus puts its feet against the fundus of the uterus and pushes to effect its own delivery."

In fact the foetus and even the fertilized egg, dictates the activity of the oviduct and uterus throughout pregnancy until the time it is sufficiently mature and ready for postnatal existence.

A probable scheme of endocrinological control of parturition is depicted in figure 7.16. The author fully realises that the field is very active and known differences exit between species so that the hypothesis is likely to be modified in the near future. However, as far as present knowledge goes, the sequence of events taking place which leads to the expulsion of foetus and placenta at the appointed time can be summarised as follows :

Towards the end of the gestation period, the foetal brain is stimulated to produce a releasing factor which stimulates the foetal pituitary to produce adreno-cortico-tropic hormones (ACTH). This strengthens the effect of switch from alpha-MSH production to ACTH production towards the end of gestation. What is the biological clock that stimulates the brain to produce the releasing factor ? Certain studies in rats have shown that the timing of parturition is determined by the photoperiod under which the rat has been maintained during the last two weeks of pregnancy. How far this is applicable to large animals is not known.

The increased ACTH action stimulates the production of cortisol by the foetal adrenals (This action of ACTH can manifest itself only towards the last 10-15 days of gestation).

Cortisol secreted by the foetal adrenals during late pregnancy, is believed to provide the stimulus to the subsequent foetal and maternal endocrine events which lead to parturition.

Thus, the present evidence points towards a hypothalamo-pituitary-adrenal axis controlling spontaneous delivery. The prepartum surge of gluco-corticoïd hormones from the foetal adrenal heralds the first phase of the drama of parturition.

Cortisol act on the placenta resulting in increasing levels of estrogen. It also results in lowering the levels of placental lactogen.

Placental lactogen (PL) favours the functioning of estrogen receptors in the corpus luteum during pregnancy favouring progesterone production. Withdrawal of PL will result in the diminishing of estrogen receptors in the CL with the resultant reduction in progesterone synthesis by the CL. This happens earlier than luteolysis but may contribute towards luteolysis. Increased estrogen levels will act on the estrogen receptors in the endometrium stimulating synthesis of prostaglandin F_2 α which causes destruction of CL. Thus the progesterone secretion is brought to the basal level.

Mechanisms which permit the maintenance of uterine quiescence during pregnancy and yet explosive activity at the time of parturition has been studied. The foetus has a number of mechanisms available to ensure that the myometrium remains quiescent during gestation to enable it to undergo development unhindered. Progesterone produced during pregnancy imposes a *conduction block* in the myometrium. It also inhibits oxytocin release and prevents the development of oxytocin receptors.

In the previous section we have seen that due to the influence of foetal cortisol, levels of estrogen go up, placental lactogen and progesterone go down. The high level of estrogen and lowered PL and progesterone act on the myometrium causing hypertrophy, increased prostaglandin production, and increase in prostaglandin and oxytocin receptors. The sensitivity of myometrium to PGs and oxytocin goes up. Similarly the contractality of myometrium increases.

The increased estrogen biosynthesis by the foetal placenta under the influence of foetal adrenal corticoid is a major factor in initiation of parturition. The oestradiol 17 beta produced in the foetal placenta traverses the placental barrier and acts on the estrogen receptors in the maternal endometrium and myometrium. They respond by increased production of prostaglandins if the progesterone level is low.

Action of estrogen on the maternal placenta causes disintegration of lysosomes and the released *hydrolase* enzymes cause placental separation. The release of *phospholipase* enzyme from the lysosomes during breakdown will lead to increased PG synthesis.

Progesterone blocks myometrial activity during pregnancy. Withdrawal of this 'block' is a prerequisite for parturition.

There is considerable evidence to show that oxytocin release from the neuro-hypophysis plays a physiological role in parturition. Oxytocin levels in cows increase during the expulsion stage of labour.

The final phase of parturition involves the active development of uterine contractions and the delivery of the foetus. This is possible only if the endocrine changes preceding it (as discussed earlier) have taken place. At low levels of progesterone and PL, increased levels of oestradiol 17-beta influence hypothalamus to release oxytocin. This sudden surge in the oxytocin level and estrogen influence proliferation of oxytocin receptors in myometrium and placenta and result in increased secretion of endometrial PGs. It also increases myometril contractility.

Fig. 7.6. Cow reaching parturition time. Note the relaxation of the sacro-sciatic ligaments and swelling of the vulva.

There are also certain changes in the cervix.

Relaxin is a hormone which is released from the ovary during the last few days of gestation. It causes relaxation of pelvic ligments (Fig. 7.6) and separation of pubic symphysis thus widening the birth canal. But relaxin has been also found to inhibit spontaneous myometrial activity. It can be hypothesised that relaxin provides a protective mechanism, in late pregnancy. This restrains the evolution of spontaneous activity while the conductivity and oxytocin sensitivity of the uterus are worked-up to peak. This peak of contractility is synchronised with the release of a centrally originating signal capable of over-riding relaxin inhibition, which culminates in labour and parturition. This signal might be the surge of oxytocin as relaxin was found to have no effect on oxytocin stimulated activity of the uterus.

7.7.2. Act of Parturition

Parturition process can be arbitrarily divided into three phases for the sake of convenience in explaining. The normal presentation in bovines is of outstreteched forelegs with chin of the calf resting on them (Fig. 7.7).

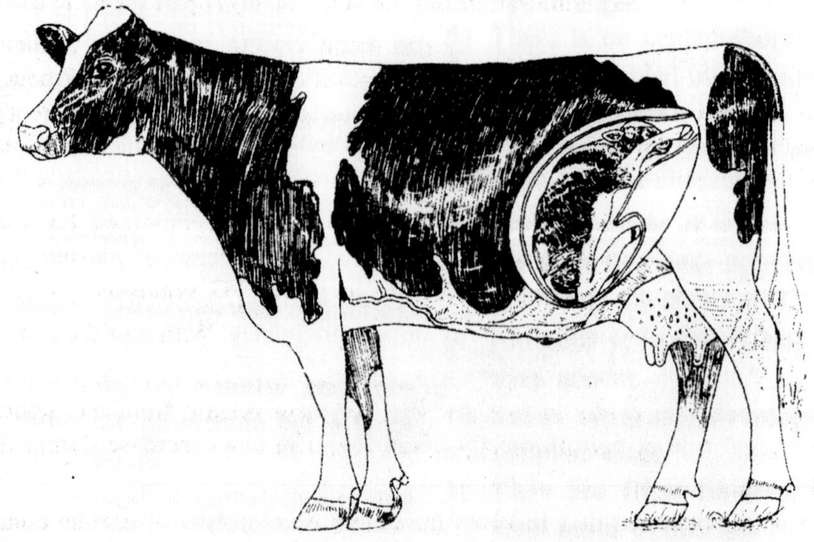

Fig. 7.7. Normal presentation of calf just before parturition:

Phase I. Preparatory Stage. This phase starts from the onset of obvious discomfort and distress in the cow. This will merge with the next phase when the first water-bag (allanto-chorion) breaks (figures 7.8 to 7.10). The duration of phase I was found to vary from 1 hour 30 minutes to 10.30 hours with a mean of 4.07 hours in Jersey-Sindhi crossbred heifers (Mathai and Raja, 1973).

Fig. 7.8-7.15. Different stage of parturition in the cow. (*Courtesy* : E. Mathai, Kerala Agricultural University).

Fig. 7.8. Parturition : Start of Phase I.

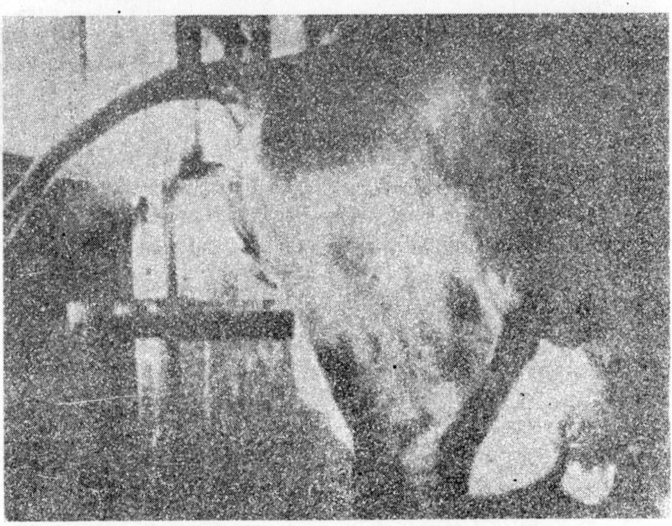

Fig. 7.9. Parturition : Phase I.

Fig. 7.10. Parturition : Phase I merging with Phase II.

Fig. 7.11. Parturition : Phase II.

Phase II. It starts with the rupture of allanto-chorion and consists of various stages of expulsion of the foetus (Figures 7.11 to 7.13). The duration of this stage varies from 35 minutes to 2 hours 45 minutes (Mean 1 hr 12 minutes) in the same crossbreds. It was observed that there was positive correlation between weight of the foetus and duration of the 2nd phase of parturition.

Phase III. It consists of the period between the birth of the calf and expulsion of placenta (Fig. 7.14 and 7.15). On an average, it takes 4 hours and 10 minutes (Range : 2 hrs 45 mts to 7 hrs 20 mts).

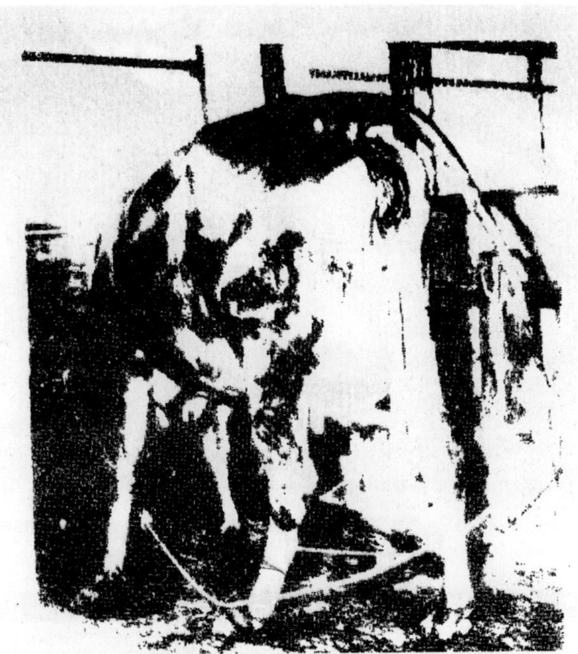

Fig. 7.12. Parturition : Phase II.

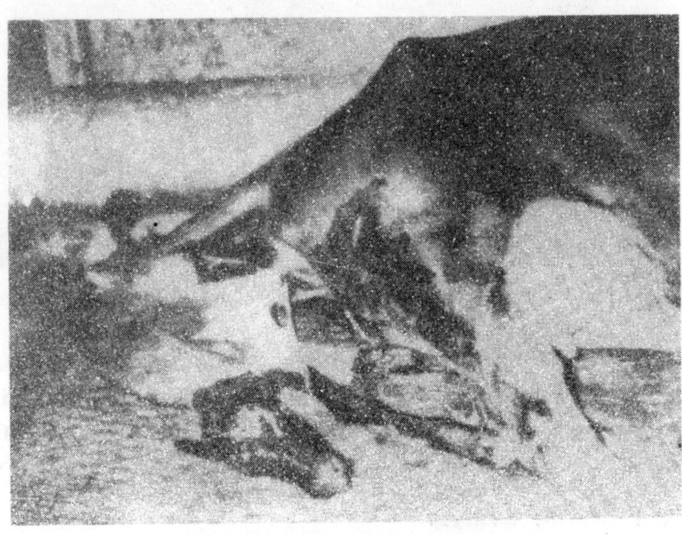

Fig. 7.13. Parturition : Phase II.

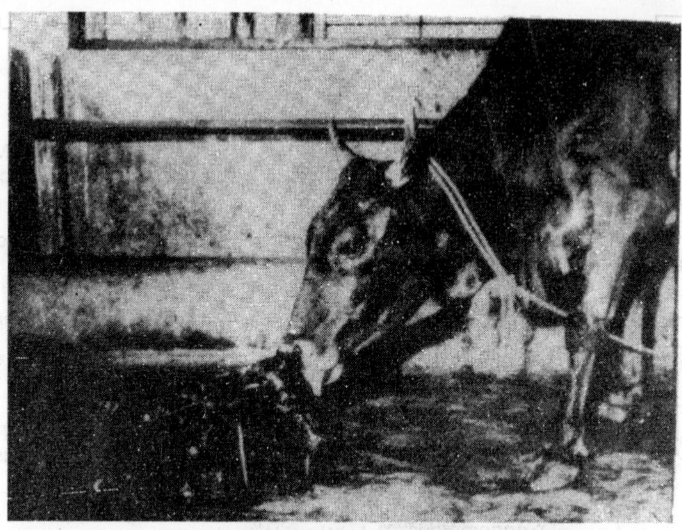

Fig. 7.14. Parturition : Phase III.

Fig. 7.15. Parturition : Completion of Phase III (expelled placenta).

7.7.3. Cow in the Post-partum Stage

The stage after calving and before the subsequent conception is called by this term. The involution of the uterus and initiation of the estrous cycles take place during this stage (Morrow *et al.*, 1969).

7.7.3.1. *Involution of the Uterus*

This is the process by which the uterus which was expanded and stretched during pregnancy involutes or shrinks to its non-gravid size and gains tonicity. This takes from 12 to 56 days. The

Morrow, D.A., Roberts. S.J. and McEntee K. (1969). A review of post-partum ovarian activity and involution of the uterus and cervix in cattle. Cornell Vet. **59** : 134.

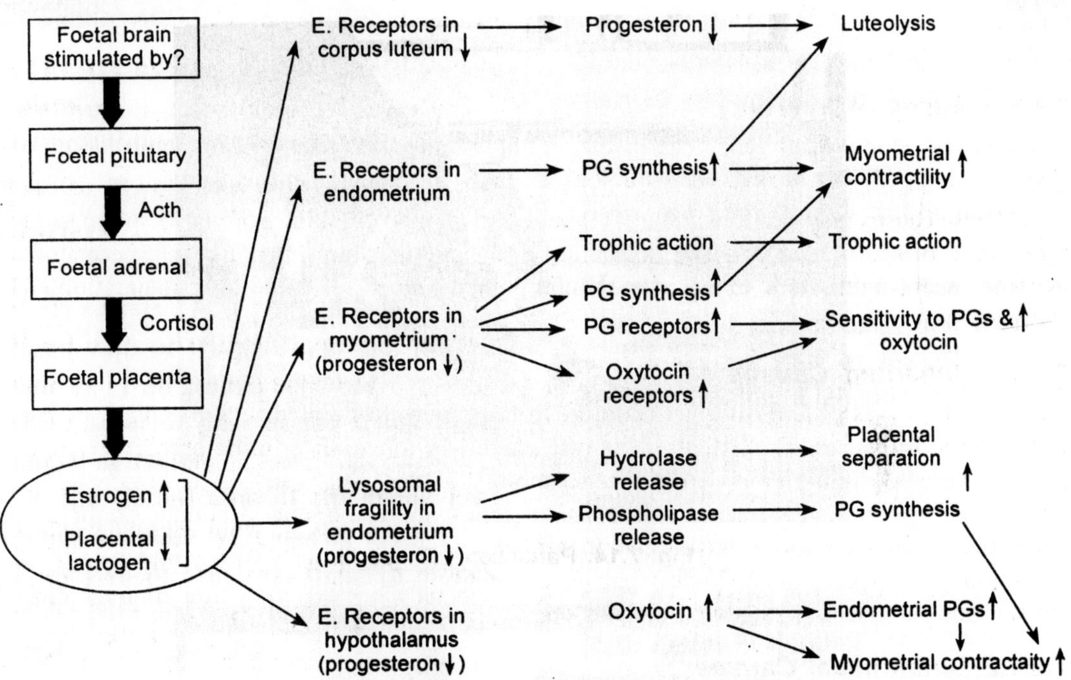

Fig. 7.16. Probable harmonal control of parturition.

length of this period assumes economic importance in that a delay to involute will delay the onset of next heat.

Involution is delayed in cows that have calved more than once (multiparous) and in cows that had complications during delivery.

The cervix shrinks and attains the normal size in about 20-24 hours after calving. Retained placenta is known to delay this.

7.7.3.2. *Subsequent Heat*

Re-establishment of estrous cycles takes one month to 2½ months. The rapidity of onset of subsequent heat is influenced by the intensity of lactation. Higher milk yield and more frequent milking are known to delay it. This is due to the action of prolactin. During the early post-partum silent heats are common (about 45 per cent) *i.e.* ovulation takes place without the overt signs of heat. On the other hand, 9 to 10 per cent cows show heat symptoms without subsequent ovulation.

7.7.3.3. *When to Re-breed*

If we can achieve a target of one calf per year, that is the ideal. But, under general conditions, an average calving interval of 14-15 months should be considered good. Therefore, one should breed dairy cows at the first estrus after 60 days post-partum. Since the gestation period of buffaloes is longer and since they show seasonality of breeding and are more prone to exhibit silent heat conditions, their calving interval is much longer compared to cattle.

7.8. INFERTILITY AND STERILITY

Infertility is the condition in which the animal is unable to reproduce successfully. While infertility refers to a temporary condition sterility means a permanent inability of the animal to reproduce.

Infertility can be of different intensities and may last for varying periods. Corrective measures can overcome infertility among cows.

7.8.1. Causes of Infertility

Lowered reproductive efficiency can be due to (1) inherited, (2) anatomical, (3) physiological, (4) psychological, (5) nutritional, (6) pathological, and (7) management causes.

These causes may result in either (*i*) failure of ovulation (*ii*) failure of viable sperm production (*iii*) failure of sperm transport and deposition at the proper time (*iv*) failure of spermatozoa and ovum to meet (*v*) failure of fertilization though gametes meet (*vi*) failure of implantation and (*vii*) abortion.

7.8.1.1. *Inherited Causes*

Many of the cases of infertility in cattle have not been pin-pointed. Often it may be due to a combination of causes. The extent of the role played by inheritance in the breeding efficiency of cattle is very negligible considering the heritability of this trait which is as low as 0.004. But, inheritance assumes significance when lethal genes are taken into account. About 27 gene-controlled lethal conditions have been identified and listed by Gilmore (1949). They include conditions like hydrocephalus, cleft palate, achondroplasia, mummification, congenital dropsy, atresia ani, anchylosis, congenital spasms etc.

7.8.1.2. *Anatomical Causes*

Many anatomical abnormalities contribute to infertility. Some of them are genetic in origin. In the females, anatomical conditions like incomplete genital tracts, double cervix, adhesion of uterus, ovarian adhesions, absence of cervix, persistent hymens etc. are observed.

Besides, there are abnormalities caused by injury or previous infection. In the bull, anatomical abnormalities like absence or under development of testicles, cryptorchidism, absence of portions of reproductive tract, deformations of reproductive tract or even limbs and feet, etc. contribute to various degrees of reduction in reproductive efficiency.

7.8.1.3. *Physiological Causes*

Most of the physiological causes of lowered fertility appear to be due to hormonal disturbances. Some of the major conditions are described below :

1. Anestrus or absence of sexual cycles in a cow may be due to infantile ovaries or due to persistent corpus luteum. In the former case follicles fail to develop and a heifer will not come in heat at all. One of the major causes of infantile ovaries is malnutrition. Besides there can be genetic causes. The second probability is the anestrus due to persistent corpus luteum. Due to certain hormonal disturbances the corpus luteum persists beyond its life expectancy in a normal cycle, thereby preventing further cycling. Sometimes the corpus luteum gets filled with a fluid (cystic corpus luteum). A common cause of persistence of the corpus futeum is retention of a portion of placenta in the uterus or infection of the uterus leading to pyometra.

2. Silent or quiscent estrus. The behavioural manifestation of heat may be very weak or imperceptible in such cases. But, there is a normal ovulation and if inseminated at the proper time these cows can conceive. Cows go through the normal ovarian changes of the estrus cycle except

Gilmore, L.O. (1949). The inheritance of functional causes of reproductive inefficiency. A review, *J. Dairy Sci.* 32 : 71.

the behavioural heat and sexual receptivity. This condition is more prevalent in water buffaloes than in dairy cows, more in summer season than in other seasons and more in heifers attaining puberty than in adult cows. Anestrus is often observed in the early post-partum period when the lactation is strong, probably due to the influence of prolactin. This is mostly a managemental problem as the dairyman misses the silent heat. In silent heats caused by hot climatic conditions in summer, especially in buffaloes, heat occurs during night which goes unnoticed. In weak heat, the symptoms of heat are subdued and the duration is very short which is not observed.

3. Nymphomania. Such cows otherwise known as 'chronic bullers' exhibit uncontrollable sexual desire due to *cystic ovarian follicles*. When unruptured follicles persist, they apparently continue to produce excess of estrogens resulting in persistent or frequently recurring estrus. The heritability of this condition is very high. The condition seems to occur more frequently in cows during their 2nd to 5th lactation than in older cows or primipara.

4. Irregular cycles. Irregular estrus cycles are an important sign of infertility. Metritis and early embryonic death are sometimes associated with irregular estrus. In cows which appear normal, estrus cycles get longer than 24 days or shortened to less than 18 days due to reasons unknown. While lengthening of cycle did not lower fertility rate, cows with cycles shorter than 18 days were found to have low conception rate.

5. Foetal Mummification. This is an inherited syndrome in which the foetus dies between 5th to 7th month of pregnancy and gets retained in the mummified form. The foetal cotyledons separate from maternal caruncles. The fluid portion is re-absorbed and the dry mummified foetus is retained indefinitely. This condition is accompanied by a persistent corpus luteum.

7.8.1.4. *Psychological Causes*

The possibilities of psychological disturbances lowering fertility rate have not been investigated much. However, general observation and limited experimental data point to the fact that mistreatment and abuse may lower the fertility and conversely gentle handling and congenial surroundings, enhance it. This is all the more important when employing artificial' insemination in which case the cow may become extremely nervous if not handled with patience and care. The nervousness on the part of cows and inexperience on the part of inseminators, both are found to cause lowered fertility. When the cow is extremely nervous, the adrenal medulla pumps in the fright-hormone epinephrine into the circulation which counteracts the action of oxytocin. Oxytocin is released at the time of mating or artificial insemination which helps in transporting sperms from the cervix to the oviduct, where fertilization takes place, by causing strong uterine contractions. When it is counteracted by epinephrine released due to fear and nervousness caused by rough handling and mistreatment, sperm transport is affected causing reduced fertility.

Mistreatment or unpleasurable experience on the part of the bull at the time of mating or semen collection may lead to psychological frigidity. Bulls may also develop a psychological inhibition against a particular surroundings, teaser or collection procedure in which case a change may overcome this problem. General mistreatment at times other than at the time of breeding has also been found to cause some lowering of fertility.

7.8.1.5. *Nutritional causes*

Cattle fed normally do not exhibit any lowered fertility due to nutritional causes because the need of nutrients for reproduction are the same as those required for maintenance, growth and milk production. Fertility is affected only when there are serious deficiencies.

Deficiency of energy and protein. Inadequate availability of energy and protein in the ration of growing animals may lead to nutritional infertility of various degree. If the restriction is only mild it will have normal fertility but at an older age. Effects of slight to moderate restriction may lead to stunted growth, infantile gonads and markedly delayed puberty. Feeding well after these changes have occurred may not reverse the situation.

This is very relevant to cattle in developing countries like India, where they depend on subsistence grazing on crop aftermaths and extremely overgrazed and denuded pastures. These effects are aggravated by drought and hot climate experienced in most of these regions.

On the other side of the spectrum, very fast growing heifers and bull calves also exhibit reduced fertility although there are some reports disproving this.

7.8.1.6. *Pathological Causes of Low Fertility*

Many pathological agents have been known to cause sterility and lowered fertility which can be extremely costly for the dairyman. Reproduction is affected only indirectly be general systemic diseases. But, some organisms cause a localised infection of the reproductive tract affecting fertility in a more direct way. These diseases may be transmitted by sexual contact or through other routes and may be caused by viruses, bacteria or protozoa. A more detailed discussion of the diseases is provided in Chapter 20. Here only their names have been mentioned.

Viral diseases

1. Catarrhal vagino cervicitis
2. Infectious pustular vulvo-vaginitis
3. Granular vulvo-vaginitis
4. Viral abortion and still-births
5. Viral orchitis
6. Fibropapillomata of the genitalia

Bacterial diseases

1. Brucellosis
2. Vibriosis
3. Leptospirosis

Protozoan diseases

1. Trichomoniasis

Besides these, certain non-specific organisms which are not normally pathogenic may also cause conditions like bovine non-specific endometritis. Some mycotic infections lowering fertility have also been identified.

7.8.1.7. *Managemental Factors Causing Lowered Fertility*

Management is a major factor that determines success of reproduction. Many decisions taken and actions carried out at the proper time go a long way in ensuring adequate fertility rate on a dairy farm. Failure in one or many may have ruinous effects on reproduction and profits. General management will affect reproduction, but here more specific aspects of management which have a direct effect on it are discussed.

(1) The most important managemental cause of lowered fertility is improper detection of heat. Some cows have silent heat or highly subdued heat. In some cases cows may have heat of

very short duration. Under hot summer conditions bovines, especially, buffaloes come into heat during the night. Such nocturnal heat may be of shorter duration and of less intensity that no tell-tale signs are left for the morning.

Detection of heat becomes more difficult when they are tied and stall-fed throughout.

(2) Breeding records are very useful managemental aids for ensuring efficiency in reproduction. Failure to keep proper records with respect to (*a*) dates the cow has come in heat (*b*) date of service (*c*) sire used (*d*) result of pregnancy diagnosis (*e*) expected date of calving (*f*) actual date of calving (*g*) health history of cow at and after claving such as dystokia, prolapse. retained placenta, metritis may lead to confusion. The whole breeding operation then will become a blind man's game and the reproductive efficiency of the herd is bound to be affected because it will be difficult to plan and organise the management to : (*i*) show the approximate date on which the cow is expected to come in heat, (*ii*) show whether the heat period is regular, (*iii*) identify cows which have previous history of breeding trouble so that special attention can be given to them, (*iv*) locate cows that may need to be treated before breeding because of reproductive disorders, (*v*) aid the veterinarians in making a proper diagnosis of the breeding trouble.

(3) Another managemental failure is due to improper time of insemination. Ovulation takes place 12 to 14 hours after the end of the heat period. But then the egg has to travel down the oviduct before it meets the sperms, which may take roughly 6 hours. Thus, if a cow is served when she first displays signs of heat, the sperms may have to remain in the reproductive tract of the cow for 30 hours or more before they encounter the descending ovum. Though many may survive this length of time, their activity tend to fall fairly quickly, and there is a risk that conception may fail to take place.

Insemination very near to or after ovulation is also not conducive to high fertility rate. The sperms must be exposed to the female reproductive tract for about 6 hours for attaining the capacity to fertilize. The ova are viable for only about 10 hours after ovulation. The maximum fertility period, therefore, is that between mid-estrus to about 6 hours after the end of estrus. Any deviation in terms of early service or late service can affect success of the breeding operation.

(4) Breeding cows in the first heat after freshening reduces conception rate. In addition, the cows do not get enough time to recoup the body reserves depleted during lactation. It is a better management procedure to let the cow pass through two estrus cycles after calving before breeding her again. A healthy cow may be bred in the third heat period. But, if the cow has history of metritis or retained placenta it will be advisable to give her sexual rest until she is completely recovered.

(5) Keeping a cow open too long after calving or delaying breeding of a heifer beyond a certain level results in lowering of fertility. Many owners due to ignorance or carelessness miss heats in early lactation and think of getting the cow bred only towards the end of lactation. Every time a cow undergoes a sexual cycle additional chance is given for some abnormal conditions to develop that may lower breeding efficiency. Therefore, the sooner a cow is rebred after she becomes normal after calving the better is the chance for conception.

CHAPTER

8

Artificial Insemination

Artificial insemination (AI) is the technique in which semen with living sperms is collected from the male animal and introduced into the female genital tract by man. This has been found to result in normal conception and produce progeny similar to what would have been under natural mating. The conception rate through AI has been satisfactory in cattle.

It was Spallanzani, the famous Italian scientist, who conducted the first recorded research in AI with dogs, even though, there are legends telling that an Arab chieftain used AI method to steal the *seed* from a prized stallion of the enemy to breed his mare. But it was Russians under the leadership of Mikhail Fyodorovich Ivanov, who put AI to extensive field use. Soon AI gained roots in Denmark and USA and from there it spread to all countries having an interest in dairying. In India, for the first time Sampath Kumaran employed this method in 1939 at the Palace Dairy Farm, Mysore. The Indian Veterinary Research Institute started scientific study of AI in 1942 under the leadership of P. Bhattacharya. Soon AI was adopted in the country especially through projects like Key Village, Community Development Blocks, National Extension Services, AI Centres, Intensive Cattle Development Programmes and Operation Flood project.

8.1. ADVANTAGES OF AI

Like any other modern technology, AI affords many advantages, but under certain unfavourable conditions it may do more harm than good. The more important of the advantages are listed below :

(*i*) ***Increased use of sires.*** AI enables an individual bull to sire a large number of progeny. This reduces the number of bulls required for breeding purpose. Consequently, a more rigid selection of males to the exclusion of inferior stock is made possible increasing the superiority of the males selected. This accelerates genetic improvement.

(*ii*) ***Progeny testing*** to evaluate young bulls can be effectively used only through the application of AI. It is the most accurate means of evaluating males which has revolutionized dairy cattle breeding in recent times.

(*iii*) Crossbreeding plans can be carried out more effectively when semen of superior sires are made use of through AI. Even bulls maintained continents away can be used for breeding.

(*iv*) ***Productive life of valuable bulls extended.*** Artificial insemination coupled with deep freezing of semen has made it possible to extend the use of bulls even years after they are dead.

(*v*) ***Prevents spread of diseases.*** AI provides a universal means of limiting the spread of venereal diseases.

(*vi*) AI gives scope of effecting certain matings which under natural conditions are incompatible due to physical disparity, breed difference or species difference.

(*vii*) ***Males carrying undesirable genes are easily detected*** because of the greater number of offsprings in AI than in natural service.

(*viii*) ***Breeding efficiency is improved.*** Regular examination of semen and frequent check on fertility helps in the detection of bulls having poor fertility. Rectovaginal technique of insemination affords opportunity to the operator to palpate the female genitalia and detect abnormalities or pathological conditions if any.

(*ix*) AI permits selected matings of outstanding bulls. It provides a means for large scale breeding work on a mass scale.

(*x*) ***Benefit to small farmers.*** The expenditure of maintaining a bull on the farm call be saved if AI service is available. Due to economic reasons a small farmer may not be in a position to maintain a superior bull. In such cases AI makes it possible to bring into service a sire of outstanding merit at a nominal cost.

(*xi*) AI is the most economical method for grading up indigenous stock with imported semen of outstanding bulls of improved breeds. Thus AI affords moving superior cattle germplasm about the world.

(*xii*) AI Centres may form a focal point of developmental activity. It brings small farmers and trained personnel into closer contact. Experience in Anand, India has shown that such AI centres can even become the inspiration for a silent social revolution.

8.2. ARTIFICIAL INSEMINATION IN INDIA AND OTHER DEVELOPING COUNTRIES

India has invested heavily in organising a country-wide AI programme for cattle and buffaloes. Another country like India which has gone headlong into introducing nationwide AI programme is Egypt. To date no systematic improvement of the stock has been made. For the funds invested, the improvement obviously has been exceedingly small. This was mainly due to :

(*i*) Lack of appreciable genetic superiority in the sires used to provide enough selection differential, and

(*ii*) Lack of a follow-up programme to ensure enough nutrition to the progeny through AI so that they can express the genetic superiority gained in terms of production.

Besides, AI has reached only to a limited extent into the remote rural areas. Although purely from the economic standpoint, the AI programmes in India and Egypt have been flops, nevertheless, they have served social and political objectives which in the long run may prove beneficial.

In contrast to the picture in these countries, AI programmes had been quite successful in Israel. Australia, Mexico, Jamacia, Brazil, Colombia, Costa Rica, Venezuela, Kenya and Uganda. One reason attributed to success in AI in these countries is that AI and other better services were intensively applied to limited areas instead of broad coverage. AI was made available to those herds only which had above average managers who appreciated and understood how they can derive the benefits from it.

The lesson is very clear. AI can be successful only where it will serve as an integral part of efficient and well managed production practices. It should also be borne in mind that AI is not a miracle which can solve fundamental problems of production. At the same time, AI should be seen as an essential tool of tremendous potential which can revolutionize dairy production if used in accompaniment of other improved husbandry methods.

It is time that in India AI programme is made more economy oriented. Instead of broad area coverage AI may be limited to planned matings in selected herds and in selected areas. It should be accompanied by sound nutritional and disease control programmes. An outstanding proof for this is the success of AMUL and Mehsana in Gujarat, India.

8.3. AI TECHNIQUE

AI has been widely adopted in dairy cattle due to the following :

(*i*) Relatively simple and practical methods of collecting semen free from urine, hair or foreign matter were evolved.

(*ii*) Successful methods were developed for diluting and storing semen. This enabled the use of a single ejaculate for large number of cows after storing for a considerable period of time.

(*iii*) Simple insemination techniques for depositing semen in the female tract were evolved.

(*iv*) These methods gave fertility rates comparable to those obtained by natural service.

8.3.1. Collection of Semen

8.3.1.1. *Artificial Vagina (AV)*

This is the universal method used for routine collection of semen. The equipment used is, in the main, similar in design to the one shown in figure 8.1 although several modifications exist. The short pattern AV is preferred because semen is ejaculated directly into the cone avoiding mixing with preputial debris which may be carried into the liner at the time of inserting the penis. This type of AV consists of a thick canvas reinforced outer rubber cylinder 28-30 cm long and an inner thin latex rubber lining rolled over on the ends and secured firmly to the outer casing. At one end of this cylinder a rubber cone is attached. At the other end of this cone a graduated test-tube for collecting semen is fastened. This test-tube and the cone are covered by an insulating muff to protect the ejaculate from temperature shock and solar radiation and also to avoid the risk of breaking the glass test tube. The space between the outer casing and inner lining of the cylinder forms a water jacket which is filled firstly with warm water and then with air to

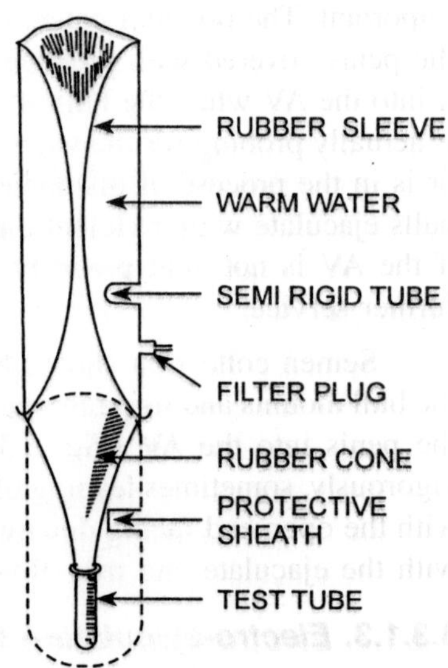

Fig. 8.1. Artificial vagina for bulls.

a temperature and pressure preferred by the species and, if known, by the individual. In other words, AV provides an environment similar to that of the vagina with a temperature and pressure simulating the female copulatory organs. The anterior part of the inner lining is smeared with sterile liquid paraffin or vaseline to provide lubrication. Prior to collection the internal temperature is adjusted slightly above body temperature using both hot and cold water. It should be about 43-45°C before leaving the laboratory as there is inevitably some heat loss before the actual collection of semen.

The bull is made to mount a 'teaser' which may be a cow in estrus, or a cow hormonally treated to exhibit signs of heat. Bulls can also be trained to mount other docile bulls or buffalo bulls and give semen in AV. The animal to be mounted is restrained in a service crate. The bull is led out to the 'mount' when the preparations for collection have been completed.

8.3.1.2. *Teasing*

Teasing or pre-collection stimulation of the bull has a marked effect on semen quality and volume. This consists essentially of exciting the bull by allowing him to see, approach, sniff at the teaser and by 'false mounting' in which ejaculation is prevented (Fig. 8.2). The bull is withheld from service until the penis is erected and accessory gland fluid starts dribbling from the prepuce. Generally 15 minutes of teasing and one or 2 false mounts will suffice. However, over-doing this procedure may be avoided as this may result in the bull passing the peak excitement period and thereby losing interest.

Fig. 8.2. Sexual preparation of the bull before semen collection.
(*Courtesy* : Abraham Mathew)

The way collection is made is also important. The operator catches the base of the penis covered with prepuce and diverts it into the AV when the bull after mounting is actually probing for the vagina. The AV should never be applied when the bull is just mounting or is in the process of dismounting. The angle at which the AV is held is also important. Some bulls ejaculate with a violent thrust resulting in sharp bending of penis and the infliction of pain if the AV is not held properly. Repetition of such painful experience may inhibit the bull for further service.

Semen collection should be done adeptly and fast enough with minimum of force. When the bull mounts and its penis starts probing for the vagina the operator holding the prepuce directs the penis into the AV (Fig. 8.3). As soon as the penis enters the AV the bulls usually thrust vigorously, sometimes leaping off the ground in that process. As the bull dismounts the AV is held with the open end facing downward so that contaminating debris and hair in the liner do not mix with the ejaculate and may flow out.

8.3.1.3. *Electro-ejaculation Method*

Not infrequently, some bulls, due to reasons such as lameness or old age, are unable to mount and give collection. If that bull is a genetically superior one, considerable loss can be prevented if the semen is collected by a method which does not involve mounting. The electro-ejaculation

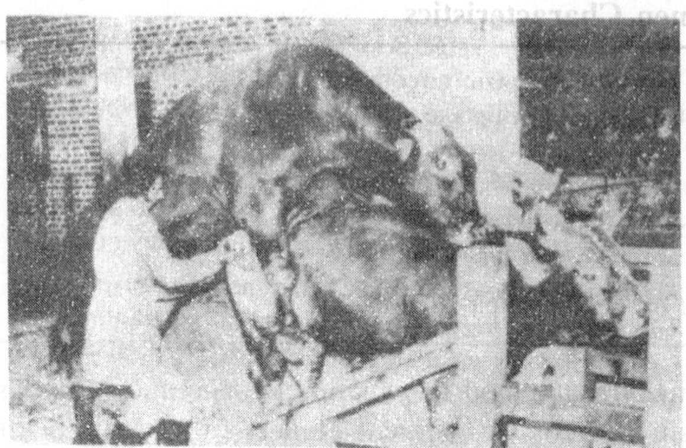

Fig. 8.3. Semen collection (*Courtesy* : Abraham Mathew)

technique originally devised for rams, but later successfully modified for cattle, is of immense use under such circumstances. Later models of electro-ejaculators employ a single rectal probe with bipolar electrodes, which is introduced into the bull's rectum. Rhythmic application of electrical stimulation will cause discharge of accessory fluids followed by semen which is collected in a graduated tube with a funnel held below the prepuce.

Although the semen collected by electro-ejaculation is healthy with normal viability and fertility, there are some indications that sperms are more susceptible to cold shock and less likely to survive freezing and thawing.

8.3.2. Evaluation of Semen

The sure test of viability of semen is to observe the fertility of inseminated females. Obviously, this is not sufficient as we need an assessment of semen quality before inseminating. We may even need the information before diluting because some samples of semen which may be considered as normal under natural service may not effect fertilization after dilution. It is also necessary to look for changes in semen which have resulted from pathological processes. A number of semen characteristics have been used for semen evaluation. Most of the methods involve estimation of sperm-cell numbers or motility and metabolism of spermatozoa.

As soon as the semen is collected, visual observation for volume and colour is made. The volume characteristic for normally collected semen is 5-6 ml. It should be creamy white in colour and free of urine, hair or other foreign matter. In some bulls 'yellowish' colour of semen may be normal.

Under routine conditions, evaluations are largely based on visual assessment of semen density and spermatozoa activity by microscopic examination of films of semen. In addition, live-dead percentages, percentages of abnormal spermatozoa and temperature shock are included periodically (Table 8.1). But in a more detailed evaluation, rate of fructolysis, methylene blue or resazurin reduction time, pH changes, hyaluronidase content, light reflecting capacity, oxygen uptake, carbon dioxide production etc. can be employed. Alterations in acrosomal cap of sperms may indicate deterioration in fertilizing capacity of sperms. Physical examination of spermatozoa with light and electron microscopes can reveal these acrosomal abberrations. The frequency of intact acrosomes of sperms in semen samples was found to be highly correlated with the fertility of bulls.

TABLE 8.1. Bull Semen Characteristics

Volume (ml)	5-6
Sperm concentration (10^8/ml)	12
Total sperms (10^9)	7
Motile sperms (per cent)	70
Morphologically normal sperms (per cent)	80
Ejaculates/week	4
Living cells needed per insemination (10^7)	1

Competitive fertilization method in which heterospermic (mixed semen of more than one bull) inseminations using breeds of different colours or animals of different blood groups as means of identifying successful fertilization has been employed by some to evaluate semen from different bulls.

Semen density may be evaluated by haemocytometer counts or by use of calibrated spectrophotometer. This measurement is required when high semen dilution rates (over 1 : 100) are contemplated. This is to ensure that the sperm concentration in the diluted semen remains above the threshold level needed for optimum fertility—say 10 million motile spermatozoa in one inseminate (one ml).

The proportion of progressively motile sperms in diluted semen sample at 37°C is measured by counting under a microscope. The minimum percentage of progressively motile sperms acceptable in a semen sample is 70.

The methylene blue or resazurin reduction tests have been widely used in USA and Scandinavia. This test is simple enough to be fitted into normal routine evaluation of semen. Semen samples having a methylene blue reduction time less than 15 minutes are acceptable.

Live and dead counts are based on the principle that as sperm cells die, they become permeable to *eosin*. Therefore. those cells which have absorbed the red colour of *eosin* can be considered as dead. *Nigrosin* is used as a background colour.

8.3.3. Semen Dilution and Diluents

The findings of Phillips and Lardy[1] in 1940 that egg-yolk can be used as an effective semen diluent contributed in no small way to extensive use of AI. The basic properties of a diluent needed for satisfactory preservation of fertility and prevention of temperature shock are as below :

(*a*) pH of the diluent similar to normal semen.

(*b*) Osmotic tension of the diluent should be isotonic with blood of cattle.

(*c*) Buffering capacity to prevent the toxic effect of end products of sperm metabolism.

(*d*) Correct balance of electrolytes and non electrolytes, cations and anions.

(*e*) Non-toxicity.

(*f*) Stability even after long periods of storage.

(*g*) Presence of nutrient materials to support the reduced metabolic activity of the spermatozoa in it.

Phillips, P.H. and Lardy, H.A. (1940). A yolk-buffer pabulum for the preservation of bull sperm. *J. Dairy Sci.* **23** : 399.

(*h*) Absence of any inhibiting effect on antibiotic additives.

(*i*) Provide lipoproteins and/or lecithin to protect sperm cells against low-temperature shock.

(*j*) Provide a source of reducing substances to protect the sulf-hydryl-containing enzymes.

(*k*) Should be sterile and free from harmful substances.

Besides, from the practical point of view, the diluent should be easily available, cheap, easy to prepare and permit the microscopic examination of the spermatozoa in a clear field.

Before 1940, the Russians developed many diluents which contained glucose and had the same pH and osmotic pressure as fresh semen. But since 1940 when Phillips discovered the beneficial effects of egg-yolk, the extenders containing this ingredient replaced the previous ones. Now, the commonly used diluents contain either egg-yolk or milk as the essential component. The protection against cold shock is provided by the lipoproteins and lecithin in egg-yolk and casein-the milk protein in milk.

Tris-sodium citrate is added to the egg-yolk to provide adequate buffering capacity, Homogenised milk has the intrinsic buffering capacity and there is no need of any addition of buffering agents to it. In addition to cold-shock prevention and buffering, egg yolk and milk contain certain substrates which will support spermatozoan metabolism. These include glucose, vitamins, enzymes, proteins, fats and minerals. The beneficial effects of these constituents have not been fully assessed.

8.3.3.1. *Preparation of Egg-yolk Diluents*

The original diluent of Phillips and Lardy had phosphate as the buffer and had the following composition :

KH_2PO_4	0.2 g
$Na_2HPO_4.12H_2O$	2.0 g
Glass distilled water	100 ml
Egg-yolk	100 ml

The egg is cleaned with soap, washed with water and swabbed with Spirit. Using a pointed scissors the egg shell is cracked and cut around in the middle. With one half of the shell holding the yolk. the other half is removed with the excess white flowing out into a beaker kept below. The yolk is then transferred into a beaker containing distilled water at 50°C to make traces of albumen adhering to the yolk, clot. The water is poured off and the egg-yolk transferred to a clean filter paper. By rolling gently back and forth the yolk is dried. Then the yolk-membrane is ruptured and the contents are poured into a sterile measuring cylinder, the membrane remaining on the filter paper. To 100 ml of yolk prepared as above 100 ml of phosphate buffer is added. The phosphate buffer solution is prepared in boiled glass distilled water by adding 0.2 g of KH_2PO_4 and 2.0 g of $Na_2HPO_4.12H_2O$ per 100 ml. Strict precautions to prevent contamination is to be exercised throughout.

Subsequently, phosphate buffer was replaced by sodium citrate. This was because phosphate diluent made diluted semen opaque due to the presence of large fat globules thus rendering microscopic examination of semen sample for motility of individual spermatozoa impossible. Use of 3.6 per cent sodium citrate solution in place of phosphate buffer gave an equally satisfactory diluent, but, with the added advantage that the fluid was clear enough for microscopic assessment of sperm motility.

Afterwards the proportion of sodium citrate was reduced to 2.9 per cent on the plea that 3.6 per cent buffer is not isotonic with bovine spermatozoa. However, these two concentrations of citrate buffer have given equally good conception rates. Attempts have also been made to lower the proportion of egg-yolk. Fertility was not significantly impaired by reducing it below 50 per cent. Extension with egg-yolk up to 33.3 per cent or even 20 per cent is in practice at certain stations at present with no apparent reduction in fertility.

The detrimental effects of bacteria on sperms and subsequent fertility have been recognised for long. It is almost impossible to prevent contamination under mass-scale semen handling. The first attempt to overcome this difficulty was by adding sulfonilamide. Later this was replaced by antibiotics like penicillin, streptomycin, neomycin etc. A combination of pencillin and streptomycin is popularly used at present. At a later stage when drug-resistant varieties of organisms pose serious problems, neomycin or a combination of lincomycin and spectinomycin can be used equally effectively. Thus an ideal egg-yolk semen diluent as per present day standards should have the following composition :

Sodium citrate	2.9 g
Glass distilled water	100 ml
Egg-yolk	100 ml
Streptomycin	500-1000 μg/ml
Penicillin	500-1000 I.U./ml

There had been many attempts to improve and modify the above extender. Cornell University workers Foote and Bratton[2] formulated a diluent termed **Cornell University Extender (CUE)** which permitted longer storage at 5°C. The composition of CUE is given in Table 8.2.

This or other similar diluents have pH adjustments for self carbonation which will keep down the metabolic activity of spermatozoa.

TABLE 8.2. Composition of CUE

Ingredients	*Composition*
Buffers :	
Sodium citrate dehydrate (g)	14.5
Sodium bi-carbonate (g)	2.1
Potassium chloride (g)	0.4
Glucose (g)	3.0
Sulfanilamide (g)	3.0
Glycene (g)	9.37
Citric acid (g)	0.87
Distilled water (final volume) (ml)	1000
Extenders :	
Buffer (per cent by volume)	80
Egg-yolk (per cent by volume)	20
Penicillin (I.U./ml)	1000
Dihydrostreptomycin (μg/ml)	1000

Foote, R.H. and Bratton, R.W. (1960). *J. Dairy Sci.*, **43** : 1322.

8.3.3.2. *Milk Diluents*

Diluents consisting of whole milk, homogenised milk, skim milk, reconstituted milk powder or milk whey are also commonly used for cattle and buffalo semen. Fresh milk has an adverse effect on the spermatozoa survival probably due to the activity of *lactenin*, an anti-bacterial substance normally present in the albumin fraction of milk. Heating at a temperature of 92-95°C for 10 minutes and allowing to cool before adding semen reduces the sperm-killing property of milk to acceptable levels. Field result based on 60-90 day non return rates indicated that over-all conception rates with homogenised milk were equal to, or superior to, those with egg-yolk citrate. In these cases, inseminations were done on the same day or the following day of collection. There are some indications that fertility is lowered considerably from the third day onward in milk diluents.

8.3.4. *Storage of Semen at 5°C*

Most of the semen used in developing countries is maintained at refrigerator temperatures (4°-5°) *i.e.,* chilled semen.

Soon after collection and removal of a small quantity for evaluation, semen vial is stoppered and placed in a water bath at 25°C. The diluent is also brought to 25°C. After gently mixing the semen with the diluent, the diluted semen is preserved in the refrigerator at 4°-5°C.

Semen handling should be as aseptic as possible. Ideally, all glassware should be sterilized and kept stoppered with cotton wool plugs and mouths of containers flamed on a spirit lamp before opening and transferring fluids.

Undiluted spermatozoa have high metabolic activity and are very susceptible to cold shock. Therefore, the time between collection and dilution should be the minimum possible. Relatively rapid cooling can be effected safely after dilution, as the diluters in common use have a protective action against cold shock.

8.3.5. Storage at Room Temperature

Expanding use of AI in developing nations, where refrigeration facilities may be lacking, has kindled an interest in developing techniques for storing semen at room temperature.

Van Demark and Sharma developed the first successful technique for ambient temperature preservation of semen and termed it the "Illini Variable temperature" (IVT) technique and, the medium, IVT-medium. It is based on the observation that bubbling of carbon dioxide gas through an egg-yolk citrate medium markedly reduced sperm activity (metabolism and motility). It was found that fertility was maintained up to 6 days. This medium consists of a 2 per cent solution of sodium citrate dihydrate with small amounts of potassium chloride, sodium bicarbonate, glucose and sulfonilamide. This solution is gassed with CO_2 until a pH of 6.35 is obtained. Egg-yolk is added at a rate of 10 per cent and 1000 IU of penicillin and 1000 μg of dihydrostreptomycin each for every ml is added. The diluted semen is stored in sealed ampoules at room temperature in the dark.

8.3.5.1. *Caprogen*

New Zealand is a country where great attention was given to development of ambient temperature extenders. *Caprogen* is the outcome of this effort. It incorporates caproic acid in an IVT-CUE type of extender followed by gassing with nitrogen. It has been proved effective in maintaining maximum fertility for a 24 hour period at ambient temperature.

CME extender is another effort in this direction in which the amount of perishable egg-yolk is reduced to as little as 0.5 per cent. Instead 15 per cent milk is added as a nutrient source. Diluting with CME has been found to preserve fertility at good level for 4 or more days at room temperature.

8.3.6. Diluents for Buffalo Semen

One of the reasons for slow development of AI in buffaloes was the lack of an effective semen diluent. Egg-yolk citrate, CUE, IVT and other diluents developed for cattle were not ideally suited for extending buffalo semen. So far, AI in buffaloes almost invariably provided very low conception rates, the figure is frequently below 30 per cent on first insemination.

Of the egg-yolk containing diluents tried, Kampschmidt's diluent with the following composition was generally found to be better :

Dextrose (Analar) solution (5 per cent) ... 4 parts
Sodium bicarbonate (Analar) Solution (1.3 per cent) ... 1 part
Sulphamezathine (ICI) Solution (2 per cent) ... 1 part
Egg-yolk ... 1 part

The solutions are first prepared separately and autoclaved at 7 kg pressure for 20 minutes. They are then mixed in the proportions mentioned above. Egg-yolk is added last and the pH is adjusted to 6.8.

Greater success has been reported with milk diluents from Pakistan. In India also milk when used as a buffalo semen extender gave equal or better results as EYC. Girish Mohan (1974) found that low fat cow milk (heated to 92-93°C in a water bath for 20 minutes, cooled in a refrigerator overnight and the fat pelicle removed by filtering through cotton plug and with usual antibiotics) gave the best results among many diluents compared.

Skimmed milk alone or with other ingredients were suggested by others. Two such diluents are listed in Table 8.3.

TABLE 8.3. Composition of Modified Skimmed Milk Diluent for Buffalo Semen

Ingredient	*Diluent I*	*Diluent II*
Heated skimmed milk	50 parts	60 parts
4 per cent glycene solution	25 parts	
Egg-yolk	25 parts	20 parts
5 per cent fructose solution		20 parts

Source : Tomer and Desai, 1961, *Indian J. Dairy Sci.* **14** : 61.

Work reported from National Dairy Research Institute, Karnal indicate that cow's milk whey is a better buffalo semen extender. Their procedure involves precipitating 100 ml cow's skimmed milk with 3 ml of 10 per cent citric acid (enough to make the pH 4.3), filtering through cotton plug and adjusting the pH of the filtrate (whey) to 6.9 by 10 per cent NaOH. The whey is heated in a water bath at 92°C for 20 minutes, cooled in a refrigerator overnight and filtered. Antibiotics like penicillin and streptomycin are added at usual levels. They have developed a ready to mix dry powder formula consisting of 9 grams of skimmed milk powder and 0.3 grams of citric acid. This had to be diluted in 100 ml distilled water and filtered and the filtrate treated in the same manner as in the previous case.

In spite of moderately satisfactory results from milk and whey diluents, it cannot be told that there has been a breakthrough. Development of a suitable buffalo semen extender which can stand the test of extended field application is yet to take place

8.4. THE INSEMINATION TECHNIQUE

Maximal fertility can be obtained by depositing viable healthy sperms in sufficient number at the proper site in the female genital tract at the optimum time in relation to ovulation. Generally two techniques are employed in cattle and buffaloes. namely, the speculum method and the rectovaginal method.

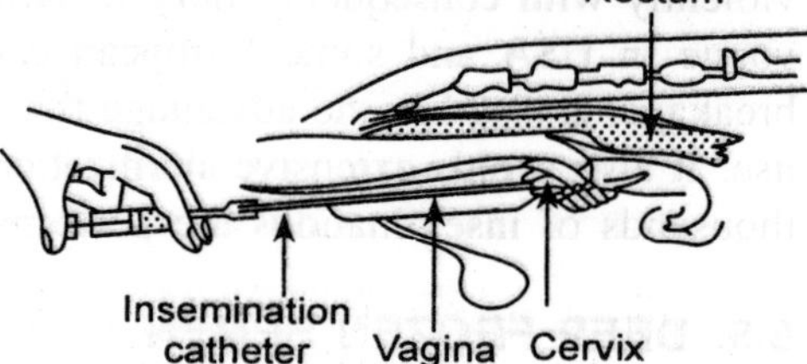

Fig. 8.4. Recto-viginal techniques of inseminating a cow.

In the **speculum method**, the vagina is held open by a lubricated sterile vaginal speculum and semen is deposited in the uterus through open cervix by means of an inseminating pipette by visual location of cervix. This method may be easy for a relatively less trained operator. But the main disadvantage is that it is not practicable to effectively sterilize the apparatus used after each insemination to prevent spread of diseases. It is also impossible to carry that many sterilized vaginal specula for each insemination when the AI man has to visit distant villages.

The **recto-vaginal technique** is relatively simpler and prevents spread of diseases (Fig. 8.4). A sterilized thick-walled glass pipette (Fig. 8.5) attached to a 2 ml plastic syringe by means of a short length of thick rubber tubing forms the apparatus. About one ml of diluted semen is drawn into the glass pipette by means of the syringe. A gloved hand is inserted into the rectum of the cow and the uterus is held in the hand through the rectal wall. Care should be taken to hold around the cervix and pull it forward to avoid tipping of the anterior opening of the cervix. The

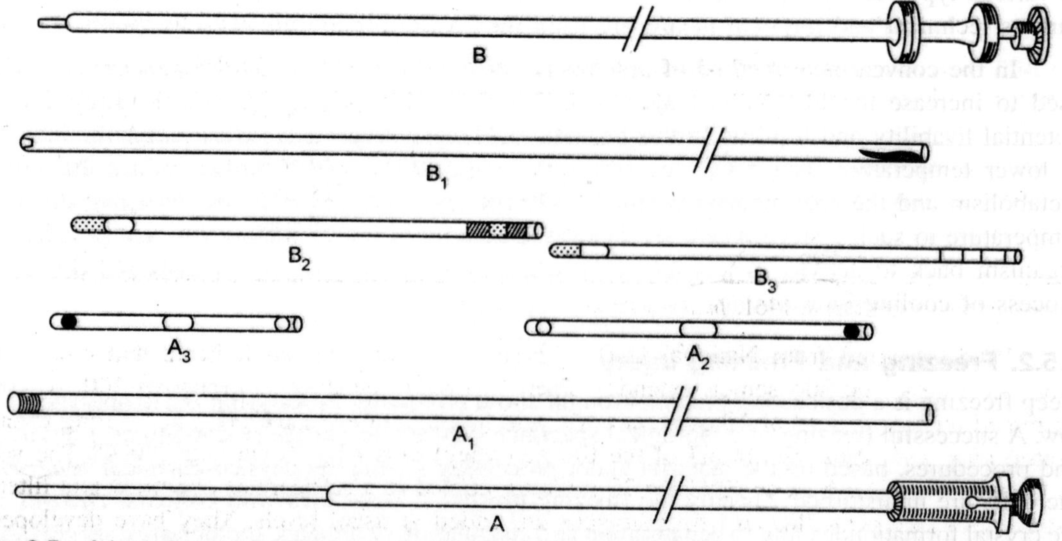

Fig. 8.5. A.I. equipment for field use. A and B—A.I. syringes, Landshut and French type. A and B—disposable sheeth, A_2—Plastic straws 0.4 ml. B_2—Plastic straws 0.5 ml. A_3 and B_3—Plastic straws 1/4 ml, land-shut and French type.

pipette is then introduced into the vestibule without touching the vulval lips. It is pushed forward gently along the dorsal surface avoiding the urethral orifice until it reaches the cervix. The pipette is manipulated to penetrate the external os. Applying slight forward pressure the cervix is manipulated until the tip of the pipette is felt as it emerges from cervical canal into the softer tissue of the body of the uterus. Keeping the pipette in this position the semen is injected inside the uterus very slowly. The pipette is slowly withdrawn after this.

One hazard encountered in this method is the breaking of the pipette when the cow struggles violently with consequent injury to sensitive organs. Use of single use plastic catheters now in vogue in USA and some European countries have removed the fear of injury due to pipette breakage. It also has the advantage that it does not contribute to spread of diseases, being single use. It also avoids extensive sterilization procedures. But, it is expensive especially when many thousands of inseminations are performed each year at a centre.

8.5. DEEP FROZEN SEMEN

8.5.1. Introduction

The concept of preserving living cells by ultra-low temperature has been existing since 18th century but it was not until 1949, when the remarkable properties of glycerol in deep freezing cells were found. a practical answer to freezing technique was evolved. Thereafter, a lot of work has been done in the field which yielded considerable changes and developments in the processing. storage. transportation and utilization of many biological specimens. The dimensions of animal breeding have been considerably widened by the technological developments in the deep freezing system of semen preservation and the potential the frozen semen offers for long term preservation and for its worldwide use. Suspended animation as a method of extending life span is an old concept. The storage of sperm in the epididymis is a suggestive example of suspended animation, being advantageously used for prolonging the life span of sperms in nature. Micro-organisms of various types assume quiescent state by which they extend their longevity of life. Artificially, various technical and physical means are used for prolonging the life span of sperms in vitro.

In the conventional method of processing semen, the effect of low temperature (4-5°C) is used to increase the life span of sperms to 3-4 days. without any appreciable damage to its potential livability and fertility. In this process, cold temperature is a partial metabolic inhibitor. A lower temperature than 5°C, *e.g.*, sub zero temperatures should further reduce the rate of metabolism and the spermatozoa would be without any visible sign of life. Bringing down the temperature to such a stage where the metabolic activity is, hardly measurable and to revive the organism back to activity is possible only when the cell structure is not damaged during the process of cooling or warming.

8.5.2. Freezing and Freezing Injury

Deep freezing is a desiccating process brought about physically by lowering the temperature ultra low. A successful freezing of a biological specimen depends on certain preconditions, precautions and procedures. based on the material under processing so that the physico-chemical integrity of the cells are undisturbed. Usually the freezing process is associated with colloidal changes and ice crystal formation leading to denaturation and coagulation of proteins, protoplasmic precipitation, enzyme release and structural derangement. The mechanism of such changes, happening during freezing is not fully understood. The velocity of molecular motion is reduced by lowering the temperature of a viable material until it ceases at the temperature of absolute zero, –273°C, and a satisfactory freezing method depends on how best the associated damaging changes are prevented to the maximum extent up to the lowest temperature required.

In general terms the causes of freezing injury can be grouped into two :

(1) Those associated with temperature trauma or cold shock.

(2) Those associated with ice crystal formation.

It is assumed that during freezing the spermatozoa are subjected to cold shock effect much in the same way as happens when semen is cooled fast at above zero temperatures. Cold shock is an effect of temperature trauma where the integrity of sperm membrane is disturbed with the resultant changes in morphology, activity and structural make up. Spermatozoa of all species do not react in the same way when subjected to cold shock: certain species has a better resistance. Cold shock and its attendant problems can be overcome to a certain extent by providing protective ingredients and selecting the right cooling rate.

On cooling as the temperature steadily descends to little below the freezing point of the diluted semen, the transition of the physical phase takes place. During this process heat is generated, the quantity of which depends on the amount of water converted to ice. The temperature remains there as long as the physical transition is over.

After this the temperature again descends fast. During the physical change of the media, no doubt, the sperms are put under high stress owing to increase of electrolyte level and osmotic pressure as the transition of water to crystalline ice results in an increasing concentration of salts, proteins, fats and other dissolved substances in the remaining fluid. The exposure of sperms to excessive concentration of electrolytes and osmotic pressure along with the mechanical injury due to extracellular and intracellular ice formation as well as changes of volume are considered to be the primary causes of freezing injury. The extent of damage depends also on the type and size of crystal formation. The lowest temperature reached, the cooling rate and the nature and concentration of solute mostly determine the type of crystals. For example, in slow cooling crystalline formation with hexagonal symmetry is obtained. Rapid cooling gives rise to irregularly constructed arborescent structures or spherulitic crystalline formation. In the case of very high solute concentration no water freezes out and the entire solution solidifies in the amorphous state. At a temperature below –130°C water molecules do not possess enough energy to assume the crystalline state. Substantial reduction of damages is possible by addition of certain protective substances to the extender, selecting a suitable freezing rate, etc.

8.5.2.1. *Cryoprotectant*

The main ingredient of the diluent protecting the spermatozoa from freezing injury is glycerol; the sugar and egg-yolk also play a part. How glycerol exerts its action is still not known clearly. Glycerol is long known to have the ability to 'protect proteins from denaturation caused by temperature change. It might be that glycerol, because of its electrolyte and water binding capability does not allow the excessive rise of electrolyte during freezing and crystallisation. Being an effective solute of salts it can do so; consequently preventing the lipoprotein denaturation and colloidal changes. Further, the glycerol modifies the type of ice formation, most probably because of its water binding capability and prevents any mechanical damage. A view exists that part of the beneficial action of glycerol is by penetrating the spermatozoa and replacing part of its free water along with extrusion of electrolyte, there by reducing the risk of intracelluar ice crystal formation as well as concentration of electrolytes. Substances of similar nature as glycerol, however, are not so good as glycerol itself. Egg-yolk or milk also reduces the freezing damage and cold shock, most probably by keeping again the integrity of lipoprotein associated with cell membrane and thus preventing loss of essential ingredients from the cell.

The crystal formation depends on the rapidity of cooling. If cooling is ultra rapid no crystals are formed and the solution freezes into an amorphous form. On fast cooling, the period of physical transition—where destruction of cells is highest is reduced to the minimum and so the crystallisation zone is by-passed by bringing the cells directly to sub-freezing temperatures known as vitrification zone. At this stage the material assumes a non-crystalline glass-like or vitrious form. But on fast cooling the risk of temperature shock is more. Extender combinations and processing procedures are shown to have some effect on the rate of cooling.

8.5.3. Extenders for Frozen Semen

Summarising most generally, the functions of semen extenders are :

(1) To increase the volume of semen.

(2) To increase the livability of sperm in vitro.

The fundamental principle of any formulation used as diluent of semen is that it should not prove harmful to sperm and should provide protective factors, metabolites and buffer salts to enhance the life span and activity of sperms. In the conventional method of processing milk and yolk citrate extenders with few additives provide a suitable easily prepared diluting media. It has the basic qualities of preserving and extending semen, such as correct osmotic tension, pH, buffering capacity, suitable balance of electrolytes and non-electrolytes, non-toxicity, presence of metabolites and protective factors. In the method of production of frozen semen, the semen as well as its media are subjected to cooling well below zero temperature which needs additional precautions. Therefore the diluents of frozen semen should provide sperms low temperature resistance in order to make them capable to withstand the freezing stress and effects of long storage, in addition to the basic factors. The attempt for formulating an extender which prevents freezing injury sufficiently high so as to extend semen at the rate followed for chilled semen, in respect of total spermatozoa per dose, has not yet been completely successful.

The number of total spermatozoa per dose is always kept high in frozen semen processing than that of chilled semen. High interaction exists between the ingredients of diluent, and the diluent and processing procedures. Therefore, what is considered optimum in one set of condition need not be optimum under other conditions.

8.5.3.1. *Formulation of Satisfactory Diluent*

The commonly used extenders for frozen semen are glycerol-egg-yolk-citrate, glycerol-milk-citrate, *'Tris'* diluents and commercial diluents like *Laciphose*. For pelletting, lactose-glycerol diluent is used. Egg-yolk and milk are being used as the basic materials into which other ingredients are added. Attempts to eliminate toxic materials completely or substituting them with synthetic materials have not yet been successful. The following fundamental principles are to be borne in mind in formulating a satisfactory diluent :

(1) Exclusion of toxic materials.

(2) Optimum concentration of different ingredients which should provide correct osmotic tension and electrolyte level in a given condition.

(3) Inclusion of protective factors.
(4) Control/Elimination of pathogens.
(5) Addition of fertility promoters.

These principles should be applied in connection with the number of sperms per dose. High extension may produce leaching of certain essential materials from spermatozoa and low dilution results in rapid accumulation of toxic waste products: both are harmful.

The concentration of buffer salts is commonly adjusted to give an initial pH value 6.8 to 7.0 in the final diluent. The best buffer solution is one which maintains a pH of 6.5 to 7.0 at the lowest electrolyte concentration. Any given isotonic buffer salt solution is improved by mixing with isotonic sugar solution. As the proportion of sugar solution increases in such a mixture the concentration of electrolytes is reduced and the sperm viability increased. The commonly used buffers in frozen semen diluents are sodium citrate and '*Tris*' (hydioxymethyle) buffer. Fructose is the sugar of choice for bull semen, and serves in the diluent as an additional energy source and regulator of the electrolytes and osmotic pressure. In freezing semen, sulpha-compounds are not used as antimicrobial agent because, while effecting the lowering of temperatures to ultra low levels, its concentration increases relatively to a harmful level. Among antibiotics. penicillin and streptomycin have the best compatibility with semen and hence are used extensively. Wide spectrum antibiotics are injurious to sperms even in minute quantities. But polymixin is reported to be effective in combination with other antimicrobial agents.

8.5.4. Processing Methodology

No processing method can increase the quality of sperm produced from the bull, whereas it can decrease the quality by faulty processing methods. The ultimate objective of any processing system is to handle the semen without reducing its inherent fertilizing power as well as to extend and get the maximum number of pregnancies in a most convenient, desirable and cheap way. The processing procedures vary with the choice of extender, packaging system and to some extent, with local resources and conditions.

8.5.4.1. *Extension of Semen*

Semen is partially or fully diluted with extender before cooling to 4-5°C. The subject of sperm numbers consistent with optimum fertility has been studied since long. The problem of sperm numbers per dose giving optimum fertility should be tackled considering processing procedures, extenders, local conditions, fertility rate of the bull, storage temperature and volume per dose. Concentration from 12 to 60 million total sperms per dose are in use but 20 to 30 million is a generally acceptable concentration. The method largely accepted for dilution of semen for freezing is in two phases, the first being at 30-32°C in an extender containing no glycerol and the second after cooling the partially diluted semen to 5°C with glycerol extender. Glycerol is considered to be best mixed at 4-5°C in 3-4 steps to avoid 'osmotic shock'. However variation from this standard procedure is not uncommon and depends on the extender composition, processing routine, etc. The semen after collection is subjected to initial evaluation. consisting of volume reading, motility test and assessment of sperm concentration, whereby the sub-standard samples are discarded. The selected samples are diluted with almost an equal quantity of diluent poured slowly and gently through the sides into the collection vial and then emptying the whole with equal caution into the premeasured diluent. Soon after the initial dilution the semen is cooled gradually to 5°C in about 1 to 1½ hours.

8.5.4.2. *Glycerolization*

The process of dilution with glycerol containing fraction of the extender is called glycerolization. The quantity of glycerol varies with extender composition and freezing condition. Generally 6-8 per cent final concentration of glycerol is considered to be the optimum for bull semen. The glycerol diluent and the pre-diluted semen should be at equal temperatures while mixing. The glycerol has a tendency to stratify and so it should be mixed thoroughly before use.

8.5.4.3. *Equilibration*

It is found that the semen stored at 4-5°C for a few hours before freezing gives better post-thaw motility and fertility. This pre-freeze holding time is termed as equilibration period. It was believed that during this period the sperm cells become permeated with glycerol and an ionic and osmotic equilibrium is established between the sperm cells and the media. Later it has been shown that addition of glycerol to semen stored at 5°C for several hours, immediately before freezing gives similar results as that in which glycerol is added at 5°C and then stored for the same period, This indicates that equilibration is not necessarily with glycerol and spermatazoa alone but can be with other ingredients and temperature as well. In any case this period allows spermatozoa to acquire better resistance to freezing stress. There are conflicting reports regarding the length of equilibration period, which varies from one minute to several hours, and its effect on spermatozoal survival rates. However, 4-6 hours of equilibration is generally practised.

8.5.4.4. *Packaging*

Packaging frozen semen in single dose containers is universally accepted. Semen is packaged commercially in various types of plastic straws (Fig. 8.5), glass ampoules and pellets. Out of these the plastic straws are gaining popularity. Other methods of packaging include pipettes, plastic vials and capsules.

Straws are plastic tubes made in various lengths and diameters with plugs of different types and materials. Glass ampoules were under common use but are getting replaced by plastic straws because of the following reasons :

1. Ease of transport in respect of weight, size and unbreakability.
2. Straws are smaller in volume than ampoules. So storage is economical.
3. Sperm loss is negligible in straws compared to ampoules during insemination process.

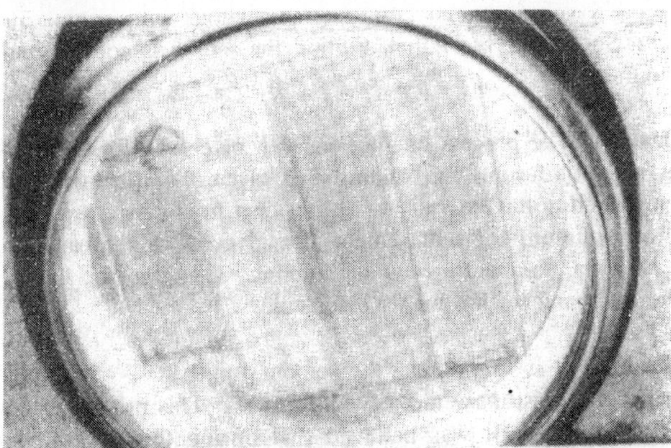

Fig. 8.6. Horizontal static vapour freezing with ½ ml pailletes.

4. Higher recovery rates can be expected in straws as the surface area in comparison to volume is more in straws that) in ampoules and a uniform freezing and thawing can be expected in the former.
5. Straws allow better handling in all phases of semen processing and utilization.

Originally 1.2 ml capacity straws were in use. The present practice is to use finer and/or shorter straws (0.5 ml and 0.25 ml) which have the additional advantage of higher storage potential while maintaining efficiency. Identification of straws is no problem. Currently 'French' straws (paillettes), 'Landshut' straws (mini-tub), Canadian straws and US or Continental straws are in use; the difference between them is in minor details like form and kind of plug, length and size etc. (Fig. 8.5).

8.5.4.5. *Freezing Methods*

The freezing procedure has been developed over years. Originally solid carbon dioxide (dry ice)—alcohol bath, (–79°C) was used as coolant. Slower rate of cooling was recommended especially upto –32°C, varying from 1 to 3°C per minute, and then cooling fast to –79°C. The procedure usually followed was to put dry ice in a wire mesh kept perpendicularly in an alcohol chest. The semen to be frozen is packaged and placed in the alcohol. The storage was done either in dry ice—alcohol bath or in liquid nitrogen. Later, automatic equipment which provided facilities of cooling by liquid nitrogen vapour at the rate required was developed.

Now the common method of freezing semen is to cool the straws fast in static liquid nitrogen vapour (Fig. 8.6). Apart from being easier, efficient and less time consuming, this procedure allows freezing of more doses at a time. Wide mouthed liquid nitrogen containers of different types and capacities are used for this purpose. A metal grill is placed at a depth of 30 or more cms below the top brim of a large container. While keeping the level of liquid nitrogen at the grill, straws racked horizontally are placed over the grill which allows cooling of semen in about 7-8 minutes to sub-freezing temperatures. The freezing rate at the critical temperature range will be very fast; the rate depending on the vapour depth in the tanker, liquid nitrogen level in relation to the straws, amount of possible mixing between vapour and air in the room, type and/or number of straws frozen at a time, volume of semen, type of diluent and initial vapour temperature. For 1/2 ml paillettes an initial vapour temperature ranging from –180 to –160°C is considered satisfactory whereas for 1/4 ml straws around –130°C is usually recommended. Ultra rapid freezing like putting the straws directly in the liquid nitrogen seems to be injurious.

There can be many variations of the procedure explained above like vertical freezing, freezing with baskets, wooden ramps or cassettes etc. depending upon the convenience and routine of the laboratory.

The lowest temperature under which the bull spermatozoa is reported to survive is that of liquid helium (–269°C). Using liquid nitrogen for freezing and storage has many advantages over solid carbon dioxide alcohol bath.

1. Solid carbon dioxide maintains a temperature –79°C whereas the boiling point of liquid nitrogen is –190°C. It has been found that keeping semen in the liquid nitrogen, increases the storage potential.
2. In dry ice chest the cold is produced in one area which is required to be conducted to the semen. It can be done satisfactorily but is inconvenient.

3. With faster freezing procedure which is better than the conventional slow freezing, liquid nitrogen vapour gives a very convenient and easy opportunity to freeze.
4. The fluctuation of temperature of semen doses is more when stored in dry ice than in liquid nitrogen.
5. Being a liquid, handling is easier in liquid nitrogen.

8.5.4.6. *Pellets*

Pellets are unpacked semen doses frozen in the from of a drop. This can be done either directly over the dry ice block or over liquid nitrogen vapour. Usually, 0.1 ml semen portion is used for a pellet which needs further dilution in a buffer solution on thawing. The initial extender does not contain buffer salts but only sugar (usually lactose) glycerol and egg yolk.

Pelletting is quicker. It needs only less storage space and production cost. But it has many disadvantages.

1. Pellets do not provide a uniform handling possibility.
2. Identification of pellets is difficult.
3. Since it is unpacked, the contamination possibilities are more.
4. Field handling is difficult; need dilution in the field with special thawing solutions. Wastage of semen during insemination is more.
5. Reports on fertility rate of pellets in comparison to straws are conflicting.
6. As the pellets are unpacked, they may get contaminated with sperms from other pellets of any bull while in storage and handling.
7. On handling the pellets may stick on to the forceps.

8.5.4.7. *Freezing of Buffalo Bull Semen*

Buffalo semen is different from bull semen in its response to extension and use in artificial insemination under any method of preservation. This suggests that extension and preservation of buffalo spermatazoa cannot be ideally carried out in medias and methods recognised as suitable for bull sperms. Most of the extenders used for bull semen have been tested for buffalo semen with varying results. Apparently extension and cryoprotection need more research in the case of buffalo semen. The practicability of freezing buffalo semen has been demonstrated unquestionably. But the reported results of most of the works are not backed with sufficient data on fertility. Techniques in freezing buffalo semen are mostly adaptation of the methodology followed for bull semen. '*Tris*', 'Citric acid whey' and Yolk citrate-glycerol diluents are being used now and a uniform procedure and large scale comparative results are awaited. The difference among buffalo bulls in producing preservable semen is more than that among bulls.

8.5.5. Handling of Frozen Semen

The basic principle of any cold storage is that the cool chain must not be broken. Even at frozen state the temperature fluctuations adversely affect the fertilizing power of spermatozoa. Some physical changes must be assumed to take place even at ultra low temperature which is influenced by storage medium, composition of diluent and freezing methods. For maintaining satisfactory fertility the storage temperature should be constant. Even within ultra low temperature, a rise causes expansion and a reduction causes shrinkage. Frequent small temperature fluctuations also should be avoided as far as possible. Bigger and long lasting changes can cause considerable damage to spermatozoa. Long term storage apparently has little effect on fertility. There is no clear evidence to show that semen properly stored will not retain its fertility indefinitely.

8.5.5.1. *Storage and Despatch*

Soon after freezing semen in the vapour, the doses are collected carefully in a goblet and plunged in the liquid nitrogen for storage (Fig. 8.7). It is not advisable to fill the goblet to its maximum capacity as this will reduce the liquid nitrogen volume to a very low level. When a standard big goblet is filled completely only about 18 per cent of its total volume is available for liquid nitrogen which on handling may evaporate away resulting in high fluctuation of temperature on straws. The storage may be planned in such a way that :

1. Transfer of straws between goblets is reduced to the minimum.
2. Semen of bulls expected to be in high demand and fast moving is easily accessible.
3. Doses meant for long storage are always kept at the bottom layer of the refrigerator.

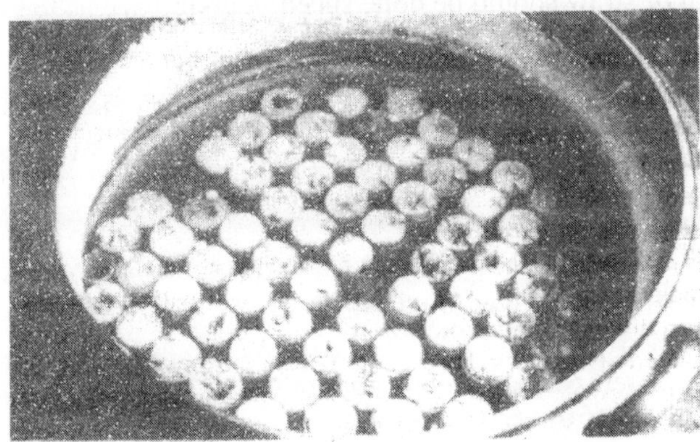

Fig. 8.7. Semen storage in large liquid nitrogen refrigerators.

Despatch of semen is made in answer to the demands. So transferring of semen from one goblet to another is unavoidable. Care should be taken to see that both goblets are filled with the coolant before the transfer is made and the operation is made quickly.

8.5.5.2. *Thawing of Frozen Semen*

Transition from the vitrious state in thawing is equally crucial as that in cooling. To avoid the crystallisation zone, thawing should be rapid. Adverse effects of temperature shock on rapid warming of semen is far less or even negligible when compared to fast cooling. But warming should be progressive and uniform at all stages. The spermatazoa do not withstand refreezing or repeated cooling and warming. If the semen is not thawed evenly and progressively the thawed part of the sample may get refrozen. The method commonly employed for thawing of frozen semen is to plunge the dose in water. The time allowed for this operation depends on the temperature of the water and the packaging system. Opinions differ as to the temperature of water in which semen is to be thawed. Data comprising this, followed by fertility tests, are not available readily. Common practice is to thaw the straws in 37-40°C for 1/2 to 1 minute eventhough a temperature upto 75°C for 12 seconds is reported to be advantageous. Air-thaw is also being practised to a limited extent. An extensive comparative study between air-thaw and water-thaw is not available followed by fertility test. The efficiency of cold insemination where the semen is expressed without thawing is not known clearly. However, a universally adopted system of thawing is not existing. Convenience and routine make the rule.

8.5.5.3. Field Application

Semen should be properly handled by the inseminators, once it is transferred to them. To expose semen outside for a long period is harmful to spermatozoa. Cumulative exposure effect on various occasions for short periods is less harmful than the changes caused by exposing it at one time, for the total length of time. The straws should be quickly taken out, from the refrigerator in about 10-12 seconds. In case it cannot be done, the canister should be lowered and then lifted again for withdrawal of semen dose. In this process other semen doses should not be touched, as far as possible. The people handling frozen semen should have a clear understanding of the following points.

1. Transfers increase the possibility of temperature damage.
2. Withdrawal of straw should be done speedily giving adequate low temperature protection for the remaining doses.
3. The liquid nitrogen level should be checked routinely at regular intervals and replenished whenever needed.
4. Semen doses should not be taken out and replaced for any reasons.
5. A uniform thawing procedure is advantageous.
6. Semen will not tolerate refreezing.
7. The post-thaw survival rate of semen is poor. Therefore thawed semen should be used as soon as possible.
8. Fluctuation of temperature below and above zero temperature should be avoided to the maximum.
9. While inseminating, the whole dose should be expelled at the right site of the female genetalia. This is more applicable when using 1/4 ml doses.

8.5.5.4. *Handling Liquid Nitrogen*

Nitrogen is an inert gas. But in liquid state it is harmful owing to its extremely low temperature. Further, on evaporation it produces about 700 times its volume of gaseous nitrogen which naturally reduces the oxygen content in the air. So accidental contact of the skin or the eyes with liquid nitrogen or the cold gas issuing from it may cause injury similar to burn. Working with liquid nitrogen in a confined area, without adequate ventilation can cause asphyxiation.

Overflow, splashing and violent boiling should be avoided. The containers and refrigerators should be cooled slowly. Objects which are ordinarily soft and pliable become very hard and brittle on contact with liquid nitrogen.

The practice of bowing head into the container for identification of straws is dangerous. One or two complete breaths of nitrogen vapour is enough to cause unconsciousness. The extremely low temperature is likely to cause injury to the conjunctiva of the eye. In cases of accidental contact, the region should be flooded with cold water. A cold compressor bandage is also beneficial.

8.5.5.5. Care and Maintenance of the Containers/Refrigerators

Extreme care should be exercised in handling liquid nitrogen containers. The cryogenic equipment are specific and they are meant to store only the particular liquid gas for which they are made.

Welding, piercing the container wall, plugging it tightly while holding liquid nitrogen (except for pressure type) are dangerous and may cause permanent damage. For transport, a wooden crate or similar protective devices may be used. Rough handling or tipping the refrigerator on its side, will lead to cracking of the inside wall. It will render the refrigerator useless without any apparent external damage. Charging of a warm refrigerator/container with liquid gas should not be rapid. Otherwise the longevity of the equipment may be adversely affected and also, due to boiling and possible splashing of liquid nitrogen the people engaged for work may get cold burn. Frequent cooling and warming should be avoided for the safety of the containers. So it is advisable not to empty them completely if they are meant for regular use. During long term use of a container/refrigerator, especially one with a wide mouth, oxygen condensation may result and also ice and dirt may accumulate.

Assessment of Liquid Nitrogen Level in the Refrigerators

The liquid nitrogen retention capacity of the refrigerators varies; very often it becomes necessary to check the nitrogen level of a container. If the mouth is large, visual examination will serve the purpose but in narrow-necked containers this method is impracticable. Accurate measurement of the volume of liquid nitrogen can be made by dipping a graduated slender stick of low thermal conductivity into the liquid nitrogen till it touches the bottom of the refrigerator. Keeping it undisturbed for 5-10 seconds the stick is withdrawn and waved in the air. Frost will appear in the wetted section. Once the length of frosted section is read the same can be converted into volume as the diameter of the refrigerator is known.

1. Use of a hollow tube should be avoided for measuring.
2. At least half the length of the straws in storage always should be below the level of liquid nitrogen. Further lowering of liquid nitrogen level will lead to decline in post-thaw motility and fertility of semen.

8.5.6. Biological Advantages of Frozen Semen

Artificial insemination with frozen semen is one of the most advanced and sophisticated techniques in livestock breeding and its impact on rapid improvement of cattle is momentous by virtue of its long term storage potential, unaffected quality on transportation and the scope it offers for spreading the valuable genotypes. Frozen semen presents profoundly flexible tool in performing nominated services, progeny testing, areawise programme of breeding, shifting of bulls or breed from one area to another etc. So, also frozen semen programme will be best suited on occasions requiring 'layoff' of bulls where instead of bulls semen can be kept waiting.

The following are the major advantages :

1. The most important advantage of frozen semen is its storage potential extending over several years. The superior genetic material can be stored for the future legitimate and planned programme.
2. Semen frozen and stored is an insurance against loss by death of the donor bull or the damaging injury to the breeding life of the animal.
3. On occasions where the use of a bull is limited to a few cows only, the bull can be disposed of if sufficient quantity of semen is stored in frozen state.
4. The frozen semen retains its full quality during transportation.
5. Frozen semen offers enough scope for planned mating programme, nominated service programme, etc.

6. The capacity to being uninfluenced over long distance shipment is made use of for wider scale use of superior bulls around the world.
7. By the introduction of frozen semen the progeny testing programmes are additionally benefited by enhancing the efficiency and accuracy, testing large number of bulls at a time and studying the superiority of bulls under different climatic zones.
8. Uninterrupted supply of semen to the breeding units is made possible by keeping a buffer stock of semen in a semen bank.
9. Possibility of establishing large bull stations is permitted by frozen semen programme instead of many small bull stations.
10. 95 per cent or more utilization of processed semen is possible with frozen preserved semen.
11. Maximum utility of bulls can be made use of.

8.5.6.1. *Disadvantages*

1. The frozen semen use may help to perpetuate certain undesired genetic characters like spastic paresis, spondylosis, etc. as they may be exhibited after 3-4 years of age. By this time if the bull is dead, the semen collected earlier will be continually used.
2. Careless processing and storing may result in spreading infectious and contagious diseases in different parts of the world and at different times, for a very long period.

CHAPTER

9

Management of Reproduction

Without reproduction there will be no lactation nor replacements for the future. Therefore, the maintenance of a high fertility rate is basic to the success of any dairy farming enterprise and its viability is a business. For high lifetime production and overall return per cow, a long life and regular reproduction are essential.

Careful attention to a number of details in management is the key word for efficient reproduction.

In any cattle population, some cows fail to conceive at the first service or subsequent services, abort or produce dead or weak calves incapable of survival. Consequently, the average calving interval is lengthened which tends to decrease the average daily milk production. Besides, less number of calves are produced for replacement. The economic implication of this problem is enormous and runs into many millions of rupees.

9.1. Reproduction Parameters

Calving Rate. The best evidence of fertility of a cow is a successful reproduction—a calf at foot. Therefore, the calving rate measures the number of cows calving out of total cows (%) during a specific period. This depends on actual service conception rate and embryo/foetus survivability.

Services per Conception. A cow may need one or more services before she is settled. When the semen is of high fertility and the cow is inseminated at proper time, the number of services required for conception gives a quantitative measure of the fertility of the cow.

Non-Return Rate. In artificial insemination programmes, a feed-back of information with respect to successful conception is very important for policy making such as needed to cull bulls of low fertility, modifications in semen handling and storage. improvement in insemination techniques etc. Therefore we cannot wait until information on conception rate or calving rate is forthcoming. This measure is based on the assumption that if a cow after an initial insemination is not brought again for another insemination within 60 to 90 days the cow is pregnant. This method is not very valid in India and similar countries because many of the owners who do not have confidence in AI and AI technicians. switch over to natural service on failure of the first insemination. In this method, such cows will be considered as conceived in the first insemination. Similarly, culled, sold and dead cows which also do not return for a second insemination are counted as conceived. On the other side, about 3-4 per cent of pregnant cows show symptoms of estrus and are rebred.

Calving Intervals. The frequency of calving or the interval between calvings is regulated by the management decision on how soon after calving a cow is rebred as well as the success of the first or subsequent services. A calving interval of 12 months for cows and 13 months for first calf heifers have been found to be optimum. However, in higher producing cows of good persistency calving interval up to 15 months will not affect production.

9.2. ASSESSING HERD REPRODUCTION STATUS

Breeding efficiency of individual cows can be calculated with the help of formulae developed for the purpose. The one used by Wilcox *et al.* (1957), which has been widely used, is based on the differences in the observed and desired intercalving period of cows. Tomar (1967) has, however, pointed out that in view of the great economic importance and wide variability of age at first calving in Zebu cattle and buffaloes, it too should be considered in calculating breeding efficiency. Earlier Singh and Desai (1962) and Acharya (1966) too suggested using calving interval and age at first calving. The formulae suggested by Tomar (1967) are given below. The breeding efficiency of the individual towards the desired standard is expressed in terms of percentage :

For cows
$$BE = \frac{[n(365) + 1020]}{AC + CI} \times 100$$

For buffaloes
$$BE = \frac{[n(365) + 1040]}{AC + CI} \times 100$$

where, BE—breeding efficiency, *n*—number of calving intervals (author suggests, considering the first two calving intervals out of the possible four in Zebu cattle), 365—desired calving interval in days, 1020—desired age at first calving in days in cows, 1040—desired age at first calving in days, in buffaloes, AC—actual age at first calving in days, and CI—Sum of actual calving intervals studied in days.

Several other parameters have also been suggested for estimation of breeding efficiency of dairy bovines : value index, a fraction representing the ratio of cow's age to the number of calves born, percentage of cows becoming pregnant at first service, number of months per calf, herd index, non-returns to first service, calving interval, age at maturity, number of services per

---

Wilcox, C.J., Pfan, K.G. and Bartlett, J.W. 1957. An investigation of the inheritance of the female reproductive performance and longevity and their relationships within a Holstein-Friesian herd. *J. Dairy Sci.* **40** : 942-47.

Tomar, N.S. 1967. A note on method of working out breeding efficiency of Zebu cows and buffaloes. *Indian Dairyman*. **19** : 345-346.

conception, number of days from first service to conception, regularity of occurrence of heat and strength and length of heat period.

In Soviet Union the index of fertility of cows, T, is calculated by the formula suggested by Dokha (Earnst *et al.*, 1977) which is as follows :

$$T = 100 - (K + 2i)$$

where, i is calving interval in months, and K is age at first calving in months.

Per cent herd reproduction status (HRS) calculated by the formula suggested by Britt and Ulberg (1970) is another good measure of reproductive efficiency of dairy-herds :

$$\text{HRS Index} = 100\ \frac{\text{ADO} \times \text{PC} \times 1.75}{\text{TC}}$$

where ADO is the average days open of problem cows. PC is the number of problem cows with days open above 100 and below 200 days. TC is the total number of cows in the herd, and 1.75 is the optimum number of services per conception.

The upper limit of HRS Index is 100 which means the herd has a 100 per cent reproductive efficiency. Anand and Balaine (1981) studied the relative effectiveness of various measures of herd reproduction rate on purebred zebu, purebred exotic and crosses between the two and concluded that the HRS Index is one of the most efficient measures of reproductive efficiency for dairy herds (Table 9.1) under Indian (Semi-arid) conditions, which could be true in similar localities in other tropical zones.

However, in view of the difficulty in obtaining accurate information on all aspects of reproduction in developing tropical countries (especially from villages) and since age at first calving and calving intervals, to a great measure are the ultimate consequences of good or bad reproductive history, a formula incorporating both of them is desirable.

9.3. REPRODUCTION MANAGEMENT—KEY FACTORS

The key factors involved in reproduction management of cattle and buffaloes are : prompt detection of oestrus, breeding at the appropriate time of oestrus, use of positively good quality semen or sire, allowing sufficient service period and dry period. regular checking of infectious as well as non-infectious reproductive disorders, hygiene at the time of calving, prompt veterinary attendance to all cases of abnormal calvings, check over quality and bulkiness of rations during the pre, pere, and post-natal periods and, last but not the least, close and regular observation and accurate record keeping.

Observations and Records. Two aspects of good management that are generally neglected are regular and careful observation of the animals in the herd and keeping of production and reproduction records. This is especially so in villages. Noting what is happening with the animals in the herd and recording the same so that future events can be predicted can greatly improve

Earnest, L.K., Beguchev, A.P. and Levantin, D.L. 1977. *Cattle Husbandry* (Russian). Kolos Publishers, Moscow. 1977. p. 179.

Brit, J.H. and Ulberg, L.C. 1970. Changes in reproductive performance in dairy herds using HRS system. *J. Dairy Sci.*, **53** : 752-58.

Anand, G.R. and Balaine, D.S. 1981. Comparative evaluation of herd reproductive status of some exotic pure-breds, their cross-breds with Hariana and Hariana pure-bred dairy cattle. *Indian J. Anim. Sci.* **51** : 271-274.

reproductive efficiency. Records help in keeping track of dates of service. pregnancy check up and calving date.

Observing milch cows and buffaloes while milking and while leaving the milking barn after morning and evening milking can greatly improve heat detection and consequent improvement of conception rate, besides locating cows with abnormal discharges through genitals. In an artificial insemination programme as well as in natural hand mating, much of the success of breeding cows depends on the herdsman's efficiency in spotting cows in oestrus. Good records and observations can also help locate open cows not returning to heat. Besides, records can be valuable not only to the herdsman but also to the veterinarian and others who can use them to trace trouble before they become too costly.

TABLE 9.1. Mean* (± S.E.) Values of Herd Reproductive Status determined by Various Methods in Different Genetic Groups of Dairy Cattle (Anand and Balaine, 1981)

Genetic groups	*% Conceived by first service*	*Stadard AFC×100 / Observed AFC*	*Standard CI×100 / Observed CI*	*% Reproductive culls*	*HRS index*
HF	52.76[b] ± 0.39 (127)	29.37[a] ± 0.21 (133)	90.80[a] ± 0.57 (50)	28.57[b] ± 0.36 (126)	64.76[a] ± 0.38 (126)
J	50.00[b] ± 1.25 (40)	94.28[a] ± 0.22 (107)	80.52[ab] ± 0.54 (74)	29.03[b] ± 0.48 (93)	62.39[ab] ± 0.52 (93)
H	43.04[c] ± 0.63 (79)	48.85[e] ± 0.43 (117)	67.84[c] ± 0.45 (103)	51.45[a] ± 0.48 (103)	22.00[f] ± 0.40 (103)
½F	32.35[d] ± 0.46 (102)	85.16[b] ± 0.35 (102)	82.13[ab] ± 0.44 (88)	27.47[b] ± 0.48 (93)	68.08[a] ± 0.50 (93)
½B	21.67[e] ± 0.69 (60)	78.70[c] ± 0.69 (59)	85.05[ab] ± 0.69 (52)	13.46[e] ± 0.66 (52)	59.01[b] ± 0.95 (52)
½J	30.95[d] ± 1.02 (42)	93.65[a] ± 0.70 (35)	88.17[a] ± 1.11 (29)	8.67[cd] ± 0.93 (29)	45.48[d] ± 1.72 (29)
½R	6.67[f] ± 0.83 (30)	71.40[d] ± 1.67 (27)	88.26[a] ± 1.14 (28)	11.54[c] ± 1.51 (26)	52.08[e] ± 1.92 (26)
½JH	64.38[a] ± 0.66 (73)	74.56[d] ± 0.59 (74)	78.29[b] ± 1.08 (38)	25.42[b] ± 0.74 (59)	39.52[e] ± 0.83 (59)
Pooled for	41.41 ± 0.09 (553)	74.92 ± 0.07 (654)	79.62 ± 0.09 (462)	27.30 ± 0.07 (581)	56.59[a] ± 0.08 (581)

Note. 1. HF = Hotstein Friesian, J = Jerrsy; B = Brown Swiss; H = Hariana.
2. Number of observations are in parenthesis.
3. *Mean values having different superscripts differed significantly ($P < 0.05$).

Increasing Useful Life. Attempts to increase the useful life must invariably include proper raising of heifer calves (*see* Chapter 13) so that they reach puberty and a safe calving size at an early age. Age and size at the time of first calving as well as the ultimate size of the cow seem to depend on the adequacy of feeding that permits good growth uptill calving and growth and production after first calving.

Among adult cows, a calf per year per cow should be the ultimate aim, though among dairy bovines reared in tropics, it can rarely be achieved among majority of the herds. However, it is not impossible to attain a 13-15 month intercalving period even among buffaloes (Gupta *et al.*, 1981) with a service period of 2 to 2½ months and a dry period of similar duration.

Silent heat is one of the greatest unsolved impediments in the efficient breeding of bovines. It occurs more frequently among first calvers than in older cows, in early lactation than in later lactations (Salisbury *et al.*, 1978), and in buffaloes during hot season (Sastry and Pal, 1982). A combination of oestrus detection methods (Barua, K.C., 1982) may be necessary for locating such animals.

The twin objectives of reproduction: initiation of lactation and giving birth to future cows. can be achieved by careful attention to a number of management factors discussed above. Shorter age at first calving, long herd life, regular calving and low calf losses depend greatly on good husbandry practices.

Overcoming summer sub-fertility in buffaloes. There are some reproductive problems peculiar to the water buffalo, namely, marked unfavourable influence of hot season on sexual activity, long age at first calving and long intercalving period (Cockrill, 1977). Measures to provide shelter from direct sun's rays and from hot air currents by wet screens, showers, wallows and fans during hot months are helpful. Provision of teaser bulls can, to some extent, help in locating buffaloes exhibiting silent or weak heat (which is often the case in summer). Buffalo males used for breeding as well as teasers must also be kept in cool sheds and splashed with cold water 2 or 3 times daily or allowed regular wallowing periods, which helps in maintenance of good libido and semen quality in them. Further, a great control of reproductive disorders (uterine pathology, cysts, early embryonic mortality etc.) is necessary in buffaloes than in cattle, especially during the hot season.

9.4. ROLE OF DETECTION OF HEAT IN HERD FERTILITY

9.4.1. Detection of Heat in Cows

Artificial insemination in developing regions has not been a major success mainly because of unsatisfactory detection of heat. Herd management practices account for some of the difficulty. '*Standing heat* is the best and simple indication of estrus in cows. But, this behaviour will be manifested only if enough of loafing area is provided to the cows. Under stall feeding systems where cows are tied throughout or if they are crowded into small lots there will be little opportunity to exhibit cardinal breeding behaviours. If the loafing area is cemented and remain wet and slippery, cows may refrain from mounting for fear of falling down.

Stimulation of movement among cows by walking among the animals or driving a truck helps in estrus detection as the cows tend to move about and exhibit mounting behaviour. Feeding time should be avoided for heat detection as the animal's attention is turned towards feeding and not sex-play. Detection work may be carried out twice daily, once early in the morning and once in the evening. Reference should always be made to the breeding history of the cow.

Gupta, L.R., Sastry, N.S.R., Yadav, R.S. and Jaiswal, U.C. 1982. Observation on calving interval of Murrah buffaloes belonging to a medium sized herd. *HAU J. Res.* **10** : 442.

Sastry, N.S.R. and Pal, R.N. 1982. *Dairy Farming*. Directorate of Publications, Haryana Agril. University, Hissar, p. 59.

Barua, K.C., 1982. Relative efficacy of different methods of heat detection in buffaloes (Personal Communications).

Cockrill, W.R. 1977. *Water Buffalo*. FAO Animal Health & Production Paper No. 13, FAO, Rome.

The occurrence of a bloody discharge on the vulva or tail of a cow indicates that she was in heat about two days previously and should be watched carefully for heat on the 17th to 19th day after this. It will be a good managerial procedure to maintain heat *expectancy charts* for the herd.

Signs of heat. Standing to be ridden by other cows represents the surest sign of estrus. But there are many other behavioural changes which also indicate estrus. In the dairy cows, symptoms at the early stages of heat are not pronounced, but they will show activities like smelling other cows, attempting to mount other cows and bellowing. They will be restless and their vulva moist, red and slightly swollen. After a time lapse of 6 to 8 hours, the heat will become more pronounced. The cow will stand still to be mounted by other cows or bulls. The period during which this sexual behaviour is manifested, is termed 'standing heat'. This extends for 14-16 hours and shows other symptoms like bellowing, nervousness, anorexia, reduction in milk yield, butting, moist and red vulva, riding other cows, clear mucous discharge from the vagina and dilated pupils of the eyes. Besides, it will be the first cow to rise in the morning. Even after 'standing heat' period, there will be a clear mucous discharge from the genitalia for another 8 to 10 hours. But the cow will not stand quietly to be mounted.

With the introduction of AI, additional responsibility has been placed on the husbands-man especially with respect to heat detection. As a result, many new techniques and innovations have been evolved recently.

Crystallization patterns of cervical mucous has been one of them. Cervical mucus during estrus will crystallize into long crystals in a typical fern-like pattern.

There are many estrus-detecting devices available. They are usually attached to the tail or rump of the cow. Mounting causes these devices to discharge a coloured fluid which can be observed afterwards even from a distance.

A New Zealand firm (Frank Pavouir Ltd., Mahana Road, Hamilton) has been marketing a gadget termed 'Chin Ball Mating Device' for heat detection. It works on the same principle of a ball point pen and is fixed by means of a halter below the chin of the teaser bull. When the checking animal mounts the cow in heat, the dye exuded round a spring-loaded ball of the device marking the back of the cow.

Russian workers have developed an instrument basically consisting of an ohm meter and electrodes. When applied to the mucous membrane of the vagina, the resistance indicated on the ohm mater shows whether the cow is in heat.

Certain other methods like use of pedometers or measuring vaginal temperature have been suggested. The former works on the principle that activity and movement of the cow increases on the day of heat and this can be detected by means of a pedometer tied to the leg of the cow. The latter depends on the fact that the vaginal temperature is elevated by about 1°C on the day of heat. Both methods are not very practicable.

The methods described above may not be fully acceptable or feasible in developing countries due to technological or economic reasons. In these countries close observation of signs of heat, especially 'standing heat' remains the most practicable method of heat detection. The same is true to some extent in developed countries also. In large farms this can be supplemented with a bull-parade using a teaser bull.

A teaser may be a vasectomised bull or a bull in which penis has been amputated and the urethra exteriorized. An intact bull can also be used by hanging a thick cloth or gunny bag curtain in front of the penis preventing entry of penis and mating. Special care should be taken to prevent spread of diseases by teasers. Vasectomised bull is more harmful in this regard.

Right Time for Breeding Cows

Success of breeding either naturally or artificially depends to a great extent on whether it is done at the right time or not. The best time to breed cattle is form the middle of standing heat and six hours following that. Three or four hours preceding and succeeding this excellent period has also given good results. As a routine practice, if a cow is seen showing signs of early heat in the morning, it may be inseminated in the evening. If such signs are first manifested in the evening, the cow may be bred next day morning.

A cow is expected to show estrus in 30-40 days after calving. Cows that fail to show heat even after 50 days have generally some problems and need examination. It may be due to infection or malnutrition and remedial measures may be taken accordingly.

9.5. PACKAGE OF PRACTICES FOR IMPROVING REPRODUCTIVE EFFICIENCY

Reproductive efficiency in a herd, as discussed already, depends to a large extent on effective management. Slackness in management of reproduction can be ruinous as reproductive inefficiency is considered the largest single cause of loss in livestock production.

A manager has a very important role in ensuring reproductive efficiency in his herd. He must directly supervise reproduction in his herd and should practise the following rules to ensure high breeding efficiency :

(*i*) Keep accurate breeding records of dates of heat, service and calving. Use records in predicting the probable dates of heat. Observe those cows expected to come in heat on a day both morning and evening. This should also include those cows having expected dates of heat 2 days earlier or later.

(*ii*) In large dairy farms, use teaser bulls for heat detection.

(*iii*) Know the complete breeding history of cow, her past performance and difficulties.

(*iv*) The manager should look for irregular estrous cycles or abnormal vaginal discharges. These are signs of possible infection of reproductive tract. He should get such animals examined by a competent gynaecologist and treated.

(*v*) Cows with retained placenta should be examined and treated within 3 days of parturition. These cows should be got examined prior to next breeding to ensure that their reproductive tract is free from infection and has undergone involution to attain normal size. In cases of retained placenta with a purulent fetid discharge, a full course of antibiotic treatment to help them get rid of the infection may be carried out under veterinary supervision.

(*vi*) A cow needs gynaecological examination after a retained placenta or uterine infection when (*a*) an abnormal discharge is observed during heat period (*b*) 30-50 days post-partum to ensure that there is proper involution.

(*vii*) A manager should examine a cow 24 to 36 hours after service for metestrus bleeding. If it occurs under 24 hours after service, the cows were bred too late. If it occurs over

36 hours after service, they were bred too early during estrus. This will help in pinpointing causes of failure of conception.

(*viii*) Get cows checked for pregnancy 45 to 60 days after service so that if they are non-pregnant, steps can be taken to re-breed them at the earliest opportunity.

(*ix*) If the conception rate under A.I. is lower than under natural service, time of insemination, insemination technique and quality of semen must be checked.

(*x*) Follow a programme of disease prevention, test and vaccination for diseases affecting reproduction. Short irregular intervals of heat frequently indicate cystic follicles; whereas, both short and long irregular intervals point to missed heats. This may be due to infection, under-nutrition or any other stress.

(*xi*) Buy heifers and not adult cows from healthy herds and test them before putting them in your herd.

(*xii*) Practice a general sanitation programme.

(*xiii*) Supply adequate nutrition.

(*xiv*) Provide suitable shelter management.

(*xv*) Detect silent or mild heat especially in buffaloes, using a teaser.

(*xvi*) Have cows give birth in isolation under sanitary conditions, preferably in a calving pen which has been cleaned and sterilized before the cow is moved in. After calving, the room should be cleaned up and sterilized.

CHAPTER

10

Lactation

Lactation is the most economically important physiological activity of the dairy animals. Production of milk by the mammary glands of mammals for the nourishment of their young for varied periods of post-natal life is known as lactation. This is characteristic of all mammals, but in dairy cows and buffaloes due to years of selective breeding, lactation capacity has been raised in great excess of that needed by the calves.

Mammary glands are specialized skin glands in female mammals. They are classified as exocrine glands because they carry their secretion to an epithelial surface.

In dairy animals it is a highly specialized biological 'factory' in which various inputs in terms of feed nutrients, transported to it through circulation are utilized to synthesize milk. This needs a large number of secretory cells, a high rate of metabolism and a very effective organisation of circulatory, nervous and endocrine systems. Its growth, development and functioning are highly integrated with the reproductive processes.

A cow weighing 400 kg and producing 6,000 kg of milk in a lactation secretes in its milk mere than 4 times as much dry matter as is present in her entire body. Some record cows have produced in the year's milk, over 7 times the organic matter of their own bodies and during a lifetime have secreted organic matter equivalent to about thirty six times that present in their own tissues.

10.1. UDDER

The udder of the cow consists of four mammary glands, weighing a total of 10-30 kg exclusive of milk. Being a skin gland, udder is connected to the abdominal cavity only through the inguinal canal. The shape of the udder is something similar to the handle of a tea cup.

In cattle and buffaloes the udder is composed of 4 separate mammary glands. These are joined closely but divided through fibrous membranes so that there is no communication between them except through blood and lymph vessels and nerves. The right and left halves of the udder show clear demarcation. On the other hand the front and rear quarters do not show separation externally (Fig. 10.1).

An ideal udder is the one strongly attached, well balanced, capacious and of fine texture indicating heavy production and a long period of usefulness. On side view, the floor of the udder should be level. Therefore udder should extend well anteriorly and well attached to the abdominal wall. Viewed from behind, the hind udder should be high, wide, fairly uniform in width from top to floor. Attachment at the rear should be high and wide.

The two posterior mammary glands are bigger and contribute about 60 per cent of the milk yield.

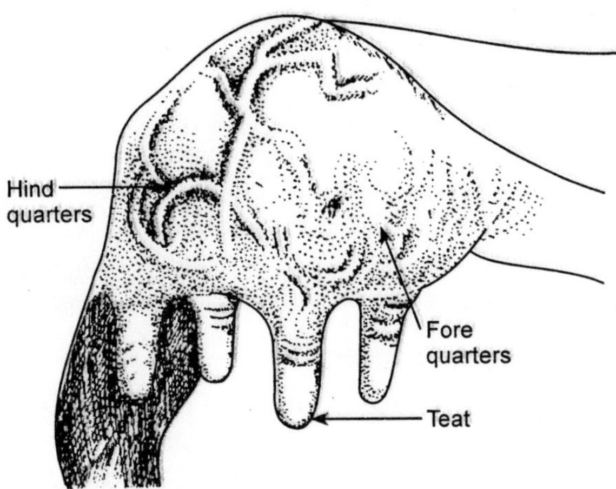

Fig. 10.1. A well balanced good udder showing the four quarters and teats.

Milk from each mammary gland are emptied through the corresponding teat. Conformation and size of teats materially affect milking operation. Ideal teats should be of uniform size, of medium length and diameter, cylindrical, squarely placed under each quarter and well spaced from each other. There should be enough tension on the sphincter muscle at the tip of teat around the orifice to allow easy milking while preventing leakage of milk and entry of infection causing mastitis.

10.1.1. Internal Structure

The mammary gland is made up of secretory cells which are arranged into hollow spherical structures called alveoli, a duct system that drains from alveoli as well as serves for storage of milk, the supportive structures and blood, lymph and nerve supplies (Fig. 10.2). *Alveoli* are composed of a single layer of epithelial cells (Fig. 10.3). These specialised cells take nutrients

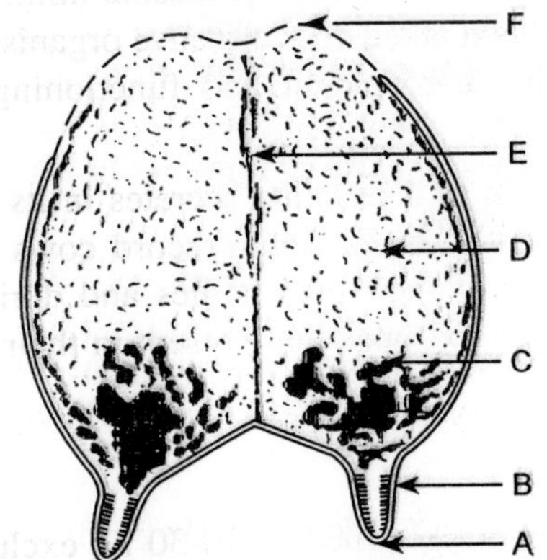

A—Teat opening, B—Teat cistern, C—Gland cistern, D—Secretory tissue, E—Median suspensory ligament F—Top of udder with fatty tissue and lymph glands.

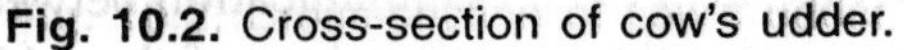

Fig. 10.2. Cross-section of cow's udder.

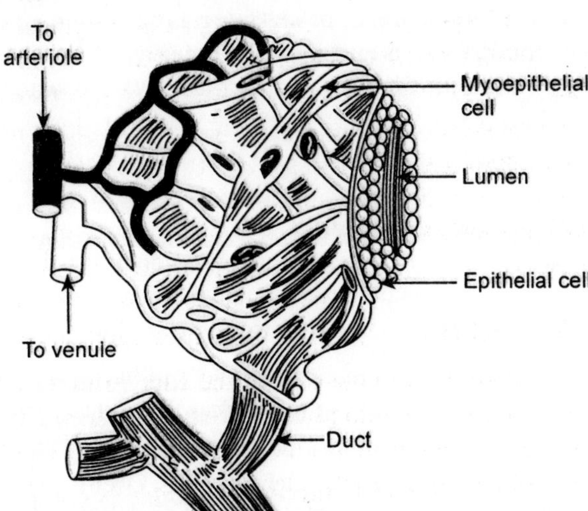

Fig. 10.3. An alveolus (diagrammatic).

from blood, transform them into milk constituents and discharge the formed milk into the lumen of each alveolus.

The organisation of the secreatory tissue in the udder is known as lobule-alveolar type in which several alveoli are held together to form lobules and many lobules are enveloped together by bands of connective tissue to form lobes. A capillary network of blood vessels surround each alveolus. The alveolus is also surrounded by myo-epithelial cells which contract in response to oxytocin secreted by posterior pituitary triggered by suckling stimulus. This forces the milk out of the lumen of alveoli into the ducts and milk cisterns in gland and teats.

10.1.2. Duct System

Small ducts from various alveoli join to form bigger ducts and similar to the tributaries of a river, these ducts join to form still bigger ducts and so on to form 12 to 48 major mammary ducts which end in the *gland cistern* (Fig. 10.2). The gland cistern is a store for milk to a limited extent. The teat cistern is situated immediately below the gland cistern. The upper portions of the teat cistern contain some small areas of accessory secretory tissue. Irritation caused to these areas due to improper milking can spread an existing localised infection to the rest of the mammary gland. When machine milking is practised, if the cups are not removed promptly from the teats when milking is finished, the teat cups crawl up the teats causing injury and irritation to the inner lining tissue.

The teat cistern opens into the exterior through the streak canal at the base of the teat. This small canal is 8-10 mm long with a series of epithelial folds that close the canal when milk is not drawn out. The teats are also closed by involuntary sphincter muscle.

10.1.3. Supporting Structures

The attachment of the mammary gland to the abdominal wall is important. Loosely attached udders may 'break down' and become prone to injury and mastitis.

The lateral suspensory ligaments and median suspensory ligaments join to form a sling like structure of connective tissue holding the mammary secretory tissue with proper support.

The median suspensory legament which is the principal support and attachment of the udder is situated in the mid-line of the body, This consists of elastic fibres and arising from the pelvic arch and mid-line of the abdominal wall, it extends between the mammary glands on either side and joins the lateral suspensory ligament. The lateral ligments are made of non-elastic fibres and arise from the tendons well above the posterior end of the udder.

10.1.4. Blood Supply

The two external pubic arteries penetrate the abdominal wall through the inguinal canal to become mammary arteries which divide into a series of smaller arteries to supply oxygenated blood to the udder tissue through a capillary network.

Venules originate from the capillaries and interlink with each other to form veins that will take the venous blood back from the udder. These veins join together at the base of the udder to become the venous circle (Figure 10.4). There are 3 major routes through which venous blood is drained from the udder (venous circle). The most obvious ones are the milk veins. These subcutaneous veins which emerge at the anterior edge of the udder and run forward in a tortuous course, penetrate the thoracic cavity at the 'milk wells'. When the cow is standing, most of the

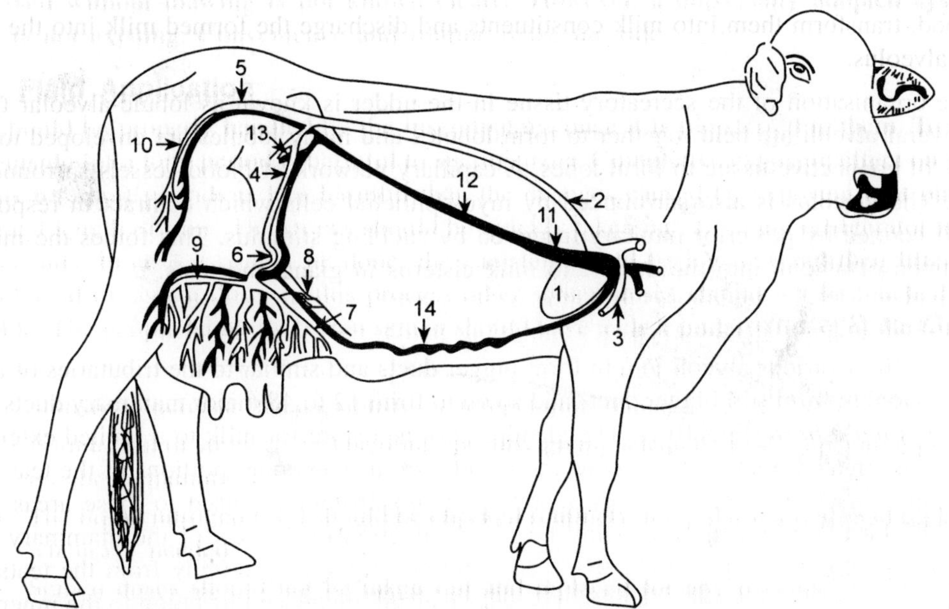

1. heart. 2. abdominal artery. 3. aorta. 4. external puedenadal artery and vein. 5. internal iliac artery and vein. 6. sigmoid flexure of the external puedenadal artery and vein. 7. cranial mammary artery. 8. subcutaneous abdominal artery. 9. caudal mammary artery. 10. perineal artery and vein. 11. anterior vena cava. 12. posterior vena cava. 13. external iliac artery and vein. 14. subcutaneous abdominal vein.

Fig. 10.4. Blood Circulation of the mammary gland.

blood from the udder takes this route. But when the cow lies down, the two external pudic veins become the major venous route from the udder. The third route is through the perineal veins which enter the body cavity through the pelvic arch. This route does not carry as much blood as the other two.

It has been calculated that it takes about 500 litres of blood flowing through the udder to form 1 litre of milk. In the first months of lactation, there is a very high turn-over of blood through the udder. Corresponding to the lactation curve, the rate of blood flow through the udder also declines.

The thin walled ducts and the supra mammary lymph glands form the lymphatic system of the udder. The lymph (a colourless tissue fluid) from the udder is collected and eventually discharged into the blood circulation by the thoracic lymph duct at the anterior vena cava, the lymphatic vessels collect the blood serum that filtrates out of the blood vessels and conveys it back to the blood circulation. The lymph nodes are also important centres for disease resistance. When the udder is infected, the lymph nodes react by increasing the output of lymphocytes which reach the udder through circulation and combat infection. The lymph nodes also act as filters removing bacteria and other foreign bodies.

10.1.5. Nervous Supply

The udder is supplied with sensory nerves (afferent) that carry impulses form touch, temperature and pain receptors to the central nervous system (Fig. 10.5). The efferent sympathetic nerve fibres

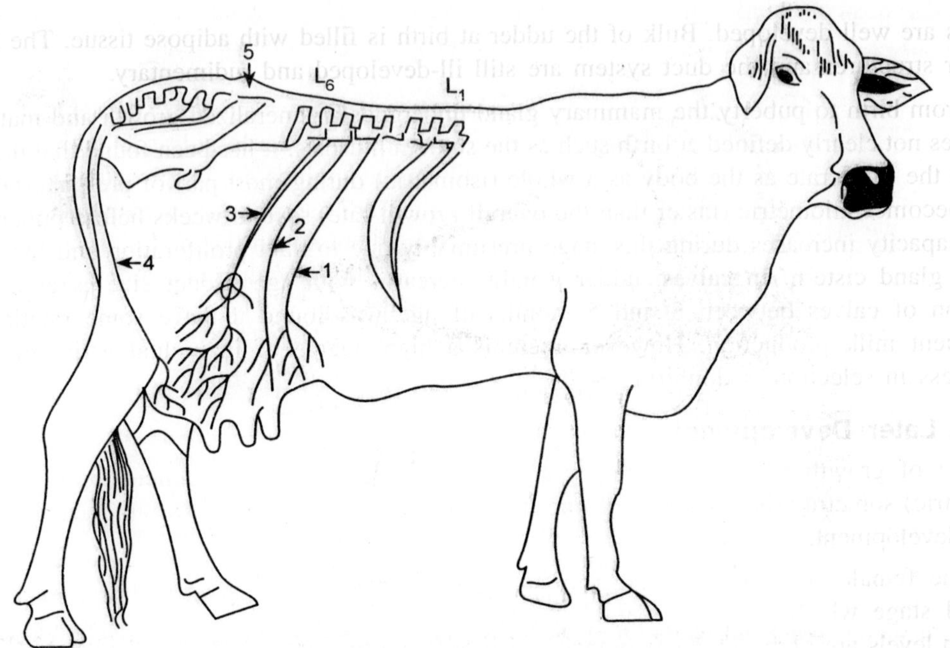

Fig. 10.5. Nerve supply of the mammary gland—L_1 first Lumbar vertebra, L_6—sixth vertebra nerve, S-sacrum. 1. first lumbar nerve, 2. second lumbar nerve, 3. inguinal nerve. 4. perineal nerve.

regulate blood flow and innervate the smooth muscles around the mammary ducts and teat sphincter. When the cow is subjected to stress or excitement, the neurohormone adrenaline (epinephrine) is discharged by the sympathetic system causing vasoconstriction including in the udder which, if prolonged, may result in reduced milk production.

Suckling is a reflex action in which the stimulii of calf licking, sucking and cuddling at teats and udder initiate impulses that evoke the discharge of oxytocin from the posterior pituitary and other hormones from the anterior pituitary. This reflex can be conditioned against other stimuli like washing with warm water, wiping and massaging with a towel, feeding concentrates etc. in weaned cows. There are evidences to prove that the afferent stimuli in the let down reflex might be carried by mechanisms other than direct innervation. Surgical section of nerves to the udder in the cow does not significantly affect udder development or milk yields. Transplanting an udder to the neck region after complete denervation in a goat did not reduce milk production (The blood supply was restored through carotid artery and jugular vein).

10.2. DEVELOPMENT OF THE MAMMARY GLAND

10.2.1. Early Development

Early in foetal life two milk lines appear on the abdomen. Soon four epidermal thickenings known as 'mammary buds' are formed on the milk lines in the inguinal region. These mammary buds become focal points for differentiation and further development of mammary structures. When the foetus is 20-30 cm in length, the embryonic teat cistern and streak canal are well defined.

At birth the teats are well formed except for the smooth muscle sphincter and smooth muscle cells surrounding the teat cistern. The connective stroma and the blood and lymph vascular systems are well developed. Bulk of the udder at birth is filled with adipose tissue. The lobulo-alveolar structures and the duct system are still ill-developed and rudimentary.

From birth to puberty the mammary gland undergoes a generalized growth and maturation of tissues not clearly defined at birth such as the smooth muscles. It has been found that the udder grew at the same rate as the body as a whole (isometric) during most part of pre-pubertal stage. But it becomes allometric (faster than the overall growth rate) several weeks before puberty. The udder capacity increases during this stage presumably due to duct proliferation and increase in size of gland cistern. In calves, udder weight increases with age. Udder size as assessed by palpation of calves between 3 and 5 months of age was found to have some relation with subsequent milk production. However, there is a high degree of individual variation and its usefulness in selection is doubtful.

10.2.2. Later Development

The rate of growth of the mammary gland becomes faster than the general body structures (allometric) sometime before onset of puberty. But, with puberty there is a rapid acceleration of udder development.

The female sex hormone, estrogen is the major growth inducing agent during the post pubertal stage which develops mainly the duct system. During the estrous cycles. when the estrogen levels are high, the ducts branch and proliferate rapidly, but regress slowly during other stages. The regression taking place is always less than the proliferation that has occurred, so that, with every cycle more ducts are added. In calves, there is a rapid growth of adipose tissue and the development of the connective tissue network is slower than that of the ducts.

Maximum development of the udder takes place during pregnancy and early days of lactation. Progesterone, in the presence of estrogen is primarily responsible for the development of the alveoli. In cattle estrogen produces a development of ducts and partially, alveoli. But, the full development of the udder during pregnancy, depends on estrogen. progesterone and pituitary hormones.

10.3. HORMONAL CONTROL OF MAMMARY GLAND DEVELOPMENT, INITIATION AND MAINTENANCE OF LACTATION

10.3.1. Initiation of Lactation

Cell proliferation and cell development are completed several weeks prior to parturition and the onset of copious milk production. The development and lactogenesis are primarily dependent on (*i*) *prolactin*, (*ii*) *corticosterone* or *cortisol*, and (*iii*) *insulin.*

The whole process involves a hectic cell proliferation and differentiation to form an array of secretory cells. The cells also should acquire the necessary enzyme complements for synthesis of milk components. Finally, the whole process should be adequately controlled and orchaestrated.

Prolactin during pregnancy sensitizes mammary secretory cells to insulin, which causes a high rate of mitotic cell division. After cell proliferation prolactin plays a central role in RNA and protein synthesis. The formation of extensive endoplasmic reticulum characteristic of secretory cells is dependent on insulin and glucocorticoids. Besides, they regulate the production and functional activity of many enzymes involved in synthesis of milk. Differentiated secretory cells of the udder are highly dependent on insulin for survival. Apart from acting on prolactin sensitized

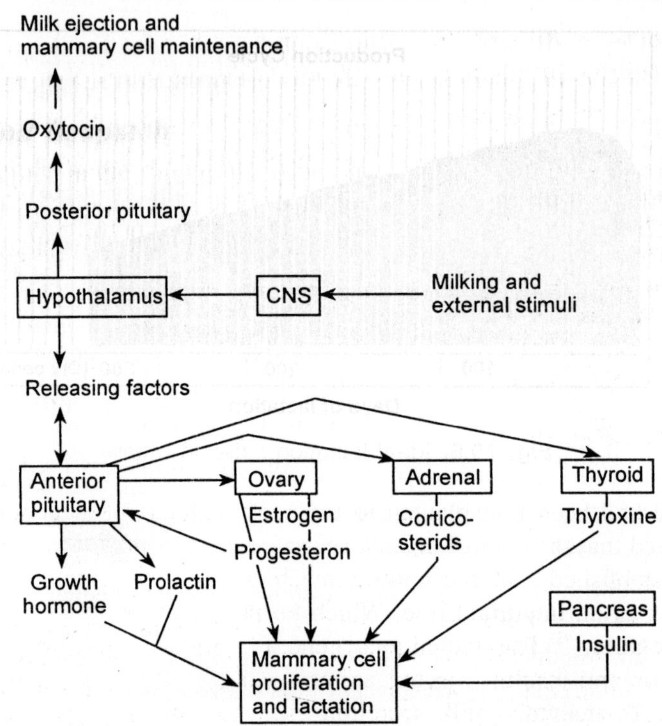

secretory cells to cause a wave of cell proliferation, insulin appears to have several specific functions along with other hormones like prolactin and glucocorticoids in biosynthesis of milk.

The precise hormonal mechanisms concerned with the initiation of milk secretion are less well established. In the pregnant rats, the lactogenic activity of pituitary prolactin and placental lactogen is blocked by a direct action of progesterone on the alveolar cells. The progesterone level falls with the onset of parturition which permits the lactogenic effects of prolactin and other hormones. The inhibitory effect is partly due to blocking the synthesis of alpha-lactalbumin, thereby inhibiting the synthesis of lactose. Thus, the decline in progesterone prior to calving "triggers" the synthesis of one or more key enzymes which heralds the initiation of lactation.

10.3.2. Maintenance of Lactation

In dairy cattle, the onset of secretory activity of the mammary gland is closely associated with parturition. The secretion formed in the first 2 to 3 days is called colostrum which differs in composition from normal milk. Soon after parturition there is a rapid increase in secretory activity and the colostral type of secretion changes to the formation of normal milk. The secretory activity goes on increasing for a while until it peaks around 4 to 6 weeks of lactation. After this, it gradually declines (Fig. 10.6). The term 'persistency' refers to the degree of maintenance of milk production. The reduction in milk yield during a month is dependent on the yield during the preceding month and it can be calculated as a percentage.

10.3.3. Control of Milk Secretion

Milk secretion involves both the intracellular synthesis of milk constituents and their passage into the alveolar lumen. How is it controlled ? Dramatic experiments by Jim Linzell in goats in which

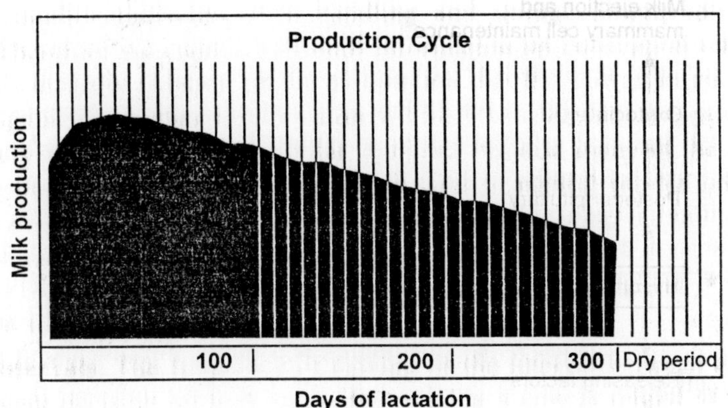

Fig. 10.6. Ideal lactation cycle in a cow.

the mammary gland had been transplanted to the neck without subsequent loss of its ability to secrete milk, indicated that the control of milk secretion is other than through the nervous system. It has been thus established that the control mechanisms are hormonal. In fact, the anterior pituitary gland plays an all important role. Which hormones of the hypophyses are concerned in maintenance of lactation ? The techniques generally involve *in vivo* studies with normal, endocrinectomised or anti-insulin injected animals and short term *in vitro* studies with tissue slices and cultures. To maintain milk secretion even partially in hypophysectomised animals, cortisol and prolactin are necessary. Somato-trophic hormone (STH) appears to improve the effects of cortisol and prolactin. Still the lactation will be far below normal in intact animals. Addition of other hormones like insulin, tri-iodothyronine and estrogens cause some more enhancement. Treatment with these hormones in addition to prolactin, cortisol and STH does not completely restore lactation. It is not yet clear whether insulin, estrogens and tri-iodothyronine have a direct effect on milk secretion or an indirect one through enhancing general metabolism of the cow.

10.3.4. Milk Removal

This involves two processes. One is the passive withdrawal of milk from the cisterns and ducts. The other one is the active oxytocin induced ejection of milk by the contraction of myoepithelial cells around the alveoli.

Milk removal, including operation of an efficient milk ejection reflex is a must for maintaining the number of secretory cells, their metabolic activity and intense lactation. After complete hypophysectomy, oxytocin must be regularly administered before milking to facilitate milk ejection, lest the mammary secretory tissue degenerates due to accumulation of milk in the glands. After milk ejection, milk should be removed by suckling or milking out to maintain lactation as milking stimuli without accompanying milk removal will not maintain lactation. Thus, regular complete milking at least twice (at 12 hr intervals) daily is a must for maximal milk secretion in cows.

10.4. MILK SYNTHESIS

Milk, the normal secretion of the mammary gland is an emulsion of fat in a watery solution of sugar and mineral salts, and with protein in a colloidal suspension. The principal constituents of cow milk are approximately as given below :

Constituents	Percentage
Water	87.00
Dry matter	13.00
Fat	4.00
Protein	3.55
Sugar	4.80
Ash	0.65
	100.00

The rate at which the secretory cells of the mammary gland uptake nutrients from the blood circulation and synthesize various milk constituents is a major limiting factor in controlling the level of milk production.

10.4.1. Alveolar Secretory Cell

Fully developed secretory cells in the alveoli have very prominent nuclei, mitochondria, golgi apparatus and an extensive rough endoplasmic reticulum.

Unlike the nuclei of gametes, the nuclei of secretory cells transmit the coded genetic information in deoxy-ribose nucleic acid (DNA) for the synthesis of milk proteins and some enzymes which direct biochemical activities in the cell. DNA (gene) transcribe the genetic information to messenger-RNA (Ribo-nucleic-acid). The messenger-RNA then leaves through the pores of nuclear membrane to the endoplasmic reticulum.

Messenger-RNA leaving the nucleus and entering the endo-plasmic network of small canals arranges the 'template'-RNA in such a way that specifies the amino acid sequence for the synthesis of specific proteins and enzymes. The inner surfaces of some of the canals of the endoplasmic reticulum are studded with these 'template' RNA which are known as ribosomes. This being the organelle where amino acids are arranged in proper sequence, is the site of protein synthesis (see chapter 14, Fig. 14.10).

Other major organelles of the acinar cell are the *lysosomes*. These are membrane bound particles which contain the enzymes which when released from within the lysosomal membranes cause degradation of the cell. During involution of the mammary gland in late lactation and when affected by mastitis, lysosomes are particularly active.

Mitochondria are the 'power house' of the cell because they store and provide energy for various cellular activities. Therefore, all metabolically active tissues have numerous mitochondria. Alveolar secretory cell being one of the most active ones has a large number of them during lactation. But, their number diminishes drastically during the non-lactating state.

The *Plasma membrane* envelops the cells and other membranes cover the organelles. These membranes have a very important and active role especially by selective and active transport of various chemicals into the cell and into various compartments within the cell. Due to this, some nutrients from blood enter the cells through the plasma membrane with ease and even get concentrated. Others are excluded.

10.4.2. Biosynthesis of Milk

Some of the constituents of milk appear to be transferred directly from blood without change. Others are synthesised by the secretory alveolar cells from **blood precursors** and secreted into milk (Fig. 10.7). Table 10.1 given below indicates the milk constituents and their precursors.

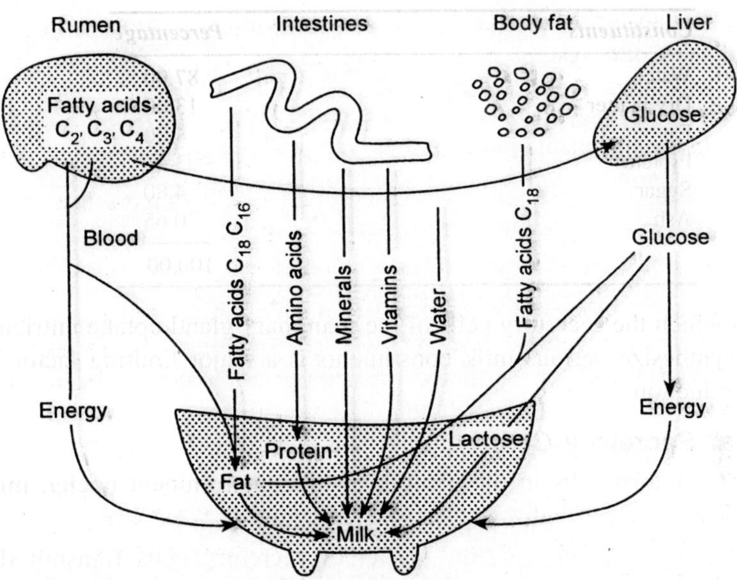

Fig. 10.7. Synthesis of various milk constituents in the uddar.

TABLE 10.1. Milk Constituents and their Blood Precursors

Milk constituents	*Blood precursors*
A. Transferred without change	
1. Water	Water
2. Minerals	Minerals
3. Vitamins	Vitamins
4. Fat	
Long chain fatty acids	Long chain fatty acids
5. Protein	
γ globulins	γ globulins
γ casein	γ casein
B. Secreted after synthesis in alveolar cells	
1. Carbohydrates	
Lactose	Glucose
2. Fat	
Short chain fatty acids	Acetate and beta hydroxybutyrate (BHBA)
3. Protein	
α casein	Free amino acids
β casein	Free amino acids
κ casein	Free amino acids
α lactabumin	Free amino acids
β lactoglobulin	Free amino acids

10.4.3. Milk Protein Biosynthesis

Milk proteins are synthesized from free amino acids most of which are derived through the blood supply. The cells require all of the essential amino acids and most of the non-essential amino acids in this way. The absorption process of amino acids through the cell membrane is not well understood. More than one system may be involved and all are likely to be active transport (energy-requiring) processes. Amino acids uptake and protein content of milk are affected by the dietary levels of these and nutritional status of animals. Heavy grain feeding to cows increased amino acid levels and milk concentration of casein and albumin in cows.

The primary proteins in milk are alpha-casein, beta-casein, K-easein, alpha-lactalbumin and beta-lactoglobulin. They make up about 90 per cent of the protein of milk and are not found in blood. Therefore, they have to be synthesized by the mammary secretory cells from amino acids.

In general terms, the biochemical mechanism involved in milk protein synthesis are similar to those in other mammalian cells. The synthesis of these proteins in milk is under genetic control, the coded information for the amino acid composition and arrangement being present in the deoxyribonucleic acid (DNA) of each cell. The general process involves, as in other cases, transcription of the coded information from DNA in messenger ribonucleic acid (mRNA) and its translation at the ribosomal level in the amino acid sequence in the polypeptide chains. Most of the protein synthesis takes place in the rough endoplasmic reticulum (RER) rich in bound ribosomes. They undergo further post-translational modifications in their secretory pathway through the Golgi apparatus and secretory vesicles before ejection into the lumen with other milk components. The secretory vesicles bleb off the distal face of Golgi cisternae and move to the apical region of the cell to fuse with the plasma membrane to replace that lost with the secretion of the fat globules. The contents of the vesicles include casein micelles, other protein and non-protein constituents of milk. They are discharged into the lumen of alveoli. Some other milk proteins like the γ globulins (antibodies), milk serum albumin and the α casein complex are transferred directly from blood without change. Synthesis of these proteins does not appear to take place in the alveolar cells.

10.4.4. Lactose Synthesis

Lactose is the disaccharide specific for milk. It is a combination of one molecule of galactose and one molecule of glucose. It is primarily synthesised from blood glucose. Two molecules of glucose are necessary to form one molecule of lactose.

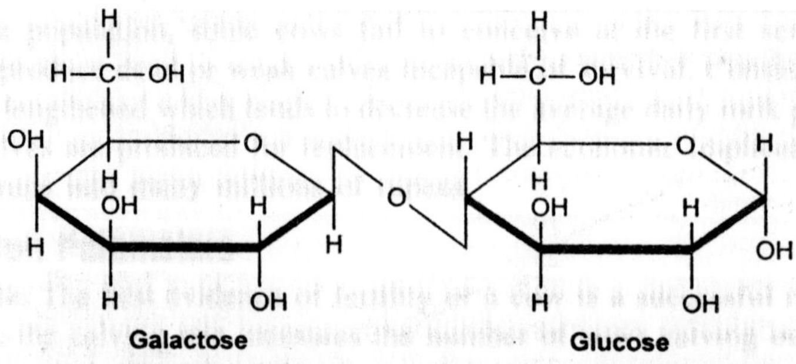

Galactose **Glucose**

The uptake of glucose into the secretory cell is therefore a limiting factor in lactose synthesis. Low blood concentration of glucose invariably causes a reduction in milk production.

For the synthesis of lactose from glucose, a part of the glucose has to be first converted to galactose. The metabolic pathway involved in this is given below :

Glucose + *ATP ⟶ Glucose-6-phosphate + *ADP

Glucose-6-phosphate ⟶ Glucose-1-phosphate

Glucose-1-phosphate + *UTP ⟶ *UDP-glucose + phosphate
UDP glucose ⟶ UDP galactose.

One such galactose molecule condenses with a glucose molecule to form one molecule of lactose.

UDP-galactose + Glucose-1-phosphate ⟶ Lactose-1-phosphate + UDP
Lactose-1-Phosphate ⟶ Lactose + Phosphate

ATP—Adinosine triphosphate, ADP—Adinosine diphosphate
UTP—Uridine triphosphate, UDP—Uridine diphosphate

The condensation reaction is catalyzed by an enzyme, *lactose synthetase*. This enzyme has 2 subunits, one of which is a milk protein—alpha-lactalbumin. Previously it was thought that the major proteins of milk are simply export proteins meant for the nourishment of the calf. Recent discoveries in the field have changed this concept as the enzymatic activity of some of these protein fractions are now well understood.

10.4.5. Synthesis of Milk Fat

Milk fat is a mixture of different triglycerides. Approximately 50 per cent of these are short chain fatty acids (C_4—C_{14}). The remaining half comprise of long chain (C_{16}—C_{20}) ones.

The primary blood precursors of the fatty acids in milk triglycerides of bovines appear to be acetate, beta-hydroxybutyrates (BHBA), chylomicra and triglycerides of low density blood lipoporteins. The long chain fatty acids of the cow's milk are mostly of direct origin from the diet. The unsaturated long chain fatty acids in the diet of cows from plant origin get hydrogenated (saturated) in the rumen. This process is not 100 per cent complete.

The precursors of milk fatty acid have been tabulated in table 10.2.

TABLE 10.2. Milk Fatty Acids and Their Precursors

Fatty acids	*Mole per cent in milk fat*	*g/100 g milk fatty acid from*	
		Acetate + BHBA	*Blood triglycerides*
C_4	10	4.0	—
C_6	4	2.1	—
C_8	2	1.3	—
C_{10}	5	4.0	—
C_{12}	3	2.8	—
C_{14}	10	5.2	5.2
C_{16}	25	14.1	14.1
$C_{16}:1$	3	—	3.6
C_{18}	8	—	10.5
$C_{18}:1$	27	—	35.0
$C_{18}:2$	3	—	3.5

Adapted from Smith, N.E. (1970), Ph.D. Thesis, University of California, Davis.

In cows short chain fatty acids are synthesized via the melonyl 1-COA and chain elongation (BHBA + acetyl COA pathways). Acetate is 2-carbon unit whereas BHBA is a 4-carbon molecule, and both molecules are derived from the fermentation of carbohydrates in the rumen to volatile fatty acids. In this reaction the chain length is increased by 2 carbon atoms at a time because acetate molecules are added stepwise. In the case of BHBA, the 4-carbon molecule is cleaved to 2-carbon units and utilized similar to acetate.

Acetic acid is the major precursor for milk fat synthesis compared to BHBA. Besides, acetate is utilized by the mammary cells for energy purposes. Thus, the production of actetate in the rumen in adequate quantities is the basic necessity for optimal milk fat synthesis in particular and milk production in general.

Secretion of milk fat involves the fat droplet being enveloped by the plasma membrane as it protrudes into the alveolar lumen. The apical plasma membrane of the secreting cell does not rupture, but rather fuses under the protruding droplet resulting in a 'pinching off effect. Often a small amount of cytoplasm of the cell including mitochondria, ribosomes and vacuoles may be included within the fat globules. The protein of the plasma membrane covering the fat-droplet after it is 'pinched off' into the alveolar lumen is termed the milk fat globule membrane (MFGM).

10.4.6. Water and Minerals in Milk

The major mineral components of milk are calcium, phosphorus, sodium, potassium, chlorine and magnesium. The buffering capacity of milk is mainly due to the citrates, phosphates, and bicarbonates of calcium and magnesium and the proteins.

In healthy cows without any udder damage there is a constant percentage of sodium, potassium, chlorine and lactose in milk. The constant amount is characteristic of each individual animal. They are responsible for keeping the osmotic balance in milk. Water moves into milk to maintain osmotic equilibrium with blood. Thus, the secretion of lactose and transfer of sodium, potassium and chlorine into milk determines the volume of milk produced. There is an inverse relationship between lactose on one side and sodium and potassium on the other.

Minerals and water are only transferred from blood to milk and the secretory cells do not change them. But, it is not conclusively known whether there is any selective uptake and transfer of these from blood to milk. There are some observations indicating that the epithelial cells can discharge minerals selectively into blood or milk indicating an active transport system.

10.4.7. Vitamins

The alveolar cells do not possess the ability to synthesize vitamins. Therefore the vitamins present in milk are transferred from blood generally in proportion to their presence in blood. Thus, the vitamin content of milk can be increased by increasing the vitamin content of blood by dietary or other means. Increasing the carotene or vitamin A content of the diet, as occurs during grazing can cause a 10 fold increase in vitamin A content of milk. Similarly, exposing cows to sunlight or feeding hay cured in sunlight can increase the vitamin D content of milk. Since rumen microflora synthesize all the B-vitamins, their concentration in diet does not change that in milk. The only exception to this is riboflavin. By changing from a straw and concentrate diet to a predominantly green forage feeding regime, the riboflavin content of milk increases.

10.5. MILK EJECTION AND MILKING

Milk secreted by the alveolar cells in the udder cannot be removed until the myoepithelial cells surrounding them contract forcing the milk from the alveoli and smaller ducts in to the major ducts and cisterns. This is brought about by the neurohumoral reflex known as milk ejection or

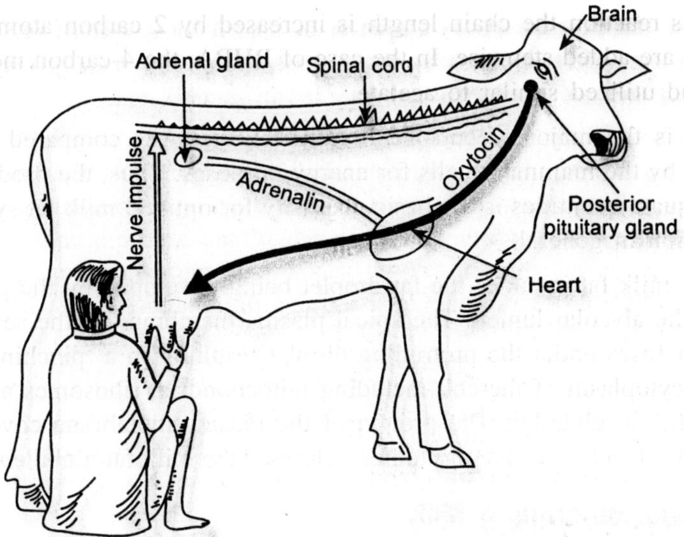

Fig. 10.8. Milk ejection.

in common parlance 'milk let-down'. (Fig. 10.8). This reflex is activated by nervous stimuli that the cow associates with the suckling or milking process, such as suckling by the calf or palpation by the milker. The nerve impulses go up the spinal cord to the superoptic and paraventricular nuclei of the hypothalamus and from there to the posterior pituitary. These sensory stimuli cause the liberation of the neurohormone oxytocin and to a lesser extent vasopressin into the blood stream. These hormones travel via blood to the mammary gland and there bring about contraction of the myoepithelial cells surrounding the alveoli. Considerable attention has been focused on the various visual and physical stimuli and conditioned reflexes that will evoke oxytocin release and consequent milk ejection. These studies have indicated that the best stimulus for triggering the reflex is manipulation of the teats during suckling or milking. The presentation of the calf to its mother and allowing her to lick up the calf, noises associated with milking, feeding concentrates prior to milking, the presence of the milker and coitus are other stimuli. Various pre-milking udder preparations like massage, water hosing etc. have been found to help let-down (Fig. 10.9).

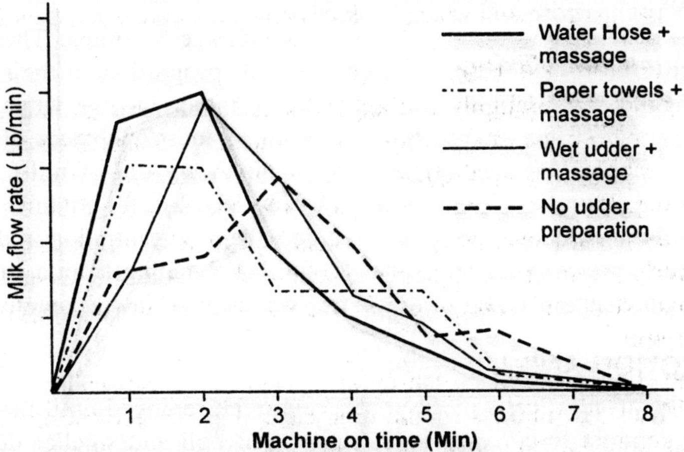

Fig. 10.9. Comparison of milk flow rates due to various pre-milking udder preparations.

Oxytocin and vasopressin are synthesized in the hypothalamus and are collected in the posterior pituitary. They are released from there when the proper sensory impulse is transmitted. Oxytocin is 5 to 6 times more active than vasopressin in milk ejection. Apart from contraction of myoepithelium surrounding alveoli, it causes relaxation of the smooth muscles of large ducts and the cisterns. This results in increased capacity of these structures to accommodate more milk.

Intramammary pressure rises as a result of milk ejection within 20 to 60 seconds. The calf or the milker has only to overcome the resistance of the teat sphincter to remove the milk. Effective levels of oxytocin remain in blood from 6 to 8 minutes. Thus it is important to milk out fast during the interval when oxytocin maintains intramammary pressure.

Milk ejection reflex is inhibited by adrinaline. Fright, pain or emotional disturbances can cause release of adrinaline from adrenal medulla. Other practices like washing the udder with very cold water can also cause contraction of blood vessels. All practices that cause vasoconstriction in the udder reduce the entry of oxytocin to the gland and thus prevent milk ejection. Therefore all care should be taken to prevent all external unpleasant events at the time of milking.

10.5.1. Management of Milking

Milking is the key operation on a dairy farm having a direct effect on the income from the enterprise. It is an art requiring experience and skill. It should be conducted at regular intervals, quickly, cleanly, gently, quietly and completely. Cows remaining comfortable yield more milk. Maintenance of clean conditions in the milking parlour and while milking will ensure udder health and production of clean milk that remains wholesome for longer time. The withdrawal of milk should be completed within 5 to 7 minutes so that the milk is removed completely when the effect of oxytocin is present. Complete milking should be practised lest the residual milk may cause regression of the secretory tissue thus reducing milk yield and also serve as a nidus for mastitis causing organisms.

Milking is a labour intensive operation requiring 50-60 per cent of the total man-hours on the dairy farm. One milker, depending on his skill may milk 10 to 15 cows including activities connected with milking like cleaning the cows, udder and feeding concentrates. Machine milking and automatic udder washing equipments may make the chore associated with milking easier and less time consuming. But, in a developing country like India where providing labour is a major social objective for any enterprise and where labour is relatively cheap, hand milking is encouraged.

10.5.1.1. *Preparation for Milking*

The milking barn should be thoroughly washed and scrubbed after each milking so that the barn will be clean and dry before the next milking is commenced. Dusty feed and silage should be avoided in the milking barn. The hind quarters, thighs and udders of cows should be brushed or washed before milking. If there is more hair growth in the udder region, it should be clipped periodically. Buffaloes should invariably be washed before milking. Just before milking (after suckling by calf if early weaning is not practised), the udder should be wiped with a cloth dipped and squeezed in some weak antiseptic solution. In cold weather, warm antiseptic solution should be used for the purpose.

Apart from the cows and udders, the milkers, their hands especially and the milking pails and cans should be clean. The milkers should wear clean dress and cover their heads with suitable caps, lest loose hair may fall in milk. Their nails should be periodically trimmed and hands cleaned and disinfected between each milking by dipping in an antiseptic solution.

10.5.1.2. *Hand Milking*

Cows are generally milked from the left side. There is no hard and fast order of teats while milking, the engorgement of teats due to letdown should be the guideline.

The first few streams of milk from each teat is milked into the strip cup to see clues of mastitis in the form of clots or streaks of blood. This also helps in getting rid of the milk in the teat canal which has high count of bacteria in it.

Full hand milking and stripping are the two commonly used methods. In full hand milking, the teat is closed at the base between the thumb and index finger and the milk trapped in the teat sinus is forced out by pressing with the other fingers on the teat (Fig. 10.10). This process is repeated in quick succession. By maintaining a quick succession of alternate compressions and relaxations, the alternate streams of milk from the two teats sound like one continuous stream. Many milkers tend to bend their thumb in, against the teat while milking. This practice should be avoided as it injures the teat tissues.

Fig. 10.10. Full-hand milking. Note the design of the milking pail which reduces contamination of milk from falling dirt. (*Courtesy* : Abraham Mathew).

Cows with large teats and she-buffaloes are milked by full-hand method, but stripping has to be adopted for cows with small teats for obvious reasons. Full hand method is superior to stripping as it is faster and also simulates natural suckling process by the calf. Stripping consists of firmly seizing the teat at its base between the thumb and index finger and drawing them down the entire length of the teat pressing the teat so that milk flows out in a stream. Stripping causes more irritation and teat injury due to repeated sliding of the fingers. In spite of this, when all milk that is available is drawn out by full-hand milking a few strippings are done. It will milk the cow completely; the last drawn milk thus is termed strippings and is richer in milk-fat.

In many cases milkers resort to wet hand milking by moistening the teats with milk, water or even saliva. This makes the teats dry and chafed, cracks and sores appear and milking becomes painful to the cows. Besides, this practice is most unhygienic and not conducive to clean milk production. The hands should be clean and dry while milking.

Immediately on noticing cracks and sores on teats some antiseptic cream should be smeared on them after milking.

10.5.1.3. *Machine Milking*

Milking machines popularly used now-a-days in most of the western and industrialised countries use alternating negative and atmospheric pressures with the help of a double chambered teat-cup assembly (Fig. 10.11).

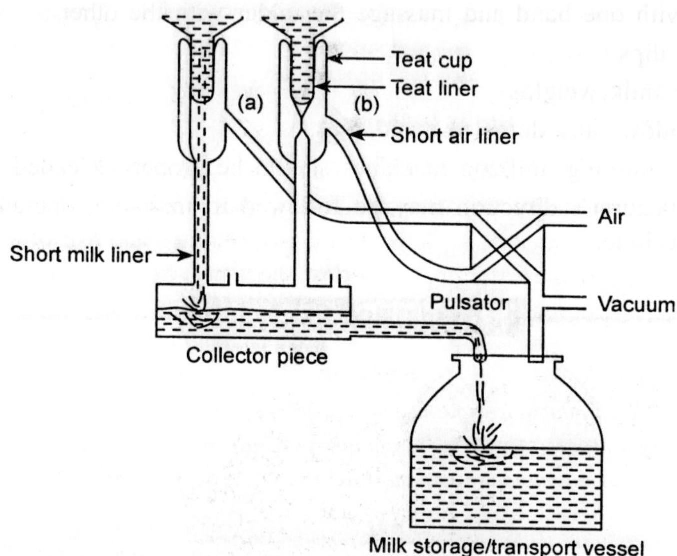

Fig. 10.11. Schematic diagram showing the principle of machine milking. Due to alternating negative (*a*) and positive (*b*) fluctuations of specific magnitudes, caused in the pressure between the teat cup and the teat liner by the pulsator, the natural suckling action is simulated on the teats. Milk from each quarter is collected through the respective milk liners into the collector piece from where it is carried out to the storage/transport vessel by vacuum suction action.

A continuous partial vacuum is maintained inside an inflatable rubber tube, *the teatcup liner*, into which the teat is inserted. A partial vacuum and normal atmospheric pressure are alternated in the space between the rubber liner and the metal shell of the teat-cup by means of a device known as 'pulsator'. When the negative pressure is applied between the liner and the shell milk flows form the teat. When atmospheric pressure enters the chamber, the rubber inflation (liner) collapses with the result that the teat is compressed and massaged. This forces the lymph and blood to flow out of the teat. A continuous vacuum alone would have caused congestion and irritation of the teats.

A number of factors influence the efficiency of machine milking like vacuum level, *pulsation rate* (the number of cycles of alternating vacuum and normal atmospheric pressure per minute), and the pulsation ratio (the proportion of time of vacuum and atmospheric pressure applied).

Milking machines vary widely in their pulsation rate, as high as 120 or as low as 40 cycles per minute. Pulsation ratio or the ratio of expansion time and collapse time is kept within the limits of 1:1 and 2.5:1.

Ideal practices in Machine Milking

1. Get the cow ready by wiping and massaging the udder and teats for 1/2 to 1 minute. Use a cloth dipped and squeezed out of warm mild antiseptic lotion.
2. Use the strip cup and draw out the foremilk into it from each quarter.

3. Put on the teat-cups promptly.
4. It is a good practice to use a timer to avoid overuse of the machine. Remove teat cups promptly breaking vacuum first.
5. Strip by machine for a few seconds. At the end of the timed period pull down on the teat cups with one hand and massage the udder with the other.
6. Apply teat dips.
7. Record the milk weights
8. Screen windows and doors to avoid flies.
9. After every milking, milking machines should be properly cleaned.
10. The manufacturer's direction may be followed to maintain, operate and service the milking machine.

FURTHER READING

Section III

Hafez, E.S.E. (ed.). 1974. *Reproduction in farm animals.* 3rd ed. Lea and Febiger, Philadelphia.

Laing, J.A. 1975. *Fertility and infertility in domestic animals.* Bailliere Tindall and Cox, London.

Luktuke, S.N. 1978. *Artificial insemination and sterility.* In : *Hand book of animal husbandry.* I.C.A.R., New Delhi.

McDonald. L.E. 1975. *Veterinary endocrinology and reproduction.* Lea and Febiger, Philadelphia.

McDonald, R.E. 1972. *Improvement of livestock production in warm climates.* W.H. Freeman, San Francisco.

Perry, E.J. 1965. *Artificial insemination of farm animals.* Oxford & IBH Publishing Co., New Delhi.

Salisbury, G.W. and Van Demark, N.L. 1961. *Physiology of reproduction and artificial insemination of cattle.* W.H. Freeman & Co., San Francisco.

Breasile, J.E. 1971. *Lactation.* In : *Text book of Veterinary Physiology.* Lee and Febiger, Philadelphia.

Courie, A.T. and Tindal, J.S. 1971. *The Physiology of lactation.* Edward Arnold (Publ.) Ltd., London.

Falconer, I.R. 1971. (Ed.) *Lactation.* Butterworths, London.

Ganguli, N.C. 1974. *Milk Proteins.* Indian Council of Agric. Res., New Delhi.

Jacobson, N.L. 1970. *The Mammary Gland and Lactation.* In : Duke's *Physiology of Domestic Animals.* (ed. Melvin J. Swenson). Comstock Publishing Associates, Cornell Univ. Press, Ithaca.

Jones, E.A. 1969. *Recent developments in the biochemistry of the mammary gland.* J. Dairy Res. **36** : 145.

Kon, S.K. and Cowie, A.T. 1961. (Ed.) *The Mamary Gland, Milk and its Secretion.* Academic Press, New York.

Larson, B.L. and Smith, V.R. 1974. (Ed.) *Lactation, a Comprehensive Treatise*, Academic Press, New York.

Patton, S. 1969. *Milk. Sci. Amer.* **221** : 58.

Reynolds, M. and Folley, S.J. (Eds.) 1969. *Lactogenesis : The Initiation of Milk Secretion at Parturition.* Philadelphia, University of Pennsylvania Press.

Schmidt, G.H. 1971. *Biology of Lactation.* W.H. Freeman and Co., San Francisco.

Smith, V.R. 1959. *Physiology of Lactation.* 5th ed. Iowa State Univ. Press Ames, Iowa.

Lactation

Lactation is the most economically important physiological activity of the dairy animals. Production of milk by the mammary glands of mammals for the nourishment of their young for varied periods of post natal life is known as lactation. This is characteristic of all mammals. In dairy cows and buffaloes, due to years of selective breeding, lactation capacity has been improved in excess to that needed by the calves.

Mammary glands are specialized skin glands of female mammals. They are classified as exocrine glands because they carry their secretion to an epithelial surface.

In dairy animals it is a highly specialized biological 'factory' in which various nutrients in terms of feed nutrients transported to it through circulation are utilized to synthesize milk. This needs a large number of secretory cells, a high rate of metabolism and a very effective coordination of circulatory, nervous and endocrine systems. Its growth, development and functioning are highly integrated with the reproductive processes.

A cow weighing 400 kg and producing 6,000 kg of milk in a lactation secretes in its milk more than 4 times as much dry matter as is present in her entire body. Some cows are known to have produced in the year's milk, over 7 times the organic matter of their own bodies and during a lifetime have secreted organic matter equivalent to about thirty times that present in their own tissues.

10.1. UDDER

The udder of the cow consists of four mammary glands, designed as a unit of [illegible] of milk. Being a skin gland, udder is connected to the abdominal cavity only through the inguinal canal. The shape of the udder is something similar to the handle of a tea cup.

In cattle and buffaloes the udder is composed of 4 separate mammary glands, joined closely but divided through fibrous membranes so that there is no connection between them except through blood and lymph vessels and nerves.

Section IV : HERD REPLACEMENTS

CHAPTER

11

Growth

Life is a whirlpool in which the body is maintained in a dynamic equilibrium with constant movement of material into and out of the body. There is a process of breaking down (catabolism) going on concurrently with a process of building up (anabolism) throughout life. When anabolism exceeds catabolism growth takes place. Growth and development of an organism from a single celled zygote to an adult animal is a spectacular biological event which can marvel the wonders of astronomical systems.

Growth is of great practical significance to the farmer because it will determine when a calf will attain adulthood and produce milk. Even though all of us understand what is growth in general terms, it has been difficult to give a precise definition of growth. Brody (1945) defined growth as "relatively irreversible time change in magnitude of the measured dimension or function".

Growth involves three processes :

(1) An increase in the number of cells,

(2) An increase in the size of cells,

(3) Incorporation of materials inside cells.

When more of protoplasm is added by reproduction than destroyed, a net accumulation of body substance takes place. The permanent increase in protoplasmic mass can be termed as growth.

Growth starts from conception. But, for convenience prenatal growth is dealt as part of reproduction and only postnatal growth is discussed here. At the same time it has to be borne in mind that prenatal growth has an important bearing on the course of growth after birth.

"As an animal grows up, two things happen : (*i*) it increase in weight until mature size is reached; this we call growth and (*ii*) it changes in its body conformation and shape and its various functions and faculties come into being; this we call development." (Hammond, 1940).

A striking feature of growth is that there is a close similarity between the growth curves of different species of animals. Not only that the curves of growth of individuals of different species of animals have the same S-shape, but populations also follow a similar growth pattern (Fig. 11.1 and 11.2).

11.1. MEASURING GROWTH

Most commonly employed method of measuring growth is by studying an increase in weight during a definite period. Such time related increase in weight has serious drawbacks as a measure of growth. An animal may add to its weight by retention of water or by deposition of fat. These do not represent an increase in the structural tissues and organs and cannot be counted as true growth. Such a defect can be remedied partially by including size measures such as height,

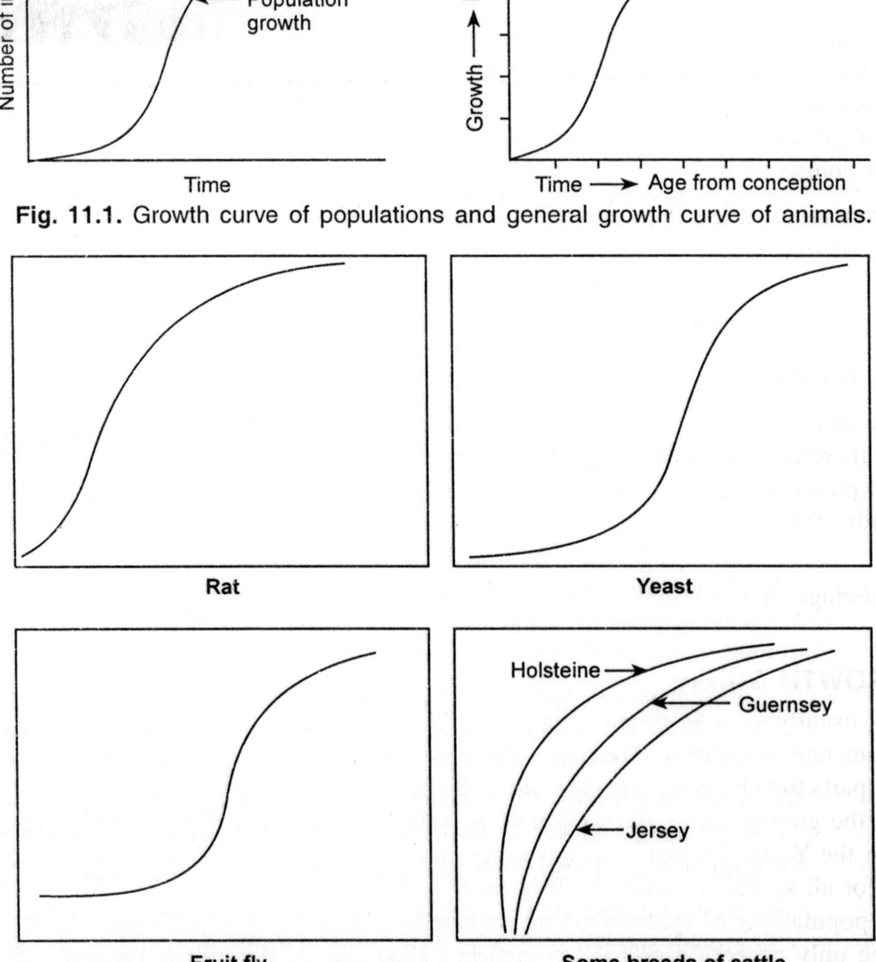

Fig. 11.1. Growth curve of populations and general growth curve of animals.

Fig. 11.2. Approximate shapes of growth curves in different species. (Adapted from Missouri Agr. Expt. Sta. Res. Bull. 96, 1926 and 97, 1927).

length, heart girth etc. These size measures crudely represent skeletal growth. They alone cannot be representative of true growth because an animal receiving insufficient energy and protein may not have any muscular growth, but may show an increase in size due to skeletal growth. A combination of body weight and body measurements remove some of the defects in each other and therefore much more useful than either alone.

Weight increase related to time may be represented in the following ways.

1. Absolute gain in weight per unit time, *e.g.*, grams per day.
2. Percentage increase in weight per unit time.
3. Cumulative weight at a given time.

Average absolute growth rate is calculated by :

$$\frac{\text{Gain in body weight over a period}}{\text{Duration during which the gain took place}}$$

Such an average value of growth rate is meaningful only when it applies to a short interval as the *velocity* of growth is different in different stages of development of an individual. Taking this into account the concept of *instantaneous growth* rate has been developed in which *dt* (interval of growth) is so short that there will not be any change in the velocity of growth rate during the course.

Percentage increase in growth rate (relative growth rate) is the gain in weight expressed as a function of initial body weight

$$\text{Percentage increase in weight} = \frac{(W_2 - W_1) \times 100}{W_1}$$

where W_1 is initial weight, W_2 is final weight.

Here also, if the interval is long, the value of W_1 is likely to be very small in relation to gain and therefore, the percentage growth rate is likely to be highly exaggerated. A simple method of partially rectifying the situation is to use the average of initial and final weights as the base (Brody, 1965) to obtain :

$$\text{Percentage increase in weight or relative growth rate} = \frac{(W_2 - W_1) \times 100}{\frac{1}{2}(W_1 + W_2)}$$

11.2. GROWTH CURVE

Growth is usually slow in infant stage, but picks up speed during adolescent stage and reaches a maximum rate at puberty. Thereafter there is a slowing down and may eventually stop. This pattern imparts the characteristic sigmoid or S shape to the growth curve (Figures 11.1 and 11.2). To obtain the growth curve, the weight of an animal is measured at regular intervals of time and plotted on the Y-axis against corresponding time intervals on the X-axis. The growth curves so obtained for all species of animals show more or less the general S-shape. Not only that, growth curves of populations of unicellular and multicellular plants and animals also exhibit the sigmoid shape. The only exception perhaps is man in which species there is an exceptionally prolonged initial juvenile period. These curves are similar because multicellular organisms represent a population of cells. So also do a colony of micro-organisms.

The growth curve gets its characteristic S-shape due to the interaction of 2 sets of opposing forces, forces which favour growth (growth-accelerating forces) and those which, inhibit growth (growth-retarding forces). The growth acceleration forces arise from the innate property of living matter for reproduction which is charactertized by (*i*) cell multiplication, (*ii*) cell hypertrophy, and (*iii*) incorporation of material into cells from outside. The retardation forces include (*i*) exhaustion of nutrient supply, (*ii*) lack of space, and (*iii*) accumulation of waste products which increasingly prove toxic when the concentration increases.

In the early stages of life, until one third to one half of mature weight is reached, growth acceleration forces will be stronger than the retarding ones. During this phase, the slope of the growth curve will be steep (increasing). After a period of acceleration, a stage will be reached when no further acceleration can take place. The opposing forces soon become equal and gradually more powerful to cause retardation of growth. The slope of the growth curve will become flatter (diminishing). The point of inflection on the growth curve in which acceleration of growth terminates and retardation begins coincides with puberty and is therefore termed pubertal inflection. This is the point at which growth rate is at its maximum and mortality at its lowest. This can serve as a landmark of reference for the determination of age equivalents in different species and breeds of animals and populations.

11.3. FACTORS AFFECTING GROWTH

As in any other trait, growth is also controlled by two sets of factors—genetic and environmental. The potential limits of growth for each individual are set by genetic control. The 'genetic codes' found in the DNA in every cell control the hormonal inter-relationship and activities and through them regulate growth and other 'physiological processes'. The realization of genetic potential depends upon the environment and its interaction with genotype. Environment includes (*i*) nutrition, (*ii*) climate, (*iii*) management, and (*iv*) diseases.

11.4. GENETIC CONTROL OF GROWTH

The DNA coding dictates the kind of function and the level of function of endocrine glands. Thus the kind of hormone, its level and the sequence and time of action has been coded in DNA With recent advances in biochemical genetics we know that the phenotypic characteristics of an individual are only manifestations of the biochemical coding which is present in the DNA of the genes interacted by environment (see Chapter 14). Growth is by and large a quantitative trait in which a large number of pairs of genes, each having a small individual effect while the overall effect of these being large, control the hereditary mechanism. To a lesser extent, qualitative gene action is also evident in which a single pair of genes or a limited number of genes may have a large effect on size and growth rate of an individual. A good example for this is dwarfism.

Only fragmentary information is available at present to explain the details of the physiological pathways of gene action. Most of the available evidence indicates that genes control growth through controlling the endocrine system. Evidences are also available to indicate that there are genetic differences in the sensitivity of certain target organs to hormones. Some evidence suggests that increased efficiency of gain and growth rate due to hybrid vigour results from more efficient enzyme systems involved in metabolism.

It is also known how DNA programme and dictate various details of growth and development of the organism.

11.5. HORMONES IN GROWTH

Nervous and endocrine are the two main co-ordinating systems in the body. These two work in unison to regulate various life processes. The central nervous system co-ordinates and controls both. It is now well known that releasing hormones from the hypothalamus of the brain control the pituitary hormones. Pituitary in turn produces a series of trophic hormones which control the activity of other endocrine glands. This three-tier system of endocrine action has a profound influence on growth and other metabolic processes. Physiological actions of a hormone cannot be singled out since many, if not all, hormones produce a response by modifying the action of or interacting with other endocrine glands. Thus, it is only for convenience that each endocrine gland is separately treated in this discussion and for brevity only the ones having major and obvious effect on growth have been dealt.

11.5.1. Anterior Pituitary Hormones

All the hormones of anterior pituitary are directly or indirectly involved in the growth process. But the action of many of them are specific on certain target organs or tissues. For example thyrotropin, TSH, causes an increase in size and functional activity of the thyroid gland. ACTH a similar effect on adrenal cortex and the gonadotropins (FSH, LH and Prolactin) the gonads. Growth Hormone (GH) on the other hand has a more generalised effect of growth promotion.

11.5.1.1. *Growth Hormone (GH)*

This hormone is also known as somatotropin (STH). The most obvious function of this hormone is a generalised growth promotion in the young animals. GH stimulates the formation of both the soft and bone tissues of the body. It influences protein, carbohydrate and fat metabolism and through that regulates growth. The overall effect is an economy in the use of protein and energy thus encouraging the body to retain these building blocks for growth or for energy. Nitrogen retention by the body is increased by GH. It also increases the permeability of cells to amino-acids thereby encouraging protein synthesis. The effect on carbohydrate metabolism is more vague and indirect. The metabolic effect of GH is by no means isolated, there being many interactions with many other hormones, factors and enzyme systems. Consequently, it is difficult to assign a specific role to GH. What can be concluded is only that GH has a major growth promoting effect through its regulatory effects on the metabolism of protein, carbohydrates and fats.

In domestic animals, growth of long bones is stimulated by GH before the closure of the epiphyseal lines of long bones soon after puberty. Puberty followed by closure of epiphyseal lines signals the stopping of skeletal growth. Once these lines are closed, GH cannot cause skeletal growth.

The fact that the levels of GH release in the growing as well as the adults remain at a constant rate is a surprise. However, the adults have less GH per unit body weight than young animals.

11.5.2. The Thyroid Gland

Thyroxine, the hormone secreted by the thyroid gland, has a very wide range of functions influencing the activity of almost all cells in the body. In general, thyroxine regulates growth and development and oxidative metabolism. The activity of the thyroid gland is regulated by TSH of anterior pituitary which in turn is controlled by the TSH-releasing factor from the hypothalamus.

The principal action of thyroxin is stimulation of oxygen utilization and consequential increase in basal metabolic level. Thyroxine causes an increase in the number of mitochondria per cell. Mitochondria are the metabolic workshops and an increase in their number means there is more metabolic machinery per cell. Thyroxine also causes an increased permeability of mitochondrial membrane and its swelling. This increased permeability is thought to increase metabolic activity through increased facility for phosphorelation. Because of the generalised effect of the hormone, it is evident that any deficiency or excess of the hormone will result in a multiplicity of symptoms among which an important one is on growth. There are also some indications that thyroxine stimulates protein synthesis and therefore growth. An animal with deficient thyroid function (or thyroidectomised young animal) will not reach normal adult size.

The general view held is that the action of thyroxin is "permissive". It is thought that the growth promoting effect of GH and other factors can manifest fully only in the presence of thyroxine.

11.5.3. Adrenal Cortex

The adrenals are composite glands which contain two distinctly different parts—medulla and the cortex. They are different in their embriological origin, chemical nature of hormones and functions.

The cortex, the outer layer, plays an essential role is the general metabolism of the animal. The adrenal cortex of cattle secretes at least 29 different steroid hormones. These include (*i*) *glucocorticoids* (those influencing carbohydrate, protein and fat metabolism), (*ii*) *mineralocorticoids* (those regulating electrolyte and water metabolism) and some sex hormones like progestagens, estrogens and androgens.

Adrenalectomy has been shown to cause cessation of growth and loss in body weight in addition to specific disturbance symptoms on digestive, circulatory, excretory and immunological systems and body tissue. Characteristic symptoms of loss of appetite and gastrointestinal disturbances, extreme muscular weakness, hypoglycaemia, haemo-concentration, increased excretion of sodium, chloride and bicarbonate and decreased excretion of potassium, reduced blood pressure, and reduced body temperature are manifested.

Adrenal cortex can influence growth through more than one way. The whole functional interrelationships of metabolic pathways essential for normal functioning and growth go haywire if there is inadequacy of adrenal cortical function.

11.5.4. Pancreatic Hormones

Pancreas produces two hormones—insulin and glucagon. Their most obvious effect is on glucose metabolism acting more or less in a reciprocal way. Insulin helps in the transport of glucose from outside the cells to inside. It is also necessary to stimulate some of the enzyme systems inside the cells. In insulin deficiency, the availability of energy to cells is reduced thereby curbing tissue synthesis. Decreased carbohydrate utilization leads to increased catabolism of fats and proteins. This will lead to loss of body weight.

11.5.5. Gonadal Hormones

Androgens secreted by testis have a growth promoting effect. They bring about their stimulatory effect on growth through influencing protein synthesis. Administration of androgens increases nitrogen retention, protein synthesis and muscular growth. The efficiency of weight gain is improved by them. The composition of the body is also affected. The carcass of an entire male will have more of muscles than fat deposition compared to castrated males.

Estrogens and progestagens are two sets of hormones produced by the ovary in the females. The effect of estrogens upon growth in diary cattle is different from that in other species. They generally produce nitrogen retention and protein synthesis, increased appetite, increased rate of gain and feed efficiency.

Progesterone, the pregnancy hormone does not cause an increase in appetite or food consumption or any appreciable growth in cattle.

11.6. NUTRITION AND GROWTH

The basic building blocks for growth are obtained by the animal form the environment through nutrition.

The foetus is adequately provided with nutrients during gestation by drawing on the body reserves of the mother. The foetus gets priority over the mother's system for nutrients to some extent. But if the mother's diet is severely deficient during gestation, calves with sub-normal birth weight and vigour are born. Lack of minerals and vitamins seriously lower the vitality of the calf without lowering the body weight significantly. Low vigour and vitality are usually accompanied by a heavy incidence of calf mortality. The effect of prenatal growth rate is transient and if fed well after birth; calves born with sub-normal birth weight can attain mature size without much difficulty.

On the other hand, postnatal feeding has a significant effect on growth and attainment of mature size. The effects of under-feeding on adult size of the animal will vary depending on (*i*) the severity of the nutrient restriction, (*ii*) the kind of restriction (energy, protein, vitamins or minerals), (*iii*) length of under-feeding, and (*iv*) the age at which feed restriction occurred.

Brookes and Vincent (1950) observed that feeding at a high plane of nutrition during early life when potential growth rate was high followed by a moderate plane in later life was the most profitable feeding system as compared to 'High-High', 'Moderate-High' and 'Moderate-Moderate' planes of feeding. Even though the 'High-High' group attained the adult size in minimum period, they were the most expensive because the rapid growth rate necessitated a high rate of feeding of expensive concentrated foods.

Research findings differ with regard to whether under-feeding can permanently stunt an animal. Severe under-feeding in early life may prevent an animal from attaining its normal mature size. It is known that after periods of retardation, if the animals are fed at a higher plane, they exhibit higher level of intake and compensatory growth (Tayler *et at.*, 1957). Restriction of feed results in depletion of fat reserves which encourages greater removal of circulating VFA from blood. This, in turn, will stimulate more feed intake and growth. Animals under realimentation and compensatory growth have improved feed efficiency partly because of the greater live-weight gain relative to body weight and partly due to a lowered maintenance requirement in restricted animals.

Brookes, A.J. and Vincent L.S. (1950). *J. Royal Agr. Soc.* **3** : 99.

Tayler, J.C., Alder F.E. and Rudman, J.E. (1957). *Nature,* **179** : 197.

Various studies have indicated that nutrition can be judiciously utilized to control growth during various stages of life. Under certain circumstances it may be advantageous to have a higher plane of feeding and faster growth rate in early life followed by a moderate plane of feeding and average growth rate through feeding roughages in later growing period. Under average conditions existing in tropical and subtropical regions of the globe, it will be advantageous to feed at a higher plane in the first 6-8 months of a dairy calf followed by a moderate feeding of reducing proportion of concentrates and increasing proportion of roughages until maturity.

11.7. CLIMATE AND GROWTH

Growth and feed efficiency are affected adversely at both climatic extremes. At temperatures below optimum the animal will have to consume more feed merely to keep warm. In contrast, under stressfully hot conditions, the feed intake is lowered in an attempt to reduce heat production.

The climatic influence on productivity and growth has two components—(*i*) the direct effects, and (*ii*) the indirect effects. The direct action of climatic excesses on the animal's physiology affects the nervous, endocrine, circulatory, respiratory, alimentary and other systems with an adverse effect on feed intake, feed utilization and growth. The indirect effects of climate is on the feed supply, nutritive value of feeds, incidence of parasitism and diseases.

Climate has four major physical components, namely, (*i*) ambient temperature (*ii*) humidity (*iii*) solar radiation and (*iv*) wind velocity.

There is great variation among the various breeds of cattle with respect to their heat tolerance. In general *Bos taurus* cattle are cold adapted whereas *Bos indicus* breeds have more capacity to withstand and perform under hot climatic conditions.

11.7.1. Modification of the Environment

It is but natural that the effect of various modifications of the microclimate around the animal vary significantly between situation to situation and animal to animal. But, in general, shade (protection from solar radiation), shelter (protection against rain, wind and snow), increased wind velocity, evaporative cooling etc. have been found to increase growth rate.

11.8. MANAGEMENT AND GROWTH

It is certain that various managemental factors like systems of housing, pen size and building design, feeding and watering systems, weaning and calf management, sanitation and disease control etc. have a definite effect on growth. Most of these effects have not been quantified so far. So it is mostly left to experience and managerial skill to chalk out an appropriate management plan for each situation.

Bovines in the growing stages are susceptible to many diseases which invariably produce a set back in growth. Depending on the severity of the disease or its duration, the effect on growth may be transient or of a more permanent nature. In the former case compensatory growth that follows after recovery will take care of the loss in growth during the course of disease. If the intensity of the disease had been severe, there is possibility of a more permanent stunting of growth. Similarly, in many chronic diseases of longer duration, the animal fails to grow satisfactorily and remains stunted.

11.9. PHYSIOLOGICAL FACTORS

Occurrence of physiological events like pregnancy, calving and lactation and the age at which these take place have a major influence on growth. The combined effect of plane of nutrition and age of calving is shown in Fig. 11.3.

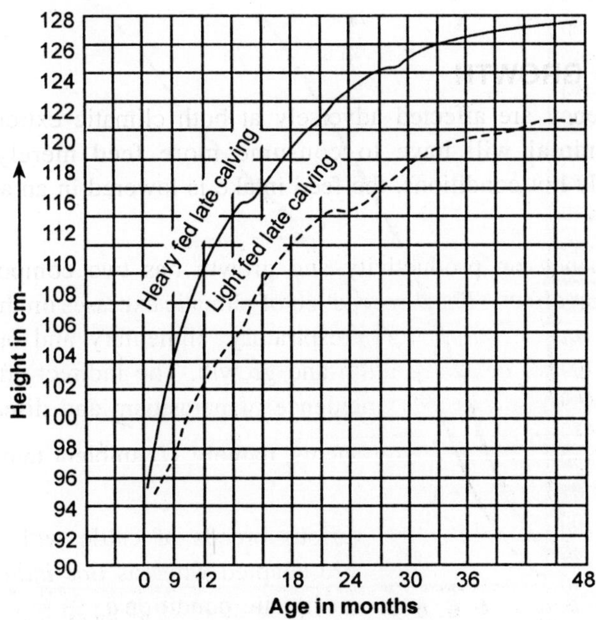

Fig. 11.3. The combined effected of light feeding and early calving upon skeletal growth of Jerseys.

11.9.1. Breed Differences

There exist vast differences between breeds with respect to growth rate under different environmental conditions. There are also vast breed differences in the mature size attained. It can be generalized that the inherent rate of growth is directly related to mature body size. The general feeling that beef cattle grow faster than dairy cattle may not be correct. In some studies Holsteins were found to grow faster than the Herefords.

11.9.2. Effect of Sex

Efficiency of growth is greatest in bulls and least in heifers. While androgens have a growth promoting effect, estrogens have been found to depress it. Weight gain and feed conversion is more in males and most of the gain is in lean meat.

11.9.3. *Growth Rates

At younger ages, the body weight increased at a higher rate with increasing Friesien inheritance. But the body weight curves at 1 and 2 years of age indicate lack of adaptability in animals

*B.G. Katpatal (1977). *Dairy cattle crossbreeding in India* : 1 Growth and development of crossbred. World Anim. Rev. 22 : 15.

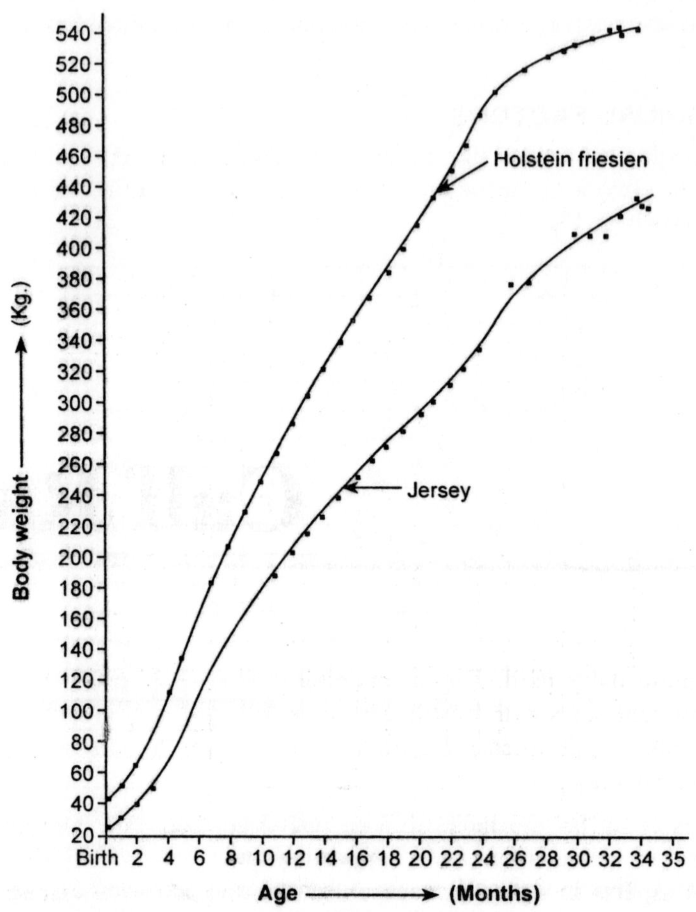

Fig. 11.4. Actual growth curve of Holsein Friestein and Jersey heifers.

containing more than 3/4 Friesien breeding. It is interesting to note that the ranking of the breed groups for both MPPA (most probable producing ability) and body weight at 1 and 2 years of age coincided with each other.

Anon (1966). *A comparison of the growth of different types of cattle for beef production.* Report of Major Beef Res. Project., Royal Smithfield Club, London.

CHAPTER

12

Calf Raising

Calves form the future dairy herd. For maintaining efficiency of production it is necessary to replace about 20 per cent cows with freshly calved heifer-cows. Heifers raised in one's own farm forms the most reliable replacements. Besides, the sale of heifers form a major component of income in the dairy business.

Raising calves is by far the most difficult operation in a dairy farming enterprise which requires a great deal of management skill, application and constant attention. Yet, this is one of the most neglected aspects in many farms as is reflected by a mortality rate of 30-35 per cent among calves in many farms. The problem is more acute in the case of buffalo calves. Although the magnitude of losses due to calf mortality, stunted growth and unthriftiness have not been assessed, it can only be assumed to be enormous in the light of available evidence.

12.1. PERI-PARTUM CARE

12.1.1. Feeding and Management of the Cow During Pregnancy

Care of the calf starts in the womb of the cow where it is a foetus. In the first 6 months of pregnancy, the additional nutrient needs of the cow due to foetal growth is negligible. But, during the last 2-3 months of gestation the foetus grows rapidly for which it draws a significant quantity of nutrients from the mother's system. This has to be replenished by additional feeding to the cow.

Unfortunately, feeding of the cow during this period is often neglected because generally at this stage the cow is dry. Good quality fodder and concentrates are fed to milch cows whereas the dry cows in advanced gestation get only straw and inferior fodder. This will have adverse consequence on the development of the calf. A cow should be fed well-balanced ration during this period for :

(*i*) maintenance of the cow,
(*ii*) growth of the foetus,

(*iii*) production of colostrum with high nutritive value which in turn will influence the growth of the calf, and

(*iv*) sufficient reserves of nutrients in the body of the cow for the ensuing lactation.

12.1.2. Care of Calf and Cow at Birth

A few days before the probable date of calving, cows should be transferred to individual calving pens. Ample amount of drinking water, laxative feed and generous supply of bedding may be provided. The calving pens should be scrupulously cleaned and sterilized before bringing in the cow. Daily fresh bedding may be provided.

12.1.3. Calving Pen

Individual loose boxes or stalls are used for calving. Such pens will have the following advantages :

(*i*) Pregnant cows get better protection. They will not be disturbed by other cows.

(*ii*) Special watch and attention can be given.

(*iii*) These pens can be disinfected thereby preventing chances of infection gaining entry at the time of calving. Chances of contamination of the general barn by infected cows and healthy cows contracting diseases of the genital tract are minimised.

(*iv*) Special arrangements can be made for preventing injuries to the newborn calf and mother. Also arrangements for first aid can be properly provided.

Pregnant cows may be transferred to calving pens 1 to 2 weeks before the expected calving date. The number of calving pens required on a farm depends on the number of breedable cows and heifers; generally 5 per cent of this number. The maternity pens should be near to the living quarters of the manager. Small farmers who have only one or two animals must tie the cow in advanced pregnancy separately to a peg in a clean area where the animals will not be disturbed. If the ground is having clean grass cover there is no need of a bedding. Otherwise a straw bedding may be provided.

In farms where abortions and calf diseases are problems, the calving pen should be sanitized and sterilized with more care. The floor and walls should be scraped, scrubbed with 4 per cent washing soda in hot water and disinfected before the cows are brought in.

12.1.4. Signs of Approaching Parturition

Knowledge of signs of approaching calving is beneficial to the farmer to plan his operations and put the animal in the calving pen. A cow which will calve soon will show the following signs :

(*i*) Will leave the herd and seek isolation.

(*ii*) Loss of appetite, distress and withdrawal from the environment.

(*iii*) Distention of teats and udder. Considerable milk appears in the udder and there may be dripping of milk.

(*iv*) Relaxation of pelvic ligaments. One day before calving, the ligaments on the sides of the tail-head is loosened so that hollows appear on either side of the backbone and the tail-head is raised and the quarters are dropped (*see* Fig. 7.6).

(*v*) The vulva becomes enlarged and flabby.

(*vi*) The labour pains will give distress. The animal will be restless and will pace about often trying to kick or scratch the flank region.

Parturition is divided into 3 stages :

I. The preparatory stage consisting of contractions of the uterine walls and dilatation of the cervical canal.

II. Active expulsive effort by the uterine walls and abdominal muscles driving the foetus out through the birth canal.

III. Expulsion of the foetal membranes. Actual duration of time has not been accurately determined and the estimates by different authors vary. The difficulty lies in the failure to determine the line of demarcation between physiological and pathological birth.

A period of 12 hours may elapse after the commencement of first stage until the calf is dropped. If a change is noticed in the morning, if she has not calved by mid-afternoon the cow may be examined. Similarly if the restlessness first occurs in the afternoon the cow ought to have calved by next morning.

In the event that the cow does not calve in the expected time, she should be examined. The examination should be short and simple—after washing thoroughly, the arm should be inserted into the vagina. If the cervix is closed or partly closed, the cow should be left alone until definite signs of straining are seen. If the cervix is open and a normal presentation of head and forelegs seen, delay may be safely ignored and the cow left until she is obviously straining fruitlessly. When the presentation of the calf is normal, the two fore feet followed by the muzzle will appear after the water-bag has burst. Any sign of abnormal presentation should mean immediate veterinary assistance. Under no circumstances should a farmer attempt to remove a calf unless he has received veterinary guidance.

Watch for expulsion of 'after-birth' (placenta). The cow should not be allowed to eat the placenta as it will cause indigestion. If after-birth is unduly delayed, veterinary help should be called for. With the expulsion of placenta, parturition is complete.

12.1.5. Management Immediately after Calving

Wash the udder and hind quarters with lukewarm water containing an antiseptic and dry with a clean towel. Cows may be milked to relieve the pressure from the udder. If weaning at birth is not practised, the cow and the calf may be allowed to remain in the calving pen for 10 days. But if weaning at birth is practised, the calf may be immediately removed. In herds in which maternal instinct is more and weaning is a problem, it is a better practice to cover the eyes of the cow before the calf is removed.

12.1.6. Attending to the New Born Calf

The cow does a neat job of licking the calf dry and stimulating its circulation and respiration with hard licking. However, sometime primiparous cows may be nervous and inexperienced, or it may be that the cow is exhausted after a prolonged labour. Under such circumstances, the husbandsman should assist by removing the phlegm (mucus) from the nostrils of the newborn calf and wipe it dry with a clean towel. He should also massage the calf vigorously for some time with a handful of straw rolled into a ball.

Sometimes the respiratory passage of the calf may be blocked by the mucus which may cause death due to choking if unattended. On noticing such a condition, immediate attention is needed as described below :

1. Hold the calf head down by lifting it holding the hock. The phlegm may flow off.
2. Take a twig of hay or grass and tickle inside its nostrils. The calf will sneeze and expel the mucus.
3. If the above methods fail, little time is left to lose. The attending person should apply his mouth to the nostrils of the animal and suck out the mucus. After that he should blow in his expired air through the calf's nostrils closing its mouth. Carbon dioxide in the expired air which has been blown-in the lungs of the calf will act as a respiratory stimulant to initiate respiration. This should be followed with intermittent pressing and releasing of pressure on the chest wall of the calf to give artificial respiration.

During winter season in North India and other colder parts of the country, the calves need some provision of warmth. If the calving takes place on a sunny winter day just allowing it to bask in the sun is all that is needed. Otherwise the calf may be shifted to a protected pen with plenty of dry bedding.

Attending to the navel of young ones is very important as infections gaining entry through the navel at the time of calving can cause serious ailments like navel ill, joint ill etc. which can debilitate and cause mortality. The calf's navel should be painted with antiseptics like tincture of benzoin or tincture of iodine soon after birth to prevent this. If the unbilical chord is not broken, a ligature may be put 2-3 cm away from the body with a sterile thread and cut 1 cm distal to the ligature with a clean sterile scissors. A small amount of antiseptic lotion may be painted at the cut end and protected from flies and crows.

12.1.7. Colustrum Feeding

The value of colostrum feeding to the calf can never be over-emphasised. It contains large amounts of gamma-globulins which are nothing but anti-bodies produced by the cow against antigens encountered during her life including those against many disease producing organisms. Absorption of these antibodies provides the calf with an *umbrella* of passive immunity against many diseases in early life.

TABLE 12.1. Composition of Colostrum and Milk

Constituents	*Colostrum (%)*	*Milk (%)*
Total Solids	28.30	12.86
Ash	1.58	0.72
Fat	0.15 to 12.0	4.00
Lactore	2.50	4.80
Casein	4.76	2.80
Albumin	1.50	0.54
Globulin	15.06	—
Total protein	21.32	3.34

Source : Petersen, W.E. 1950. Dairy Science, 2nd Ed. J.B. Lippincolt Co. Philadelphia.

Besides, colustrum is a highly fortified source of nutrients having 7 times the protein and twice the total solids of normal milk. Thus, it gives an early boost in protein and solids intake. The average composition of colostrum and normal milk of cows is given in Table 12.1.

It also contains a higher than normal content of vitamins and minerals. A calf when born has little or no antibodies and vitamin-A, which are essential to combat diseases. Ingestion of these through colostrum substantially increases the calve's survivability.

Quality of colostrum vary from cow to cow depending on the diseases to which she has been exposed by natural infection or vaccination. While the former is uncontrollable and undesirable the latter one can be beneficially exploited.

Vaccinate dry cows. As far as possible, vaccination should be concentrated to the dry period. Only exceptions should be those which may cause abortion or those which are to be done in a different time of the year. Giving the vaccination during dry period helps to increase the quantity and quality of γ-globulins in colostrum. In turn, colostrum provides more protection to newborn calves. This procedure also allows the cows to undergo the stress of responding to a vaccine at a time when it does not sacrifice milk production. Besides, the cows get protection against diseases at times when they need it most, *namely*, late gestation, parturition time and early lactation when vitality is likely to be low.

Colostrum of mature cows best. For the basic reasons stated before, colostrum of mature cows is of higher quality than that produced by heifers. Older cows have a greater chance to be exposed to many infections, many of them repeatedly. This in turn stimulate them to produce greater quantities of γ-globulins effective against larger number of diseases.

12.1.8. Time of Feeding Colostrum

Of equal importance to quality of colostrum is the time at which the baby calf receives its first colostrum. To be of any value to the calf for passive immunity, the γ-globulins must be absorbed as such across its intestinal wall into the blood stream without being broken down into the constituent peptides or amino acids. Once broken down, the constituent amino acids of γ-globulin enter the general pool of amino acids and will serve only the purposes to which any protein is generally put to. The intestinal wall of the calf will allow the globulins to pass from inside the intestine to the blood stream for only a short period of time after the calf is born. This permeability is rapidly lost after the first few hours of life. Many studies have shown that these globulins pass across the gut wall at the most rapid rates during the first 1-2 hours of life. Taking this into view, it will be highly useful to feed colostrum in the first 15-30 minutes followed by a second dose in approximately 10-12 hours. (See figure 12.1).

Quantity of colostrum to be fed. A calf should receive colostrum during the first 3 days. Quantity of colostrum to be fed to calves depends on their body weight. One recommendation is given below :

	Colostrum as per cent of body weight
15-30 mts of life	5-8
10-12 hrs. of life	5-8
2nd day	10
3rd day	10

In most of the Indian recommendations it is advised to feed colostrum at a rate of 1/10 of body weight for five days. In conditions where there are no facilities to weigh the calf a quantity of 3 kg colostrum per day was recommended by Razdan and co-workers. I.A.R.I. workers on the other hand recommended a daily allowance of 2.5 kg colostrum per calf for the first 3 days.

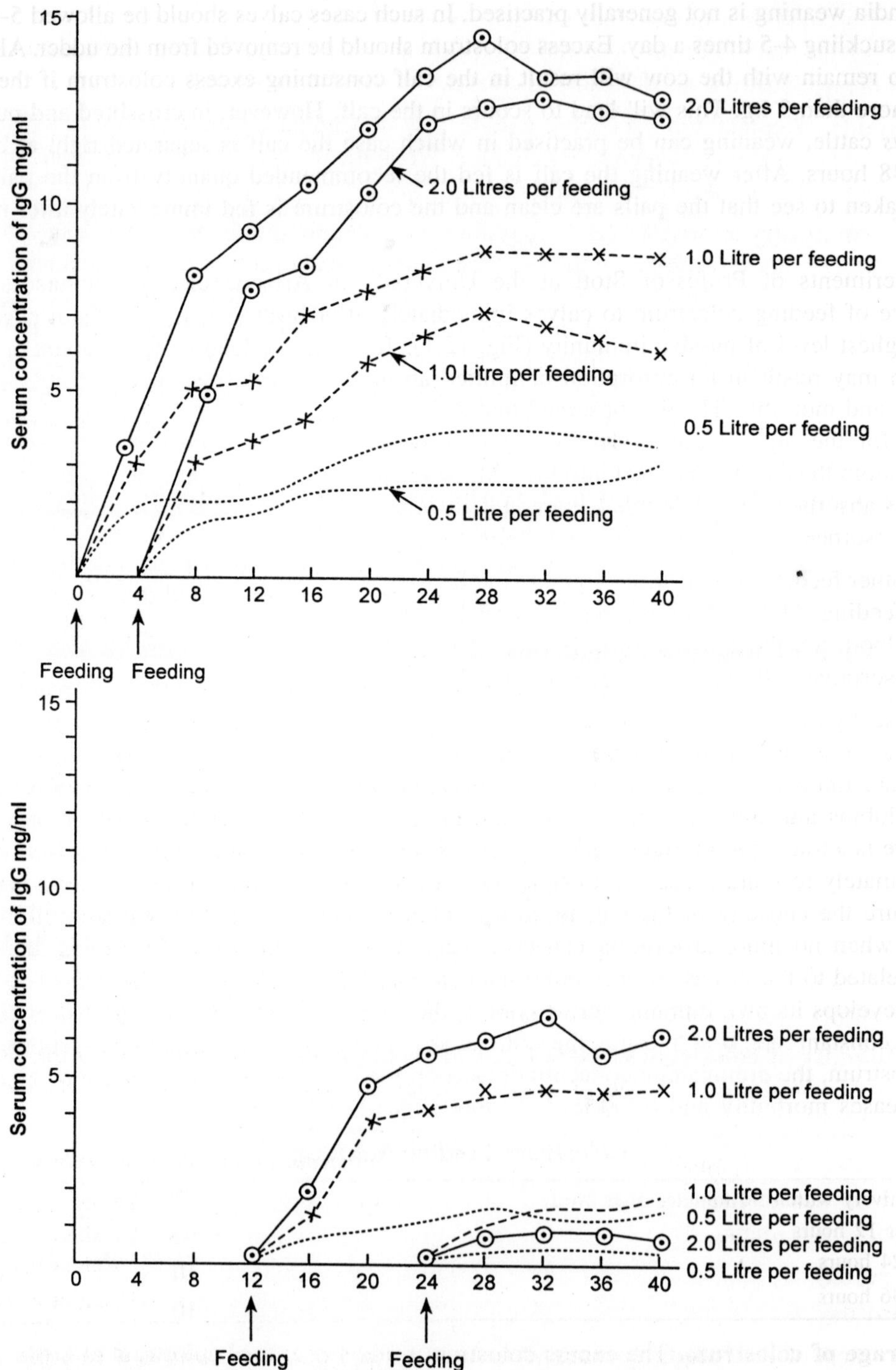

Fig. 12.1. Effect of feeding colostrum at three different levels at 0 hour; 4 hour, 12 hours and 24 hours from birth on immunoglobin (IgG) level in calves' blood. (Adapted from Stott, G.H., 1979; *J. Diary Sci.* **62** : 1766).

In India weaning is not generally practised. In such cases calves should be allowed 5-minute period of suckling 4-5 times a day. Excess colostrum should be removed from the udder. Allowing the calf to remain with the cow will result in the calf consuming excess colostrum if the cow's yield is more than 3 kg. This will lead to scours in the calf. However, in crossbred and purebred *Bos taurus* cattle, weaning can be practised in which case the calf is separated right at birth or after 24-48 hours. After weaning the calf is fed the recommended quantity from the pail. Care must be taken to see that the pails are clean and the colostrum is fed immediately after milking out.

Experiments of Professor Stott at the University of Arizona clearly demonstrated the importance of feeding colostrum to calves immediately after birth to gain maximum absorption for the highest level of passive immunity (Fig. 12.1). To delay the feeding to 12 hours or longer after birth may result in no absorption or limited absorption with a resultant high incidence of morbidity and mortality. He also observed that 2 litres of colostrum in the first feeding are quite optimum for the highest colostral antibody absorption. It can be noticed that the amount of immunoglobin the calves absorbed into their blood relates to the amount of colostrum fed. Calves fed 2 litres absorbed almost double the amount absorbed by those fed 1 litre, and calves receiving 1/2 litre absorbed only 25 per cent of the same.

Another feeding of colostrum equal to the first feeding was given to all calves 12 hours after the first feeding. Only calves receiving 1 and 1/2 litre responded with a rise in immunoglobin level in blood; not those receiving 2 litres. Apparently 2 litres in the 1st feeding was enough to satiate absorptive cells in the intestine before they closed to further immunoglobin uptake.

These figures also demonstrate clearly the diminishing absorptive response of calves to immunoglobin as they grow older (from 4 hrs. to 24 hrs.). The absorptive cells lining the small intestine are immature at birth. In this stage they indiscriminately take up large molecules like immunoglobins and transport them to the general circulation. As the calf grows older hour by hour, there is a transition of epithelial cells of small intestine from the immature type that absorbs indiscriminately to mature cells that cannot take up large protein molecules. As more and more cells mature the capacity of the calf to absorb immunoglobins diminishes proportionately until 'closure' when no more absorption can take place. Concentration of antibodies at 'closure' is directly related to the disease resistance of the calf. Until about 24 days to 1 month of age when the calf develops its own immunological system, the blood content of immunoglobulins diminish at fairly a constant rate. If at 'closure' the calf had absorbed only a small amount of immunoglobins from colostrum, the diminishing concentrations soon put the calf into a critical immune position. This increases morbidity and often leads to mortality of the calves.

Colostrum Feeding Schedule

1. Positively within 30 minutes after birth	2 litres
2. After 12 hours	1/20th of body wt.
3. At 24 hours	1/20th of body wt.
4. At 36 hours	1/20th of body wt.

Storage of colostrum. The excess colostrum milked out from cows is a valuable food. It can be mixed with normal milk and fed to other calves. It can also be stored. Storing colostrum affords facilities to feed colostrum to calves whose mother died immediately after calving or failed to produce enough colostrum due to mastitis or other diseases. It also makes it possible to feed colostrum from mature cows to calves of heifers. Storage can be done through refrigeration in small herds. It can also be frozen and stored indefinitely. It is also possible to ferment colostrum and store during winter season. Plastic containers are filled to the brim with colostrum and stored completely air-tight. Fermentation takes place due to naturally occurring organisms. These containers can be opened 5-7 days afterwards for feeding.

12.2. CALF FEEDING FROM DAY 3 TO 90 DAYS AGE

Feeding is the most important factor in the management of the young calf. After about 3 days the composition of the secretion gradually changes from that of colostrum to normal milk.

The reticulo-rumen, although has got capacity of about 2 litres, is non-functional at birth. Therefore, for all practical purposes of feeding the calf should be treated as a non-ruminant. This requires consideration of the following points while computing rations for a calf :

(*a*) They are not equipped to utilize cellulose. Cellulose is broken down by the enzyme cellulase elaborated by micro-organisms present in large numbers in the rumen of older animals. Since such microbial populations are established only slowly, the calf cannot utilize roughages containing large proportions of cellulose to any great extent. However, feeding small quantities of good quality hay and other roughages will help in early establishment of microbial populations in the rumen and early rumen development.

(*b*) They require well balanced high quality proteins with all the essential amino acids in the required quantity in the ration. This is so because the rumen microbial protein of high biological value is not available to them in any significant quantity.

(*c*) The calves have little capacity to utilize non-protein nitrogenous compounds and therefore substances like urea should not be included in their ration.

(*d*) Due to the same reason, B-complex vitamins also are dietary essentials for calves in addition to vitamins A and D.

In the calf, a canal called the oesophageal groove exists in the reticulum connecting the oesophagus with the omasum. During nursing and milk feeding, the sides of the groove are raised by reflex action to form a tunnel through which milk passes from oesophagus to omasum by-passing the rumen and reticulum. This continues to function even after considerable development of the rumen if milk feeding is continued. Avoiding rumen and reticulum improves the efficiency of utilization of milk. In case of high quality feeds like milk, the bacterial fermentation in the rumen is more wasteful of energy and protein than the enzymatic digestion in the abomasum and small intestine.

12.2.1. Nurse-cow Method

The food intended for calf by Nature is milk and the process nursing. In the case of low producing cows, the milk produced by them is just sufficient for their calves and the calves can be allowed to remain with their mothers drinking at will. This is indeed the method adopted in raising male calves of draft breeds. But, in the case of milch breeds, the purpose is to produce milk for human consumption and therefore nursing has to be regulated. Moreover, the cows of milch breeds produce much more milk than to meet the requirement of the calf. If the calf is allowed to nurse it will overfeed with resultant indigestion and scouring.

Under such circumstances, procedures like leaving one or two teats for the calf while milking out the other, or allowing the calf to suckle for a limited period after let down etc. can be practised. Another procedure is to allow 3 or 4 calves to a nurse cow. The mucous at the time of calving is applied to all the calves to be adopted by the cow and placed before her to lick and dry. This will help in solving problems of rejection. The number of calves attached to a cow for nursing should be according to her production. Calves should be fed additional green fodder, hay and concentrates. Calves can be raised up to 6 months on such feeding or they may be weaned at 2 or 3 months. Using a nurse cow is the easiest way of raising calves and demand the least management skill.

12.2.2. Whole Milk Feeding after Weaning

In this system, calves are weaned right at birth. They are taught to drink milk from a pail (Fig. 12.2) or the milk may be fed from a nipple pail (Fig. 12.3). A feeding schedule which has been found very successful on many university and government farms in India is given in Table 12.2.

TABLE 12.2. A Proven Whole Milk Feeding Schedule

Week	*Whole milk*	*Legume hay*	*Calf starter*
4th to 7th day	1/10 of birth weight	*ad lib.*	*ad lib.*
2-8	1/10 of body weight	-do-	-do-
9	1/10 of body wt—2 kg	-do-	-do-
10	1/10 of body wt—4 kg	-do-	-do-

The calves are weighed every week and the quantity of milk to be fed is adjusted accordingly. Leguminous hay like lucerne hay or berseem hay is made available *ad libitum* from the 4th day onwards; so also a highly palatable and nutritious calf starter.

However, it may not be possible for every farmer to weigh the calves periodically. In such a case the farmer can weigh the calf at birth on a spring balance and use the birth weight alone as a basis for feeding them. The general guideline is to feed approximately 8, 9, 10, 8 and 5 per cent of the birth weight during 1st, 2nd, 3rd, 4th and 5th week respectively. A feeding schedule suggested by Illinois workers (1968) on this basis is given in Table 12.3.

Fig. 12.2. A weaned buffalo calf is trained to drink milk from the pail. (Courtesy : N.D.R.I., Karnal)

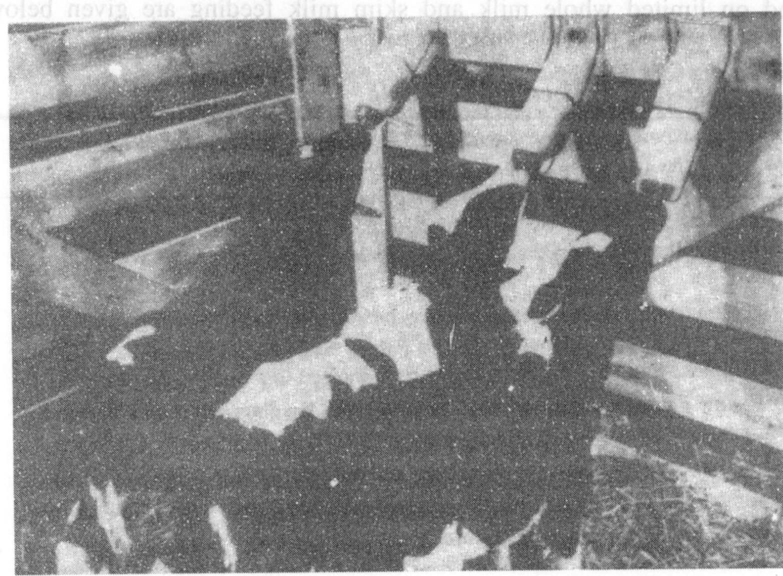

Fig. 12.3. Calves drinking milk from feeding bottles with nipples.

TABLE 12.3. Illinois Whole Milk Feeding Schedule

Feeding category	*Birth weight (kg)*	*Weeks* —(kg whole milk/day) 1	2	3	4	5	*Total Milk (kg)*
1	23-29.5	2.25	2.50	2.70	2.25	1.80	80.00
2	30-33.0	2.50	2.70	3.00	2.70	1.80	90.00
3	33.5-37.5	2.70	3.00	3.50	3.00	1.80	100.00
4	38.0-42.0	3.00	3.50	4.00	3.50	2.25	115.00
5	42.5-46.5	3.50	4.00	4.50	3.50	2.25	125.00
6	47.0-51.0	4.00	4.50	5.00	4.00	2.25	140.00
7	51.5 and above	4.50	5.00	5.50	4.50	2.25	150.00

Source : Based on Fryman L.R., Harshbarger, K.E., Muekling, A.J. (1968). *Feeding, managing and housing dairy calves.* Ext. Circ. 992. College of Agriculture, University of Illinois.

Those farmers who do not have any means of finding out the birth weight of calves may judge whether the calf is large, medium or small and accordingly feed as per categories 6, 4 and 2 in the case of large breeds like Holstein Friesien, Brown Swiss and their crosses, and large buffalo breeds. But in case of smaller breeds like Jersey, Jersey crossbreds and most of the Bos Indicus breeds, large, medium and small calves may be fed according to categories 3, 2 and 1.

12.2.3. Raising Calves on Limited Whole Milk and Calf Starter

By far butter fat is the costliest component of milk. Therefore its saving in calf rearing will economize the operation and save butter fat for human consumption.

As early as 1904 Otis and co-workers reported gains of 803 g per day in calves fed skim milk. The change from whole milk to skim milk should be gradually made when the calf is 2 to 4 weeks old. Skim milk is fed for a period of 2 to 4 months. Some of the recommended feeding schedules based on limited whole milk and skim milk feeding are given below (Table 12.4 to 12.6).

TABLE 12.4. Suggested Calf Feeding Schedule

Age	*Colostrum/ Whole milk (kg)*	*Skim milk (kg)*	*Hay/Green*	*Starter* (kg)*
1-5 days	3.00	—	—	—
6-7 days	2.75	—	*Ad. lib.*	—
8-14 days	3.25	—	"	—
15-21 days	2.75	1.00	"	0.10
22-28 days	1.75	2.00	"	0.20
29-35 days	1.00	3.00	"	0.30
36-42 days	0.50	3.50	"	0.50
43-56 days	—	3.50	"	1.00
57-84 days	—	2.50	"	1.25
85-112 days	—	0.50	"	1.50
113-140 days	—	—	"	1.75
141-182 days	—	—	"	2.00
			DCP per cent	TDN per cent
*upto 56 days			24	75
57-182 days			21	72

Source : Razdan *et al.* (1965).

TABLE 12.5. Recommended Feeding Schedule for Calves of Various Ages

Age of calf	*Whole Milk and Skim Milk Fed*	
	Whole milk (kg)	*Skim milk (kg)*
1-3 days	2.50 (colostrum)	—
4-7 days	2.50	—
2nd week	3.00	—
3rd week	3.25	—
4th week	3.00	—
5th week	1.50	1.00
6th week	—	2.50
7th week	—	2.00
8th week	—	1.75
9th week (upto 60th day)	—	1.25

In addition to the above calf starter and good quality hay be fed to various breeds of calves as indicated in Table 12.6.

In most of the recommendations, a total of 100-150 kg whole milk and 250-300 kg skimmed milk are required for raising one calf.

Skim milk is obviously deficient in energy. Attempts have been made to supplement this deficiency by incorporating grain mixture, sucrose, glucose, dextrins, jaggery, oils, lard and colostral fat in the diet of calves. But calves do not seem to have much capacity to utilize starch and sucrose before 4-6 weeks of age and incorporation of these in large quantities cause diarrhoea.

TABLE 12.6. Recommended Feeding Schedules for Calves of Breeds Generally Found in India (From Various N.D.R.I. Publications)

Breed :	*Haryana*		*Murrah Buffalo*		*Jersery Crossbreds*		*Holstein and Brown Swiss crossbred*	
	Calf starter (kg)	*Hay (kg)*	*Calf starter (kg)*	*Hay (kg)*	*Calf starter (kg)*	*Hay (kg)*	*Calf starter (kg)*	*Hay (kg)*
2nd week	0.50	0.20	0.05	0.25	0.05	0.20	0.10	0.30
3rd week	0.10	0.35	0.10	0.35	0.10	0.35	0.30	0.50
4th week	0.20	0.40	0.30	0.50	0.20	0.40	0.40	0.60
5th week	0.35	0.50	0.40	0.55	0.35	0.50	0.50	0.70
6th week	0.55	0.55	0.60	0.60	0.50	0.55	0.65	0.75
7th week	0.60	0.60	0.70	0.70	0.60	0.60	0.80	0.85
8th week	0.70	0.65	0.80	0.80	0.70	0.65	1.00	1.00
9th week	0.80	0.75	1.00	1.00	0.80	0.75	1.20	1.10
10th week	0.90	0.85	1.20	1.10	0.90	0.85	1.30	1.20
11th week	1.00	1.00	1.30	1.20	1.20	1.10	1.40	1.30
12th week	1.30	1.20	1.40	1.40	1.30	1.30	1.50	1.50
13th week	1.50	1.50	1.70	1.90	1.60	1.70	2.00	2.00

Jaggery is a product cheaply available throughout India. A Feeding schedule in which skim milk partially replaces whole milk from the 5th day and jaggery is introduced from the second week has been found to produce a growth rate similar to whole milk feeding schedule. Jaggery is to be introduced from the second week of age in very small amount initially to avoid incidence of diarrhoea. The details of such a feeding schedule are given in Table 12.7.

TABLE 12.7. Calf Feeding Schedule with Limited Whole Milk, Skim Milk and Jaggery

Age days	*Colostrum*	*Whole milk (As a proportion of body weight)*	*Skim milk*	*Calf starter g*	*Roughage*
0-4	1/10	—	—	—	—
5-12	—	1/20	1/20	To be induced	*Ad lib.*
13-20	—	1/40	3/40	50-100	-do-
21-30	—	—	1/12	250	-do-
31-40	—	—	1/12	500	-do-
41-60	—	—	1/15	750	-do-
61-90	—	—	1/20	800	-do-
91-120	—	—	—	1,000	-do-

Amount of Jaggery Fed Daily per Calf

Age (weeks)	*Jaggery fed (g)*	*Age weeks*	*Jaggery fed (g)*
1	—	9	170
2	30	10	180
3	50	11	200
4	70	12	215
5	85	13	240
6	110	—	—
7	125	—	—
8	150	—	—

12.2.4. Colostrum as Substitute for Whole Milk

When colostrum is fed to calves on the same dry matter basis as that of whole milk, they obtain similar growth rate. Apart from the needs of colostrum for its own calf, a cow produces 20-40 kg extra colostrum with an average dry matter content of 18 per cent. This colostrum could be preserved by various means and later fed to calves by mixing 2 parts colostrum to 1 part water. Colostrum can be preserved by freezing or by natural fermentation. This can make a saving of ₹ 50 to ₹ 100 per calf.

12.2.5. Milk Replacers and Calf Starters

Milk replacers or milk substitutes, consist basically of skim milk powder and lard or vegetable fat although a proportion of butter milk powder and whey powder is often included. A small proportion of glucose, soyabean flour and cereal flour may also be added together with certain minerals and vitamins. If good quality milk replacer is used, there is no need for feeding any whole milk after the colostrum feeding period.

General Characteristics of good quality milk replacers

1. Contains minimum of 50 per cent spray dried skim milk powder.
2. Contains 10-15 per cent stabilised high quality fat, mainly lard homogenised into skim milk or butter milk before spray drying.
3. Supplemented with Vitamin A, E and B_{12}.
4. Antibiotic feed additives incorporated.
5. Minimum of 22 per cent and upto 25 per cent protein containing no poorly digested protein.
6. Contains little or no starch or fibre.
7. Must flow well as a powder for automatic feeding equipment.
8. Must be readily dispersible in warm water.

Milk replacers should be mixed in the correct proportions of warm water since dilutions which are too weak or too strong tend to create digestive problems.

Suggested mixing ratios :

Milk replacer (kg) : Water (litre) :: 1 : 8

A good milk replacer with the following composition resulted in growth rate and condition similar to whole milk feeding (Williams and Knodt, 1949).

Spray dried skimmed milk powder	50 parts
Dried whey	10 parts
Non-milk source	40 parts

A good milk replacer suggested by Ohio workers is as follows :

Items	*Amount in kg*
Dried skim milk	70
Dried whey	18
Lecithin	2
Animal fat	10
Dicalcium phoshate	1.7
$CuSO_4$	+
$FESO_4$	+
$MNSO_4$	+
$CoSO_4$	+
Antibiotic	+

+ = trace amounts. (Porterfield, R.A. McGrew, C.D. and Hibbs, J.W., *Raising dairy herd replacements*, Ext. Bul. 514, College of Agriculture, Ohio State University, 1969.)

12.2.6. Partial Milk Replacers

These are certain feed combinations which when combined with a limited whole milk feeding give good results. The difference is that these do not contain high proportions of milk or skim milk powder. An example of a partial milk replacer and a feeding schedule incorporating it are given in Table 12.8 and 12.9.

TABLE 12.8. Partial Milk Replacer and Calf Starter

Partial milk replacer	***kg.***
Wheat	10
Fish meal	12
Linseed meal	40
Milk	13
Coconut oil	10
Butyric acid	0.3
Citric acid	1.5
Mineral mixture	3.0
Antibiotic	0.2
Calf starter	
Maize	35
Groundnut cake	33
Gram	10
Wheat bran	20
Mineral mixture	3
Rovimix (Vitamins)	15 g/100 kg

Williams, J.B. and Knodt, C.B. (1949). *The value of milk replacements in the ration of dairy calves.* J. Dairy Sci., **32** : 986.

TABLE 12.9. Feeding Schedule

Age (days)	*Body wt. (kg)*	*Whole milk (kg)*	*Milk replacer (g)*	*Calf starter (g)*	*Green folder*
0-5	—	1/10 of body wt. (colostrum)	—	—	*ad lib.*
6-9	—	1/10 of body wt.	—	—	"
10-13	—	-do-	100	—	"
14-17	—	0.5	175	—	"
18-21	—	1	250	—	"
22-25	—	1.5	325	—	"
26-29	35	2	375	—	"
30-33	—	2.5	450	—	"
	40	1	450	600	"
	45	1	500	600	"
	50	1	600	800	"
	55	1	600	800	"
	60	1	700	1,000	"
	65	1	750	1,000	"
	70	1	800	1,200	"
	75	1	850	1,200	"
	80	—	—	1,300	"
	90	—	—	1,500	"
	110	—	—	1,600	"
	120	—	—	1,700	"
	130	—	—	1,700	"
	140	—	—	1,800	"
	150	—	—	1,800	"

Source : Arora, S.P., Bajpai, L.D. and Dave, B.K. (1973). Raising cross-bred and buffalo calves on milk replacer. *Indian J. Anim. Sci.* **43** : 462.

12.3. CALF STARTERS

They are the first dry concentrate mixture fed to calves. Calves start eating small amounts of dry starter from the 2nd week of life. To train them to eat starter mix, the following procedures may be useful :

(*i*) When they complete drinking milk from the pail put some starter mix at the bottom of the vessel. The calves will lick it dry. Gradually increase the quantity.

(*ii*) Rub a small amount of the concentrate on the tongue and lips of the calf which will induce it to eat more.

Care may be taken to clean the manger or vessel in which starter mix is fed to calves daily after which fresh concentrate mix is placed in amounts that the calves will just consume. Calves receiving limited whole milk and a calf starter grew at a slower rate in early calfhood than those receiving conventional quota of whole milk. But when they grew older, they compensated by growing faster and at one year of age had similar live-weight as conventionally milk fed calves.

A calf starter should be highly palatable. It should be high in energy (75 per cent TDN) and contain 14 to 16 per cent digestible crude protein.

Calf starters may be fed on a free-choice basis until the calf starts consuming about 1 to 1½ kg of the starter mix a day after which the amount may be restricted. Calves generally reach this stage by 2½ to 3 months of age. Milk feeding can be discontinued earliest when the calf is consuming 0.4 to 0.5 kg concentrate per day depending on the breed.

A great variety of calf starters are available. The constituents of calf starters may be altered accordingly to the availability of the feeds in the region and the cost. The composition of some of the proven calf starters are given in Tables 12.10 to 12.14.

TABLE 12.10. A Calf Starter used at Haryana Agricultural University, Hissar

Ingredients	*Parts*
Maize	42
Ground nut cake	35
Wheat bran/Rice polish	10
Fish meal	10
Mineral mixture	2
Salt	1

Source : Dr. S.S. Khirwar, Dept. of Animal Nutrition, Haryana Agric. Univ. Hissar (Personal Communication)

TABLE 12.11. A Complete Calf Starter Recommended by Iowa State University

*Constituents**	*Parts*
Dehydrated lucerne	20
Coarse ground shelled maize	32
Coarse ground oats	20
Molasses	10
Soyabean meal	15
Steamed bone meal or dicalcium phosphate	2
Trace mineralized salt	1

*Add per kg starter 4400 I.U. Vitamin A.
660 I.U. Vitamin D.
22 mg. chlortetracycline
(Auromysin) or Oxytetracyclin (Teramycin)

Source : Nelson. K. (1970). Calf survival-major link in herd improvement. New England Holstain Bull, May-June 1970.

TABLE 12.12. Calf Starter used at Indian Veterinary Research Institute

Ingredients	*Parts*
Barley (crushed)	50
Ground nut cake	30
Fish meal/Dried skim milk/Meat meal	10
Mineral mixture	2

To every 100 kg of the above mixture the following may be added :
(*i*) Molasses—5-10 kg.
(*ii*) Rovimix—10 g
(*iii*) Salt—0.5 kg.
(*iv*) Aurofac (chlor tetracycline)—20 g

Source : Rajnhan, S.K. (1977). *Animal Nutrition and Feeding Practices in India.* Vikas Publisbing House, New Delhi, page. 213.

TABLE 12.13. A calf starter used in National Dairy Research Institute along with a partial milk replacer

Ingredients	*Parts*
Maize	35
Ground nut cake	33
Gram	10
Wheat bran	20
Mineral mixture	3
Rovimix	15 g/100 kg

Arora C.L. (1973)

TABLE 12.14. A calf starter used at Haryana Agricultural University

Ingredients	*Parts*
Maize (crushed)	35
Ground nut cake	10
Wheat bran	35
Barley	17
Mineral mixture	2
Salt	1
Rovimix	20 g/100 kg

Source : D.N. Sharma (1977).

12.3.1. Feeding Good Quality Hay

Feeding hay encourages early rumen development which will mean that the calf can be fed as a ruminant from an early stage. This will economize calf raising. Calves start nibbling at hay and fodder from 5 to 7 days of life. They should be given soft textured, high-quality hay (preferably leguminous) on a free choice basis. Soiled hay should be removed and always fresh hay fed. Hay left over in the calve's manger can be fed to older animals.

12.3.2. Weaning

Making the calf independent of its mother is known as weaning. Under early weaning system, the cow is not allowed to suckle its calf. Instead, the cow is completely milked out and required quantity of whole milk, skim milk, milk replacer etc. are fed to the calf.

Advantages

1. Adequate measured quantity of milk can be fed to the calves preventing over-feeding and under-feeding.
2. Exact amount of milk produced by the cow is known. This is essential for maintaining proper production records for breeding and improvement. Knowing exact milk production of a cow will also help feed her properly.
3. Saves milk for human consumption as calves can be fed milk replacers.
4. Ensures clean milk production.
5. Prevents injury to the teats and chances of infection gaining entry through such wounds causing mastitis.

Disadvantages

In *Bos indicus* breeds and buffaloes maternal instinct is very strong and weaning is sometimes problematic. In these animals, weaning may cause reduced milk yield, early drying off, and temperamental problems. If weaning is done from the first calving and the calf is removed immediately after parturition with the cow blindfolded, such problems can be minimised.

Procedure

1. Immediately after calving blindfold the cow and remove the calf.
2. Dry the calf with a clean towel and massage it.
3. Attend to the navel of the calf.
4. Weigh the calf.
5. Milk out colostrum and immediately feed to the calf 1/20th of its birth weight. Follow it up with an equal dose of colostrum after 8-10 hours.

12.3.3. Teaching Calf to Drink

Colostrum or milk is poured into a pail kept scrupulously clean. It may be kept raised on a firm stand at a level convenient for the calf to drink. The calf is brought to the pail and its nose into contact with the milk. The husbandman should dip his fingers (after cleaning) in milk and allow the calf to suck at it (Fig. 12.2). He should then slowly lower his hand dipping it in the milk. When the calf takes in one or two mouthfuls remove the fingers. This procedure may be repeated whenever the calf stops drinking and lifts its head. Patience is needed on the part of the trainer.

Where labour is costly, the weaned calves may be fed from nipple pails (Fig. 12.3). It is a pail equipped with one or more rubber nipples or bottles with nipples from which calves suck milk. These nipples should be removed once a day, washed thoroughly and sterilized.

The following points are important in weaning :

1. Feeding colostrum as per recommendations given earlier.
2. Feed freshly drawn milk.
3. It is always necessary to feed wholesome and clean milk avoiding contamination.
4. Feed milk at calve's body temperature.
5. Whenever there is large amount of foam on the milk or separated milk, it should be removed by drawing a paddle on the surface or by filtering through a clean cloth. Foam causes the calves to take in entrapped air which may lead to bloating.
6. Stanchion calves at milk feeding to prevent spilling.
7. Weigh proper amount of milk for each calf.
8. Clean and sterilize pails and other utensils daily.
9. Whenever there is incidence of calf scour on the farm, milk and water may be boiled, cooled to body temperature and fed.
10. Calf-starters and good quality hay should be introduced from 16th day onwards.
11. Deworming should be done with an anthelminthic drug at proper time.
12. Special attention should be given to vitamins and minerals in the ration of calves.

12.4. FEEDING THE CALF FROM 3 MONTHS TO 6 MONTHS

The calf at 3 months is generally weaned from milk feeding. This is a period during which calves are most often neglected. Too often when milk feeding ceases, the calf is put entirely on roughages. But it is not yet a fully developed ruminant in its ability to utilize large quantity of roughages. Therefore, it will need some concentrates along with roughages to make normal growth until it is 10 to 12 months old. But if good quality legume hay or good quality green fodder is available, the calves can be raised entirely on them from 6 months onwards.

Once the calf has reached 2 to 3 months age, the most risky period of calf mortality and the most expensive feeding stage have passed. Hereafter, the emphasis would be to achieve the required growth rate with maximum economy. The expensive starter ration comprising of many ingredients can be gradually shifted to a concentrate mixture having fewer ingredients.

If plenty of good quality leguminous fodder is available the calf may be fed a simple concentrate mixture having 10-11 per cent digestible crude protein (DCP) and 75 per cent digestible nutrients (TDN). If the quality of the fodder is poor or if non-leguminous fodder is available then the calves should get a concentrate mixture having 12-14 per cent DCP and 75 per cent TDN. If only straw is available they may require 16-18 per cent DCP in the concentrate mixture. Calves in this age group may be fed between 0.75 to 1.5 kg concentrate mixture per day depending on their body weight and condition. For a calf in normal condition 1 kg concentrate mixture per day can be fed for every 100 kg body weight. This amount may be increased if condition of the calf is poor.

When leguminous fodder or hay is available requirements of calcium and trace minerals are generally met. So the only minerals that are to be added are phosphorus, sodium and chloride. By incorporating an ingredient rich in phosphorus like wheat bran in the concentrate mixture or a crop rich in phosphorus like cowpea in the roughage ration the deficiency of phosphorus can be easily remedied. One per cent common salt may be added to concentrate to supplement sodium and chloride. When enough of fodder is not available or when non-leguminous fodder is fed, the concentrate mixture should include 1 per cent each of dicalcium phosphate and trace mineralized salt.

12.5. CALF HOUSING

The type of housing used for calf rearing vary from situation to situation. There is great diversity of opinion whether calves should be housed in individual pens or in groups. In India, calves are generally reared in groups in ordinary stall barns or often in the same house along with adult cattle. Even in organised farms situation is not much different. The high level of calf mortality amounting between 30-40 per cent on many farms can be attributed to this complacent attitude towards housing and management of calves.

Experiments have indicated that under prevailing unhygienic conditions it is better to keep calves in individual pens for at least 1 month and if possible for 3 months.

A good method of housing for calves is to convert a barn in which individual calf-hutches are placed. In the Government Livestock Farm, Hissar installation of such calf-hutches reduced calf mortality among pure-bred Jersey and Holstein Friesian calves from 30 to 3 per cent. Calves are reared in the individual calf hutches until they are one month old. Afterwards they are transferred to larger hutches which can hold 5 calves where they remain up to 3 months of age

Fig. 12.4. Calf-hutch for groups of calves from 1 month to 3 months of age.

(Fig 12.4). Then they will be shifted to a loose house with straw bedding. These calf hutches are made out of angle-iron and strong wire meshes with slotted wooden floors. They are kept in double rows with a central passage.

In an experiment individual calf housings were made by erecting half walls of 1.6 m height in a barn. A lime mixed dried-dung bedding was provided. This was found to be advantageous in increasing growth rate. It also reduced the incidence of diseases like pneumonea, calf scours, navel ill, mange and tick infection as well as mortality in comparison to group housed calves.

Another efficient but simple form of calf housing for colder regions make use of the warmth and insulation provided by baled straw or hay. It is a simple construction with corrugated iron or asbestos cement roof on pillars. The walls are made out of bales of hay or straw allowing sufficient ventilation. The floor may be of concrete cement or tiles and a bedding of straw may be given.

12.6. CALF DISEASES AND MORTALITY

Mortality. Incidence of mortality of young calves is much higher in India and other developing countries in comparison to advanced countries. For example, in England and Wales, the calf mortality rate is as low as 6 per cent up to the age of 6 months. In contrast, the mortality rate among calves vary from 10 to 86.6 per cent in various farms in India. In most of the investigations, the mortality rate was found to be higher in buffalo calves than in cow calves (Ram, 1955).

Among buffalo calves the mortality rate is higher in male calves. On the other hand there is more mortality among female calves in cattle. The higher rate of survival of bull calves and female buffalo calves have a definite bearing on their future utility and the preferential treatment received from the owners.

There also seems to be a seasonal variation in calf mortality. In the Government Livestock Farm, Hissar, the buffalo calf mortality was higher from September to February (Verma, 1970).

Ram, T. (1955). Paper presented at the 5th Disease Investigation Officers Conference, Ranchi, 1955.

Verma, P.C. (1970). Studies on the pathology of buffalo calf mortality with special reference to its etiology. M.Sc. Thesis, Haryana Agric. University, Hissar.

Calf mortality is relatively high during the first 3 months of age (60 per cent), thereafter it tends to decrease with advancing age.

12.6.1. Calf Diseases

The important causes of mortality among calves are :

(*i*) Gastro-enteritis

(*ii*) Pneumonea

(*iii*) Pyosepticaemea including navel ill and joint ill

(*iv*) Diseases acquired from or common to older animals.

12.6.2. Gastro Enteritis

Gastro-enteritis in calves may be due to

1. *Escherichia coli*
2. Saccharolytic organisms
3. *Ascaris vitulorum*
4. Coccidia and
5. Other miscellaneous organisms.

12.6.2.1. *Escherichia coli*

Escherichia coli is by far the most important pathogenic organism causing calfhood diseases. In nature these are ubiquitous and in adult animals they thrive causing no apparent ill effects in their alimentary tract. But in calves under certain circumstances they assume a pathogenic role.

In acute cases there is septicaemea and death, but in milder cases they produce enteritis or a combination of enteritis and pneumonia. Lack of proper colostrum feeding with little or low levels of γ-globulins in the calve's circulation is the major cause of *E. coli* septicaemea in calves. These cases are highly fatal.

In localised intestinal cases, diarrhoea is the main symptom with consequent dehydration. If unattended the calves become prostrate and die mostly from dehydration.

The main predisposing cause for diarrhoea by *E. coli* is the **dysbacteriosis** in the alimentary canal of the calf when the balance between saccharo-proteolytic organisms (including *E. coli*) and saccharolytic organisms (mainly Lactobacilli) is disturbed. The dominance of *E coli* causes a putrificative type of diarrhoea whereas that of Lactobacilli causes a fermentative type of diarrhoea. The former type is predisposed by overloading the duodenum and small intestine with excessive amounts of proteins and peptides. For example, if a calf is raised solely on skim milk, the excessive protein content in the intestine is likely to cause diarrhoea. Milk substitutes contain a large quantity of skim milk. If in the preparation of skim milk powder, denaturation of whey proteins takes place to a considerable degree, milk will not clot properly in the stomach. This will cause undigested protein to pass in the chyme to the duodenum which forms a suitable substrate for *E. coli.*

E. coli infections can be prevented to a great extent by proper management and by using proper method of skim milk powder manufacture. The following managemental practices may be adopted to reduce the incidence of *E. coli* diarrhoea.

(*i*) Colostrum from the first milking should be given within ½ to 1 hour of parturition at a rate of 6-8 per cent of the calf's birth weight to be followed by another 5 per cent after 10-12 hours. Continue colostrum feeding up to 3rd day of life at a rate of 10 per cent of birth weight per day.

(*ii*) Pre-partum milking may not be practised as the colostrum produced after calving in such cows will have low concentration of γ-globulins.

(*iii*) If milk substitutes are fed, always take care to use one incorporating high quality skim milk powder.

12.6.2.2. *Saccharolytic Organisms*

The neonatal calves can utilize, among sugars, only lactose and glucose. Even these in excess can cause diarrhoea. Upto the age of one month, the calf practically does not possess the capacity to digest sucrose, starch, maltose and dextrin. Feeding of these in sufficiently large level will result in the proliferation of saccharolytic microflora which produce lactic, acetic and other organic acids which may cause an irritant effect on the intestinal wall. This may stimulate hyper-peristalsis resulting in a fermentative diarrhoea. To prevent such a condition, the following guidelines may be followed :

(1) It is advisable not to feed starch, sucrose, dextrins or maltose before the calves are one month old, except in small amounts to train them.

(2) A calf may not be fed more than 200 g lactose or glucose per day.

12.6.2.3. *Parasitic Diarrhoea*

In tropical and subtropical regions, ascariasis caused by *Ascaris vitulorum* is a major health problem in calves. Gastro-enteritis and diarrhoea are caused by adult worms when they are in large numbers in the abomasum and intestines. The disease is characterized by a progressive loss of condition, unthriftiness and the coat becomes shaggy and skin 'hide bound' and thick. Generally diarrhoea is present, but not invariably. If large number of larvae migrate through lungs, it may weaken the organ and cause cough and secondary bacterial infection leading to pneumonea. In severe cases, the calf will be anaemic and pot bellied. Often the calf will have a vacant look, unthriftiness and when they become prostrate it may be too late to save them.

Under natural conditions, pre-natal infection is the main source of adult parasites. Mature worms will be present only during the first 4 to 5 months of calfhood. Most probably *A. vitulorum* larvae remain 'dormant' in the tissues of the adult cow until mobilised by pregnancy when they migrate across the uterine wall and placenta to the foetus. In the foetus they migrate through liver and lungs and reach the bowels of the calf (when it is born) through the trachea.

Another group of round worms causing diarrhoea in calves belongs to the family *Trichostrongylidae.*

Prevention of parasitic diarrhoea can be afforded through the following means :

(*i*) Parasites take the upper hand and cause disease conditions only when the animals are under-nourished and their vitality is low. Therefore, the first preventive measure will be ensuring adequate nutrition.

(*ii*) Prevention of overcrowding of calves.

(*iii*) Calves should never be housed along with adult animals.

(*iv*) Prophylactic use of drugs like piperazine adipate and phinothiazine. Piperazine adipate may be given at the rate of 1 oz per calf on 3rd day and 7th day. Two or 3 medications with phenothiazine at a rate of 15-20 g per calf can control infestation by Trichostrongyl worms. Drugs like Albendazole, Fenbendazole and Mebendazole are very effective and less toxic to the calves.

12.6.2.4. *Coccidiosis*

Coccidiosis in calves is more prevalent than is generally realized. Young calves under 6 months of age are susceptible in whom coccidiosis results in stunted growth, general morbidity and even mortality.

The disease is caused by a protozoan parasite Eimeria. Of about a dozen species known from cattle and buffaloes, *Eimeria bovis* and *E. zurnii* are the most pathogenic.

Diarrhoea, dysentery with foul smelling and mucous and blood mixed faeces, arched-back-posture, anaemia, weakness, emaciation and loss of appetite are some of the major symptoms. Overcrowding of young calves, mixing of calves of various ages, lack of proper sanitation, stress, concurrent gastro-intestinal parasitic infections and other diseases etc. are the major predisposing causes.

Cattle harbouring these parasites pass large number of oocysts which under favourable conditions of warmth and moisture become infective in 2 days. Therefore, prevention can be achieved by scrubbing, cleaning and disinfecting calf houses on alternate days. Care should also be taken not to overcrowd calves. Similarly, young calves should not be introduced into a pen already occupied by a group of calves of various ages.

12.6.2.5. *Other Miscellaneous Organisms causing Diarrhoea*

A number of other organisms were found to be associated with diarrhoea in calves in certain cases. *Salmonella dublin* and *S. typhi murium*, for instance, have been known to cause septicaemea and gastro-enteritis in calves. In most of the cases, prevention lies in the identification and removal of the source of infection.

12.6.3. Management Measures to Prevent Calf Diarrhoea

1. Overfeeding of calves should be avoided.
2. Purchased calves should be isolated.
3. On arrival of purchased calves, milk should not be fed for fear of overfeeding. They may be given 3-5 per cent glucose solution to quench their thirst.
4. Milk or freshly reconstituted milk should be fed at body temperature (38°C).
5. When milk substitutes are fed, the quality of ingredients, especially skim milk powder should be given great emphasis.
6. Colostrum in proper doses should be fed at appropriate time.
7. Calves should be fed twice a day at regular intervals.
8. Calves may be dewormed at appropriate intervals.
9. If diarrhoea is prevalent among calves, one meal may be withheld from the affected ones. Afterwards, the daily allowance of milk for the calf may be curtailed to 60-80 per cent of the normal until the calf recovers. Once the consistency of the dung is normal, the milk ration can be gradually increased to normal levels.

12.6.4. Pneumonea

Generally, pneumonea occurs in the winter months. Chilling due to cold and wet floors, cold drafts and drenching in rain are the major predisposing causes. Stress and exposure due to transportation also precipitate this disease. Viruses, Corynebacterium, Salmonella and *E coli* and some other organisms have been identified as the causal agents in pneumonea. Often there is a primary infection by viruses followed by secondary bacterial attacks. Passive immunity gained through colostrum appears to protect the calves up to one month of age against pneumonea and only in calves above one month this condition is frequent. Non-availability of proper amount of colostrum can cause pneumonea at an earlier age.

Incidence of pneumonea can be reduced significantly by :

(*i*) individual calf pens with slotted floors,

(*ii*) providing bedding in calf pens, and

(*iii*) feeding proper amount of colostrum.

12.6.5. Septicaemea, Naval Ill and Joint Ill

In the first few days of life of the calf some of the most ubiquitous organisms, which are generally non-pathogenic otherwise, assume a pathogenic role under certain conditions and cause serious infections. A number of organisms are associated with this condition. The important among them are *Escherichia coli*, Citrobactor, Streptococci, Klebsiella, Proteus, a species of Mycoplasma and an organism of *Psittacosis-lymphogranuloma venerium* group (PLV).

Some of the major predisposing causes are :

1. Improper colostrum feeding.
2. Contamination of the umbilicus.
3. Unsanitary premises.
4. Bringing calves into an environment to which their mothers are not exposed. Cow may not be able to transfer antibodies against strains of organisms present in the new environment to which she has not been exposed.
5. Deficiency of vitamin A and minerals in the diets of cows when pregnant and
6. Faulty feeding of calves.

Infection may gain entry through the umbilicus or mouth. In acute cases seen early in life, symptoms of septicaemea like fever, loss of appetite, depression, clay coloured diarrhoea which soon becomes watery and blood stained etc. may be seen. In the umbilical infections the umbilicus will be generally swollen. In acute cases, the calves mostly die. In older calves the disease is milder and the animals recover. But, a majority of them develop navel ill (an abscess of the umbilicus), joint ill (swelling of the joints), pneumonea or combinations of some or all. The preventive measures can be :

1. Feeding of colostrum immediately after birth in proper amount.
2. Dressing naval chords with an antiseptic like tincture of iodine or tincture of benzoin.
3. Disinfection of calf premises.

12.6.6. Diseases Acquired from Older Animals

In addition to the diseases discussed above, calves acquire diseases from in-contact adult animals or they can be attacked by diseases which are common for young and old alike. Tuberculosis,

Johne's disease and brucellosis are transmitted by the cows to their calves through milk or contact. Milk from mastitic cows also causes mild to severe infections in calves.

Calves are susceptible to diseases like foot and mouth, rinderpest, haemorrhagic septicaemea, mucosal disease, black quarter, thilariasis etc., which may be more fatal in the calves than in adults. In addition, they can also suffer from bloat, rickets, *degnala disease* and other poisonings.

12.6.7. External Parasites

Under conditions of malnutrition, unhealthy and unsanitary housing conditions and in certain regions with certain environmental conditions, external parasites become a major menace. Ring worm, mange, ox warble flies, ticks, lice etc. are some of the important external parasities.

These can be checked by periodic spraying of calves and pens with suitable insecticides.

12.7. GENERAL MANAGEMENT

Every calf born on the farm must be identified immediately after birth by tattooing or tagging and also branded by around one year (see Chapter 22). Some other routine and necessary calf management practices like dehorning, castration, removing extra teats etc., to be carried out at various ages are also discussed in chapter 22.

CHAPTER 13

Heifer Management

Heifer is a female animal from one year of age upto first calving according to the definition of the British Society of Animal Production. Thus heifers are future cows of the herd. It is time to realise that the productive performance of cows depends on the care and attention given to them during their heiferhood. But unfortunately the care of heifers is neglected and in fact, is considered an unproductive burden.

13.1. PRESENT STATE OF AFFAIRS

Age at first calving of Zebu cattle ranges from 38 to 50 months and even much later in buffaloes depending on breed, level of feeding and management status as compared to around 30 months in *Bos taurus* cows (see Chapter 16). How far low standards of feeding and management are responsible for this, apart from the often quoted lower genetic potential for this trait, is a moot point. Suffice to say that heifers are the most neglected stock on the Indian farmer's household.

A common problem in dairy heifer management in general and buffalo heifer management in particular is under-feeding due to various reasons—diversion of scarcely available better feeds and fodders to milch stock, heavy parasitic load, improper housing that makes shy calves remain away from feeds, overcrowding, ignorance of the importance of heifer care etc.

After weaning (actually in India, after the cow/buffalo dries off) farmers leave the female calves to fend for themselves on poor grazing and dry fodders. Hardly any concentrate feeds are given as it is generally considered to be "uneconomical". Unfortunately it is at this age, 6 to 12 or 15 months, the highest growth potential exists (see Chapter 11). The farmer takes greater interest in heifer as she matures, when its growth potential has already slackened. A survey conducted in Haryana (Sharma, 1979) has revealed that the buffalo heifers receive adequate (or even excess) quantities of DCP and TDN only after 2 years of age, upto which time they would have remained stunted due to inadequacy of nutrients (Table 13.1). In crossbreds too feeding cost come lowest around six months of age around weaning (Table 13.2). Rearing costs also are higher at later ages.

TABLE 13.1. An Appraisal of Feeding Practices of Buffalo Heifers in Haryana Villages (Sharma, 1979)

Age of heifers (months)	*DCP (kg)*		*TDN (kg)*		*Body weight (kg)*
	Requirements	*Supplied*	*Requirements*	*Supplied*	
0-3	0.22	0.06	1.30	0.42	65.1 (13)
3-6	0.25	0.14	1.90	0.87	77.1 (9)
6-12	0.28	0.24	1.90	1.78	92.7 (23)
12-24	0.40	0.46	3.00	3.94	213.4 (15)
24-1st service	0.48	0.61	5.00	4.59	392.9 (13)

Note. Figures in parenthesis indicate number of animals actually measured to get the estimate.

TABLE 13.2. Cost in Rupees per Kilogram Gain in Body Weight

Genetic Group	*0-70 days*	*71-180 days*	*181-365 days*	*1 Year to AFC*
Halfbreds				
½F ½H	15.61	3.62	5.08	10.73
½B ½H	22.60	3.74	5.35	11.65
½J ½H	17.97	3.53	5.30	9.35
Triple Crosses				
½F ¼B ¼H	16.52	3.76	5.10	7.25
½F ¼J ¼H	16.57	3.39	5.78	7.78
½B ¼F ¼H	20.91	4.52	5.23	7.80
½J ¼F ¼H	18.15	4.52	5.23	7.80

Note. F = Friesian; B = Brown Swiss; J = Jersery; H = Haryana.

Source : Chopra, S.C. and Baalaine, D.S. 1978. Proc. National Conference on Crossbreeding. April 27-28, 1978, NDDB. Anand, p. 299.

Such a feeding practice is not only unsound and uneconomical, but also results in fatty heifers (due to overfeeding of nutrients at later ages) which condition interferes with efficient reproduction.

If this is the practice followed by renowned cattle breeder-farmers in the home tract of the famous Murrah buffalo, the situation cannot be much different, if not worse, in other parts of the country.

On some institutional farms like those of NDRI, HAU, PAU etc. it has been proved beyond doubt that, even in Zebu and buffalo heifers, age at first calving can be reduced by following scientific feeding and management practices. Cost of rearing heifers upto age at first calving on NDRI farm is shown in Table 13.3.

Aspects of reducing age at first calving in dairy heifers through proper feeding and management practices are discussed in the succeeding sections of this chapter.

13.2. AIMS AND OPTIONS OF HEIFER REARING

The main aims of management of calves during the post weaning period is to produce large, fast growing heifers at minimum cost. Feeding and management of heifers, though neglected badly,

Sharma, V.B. 1979. *Economic level of concentrate feeding for buffalo heifers.* M.Sc. Thesis, Haryana Agricultural University, Hissar.

TABLE 13.3. Cost of Raising Calves to the Age at First Calving (Rs. per head) for Different Breeds of Cattle and Buffaloes at NDRI, Karnal (1975-76)

Components of cost	*Breed*											
	PUREBRED (Zebu)				*PUREBRED (Exotic)*			*CROSSBRED CATTLE*				
	Thar-parkar (3)	*Sahiwal (18)*	*Red Sindhi (2)*	*Purebred All (23)*	*Jersery (2)*	*Holstein (3)*	*All (5)*	*Brown Swiss (116)*	*Jersery (18)*	*Holstein (65)*	*All (199)*	*Murrah buffalo (15)*
Feed cost	3,038.98	2,960.75	2,785.44	2,955.71	2,099.00	2,597.45	2,398.07	2,651.02	2,269.78	2,690.37	2,629.39	3,569.99
Labour cost	499.78	489.32	465.76	488.64	345.55	429.37	395.84	423.14	369.08	419.33	417.01	485.16
Supervision cost	107.87	105.14	99.13	104.97	68.56	89.85	81.33	88.26	83.31	87.29	86.67	104.08
Vety & Medicinal	18.65	3.65	6.35	5.84	22.38	19.19	20.45	7.64	5.78	6.58	7.13	32.10
Misc. cost	390.72	380.81	359.21	380.21	248.32	325.44	294.59	319.68	269.76	316.16	314.24	376.96
Gross cost	4,056.00	3,939.67	3,715.72	3,935.37	2,783.81	3,461.30	3,190.30	3,489.74	2,997.71	3,519.73	3,454.44	4,568.29
Net cost	3,974.20	3,859.94	3,640.55	3,855.77	2,731.82	3,393.03	3,128.54	3,422.81	2,941.23	3,453.53	3,388.69	4,495.61
Age at first calving (days)	1,221	1,190	1,122	1,188	776	1,019	894	999	843	988	982	1,178

Note. Figures in parentheses indicate the number of calves reared.
Source : Dairy Handbook, 1977, NDRI, Karnal, p. 116.

are easier and less costly in the period from weaning to first calving than during any other period in the life of the female bovine.

The main purpose of rearing heifers is to take them into the producing herd in place of cows that might have been dead or culled due to old age, disease, low production etc. The rate of such replacements can vary from 20-25% (Etgen and Reaves, 1978) depending on several herd management factors. In US it becomes mandatory that only such heifers are reared that would eventually become good cows *i.e.*, daughters of high yielding dams or proven sires etc. But in this country, since no such records are available, the farmers have to go by physical condition of the animal, freedom from disease and physical defects, dam's yield (whenever possible) and by the word of the seller. Size of the heifer in relation to her age (judged by teeth and other means) can be a very valuable guideline. Also if. the date of birth is known, the animal can be weighed on a test day or on the day of purchase and the daily weight gain estimated by dividing the weight by age in days. This method is used very effectively in selecting breeding stock in the Federal Republic of Germany and other western countries.

The farmer or manager has two options for obtaining herd replacements, namely, purchasing or rearing them himself. The decision to go for one or the other option has to be taken after considering carefully the costs involved, the facilities available on the farm, availability of good stock for the purpose, health situation outside the farm and so on.

Rearing heifers on farm permits implementing a general plan for genetic progress in the herd (progeny testing of herd sires), removes or reduces chances of bringing in disease on the farm and provides chances of getting more profits by sale of surplus full-grown heifers. Its main drawback is that it leads to investments without profits for 2-3 years. Purchasing is beneficial especially when there are sources where genetically superior stock can be had and this latter practice does permit ready returns from the money invested on the purchase. The decision is that of the farmer/manager.

It is difficult to suggest a possible replacement rate under village conditions because of the very small size of herds and general lack of facilities for scientific rearing of heifers. As mentioned earlier, heifer rearing in the villages mainly comprises of leaving the weaner calves to fend for themselves. However, on organized farms the number of replacements to be reared can be reckoned from the herd's culling and disposal rates, and mortality rate among growing stock.

Possible sources for purchase of heifers in this country include-government dairy farms, milk colonies, private dairy farms, university and institutional farms and scientifically run *"Goshalas"*, *"Gosadans"* and *"Panjarpoles"*. However, the most accessible sources are the villages covered by Key Village Schemes, Intensive Cattle Development Projects and native tracts of recognized dairy and dual purpose breeds of cattle and buffaloes.

13.3. REARING SYSTEMS

After weaning, female calves can be reared sucessfully on green fodders alone, as by this time the rumen functioning would have been thoroughly established. In western countries there is a heifer feeding practice known as "turning on pasture", pasturage being the major, if not the sole,

---

Etgen, W.M. & Reaves, P.M. 1978. *Dairy cattle feeding and management.* 6th Ed. John Wiley & Sons, New York. p. 326.

nutrient source. But, in India, the pasture (actually natural grasslands) quality is perpetually poor due to overgrazing and non-maintenance. During summer, grazing has perhaps practically no feed value. Whatever the animals get from grazing on dry lignified stubble may not be worth the energy spent on the affair.

In case of buffalo heifers, rearing by grazing was found to be less efficient than stall feeding individually but superior to stall feeding in groups (Table 13.4). Grazing under village ,conditions seems economical, though it takes longer to attain breeding stage, because no cost is imputed to labour; family members-wife and children-perform this task free. This study shows that stall feeding heifers individually, though may mean higher per day cost, will result not only in earlier maturity but also lower cost to attain the same (₹ 4,091.60 against ₹ 4,175.50 in grazing and ₹ 5,229.00 in group stall feeding). However, it may be, pointed out that grazing cannot always be as good as in this study; there were unusually good rains during this study period. The quality of grazing varies greatly from region to region and season to season.

TABLE 13.4. Performance of Young and Older Buffalo Heifers under three feeding systems (From Yadav, *et al.*, 1985)

Parameter	*Grazing*	*Stall feeding individually*	*Stall feeding in groups*
1. Daily weight gain (g) (6 to 18 months)	364	385	214
2. Daily weight gain (g) (18 to 30 months)	474	513	441
3. Dry matter intake per kg gain (kg)	17.5	11.7	15.4
4. Cost/kg gain (₹)	11.93	11.69	14.94
5. Per day cost upto* 350 kg body weight (₹)	4.68	5.18	5.30
6. Per day cost minus labour cost*	3.20	4.78	4.92
7. Age of attaining 350 kg body weight (days)	893	790	986

*Extrapolated values.

Even in the study just mentioned, more is the fall (12-15%) in the intensity of growth during hot dry and cold seasons under all systems especially in grazing. The reason for this was considered to be the fact that herbage contained low levels of crude protein and high levels of dry matter especially crude fibre in hot dry season and low levels of all the three components during winter. This would lead to the ultimate conclusion that grazing needs concentrate supplementation in summer.

Comparative merits of individual and group feeding of cattle has been a controversial subject amongst nutritionists and husbandmen for years. Perhaps every report in this regard has to be considered in the light of the quality and quantity of feeding available. A critical appraisal of the review of several reports on the topic by Yadav (1984) revealed that there was either no difference in growth rate of calves fed individually and in groups or the latter were superior in case of practically all quoted reports from temperate regions. On the other hand, in majority of reports emanating from tropical regions including India (Arora, 1978; Bose and Thomas, 1979;

Yadav, R.S., Sastry, N.S.R. and Aggarwal, C.K. 1985, *Indian J. Anim. Prod. Mgmt.* **1**: 19-27.
Yadav. R.S. 1984. *Effect of three rearing systems on growth performance of female buffalo calves.* Ph.D. Thesis, Haryana Agricultural University, Hissar.
Arora, S.P. 1978. *Calf raising.* NDRI Pub. No. 135. NDRI, Karnal.
Bose, A.M. and Thomas, C.K. 1979. *Indian J Anim. Sci.* **49**: 419-422.

Sastry *et al.*, 1981; Singh *et al.*, 1985), individual feeding was invariably found to be a superior system. This indicates that when feeding is adequate (as in western countries) both rearing systems are similar, while the individual feeding system is desirable when there are constraints to feed availability, both quality and quantity-wise, as in India.

Out-door housing and natural grazing are generally, but not necessarily, economical propositions in warm countries (Williamson and Payne, 1977). But it is a universally agreed fact that the grazing, unlike in temperate regions, cannot effectively sustain cattle and buffaloes in tropical countries. Grazing bovines tend to lose weight during most part of the year, especially during dry summer.

However, grazing animals supplemented with some concentrates allows maintenance of live weights and even gains unlike in those kept solely on grazing (Mahajan *et al.*, 1976) though when herbage stand is good no such differences may be discernible (Gupta *et al.*, 1970). Ranjhan (1977) reported that when calves are maintained on part monsoon or autumn pasture with 1.0 to 1.5 kg of supplemental concentrate mixture, a rate of growth as high as 0.50 to 0.55 kg per day can be maintained in Haryana calves over short periods. When growing animals are maintained exclusively on pasture without any concentrate or roughage supplementation during autumn season, the younger calves have low gains (0.25 kg/day) as compared to 0.4 kg/day for older ones. However, after December, all the animals generally start losing weight, possibly due to inadequate dry matter consumption. This was essentially the finding of Yadav *et al.* (1985) too in case of Murrah heifers.

What system of heifer rearing can be recommended to the Indian farmers ? Based on the reports discussed above, perhaps grazing on village grazing lands, field bunds, tank and canal embankments and other wastelands coupled with one to one and half kilo concentrates (groundnut cake/grain/bran with mineral mixture) especially from March to July offers the best solution.

On organized farms, maintaining heifers on good quality green fodders, should be opted for. Some sort of concentrate supplementation is necessary, though, during dry summer and cold months in the north and from September/October to next monsoon in other regions.

13.4. HEIFER MANAGEMENT

In the very outset the reader is advised to acquaint him/herself with the general principles of feeding (Chapter 6). reproduction (Chapter 7), housing (Chapter 19) and health control (Chapter 20) of heifers described elsewhere in the respective sections of this book. Aspects of growth in calves are described in Chapter 11 and recorded normal growth patterns of different breeds of cattle and buffaloes are given in section 16.3. The aim is to achieve growth rates higher than those reported averages and never below average performance.

Sastry, N.S.R.,Yadav, RS., Sharma, V.B. and Juneja. I.J. 1981. *Haryana Agric. Univ. J. Res.* **11** : 452-457.

Singh, M.P., Kakri, D.B., Mehla, O.P., Singh, R.A. and Sastry, N.S.R. 1985. *Indian J Anim. Prod. Mgmt.* **1** : 160.

Williamson, G. and Payne, W.J.A. 1977. *An introduction to animal husbandry in the tropics*. ELBS & Longmans, London, p. 269.

Mahajan, J.M., Chauhan. D.S. and Tomar, V.P.S. 1976. *Indian J. Anim. Res.* **10** : 90-92.

Gupta, B.S., Ranjhan, S.K. and Chhabra, S.S. 1970. *Indian J. Anim. Prod.* 1 : 82-85.

Ranjhan. S.K. 1977. *Animal nutrition and feeding practices in India*. Vikas Pub. House, New Delhi, pp. 216-217.

Yadav, *et al.*, 1985. *loc. cit.*

Apart from the above mentioned generalities, certain aspects of management specific to heifers are discussed in this section. Care of heifers during summer is of great importance in the tropics as animals grow poorly or even lose weight during this period largely due to fall in status of feeding. This superimposed by climatic effects can affect the tempo of growth in calves. The case of buffalo heifers reared at HAU Farm (Table 13.5) illustrates this point. Fastest growth rates were obtained during berseem season (743 g/day). The most critical period is the hot season during which there is greater climatic stress as well as shortage of good quality greens resulting in poor growth (364 g/day). However, if legume hay is fed during such periods, higher growth rates (624 g/day) can still be obtained (Table 13.5). In a recent study (Juneja, 1985), poor growth during the hot humid period despite adequate feeding, in case of heifers of purebred breeds was recorded.

TABLE 13.5. Growth Performance of Murrah Buffalo Heifers (6-18 months old) at Haryana Agricultural University Farm during Different Seasons of the Year on a Predominantly Roughage Based Ration

Months	*Number of animal studied*	*Type of fodder available*	*Mean max. ambient temperature (°C)*	*Mean min. ambient temperature (°C)*	*Mean relative humidity (%)*	*Mean daily wt. gain (g)*	*Mean daily dry matter intake (kg)*	*Dry matter intake per kg gain in wt. (kg)*
1. ½ March, April ½ May	20	Maturing green barseem	34.6	18.7	72.6	456	3.4	7.3
2. ½ May, June, ½ July	20	Berseem hay	37.3	24.1	81.0	624	3.7	6.1
3. April, May, June	12	Maize and Jowar	28.0	24.5	79.0	428	4.2	9.2
4. July, Aug., Sept.	20	Mature Jowar	34.8	22.1	93.0	364	4.8	13.7
5. Dec., Jan., Feb., March	20	Green Oats, Oats + Berseem and Berseem Green	25.4	7.2	82.0	743	5.4	7.7

1. Compiled from the original reports of Tripathi *et al.*, 1972 (3) Dall, 1980 (1&2), Grewal *et al.*, 1982 (5) and Sastry *et al.*, 1980 (4).

Thus along with better feeding, heifers should be protected from thermal stress by (*i*) showering or splashing cold water on animal 2-3 times during the hottest part of the day (Tripathi *et al.*, 1972; Thomas *et al.*, 1975); provision of cooled drinking water (Radadia *et al.*, 1980) and (*ii*) protecting under shade during day and keeping in the open at night (Karti *et al.*, 1983).

Tripathi, V.N., Thomas, C.K., Sastry, N.S.R., Pal, R.N. and Gupta, L.R. (1980). *Indian J. Anim. Sci.* **42**: 742-745.
Dall, J.S. (1980). *Studies on evolving an economic system of raising buffalo heifers.* M.Sc. Thesis. Haryana Agric. Univ., Hissar.
Grewal, S.S., Sastry, N.S.R and Yadav, R.S. (1982). *Indian J. Anim. Sci.*
Sastry, N.S.R, Yadav, R.S., Sharma, V.B. and Juneja, I.S. (1980). *Haryana Agric. Univ. J. Res.*
Juneja, I.J. (1985). *Personal Communication.*
Radadia, N.S., Pal, R.N., Sastry, N.S.R. and Juneja. I.J. (1980). *Haryana Agric. Univ. J. Res.* **10** : 152-156.
Thomas, C.K., Tripathi, V.N. and Sastry, N.S.R. (1975). *Indian J. Anim. Sci.* **45** : 823-826.

Housing heifers in loose houses (see Chapter 19) was found to be most efficient in all seasons (Singh *et al.*, 1985). The village practice of housing animals inside during day and in the open during night in summer and *vice versa* in winter was found to be only marginally inferior to loose housing. Buffalo heifers protected through thermal amelioration/extra shelter and water sprinking during summers calved three months earlier than comparable but unprotected ones (Table 13.6).

TABLE 13.6. First Lactation Productive and Reproductive Performance of Buffaloes under Two Summer Management Regimes during Calfhood

Variables	*Protected during 6-18 months of age*	*Unprotected during 6-18 months of age*	*Difference in favour of protected group*
Daily weight gain during 6-18 months of age coinciding with summer season (kg)	0.426 ± 0.011 (10)	0.397 ± 0.047 (10)	+ 0.029 (10)
Age at first calving (days)	1,771 ± 72 (10)	1,882 ± 68 (10)	– 111
1st lactation milk yield (kg)	1,715.9 ± 252.7 (9)	1,712.9 ± 156.4 (9)	+ 3.0
1st lactation length (days)	301 ± 25 (9)	332 ± 20 (9)	– 31
Daily milk yield during 1st lactation (kg)	5.63 ± 0.48 (9)	5.07 ± 0.81 (9)	+ 0.56
300 day milk yield of 1st lactation (kg)	1,710.0 ± 50.0 (9)	1,523.0 ± 87.3 (9)	187*
1st dry period (days)	125 ± 14 (9)	222 ± 62 (8)	– 97
1st calving interval (days)	427 ± 30 (9)	514 ± 62 (8)	– 87

Note. Figures in parentheses are number of animals.
*Significant at 5% level of probability.
Source : Sastry *et al.* (1981).

The deworming practices should be continued during hot season even in older heifers. Medication against fasciolasis is also a must in high humid marshy areas. Further, monthly spraying with proper insecticides against ticks is highly necessary during hot season. Not only the animals but the floors, walls and roof must also be sprayed. Calves regularly dewormed and sprayed against ticks are known to show 10% higher growth rates.

13.5. CARE OF FULL-GROWN HEIFERS

13.5.1. Breeding

The aim of all heifer growing plans is to mate her as early as possible so that she calves early. The time of breeding heifers has to be decided on the basis of their body size rather than age.

Tripathi, V.N., Thomas, C.K., Sastry, N.S.R., Pal, R.N. and Gupta, L.R. (1972). *Indian J. Anim. Sci.* **42** : 745-748.

Karki, D.D., Singh. M.P., Sastry, N.S.R. and Grewal, S.S. (1983). *Indian J. Anim. Sci.* **53** : 1079-1081.

Singh, M.P. Karki, D.B., Mehla. O.P., Singh, R.A. and Sastry, N.S.R. (1985). *Indian J. Anim. Sci.* **55** : 462.

It is generally considered that attainment of around 60% of the mature body weight is the stage at which the heifer should be bred. The aim should be to see that Zebu and buffalo heifers attain this body weight by at least two years of age so that their age at first calving is below three years. However, such a satisfactory growth is rarely attained though not impossible in tropics. Exotic and crossbred heifers can attain such weights by 20 months of age with proper care.

It may be better to get heifers examined per rectum for soundness of reproductive tract. Heifers should be mated only with semen of proven or certified sires.

13.5.2. Gestation

The demands from the growing foetus on the mother has been greatly exaggerated. The common belief is that the developing foetus causes too severe a drain of nutrients from the heifer's system to cause a retardation of growth. But scientific evidence does not contribute to this view. In fact late pregnancy is a period of rapid growth when the output of growth hormone from the anterior pituitary is at one of its peaks (The other peak is during early calfhood). The dry matter content of the entire foetal material is only 10-12 kg which is equivalent to only 50 to 60 kg milk on dry matter basis. Gestation, therefore, can have only very little influence in retarding growth. Experimental evidence indicates that pregnant and non-pregnant heifers under similar feeding grew at the same rate.

But during pregnancy the heifer can be made to grow rapidly to attain mature weight and to have sufficient reserves for the ensuing lactation. It is futile to grow calves too rapidly making them over conditioned and too big at an early age. It will mean feeding a larger heifer for a longer time which will need more nutrients for maintenance and growth. Instead, the heifer may be fed mostly on roughages and allowed to remain lean (not stunted) until pregnancy. During the last half of pregnancy it can be fed at a higher plane to achieve rapid growth. Growth hormone has an important role to play in the development of alveoli and ducts in the mammary gland. During late pregnancy if a high plane of nutrition is given, it, coupled with high level of growth hormone, is likely to cause maximum development of ducts and alveoli in the heifer's udder.

13.5.3. Lactation

While gestation is of minor consequence to growth of heifers, the ensuing lactation has a major effect on it. A cow producing 10 kg milk secretes about 1.4 to 1.5 kg dry matter, in milk. A normally growing calf adds only 0.35 to 0.4 kg dry matter a day which is 1/3 to 1/4 the amount of dry matter secreted by a cow in her milk daily. This implies that the heifer has to be fed heavily for sustaining milk production in addition to growth demands. In early lactation, the heifer may mobilise her body reserves and even lose weight to maintain lactation. This implies that the heifers should have enough body reserves to tide over the period of heavy demand for nutrients in early lactation.

In the feed manger, during the last fortnight of gestation, the heifers should receive larger amounts of concentrates. This will not only help build up reserves, but also give time to make a gradual shift to high concentrate feeding which is necessary during lactation. Such a practice will allow gradual change in rumen microbial population to suit heavy concentrate feeding. Such feeding is termed 'challenge feeding'.

Initiation of lactation in heifers can sometimes be accompanied by oedematous swellings around the udder especially on the under surface of the belly in front of the udder. This can be a source of pain to the animal and hence need be attended. Smearing of glycerine and magsulph paste is the standard remedy. In difficult cases veterinary help may be sought.

Introduction of heifers to the routines of the milking parlour is also important, as the animal may get nervous if taken to the parlour all of a sudden. It may be a good idea to let pregnant heifers in later stages into the milking parlour and go through the routine. This will give an opportunity for them to get adapted to the milking routine.

FURTHER READING

Section IV

Brody, S. 1968. *Bioenergetics and Growth.* Hafner Publ. Co., Inc. New York.

Buttery. P.J. and Lindsay. D.B. 1980. *Protein Deposition in Animals.* Butterworths, London.

Davis, R.F. 1965. *Modern Dairy Cattle Management.* Prentice-Hall of India (Pvt.) Ltd. New Delhi.

Diggens, R.V., Bundy. C.E. and Christensen. V.W. 1979. *Dairy Production.* 4th ed. Prentice-Hall. Inc., Englewood Cliffs. New Jersey.

Etgerr. W.M. and Reaves. P.M. 1978. Dairy cattle feeding and management. 6th ed. Jonn Wiley & Sons, New York.

Goss. R.J. 1978. *The Physiology of Growth.* Academic Press, New York.

Hafez, E.S.E. and Dyer, L.A. 1969. eds. *Animal Growth and Nutrition.* Lea and Febiger, Philadeipha,

Hammond, J. 1960. *Farm Animals, Their Growth, Breeding and Inheritance.* 3rd ed. Edward Arnold. London.

Juergenson, E.M. and Mortenson, W.P. 1966. *Approved Practices in Dairying.* Oxford & IBH Publ. Co., New Delhi.

Kleiber, M. 1961. *The fire of life.* John Wiley & Sons. New York.

Maynard, L.A. and Loosli, J.K. 1973. *Growth,* In : *Animal Nutrition.* pp 435. Tata McGraw-Hill Publ. Co, Ltd. Bombay.

Roy, J.H.B. 1980. *The Calf.* 4th ed. Butterworths, London.

Schmidt, G.H. and Van Vleck, L.D. 1974. *Principles of Dairy Science.* W.H. Freeman and Co. San Francisco.

Warner, R.G. 1970. Raising dairy replacements from birth to weaning and from weaning to freshening. *Hoard's Dairyman.* **115**: 976, 1029.

Williamson, G. and Payne, W.J.A. 1977. *An introduction to animal husbandry in the tropics.* ELBS & Longmans, London.

Section V : GENETIC IMPROVEMENT

CHAPTER

14

Material Basis of Inheritance

Formulation of selection plans is possible only on the basis of a clear knowledge about the mechanisms involved in the formation, manifestation and transmission from generation to generation of selection characters. Genetics, the science that teaches the laws of heredity and variability in living beings, gives us such a knowledge.

Generally, by heredity we mean the capability of the animal to transmit its own characters to the offspring. Actually it is not the characters that are transmitted, but the genetic factors, which determine the former. The various characters develop and manifest themselves as the animal grows as a result of specific biochemical processes.

Heredity is that nature of all living beings to repeat in a series of generations similar type of biosynthesis and body metabolism as a whole. These, in their own turn, cause and determine the structural and functional attributes of different types of body tissues as the animal develops.

14.1. CELL IN REPRODUCTION

All the manifestations of heredity are linked with the fundamental unit of any living being, namely, the cell (Fig. 14.1). The complex structure of the cell can be known by cytological studies with the help of electron-microscope that magnifies the image of objects by several thousand times. The cell is enclosed by a semipermeable cell membrane through which metabolic exchanges take place. Its thickness is hardly in millionth parts of a millimeter.

The cell comprises of two major parts—the nucleus and the cytoplasm. The nucleus is enclosed by a double membrane. The cytoplasm is interspersed by a network of large number of microcanals which serve as connections between different parts of the cell. In the nucleus as well as in the cytoplasm exist several structures which are also enclosed within membranes.

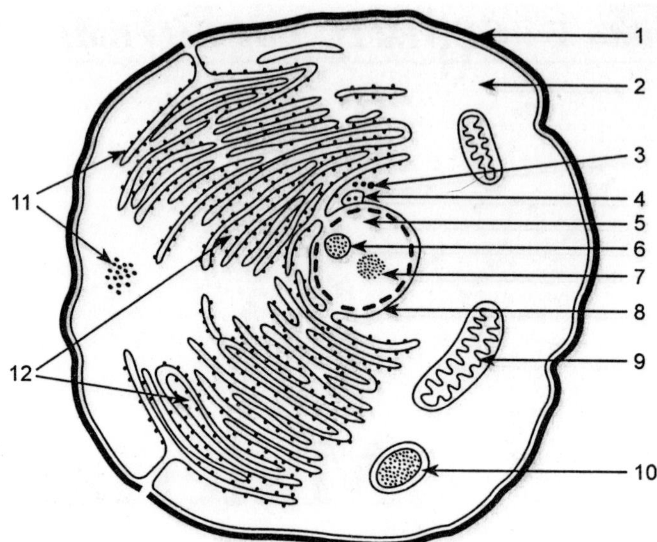

Fig. 14.1. Diagrammatic representation of the structure of an animal cell. 1. Cell wall; 2. Cytoplasm; 3. Golgi bodies; 4. Centrosome with centrioles; 5. Nucleus; 6. Nucleolus; 7. Chromosomes; 8. Nuclear membrane; 9. Mitochondria; 10. Lysosomes; 11. Ribosomes; 12. Microcanal network.

These structures perform specific functions connected with the life activity of the cell. Mitochondria located in the cytoplasm can be termed as power stations of the cell. It is in these structures energy required for the cell is accumulated. These are centres of enzyme activity in oxidative metabolism. Ribosomes located along the surface of microcanals are concerned with synthesis of protein. On the other hand, protein disintegration takes place in the lysosomes. The function of Golgi bodies contained in animal cell, is not yet precisely known. The centrosome with centrioles functions in cell division. Chromosomes are located in the nucleus and are of most interest to geneticists.

14.2. CHROMOSOMES

The chromosomes are the carriers of genetic material. All somatic or body cells of animals contain two strands of each chromosome—diploid number of chromosomes. On the other hand, sex cells—sperm and ovum, contain only single strand of each chromosome—haploid number of chromosomes. The haploid number is denoted as n, while the diploid number as $2n$. The number of chromosomes is constant for a given species.

Chromosomes vary considerably in their size, shape and location of centromeres, both between and within species. Centromere is the point at which, before complete cell division, a dividing chromosome is still attached. Recent investigations in cytogenetics permitted identification of each pair of chromosomes of cattle by a specific and reproducible banding pattern when stained by means of different techniques. This banding pattern is similar for both members of a homologous pair, which fact increases the accuracy of identifying every individual pair.

At the metaphase stage of cell division each chromosome appears doubled and is united at the centromere (Fig. 14.2). Chromosomes can be classified according to the position of centromere as metacentric (with a median centromere), sub-metacentric (with a sub-median centromere) and

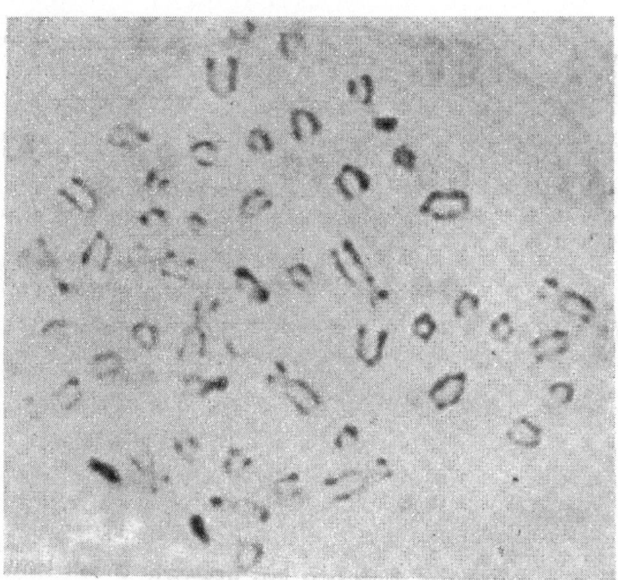

Fig. 14.2. Photomicrograph of metaphase spread from cultured lymphocytes of male buffalo (*Bos bubalis*, $2n$ = 50). Note that at metaphase stage the chromosomes appear doubled and connected at the centnomere. (*Photo Courtesy* : Dr. Vijay Kumar, K.A.U., Trichur).

acrocentric (with a terminal centromere) chromosomes. The karyotypes (systematic arrangement of chromosomes) for individuals of several species have been determined. Karyotype of male *Bos indicus* × *Bos taurus* crossbred cattle is shown in Fig. 14.3. Crossbred of this type shows karyotype typical of *Bos taurus* cattle. Karyotype of Murrah buffalo male is shown in Fig. 14.4.

Body cells of cattle contain $2n$ = 60 chromosomes, while the ova and sperm cells of cattle contain n = 30 chromosomes. In case of buffaloes the information on number of chromosomes is rather puzzling. According to I.L. Mason, the Swamp buffalo (SE Asian) has a diploid number

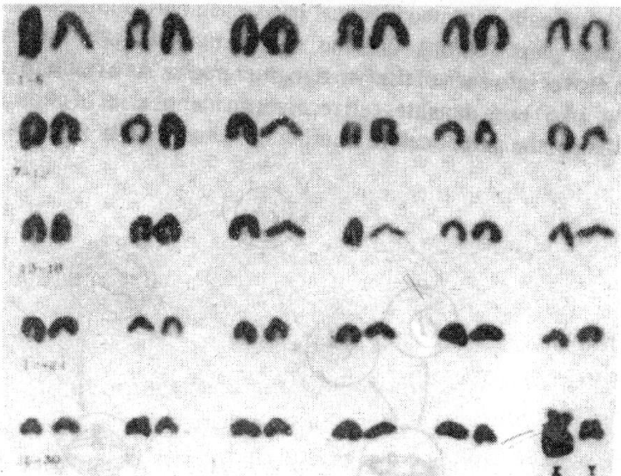

Fig. 14.3. Photomicrograph of chromosomes from lymphocytes of *Bos tauras* × *Bos indicus* male cattle arranged in a karyotype. Such crossbred cattle show karyotype typical of *Bos taurus* cattle. (*Photo Courtesy* : Dr. Vijay Kumar, K.A.U., Trichur).

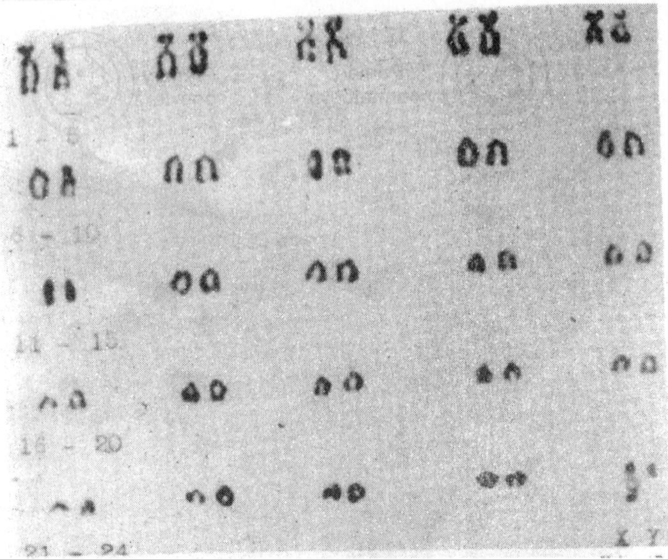

Fig. 14.4. Photomicrograph of chromosomes from lymphocytes of male buffalo (*Bos bubalis*) arranged in a karyotype. (*Photo Courtesy* : Dr. Vijay Kumar. K.A.U., Trichur).

of 48 chromosomes and the River type (Indian)—50, but the amount of chromosomal material is similar in the two types. In the latter case there are 5 pairs of metacentric or submetacentric chromosomes and 25 pairs (including the sex chromosomes) of acrocentric chromosomes. This gives the 60 chromosomes typical of *Bovidae*.

14.3. CELL DIVISION

The body cells form by a process of cell division. During cell division the nucleus also gets divided. Just before the cell division, all chromosomes (*i.e.* both members of each chromosome pair) split longitudinally into two identical parts, each part comprising of a full compliment of all chromosome pairs. During division of the nucleus, one identical part each of the chromosome moves into each of the two daughter nuclei. As a result of such a division, known as mitosis (Fig. 14.5) each daughter cell receives an identical set of chromosomes that is exactly the same as that of the parent cell in number (2*n*) and quality (Fig. 14.5).

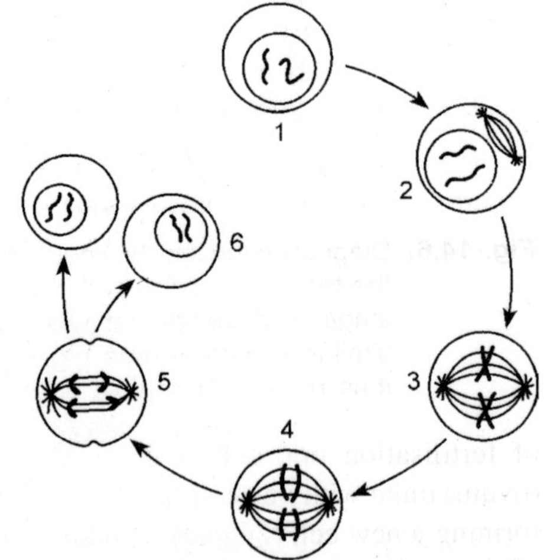

Fig. 14.5. Diagram showing the duplication of chromosomes during mitosis. Note that the daughter cells (6) resulting from mitotic cell division contain the same number (2*n*) of chromosomes as that of the mother cell (1). The various stages of cell mitosis can be seen in figures 2 to 5, 3 being the metaphase stage.

But in case of sex cells, by a process of reduction division—known as meiosis (Fig. 14.6) the number of chromosomes in the daughter cells get halved (*n* chromosomes) on cell division. At the time

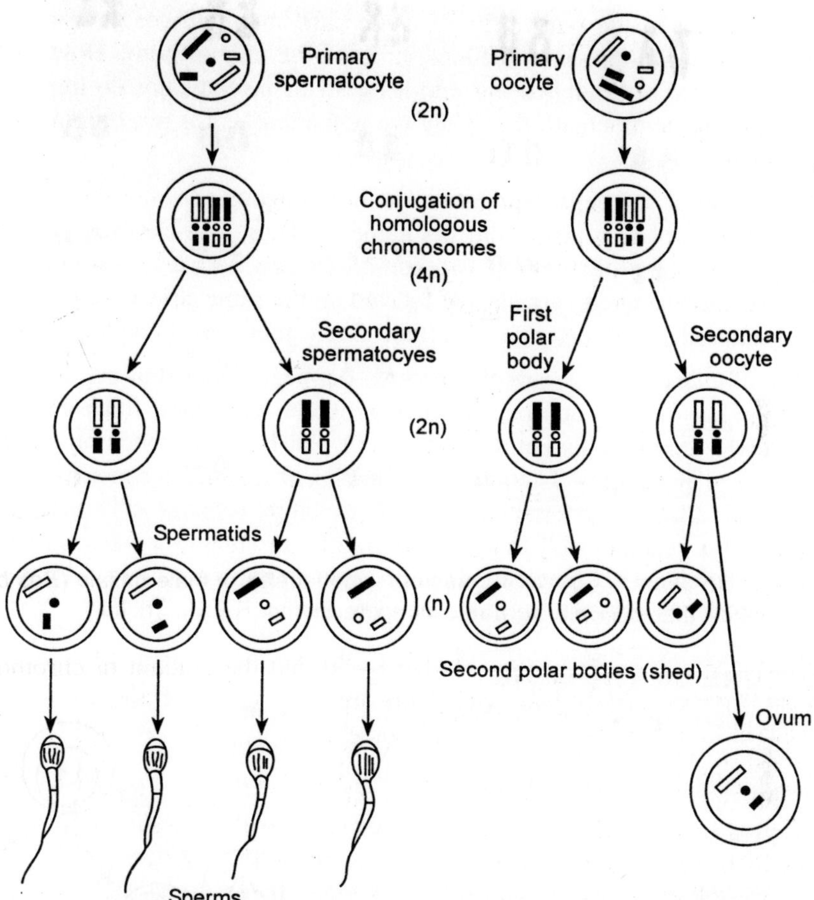

Fig. 14.6. Diagram showing the reduction division or meiosis. This type of cell division takes place in the testis and ovary in the formation of gamete; or sex cells *i.e.* sperms and. ova. At a given stage of development two cell divisions take place in rapid succession. One of these is similar to mitosis (see Fig. 14.5). In the other, chromosomes do not duplicate themselves, thus producing haploid cells-spermatids and ovum.

of fertilisation one cell (with n or haploid number of chromosomes) from the female parent (ovum) unite with one cell from the male parent (sperm), also with n or haploid set of chromosomes, forming a new cell (zygote) with $2n$ or diploid set of chromosomes. Thus, the number characteristic of the species is restored. That is how cells of an individual animal contain half the number of chromosomes (so genes) from its male parent (sire) and the other half from its female parent (dam).

14.4. GENES

The elementary unit of heredity is known as gene. Genes are located on chromosomes. The importance of genes is such that they transmit genetic information from cell to cell, from parent to offspring and take active part in the realisation of this genetic information. Regulation of the activity of cells in relation to the conditions of internal and external environment of the organisms, takes place through the agency of genes. Each chromosome carries a large number of genes. The

60 chromosomes present in the body cells of cattle carry hundreds of thousands of genes that control every stage of development and functioning of the animal body. However, as stated by J. Lush (1961), genes are not independent entities unto themselves nor do they, like 'constants' of physics, act similarly in all situations. "They are potentialities for reacting in a given way with a given environment."

During formation of sex cells, along with the homologous chromosomes the corresponding genes carried on them will also move into the gametes. Thus, like chromosomes the genes also segregate purely by chance probability. If two non-allelic genes (that is—two different gene pairs that influence different characters) are closely located on the same chromosome, in all probability, they will be transmitted to the offspring together. Such genes are known as linked genes.

Every gene of every chromosome has its own duplicate located at the same point or locus, on the second member of the same pair of analogous chromosomes. Such genes located at the same loci of the two homologous chromosomes and that have influence on the same character are called alleles or allelomorphs. The number of alleles of a particular gene can be two or more. In the latter case the alleles might have formed by mutation (change of structure and so action) of the original or wild gene, as well as by a mutation of an already mutated gene. Mutations are caused by various known (like nuclear radiation) and unknown factors. Mutation brings such changes in the genes that become heritable henceforth.

14.5. STRUCTURE OF THE GENE

The gene, as is understood now, is biochemically a part of the molecule of desoxy ribo-nucleic acid (DNA). A molecule of DNA is the fundamental material carrier of genetic information.

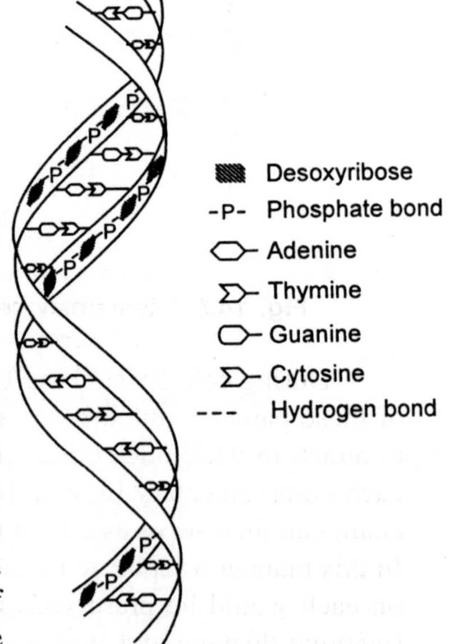

Fig. 14.7. The double helix structure of gene as proposed by Watson and Crick.

DNA is found in the chromosome in association with protein, from which it may be separated by chemical methods. It is made up, according to Watson and Crick, of long chains of nucleotides. Each nucleotide consists of nitrogen containing base (a pyramidine or a purine comparable to an amino acid in size), a pentose (five-carbon) sugar and phosphoric acid. In DNA, the sugar is deoxyribose; the purines are of only two kinds (*adenine* and *guanine*) and pyramidines are also of only two kinds (*thymine* and *cytosine*). One of these bases is attached to a sugar group and the latter are joined together by phosphate linkages as shown in Fig. 14.7. There may be from three to four thousand such units in a DNA molecule. Watson and Crick proposed a model in which such chains may produce copies of themselves. The principal features are as follows : (1) The structural unit of DNA consists of two long chains of nucleotides coiled in a helix around a common axis (Fig. 14.7). (2) The two chains are complementary to each other in such a way that a purine on one is always paired with a pyrimidine on the other, the two being linked by hydrogen bonds. (3) *Adenine* always pairs with *thymine* and *guanine* with *cytosine*.

Lush, J.L. (1961). Additional sources of germ plasm: Large Animals. In: *Germplasm resources*. Radha Krishna Prakashan, Delhi.

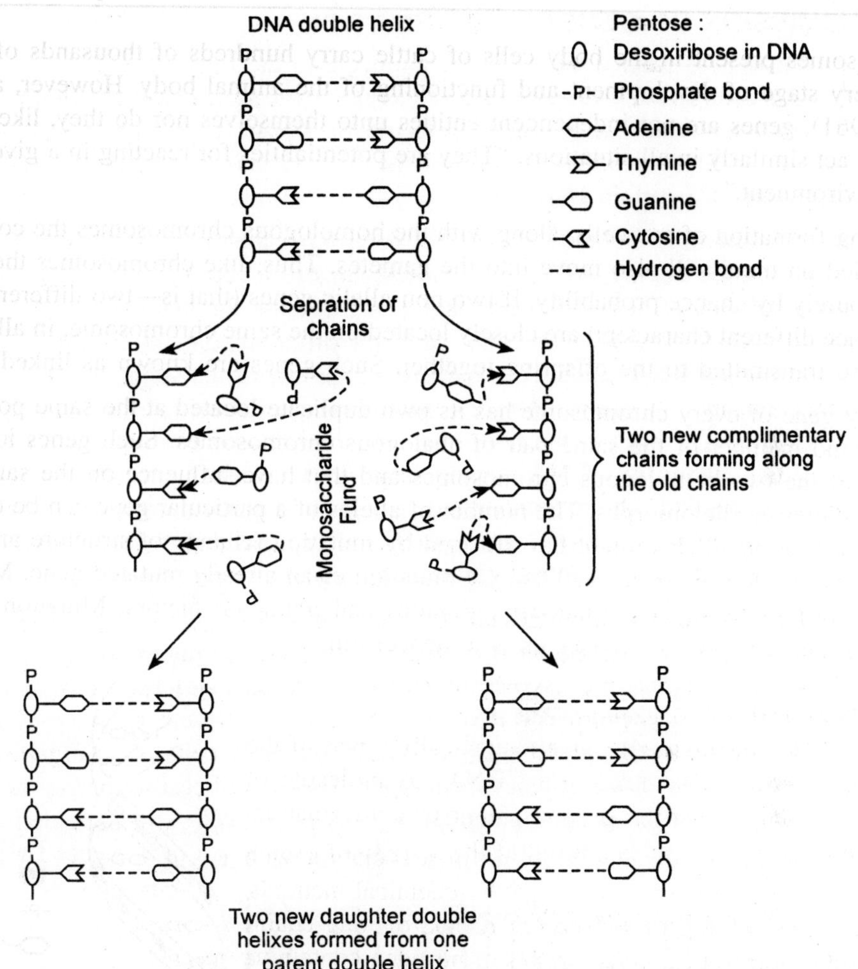

Fig. 14.8. Diagramatic representation of the duplication of the DNA molecule (gene).

During cell division the DNA spiral splits into two identical halves resulting in the splitting of these pairs of amine bases along the hydrogen bonds (Fig. 14.8). Thus, each half will be free to attach to itself nucleotides or their precursors from the environment within the cell to replace each complementary base that was separated from it in the unwinding and bond breaking. Each chain can then serve as a mould or template on which its complement can be built up (Fig. 14.8). In this manner a replicate of each of the complementary chain is produced. The order of the bases on each would be preserved, since each strand, because of the complementary pairing relations (*adenine-thymine* and *guanine-cytosine*) specifies the order on the other strand.

This helps in maintaining the constancy of number as well as identity of chromosomes during mitotic cell division. This results in true and complete transfer of genetic information from parent cells to daughter cells. Without this relationship, similarity of all processes taking place in the organism would not have been possible.

Thus the capability of DNA molecule of 'self-copying' is the essence of transmission of genetic information from cell to cell, from parent to offspring.

14.6. TRANSMISSION OF GENETIC INFORMATION

The genetic information is carried by the DNA molecule in the form of difference in the order of nucleotide base pairings (*adenine-thymine* and *guanine-cytosine*). There are thousands (15,000 on an average) of nucleotides in a single molecule of DNA. A part of the DNA molecule containing 1000 to 1500 nucleotides could well be a gene carrying genetic information that controls development of a certain character.

The genetic information included in the DNA determines the order of amino acids in protein molecules, as every three nucleotides (triplet) corresponds to a certain amino acid. Inclusion of different amino acids in a protein molecule takes place through the medium of ribonucleic acid (RNA). Ribonucleic acid, present in all cells, performs different roles in protein synthesis. The messenger or information RNA (*m*RNA) is synthesised on one of the spirals of DNA (Fig. 14.9) and then diffuses into the cytoplasm. Here the *m*RNA attaches itself to ribosomes (Fig. 14.10). The transfer RNA (*t*RNA) carries to the ribosomes amino acids and arranges them in accordance with the order of triplets on the *m*RNA. Amino acids are the building 'bricks' of proteins. The character of a protein depends on the number and arrangement of amino acids in it. The 20 amino acids, about which we know now, form a great number of proteins that form the basis of all living organisms. The various stages of protein synthesis are shown in Fig. 14.10.

During cell division every daughter cell gets identical sets of genes. The combined essence of the genetic information contained on the genes located on the chromosomes constitute the genotype of that cell. Whereas the essence of all the characters and functions of a cell is its phenotype. In course of succeeding cell divisions, cells with different phenotypes form nerve cells, muscle cells, pigment cells, gland cells, so on. All the body cells of an organism have the same genotype, though they differ in their phenotypes. This is so because during the process of development only a part of the genetic potential of cells is realised. The characteristic features of a cell, that is, its phenotype is a result of the activity of only a part of its genes (Aisner, 1976).

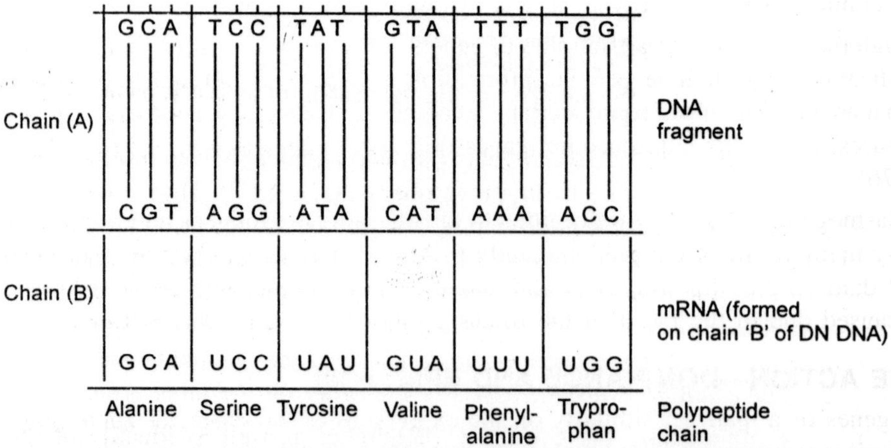

Fig. 14.9. Schematic diagram showing that order of amino acids in a polypeptide chain is determined by the order of the triplets in mRNA, which in turn depends on the order of triplets in the DNA molecules. Note that in RNA molecules Uracil (U) replaces the nucleotid thymine (T). (Diagram redrawn after Taranov, 1977).

Aisner, F.F. (1976). *Exploitation of selection traits in cattle husbandry* (Russian). Kolos, Moscow.

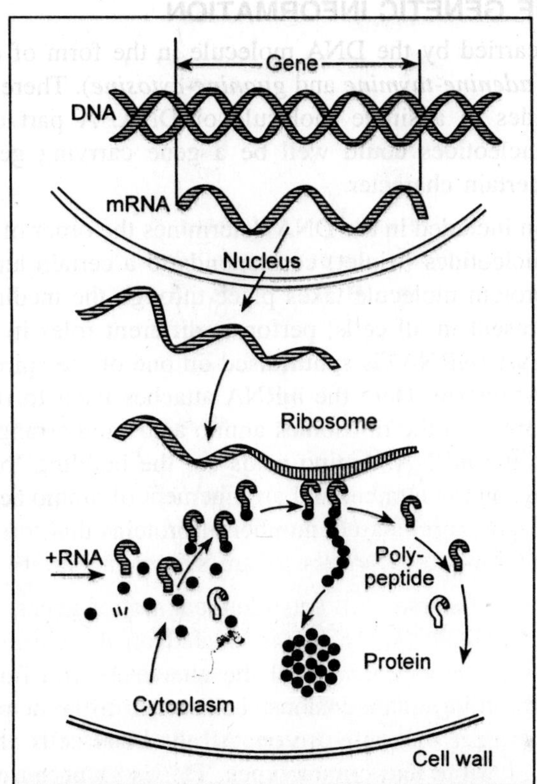

Fig. 14.10. Diagram showing the involvement of the gene in protein synthesis in cells.

Hence, the phenotype of a cell can be termed as the result of realisation of part of the genetic information contained in its genotype; it acts as one united system.

Although the main role in transmission of genetic information is played by the chromosomes, there are indications that definite genetic information is transmitted through protoplasm as well. Structures that are capable of self reproduction, called plasmogenes, have been found in protoplasm. These structures, unlike the chromosomal genes drift into daughter cells in a disorderly fashion (Aisner, 1976).

The plasmogenes are mainly transmitted to the offspring through ovum. Thus, the characters controlled by them are inherited predominantly from the dam. This causes an additional genetic influence of dam on the offspring, over and above the well known phenotypic maternal effects, which are caused due to the fact that the foetus grows in the body of its mother.

14.7. GENE ACTION—DOMINANCE AND EPISTASIS

If both the genes of a ·pair act similarly on the expression of the character controlled by them, such an organism is known as homozygous for that allele or character. In case the two alleles act differently on the character controlled by them the animal is referred to as heterozygous for the given character or gene pair. As a rule one member of a pair of the genes is more active or dominant and suppresses the activity of the other member of the pair—known as recessive gene.

Gene activity of this nature can be explained by the polled or horned conditions in cattle (Fig. 14.11). Polled condition or hornlessness is dominant over horned condition and is influenced

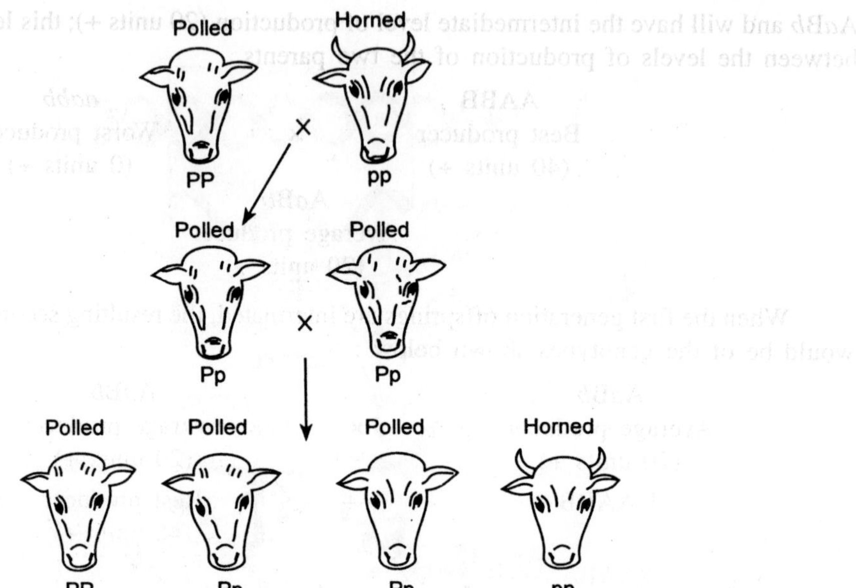

Fig. 14.11. Pattern of inheritance of polled condition in cattle, a character influenced by a single pair of genes.

by a single pair of genes. If this dominant gene can be denoted by 'P', and its recessive counterpart by '*p*', then the genotype or gene make-up of various individuals upto the second generation will be as shown in figure 14.11. Note that homozygous 'PP' and heterozygon 'P*p*' individuals have the same external appearance or phenotype. Such type of characters that are expressed as sharply distinct classes (horned or polled) are known as qualitative characters. Sometimes dominance of one gene over its allelic pair can be incomplete.

Sometimes a pair of genes may mask the presence and expression of another pair of genes. Such a type of gene action is known as epistasis.

14.8. ADDITIVE GENE ACTION

Most of the economic characters of dairy cattle like milk yield, body weight, efficiency of feed conversion etc. are inherited in a different manner. The expression of such characters range from one extreme to the other, with no clear cut and definable natural classes. Such characters, known as quantitative or continuous characters, are due to many genes (each having a small effect) rather than due to one or two genes (each with a distinct large effect) as in the case of qualitative or discontinuous characters.

The genes that control quantitative characters are known as multiple factors or multiple genes. The expression of quantitative characters is due to additive nature of the action of multiple genes. Multiple genes, generally, do not show complete dominance.

This type of gene action can be illustrated by the following hypothetical example. Let us presume that one member of the gene pair 'A' or 'B' each cause increase in production by 10 units, and genes '*a*' and '*b*' do not cause such an increase in production. Then the best producing animal will be AABB (40 units +) and the worst producer— *aabb* (0 units +). When the best animal is mated to the worst animal, the first generation offspring (F_1) will all be having genotype

A*a*B*b* and will have the intermediate level of production (20 units +); this level being intermediate between the levels of production of the two parents.

AABB		*aabb*
Best producer	×	Worst producer
(40 units +)		(0 units +)
	A*a*B*b*	
	Average producer	
	(20 units +)	

When the first generation offsprings are intermated, the resulting second generation offsprings would be of the genotypes shown below :

A*a*B*b*		A*a*B*b*
Average producer	×	Average producer
(20 units +)		(20 units +)
1 AABB	—	Best producer = 1 (40 units +)
2AAB*b* 2A*a*BB	—	Best producer = 4 (30 units +)
1 AA*bb*) 4 A*a*B*b* 1 *aa*BB)	—	Average producers = 6 (20 units +)
2 A*abb* 2 *aa*B*b*)	— —	Poor producers = 4 (10 units +)
1 *aabb*	—	Worst producer = 1 (0 units +)

Though actually nine different groups of genotypes are possible, these can be grouped into five production groups (best, fair, average, poor and worst) according to the number of "good" and "bad" genes. The continuous nature of variation in production of second generation offsprings is shown in Figure 14.12.

It should be kept in mind that to simplify the illustration only two pairs of genes were considered. But actually economic characters of dairy cattle are controlled by a large (unknown) number of genes. The number of possible genotypes that can occur when a large number of pairs of genes arc involved are infinite. The continuity and overlapping between phenotypic classes is more gradual with such a large number of genes, so much so that their distribution fits a normal curve (see Section 15.6)

Frequencies of different levels of milk production in 350 lactations of 70 buffaloes belonging to the herd of Haryana Agricultural University are plotted on a graph in Fig. 14.13. The typical bell shaped normal curve resulted. If sufficiently large number of observations are plotted in the same fashion the curve will be smoother and more symmetrical.

14.9. HOW GENES FUNCTION

We have said in the previous section that all the body cells of an individual contain a similar component of all the genes the individual possesses. Thus, the genetic make-up or the genotype

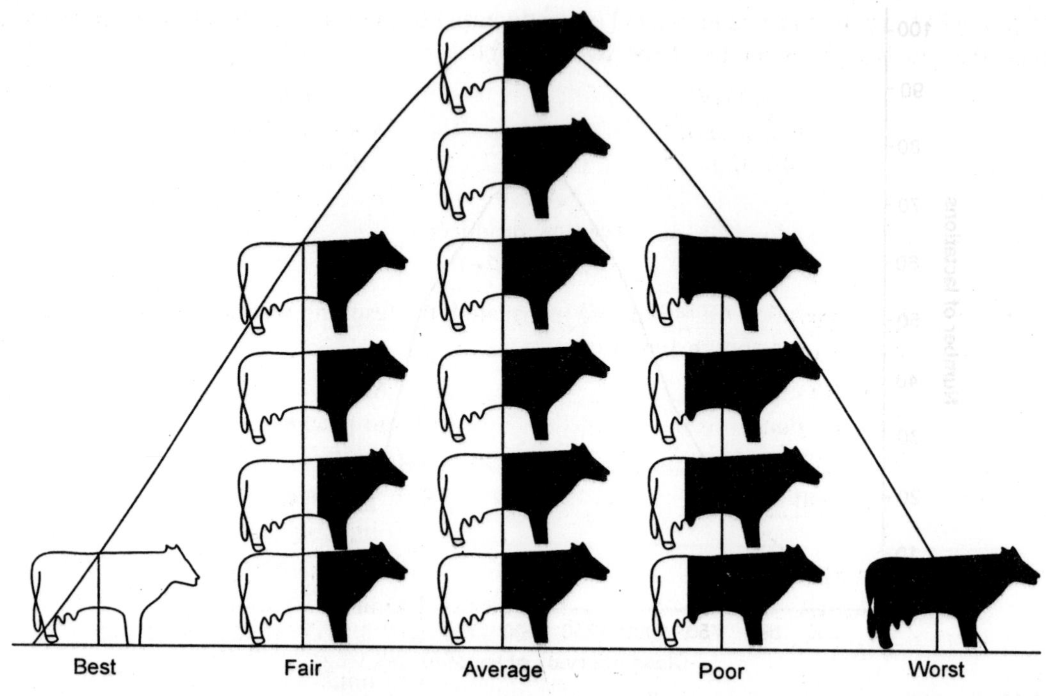

Fig. 14.12. Chart showing the continuous nature of increase in milk production of cows from worst to the best with no clear distinctions between succeeding groups.

of all the cells is the same. However, all the genes may not be active at a given stage of development of the individual. Further, the complete genetic information contained in the cell may not be expressed.

According to Jacob and Monod (1961), whose theory is most widely accepted in microorganisms, there are two types of genes, *viz.*, the ***structural gene*** which synthesises various proteins and ***regulator gene***, which does not directly synthesise protein but actively controls the activity of the structural gene. The ***regulator gene***, which may be located on a different chromosome as the structural gene produces a protenous repressor substance that prevents the structural gene from synthesising *m*RNA and thus, from protein synthesis. There is some evidence to indicate that in larger animals too the genes function in an identical manner. However, what triggers the regulator gene in controlling the protein synthesis by the structural gene is not yet clearly established.

Genes synthesise enzymes (proteins) that participate in thousands of biochemical reactions taking place in the animal body. Most of the biochemical reactions of animal body are of consecutive or chain type. Each compound is acted upon by its specific enzyme resulting in the formation of a second compound. The latter is acted upon by its own specific enzyme resulting in a third compound. The reaction continues in this fashion. Absence of an enzyme causes disruption in the chain of reactions resulting in the non-formation of all the compounds beyond the stage at which the enzyme is missing.

Jacob, F. and Monod, J. (1961). *J. Molecular Biol.* **3** : 318.

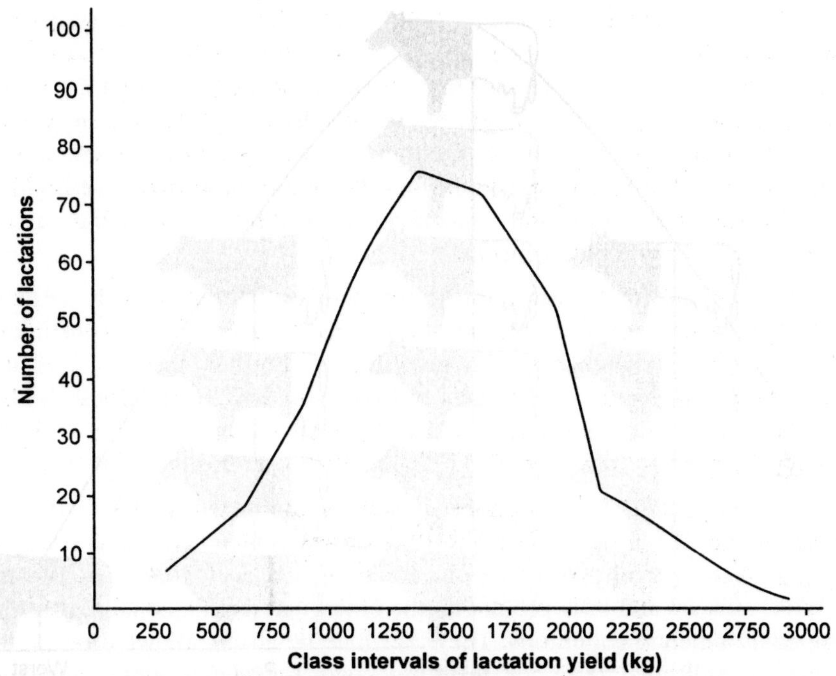

Fig. 14.13. Levels of lactation yields and their frequencies in Murrah buffaloes were plotted in this graph. Note the resulting bell-shaped normal curve.

Our knowledge of functioning of genes through the media of enzymes produced by them is mostly obtained by study of various hereditary metabolic disorders in humans and animals. Most of the lethal effects are also considered to be due to non or disfunction or lack of enzyme systems due to mutation of the concerned gene.

One of the most common and widespread metabolic disorders known in animals is albinism (or partial albinism). The condition is caused by total absence or insufficiency of enzyme tyrosinase that participates in the formation of melanin pigment from the amino acid tyrosine.

14.10. GENES AND METABOLISM

In literature pertaining to dairy cattle breeding, we often find expressions like heritability of milk yield, fat content in milk and so on. As a matter of fact, such expressions are not entirely correct because it is not milk yield or fat content that is inherited, but the units of genetic information (that cause the development and expression of these characters), that are inherited.

Every animal develops out of a fertilized ovum—zygote, by a process of successive cell divisions. The diverse processes that take place in the animal body are determined by genes. By controlling protein synthesis, genes control growth of cells, leading to the formation of different tissues and organs. On tissues and organs depend the nature and speed of metabolism taking place in the organism.

And the productivity of animals—growth performance, milk production, etc., is the direct end product of the metabolic processes (digestion, assimilation and utilisation of nutrients) taking place in the body. An animal with superior metabolism, thus will be a superior producer. Evidence in support of this can be seen in literature.

The author's (Sastry and Kravtchenko, 1980) study showed that calves in whose blood levels of haemoglobin, haematocrit, urea nitrogen, total serum proteins and activity of aspartic amino-transpherase were slightly higher at 9 months of age turned out to be the fastest growers. Such calves attained the highest body weight by 18 months of age. Calves in whose blood the levels of the said 5 metabolites were either too low or too high than average were inferior as regards their growth rate. Further, it was found out that these parameters are highly repeatable and heritable at the ages mentioned, thus indicating a possibility of selecting calves for better growth potential based on their blood profile.

Workers from Institute for Research on Animal Disease, Compton, U.K. (Rowlands *et al.*, 1974 and Payne, 1973) have reported that concentrations of certain blood constituents are both heritable and have strong relationship with growth rate. Further, there is evidence to show individuality of blood constituents which tends to persist for long times. These findings suggest that not only calves might be screened early in life for indications of rapid growth rate, but also that animals might be bred to have the blood characteristics indicative of this quality.

Bandarenko *et al.* (1974) from a series of studies conducted at the Institute of General Genetics of the Academy of Science of U.S.S.R. have observed that there exists strong relationship between levels of glucose, total protein and its fractions and acetone bodies on one hand and production of milk, milk fat and milk protein on the other, that remain consistent for long periods under changing environmental conditions. They concluded that this allows utilisation of some of these parameters as additional tests for selecting cows for higher yields and superior milk composition. Taranov (1976) and his associates, at All Union Institute of Animal Science in the Soviet Union have obtained considerable evidence to prove the gene-metabolism-production axis in farm animals in a series of studies conducted over the past 36 years.

The realisation of the genotype into phenotype takes place as a result of gradual unfolding of the genetic information during the process of ontogenesis—growth and development of the animal. However, the action of the genes at different stages of growth is not the same.

14.11. REALISATION OF GENOTYPE

Given below is a modified version of the scheme suggested by Soviet geneticist and breeder, F.F. Aisner, that explains the process of realisation of genetic information in farm animals (Fig. 14.14). This needs some further explanation.

The growth period of an animal can be divided into two phases—morphogenetic and post-morphogenetic (Fig. 14.14). The first phase coincides with the prenatal or embryonal growth. During this period, even by the first few divisions of the zygote, form rudiments of various tissues and organs of the embryo. It is during this stage that embryonal cells get differentiated.

The embryo which grows in the uterus, is protected to a considerable extent against external influences. This is the reason why the genetically determined growth programme is more or less

Sastry, N.S.R. and Kravtchenko, N.A. (1980). *Indian J. Anim. Gen., Breed.* **2** : 21.

Rowlands, G.J., Payne, J.M., Sally M. Dew, Manston, R. (1974). *J. Agric Sci. Comb.* **82** : 473.

Payne, J.M. (1973). *British Cattle Breeder's Club* **28** : 55.

Bandarenko, G.A., Varonova, V.N., Guseva, N.T. and Suchova, E.C. (1974). Peculiarities of intermediary metabolism of cows with different levels of milk production, fat and protein contents in milk. In : *Investigations into genetics, immunogenetics and selection of farm animals.* (Russian). Navuka Publisbers, Moscow.

Taranov, M.T. (1976). *Biochemistry and productivity of cattle* (Russian). Kolos Publishers, Moscow.

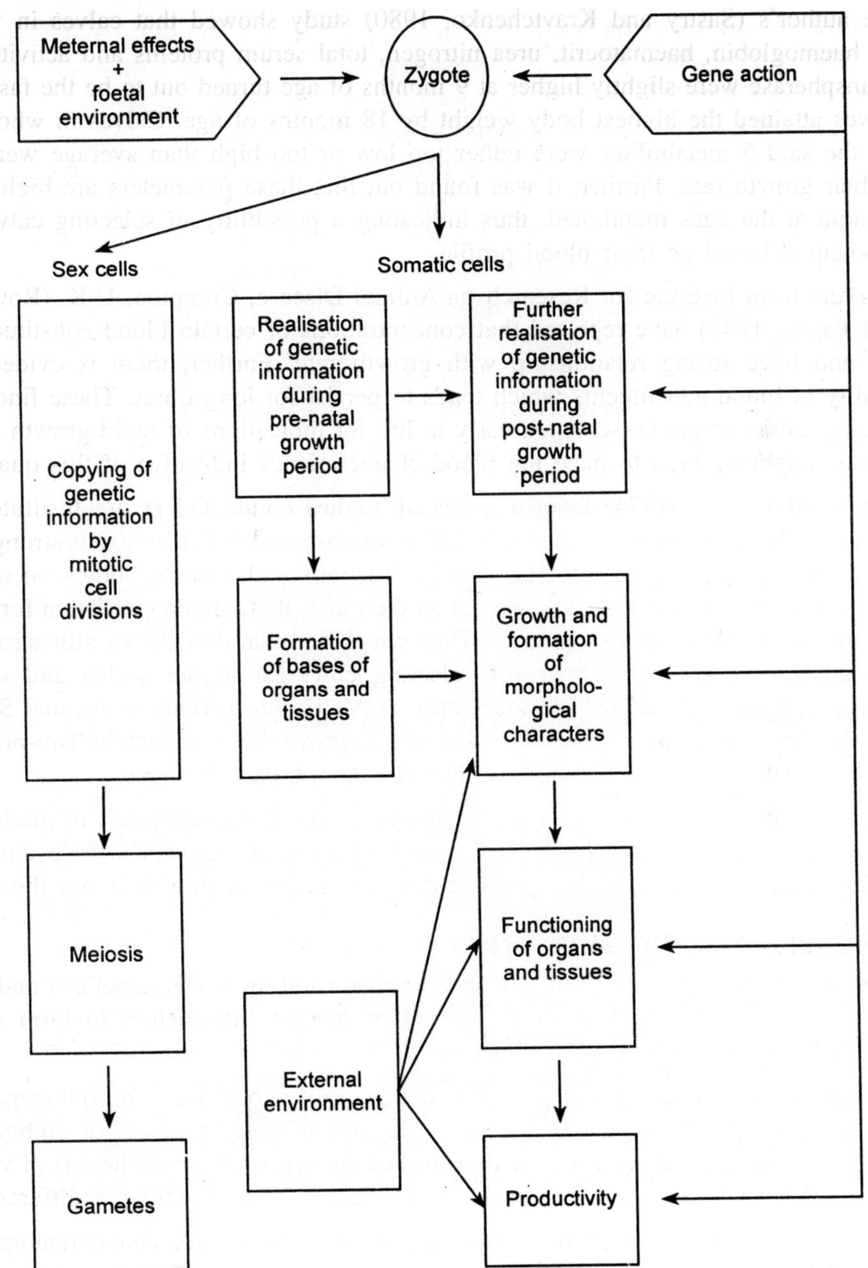

Fig. 14.14. Schematic diagram showing the mode of realisation of genotype of animals during the period of growth and afterwards (Redrawn from F.F, Eisner. 1976).

clearly realised. However, the maternal organism too, by controlling the embryo's nourishment, influences the growth of the embryo to some extent.

Towards the end of the morphogenetic period of growth the animal already acquires characteristic features of the adult animal. Formation of the basic features of metabolism takes place at this stage.

Diversity of characters formed during the prenatal growth period is much less dependent on external factors as compared to that of postnatal period. After birth, influençe of external factors increases considerably. Not only the external factors influence the development of different characters at each stage of growth, but they also influence realisation of genetic information in the succeeding stages. As a matter of fact, formation of various characters at different stages of growth is made possible only by virtue of interaction with the environment of genetic information already realised in the preceding stage. Thus, result of interactions with and influences of external factors at succeeding stages of growth somewhat get accumulated in the phenotype of the animal.

Most of the economic characters of dairy cattle and other farm animals are result of realisation of several bits (types) of genetic information at various stages of growth. That is so because each economic character depends on functioning of more than one organ and system that attain functional status at different stages of growth of the animal. Milk production alone demands the active participation of the udder, liver, digestive system, circulatory system, endocrine system, and so on. It should be further remembered that productivity depends on the genetic make-up of the animal, which acts via determining the intensity and nature of metabolism as well as other life activities of the animal.

Thus, in productivity, an animal that was reared under poor feeding conditions during earlier stages of growth lags behind another animal reared under proper conditions, even if both the animals have the same genotype. The reverse is also true. The best genetic information, if not realised completely in time (due to unfavourable environment) may not allow development of the full production potential of the animal, even if ideal conditions of feeding and management are provided at a later stage.

From the foregoing discussion one point becomes apparent. By evaluating the phenotype of an animal we cannot know its exact genotype. In other words phenotype reflects just one version (not necessarily the best) of genotype. Hence, phenotype evaluation of animals does not always give us precise idea of their genetically controlled production potential.

The genetic worth of an animal is best estimated by evaluating the quality of its progeny. It is necessary here to remember that each parent gives to the next generation only half of its chromosomes. Hence, the genetic information transmitted to the offspring is never a complete reflection on the genotype of its parents. Transmission of genetic information from generation to generation is also dependent on the transformations (segregation, crossing over, linkage, mutation etc.) that it undergoes during this process. These transformations in their turn cause genetic variability of characters. However, until more precise methods become available, we have to relay on the study of production characters of animals as a means of evaluating their genotype.

CHAPTER

15

Population Genetics

Population genetics deals with large number of accurate facts that are connected with the study of variability, heritability and repeatability of characters in a population of animals. A population is a large group of animals living in the same general area and breeding freely among themselves for several generations, as a result of which a definite homogeneity of their genotype is established.

The study of genotypes of dairy cattle is a complex process because of small number of offspring per animal, long interval between generations, differences in the environment in which animals of different generations live and produce, etc. At present the study of inheritance of quantitative characteristics of dairy cattle is based principally on three methods : 1. Experimental study of twins, especially monozygotic twins; 2. Experimental crossing of animals of different groups of breeds having diverse types of inheritance and expression of traits; and 3. Analysis of masses of data on performance of cows undergoing different types of breeding in different herds. The method last mentioned involves analysis of populations using the principles of biometry (see Section 15.6) for estimation of genetic parameters of populations, namely—variation, heritability, repeatability and inter-relationships between traits.

15.1. GENES IN POPULATIONS

Every population has its own genes that constitute the population's gene fund. The peculiarities of distribution of genes and their combinations among the individuals within the population give rise to the latter's genetic structure.

The breeder is actually dealing with a large number of genes among animals and populations of animals in which he wants to bring about genetic improvement. Appraising action of individual genes seems to be an impossible task under such circumstances. Fortunately, mathematical and statistical tools have come to our rescue to characterise and give a rational explanation to the dynamics of such multi-gene actions in populations.

The degree to which a particular gene is abundant or rare in a population as compared to other alleles is referred to as gene-frequency. The frequency of genes in population varies from zero to one.

The Hardy-Weinberg law states that in large randomly mating populations, if (1) the frequency of one of the two alleles is equal to "*a*", (2) the frequency of the other is equal to "*b*" and (3) the sum of the frequencies "*a* + *b*" equals one; the offsprings of the three genotypes will occur in a definite ratio, or will be in equilibrium in the next generation at the frequencies of a^2 $2ab$ and b^2. If "P" and "*p*" are the two allelic genes at a locus, then, after one generation of random mating, the frequency of occurrence of the genotypes will be :

Number of homozygous PP individuals = a^2
Number of heterozygous P*p* individuals = $2ab$
Number of homozygous *pp* individuals = b^2

One of the major prerequisites for the operations of Hardy-Weinberg law is that the population should be under random mating. Random mating in a population for a certain character means that no discrimination is shown to any genotype for mating and that matings take place at random, irrespective of the genotype to which the individuals belong.

In a large random mating closed population, there is very little likelihood of change in gene-frequencies. The gene-frequencies in a population can be altered by the following reasons :

1. Random genetic drift is possible in the case of small populations in which purely due to reasons of chance, gene-frequencies may change.
2. *Migrations :* When animals from outside are allowed to mate with animals in a population, the frequency of genes are likely to be disturbed and altered. A good example can be crossbreeding of zebu cattle with European breeds, by which frequency of genes for milk production are increased in the former.
3. *Mutations,* in which new alleles are produced which differ from the original genes in their effects on expression of the particular characters is another factor, which will alter gene-frequency.
4. *Selection* of animals in which certain favoured animals are allowed to reproduce more than certain others also alter gene-frequency. Here individuals of certain genotype are retained for breeding purpose in larger numbers than others.

However, while analysing the genetic structure of any population we deal neither with genes nor with genotypes directly. We study the characters controlled by them. One of the basic parameters used in studying selection traits in population is variation. Variation of characters is dependent on the differences in its expression by each of the individuals of the population.

15.2. VARIATION

The kaleidoscope of Nature is so resourceful in ever evolving newer patterns of life, so that, not even two blades of grass, no two pigs, no two cows, nor two men (except identical twins) ever lived, living or will live have been or will be alike. Thus, variation can be observed in all the quantitative traits of animals like height, weight, growth rate, feed efficiency, milk, wool, meat and egg production, speed, stamina for work, quality of meat and wool, etc., disease resistance and heat tolerance. Variation refers to the measurable difference of a trait between individuals. It is the raw material with which the breeder has to work as we are not yet equipped to make selection at the gene level. "Variation is at once the hope and despair of the breeder". If there were no variation, there would be no scope for selection as two individuals would look and perform exactly alike. Thus, variation gives us hope that progress, by genetic improvements, is possible through selection. It is the despair because an improved animal may be carrying genes for inferior production and mediocrity and, therefore in subsequent regroupings of genes in the reproductive process, may produce mediocre and inferior animals.

An idea about the extent of variation in the production traits among various breeds and genetic groups can be had from the discussions on this aspect presented in Chapter 16.

When we analyse variation, we find it has three essential components, namely, hereditary, environmental and interaction between the two.

The total phenotypic variation can be partitioned as shown below :

$$\sigma_P^2 = \sigma_G^2 + \sigma_E^2 + \sigma_{GE}^2$$

where σ_G^2 = Genetic variation, σ_E^2 = Variation due to environment and σ_{GE}^2 = variation due to interaction between genotype and environment.

15.2.1. Genetic Variation

The hereditary portion of the phenotypic variation *i.e.,* hereditary or genetic variation is due to difference in the genetic make up between individuals, symbolised as σ_G^2.

Genetic variation of characters is due to influence of different types of gene action on the character. The genetic variation thus, can be further partitioned into additive and epistatic as well as variation caused by dominance and superdominance.

Now we can write down the various components of genetic variation of characters as follows

$$\sigma_G^2 = (\sigma_a^2 + \sigma_e^2 + \sigma_d^2)$$

Genetic variation due to non-additive factors has rather a limited place in dairy cattle breeding : such as the 'heterosis' effect obtained in crossbreeding (*see* under crossbreeding). However, the main basis for selection of dairy cattle is additive genetic variation.

There are two forms of genetic variability : (*a*) those caused by mutation, and (*b*) those caused by recombination of genes. Mutation of chromosome include loss of a part of a chromosome, equal or unequal exchanges of broken parts between chromosomes—crossing over. Gene mutations are caused by changes in the chemical structures of one or more genes. There are known and unknown factors that cause mutations. Some of the identified causes include—radiation-activity, certain chemicals, intoxication caused by certain pathogenic agents etc.

Mutation can take place in somatic as well as sex cells. Mutations occurring in the somatic cells of animals are expressed in them individually, but are not transmitted to the next generation. Consequently, such a variability is not genetically transmissible.

Mutations occurring in the gamets of parents are however, inherited. As the animal carrying a mutant gene grows, there occurs a deviation in the normal synthesis of proteins. This causes changes in the structure of organs, tissue and systems and so changes in metabolism. If these deviations are considerable the animal will generally be unable to live and dies in early stages of growth. Almost all the mutations are harmful and mutations that can be selected for their utility are very rare.

Of greater importance to selection of farm animal is the variability caused by recombinations of chromosomes and genes as a result of which genetic information of parents undergoes redistribution and recombination among the offsprings.

Recombination of genetic information takes place during sexual reproduction—during, gamete formation, to be more precise. Gametes or sex cells as mentioned earlier, form as a result of a special type of cell division called meiosis (*see* Fig. 14.6). Contrary to mitosis, meiosis comprises of two successive divisions. As a result of this, after doubling of the set of chromosomes (4*n*) their number is reduced by four times until it becomes '*n*'. During meiosis, homologous chromosomes, out of which one was received from female parent and the other from the male, represented in black and white in Fig. 14.6, undergo segregation and recombination forming different combinations. The number of such possible combinations are equal to 2^n, where n is the haploid number of chromosomes. In cattle this is 2^{30} *i.e.*, 1073741724. This number is more than the existing number of heads of cattle in the whole world at present.

A chromosomal aberration known as crossing-over may take place during meiosis. This comprises of mutual exchange of parts between two homologous chromosomes during conjugation *i.e.* when they are arrayed close and parallel to each other. As a result of crossing over one set of homologous chromosomes of parents form a pair that is identical to parental chromosomes and another pair of crossed-over chromosomes. Crossing-over can take place independently in several chromosomes simultaneously during spermatogenesis as well as oogenesis.

During meiosis, apart from exchange of entire pieces of chromosomes along with the genes carried on them, exchange can take place between different parts of gene itself. An extreme case being exchange of different nucleotides of a gene. Such exchanges are called internal recombination which is another source of variability in the genetic material.

Thus, meiosis is directed at increasing genetic variation, while mitosis prevents changes or rather maintains copies of the same genetic material in all somatic cells. As a result of meiosis genetic diversity increases, thanks to the recombination of maternal and paternal chromosomes and their reorganisations.

The afore-mentioned processes cause great variability of gametes and thus scope for natural selection even at the gamete level. A far greater selection pressure is on the male gametes—spermatozoids. Out of millions and millions of sperms just one fertilizes the ovum. The process of spermatogenesis itself is fraught with great diversity—each primary spermatocyte give off 4 different spermatozoa, while in oogenesis one primary oocyte forms just one ovum (Fig. 14.6).

Thus, even during gametogenesis, genetic variability is created, which continues in later stages after formation of zygote. As can be seen, the animal selector is concerned with only that part of genetic diversity (material) obtaining in the individual animals, after they undergo a rather strict natural selection upto fertilization. It is this part of genetic diversity (material) among individual animals in a population that determine, along with the environmental influence, the phenotypic variation of characters.

15.2.2. Environmental Variation

Superimposed on genetic variance is the variation caused by the environment in which the individual grows up. This is known as environmental variation and designated by σ_E^2. This results from the influences of maternal environment, nutrition, management climate, disease incidence, etc.

On variability of qualitative characters environmental conditions do not have any significant influence. Hence, generally the phenotypic variation of such characters reflects truly the degree of diversity among individuals that is due to gene action. The heritability of such characters is very high, often nearing unity.

In case of all economic characters, that are quantitative in nature, the relation between genetic and phenotypic variation is entirely different. Quantitative characters are determined by a large number of genes and the environmental conditions have a significant and, at times, deciding influence on them.

The genetic variation is expressed only as the environmental conditions permit its expression. However, the environment does not directly change the hereditary make-up of an individual. Animals of high genetic potential if reared under poor conditions may be stunted and not develop to the scope of their inherent capacities, but these animals are still capable of producing offsprings of good genetic potential. Transmission of genes from parent to offspring are not affected by the poor growth and stunted nature of the parent.

15.2.3. Genetic × Environment Variation

Even though environment does not directly influence the germplasm, it most certainly has a marked influence on life and causes gradual change in germplasm. Such environmental influences perpetuate "natural selection" among plants and animals causing establishment of individuals in environments most suited to them. Farm animals of a certain genotype may perform much better in a certain environment than they do in another, where other genotypes may do well.

The best possible inheritance will not give rise to a superior dairy herd unless proper environment (feeding, management and disease control) is provided, so that the cows can attain the limits determined by their inheritance. Also, the best possible environment will not make superior cows out of animals having poor genetic production potential. The limit to animal performance is set by the animal's own heredity, and the best possible environment will not make the individual exceed its own genetic potential. There are probably physiological limitations set by inheritance.

Whenever breeds of *Bos taurus* cattle were imported into tropical zones a marked reduction in their performance has been observed. None of the breeds of *Bos indicus* cattle attained the level of performance of developed dairy breeds in tropics in spite of providing the best environment.

Data from India show that when the proportion of inheritance from European breeds reaches 3/4 or 7/8 level in crosses, their performance is lower than that for crosses with 1/2 or 5/8 exotic breeds inheritance. Pure European breeds have not performed as well as 1/2 or 5/8 crosses in several experiments.

The breeder should always consider interplay between heredity and environment. Even though environmental variation is not transmitted, the breeder must strive to provide a favourable environment, which will permit the attainment of the cows' inherent potentialities.

The extent of interaction between genotype and environment is difficult to determine in case of dairy cattle because of the complicating fluctuations in climate, feeding and managemental practices. Undoubtedly, the genotype-environment interactions are important from economics point of view. In addition, it becomes obvious because of these interactions, that not only cows should be selected for suitability to the environment in which they are to be kept, but also that cows should be kept in a kind of environment that allows them to produce according to their genetic potential.

15.3. HERITABILITY OF TRAITS

J. Lush, the great American livestock breeder and father of large animal population genetics, first coined the term "heritability", represented by the symbol—h^2. He (Lush, 1945) pointed out that heritability estimates concern with differences between individuals or groups of individuals and not with their absolute values. Thus, heritability refers to that portion of the phenotypic variance in a population for a particular trait that is due to heredity.

What does it mean, for example, if the coefficient of heritability of milk yield of cows is 0.3 ? It means, that if a group of cows were selected, whose average milk yield is higher than the average for the herd to which they belong by 800 kg, then their daughters can reasonabiy be expected to be superior to their own (daughters) herdmates not by 800 kg but by 240 kg (800 × 0.3 = 240). This expectation is based on heredity. The rest of the superiority over and above this may be the effect of environment.

Lush further pointed out that heritability can be used in either a broader or a narrower sense. Heritability in a broader sense is that portion of the phenotypic variation that is due to all genetic factors—

$$h^2 \text{ in a broader sense} = \frac{\sigma_G^2}{\sigma_P^2} = \frac{\sigma_A^2 + \sigma_D^2 + \sigma_E^2}{\sigma_A^2 + \sigma_D^2 + \sigma_E^2 + \sigma_{En}^2}$$

where σ_D^2 is genetic variation comprising of variance due to additive gene action (σ_A^2), dominance (σ_D^2) and epistasis (σ_E^2), σ_P^2 is phenotypic variance and σ_{En}^2 is environmental variation.

Whereas heritability in a narrower sense is that portion of phenotypic variation that is due to additive gene action, h^2 in a narrower sense $= \frac{\sigma_A}{\sigma_P}$.

This is roughly the same as the percentage of genetic progress made in succeeding generations when superior animals are selected as parents. Almost all the methods of computing h^2 (*see* Section 15.6) include only a little more than additive portion of genetic variation, though this can vary depending on the method of computation.

Coefficient of heritability is generally calculated based on the degree of likeness or resemblance or similarity for traits under question, between related individuals (parent-offspring or sibs) in comparison with non-related individuals. The heritability is expressed either as a percentage (0 to 100 per cent) or as a fraction of unity (0 to 1). A negative value can be due to extremely low genetic variance and/or high sampling variation of the trait.

Genetic differences among animals in respect of a particular characteristic depend on a number of factors (previous selection, method of breeding—inbreeding, out-breeding, crossbreeding etc.) and these are different in different herds. Hence, coefficient of heritability of a characteristic is different for each herd of animals. Apart from this, the degree of genetic diversity in the herds can vary from time to time and so the h^2 tells about the genetic situation of the herds only at the time of determination of the coefficient.

Also, factors such as breed, herd, season, husbandry status and other environmental factors can influence the magnitude of the h^2 estimate by influencing the denominator of the formula,

---

Lush, J.L. (1945). *Animal breeding plans.* Iowa State College Press, Ames, Iowa.

namely phenotypic variation. Method of computation of the h^2 estimate itself can cause variation in its magnitude. This is actually the reason for the considerable diversity in reported values of h^2 of the same trait, but on different farms, breeds, locations and periods (Table 15.1).

Hence coefficient of heritability should not be taken as a static but a dynamic parameter. It is a specific parameter of the genetic diversity for a specific character among animals within a specific herd, at a specific time and place, and under specific conditions. Thus, while planning selection programme in a particular herd, it is appropriate to use the h^2 estimated for that herd only.

Some traits are more heritable than others. Heritability of lactation milk yield varies from 0.20 to 0.50 in most of the herds with a mean value of 0.37 (Table 15.1). This means that 37 per cent of the differences between selected animals and herd average (this estimate is known as *selection differential*) could be due to genetic factors.

The reproduction traits, on the other hand, have very low heritability (Table 15.2 and 15.3). This means that most of the variation for such traits is due to environmental factors. Improvement in such traits can by far be achieved by improving the nutrition and management of stock. Actually there are reports indicating that heritability of some components of calving interval. namely, service period and dry period is very close to zero.

TABLE 15.1. Heritability values (h^2) of Lactation Yield and Length of Dairy Bovines

Breed	*Number of reports considered*	*Lactation yield*			*Lactation length*		
		Range of reported values		*Probable average value*	*Range of reported values*		*Probable average value*
		Lowest	*Highest*		*Lowest*	*Highest*	
Red Sindhi	5	0.32	0.37	0.34	0.04	0.43	0.27
Sahiwal	10	0.15	0.58	0.37	0.15	0.67	0.33
Tharparkar	11	0.15	0.63	0.37	0.08	0.43	0.29
Gir	3	0.20	0.67	0.40			
Hariana	17	0.04	0.58	0.33	0.22	0.61	0.28
Murrah buffalo	4	0.04	0.33	0.29	0.07	0.16	0.11
Zebu × *B. tilurus* (different grades)	9	0.02	0.76	0.34	0.08	0.17	0.31
Probable overall average :				0.37			0.30
Probable overall range :				0.20 to 0.54			0.10 to 0.50

Note. The data have been compiled from too large a number of reports to be listed here. The probable average values were arrived at by crude computations, but with due consideration to the number of animals purported to have been studied by the authors.

Low h^2 values of fertility traits among cows should be considered minutely. It may be due to the fact that fertility has always been under strong control of natural selection and in this process individuals with poor fertility are constantly eliminated from the population, which situation leads to greater stability of genotype for this characteristic. This reflects on the coefficient for heritability.

Growth traits of dairy bovines are moderate to highly heritable (Table 15.4), which gives sufficient scope for improvement of such traits through selective breeding. This is important since a higher rate of growth means an earlier age at first calving.

TABLE 15.2. Heritability Values of Age at First Calving and Calving Interval of Dairy Bovines

Breed	*Number of reports considered*	*Age at first calving*			*Calving interval*		
		Range of reported values		*Probable average value*	*Range of reported values*		*Probable average value*
		Lowest	*Highest*		*Lowest*	*Highest*	
Red Sindhi	5	–0.09	1.39	0.16	–6.08	0.88	0.01
Sahiwal	5	0.14	0.56	0.27	–0.45	0.09	–0.07
Tharparkar	7	–0.44	0.48	0.07	–0.31	0.27	–0.06
Gir	3	–1.24	0.13	–0.32	–0.37	0.12	–0.05
Hariana	9	–0.31	0.54	0.18	0.09	0.43	0.27
Murrah buffalo	8	–0.02	0.77	0.27	–1.40	0.40	–0.09
Crossbreds (different grades)	4	0.13	0.63	0.31	.0.00	0.29	0.16
Probable overall average				0.16			0.03
Probable overall range				–0.20 to 0.46			–0.24 to 0.32

Note. Same as for Table 15.1.

TABLE 15.3. A Compilation of Several Literature Reports on Heritability and Repeatability Values of Reproduction Traits in Large Cattle Herds (after Maijala, 1976)

Reproduction trait	*Heritability (%)*	*Repeatability (%)*
Length of calving interval	1.7	13.1
Number of services per conception	3.2	7.6
Conception/calving rate to 1st service	5.1	4.8
Conception/calving rate to 2nd service	0.4	3.1
Days from calving to first oestrus	2.3	21.8
Regularity of heat periods	11.0	18.0
Strength of heat	21.0	29.0
Length of heat period		28.4
Final conception rate to AI	2.7	
Pathology in reproductive trait	5.0-13.5	

Source : Maijala, K. (1976). *World Rev. Anim.* 12 : 69.

When the heritability of a particular trait is high, the correlation between an individual's genotype and phenotype is likely to be high and selection based on individual's own phenotype is also likely to be effective. It also indicates that the greater part of variation is due to additive gene action and hence mating best to best (as regards their own performance) should result in production of the desired progeny.

On the other hand, low heritability values indicate—(*i*) that correlation between genotype and phenotype of individual is low; (*ii*) that an individual found superior by its own phenotype may not produce desirable offspring; (*iii*) that for surer progress by selection greater attention should be paid to the phenotype of collateral relatives and progeny of the individual under consideration; and (*iv*) that the additive gene action may be of low importance and non-additive gene action (dominance, super-dominance and epistasis) may be important.

TABLE 15.4. Reported values of coefficient of heritability of some growth parameters of dairy animals

Parameters	*Baffalo (B. bubalis)*	*Sahiwal (B. indicus)*	*Crossbred (F_1) (B. taurus × B. indicus)*
Weight at birth	0.74, 0.21	0.23, 0.21, 0.39	0.61, 0.60
Weight at 6 months	0.43	0.73, 0.61	0.62, 0.57
Weight at 12 months	0.72	0.70, 1.05	0.66, 1.05
Weight at 24 months	0.43	0.79, 0.28	0.85, 0.57
Weight at first calving	0.23	0.18	0.21
Authors	Nagarcenkar (1978) and Johri and Bhat (1978)	Taneja and Bhat (1978a) and Taneja *et al.* (1979)	Taneja and Bhat (1970b) and Taneja *et al.* (1978).

Heritability estimates are used in animal breeding work for a variety of purposes—in construction of selection indices of bulls; for determining genetic progress obtained through selection in herds; to find out in which herds and under what circumstances information on pedigree and individuality can be or cannot be used in determining the breeding worth of animals; for analysis of reciprocal crossings (crossing males of one breed to females of another breed and *vice-versa*); comparison of monozygotic twins with dizygotic ones and full-sibs (both parents common) with half-sibs (one parent common); so on.

15.4. REPEATABILITY OF TRAITS

Repeatability is the form of realisation of genetic information within one generation or group of animals at different ages or periods or conditions of environment. In other words, repeatability estimates refer to the expression of the same trait, such as milk production, at different times in the life of the same individuals. There is no new segregation or independent assortment of genes during the life time of animals, but the same genotype is expressed differently at different stages of life.

Another definition of repeatability estimate is that it is the fraction of differences or variation between single records of the same individuals that are likely to occur in future records of the same individuals. This portion of variance includes additive, epistatic and dominance genetic components, as the gene or gene combinations do not change when they influence the successive expressions of the trait in the same individuals, though it may also include a portion of the same permanent environmental influences. Thus, repeatability gives the upper limit of heritability and thus can be as large as coefficient of heritability in its broader sense :

Johri, D.S. and Bhat, P.N. (1978). *Indian J. Anim. Sci.* **40** : 79.

Nagarcenkar, R. (1978). Riverine buffaloes of India and possibilities of genetic improvement *vis-a-vis* cattle. *FAO Animal Production and Health Paper No. 13.* FAO Rome.

Taneja, V.K. and Bhat, P.N. (1970a). *Indian J. Anim. Sci.* **41** : 897.

Taneja, V.K. and Bhat, P.N. (1970b). *Indian J. Anim. Prod.* **1** : 64.

Taneja, V.K., Bhat, P.N. and Gars, R.C. (1978). *Indian J. Dairy Sci.* **31** : 191.

Taneja, V.K., Basu, S.B., Bhatnager, D.S. and Rao, V.P. 1979. *Indian J. Dairy Sci.* **32** : 326.

$$\frac{\sigma_A^2 + \sigma_D^2 + \sigma_E^2 + \sigma_{Enp}^2}{\sigma_A^2 + \sigma_D^2 + \sigma_E^2 + \sigma_{Enp}^2 + \sigma_{Ent}^2}$$

where, σ_{Enp}^2 is that part of phenotypic variation that is due to permanent environmental effects and σ_{Ent}^2 is variation due to temporary environmental effects ($\sigma_{Enp}^2 + \sigma_{Ent}^2 = \sigma_{En}^2$ = environmental variation). Note that σ_{Enp}^2 (variation due to permanent environmental effects) is the only additional value in the numerator as compared to heritability in broader sense (*see* Section 15.3).

Repeatability is estimated by finding out correlation between levels of one and the same trait of the same animals at different periods or ages (body weight at different ages, milk yield in succeeding lactations, etc.) or by analysis of variance method (*see* Section 15.6). The repeatability estimates can vary from 0 to 1.

Lactation milk yield is moderate to highly repeatable (0.48), the differences in this respect between different breeds being small (Table 15.5). The reproduction traits have generally low repeatability (Table 15.3 and 15.6).

TABLE 15.5. Repeatability Values of Lactation Milk Yield of Diary Animals

Breeds	*Numbers of reports considered*	*Probable average value*
Red Sindhi	4	0.50
Sahiwal	3	0.54
Tharparkar	3	0.48
Gir	3	0.55
Hariana	4	0.60
Murrah buffalo	9	0.33
Other buffaloes	6	0.52
Probable overall average :	0.48	
Probable overall range :	0.34 to 0.62	

Note. Same as for table 15.1.

TABLE 15.6. Repeatability Values of Calving Interval and Its components in Dairy Animals

	Lactation length	*Service period*	*Dry period*	*Calving interval*
Cows (6) :				
Probable average :	0.28	0.10	0.21	0.29
Probable range :	0.15 to 0.41		(Single reports)	0.09 to 0.49
Buffaloes (13)				
Probable average :	0.29			0.11
Probable range :	0.15 to 0.43			0.05 to 0.17

Note : Figures in brackets are number of reports considered.
Source : Same as for Table 15.1.

Coefficient of repeatability determined by calculating coefficient of correlation between successive expressions of the same trait depends to a great extent on the stability of external factors in which the animals are located in succeeding years. The more stable the environment is, the higher will be the coefficient of repeatability. It was also established that the level of repeatability of one and the same character varies from period to period. Low repeatability indicated considerable changes in the external factors *i.e.,* feeding and managemental condition on the farms which has a direct bearing on the productivity. Hence, it is very necessary to reduce the variation in feeding and managemental conditions to the minimum, if a higher effectiveness of selection programmes is desired. By controlling the feeding and managemental conditions the accuracy of evaluation of cows can be increased considerably.

A knowledge of repeatability estimates for various traits may be used in selection for future performance. When repeatability of a trait is high, culling on the basis of the first records should be effective in improving the overall performance of the herd in the next year. In addition, offspring from superior individuals in the herd can be given preference when selection is made for replacement stock.

Repeatability estimates also give an indication of how many records should be obtained on an individual before it can be culled from the herd. Repeatability value can be used for selection of the same individuals for future production and for computing the most probable producing ability of an individual in its life time. Such computed values of life time performance are of value in selecting for traits with low repeatability estimates.

15.5. CORRELATIONS BETWEEN SELECTION TRAITS

Most of the selection traits of dairy bovines are inter-related. The nature of their inter-relationships are diverse and are dependent on the condition of development of the population. Relationship between production characteristics can be of straight line type or curvilinear type; the latter tendency cannot be completely expressed by coefficient of correlation.

Charles Darwin wrote that *"with changes in one part of the organism consequent to selection (natural or artificial), other parts of the organism will also change."* This fact is always considered by the selectors either directly or indirectly. The importance of correlation analysis (*see* Section 15.6) is of much help in carrying out selection simultaneously for several important characters of animals. It is well known, however, that dividing selection pressure among the several characters reduces the rate of genetic progress of the animals. Hence it is necessary to know the direction and degree of relationship between different characters while selection is carried out for one of them. In case the correlation is positive selection can be carried out for this one character. If two important characters are closely and negatively correlated selection will be complicated. It will be necessary to carry out selection for simultaneous improvement of all characteristics. This causes change not only in the degree of the development of the character but also on nature and level of relationship between them.

Lactation milk production and factors influencing the same are of great importance to dairy producers and breeders alike. Garcha and Tiwana (1981) reported that, in Murrah buffaloes, while phenotypic correlations between lactation yield and lactation length and between these two traits and persistency and peak yield were positive and considerable (0.12 to 0.58), such correlation between persistency and peak yield were not (Table 15.7). More or less similar findings (Table 15.8) were reported in case of Jersey × Hariana crossbreds (Koley *et al.*, 1979) and in other breeds (Dhoke and Johar, 1977; Deshpande and Singh, 1977 and others).

However, genetic correlation between the same traits reported by Koley *et al.* (1979) were not significant. The relationship between milk yield and fat content are negative.

Garcha, D.S. and Tiwana, M.S. (1981). *Indian J. Anim. Sci.* **51** : 103.
Koley, N., Chaudhary, G. and Mitra, D.K. (1979). *Indian J. Dairy Sci.* **33** : 301.
Dhoke, M.K. and Johar, K.S. (1977). *Indian vet. J.* **54** : 587.
Deshpande, K.S. and Singh B.P. (1977). *Indian vet. J.* **54** : 814.

Some reproduction traits like dry period, service period, calving interval, breeding efficiency have quite diverse relationships with lactation performance in cattle (Brijendra Singh *et al.,* 1980) and buffaloes (Bhat, 1978a).

Reproductive efficiency has either low or negative phenotypic and genetic correlations with lactation parameters (Tables 15.9 and 15.10). The correlations between service period and dry period and between these two parameters and calving intervals are positive (Table 15.9) as the former determine the magnitude of the latter (Bhat, 1978a).

Thus, longer dry period, service period and calving interval are somewhat associated with higher milk production and *vice-versa* (Table 15.8 and 15.9). Production and reproduction parameters of first lactation are also genetically associated with life time performance (Table 15.10). Correlations given in this table indicate that early first calvers have more number of calvings and higher life time milk production. Heavier animals at first calving (better growers) too have longer herd life and produce more milk in their life time than lighter animals. Higher production and reproduction traits in first lactation seem to be maintained even during the life time (Sahota and Gill, 1980).

Growth parameters—body weights and daily weight gains at various ages are moderate to highly correlated between themselves (Table 15.11 and 15.12.) both phenotypically and genetically (Daya Singh *et al.*, 1968., Bhat, 1978b; and others).

TABLE 15.7. Phenotypic Correlation among Persistence, Peak Yield, Days to Reach Peak Yield, Lactation Length, and Lactation Yield (Garcha and Tiwana, 1981)

Trait	*Peak yield*	*Days to reach peak yield*	*Lactation length*	*Lactation yield*
Persistency	–0.038 ± 0.033	0.177 ± 0.035	0.353 ± 0.032	0.432 ± 0.030
Peak yield		0.032 ± 0.033	0.119 ± 0.033	0.579 ± 0.026
Days to reach peak yield			0.173 ± 0.033	0.156 ± 0.033
Lactation length				0.534 ± 0.028

Garch and Tiwana (1981). *loc. cit.*

TABLE 15.8. Correlations of Persistency Index with Production and Reproduction Traits in Jersey-Hariana crossbred (F_2) Cows (Koley *et al.*, 1979)

Traits	*Genetic Correlation*	*Phenotypic Correlation*
Total yield	0.258	0.157
300-day yield	0.102	0.131
Peak yield	0.034	– 0.005
Lactation period	0.823	0.115
Dry period	0.503	0.026
Service period	1.134	0.130
Calving interval	1.066	0.131

Brijendra Singh, Bhat P.N. and Satish Kumar (1980). *Indian, J. Anim. Sci.* **50** : 915.

Bhat, P.N. (1978a). Genetic parameters of milk production and scope of increasing milk production in buffaloes *vis-a-vis* cattle. *FAO Animol Production and Health paper No. 13.* FAO, Rome. p. 129.

Daya Singh, Acharya, R.M. and Sunderesan, D. (1968). *J. Res.* (PAU). **5** : 555.

Bhat, P.N. (1978b). Cross breeding in hilly and high rainfall areas. Paper presented at National Seminar on *Progress in Research on Animal Genetics and Breeding.* NDRI, Karnal, 12-15 June, 1978.

TABLE 15.9. Genetic and Phenotypic Correlations among Some First Lactation Production and Reproduction Traits in Sahiwal Cattle

Trait	*Breeding efficeincy*	*Weight at first calving*	*Age at first calving*	*First service period*	*First dry period*	*First calving interval*	*First lactation length*	*First lactation milk yield*	*Milk per day of calving interval*
Breeding efficiency		0.41	0.15	–0.64	–0.52	–0.96	–0.34	0.22	0.36
Weight at first calving	0.03		–0.18	–0.26	–0.01	–0.08	–0.15	–0.14	0.09
Age at first calving	–0.04	0.25		0.21	0.27	0.30	0.26	–0.43	0.37
First service period	–0.44	–0.09	–0.01		0.42	0.89	–0.71	–0.27	–0.24
First dry period	–0.36	–0.08	–0.01	0.77		0.46	–1.10	–0.68	–0.67
First calving interval	–0.47	0.07	0.01	0.96	0.78		–0.81	–0.48	–0.39
First lactation length	–0.23	–0.02	0.01	0.38	–0.17	0.41		0.80	0.93
First lactation milk yield	–0.21	0.10	0.01	0.24	–0.16	0.25	0.68		–1.03
Milk yield per day of calving interval	0.05	0.17	0.01	–0.30	–0.60	–0.30	0.41	0.79	

Note. 1. Values above the diagonal are genetic correlations and values .below the diagonal are phenotypic correlations.

2. According to the author all the values falling between 0.08 and 0.14 are statistically significant at 5 per cent and values above 0.14 at 1 per cent levels of probability.

Source : Brijendra Singh *et al.* (1980).

TABLE 15.10. Genetic Correlations between First lactation and Life Time Production and Reproduction Traits of Friesian × Sahiwal crossbreds

First lactation traits	*Life time production and reproduction traits*						
	Productive life	*Longevity*	*No. of calvings*	*Total life time milk*	*Milk per day of productive life*	*Milk per day of life in the herd*	*Per cent days in milk*
Age at first calving	0.287	0.629	–0.739	–0.268	–0.130	0.146	>1
Weight at first calving	–0.133	–0.032	0.294	0.237	–0.241	0.093	0.540
First lactation milk yield	–0.444	–0.410	–0.283	– >1	0.446	0.249	>1
305 days milk yield	0.146	0.223	0.102	–0.121	0.341	0.158	0.892
Lactation length	0.400	0.294	0.447	–0.094	0.287	0.500	>1
Dry period	0.650	0.676	0.439	0.044	–0.629	0.012	–0.562
Service period	0.306	0.251	0.248	–0.212	–0.111	0.223	–0.061
Milk/day of first lactation	–0.253	–0.230	–0.225	–0.289	0.029	–0.093	0.102
Milk/day of first calving interval	0.340	0.433	0.146	0.158	0.240	0.144	0.884

Source : Sahota and Gill (1980). *loc. cit.*

TABLE 15.11. Phenotypic and Genetic Correlation between Various Traits of Growth and Production in Sahiwal Cattle

Traits	*Traits*								
	1	*2*	*3*	*4*	*5*	*6*	*7*	*8*	*9*
1. Birth weight		0.38**	0.14**	0.13**	0.18**	0.10**	–0.09	0.00	0.01
2. Weight at 8 weeks	0.62**		0.6 *	0.43**	0.36**	0.14**	–0.16**	–0.03	0.02
3. Weight at 26 weeks	0.45**	0.87**		0.74**	0.52**	0.17**	–0.21**	–0.04	0.00
4. Weight at 52 weeks	0.48**	0.72**	0.86**		0.65**	0.24**	–0.17**	0.02	0.00
5. Weight at 24 months	0.51**	0.64**	0.78**	>1		0.28**	–0.22**	0.03	0.00
6. Weight at 1st calving	>1	0.17	0.22	0.46**	0.40		0.16**	0.18**	0.00
7. Age at 1st calving	–0.04	–0.74**	–0.69**	–0.62**	–0.50**	–0.03		0.13**	0.10*
8. 1st lactation yield	0.33	–0.32	–0.38*	–0.05	–0.00	0.17	0.38		0.66**
9. 1st lactation length	0.54*	–0.09	0.11	0.20	0.26	0.20	0.46	0.74**	

*Significant at 5 per cent level. **Significant at 1 per cent level.
Upper diagonal indicates phenotypic correlations and lower diagonal indicates genetic correlations.
Source : Taneja *et al.* (1978). *loc. cit.*

TABLE 15.12. Phenotypic and Genetic Correlations among Various Economic Traits of Indian Buffaloes

Traits	*Birth weight*	*Weight at one year*	*Daily weight gain during first year*	*Age at first calving*	*First calving interval*	*First lactation milk yield*	*First lactation length*
Birth weight		0.11**	0.01	0.03	0.03	0.03	0.04
Weight at one year	0.25**		0.98**	–0.21**	–0.08	0.06	0.03
Daily weight gain during 1st year	0.17	0.92**		–0.20**	0.08	–0.05**	–0.04**
Age at lint calving	0.28	–0.21	–0.16		–0.00	0.08*	–0.02
First calving interval	0.45	0.67**	0.49	>1		0.36**	0.59**
First lactation milk yield	0.37	0.24	0.12	0.41*	0.23		0.63**
First lactation length	0.76**	0.30*	0.32	0.33**	0.12	0.46	

Note. 1. * P< 0.05 and ** P< 0.01.
2. Above diagonal are phenotypic correlations and below diagonal are genetic correlations.
Source : Bhat (1978) *loc. cit.*

While growth rate is naturally negatively related to age at first calving, the latter is positively related (Table 15.12) to first lactation production (Singh *et al.*, 1968, Brijendra Singh *et al.*, 1980; Bhat, 1978; Sikka and Taneja, 1979), though some authors questioned the actual practical beneficial effect of increasing age at first calving on first lactation yield (notably Chawla and Mishra, 1978).

Daya Singh *et al.* (1968). *loc. cit.*
Brijendra Singh *et al.* (1980). *loc. cit.*
Bhat (1978). *loc. cit.*
Sahiwal crossbreds in northern India. *Proc. National Seminar on Crossbreeding.* 27-29, Apr. 1978, NDDB, Anand. Chawla, D.S. and Mishra, R.R.(1978). Relative performance of grades of Brown Swiss × Sahiwal and Friesian.

The essence of all the reports on the topic suggests that if age at first calving is reduced by selection, the first lactation yield is also likely to be affected adversely. However, earlier age at first calving increases the total life time production and so the net effect of reducing age at first calving is still advantageous.

15.6. BIOMETRICAL METHODS USED IN DAIRY BOVINE BREEDING

Biometry means the application of statistical methods for analysing group characteristics in biology. Biometry aims at analysing the diversity between and within groups or populations of animals or plants.

15.6.1. Sampling

The biometrical methods are based on the theory of probability. Let us presume that there is a population of 100,000 dairy cows in a region. This whole population is the total entity. There are no conceivable means of studying all the individuals of such a large group of animals. Hence, for a study of the general nature of this population or nature of certain traits in this population, a part of this population, *i.e.,* a smaller group out of this population is selected for closer study. This act is known as sampling. The investigator studies the nature of the selected sample group of cows and then he generalises the results so obtained from the study of the sample to that of the whole population.

The sample should be so selected that it reflects the nature of the population effectively. If sampling is done for study of say, milk yield, the sample must contain animals of different production levels proportionate to their number in the population. There should be cows with high, medium as well as low yields in the sample and not just high yielders alone or low yielders alone.

Such an unbiased sample can be drawn only on the principle of probability. Samples drawn from population with no other criterion except chance or probability are known to have been drawn randomly. Randomisation of drawing samples can be best done by referring to Random Numbers. Most books on statistics (*see* Snedecor and Cochron, 1976) give Random Number Tables which can be made use of by the dairy cattle breeders.

Random Number Tables contain a large number of figures arranged in 1 to 5 digit columns. The order of occurrence of any given figure in each digit is determined purely by chance.

After obtaining a sample the next step will be to estimate statistical values that tell about the nature of the sample as regards a particular character. All the calculated statistical values are likely to have some error due to the nature of distribution of sample means in relation to that of the population. Hence, generally standard errors are also calculated for each statistical value. The standard error gives an idea about the accuracy with which the sample mean represents the population mean.

Described below are some common methods of estimating various statistical values that reflect the nature of characteristics in populations. Only some of such methods of computations that are commonly used in programmes of selection and breeding of dairy cattle are described. It should be kept in mind that there could be other methods of estimations as well.

15.6.2. Estimation of Mean—($\overline{X}$)

The fundamental statistical value that tells about the average level of expression of a given quantitative character is the 'mean', which is generally represented by the symbol '$\overline{X}$', whereas 'μ' is the population mean. Means can be of three types but the most common one used in breeding work is the arithmetic mean.

The direct method of calculating the arithmetic mean comprises of the summing up of all the variants and then dividing the sum by the number of variants (observations) summed up.

$$\overline{X} = \frac{x_1 + x_2 + x_3 + \ldots + x_n}{n} = \frac{\Sigma x}{n} \qquad \ldots(1)$$

where x_1, ...x_n are the values of different variants, n is the number of variants and Σ is the symbol used to indicate summation.

Let us work out an example. The milk yields in 50 randomly selected lactations of Murrah buffaloes belonging to the HAU farm are given in Table 15.13. We calculate the arithmetic mean from this data.

TABLE 15.13. Milk Yield in 50 Randomly Selected Lactations of Murrah Buffaloes

1413	1890	1399	1320	691
1903	1350	1745	931	589
1606	1389	1925	863	1776
1620	1042	1563	652	1627
1015	1486	1911	2133	1566
1270	1103	751	1695	1613
2070	669	1177	1540	1270
1692	1017	1546	1670	740
1423	1235	1374	1622	734
1610	1218	1326	835	1194
				Total 67791

The first step is to sum up all the 50 observations, *i.e.*, 1413 + 1903 + 1606 + ... + 740 + 734 + 1194. By so summing up we obtain 67791. The second step is to divide this sum by 50, which is the number of observations.

$$\frac{67791}{50} = 1355.8 \text{ or } 1356.$$

Thus we obtain that the average lactation milk yield of Murrah buffaloes in this particular herd is 1356 kg.

15.6.3. Weighted Mean $(\overline{X})_w$

If one estimates means from a number of samples with different sizes, one can also estimate an average of all these sample means. If all the means are summed up and the sum then divided by the number of means we can get an estimate of the average values. But it is likely to be an inaccurate value as each sample mean is calculated from samples of different sizes.

To overcome this ambiguity a weighted mean can be calculated. The principle underlying behind this is to give proper weightage to each sample mean according to respective sample size and then calculating an average of all the means. The formula for doing so is—

$$\overline{X}_w = \frac{\overline{X}_1 n_1 + \overline{X}_2 n_2 + \ldots + \overline{X}_i n_i}{n_1 + n_2 + \ldots + n_i} = \frac{\Sigma(\overline{X}_i n_i)}{\Sigma n_i} \qquad \ldots(2)$$

where ($\overline{X}_w$) is the weighted mean, $\overline{X}_1$, $\overline{X}_2$ etc. are means from 1st, 2nd etc. samples, and n_1, n_2, n_i weights (sizes) of 1st, 2nd, ... ith samples.

The arithmetic means and sizes of 5 samples drawn for estimation of lactation milk yield of Murrah buffaloes are as under :

		($\overline{X}$)	n
Sample 1	:	1340 kg	50
Sample 2	:	1420 kg	25
Sample 3	:	1560 kg	20
Sample 4	:	1250 kg	30
Sample 5	:	1410 kg	40

The weighted mean according to formula (2) is calculated below :

$$\frac{(1340 \times 50) + (1420 \times 25) + (1560 \times 20) + (1250 \times 30) + (1410 \times 40)}{50 + 25 + 20 + 30 + 40} = \frac{22760}{165}$$

$$= 1379.3 \text{ kg.}$$

Arithmetic mean of a sample tells about the mean value of the population with only a certain degree of reliability. In other words it has an error in expressing the population mean. This error, known as standard error of the mean (represented by the symbol 'm'), can be calculated by the following formula :

$$\text{Standard error of arithmetic mean} = m\overline{X} = \frac{\sigma}{\sqrt{n}} \qquad ...(3)$$

where σ is standard deviation about which we will know shortly. The error is always expressed as more or less of the mean as shown below :

$$\overline{X} \pm m$$

In our example the σ calculated as described in section 15.64 is 410.63. Then the standard error of mean is

$$m = \frac{410.63}{\sqrt{50}} = 58.08$$

We can now express the mean lactation milk yield and standard error of the same as : 1355.8 ± 58.08

15.6.4. Standard Deviation and Variance

It is very important for the dairy cattle breeder embarking on a selection programme first to know the degree of diversity among the animals in his herd with regards to each selection trait. A good measure of such diversity can be obtained by calculating standard deviation, represented by the symbol 'σ', which considers the deviation of each of the variants from the arithmetic mean. The square of standard deviation is variance.

$$\text{Standard deviation} = \sigma \text{ and variance} = \sigma^2.$$

In small samples the σ is generally calculated according to the formula :

$$\sigma = \sqrt{\frac{\Sigma(X - \overline{X})^2}{n - 1}} \qquad \ldots(4)$$

that is, squaring the deviation of each variant (X) from mean ($\overline{X}$) irrespective of their sign, dividing them by the number of variants (n) minus 1 and taking the square root of the resulting figure.

On electronic calculator σ is calculated by a simpler method according to the formula :

$$\sigma = \sqrt{\frac{\Sigma X^2 - \frac{(\Sigma X)^2}{n}}{n - 1}} \qquad \ldots(5)$$

As an example let us calculate the σ from the lactation milk yield data given in Table 15.13.

$$\Sigma X^2 = (1413)^2 + (1903)^2 + \ldots + (734)^2 + (1194)^2 = 100174657$$

$$\Sigma X = 67791$$

$$\frac{(\Sigma X)^2}{n} = 91912394 \text{ and } n = 50$$

Then

$$\sigma = \sqrt{\frac{100174657 - 91912394}{49}} = \sqrt{\frac{8262263}{49}}$$

$$= \sqrt{168618} = 410.63$$

The standard error of σ is calculated according to the formula—

$$m\ \sigma = \sigma/\sqrt{2n}$$

The standard error of σ in the above example will then be equal to—

$$m\ \sigma = \frac{410.63}{\sqrt{2 \times 50}} = 41.06$$

Now we can write $\sigma \pm m = 410.63 \pm 41.06$.

15.6.5. Normal Distribution

While talking about the additive gene action in section 14.8 it was mentioned that when expression of a given trait, say milk yield, by the individual animal is measured, we find that the variable (milk yield) so varies that there are no specific number of individuals with different levels of measurement. There is no clear cut demarcation in production levels. But they vary continuously, one level merging into the adjoining one. Such variables are said to be normally distributed. Most of the biological measurements can be considered to be normally distributed. This fact greatly aids and enables us to describe populations of biological measurements.

The characteristic feature of a normal distribution (Fig. 15.1) is that the various individuals tend to group around the mid point or the average value, thinning out gradually towards both

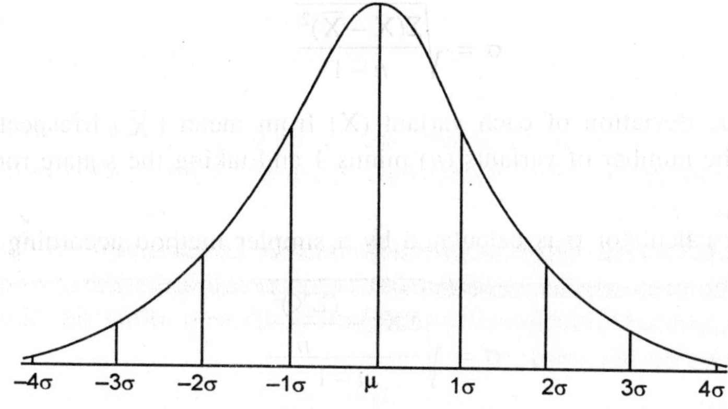

Fig. 15.1. A normal curve.

extremes as regards the level of production of any given trait. Animals with extreme high values and extreme low values are rare. This type of distribution is represented by a bell-shaped curve (Fig. 15.1). The height of the curve at a particular point represents the frequency with which occur individuals having that particular level of expression of a given trait.

Within the limits of -1σ and $+1\sigma$ on either side of the 'μ' (*see* Fig. 15.1) fall 68.26 per cent of the individuals, within the limits of -2σ and $+2\sigma$ fall 95.46 per cent individuals, and within the bounds of -3σ and $+3\sigma$ fall 99.7 per cent individuals. Thus mean and standard deviation together describe effectively the nature and extent of variation in any given population.

15.6.6. Coefficient of Variation

The coefficient of variation (represented by the symbol 'CV') is the fraction or percentage that the standard deviation is of the mean.

$$CV = \sigma/\mu \times 100 \qquad ...(6)$$

Coefficient of variation can be used to compare variations of two unrelated groups or two different traits. If CV of trait A is higher than that of trait B, it means trait A is more variable than trait B. Also if the CV of say, milk yield in herd I is higher than in herd II, it means there is a greater variation among cows of herd I as regards their milk production than among cows of herd II. In our example of lactation milk yield of Murrah buffaloes the CV of their lactation milk yield is—

$$\frac{410.63}{1355.8} \times 100 = 30.29 \text{ per cent}$$

The standard error of coefficient of variation is equal to

$$m\text{CV} = \frac{\text{CV}}{\sqrt{2 \times n}} = \frac{30.29}{\sqrt{2 \times 50}} = 3.03$$

Then the coefficient of variation = CV = 30.29 ± 3.03 per cent.

15.6.7. Coefficient of Correlation

The coefficient of correlation is referred to as *r* and it gives a measure of how two variables tend to vary together. When an increase or decrease in one variable (independent) is accompanied by an increase or decrease in the other variable (dependent), they are said to be positively correlated. They are said to be negatively correlated when increase or decrease in the independent variable is accompanied by a decrease or increase in the dependent variable. The *r* can vary from 0 to 1 in either direction *i.e.,* ± 1.

An estimate of the coefficient of correlation is used in population genetics for knowing the inter-relationship between selection traits (phenotypic correlations between traits), estimation of repeatability of traits etc.

Actually, *r* does not tell which of the two variables is dependent on the other. This cause and effect relationship between the variables has to be determined from knowledge regarding the nature of the two variables. Coefficient of correlation tells only about the direction and intensity of relationship between the variables.

Whether a particular estimate of *r* is significant or not can be ascertained by referring to tables (*see* any standard book on statistics like *Snedecor's*) in which the levels of significance at different levels of probability are given. The degree of significance of *r* depends on the size or the magnitude of the coefficient of correlation and the number of individual pairs of items used to calculate it. Coefficient of correlation is generally calculated by the formula given below :

$$r = \frac{\Sigma XY - \frac{(\Sigma X)(\Sigma Y)}{n}}{\sqrt{\Sigma X^2 - \frac{(\Sigma X)^2}{n} \cdot \frac{(\Sigma X)(\Sigma Y)^2}{n}}} \quad \ldots(7)$$

where *r* = coefficient of correlation; X = variables of one character; Y = variables of the second character, *n* = number of items for both the characters (pairs), and Σ = symbol for summation.

The values in the denominator are nothing but variances of characters X and Y [(see formula (5)], while the value in the numerator is the covariance of traits X and Y.

Let us work out an example.

TABLE 15.14. First Lactation Yields of 35 Russian Bestuzevskaya Cows and their daughters (Gofman-Kodashnikov and Larsteva, 1975)

Milk yield of dams (X)	*Milk yield of daughters (Y)*	*Milk yield of dams (X)*	*Milk yield of daughters (Y)*	*Milk yield of dams (X)*	*Milk yield of daughters (Y)*
3899	3725	3203	4918	3372	4475
3035	5422	4207	5211	5352	6037
3692	4701	4288	4336	3158	4064
3295	4622	3281	3486	3149	5363
2822	5123	3158	3483	2793	4183
5678	5143	4064	3764	3050	5288
3472	3475	2991	3158	5503	4500
3484	3401	2991	4394	5286	5046
4228	3864	3989	5678	4578	5109
5685	5315	3152	4227	5486	4050
3216	3919	4711	5000	4346	3447
4469	3208	4164	5286		

Gofman-Kodashnikov, P.B. and Larsteva, S.K. (1975). *Practical exercises in genetics* (Russian). Kolos Publishers, Moscow.

Given above (Table 15.14) are data on first lactation yield of 35 Russian Bestuzevskaya cows (X) and first lactation yields of their daughters (Y). To calculate the phenotypic correlation between first lactation yields of dams and their daughters :

$\Sigma X = 137267.00$	$\Sigma Y = 156431.00$	$\Sigma XY = 619865013.00$
$\Sigma X^2 = 566020659.00$	$\Sigma Y^2 = 720220493.00$	$(\Sigma X)(\Sigma Y) = 613508973.63$
$(\Sigma X)^2/n = 538349408.26$	$(\Sigma Y)^2 = 699161650.31$	$n = 35$

Now the coefficient of correlation is equal to—

$$r = \frac{619865013.00 - 613508973}{\sqrt{(566020659.00 - 538349408.26)(720220498.00 - 699161650.31)}}$$

$$= \frac{6356039.37}{763556.90} = 0.83$$

By referring to the table in which significant levels are given (*see* any standard textbook on statistics) we find that this particular value of coefficient of correlation is significant at 1 per cent level of probability.

The standard error of the coefficient of correlation is calculated according to the following formula :

Standard error of 'r' $$m_r = \frac{\sqrt{1 - r^2}}{n - 2} \quad \ldots(8)$$

In our example $$\frac{\sqrt{0.31}}{33} = 0.10$$

The coefficient of correlation for the data on first lactation yield of Russian Bestuzevskaya cows and their daughters can then be expressed as—

$$= r \pm m_r = 0.83 \pm 0.10$$

15.6.8. Linear Regression Coefficient

A linear regression coefficient indicates the extent of change in one trait with a unit change in another related trait. The regression coefficient is represented as 'b' or 'R' and is extensively used in calculating heritability of characters and other parameters in animal breeding. It is generally estimated in the following manner.

Suppose there are two traits—X and Y. The coefficient of regression of trait X on trait Y is equal to—

$$b_{xy} = \frac{\text{Covariance XY}}{\text{Variance Y}} \text{ which is same as } r\frac{\sigma_X}{\sigma_Y} \quad \ldots(9)$$

The coefficient of regression of trait Y on trait X is equal to

$$b_{yx} = \frac{\text{Covariance XY}}{\text{Variance X}} \text{ which is same as } r\frac{\sigma_Y}{\sigma_X} \quad \ldots(10)$$

In small samples, however, the coefficient of regression is calculated according to the following formulae

$$\text{Regression of X on Y} \quad = b_{xy} = \frac{\Sigma XY - \dfrac{(\Sigma X)(\Sigma Y)}{n}}{\Sigma Y^2 - (\Sigma Y)^2/n} \quad \text{...(11)}$$

$$\text{Regression of Y on X} \quad = b_{yx} = \frac{\Sigma XY - \dfrac{(\Sigma X)(\Sigma Y)}{n}}{\Sigma X^2 - (\Sigma X)^2/n} \quad \text{...(12)}$$

The various symbols used in the above formulae are the same as those given under formula (7).

If formula (7) is compared with formulae (11) and (12), one will notice that the values r and b are inter-related. The relation between these two estimates is such that the product of b_{xy} and b_{yx} equals the square of the coefficient of correlation

$$b_{xy} \times b_{yx} = r^2 \text{ or } r = \sqrt{b_{xy} \cdot b_{yx}}$$

Let us calculate the regression of first lactation yield (Y) of daughters on that of their dams (X) using the data on Bestuzevaskaya cows given in Table 15.14.

$$b_{yx} = \frac{\Sigma XY - \dfrac{(\Sigma X)(\Sigma Y)}{n}}{\Sigma X^2 - \dfrac{(\Sigma X)^2}{n}} = \frac{6356039.37}{27671250.74} = 0.228$$

Calculation of Heritability from Regression Coefficient

The coefficient of heritability of a trait is twice the estimate of regression of offspring on parent based on the fact that an individual gets half of its inheritance from each of the parent.

$$\text{Coefficient of heritability} = h^2 = 2b \quad \text{...(13)}$$

In our example the coefficient of heritability of first lactation yield of Bestuzevaskaya cows is

$$h^2 = 2 \times 0.228 = 0.456$$

15.6.9. Linear Regression Equation

A regression equation described the nature of relationship between two traits. When the relationship between two traits is linear the regression is expressed as follows :

$$Y = a + bX \text{ or } \overline{Y} = a + b\overline{X} \quad \text{...(14)}$$

where Y is the expected value of trait Y, constants a and b specify the nature of the straight line, and X is the actual value of trait X. The uniform rate of change in the straight line is given by the regression coefficient, b which describes how a trait Y changes for b units of change in X.

The regression equation $Y = a + bX$ can be used to predict the estimates of trait Y (dependent) for any particular value of trait X (independent) since the regression equation is fitted to suit the data within the range of study.

As an example let us obtain a linear regression equation using the data given in Table 15.14 so as to estimate the first lactation yields of daughters (Y) from the first lactation yields of their dams (X).

We know that $\overline{Y} = a + b\overline{X}$

then $a = \overline{Y} - b\overline{X}$

In our example $\overline{Y} = 4469.46$

$$\overline{X} = 3921.91$$

$$b = 0.228$$

$$4469.46 = a + (0.228 \times 3921.91)$$

then

$$a = 4469.46 - (0.228 \times 3921.91)$$
$$= 4469.46 - 1129.51$$
$$= 3339.95$$

The regression equation will then be as follows :

$$Y = 3339.95 + 0.228\ X$$

From this equation the fi:st lactation yield of daughters (Y) of Bestuzevkaya cows can be predicted if the latter's first lactation yield (X) is known. Suppose a particular cow's yield is 4000 kg. Then the predicted or likely lactation yield of her daughter will be

$$= 3339.95 + (0.228 \times 4000)$$
$$= 3339.95 + 912.00$$
$$= 4251.95$$

The regression equation is widely used in selection work for calculating expected production ability of animals, construction of selection index etc.

15.6.10. Genetic Correlation (rG)

Genetic correlation between characters indicates genetic relationship between them. The result of such a relationship is actually the change expected in one character when selection is practised for a second character. Coefficients of genetic correlation are used in calculating selection indices that are used as a means of selection of sires of desired characters.

Based on the method of path analysis S. Wright and L.N. Hazel suggested the following formula or calculating coefficient of genetic correlation

$$r_G = \frac{r_{x_D y_T} . r_{x_T y_D}}{r_{x_D y_D} . r_{x_T y_T}} \qquad ...(15)$$

where x_D is the first character of dams, y_D is the second character of dams, x_T is the first character of daughters, y_T is the second character of daughter, $r_{x_D y_T}$ is the phenotypic correlation between the first character of dams and second character of daughter, $r_{x_T y_D}$ the phenotypic correlation between the first character of daughters and second character of dams, $r_{x_D y_D}$ is the phenotypic correlation between first and second characters of dams, and $r_{x_T y_T}$ is the phenotypic correlation between the first and second characters of daughters.

When one of the phenotypic correlation in the formula (15) is negative the following formula was recommended.

$$r_G = \frac{r_{x_D y_T} + r_{x_T y_D}}{\sqrt{r_{x_D y_D} \cdot r_{x_T y_T}}} \quad ...(16)$$

15.6.11. Analysis of Variance

This technique is used in genetics and animal breeding for several purposes such as calculation of coefficient of heritability, evaluation of herd sires, etc. Analysis of variance is partitioning of the sum of squares of deviations from mean into components arising out of several sources.

Diversity of characters—their variation around arithmetic mean, depends on several factors. It often becomes necessary to find out the contribution of each factor to the total variability of a given character. For example a dairy cattle breeder may be interested to know to what extent variation in milk yield of daughters of a bull is due to the latter's genotype. When contribution of just one of the factors to the total variation is to be calculated the single factor (one way classification) analysis of variance has to be carried out. In more complex situations, contribution of two or more factors can be studied simultaneously. Such a situation is known as two or multi-factor analysis of variance.

We have already seen how standard deviation (σ) is used as a measure of variability of traits. In analysis of variance, measures like variation (σ^2) and dispersion (C) are used. Dispersion is the sum of squares of deviations of individual variants from the arithmetic mean :

$$C = \Sigma(X - \bar{X})^2$$

In single factor analysis of variance, first the general dispersion (C_z) is calculated, which is the sum of squares of deviation of all variants as influenced by all the factors that affect the trait. Secondly, dispersion due to the factor under question (C_x), *i.e.*, that fraction of the general dispersion that is due to the factor under consideration, is calculated. Then the remainder of the dispersion (C_y) is estimated, which is due to other chance factors. The dispersion calculated in the above manner are interrelated thus :

$$C_z = C_x + C_y$$

In two-factor analysis of variance contribution of two factors to the general variation is calculated. For example an investigator studying the milk production of cows in several herds in a district may be interested to find out the extent of variation in milk yield of cows due to differences between herds (factor 1) and differences between bulls used (factor 2). In such a case the interrelationship between various fractions of dispersion will be as follows :

$$C_z = C_w + C_x + C_y$$

where C_z is the general dispersion, C_w is dispersion due to factor 1, C_x is the dispersion due to factor 2, and C_y dispersion due to other chance factors *i.e.*, error.

Given below are methods of computing the analysis of variance when the data are available in one-way and two-way classification.

Computation of Analysis of Variance from One way Classification of Data

(*a*) The first step comprises of partitioning the various components of variation as shown in Table 15.15.

TABLE 15.15. Table showing how to calculate different sums of squares from one way classification of data

Source of variation	*Sum of squares*	*Mean sum of squares*
Total sum of squares (SS)	$\Sigma X^2_{ijk} - CF$	—
Between sires (SS_s)	$\frac{\Sigma T_i^2}{n} - CF$	$SS_s/S - 1 = MSS_s$
Within sires (SS_w)	$SS_t - SS_s$	$SS_w/n - s = MSS_w$
Correction factor (CF)	G^2/n	

where G = grand total of all the observations, n = total numbers of observations, S = number of sires, T_i = total of daughters of each bull, n_i = number of observation (daughters) within each sire, x_{ijk} = each individual observation irrespective of sires, Σ = sign to indicate summation of values.

When there are unequal number of daughters per sire sum of squares between sires can be calculated thus :

$$\frac{T_1^2}{n_1} + \frac{T_2^2}{n_2} + \ldots + \frac{T_s^2}{n_s}$$

which are totals of and number of daughters of bulls 1, 2, ..., s.

(*b*) Testing significance of mean sums of squares. After calculating the different sums of squares the analysis of variance table is prepared as shown below.

TABLE 15.16. Analysis of Variance for Data Considered in Table 15.15

Source of variation	*Degrees of freedom*	*Sum of squares*	*Mean sum of squares*	*F Value*
Between sires	(S – 1)	SS_s	MSS_s	MSS_s/MSS_w
Within sires	(n – S)	SS_w	MSS_w	
Total	(n – 1)	SS_t		

The magnitude of the F value enables us to learn whether the MSS_s is significant or not. The value MSS_s/MSS_w follows F-distribution with (S – 1) and (n – s) degrees of freedom. To know whether the calculated value of F is significant or not one has to see whether it is higher or lower than the tabulated value of F given in "F-distribution tables" at (S – 1) and (n – s) degrees of freedom. All standard books on statistics give table of F-distribution which may be referred to.

Given below (Table 15.17) are data pertaining to the fat per cent in milk of Murrah buffaloes which are daughters of five bulls. Carry out analysis of variance on this data.

$$\text{Correlation factor (CF)} = \frac{(234.3)^2}{40} = \frac{54896.5}{40} = 1372.4$$

$$\text{Total sum of squares } (SS_t) = \Sigma X^2_{ijk} - CF$$
$$= (5.9)^2 + 5.9^2 + 6.9^2 + \ldots + 6.9^2 + 5.5^2 + 5.9)^2 - 1372.4$$
$$= 1383.9 - 1372.4 = 11.5$$

TABLE 15.17. Data on Fat per cent in Milk of Murrah Buffaloes which are Progeny of Five Bulls

Daughter	*Sires*				
	A	*B*	*C*	*D*	*E*
1	5.9	5.2	5.2	5.0	6.2
2	5.9	5.8	5.9	5.6	5.6
3	6.9	4.9	6.6	5.7	5.7
4	5.7	5.8	6.5	5.1	5.1
5	6.0	5.3	5.8	6.2	5.8
6	6.5	5.9	6.4	5.9	6.9
7	6.0	5.9	6.3	5.7	5.5
8	7.2	6.3	5.0	5.5	5.9
Total	50.1	45.1	47.7	47.7	46.7

Total = 234.3

Between sires $(SS_s) = \frac{\Sigma T_i^2}{n_i} - CF$

$$= \frac{(50.1)^2 + (45.1)^2 + (47.7)^2 + (44.7)^2 + (46.7)^2}{8} - 1372.4$$

$$= \frac{10998.29}{8} = 13748.8 = 1374.8 - 1372.4 = 2.4$$

Within sires $(SS_w) = SS_t - SS_s = 11.5 - 2.4 = 9.1$

TABLE 15.18. Analysis of Variance

Source of variation	*Degree of freedom*	*Sum of squares*	*Mean sum of squares*	*F. Value*
Total	$(n - 1) = 39$	11.5	—	—
Between sires	$(n - 1) = 4$	2.4	0.62	2.308
Within sires	$(n - 1) = 35$	9.1	0.26	—

The tabulated value of F at 4 and 35 degrees of freedom is 2.61 and 3.83 respectively at 5 per cent and 1 per cent level or probability. Thus, the observed value of F is not significant. In other words, the variation between sires is not statistically significant.

15.6.12. Computation of Analysis of Variance from Two-way Classification of Data

When the date are classified in a two-way fashion also the various components of variance are partitioned on the same principle as in one-way classification of data. Suppose the fat content in milk of 10 buffaloes during five (I to V) successive lactations has to be tabulated. Naturally it will be tabulated under two headings, *viz.* lactations and buffalo cows and is analysed as shown below (Table 15.19).

TABLE 15.19. Data on Fat Content in Milk of 10 Buffalo Cows in Five Successive Lectations

Buffalo Cows	*Lactations*					*Totals*
	I	*II*	*III*	*IV*	*V*	
1	5.8	6.1	6.3	5.4	6.0	29.6
2	4.2	4.8	4.4	5.0	4.4	22.8
3	6.4	6.0	6.3	6.5	6.8	32.0
4	7.4	6.8	6.7	6.8	6.4	34.1
5	7.0	6.5	5.8	6.0	5.9	31.2
6	7.0	6.0	5.9	5.9	6.3	31.1
7	5.7	5.7	6.0	6.1	5.8	29.8
8	7.4	7.0	7.0	6.8	6.6	34.8
9	4.9	5.6	4.8	5.4	5.5	26.2
10	5.6	5.0	5.1	5.4	5.5	26.6
Total	61.4	59.5	58.3	59.3	59.2	297.7

$$\text{Correction factor (CF)} = \frac{(297.7)^2}{50} = 1772.51$$

$$\text{Total sum of squares } (SS_t) = \Sigma X^2_{ijk} - CF$$

$$= [(5.9)^2 + (4.2)^2 + (6.4)^2 + \ldots + (5.8)^2 + (6.6)^2 + (5.5)^2 + (5.5)^2)] - 1772.51$$

$$= 1802.33 - 1772.51 = 29.82$$

$$\text{Between buffalo cows } (SS_c) = \frac{(29.6)^2 + (29.8)^2 + \ldots + (26.2)^2 + (26.6)^2}{5} - 1772.51$$

$$= 1797.40 - 1772.51 = 24.89$$

$$\text{Between lactations } (SS_1) = \frac{(61.4)^2 + \ldots + (59.2)^2}{10} - 1772.51$$

$$= 1773.02 - 1702.51 = 0.51$$

$$\text{Error sum of squares } (SS_s) = SS_t - (SS_c + SS_w) = 29.82 - (24.89 + 0.51) = 4.42$$

TABLE 15.20. Analysis of Variance

Source of variation	*Degree of freedom*	*Sum of squares*	*Mean sum of squares*	*F. Value*
Total (SS_t)	49	29.82	—	—
Between cows (SS_c)	9	24.89	2.766	22.488**
Between lactations (SS_l)	4	0.51	0.128	1.041
Error (SS_e)	36	4.42	0.123	—

By referring to the F distribution table at appropriate degrees of freedom (9 or 4 and 36) it can be seen that variation among cows is highly significant, whereas variation among lactations is not.

Estimation of Heritability from Analysis of Variance Table

Coefficient of heritability of traits can be estimated from intra-class correlation among half-sibs by analysis of variance technique. This comprises of estimation of the various genetic parameters out of the different components of variation. For dairy cattle which produce only one offspring at a time and which have long generation intervals, one-way layout of data is useful. Each sire is mated to several dams and each dam produces one offspring (in case of dairy cattle we are interested only in the female progeny).

Given below is the technique of computing various genetic parameters from analysis of variance table :

Step 1. Estimation of variance components. This part has already been explained while discussing the computation of analysis of variance from one-way classification of data. The further steps are as follows :

(*i*) Variance due to sires = $\sigma_s^2 = \dfrac{MSS_s - MSS_w}{n_i}$

where MSS_s is mean sum of squares between sires (see Table 15.16). MSS_w is mean sum of squares within sires, n_i is number of daughters per sire. When the numbers of daughters per sire are unequal the denominator in the above formula (n_i) is calculated as shown below :

Average number of daughters per sire $n_i = \dfrac{1}{S-1}\left(N - \dfrac{\Sigma n_i^2}{N}\right)$

where S is the number of sires, n_i is number of observations (daughters) for each sire, and N is total number of observations on all sires.

(*ii*) Variance with sires $\sigma_w^2 = MSS_w$ (Table 15.16)

Step 2. Estimation of genetic and phenotypic variance. Genetic variance = σ_s^2. As the offspring inherits only half of the genetic material of one parent (sire or dam) the variance among the sires or among the dams measures only one fourth of the genetic variation and hence the need of multiplying σ_s^2 by 4.

$$\text{Phenotypic variance} = \sigma_s^2 + \sigma_w^2$$

Step 3. Estimation of heritability

$$\text{Heritability} = h^2 \frac{\text{Genetic variance}}{\text{Phenotypic variance}} = \frac{4\sigma_s^2}{\sigma_s^2 + \sigma_w^2} \quad \ldots(17)$$

Let us calculate the coefficient of heritability of fat content of milk of Murrah buffaloes from data given in Table 15.17. From the analysis of variance table (Table 15.18) we know that the MSS_s is 0.60, MSS_w is 0.26 and $n_i = 8$.

Now the variance due to sires $= \dfrac{0.60 - 0.26}{8} = 0.043$

Genetic variance $= 4 \times 0.043 = 0.172$

Phenotypic variance $= 0.043 + 0.260 + 0.303$

$$\text{Heritability} \quad = h^2 = \frac{\text{Genetic variance}}{\text{Phenotypic variance}}$$

$$= \frac{0.172}{0.303} = 0.568 \qquad \ldots(18)$$

Calculation of Repeatability from Analysis of Variance Technique

Repeatability estimates the fraction of difference between single records of individuals that are likely to occur in future records of the same individuals. Though repeatability estimate also is related (like heritability) to the genetic superiority or inferiority of the individuals, it tells about the chance that a particular individual will tend to repeat records for a particular trait throughout its life. Let us see how repeatability estimate for milk fat content of Murrah buffaloes can be computed from the data given in Table 15.19. For this purpose it is sufficient to estimate the mean sum of squares due to individual cows and the error mean sum of square. The analysis of variance will then be as follows :

TABLE 15.21. Analysis of Variance of Data Given in Table 15.19

Source of variation	*Degree of freedom*	*Sum of squares*	*Mean sum of squares*
Total	49	29.82	
Individual cows	9	24.89	2.766 = MSS_c
Error	40	4.91	0.123 = MSS_e

The value of chance variation = σ_e^2 = MSS_e = 0.123

$$\text{The value of variation among cows} = \sigma_c^2 = \frac{MSS_c - MSS_e}{k}$$

where 'k' is the number of observations per individual, which is 5 in our example.

$$= \sigma_c^2 = \frac{2.766 - 0.123}{5} = 0.529$$

When number of observations per individual are unequal, the value of 'k' (according to W.A. Becker (1968) is equal to $\frac{1}{N-1}\left(n - \frac{\Sigma n^2}{n}\right)$ where N is the number of individuals and n is total number of measurements.

$$\text{Repeatability,} \qquad R = \frac{4\sigma_c^2}{\sigma_c^2 + \sigma_e^2} = \frac{0.529}{0.529 - 0.123} = 0.801 \qquad \ldots(19)$$

Thus, we find that the fat content in milk of buffaloes is a highly repeatable trait.

CHAPTER

16

Production Traits of Dairy Bovines

Economic production traits of dairy bovines can for description purposes be classified as production traits, reproduction traits and traits concerning growth, vitality, disease resistance and heat tolerance. It must, however, be remembered that these groups of traits are expressed simultaneously and are interrelated biologically. Let us consider now the extent of variability that can be encountered in different selection traits.

16.1. MILK PRODUCTION AND ASSOCIATED TRAITS

In a dairy enterprise milk is the direct source of economic return and so milk production is the most important selection trait. Coefficient of variation of this trait varies from 15-35 per cent among various herds. There are considerable differences in level of milk production between breeds, between herds within breeds and between individuals within herds. The extent of variation among different breeds of dairy bovines found in India is shown in Table. 16.1

Cattle in the tropics are generally of the Zebu type (*Bos indicus*). While in the Indian sub-continent well developed breeds of cattle found are of pure zebu type, in some tropical countries of Latin America and Africa exist some stabilised breeds, which seem to possess some inheritance of *Bos taurus* cattle (Willianson and Payne, 1978). Apart from a select few individuals, milk yield of tropical (Zebu) breeds is much lower than that of European cattle, even under good managemental conditions (Table 16.1). The production data from tropical countries were mostly obtained when the animals are well fed and managed (on government farms). The yields may be still lower on farmers' holdings.

TABLE 16.1. Lactation Milk Yield (kg) of Various Breeds of Cattle and Buffaloes

Breeds	*Lactation yield*	*Breeds*	*Lactation yield*
Bos indicus breeds		***Bos taurus breeds* (in temperate zones)**	
Sahiwal	1548-2283	H. Friesian	4200-5633
Red Sindhi	1514-1634	Jersey	3548-4105
Tharparkar	1456-2177	Brown Swiss	3509-5603
Gir	1312-1804	Red Dane	3797-4522
Hariana	721-1436	***Bos taurus breeds* in India (tropics)**	
Ongole	613-1120	H. Friesian	2453-5256
Kankrej	960-1856	Jersery	1969-3902
Rathi	1527-1931	Brown Swiss	1792-2121
Deoni	879-1049	Red Dane	3182-4111
Kangayam	642	***Bos bubalus* (buffalo) breeds**	
Dangi	616	Murrah	1031-2565
Non-descript	376-535	Surti	1086-2095
***Bos indicus* × *Bos taurus* (F_1)**		Nagpuri	1063-1200
With H. Friesian	2242-3128	Nili-Ravi	1586-1955
With Jersey	2162-2979	Bhadawari	1100-1252
With Brown Swiss	2045-2379	No-descript	541-650

Note. The above data have been compiled from too large a number of reports to be listed here.

Buffaloes are generally better milk producers than the Zebu in South Asia (Table 16.1). Milk production of cattle breeds found in other tropical regions are more or less akin to the cattle indigenous to India (Mahadevan, 1966; and Willianson and Payne, 1978) only varying with local environmental and managemental conditions.

The milk of zebu cattle is generally richer in fat content than that of European breeds, with the exception of Jersey (Table 16.2). The fat content is highest in the milk of buffaloes. Variability of fat content is generally lower than that of milk yield within breeds.

TABLE 16.2. Butter Fat content in Milk of different Cattle and Buffaloes

Breeds	*Fat %*	*Breeds*	*Fat %*
H. Friesian	3.3-4.1	Tharparkar	4.6
Jersey	4.9-5.5	Gir	4.7
Brown Swiss	3.6-4.1	Hariana	4.4
Red Dane	3.8-4.2	Murrah buffalo	5.5-7.2
Sahiwal	4.6	Nili-Ravi buffalo	6.5-7.0
Red Sindhi	4.9	Surti buffalo	6.3-8.1

Note. Same as for Table 16.1.

Several environmental and animal factors cause considerable variation in level of milk production of dairy animals over and above the effect of genes. Some of the more important of such factors are—level of nutrition, ration quality, season, frequency of milking, age of the cow parity or number of lactation, stage of lactation, length of preceding dry period, lactation length, persistency of milk production etc.

Stage of lactation has pronounced effect on both milk yield and composition. Milk production function during a lactation is best described by the shape of the lactation curve (Figure 16.1).

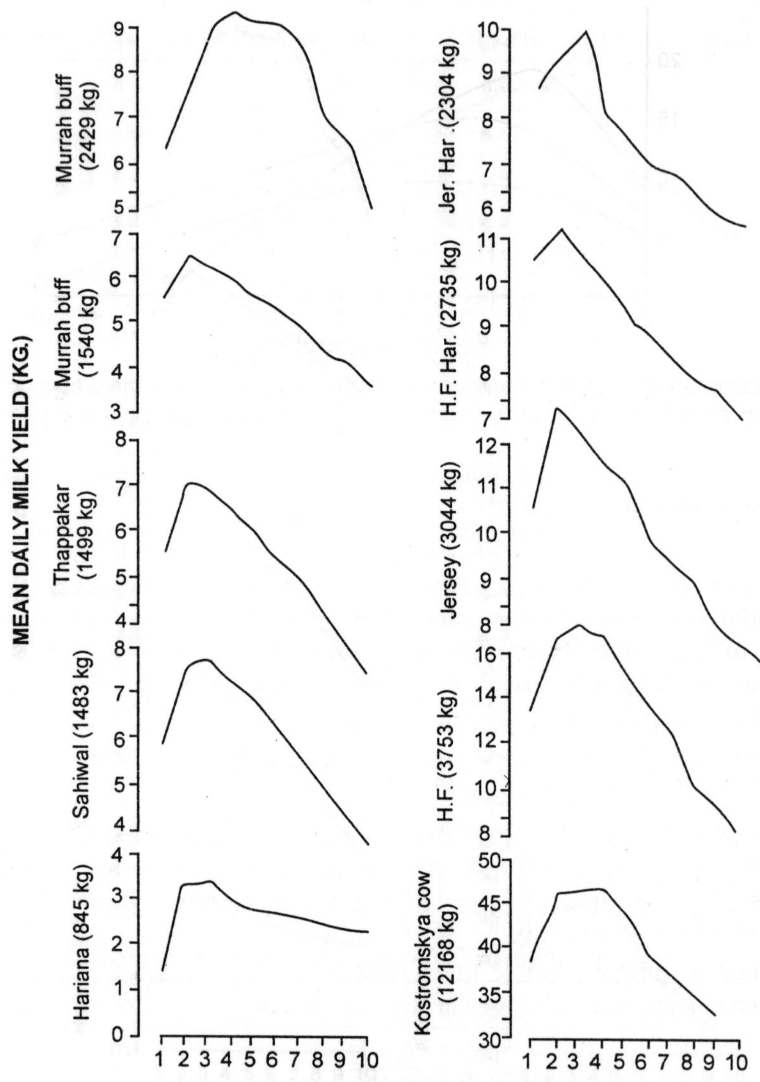

Fig. 16.1. Lactation curves drawn from uncorrected data of herds (30 to 287 observations) of Hariana (Acharya *et al.*, 1977). Sahiwal (Abhey Singh, 1988). Tharparkar (Abhey Singh, 1988) low yielding Murrah buffalo (Rhotas Singh, 1980), high yielding Murrah buffalo (Grewal and Sastry, 1988), Jersey × Hariana F_1, (Kashmiri Lal, 1975), Friesian × Hariana F_1 (Kashmiri Lal, 1975), Jersey (Juneja, 1987) and Friesian (Juneja, 1987) breeds located at Hissar and mean of seven lactations of champion Koslromskya cow (Earnst *et al.*, 1977), Note that in herds with higher potential (lactation yield) the lactation is more persistent as indicated by bell like shape of the curve.

Mean daily yield increases from the date of calving reaching a maximum by 4-6 week after parturition. After attaining this maximum or peak yield, level falls gradually until the animal dries off.

As seen in Figure 16.1, the peak yield is attained generally by the second month in buffaloes and zebu cattle as well as in crosses of the latter with *Bos taurus* breeds. The ability of the cow

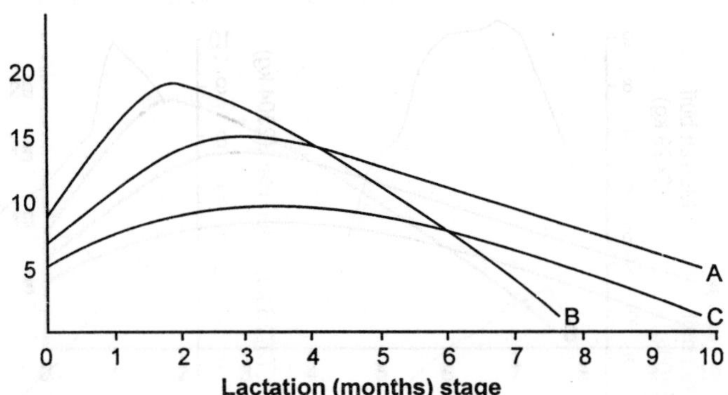

Fig. 16.2. Possible shapes of lactation curves of cows with good milk yield and good persistency (A), good peak yield but poor persistency (B) and poor peak yield but good persistency (C).

to sustain good daily yields for a longer period, *i.e.,* the slope of the descending phase of the lactation curve is known as persistency.

A cow with higher peak yield and gentler slope is the one with good persistency (Figure 16.2A.) and yields more milk in one lactation than a cow with higher peak yield but steeper fall of the descending phase of lactation curve (Figure 16.2B) or a cow with lower peak yield but gentler slope of the curve (Figure 16.2C). Persistency is more relevant to commercial milk production than just high peak yield though a combination of both is most ideal. The typical appearance of the lactation curve may be disturbed if the peak of the curve coincides with the summer season and periods of fodder scarcity. This is specially true in crossbred animals in tropics, their overall lactation yields falling as a result of this.

The persistency is higher in *Bos taurus* breeds (rate of decline of yield per month 5.3-5.7 per cent of the previous month) than in *Bos indicus* breeds (rate of decline 7.5-11 per cent) and buffaloes (Figure 16.1). The initial yield and persistency account for more than 95 per cent of the variation in lactation yield (Koley *et al.,* 1979).

Persistency is quite a variable trait (Table 16.3). The persistency index mentioned in the table can be calculated according to the formula of Mahadevan (1966) given below :

$$\text{Persistence P} = \frac{A - B}{B}$$

where A and B are respectively milk yield in the first 180 days of lactation and initial yield of the first 10 weeks of lactation. One may take A as the lactation yield in 300 days or less and B yield upto 45 or 60 days when most of the zebu breeds and crossbreds reach their peak production (See Figure 16.1). In fact several workers have followed this procedure (Balaine *et al.,* 1970; Koley *et al.,* 1979 and others).

Age of the cow and so the order of her lactation or **parity**, too affects her milk production. In European breeds of cattle the peak production is achieved by the third lactation, production decreasing in successive lactations after the third, whereas, in zebu breeds the best lactation could be the third or even the fourth (Figure 16.3). In buffaloes maximum yield may be obtained even

Balaine, D.S. (1971). *Indian J. Dairy Sci.* **24** : 25.
Koley, *et al.,* (1979). *loc. cit.*

TABLE 16.3. Persistency Indices of Some Breeds of Dairy Bovines

Breed	*Persistency index**	*Coefficient of variation*	*Authors*
Hariana	2.39	38.6	Lal, 1975
Sahiwal	3.07		Sexana and Kumar, 1960
Red Sindhi	3.96		Sharma and Bhatnagar, 1973
Brown Swiss × Sahiwal (F_1)	3.03		Sharma and Bhatnager, 1973
Brown Swiss × Hariana (F_1)	3.08	12.0	Lal, 1975
Friesian × Hariana (F_1)	3.27	11.2	Lal, 1975
Red Dane × Hariana (F_1)	3.13	17.8	Lal, 1975
Jersey × Hariana (F_1)	3.07	13.1	Lal, 1975
Jersey × Hariana (F_1)	4.44	31.0	Koley *et al.*, 1979
Murrah buffalo	1.23	13.0	Kumar and Narain, 1978

*See text.

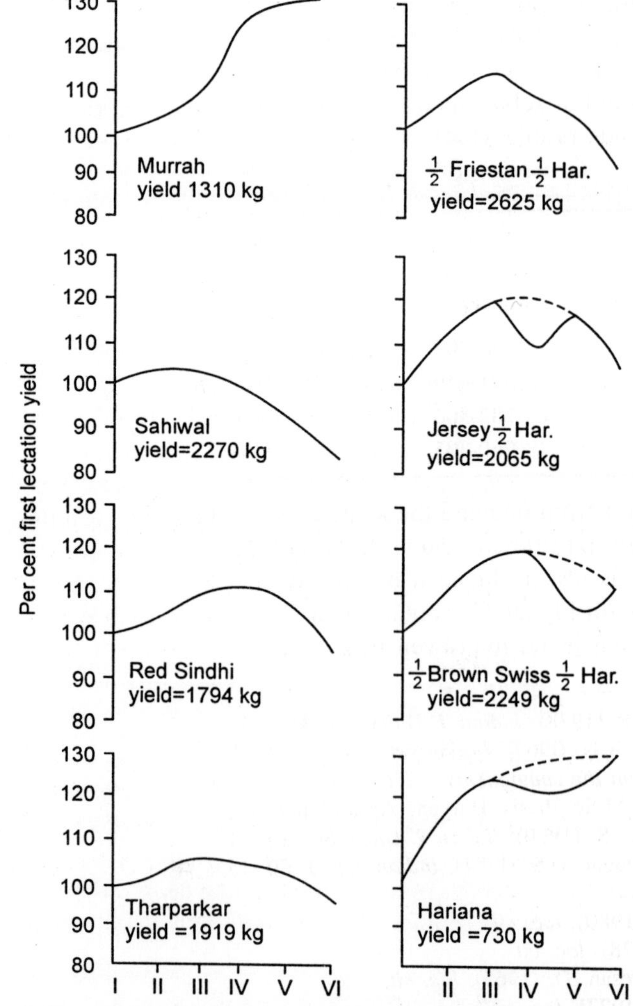

Fig. 16.3. Variation in lactation milk yield (as per cent of first lactation yield) of dairy animals from 1st to IVth lactation (drawn from data of Lal, 1975, Sharma *et al.*, 1974, Nagarcenkar, 1979).

much later (Krishna and Desai 1969). However, there is considerable diversity in the nature of this characteristic among breed groups generally reared in tropics; the best lactation can be anything from the 2nd to the 6th (Fig. 16.3).

Besides, milk yield in the first lactation is influenced by age at first calving (Figure 16.4), practice of weaning and other factors. In breeds with comparatively better growth rates, Sikka and Taneja (1979) estimated that with one day increase in age at first calving the milk yield in 300 days (or less) would increase by 0.53 kg. Similar nature was observed in late maturing breeds like Hariana (Singh *et al.,* 1968). There are others (notably Chawla and Misra, 1978) who could not find any beneficial effect of increasing age at first calving on first lactation yield in relatively high yielding cows. In fact some have reported (Daya Singh and Sundaresan, 1965; Brijinder Singh *et al.,* 1980; Bhat 1978 and others), a positive correlation between age at first calving and first lactation yield, the latter falling with decrease in the former.

The length of lactation or the lactation period (*see* Figure 16.5) also influences the milk yield considerably, there being a positive correlation between the two traits (Deshpande and Singh, 1977; Sharma *et al.,* 1978 and others), Bhat (1980) estimated that in Tharparkar cows, an increment of 4.4 to 5.6 kg in total lactation yield would result per day increase in lactation length. The lower lactation yield of zebus in comparison to *Bos taurus* cows may be to some extent, their shorter lactation periods (Table 16.4).

TABLE 16.4. Lactation Length (days) in Different Cattle and Bufalloes Recorded in India

Breeds	*Lactation length*	*Breeds*	*Lactation length*
H. Friesian	279-392	Tharparkar	274-286
Jersery	287-417	Gir	252-301
Brown Swiss	296-307	Hariana	230-251
Red Dane	391-459	Murrah buffalo	228-351
Sahiwal	282-306	Nili-Ravi buffalo	250-323
Red Sindhi	260-305	Surti buffalo	238-350

Milk composition from one and the same cow is also influenced by various factors such as stage of lactation, age, nutritional status, ration composition, season, frequency of milking etc. Composition of milk is also influenced by species and breed (Table 16.2) of the animal. Zebu cattle produce milk with higher fat content than European cattle (excepting Jersey) and buffalo milk is much more rich in fat than even that of Zebu.

Kumar, K. and Narian, P. (1978). *Indian J. Dairy Sci.* **31** : 220.
Krishna, S.T. and Desai R.N. (1969). *Indian vet. J.* **46** : 409.
Lal, K. (1975). *Studies on the components of lactation curve in Hariana purebreds and their crosses with some elite exotic breeds.* M.Sc. thesis, Haryana Agric. Univ., Hissar.
Saxena, P.N. and Kumar. S. (1960). *Indian J. Dairy Sci.* **13** : 45.
Sharma. R.C. and Bhatnagar, D.S. (I973). *Indian vet.,* J. **50** :153.
Bhat, P.N. (1978). *loc. cit.*
Brijendra Singh *et al.* (1980). *loc. cit.*
Chawla and Mishra (1978). *loc. cit.*
Daya Singh and Sundaresan. D. (1965). *loc. cit.*
Deshpande and Singh (1977). *loc. cit.*
Singh. D., Acharya, R.M. and Sundaresan, D. (1968). *Punjab Agric. Univ. J. Res.* **5** : 555.

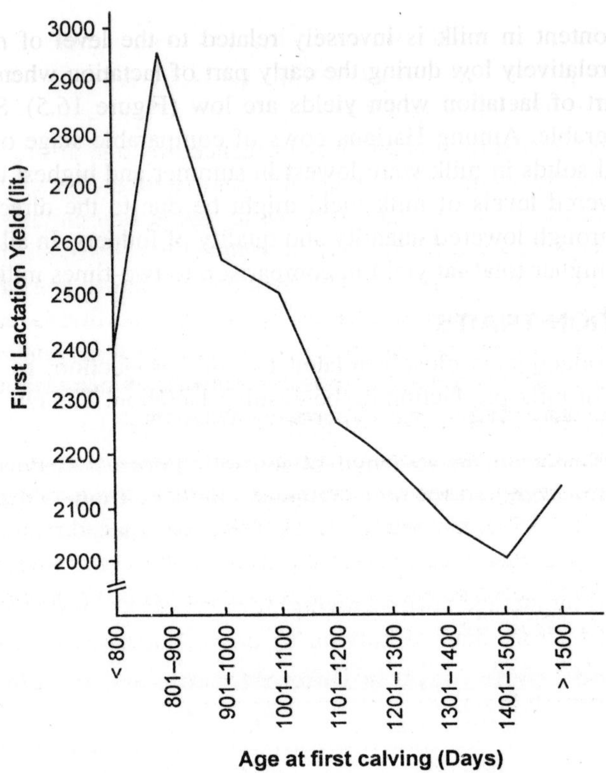

Fig. 16.4. Relationship between age at first calving and first lactation (305-day) yield in Friesian × Sahiwal crosses (Ram, 1979). Too early or too late age at first calving obviously results in lowered first lactation yields, 800-900 day period being ideal among crossbreds.

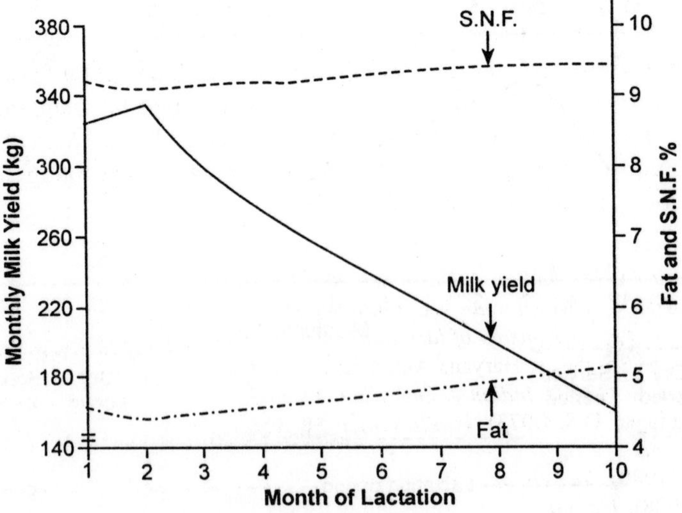

Fig. 16.5. Changes over the lactation period in monthly milk yield and fat and solids-not-fat in milk of crossbred cows (After Sutej Bhan, 1980)

Generally, fat content in milk is inversely related to the level of milk production (Ling, 1963). Fat content is relatively low during the early part of lactation when yields are higher, and higher in the later part of lactation when yields are low (Figure 16.5). Seasonal (and so feed) effect are also considerable. Among Hariana cows of comparable stage of lactation fat, protein, solids-not-fat and total solids in milk were lowest in summer and highest during winter (Kaushik *et al.*, 1979). The lowered levels of milk yield might be due to the direct effect of hot climate as well as indirectly through lowered quantity and quality of fodders. In all seasons, milking three times a day results in higher total fat yield in comparison to two-times milking (Das *et al.*, 1976).

16.2. REPRODUCTION TRAITS

In dairy bovines reproduction is closely related to milk production, as giving birth to calf is essential for initiation of milk production in the ensuing lactation. The rate of reproduction is also important for rearing herd replacements and in determining annual genetic progress in the herd, as poor reproductive efficiency increases length of generation interval. Reduced rates of reproduction can also increase the time required for progeny testing bulls by limiting the numbers of daughters available.

The traits under this category that are of interest to the breeder are : age at maturity (practically represented by the age at first calving), dry period (from date of drying to subsequent calving), service period (from date of calving to date of successful service), gestation period (from data of successful service to subsequent calving date), and calving interval (duration between two successive calvings).

For better understanding, the above parameters are pictorially represented in Figure 16.6. The reader may note that the calving interval is equal to service period + gestation period as well

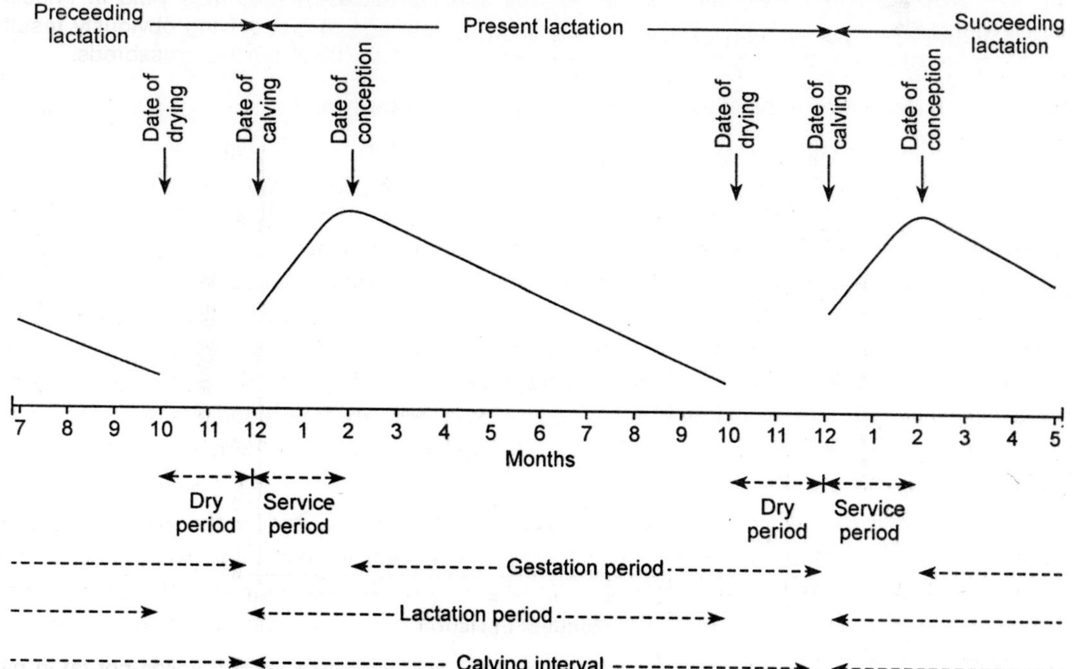

Fig. 16.6. The cycles of calvings, conceptions and dryings off of dairy bovines. Note the durations of various components of calving interval.

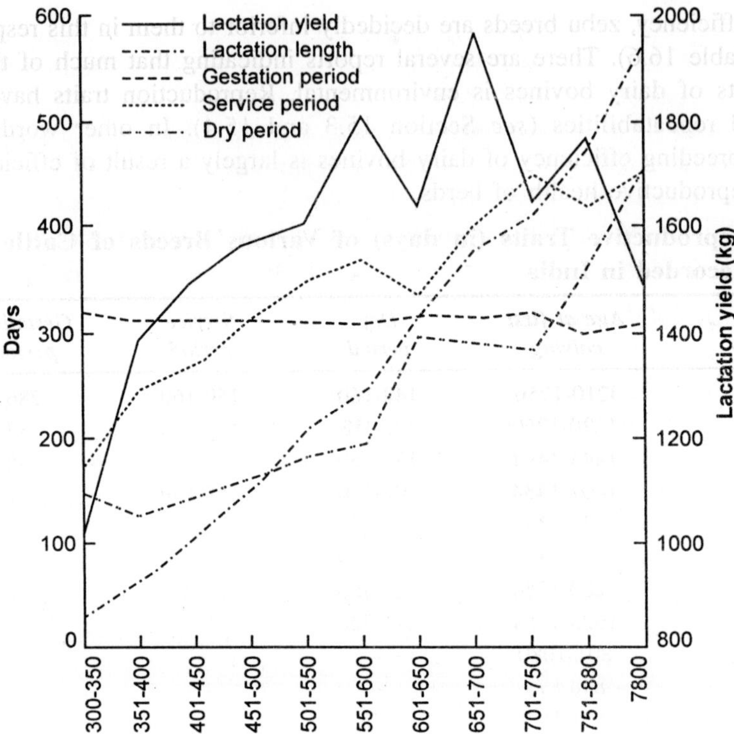

Fig. 16.7. Variation in lactation yield lactation length, gestation period, service period and dry period of Murrah buffaloes due to length of their calving interval (Sastry *et al.*, 1982).

as to lactation period + dry period. All the reproduction traits are dependent mostly on one single fact, namely, how soon the cow is conceived after calving, *i.e.*, service period, which influences other traits. Interrelationships among different reproduction traits and relationship with milk yield can be seen in figure 16.7.

Generally speaking, within breed or type, high yielders have low reproduction rates and *vice-versa* (Basu, *et al.*, 1980; Bhat *et al.*, 1980; Sastry *et al.*, 1982). In fact, significant positive genetic and phenotypic correlation between reproduction efficiency and milk yield per day of calving interval and a negative correlation with calving intervals was observed in Sahiwal cows (Brijendra Singh *et al.*, 1980). The dairy breeder has to strike a balance between the two (though achieving high yields and high reproductive efficiency should be the ultimate goal) and settle for cows that are moderately high yielders and have moderately high reproduction rates. Such cows would generally yield more per day of calving interval and so have good life time production. Such an aim is justified in tropics in view of the constraints imposed there by climate, feed and level of husbandry to milk production.

Excluding gestation period, which is a species specific trait, by far the greatest variability is seen in case of reproduction traits both between and within breeds (Table 16.5). Coefficient of correlation of such traits is rather high (30-70 per cent). While *Bos indicus* × *Bos taurus* crosses seem to be as efficient as purebred *Bos taurus* cattle (reared under similar tropical conditions)

Brijendra Singh *et al.* (1980). *lac cit.*

Sastry, N.S.R., Juneja, I.J., Yadav, R.S. and Gupta, L.R. (1982). *Haryana Agric. Univ. J. Res.* **12** : 6.

in reproductive efficiency, zebu breeds are decidedly inferior to them in this respect (Anand and Balaine, 1981, Table 16.6). There are several reports indicating that much of the variability in reproduction traits of dairy bovines is environmental. Reproduction traits have generally low heritabilities and repeatabilities (*see* Section 15.3 and 15.4). In other words, good rate of reproduction or breeding efficiency of dairy bovines is largely a result of efficient management and control of reproductive health of herds.

TABLE 16.5. Reproductive Traits (in days) of Various Breeds of Cattle and Buffaloes Recorded in India

Breed	*Age at first calving*	*Dry period*	*Service period*	*Gestation period*	*Calving interval*
Sahiwal	1210-1250	148-160	150-160	286	439-580
Red Sindhi	1220-1260	145-152	142-154	283	425-540
Tharparkar	1444-1494	138-150	138-154	288	430-456
Gir	1401-1434	207-230	155-169	281	511-630
Hariana	1731-1779	158-172	167	290	444-632
Kankrej	1400-1448	130-150	206-220	—	475-506
Ongole	1200-1350	321-380	221-240	286	490-515
Non-descript	1625-1754	271-326	271-284		571-632
H. Friesian	801-1080	—	—	275	396-425
Jersey	848-871	—	—	—	392-453
Brown Swiss	1031	175	192	287	388-490
Red Dane	899	—	—	—	454-525
B. indicus × *B. taurus*	879-1027	90-94	105-122	278-284	384-400
Murrah buffalo	1200-1381	60-200	68-218	311	370-650
Nili-Ravi buffalo	1600	—	223-236	310	445-525
Surti buffalo	1305-1340	111-343	193-236	304	461-581

Note. 1. The above data have been compiled from too large a number of reports to be listed here.
2. Data on *Bos taurus* breeds pertains to their performance in tropics.

16.3. GROWTH TRAITS

Rate of growth of dairy calves has a great bearing on their productivity in future life. It is during the growing stage that a strong anatomical and physiological foundation is laid for future performance. If superior metabolism (utilisation of feeds) is the basis of an animal's superior performance, animals showing superiority in one trait (growth), in all probability, will be superior in other traits like milk production, which also demands superior metabolism (Taranov, 1977).

There is considerable variation in the growth rates of dairy animals reared in tropics (Table 16.7), especially during early ages, which can possibly give room for early selection of outstanding animals. There are a good number of reports indicating that growth rate during early stages is related to later production traits. The most important of such relationships is the one between growth and age at first calving which is positive. Diversity in rates of body weight gain are reflected in the age at first calving in different types of cattle. Animals with superior growth rate (*e.g.*, *Bos taurus* cattle) have invariably early age at first calving and *vice-versa* (Table 16.8).

Anand, G.R. and Balaine. D.S. (1981). *Indian J. Anim. Sci.* **51** : 271.
Taranov (1977). *loc cit.*

TABLE 16.6. Average Values of Herd Reproductive Status of Purebred Zebu and European Cattle and Crosses between them as determined by Various Methods

Breed/ Genetic group	*Per cent conceived by first service*	*Standard AFC × 100 / Observed AFC*	*Standard CI × 100 / Observed CI*	*HRS Index*	*Per cent Reproductive culls*	*Per cent Non-return rate by 120 days*
H	43.0	48.9	67.8	22.0	51.5	26.2
F	52.8	91.4	90.8	64.8	28.6	44.4
J	50.0	94.3	80.5	62.4	29.0	46.2
½ F	32.4	85.2	82.1	68.1	27.5	48.4
½ B	21.7	78.7	85.1	59.0	13.5	50.0
½ R	6.7	71.4	88.3	52.1	11.5	56.8
½ J	31.0	93.7	88.2	45.5	8.7	48.6
½ Ja	64.4	74.6	78.3	39.5	25.4	38.9
Pooled	41.4	74.9	79.6	56.6	27.3	42.7

H = Hariana, F = Friesian, J = Jersey, ½F = Friesian × Hariana; ½B = Brown Swiss × Hariana; ½R = Red Dane × Hariana; ½J = Jersey × Hariana; ½Ja = Jersey × Hariana (herd different from other ½J crossbreds); AFC—Age at first calving; CI = Calving interval; HRS—Herd reproductive status index.

Source : Anand and Balaine (1981) *loc. cit*

TABLE 16.7. Mean Body Weights at Different Ages of Various Dairy Bovines

Breed	*Weight at*						
	Birth	*6 Months*	*12 Months*	*18 Months*	*24 Months*	*30 Months*	*Period of Maximum gain (12-26 weeks) per day (kg)*
	kg	*kg*	*kg*	*kg*	*kg*	*kg*	
			CATTLE				
Friesian*	38.0	176.0	297.0	400.0		536.0	0.958
Sahiwal	19.6	104.6	176.3	235.9	281.5	328.9	0.568
Sahiwal × Friesian	23.9	135.4	213.9	273.0	326.8	365.1	0.684
Hariana	18.7	113.9	178.7	238.4	277.2	334.5	0.566
Hariana × Friesian	23.9	139.9	222.6	270.3	333.2	390.3	0.689
Tharparkar	19.5	96.7	168.8	217.0	263.8	305.4	0.488
Tharparkar × Friesian	22.1	126.3	209.5	267.6	327.0	372.6	0.621
Red Sindhi	17.3	97.8	174.2	222.1	267.3	305.4	0.464
Red Sindhi × Friesian	19.8	131.6	209.1	277.7	347.7	308.9	0.668
Gir	20.3	113.3	166.1	203.5	245.5	285.8	0.540
Gir × Friesian	22.1	131.6	223.2	264.0	310.5	358.6	0.653
			BUFFALO				
Murrah	29.3	118.6	212.1	164.0	324.7	376.9	0.500
Murrah grades	29.3	118.6	211.8	265.9	325.7	382.9	0.500
Nili-Ravi	30.5	133.9	218.8	288.9	355.8	423.9	0.570
Nili-Ravi grades	29.9	123.5	205.8	267.9	339.7	400.9	0.502

*From Beguchev (1977). Performance recorded in USSR.

Source : Bhat *et al.* (1980) *loc. cit.*

Rate of gain in body weight is higher during the early ages and it gradually falls off as the animal matures. Both genetic and environmental factors cause variation in growth traits (body weight at different ages, absolute weight gain per day, week, month etc., relative growth rate, growth in body measurements). While there does not seem to be any real genetic differences in most of the growth traits among various zebu breeds (Teneja *et al.,* 1978), there are considerable differences between buffaloes, purebred zebu and zebu × European breed crosses (Table 16.7). The crossbreds are generally superior to their zebu counterparts in this respect, which in part may be due to heterosis.

Among the environmental effects, level of nutrition is the single largest factor contributing to variation in growth and hence the need to maintain all the breeding stock under similar feeding conditions. Besides, there can be variations in growth rate of dairy calves between different herds (farms), seasons and calendar years (Rao and Nagarcenkar, 1979; Bhat *et al.*, 1980), which may be due to the diversity in climate, quality and quantity of feed available and managemental conditions.

16.4. TYPE AND CONFORMATION

While there are several characters that are peculiar to individual breeds (colour, size, weight, etc.), there are certain other characteristics which are common to all dairy cows. Dairy type cattle possess typical body conformation and shape. A sum of all such physical characters are referred to as dairy types.

Dairy cows are expected to be free from excess fleshing, possess lean and long neck, light brisket and shoulder, the hind quarters looking light but broader and deeper. Their udder should be free from "meatiness" and is usually quite large, which combined with the light and narrow forequarters give the dairy cow a marked triangular appearance when viewed from side. Indeed, it is an important procedure in dairy cattle judging and physical evaluation to study the so called angularity of the cow. The top, side and front "wedges"—the triple wedges of the cow, are looked for as measures of this angularity.

Unified score cards (*see* any standard book on dairy cattle judging) have been developed in the West for evaluating the dairy type and conformation of cows and bulls. Different body parts and features are given different weightages depending on their relative importance to the formation of the overall type conformation of the cow. A cow with high type score (as judged by experienced cattlemen) is to be deemed nearer to the ideal dairy cows than a cow with lower type score. Such score card traits (general appearance, body capacity, dairy characters and mammary system) can increase accuracy of selection for milk production if they are highly heritable and closely related to milk production.

However, such typical dairy cow conformation features (score card traits) are not always clearly discernible in zebu cattle; perhaps due to lack of typical dairy characteristics in them. Attempts have been made by several workers in India to study the relationship between conformation and production traits. Considerable diversity in body measurements of zebu and buffalo breeds of India has been recorded (Table 16.9). As seen from data given in table 16.9 no clear physical character-production relationship is evident in the zebu breeds, unlike in the

Bhat *et al.* (1979) *loc. cit.*

Rao, G.N. and Nagarcenkar, R. (1979), *Indian J. Dairy Sci.* **32** : 306.

Taneja *et al.* (1978), *loc. cit.*

recognised dairy breeds of the West. Buffaloes seem to possess heavier hind quarters than zebu cattle.

TABLE 16.8. Age at First Calving of Different Breeds in Different Countries

Breed	*Country*	*Average in months*	*Authors*
Zebu, in South-Asia			
Red Sindhi	India	42.0	Amble *et al.*, 1958
Red Sindhi	India	44.4	Rajagopalan, 1952
Hariana	India	58.8	Kohli and Suri, 1957
Hariana	India	53.7	Sharma, 1951
Sinhala	Sri Lanka	42.4	Mahadevan, 1953
Zebu in Africa			
Zebu	E. Africa	42.7	Galukande *et al.*, 1962
Zebu	Uganda	51.7	Sacker and Trail, 1966
Ankole	Uganda	51.3	Sacker and Trail, 1966
Nganda	Uganda	41.0	Williams & Bunge, 1952
Local	Egypt	34.3	Ragab *et al.*, 1961
Kenana & Butana	Sudan	44.0	El Amin, 1969
Butana	Sudan	44.0	Alim, 1962
Kenana	Sudan	40.6	McLaughlim, 1955
Bos Taurus in Temperate Zones			
Jersey	Denmark	26.5	Ostegard, 1950
Red × White	Norway	32.5	Berge, 1949
Swedish breeds	Sweden	26-28	Hansson, 1941
German breeds	Germany	24-26	Lowe, 1939

Source : Anim, F.M. El. (1976). *World Rev. Anim. Prod.* **12** : 69.

While the score card traits were found to be low to moderately heritable in purebred zebu cattle, they seem to be moderate to highly heritable in *Bos indicus* × *Bos taurus* crosses (Aggarwal and Basu, 1976). Differences in all score card traits except mammary system and milk production among different genetic groups were found to be large; Jersey crosses excelling in mammary system over Brown Swiss and Holstein crosses, while Brown Swiss crosses being the best in overall type score.

Mishra *et al.* (1978) observed that heart girth, abdominal girth and escutcheon length showed a significant positive correlation with milk yield and other economic traits in crossbred and Red Sindhi cows, which are nearer to ideal dairy type. Whereas, similar values in Hariana (a draft cum moderate milch type breed) were relatively lower. However, not all workers could record such reliable relationship between type traits and production traits in zebu breeds.

The thickness of skin is another trait of significance in type evaluation. High yielding cows possess relatively thinner skin than low yielders. In fact, a significant negative correlation (–0.23 to –0.46) was observed between average lactation yield of Sahiwal and crossbred cows and thickness of their skin at different sites, *viz.* neck, dewlap, chest, abdomen and rump (Bhatnagar

Aggarwal, S.B. and Basu, S.B. (1976). *Indian J. Dairy Sci.* **29** : 243.
Mishra, R.R. (1978). Cited by Rao and Nagarcenkar (1979).

et al., 1980). Regression coefficients calculated by these authors indicated that a decrease in skin thickness at different sites by 1 mm would cause an increase in average lactation yield from 76 to 264 kg in Sahiwal cows and from 133 to 252 kg in Brown Swiss and crossbred cows.

TABLE 16.9. Some Physical Characteristics of Cows belonging to Different Breeds with different levels of Production

Characteristics	*Cattle*						*Buffalo*		
	Karn Swiss (Sahiwal × Brown Swiss)	*Sahiwal*	*Red Sindhi*	*Tharpar-kar*	*Gir*	*Hariana*	*Murrah*	*Nili-Ravi*	*Surti*
Height behind hump (cm)	127.2	124.0	116.0	122.0	128.0	133.0	133.0	136.0	124.0
Length from point of shoulder to pin bone (cm)	148.9	140.0	129.0	145.0	140.0	137.0	149.0	148.0	138.0
Girth (cm)	226.9	172.0	158.0	172.0	168.0	170.0	220.0	225.0	191.0
Length of quarter from angles of hip to pin bone (cm)	46.6	45.0	40.0	43.0	43.0	44.0	45.0	44.0	41.0
Width between angles of hip (cm)	46.1	46.0	39.0	42.0	43.0	44.0	61.0	53.0	52.0
Slope of rump *i.e.,* height at hook bone-height at pin bone (cm)	11.2	13.0	11.0	9.5	11.0	11.0	9.0	14.0	12.0
Average (estimated) body weight (kg)	475.0	408.3	340.0	385.6	386.0	353.0	431.0	454.0	408.0
Average lactation yield (kg)	3355	2270	2146	1136	1590	855	1793	1600	1500

Source : *Characteristics of Cattle and Buffalo Breeds*; ICAR., New Delhi, 1979.

Shape and size of udder and teats has been found to affect test milk yield in zebu dairy breeds (Sharma *et al.*, 1976) and crossbreds (Swaid and Sastry, 1982). The former workers' estimation of increment in milk yield with unit increase in length and width of udder in different breeds is as follows :

Breed	*Increase in test milk yield with 1 mm increase in udder length (g)*	*Increase in test milk yield with 1 mm increase in udder width (g)*
Tharparkar	82	74
Sahiwal	86	170
Crossbreds	121	117
Murrah buffalo	83	82

Texture of udder (collapsible, meaty or non-collapsible and intermediate. type), as judged by palpation before and after milking, has a bearing on the production ability of the cow. The rate of milk flow out of the udder (1.20 kg/minute) as well as test milk yield (10-12 kg) were found to be highest in case of crossbred cows possessing collapsible udders (Swaid and Sastry, 1982).

Swaid, A.H. and Sastry, N.S.R. (1982). *Indian J. Anim. Sci.* **52** : 891.

Evaluation of type conformation for predicting milk production of village cattle is worthy of consideration in India and other developing countries, where milk recording is difficult due to various reasons. However, more field data is required for standardisation of type-production relationship under such conditions.

16.5. RESISTANCE TO DISEASES AND ADVERSE CLIMATE

To thrive, grow well and produce well in tropical climate, dairy animals should be able to withstand the hot and hot-humid conditions as well as various diseases and insect vectors that abound in the tropics. This is becoming increasingly important with the introduction of *Bos taurus* inheritance in tropical cattle. Though inherited resistance is long known, there is only a limited evidence suggesting that breed, strain or individual differences exist.

It is well known that *Bos taurus* breeds, native to the temperate zones, when introduced into tropics lose some vitality and production performance. Studies, especially in India, have shown that crossbreds with ½ to 5/8 inheritance of *Bos taurus* give satisfactory performance. As the *Bos taurus* inheritance increases beyond this level there is increase in mortality rate, reduction in growth rate and reproductive performance. Much of this is presumably due to lack of resistance in *Bos taurus* cattle to hot climate, diseases and insects of tropics.

McDowell (1972) suggested that instead of genetic resistance to diseases *per se*, there may be traits with genetic association that directly influence resistance to diseases transmitted by vectors as well as to the syndrome of stress in a tropical environment. The ability of the zebu cattle to suffer less from ticks than European cattle might be their short hair coat, greater number of hair follicles per unit area, thicker skin, denser epithelial layer, greater depth of hair follicles, and perhaps the peculiar secretions from the sebaceous glands excudated through the skin which acts as a fly repellant (McDowell, 1972). Further, this ability of the cattle indigenous to India and other tropical countries to ward off fly menace could well be due to their ability to move skin voluntarily which is due to the presence of voluntary sub-cutaneous muscles.

The number one problem disease of dairy cattle, mastitis, might have predisposing anatomical features like heavy and pendulous udders, weak udder attachment, weak teat spinchter (Appleman, 1973 and Blake *et al.,* 1978), the latter having genetic basis of expression.

As regards heat tolerance of zebu cattle *vis-a-vis* European cattle, it is well known that the former's ability to dissipate body heat by evaporation through the skin surface is a factor in its favour. Cutaneous evaporation is efficient in this species as water droplets are brought to the skin surface by activity of sweat glands and other associated skin structures. Nearness of the sweat glands to the skin surface and their high density are important in determining the sweating rate and thus relatively better evaporative heat loss in cattle native to tropics (Dalal and Chopra, 1980). Govindiah *et al.* (1980) found that the density of sweat glands is highest in Red Sindhi cattle and lowest in Brown Swiss, crossbreds being intermediate in this respect.

Thus, much of the resistance of cattle to hot climate, diseases and insects does seem to be due to definite favourable anatomical features. Hence, as suggested by McDowell, the dairy cattle breeder should attempt at selecting such desirable anatomical features as a means of indirect selection for resistance to diseases and adverse climate.

McDowell, R.E. (1972) *loc. cit.*
Appleman, R.D. (1973). *J. Dairy Sci.* **56** : 411.
Blake, R.W. McDaniel, B.T. and King, R.A. (1978). *J. Dairy Sci.* **61** : 474.

16.6. DRAUGHT CAPACITY*

In developing countries draught animal power is important for agriculture and rural transport. While aiming at increasing the milk production of local cattle by introducing *Bos taurus* inheritance, the breeder should not as yet lose sight of this important trait of cattle, as farmers are in need not only of high milk producers but also good workers.

The inability of the *Bos taurus* and its crosses with zebu to perform sustained work under tropical conditions seem to be related to their low heat tolerance (Sastry *et al.,* 1970*a&b*; Georgie *et al.,* 1970; Roy *et al.,* 1972; Rao *et al.,* 1974 and Acharya *et al.,* 1979). As early as in 1947, Vaugh reported that bullocks develop drafts equivalent to 1/5th to 1/6th of their body weight. His study showed that the maximum pull the bullocks could exert varied from 49.5 to 60.5 per cent of their body weight. Mukhherjee *et al.* (1961) found that the maximal load drawn by the bullocks for 6 hours varied from 310 to 540 per cent of their body weight. They also observed an increase in pulse rate, body temperature and rate of respiration of bullocks with increased hours of work. The study of Swamy Rao (1964) showed that Hariana cross bullocks develop drafts varying from 14.5 to 24.5 per cent of their body weight with best harness. Sundar (1973) reported a decrease in speed and draft developed by a pair crossbred (Hariana × Jersey) bullocks with passage of time of testing. Devadattam and Maurya (1978) reported the optimum draft for an average pair of Hariana bullocks to be about 60 kg.

Mukherjee *et al.* (1961) have observed that variation in the pulse rate, respiration rate and body temperature of Hariana bullocks increased with increasing hours of work. Izzatnagar workers (Anon., 1957 and 1960) observed that in case of Hariana bullocks, the pulse rate of animals at rest had significant negative correlation with the maximum loads drawn by them. Furthermore, there was an indication of the respiration rate of animals at rest, too having a negative correlation with the maximum traction loads. Singh *et al.* (1968), also working on Hariana bullocks, too recorded a progressive rise in all the three physiological responses during three operations to which the bullocks were put, namely, disc ploughing, cultivating and harrowing. However, there were differences in the degree of rise of these responses between the three operations, between periods of work and between teams as well. These workers further reported a progressive fall in erythrocyte count, haematocrit and platelet count and a significant rise in total leucocyte count with the progress of work.

The Indian farmers in general have a notion that the crossbred bullocks are not as efficient as the indigenous ones in working capacity. Studies were conducted at NDRI, Karnal to compare crossbred bullocks with their Zebu counterparts (Sastry *et al.,* 1970*a&b* and Georgie *et al.,* 1970). After studying the cardio-respiratory activity, oxygen carrying capacity of blood, bicarbonate buffer system and body water distribution of crossbred and zebu bullocks the above authors concluded that the crossbred bullocks can be put to work as efficiently as Zebu bullocks, provided adequate rest is allowed between working periods during hot weather. The manifestation of distress signs in crossbreds during summer could be minimised and their working efficiency increased if they were allowed to work during early morning and late evening hours (Roy *et al.,* 1972 and Rao *et al.,* 1974).

---

*This section is reproduced from Razdan, M.N. and Sastry, N.S.R. (1984). *Draft Animal Power in India.* Talk delivered at the 1st Workshop of Animal physiological workers, 2-5-Dec. 1984, NDRI, Karnal. See Original for references.

Also, since crossbreds grow faster, at all given age they will be heavier than local zebu bullocks. In other words, crossbred bullocks can be put to work when they are as young as 2½ to 3 years, while indigenous bullocks will be ready for work only by 3½ and 4 years of age (Sastry, 1969). Crossbreds are massive beasts and hence make powerful bullocks. The notion that absence of hump in crossbreds reduced their efficiency is untenable as hump plays no role in pulling loads. It is on the neck that the yoke rests while the animals are working. Besides, crossbreds were found to have marginally greater speed and depth of furrow while ploughing (Acharya *et al.*, 1979).

Studies at Orissa University of Agriculture and Technology (Acharya *et al.*, 1979) too revealed that although crossbreds showed higher rates of respiration and pulse accompanied by an earlier manifestation of panting and salivary frothing during rainy and summer seasons, they exhibited more stamina, higher speed and greater capacity for work than non-descript indigenous bullocks. They further reported that crossbreds pulled a maximal load of 1310 kg, which was significantly higher than the 1210 kg maximal load pulled by indigenous bullocks. The crossbreds were observed to have higher carting speed than local bullocks, it fell as the load pulled and duration of pulling increased. Adkine *et al.* (1977) too found that with increased load, the speed of the bullocks is reduced.

CHAPTER

17

Selection Principles and Practices

Selection is allowing only certain of the members of a population to leave progeny for the next generation *i.e.,* only chosen males and females reproduce. In case of additive characters, selection of superior individuals means selection of genotypes possessing plus genes. This will lead to the next generation having higher frequency of such genes, and average performance of the progeny ought to be higher than the average performance of the unselected parental generation. If environmental influences, dominance and epistasis are absent, the average progeny performance is expected to equal the average breeding value of the selected individuals of the parental generation (Pirchner, 1964). Genetically, selection is a means to deviate the average of the population in a desirable direction by increasing the frequency of desirable phenotypes, and so desirable genotypes.

Culling, in simple terms, is the opposite of selection, namely, not allowing unwanted individuals to produce the next generation. Thus selection and culling are two sides of the same coin. However, animals may be culled (removed) out of the herd for non-genetic reasons too, *e.g.,* old age, incurable disease, infertility, etc.

Natural selection has been going on since ages, where animals and plants which were stronger, which had better survivability and which were in more unison with the environment around them found a better chance to reproduce; thereby, genes for such characters getting more chances to get selected—to form individuals in the subsequent generations. Indeed, the natural selection process—survival of the fittest—has been one of the cardinal principles of Charles Darwin's theory of evolution. Since domestication, man has been ever seeking superior genotypes among superior phenotypes for traits useful to him and selecting such animals to form the parental generation, *i.e.,* the best mated to the best.

17.1. PRODUCTION RECORDS AND THEIR STANDARDISATION

Before the breeder can actually decide which individuals of a herd (population) he should select and which he should cull out, he should have all possible information about the likely breeding value of individual animals as well as the genetic make-up of the herd. Such information is much more important in selecting males, as the herd sire contributes half of the genes to the next generation and hence only the best should be selected.

The various steps involved until this decisive stage in the selection process can be sequentially listed as follows :

1. Identification (marking) of individual animals.
2. Recording production and reproduction information of individual animals.
3. Standardisation of production records to eliminate extraneous sources of variation.
4. Estimation of the genetic parameters (variability, heritability, repeatability, correlations between traits) of the herd.
5. Estimation of the possible breeding value of the individuals (based on individuality, pedigree or progeny quality).
6. Selection (or culling) of the individuals based on their breeding value (selection by tandem, independent culling levels or selection index method).

Identification of stock, keeping production and reproduction records and estimation of selection parameters, discussed elsewhere in this book, may be consulted by the reader. Discussion is restricted in this section to aspects of standardisation of records before their actual usage.

There can be considerable variation in the production records of individuals arising out of different sources : herd differences, age differences, differences due to variation in length and order (age of cow) of lactation, frequency of milking, etc. For carrying out the selection process it becomes necessary to compare individuals of quite contrasting nature; daughter (young) with dams or grand dams (old), cows located in different herds and milked differently, so on.

Before estimation of selection parameters or breeding value of individuals from such records, they have to be standardised to a uniform scale to estimate effects of different factors listed above. It is a standard practice to reduce all records to a—305 day, 2X, M4 per cent record *i.e.,* 4 per cent fat corrected milk yield in a 305 day lactation, when milked twice a day and of mature equivalent cows (which may be 3rd or 4th lactation). Standardising yield on the basis of First Lactation Equivalent is also in vogue. All such correction factors are evolved based on observations on a large number of animals.

17.1.1. Fat Corrected Milk

If comparisons are to be made between different breeds, most often, the fat content also is kept at a constant 4 per cent, *i.e.,* fat corrected milk (FCM) is taken into consideration for cows. When only a single breed is involved, instead of 4 per cent the milk yield can be corrected for the average fat per cent of that breed.

The formula suggested by Gaines and Davidson (Rice *et al.,* 1957) for correcting milk yield to 4 per cent fat is

$$4 \text{ per cent FCM} = 0.4 \text{ (Total milk)} + 15 \text{ (Total Fat)}$$

Rice, V.A., Andrews, F.N., Warwick, E.J. and Legates, J.E. (1957). *Breeding and improvement of farm animals.* McGraw Hill Book Co. & Inc.—Kogakusha Co. Ltd., Tokyo, pp. 404-407.

For correcting buffalo milk yield to 7.5 per cent FCM (in small breeds like Surti and Mehsana) and 6.5 per cent (in large breeds like Murrah and Nili) respectively, the National Dairy Development Board recommended use of Nagarcenkar's factors—1.525 and 1.375 (NDDB, 1980), respectively.

17.1.2. Correction for Age

Correction of lactation yield (early or late) to mature lactation yield or correction of later lactation yield to first lactation equivalent are the common practices involved in correcting lactation yield for age (since the parity of the cow increases with age). There have also been correction factors for age based not on order of lactation but on age of cows in months.

Sharma and Aggarwal (1974) compared three methods, *viz.*, Sander's, Modified Sander's and Ratio (later lactation as a ratio of the fist) methods and found the modified Sander's version to be marginally better for Zebu breeds, crossbreds and buffaloes (Table 17.1). One may notice that these correction factors, in a broad sense, follow in magnitude the levels of yields in different lactations of the various breeds of cows. Age correction factors based on the actual age (in months) for Sahiwal cows as reported by Ram Kumar *et al.* (1974) are given in Table 17.2. Others (Amble, *et al.,* 1958 and Kumar and Narain, 1978) worked out age correction factors for other production and reproduction traits also (Table 17.3).

TABLE 17.1. Conversion Factors for Converting Later Lactation Yield to First Lactation Equivalent

Lactation number		*Tharparkar*	*Sahiwal*	*Hariana*	*Red-Sindhi*	*Crossbreds*	*Murrah buffaloes*
Sander's Method	I	1.0000	1.0000	—	1.0000	1.0000	1.0000
	II	0.9807	0.9852	—	0.9258	0.9865	0.9947
	III	0.9801	0.9953	—	0.8892	0.8780	1.0052
	IV	0.9980	1.0319	—	0.8822	0.7910	1.0325
	V	1.0366	1.0111	—	0.9036	0.7826	1.0794
	VI	1.1007	1.2166	—	0.9574	—	1.1514
Modified Sander's Method	I	1.0000	1.0000	—	1.0000	1.0000	—
	II	0.9654	0.9812	—	0.9290	0.9782	—
	III	0.9509	0.9867	—	0.8961	0.8756	—
	IV	0.9548	1.0176	—	0.8938	0.8133	—
	V	0.9774	1.0786	—	0.9217	0.8659	—
	VI	1.0218	1.1812	—	0.9860	—	—
Ratio Method (Later	I	1.0000	1.0000	1.0000	1.0000	1.0000	—
lactation as fraction	II	0.9756	0.9765	0.8560	0.9560	1.0152	—
of 1st lactation).	III	0.9452	0.9785	0.8260	0.9107	0.8489	—
	IV	0.9643	1.0214	0.8710	0.9001	0.8403	—
	V	0.9746	1.0776	0.8800	0.9183	0.8665	—
	VI	1.0267	1.1751	0.8250	1.0091	—	—

Compiled from : Sharma and Aggarwal (1974) *loc. cit.* Chhabra *et al.* (1972) *loc. cit.* and Kumar and Narain (1978) *loc. cit.*

NDDB (1980). *Breeding and feeding for milk production in Operation Flood II.* National Dairy Dev. Board, Anand.
Sharma, K.N.S. and Aggarwal, S.B. (1974). Proc. Seminar on Current Res. in Anim Breed., NDRI, Karnal, p. 160.
Amble, V.N., Krishnan, K.S. and Srivastava, J.S. (1958). *Indian vet. J.* **28** : 33.
Kumar, K. and Narain, P. (1978). *Indian J. Dairy Sci.* **31** : 220.
Ram Kumar, C., Mahabir Prasad and Saxena, P.N. (1979). *Indian J. Dairy Sci.* **27** : 62.

TABLE 17.2. Conversion Factors for Mature Yield with Age Measured in Months for Sahiwal Cows

Age in months	*Average yield in litres (Actual)*	*Relative yield in litres (Estimated)*	*Relative values*	*Correction factor*
19-24	1929	2119	100	1.561
25-30	2359	2265	107	1.461
31-36	2557	2402	113	1.376
37-42	2432	2529	119	1.309
43-48	2700	2646	125	1.251
49-54	2771	2784	130	1.202
55-60	2923	2852	135	1.160
61-66	3008	2941	139	1.125
67-72	2968	3020	142	1.096
73-78	3041	3089	146	1.071
79-84	3098	3149	149	1.051
85-90	3195	3199	151	1.034
91-96	3211	3240	153	1.021
97-102	3181	3271	153	1.021
103-108	3219	3293	155	1.005
109-114	3489	3305	156	1.001
115-120	3367	3307	156	1.001
121-126	3190	3300	156	1.003
127-132	3352	3284	155	1.008
133-138	3008	3257	154	1.016
139-144	3122	3221	152	1.027
145-150	3838	3176	150	1.042
151-156	2835	3121	147	1.060
157-162	2992	3056	144	1.083

Source : Ram Kumar *et al.* (1974) *loc. cit.*

17.1.3. Correction for Length of Lactation

There can be considerable variation in the length of lactation period, especially in Zebu cattle and buffaloes. While very short abnormal lactations (below 50 days or so) should be omitted, there is need to reduce yield in such widely varying (in duration) lactations to a standard 305-day lactation yield (Table 17.4). Further, while progeny testing bulls, it may be necessary to evaluate them based on the incomplete records of their daughters to speed up evaluation. Correction factors (mostly used in the West for *Bos taurus* cattle) have been developed for the purpose of standardising the lactation length and/or yield.

In such cases first calvers' part lactation yield can be extrapolated to 305-day yield using factors given in Table 17.5. Nagarcenkar (1978) worked out ratio estimates for obtaining lactation yields from part lactation yields in Murrah buffaloes (Table 17.6) and found them to be ideal for early assessment of daughter's production and initial selection of superior sires. Further, the lactation yield can be corrected for the age at calving, as the latter influences the former. The factors in practice in USA for well known dairy breeds are reproduced (From Kravtchenko, 1973)

TABLE 17.3. Sander's Correction Factors for Age Correction of Various Parameters

Order of lactation	*Lactation yield*	*Lactation length*	*Yield/day of lactation*	*Calving interval*	*Yield/day of calving interval*
Sahiwal					
I	1.0000	1.0000	1.0000	1.0000	1.0000
II	0.9718	1.0614	0.9137	1.0382	1.0516
III	0.9636	1.1186	0.8408	1.0667	0.8710
IV	0.7754	1.1693	0.8189	1.0837	0.8979
V	1.0078	1.2108	0.8854	1.0081	0.9468
VI	1.0653	1.2407	0.9363	1.0795	1.0340
VII	1.1563	1.2570	1.0188	1.0585	0.9682
VIII	1.2995	1.2586	0.9732	1.0265	0.9423
Murrah buffalo					
I	1.0000	1.0000	1.0000	1.0000	1.0000
II	0.9947	1.0869	0.9347	1.0801	0.8131
III	1.0052	1.1379	0.8941	1.0678	0.8402
IV	1.0325	1.1606	0.8836	1.0473	0.8821
V	1.0794	1.1711	0.8972	1.0395	0.8963
VI	1.1514	1.1888	0.9373	1.0318	0.9251
VII	1.2580	1.2364	1.0116	1.0102	1.0546
VIII	1.4180	1.3473	1.1367	0.9877	1.4367
Red-Sindhi (Pooled for 2 farms)					
I	1.0000	1.0000	1.0000	1.0000	1.0000
II	0.9420	1.0130	0.9115	1.0250	0.8867
III	0.9660	1.0285	0.8460	1.0115	0.8410
IV	0.9170	1.0470	0.8650	0.9855	0.8370
V	0.9390	1.0725	0.8730	0.9750	0.8635
VI	0.9820	1.1000	0.9025	0.9750	0.9180
VII	1.0500	1.1315	0.9650	0.9755	0.9975
VIII	1.1640	1.1695	0.9950	0.9840	1.1060

Compiled from : Amble *et al.* (1958) *loc. cit.* and Kumar and Narain (1978) *loc. cit.*

TABLE 17.4. Factors for Computing Lactation Records of Over 305 Days to a 305-Day Basis

Lactation length	*Factor*	*Lactation length*	*Factors*
305-308	1.00	337-340	0.92
309-312	0.99	341-344	0.91
313-316	0.98	345-348	0.90
317-320	0.97	349-352	0.89
321-324	0.96	353-356	0.88
325-328	0.95	357-360	0.87
329-332	0.94	361-364	0.86
333-336	0.93	365	0.85

Source : Rice *et al.* (1957) *loc. cit.*

TABLE 17.5. Factor for Converting Incomplete Milk Records to a 305 Day Basis

Days	*Age at first calving*		*Days*	*Age at first calving*	
	< 36Mo	*≥ 36Mo*		*< 36 Mo*	*≥ 36 Mo*
30	8.32	7.42	170	1.58	1.48
40	6.24	5.57	180	1.51	1.41
50	4.99	4.47	190	1.44	1.35
60	4.16	3.74	200	1.38	1.30
70	3.56	3.23	210	1.32	1.26
80	3.15	2.85	220	1.27	1.22
90	2.82	2.56	230	1.23	1.18
100	2.55	2.32	240	1.19	1.14
110	2.34	2.13	250	1.15	1.11
120	2.16	1.98	260	1.12	1.09
130	2.01	1.85	270	1.08	1.06
140	1.88	1.73	280	1.06	1.04
150	1.77	1.64	290	1.03	1.03
160	1.67	1.55	300	1.01	1.01

Source : Rice *et al.* (1957) *loc. cit.*

TABLE 17.6. Ratio Estimates of Partial Lactations of Murrah Buffaloes. (Multiply Yield in Each Period by the Corresponding Ratio Estimate to Obtain Estimate of Lactation Yield)

Sr. No.	*Partial lactation yield*	*Ratio estimates*
1	Yield upto 4 weeks	8.7486
2	Yield upto 6 weeks	5.6772
3	Yield upto 8 weeks	4.2076
4	Yield upto 10 weeks	3.3580
5	Yield upto 12 weeks	2.8096
6	Yield upto 14 weeks	2.4254
7	Yield upto 16 weeks	2.1401
8	Yield upto 18 weeks	1.9216
9	Yield upto 20 weeks	1.7474
10	Yield upto 22 weeks	1.6058
11	Yield upto 24 weeks	1.4883
12	Yield upto 26 weeks	1.3992
13	Yield upto 28 weeks	1.3122
14	Yield upto 30 weeks	1.2438
15	Yield upto 32 weeks	1.1895
16	Yield upto 34 weeks	1.1436
17	Yield upto 36 weeks	1.1051
18	Yield upto 38 weeks	1.0722
19	Yield upto 40 weeks	1.0431
20	Yield upto 42 weeks	1.0154

Source : Nagarcenkar (1978). *loc. cit.*

in Table 17.7. Perhaps these can be used in cases of crossbreds developed in tropics by crossing with one of the *Bos taurus* dairy breeds.

TABLE 17.7. Correction Coefficient for Age of Different Breeds of Cows used in U.S.A. in Progeny Testing of Bulls

Age at calving	*Breed*		
	Ayrshire, Jersey, Guerency	*Brown Swiss*	*Holstein Friesian*
1 year 6 months	1.343	1.718	1.515
2 year	1.262	1.538	1.377
2 year 6 months	1.195	1.400	1.275
3 year	1.141	1.285	1.203
3 year 6 months	1.099	1.196	1.131
4 year	1.063	1.136	1.077
4 year 6 months	1.037	1.038	1.035
5 year	1.020	1.032	1.017
5 year 6 months	1.008	1.028	1.006
6 year	1.000	1.012	1.000
6 year 6 months	1.000	1.000	1.000
7 year	1.000	1.000	1.006
...	...	...	...
16 year	1.199	1.192	1.378

Kravtchenko (1973). *lac. cit.*

17.1.4. Corrections for Frequency of Milking

Cows may be milked 2, 3 or 4 times a day; the yield being positively correlated to the number of milkings. Hence, cow's yield has to be taken as when milked twice a day, so that all the cows are comparable. In USA the following coefficients (as per data of V. Rice) are used for correcting milk yield for frequency of milkings per day (Kravtchenko, 1973) :

365 day lactation yield to 305 day lactation yield : 0.85
305 day lactation yield to 365 day lactation yield : 1.17
4 time milking yield to 3 time milking yield : 0.88
4 time milking yield to 2 time milking yield : 0.74
3 time milking yield to 4 time milking yield : 1.13
3 time milking yield to 2 time milking yield : 0.83
2 time milking yield to 4 time milking yield : 1.35
2 time milking yield to 3 time milking yield : 1.20

While comparing cow A giving 5000 litres in 365 days under 4 time milking with cow B giving 3000 litres in 305 days under 2 time milking and with cow C giving 4000 litres in 305 day under 3 time milking, we get the following standard 305 day 2-time milk production :

Cow A : 5000 litre × 0.85 × 0.74 = 3145 litres
Cow B : 3000 litre × 1.00 × 1.00 = 3000 litres
Cow C : 4000 litre × 1.00 × 0.83 = 3320 litres

While uncorrected yield shows cow A to be the best, actually it is only the second best.

Kravtcenko, N.A. (1973). *Breeding farm animals* (Russian), Kolos Publishers, Moscow.

17.1.5. Effect of Data Standardisation

Sometimes magnitude of various selection parameters (correlations, heritability etc.) can considerably increase when standardised data is used in comparison to using raw data. Maibarada and Naidenko (1971) studied coefficients of correlation between milk yields of Russian Simmental dams and their daughters before and after correction of data for various aspects. By using such corrected data the correlations have increased in magnitude by 0.11 to 0.17 (Table 17.8). This would lead to higher h^2 values, which is desirable.

Thus in dairy cattle breeding, standardisation of data is in extensive use for comparing various groups of animals of different generations, pedigrees, ages and herds. Also, as we will see soon other methods of genetic/statistical analyses are also in vogue that do not need standardisation of the original data.

TABLE 17.8. Correlation between Yield of Russian Simmental Dams and Their Daughters According to Uncorrected and Corrected data

Lactation Number	*Dam-daughter pairs*	*305-day lactation yield (kg)* Daughters	*305-day lactation yield (kg)* Their dams	*Coefficient of correlation dam-daughter*	*Increase in 'r' after correction of data*
		Based on uncorrected data			
I	459	2971 ± 43	3662 ± 77	0.12 ± 0.04	—
II	414	3723 ± 55	4221 ± 54	0.34 ± 0.04	—
III	356	4088 ± 67	5049 ± 59	0.28 ± 0.05	—
		Based on corrected data			
I	459	2971 ± 43	3215 ± 49	0.23 ± 0.04	+ 0.11
II	414	3723 ± 55	3840 ± 59	0.43 ± 0.04	+ 0.09
III	356	4088 ± 67	4274 ± 65	0.45 ± 0.04	+ 0.17

Source : Maibarada and Kravtcenko (1973) *loc. cit.*

17.2. SELECTION : ITS EFFECTS, BASIS AND METHODS

17.2.1. How Selection Causes Genetic Change

Rate of genetic change of a population due to selection is determined basically by three factors in the following way :

$$\text{Rate of genetic change} = \frac{\text{Heritability} \times \text{Selection differential}}{\text{Generation interval}}$$

The larger the values in the numerator and or smaller the value in the denominator, the faster is the rate of genetic change due to selection. In other words, heritability and selection differential should be as high as possible and generation interval as short as possible to make the maximum progress.

(1) Heritability (*see* section 15.3) represents the function of the total difference between individuals that is caused by genetic differences. It is only the genetic difference that influences genetic change. Heritability is not constant, as its value for a trait may vary under different

Maibarada, N.N. and Naidenko, I.T. (1971). Cited by Kravtchenko (1973).

environmental and other conditions in the population. It only serves to set a limit on the rate of progress. However, before starting selection for a trait we must study its heritability only to make sure that it is heritable as otherwise we get no genetic change.

(2) Generation interval represents the average age of the parents when their offspring is born. In zebu cattle, the generation interval can be 4.5 to 5 years. If the heritability of the trait is the same, we would expect to make more progress in four generations than in two. But, this advantage in shorter generation interval by selecting young animals may be often more than offset by the greater selection differential when sophisticated techniques like progeny testing are employed. All the same, efforts should be made to reduce generation interval by proper rearing of young stock and breeding heifers as early as possible.

(3) Selection differential represents the superiority of the animals selected to be parents of the next generation. This is the only factor man can influence to a marked degree in his endeavour to improve his farm animals. Unless the selected animals are better than the average of the population, no improvement can be made in the subsequent generation. If the average production of a herd is 2000 litres and that of selected individuals is 2500 litres then the selection differential would be 500 litres. There are a number of factors which influence the selection differential. They are :

(*a*) *Accuracy of selection.* The evaluation of the trait in the parental generation must be as near to its genotype as possible. If they are selected only due to environmental advantage, rather than genetic superiority, the selection will be invalidated as the superiority will not find expression in the progeny (Acquired characters due to environment are not inherited). Adequate measurement coupled with proper evaluation is necessary for ensuring accuracy.

(*b*) *Genetic variation of the population.* The greater the variation, the larger the selection differential possible. While very little can be done about this factor, the power of the genetic mechanism is that it creates variation and maintains it. While selection reduces variation among selected parents, variation is generated again among the offsprings.

(*c*) *Intensity of selection.* The more intense the selection, the larger the selection differential. It is as simple as the fact that the top five per cent of the population will have better averages than the top ten per cent and so on. The selection intensity can be more for males than for females because one male can mate and fertilise many females. On the other hand, most of the females have to be retained for production in the case of dairy cattle.

(*d*) *The number of traits under selection.* The fewer the traits under selection, the faster the progress that can be made in those traits. When selecting for only one trait, 100 per cent of the selection pressure can be devoted to that trait. However, when selecting for two traits at a time, only 71 per cent as much progress can be created in each trait; when selecting simultaneously for three traits only 58 per cent as much progress can be made in each trait as when selection is for a single trait.

Therefore, any time an additional trait is added to a selection scheme, we must consider whether we lose more than we gain in real income.

17.2.2. Basis of Selection

Genetic improvement in animals will depend upon the ability of the breeder to recognise individual animals with superior genotype (not necessarily superior phenotype). These superior animals may

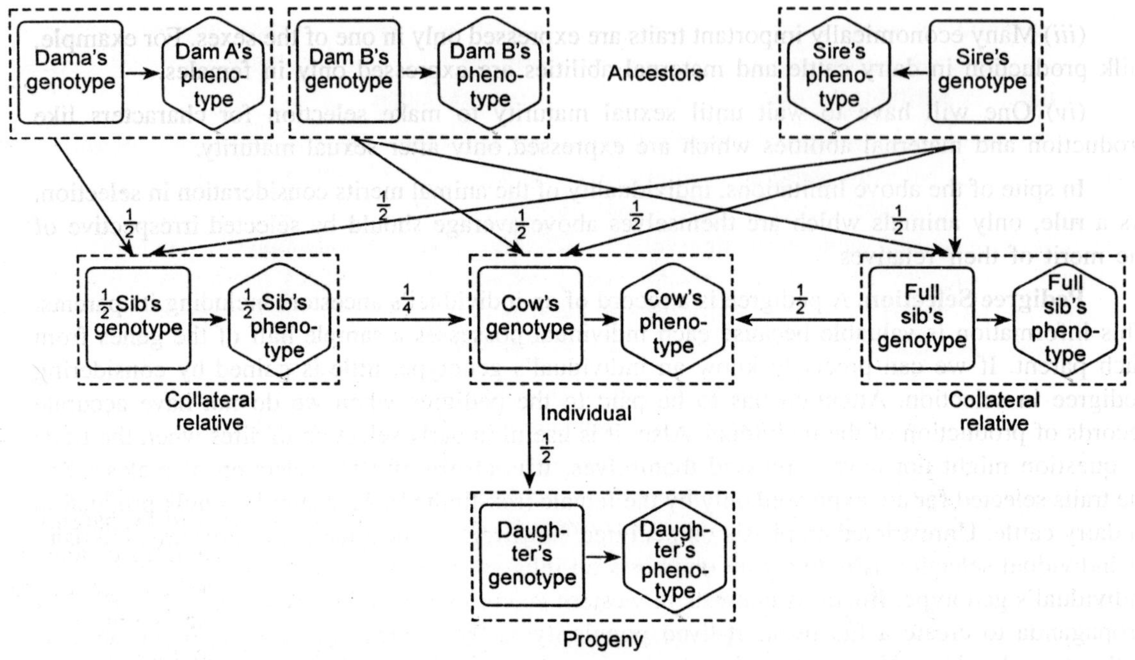

Fig. 17.1. Relationship of genotypes of various relatives to the individual's genotype.

be mated among themselves to produce offsprings with superior qualities. At present we are equipped to study the genotype of an individual only through the phenotype of its relatives. Relatives can be ancestors, progeny or collaterals (Figure 17.1). Ancestors that can be studied include sire, dam and the grand parents two each from maternal and paternal side. Progeny which can be of use in selection consists of sons and daughters. Collaterals include relatives of the same generation, *i.e.,* brothers, sisters, cousins, etc.

Individual selection means that animals are selected purely on the basis of their own phenotype. Individuality for certain traits should always be considered in a selection programme. These traits are those with high heritability which are affected to a greater degree by additive gene action. High heritability estimate indicates that the phenotype strongly reflects the genotype and that if an individual has better performance, it is most likely to possess better genes for the trait also. Undoubtedly, individual selection has played a major role in the improvement of livestock to date. It is also known as phenotypic or mass selection. Often it is combined with pedegree and/or family selection.

One of the advantages is that the traits involved in individual selection are usually available by the time breeding stock has to be selected. Besides, such a selection is based on the individuality of all animals whereas in methods like progeny test, only a few individuals are considered. However, individual selection has some disadvantages which are listed below :

(*i*) Individual selection can bring about erroneous results in traits for which heritability is low. Most of the economically important traits have low to medium heritability.

(*ii*) The easy appraisal of type tempts the breeder to emphasise on such traits in selection. But it has been generally observed that type and performance are not closely correlated. This indicates the importance of selecting separately for economic traits in livestock production.

(*iii*) Many economically important traits are expressed only in one of the sexes. For example, milk production in dairy cattle and maternal abilities are expressed only in females.

(*iv*) One will have to wait until sexual maturity to make selection for characters like production and maternal abilities which are expressed only after sexual maturity.

In spite of the above limitations, individuality of the animal merits consideration in selection. As a rule, only animals which are themselves above average should be selected irrespective of the merit of their relatives.

Pedigree Selection. A pedigree is a record of an individual's ancestors including its parents. This information is valuable because each individual possesses a sample half of the genes from each parent. If we can precisely know an individual's genotype, little is gained by considering pedigree in selection. Attention has to be paid to the pedigree when we do not have accurate records of production of the individual. Also, it is useful in early selection of sires when the traits in question might not have expressed themselves. It is also useful for selection of males when the traits selected for are expressed only by the female (sex-limited). An example is milk production in dairy cattle. Unrestricted emphasis on pedigree, however, can be a danger to progress. Intensity of individual selection is reduced by emphasis on relatives who contribute only very little to the individual's genotype. Business interests in western countries have often succeeded through mass propaganda to create a fad or short-lived popularity in the progeny of some favoured animals without much relevance to the genetic superiority contributed by such an ancestor to the individuals in question. This has often resulted in serious setback.

In a sound programme, weightage is given to a particular pedigree on the basis of the correlation of the ancestor's phenotypic merit and the breeding value of the pedigreed individual. In such an appraisal, closely related ancestors (like parents) should receive most of the emphasis. Pedigree selection is more accurate when the heritability of a trait is high. But, considering remote ancestors will not give any additional information when heritability of the trait is high. When the heritability of the trait is low, more weightage may be given to the information on more remote ancestors.

Pedigree selection has its greatest applicability in initial selection for sex-limited traits. Such selection has economic importance as the number of males to be retained can be reduced, thereby reducing expenditure. In a progeny testing programme, such preliminary selection on the basis of pedigree is of great value as progeny testing can be done only on limited individuals due to various limitations. It is extremely important that the pedigree selection is accurately done, making sure that all the potentially superior sires are given an opportunity to be progeny tested later. It is necessary to maintain elaborate pedigree trees for proper utilisation of this method.

Family Selection. Family line, traced through ancestors to an outstanding foundation ancestor after which the family has been named, has very little meaning in inheritance of character, unless, through line breeding, the relationship had been kept high to that outstanding ancestor. Therefore, family in this sense is of very little use in selection. This method is useful in case of lowly heritable and sex-limited characters. However, it would be far better for the breeder to concentrate on individual excellence and the excellence of near relatives (parents, grand-parents, sibs, half-sibs etc.).

However, in a different concept of family, it is defined to represent a group of animals having a common genetic relationship. In animal population, half-sib and full-sib families are most common (*see* Figure 17.1). They are not directly related with each other, but through a common ancestor. The average performance of the collaterals is, however, an indication of the genotype the individual might have received from one or more common ancestors. Thus, family selection in essence is another form of pedigree selection.

For a long time, animal breeders were asking whether selection should be (*i*) entirely based on individuality, (*ii*) entirely based on family, or (*iii*) based on both. Lush studied this problem in detail and came to the conclusion that the relative effectiveness of the three methods depend on the relationship or genetic correlation among members of the family (R), the genotypic variation among the members of the family (t), and the number of members in the family (n). Selection based on family average alone is

$$[1 + (n - 1)\mathrm{R}] \div \sqrt{n[1 + (n - 1)t]}$$

times more accurate than individual selection. If this factor is more than 1.0, family selection is preferred to individual selection and *vice versa*. Individual selection is more accurate when t is large and R is small. Conversely, family selection is more accurate when t is small and R is large. Information on family and on individual can both be combined to give an optimum basis for selection. Such a selection will be :

$$\sqrt{\frac{(1 + (\mathrm{R} - t)^2}{1 - t} \quad \frac{n - 1}{1 + (n - 1)t}}$$

times as effective as individual selection alone. If R and t are unequal a combination of individual selection and family selection will yield better results. If R and t are equal, the combination will be only as effective as individual selection alone.

Selection on the basis of progeny tests. In this method we evaluate the breeding value of an individual by a study of the expression of the trait in its offsprings. *"Individuality tells us what an animal seems to be, his pedigree tells us what it ought to be, but the performance of his progeny tells us what it is."*

Measure of performance of a polygenic or quantitative character is observed as an expression of the phenotypic value of the individual. Since phenotypic value is not an accurate estimate of the genotype of the animal, it does not assert much in bringing about improvement in the succeeding generation. It is essential, therefore, to estimate the breeding value of an animal for the specified trait. Progeny testing is a method of accurate estimation of such breeding value or genotype.

Working from the Mendelian principles, we can consider the factors that determine the potential of a young animal's future as follows: Every animal that is born receives (*a*) a sample half of its sire's genes, (*b*) a sample half of its dam's genes, and (*c*) a sample of the environment into which it is born and grows. Everything that happens to or affects an animal must fall into anyone of these categories.

In evaluating a sire on the basis of the progeny test, all the above factors have to be considered in order to make the progeny test as accurate and as efficient as possible. The three factors are, therefore, discussed below in more detail :

(1) Each daughter of a sire carries a sample half of his genes. For assessing the sire's genotype from such samples (daughters) the number of daughters *i.e.*, the number of samples of

his genes, must be large enough to prevent chance from leading us astray. At the same time, taking large number of daughters will be very expensive and may not be practicable. The breeder must make some decision as to how many sires to test and how many progeny must be produced to give better results from the test. The number of offsprings required for an accurate progeny test depends on the heritability of the trait. The higher the heritability the fewer the offsprings required and *vice-versa*. For milk production in cattle, from practical, economic as well as genetic considerations, this number should be between 30 and 50 daughters. The effect of heritability and number of daughters on accuracy of test is shown in Table 17.9. Considering the h^2 of milk yield to be about 0.35 around 15 daughters should give reasonably accurate evaluation. Lush suggested 10 daughters for full evaluation and 3-5-7 for preliminary evaluation. Waiting for more daughters for slight improvement in accuracy may delay the progeny testing unduly. It is very necessary that daughters must not be selected or else the sample will not be a random one and the evaluation will be faulty. The purpose of sire evaluation is not to make sire X or Y look good or bad. Rather, it is to set up a system whereby the sires having truly the best genotypes can be accurately identified.

(2) The prime purpose of sire evaluation is neither to evaluate dams nor to determine daughter-dam differences; although these may be useful bits of information. In fact, it is preferred that the mates of each sire to be evaluated be as nearly equivalent to each other as practical, so that the only difference in the genetic potential between progeny groups lies in the genetic contribution from the sires. This factor can be controlled either by design of the testing schemes so that different groups of mates tend to be equivalent or by statistical adjustment. The former is preferable.

How much of the phenotypic differences between groups of dams is reflected in the differences between their progeny groups? This is determined by the factor derived by multiplying the heritability of the trait with the genetic relationship ($h^2 \times$ R). For example, the genetic relationship, R between dam and daughter is 0.5 and the heritability of milk production is 0.25 (it can vary). The factor will be, therefore, 0.125 or 1/8. Thus, only about 1/8 of the phenotypic difference in milk production of dams (cows) is reflected in milk production differences between progeny group. Effects of yields of dams mated to a sire on their daughter's yield and the resultant effect on sire evaluation can be judged from the data on Simmental cows at Trostyanets farm in Soveiet Ukraine (Table 17.10). The bull may get a good positive evaluation if yields of daughters of dams 1 to 3 are considered and a negative evaluation of yields of daughters of dams 7 and 8 are considered.

TABLE 17.9. Degree of Accuracy of Milk Production Evaluation as Determined by the Level of Heritability and Number of Daughters Evaluated

Coefficient of heritability	*No. of daughters evaluated* 10	20	30	40	*Original author whose work was considered*
0.2	0.34	0.51	0.68	0.75	
0.3	0.45	0.62	0.76	0.83	K. Panye
0.4	0.58	0.69	0.82	0.85	N.M. Karmarenko and L.K. Ernst
0.5	0.59	0.74	0.85	0.90	

Source : Kravtchenko (1973) *loc. cit.*

TABLE 17.10. Role of Level of a Productivity of Dams of Daughters of Sire in Determining his Evaluation. The example is of the evaluation of the bull granite belonging to the Trostyanets farm in Soviet Ukraine

Sl. No.	*Yield of dams (M) of daughters of Granite*	*Yield of daughters (D) of Granite*	*Difference (D – M)*
1	3065	5330	+ 2265
2	3855	5141	+ 1285
3	4311	5369	+ 1058
4	4794	5463	+ 669
5	5226	5967	+ 741
6	5689	5269	+ 419
7	5966	5672	– 94
8	7146	6130	– 1015

Source : Kravtchenko (1973).

(3) The sample of environment each animal receives is the most important factor in its life. The best sire in the world cannot contribute an iota of environment through his genes.

The environment of an animal begins even when it is not conceived (the condition of the dam and her reproductive organs in which the foetus is carried and nurtured during gestation). It includes everything that happens to or affects it from its embryonic life, through early infancy, throughout its productive life. The environment received by any two animals cannot be the same in all details and many events that can affect the productivity of an animal are beyond the control of the farmer and often are unknown to him.

In many economic traits, the major part of the variation seen is due to environmental effect, for example, 90 per cent of the difference in milk production between contemporary herds in UK and USA has been attributed to difference in the environment of the herds. Conversely, only ten per cent of the differences between contemporary herds are due to differences in genotype.

All the same, progeny testing is no doubt the most accurate tool in the hands of breeders as it can form the corner stone for animal improvement, if extensively used.

Genetic gain (ΔG) from selection is a linear function of the accuracy of selection, selection intensity and additive genetic variance. It is expressed as :

$$\Delta G = r_{GI} \, . \, i \, . \, \sigma_g$$

where, r_{GI} corresponds to the accuracy of selection of sires and depends on the heritability of the trait and number of daughters available for each sire. If 'n' is the number of observations (daughters) and h^2 is the heritability, the r_{GI} is expressed as :

$$r_{GI} = \left(\frac{(nh^2)}{4 + (n-1)h^2} \right) 0.5$$

Progeny testing is an effective means of arriving at accurate estimates of r_{GI} and σ_g. This method is indispensable in dairy cattle breeding because of the sex limited nature of various economic traits and low heritability of most of these traits.

However, progeny testing could be a serious matter because of confusion of the picture due to the following sources of errors (Hofmeyer, 1955) : (*i*) randomly distributed source of variation, (*ii*) systemic errors—differences in yield, due to age, previous selection etc. and (*iii*) systematically distributed environmental differences between progeny groups in that they have been kept in different herds or at different periods of time or both. Of course there are means of avoiding such errors.

The time and expenses involved in progeny testing is a major prohibitive factor. The generation interval is prolonged if we decide to wait for progeny testing information on sires, thus affecting rate of genetic change. These factors must be weighed against accuracy of the progeny test before deciding the type of selection. At least four to five times more males than required must be progeny tested so that selection can be practised after getting the progeny test information. The progeny test should be accurately designed. Testing of progeny at several locations and the adoption of comparison with contemporary animals can improve its accuracy. Once an animal is proved to be superior in a progeny test, it should be extensively used through artificial insemination with frozen semen technique. If such an extensive use of the sire can be made, it will pay for the huge investment in progeny testing programme.

Effective progeny testing requires a large population. In herds of 100 to 150 dairy cattle, the most efficient progeny testing can only equal in efficiency a system based on record of their dams and other close relatives. But, larger scale progeny testing programme with AI in a population of 10,000 breeding females has been found to produce an average annual genetic improvement in milk yield equal to 1.5 to 2.0 per cent of the herd average. The corresponding figure without testing has been expected to be only 0.7 to 1.0 per cent of the mean.

The inability to get many offsprings from females is the major reason why the pathway through the dam is not of major importance for progeny testing in livestock.

In evaluating a bull by means of his progeny test, it is necessary to consider the following aspects :

1. For practical, economic as well as genetic considerations, the number of unselected daughters studied to evaluate a bull should be between 30 and 50.
2. It is preferred that the mates of each bull for evaluation be as nearly equivalent to each other as possible in production, so that the only difference in the genetic potential between progeny groups lies in the genetic contribution from the sires.
3. The sample of environment each calf receives is the most important factor in its life. The control of the environment is largely up to the individual dairyman.

There are many methods of evaluating a dairy sire based on his daughter's performance :

1. Daughter average. The average production of the daughters of a particular bull for valuable information on that bull's breeding value, provided the daughters are unselected and their number is sufficient.

$$S = D$$

where S is the breeding value index of the sire and D is the average yield of his daughters.

2. Daughter-dam comparison. This method has the disadvantage in that the dams and daughters usually are not milked during the same period and consequently some of the variations in production observed may be due to the differences in the environments under which they make their records. Besides, comparisons are usually restricted to limited range of environmental conditions in a few herds. Many times, data on daughters' production will have to be discarded due to lack of records of dams.

$$S = D - M$$

where M is the average performance of the dams of the daughters of the sire in question. Then if

D > M, the bull has '+' index
D = M, the bull is 'neutral' and
D < M, the bull has '–' index.

3. Herdmate-daughter comparisons. Herdmates of a sire's daughters are the other cows in the herd that calved in the same year and season as the sire's daughters did. This has the advantage that the records being compared are contemporary and, therefore, likely to be under the same environment.

A simple and direct formula for sire evaluation that combines the principles and techniques described above is as follows :

Three comparisons are made in this formula, *viz.*, (*i*) daughters compared to their herdmates, (*ii*) dams compared to their herdmates, and (*iii*) all herdmates compared to the breed average. These comparisons are weighted by the appropriate factors corresponding to their genetic contributions. These factors are :

(*i*) *b* is the regression of daughter's production on dams' production which increases in value as the number of daughters of the sire and the number of herds in which the daughters are located increases. For example the *b* value is only 0.15 for 10 daughters in one herd but 0.87 for 100 daughters in many herds.

(*ii*) One-eighth is the contribution of dams as discussed earlier.

(*iii*) One-tenth is the genetic difference between contemporary herds.

Predicted Daughter Superiority = *b* (Daughter average – Herdmates average) – 1/8 (dam average – Herdmates average) + 1/10 (Herdmates average – Breed average)

17.2.3. Methods of Selection

The net value of an individual is dependent on several traits. These traits usually are of varying economic importance. The heritability and the magnitude of genetic variance of the traits differ. There may be phenotypic and genetic interrelationship among the-traits. If too many traits are selected at a time, less progress will be made in the improvement of any particular trait. Hazel and Lush examined the theoretical superiority of three methods of selection involving selection for many traits. These were (*i*) the tandem method, (*ii*) the independent culling method, and (*iii*) the total score or selection index method.

In the **tandem method** of selection, only one trait at a time is considered for selection until satisfactory improvement has been brought about in that trait. Selection efforts for this trait are then relaxed and selection is done for a second trait, then a third, and so on. The improvement is dependent on the genetic correlation between the, traits. If there is a positive correlation between the traits, selection for one trait will improve the other traits also. On the other hand, if the traits are negatively correlated, selection for the second trait could undo the progress

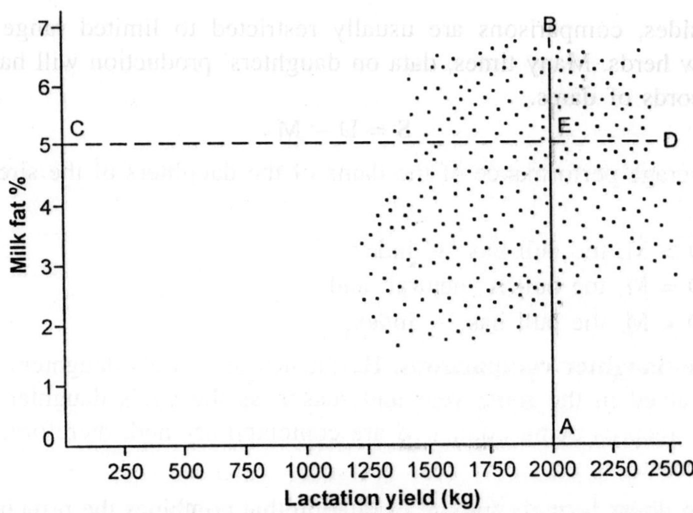

Fig. 17.2. An illustration of independent culling method of selection for milk yield and milk fat simultaneously in buffaloes. Only animals satisfying milk yield standard (2000 kg) and milk fat standard, (5%) *i.e.*, those falling between B, E, D will be selected. High yield alone (beyond A, B) or high fat alone (beyond CD) will not permit selection of a buffalo.

previously made in the earlier selection. When there are no correlations, this is the least efficient method.

In the second method, *i.e.*, **independent culling level,** selection for many traits can proceed simultaneously, but, a minimum standard is set for each trait (Figure 17.2). Any animal falling below that minimum level is culled irrespective of its meritorious performance in the other traits. The superiority of the animal in some traits will not have an opportunity to offset the weakness in other traits. Thus, such a selection will try to favour medium type animals which may just qualify in each trait above the culling mark, but, may not possess much excellence in any trait. On the other hand, outstanding animals with respect to other traits may be rejected if they fail to qualify in one trait.

Selection index or total score method was developed by Smith and Hazel which rectifies many of the drawbacks of the previous two methods. Each trait is given a weightage depending on its economic worth, heritability and interrelationship with other traits. The total of the scores given to the traits is added up to arrive at a total score. These indices are computed in such a way that they are closely correlated to the composite breeding value of the animal, as is possible with a linear joining together of traits.

The selection index method is more efficient than the other ones under most circumstances because it allows the individuals which are markedly superior in some traits to be saved for breeding purposes, even though, they may be slightly deficient in one or more other characters.

The selection index is now adopted widely by livestock breeders. This can be used for individual selection, pedigree selection, family selection or selection based on progeny testing.

17.2.4. Sire Index*

In modern times newer methods of dairy sire evaluation are being evolved steadily and in proportion to the understanding of the structure of dairy cattle population. Statistical techniques of analysis have been developed as added knowledge has been gained of factors about which, previously, little had been known and which consequently, had decreased the accuracy with which sire indices were estimated. Some of the common methods of estimation are : best predictor (BP), best linear predictor (BLP), best linear unbiased predictor (BLUP), generalised least squares estimator (GLS) and maximum likelihood (ML) estimator. Amongst all these, BLUP is the latest technique. This was primarily developed by Henderson and has been discussed by Henderson (1973, *a* & *b*), Henderson *et al.* (1957) and Searle (1974).

Harvey (1979) conducted a computer simulation study comparing the accuracy of BLUP versus simplified regressed least squares (SRLS) as methods of ranking sires. This study established that BLUP estimates were more accurate than SRLS when all the usual assumptions were met and moderate amounts of heterogeneous error variance was generated. But the SRLS estimates were more accurate when a correlation of 0.26 or higher was generated between herd and sire effects or when paternal half sib correlation existed between the sires under ranking. This indicates that, in circumstances where there has been significant sire selection in/between herds, the BLUP estimates are less accurate than SRLS estimates.

Henderson and Carter (1957) apportionated the total variation in age-adjusted milk production as : sire 7 per cent, herd 30 per cent, year-season 4 per cent, sire-herd 2 per cent, herd-year-season 14 per cent and residual 43 per cent. Thus herd, year and season variation accounts for about 50 per cent of the total variation in milk yield.

Herdmate Comparisons. Most of the recent research in sire evaluation methodology has centered on the herdmate comparison procedure. Evidence from the USA indicates that the herdmate comparison procedure gives a reasonably accurate assessment of bulls used under field conditions. The herdmate comparison is relatively easy to programme and does not have elaborate computational requirements. The herdmate comparison can be computed on a sequential basis, thus reducing computer storage requirements in contrast to procedures which require a matrix inversion process.

The herdmate comparison largely removes from the estimate of a sire's breeding value complications arising from herd, year, and season of calving variations in production. Herd-year-season yield is based on a 5-month moving average. The appropriate herdmate average for each daughter of a sire is obtained by averaging all records of daughters of other sires calving in the herd in the same month as that daughter, in the previous two months, and in the succeeding two months. In order to increase the accuracy of herdmate comparisons, the predicted difference (PD) is used for ranking sires on a standard basis. This value is the expected production deviation of a sire's progeny from their herdmates in breed-average herds. When each daughter of a bull has only one record, the PD is computed as follows :

$$\text{PD} = n/n + 20 \text{ (adj. daughter av. – breed av)}$$

where, $n/n + 20$ is an adjustment for varying number of daughters :

$$\text{adj. dau. ave. = dau. av. – 0.9 (adj. herdmate av. – breed av.).}$$

*Reproduced from, *Breeding and feeding for milk production in Operation Flood II*. National Dairy Development Board 1980 with permission. *See* references in the original.

The factor 0.9 is to correct daughters' records that reflect about 0.9 of the difference between the adjusted herdmate average and the breed average.

$$\text{AHMA} = \text{RBYSA} + \frac{1}{\text{N}+1} - (\text{HMA} - \text{RBYSA})$$

where, AHMA = Adjusted herdmate average for each lactation; RBYSA = Regional breed-year-season av., N = Number of herdmates; HMA = Herdmate average.

A regional breed-year-season average is used to minimise effects of location specific factors on yields in different regions of the country, which reduces environmental distortion, of results. It should be calculated from the 305-day, 2X, ME records of cows which have calved over a period of several years (10-15 years). The use of N/(N + 1) is to correct for the number of records. The AHMA will increase with increase in number of herdmates. When several sires are being compared, and some sires are used more often in the better herds, the bias is removed by use of the adjusted daughter average, which increases in poorer herds and decreases in better herds.

Hargrove *et al.* (1974) found that the daughter-dam method was least satisfactory. Predicted difference from herdmates showed essentially no improvement when adjusted for dams. PD from contemporaries usually gave lower correlations than PD from herdmates.

Herdmate Comparison vs. Linear Model Techniques. Cunningham (1965) pointed out that the herdmate comparison procedure amounts to a two stage process in which herd and year-season effects are removed (by expressing the records as deviations from herd-year-season means) and sires are evaluated by analysing these deviations, ignoring herds and year-seasons. As an alternative to the herdmate comparisons, Cunningham (1965) suggested the simultaneous consideration of the sire, herd and year-season effects by weighted least squares procedures. This method requires the application of a linear model requiring matrix inversion for a solution. Depending upon whether the effects are fixed, random or mixed, it may lead to a least squares analysis or maximum likelihood analysis.

Miller *et al.* (1968) compared the above three methods of sire evaluation, *i.e.*, herdmate comparison, least squares and maximum likelihood. The results for all the three methods were in close agreement, as measured by the rank correlations of sire effects. These correlations ranged from 0.94 to 0.99 based on breeding values for the same trait. Maximum likelihood, theoretically, circumvents certain biases which may affect herdmate and least squares results, but such situations were apparently of little importance in the data studied. Maximum likelihood estimates in theory have the smallest possible sampling errors. It was concluded that the three methods rank bulls in the same order and with similar precision.

Modified Contemporary Comparison. The use in dairy sire evaluation of only progeny in the first lactation having contemporaries leads to less precision than is possible if all progeny with one or more herdmates of any age are utilized. Comparing cows in first lactation with older cows that are survivors of culling for yield can be an important source of bias in dairy sire evaluations. These biases should be considered in any sire evaluation procedure. McDaniel *et al.* (1973) presented a modified contemporary comparison which uses all cows with at least one herdmate of any age but which is unbiased by culling. For first lactations, the modified contemporary average (MCAv) is calculated as follows :

$$\text{MCA}v = n_1 x_1 + w(\bar{x}_1 - bias)/(n_1 + w)$$

where, n_1 = number of contemporary first lactations;
x_1 = av. yield of contemporary first lactations;

x_1 = av. yield of later lactation herdmates;
w = weight to give to later lactation herdmates (probably 1, 2, or 3);
bias = adjustment for later lactation cows being the survivors of culling for yield or
= average within herd-year-season difference between first and later lactation cows, adjusted for genetic trend.

Thus, herdmate average adjusted for the mean bias expected from female selection would be used for cows in first lactation that did not have contemporaries. Herdmates in later lactation would be used progressively less as more contemporaries become available. Deviations of cows in first lactation from the modified contemporary average should be unbiased and have less sampling variance than deviations from only contemporaries.

The modified herdmate average (MH*m*A*v*) for cows in later lactations would be constructed in the same manner so that it would be an unbiased deviation from other later lactation survivors of culling as follows :

$$\text{MCA}v = n_1x_1 + w(\bar{x}_1 - bias)/(n_1 + w)$$

where, n_l = number of later lactation herdmates and x_1, $\bar{x}_1$, *bias* and w are the same as defined previously.

In general, the modification to the later lactation average would be quite small, as a large proportion of cows in later lactations have five or more similar aged herdmates.

The modified contemporary comparison (MCC) replaced the herdmate comparison that had become vulnerable to biases from genetic improvement over the past decade in the USA. The MCC methodology incorporated effects were either not considered or considered less effectively in the herdmate comparison. Some of these effects are genetic differences in herdmates, pedigree information, daughter distribution over herds, length of lactation, number of herdmates, number and average repeatability of herdmates sires. Norman *et al.* (1976) observed that the predicted difference by MCC had the greatest advantage over that of the herdmate comparison when bulls had fewer daughters.

$$\text{MCCPD} = \text{R (MCD)} + (1 - \text{R) GA}$$

R = repeatability of herdmate comparison;
MCD = modified contemporary deviation or the weighted difference between D and MCA*v*.
GA = group average of MCD's of bulls with similar pedigrees.

Multiplying (1 – R) GA by various constants (K) produced modifications to PD that gave less or greater emphasis to pedigree than the MCC method.

Thus, PD's with variable pedigree emphasis were calculated by :

$$\text{PD}_k \text{ (variable pedigree)} = \text{R(MCD)} + \text{K}(1 - \text{R})\text{GA}.$$

Most data available to animal breeders do not meet the usual requirements of random sampling, the problem being that the data arise either from selection experiments or breeder's herds which are undergoing selection. Consequently the usual methods are likely to yield biased estimates and predictions. Henderson (1975) presented a method known as best linear unbiased prediction for dealing with such data. The computing technique is a simple modification of least squares.

In conclusion, selection of herd sires may be carried out in a three-tier system : (*i*) First, based on pedigree performance (sons of high yielding dams and proven sires etc.), certain number of bull calves may be set apart for rearing; (*ii*) Some sort of further selection of the above group of bull calves may be carried out based on their growth, vigour, age at puberty and if possible, libido and semen quality (as no genetically superior bull is good if its semen is bad); (*iii*) Finally the ultimate young sires should be progeny tested on institutional as well as farmers herds. They may be first chosen via preliminary evaluation based on part first lactation yields of daughters, to be ultimately evaluated confirmatorily (proving) based on complete daughter's data. Semen from such proven sires should be collected substantially and stored by deep freezing for further extensive use.

17.2.5. Selection of Females

In dairy cows, often, economic and other considerations make phenotypic selection the most suited one. Records of production of the individuals are compared to the unselected average production of the herd from which the animal comes from. The more number of lactation records on which a cow's average production is calculated, the more reliable it will be as an indicator of its genetic potential for production. Therefore, more emphasis should be placed on such a value as an indicator of her producing ability. A selection standard will have to be formulated for each separately and selection should be relative rather than rigid.

The most probable producing ability of a cow represents mostly the inherent producing ability of a cow. The regression of the future records on the present records or an estimate of how far the milk yield will be repeated in the next lactation will give a reasonable idea about the genetic potential of the animal for milk production. The 'most probable producing ability' of a cow is calculated on the basis of differences between the cow's n records and the herd average as follows :

$$\frac{nr}{1+(n-1)r} \quad \frac{(\text{Cow's record} - \text{Yearly herd average})}{n}$$

where n is the number of records and r the value of repeatability.

The cows in a herd should be ranked on the basis of their expected producing abilities and a certain percentage depending on genetic and economic considerations be culled from the lowest strata.

Cows giving very low yield of milk under normal circumstances should be culled outrightly. In doubtful circumstances one more chance in the form of a second lactation can be allowed.

Pedigree selection is of limited value in selecting dairy cows. But, in certain circumstances, particularly in selecting dams of young bulls, it is worthwhile to pay attention to the information on close relatives over and above its own performance. It is futile to consider distant relatives with relationship below 0.25 to the cows.

Most of the emphasis should be given to the actual record of the cow for traits that are highly heritable. The relative's performance assumes more and more importance as the heritability of the trait become lower and lower. In predicting a cow's breeding value with respect to milk and fat production, the performance of her mother and each daughter should get about 30-35 per cent of the emphasis placed on her own phenotype. A selection index combining the information on the individual itself and its close relatives may be used.

CHAPTER 18

Methods of Breeding

Breeding or mating of animals bas two aspects : (*i*) which group/breed of bulls to be mated to which group breed of cows *i.e.,* system of mating or breeding; and (*ii*) which individual bull is to be mated to which cow or cows *i.e.,* pairing. First the breeder has to determine his breeding policy—which system of breeding he would follow on a farm or in a location. Then he should consider formulating plans for individual pairings.

Before going further into the study of breeding systems the reader should have studied the principles of selection in the previous section; selection and mating are two inseparable aspects of genetic improvement of animals.

A breeding system is one which allows the breeder to alter the genotypes in his herd (or in a locality) in a desired direction, resulting in evolution of animals with desired levels of performance. Individual pairing plans have to be decided next, depending on his subsidiary aims like—mating individual bulls to given sets of cows for progeny testing; breeding individual animals to a particular bull with which close relationship is desired; pairing related individuals like dam to son, sire to daughter, so on.

18.1. SYSTEMS OF BREEDING

Ivar Johansson (1962), the renowned Swedish cattle breeder, classified the commonly used breeding or mating systems as follows (also *see* Figures 18.1 to 18.4).

A. Mating within a Breed (Purebreds)

1. ***Random mating.*** The same probability exists for each possible mating in the population.

2. ***Inbreeding (Inbred).*** The mates are more closely related than the average relationship between the animals within the breed. (Figure 18.1).

Inbreeding for the Formation of Inbred Lines or Strains. Continued sib matings are the most effective means of the line formation. To be classed as an inbred line the individuals should

---

Johansson, I. (1962). *Genetic aspects of dairy cattle breeding.* Oliver and Boyd, London, p. 186.

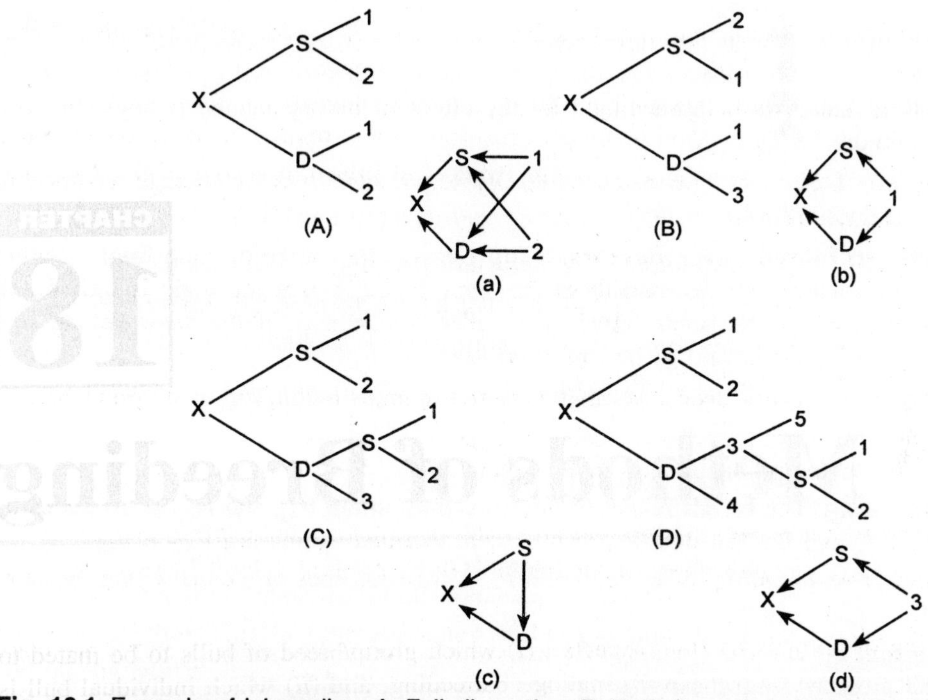

Fig. 18.1. Example of inbreeding. A. Full-sib mating. B. Half-sib mating; C. Parent-offspring mating; D. Grand parent-offspring mating; *a*, *b*, *c* and *d* are, respectively the arrow diagrams of the same: S is the male (sire) and D the female (dam) parent and X is the inbred individual.

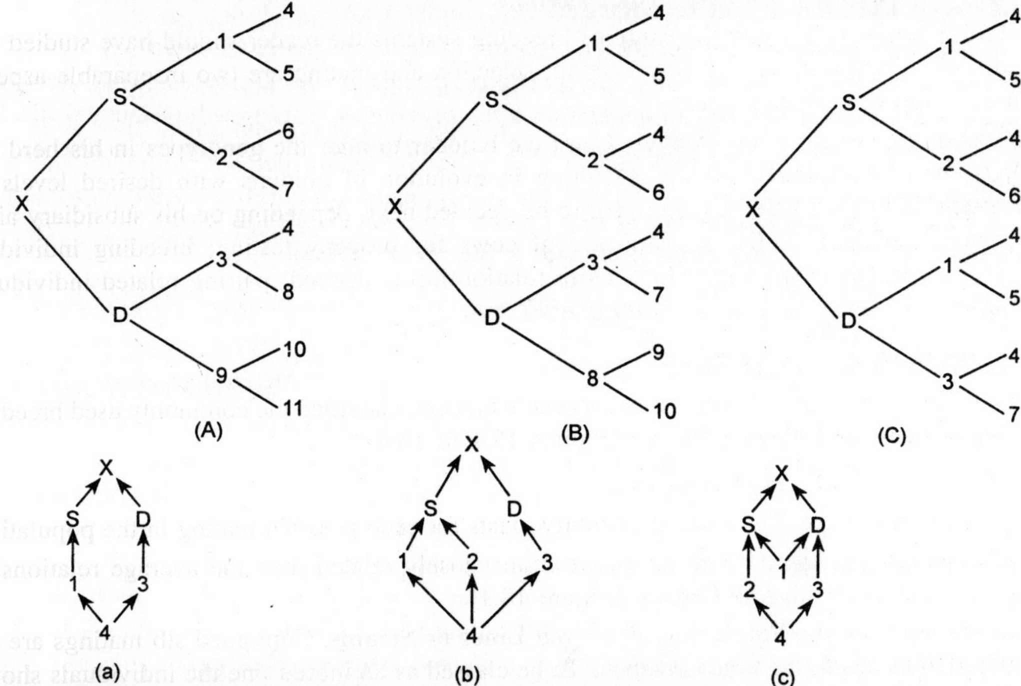

Fig. 18.2. Examples of line breeding in which relationship is maintained with a renowned male ancestor (No. 4) not so closely (A), closely (B) or very closely (C), *a*, *b* and *c* are, respectively arrow diagrams of the same. There can be other ways of linebreeding.

have reached an inbreeding coefficient of at least 37.5 per cent corresponding to two generations of full-sib mating.

Line Breeding (Linebred). Inbreeding with the effect of increasing the relationship to a certain ancestor. (Figure 18.2).

3. ***Out Breeding (Outcross).*** The mates are less closely related than the average relationship within the breed. (Figure 18.3).

Crosses between Inbred Lines (in cross) : Top Cross. The mating of male from a certain line or strain to non-related females usually of the same breed. When the male is inbred and mated to non-inbred dams of the same breed, it is called *top incross*. If the dams belong to a different breed, the offspring is said to be *top cross-bred*.

B. Cross-Breeding (cross-bred). Matings between animals belonging to different breeds. (Fig. 18.4).

1. Cross-breeding for production of an F_1 generation.

2. ***Double-crossing.*** The breeds A and B are crossed to obtain F_{AB} and F_{CD} is produced by crossing the breeds C and D. The double cross is then obtained by mating $F_{AB} \times F_{CD}$.

3. ***Rotational cross-breeding.*** (Can be used also in crossing lines or strains within the same breed).

Criss Crossing. F_1, animals (females) from a cross between the breeds A and B are back-crossed to A; the progeny from these matings are back-crossed to B; etc.

3-Breed Rotation. F_1 females from cross between A and B are mated to males from breed C; the female offspring from these matings are mated to males of breed B; etc.

4-Breed Rotation. It can be organised in a similar way.

4. ***Grading (grades).*** Continued back-crossing to the same (parental breed) *e.g.* scrub cows are mated to purebred Holstein bulls; the female offsprings are back-crossed to Holstein bulls; and such back-crossing is repeated generation after generation. This breeding method has been used extensively in areas where purebreds have been imported to improve the native cattle or mongrels.

5. ***Crossing for the foundation of new breeds.*** The Santa Gertrudis breed, for example was developed at the King Ranch from Brahmans × Shorthorn crosses.

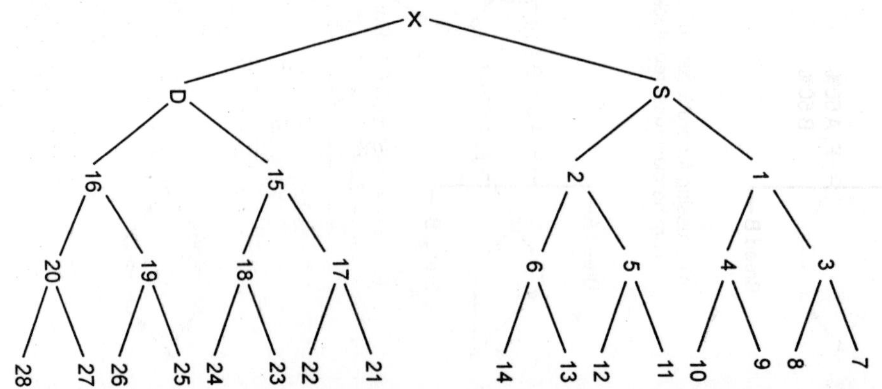

Fig. 18.3. An example of cutbreeding. There is no common ancestor upto four generations on either parent side of the individual X and there is no evidence of mating related individuals, either.

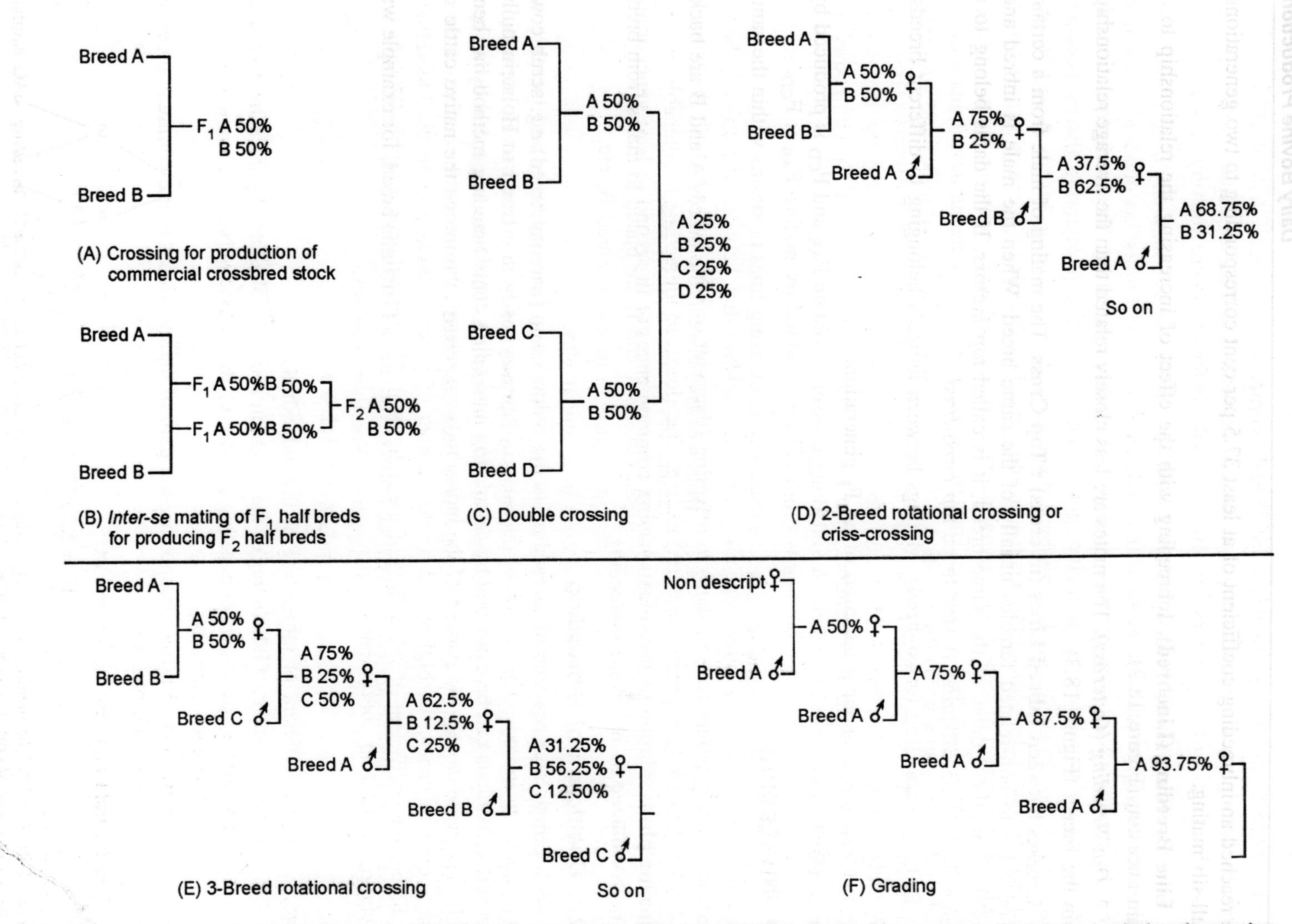

Fig. 18.4. Different system of crossbreeding : I. Crossing for production of halfbreds (F_1); II. Double-crossing; III. 2-breed rotational crossing or criss crossing; IV. 3-breed rotational crossing and V. Grading-up.

The extreme of mating unrelated individuals comprises of *species hybridization*, where individuals belonging to two different species are crossed. The best example of this type of mating is the mule, which is obtained by crossing male donkey to mare. Such hybrids are sterile.

Selective Breeding or Selective Pure Breeding. Another term commonly used in practice, refers to the mating of selected males to selected females within a breed, as opposed to *random mating*. In crossbreeding work *inter se* mating is sometimes practised. This technique refer to mating among males and females of a given genetic group, resulting in offspring also of the same genetic group. For example, the F_1 crossbred males (obtained by mating breeds A and B) are mated to F_1 females (also obtained by mating breeds A and B), obtaining F_2 generation crossbreds. The relative proportion of 'blood' of original parental breeds will be same in the animals beings *inter se* mated as well as their offspring. Though the average value of quantitative traits may be similar in F_1 and F_2 crossbreds the variation is higher in F_2. Mean lactation milk yield with their standard errors for Jersey × Sahiwal filial generations F_1, F_2 and F_3 obtained in Australia (Hayman, 1977) were 758 ± 114.9, 895 ± 150.0 and 861 ± 260.8 kg, respectively. Note the increase in standard error with each filial generation. Also F_2 individuals vary more than F_1 with respect to some physical traits.

18.2. INBREEDING

Inbreeding can be close or wide depending upon the relationship between the animals being mated, like mating parent (sire or dam) to offspring (daughter or son), full-sib (both the parents common) or half-sib (one parent common) mating. The closest of inbreeding is attainable in self-pollinating plants. Mating of individuals having common ancestor in second or later generations can be considered as not close inbreeding.

18.2.1. Coefficient of Inbreeding

The main genetic consequence of inbreeding is the reduction in heterozygosity and increase in homozygosity in inbreds. The degree of decrease in heterozygosity in a free-mating population of related individuals can be estimated by calculating inbreeding coefficient suggested by Wright (1921). The more closely the parents of the individuals are related, the more the chances of their having common genes, the higher will be the number of genes homozygous in the individual, and, higher the coefficient of inbreeding of the individual. The formula suggested by Wright for calculating inbreeding coefficient is :

$$F_x = \Sigma(½)^{n+n_1+1}(1 + F_a)$$

where, F_x = coefficient of inbreeding of the individual;

F_a = coefficient of inbreeding of the common ancestor calculated by the same formula;

n = number of arrows connecting the sire of the individual to the common ancestor via each pathway;

n_1 = number of arrows connecting the dam of the individual to the common ancestor via each pathway.

The formula can also be written as :

$$F_x = \Sigma(½)^{n+1}\ (1 + F_a)$$

Hayman, R.H. (1977). The development of the Australian Milking Zebu. In : Animals Breeding. FAO Animal Production and Health Paper No. 1, FAO, Rome.

Wright, S. (1921). Genetics **6** : 111.

if one considers '*n*' as the number of arrows connecting dam and sire via the common ancestor ($n = n + n_1$). If the ancestor in question is not inbred itself, then the value ($1 + F_a$) can be deleted from the formula.

Drawing the pedegree details of the animals in the form of an arrow diagram (Figure 18.5) helps greatly in finding out common ancestors and their exact position in the ancestry.

Since there can be several pathways (several intermediary parents) through which the individual can get genes from the common ancestor(s) all possible pathways should be considered in calculating the estimate. An individual cannot be a common ancestor if it does not appear in the sire as well as dam side of the ancestry at least once. In other words, an individual whose parents are not related cannot be an inbred.

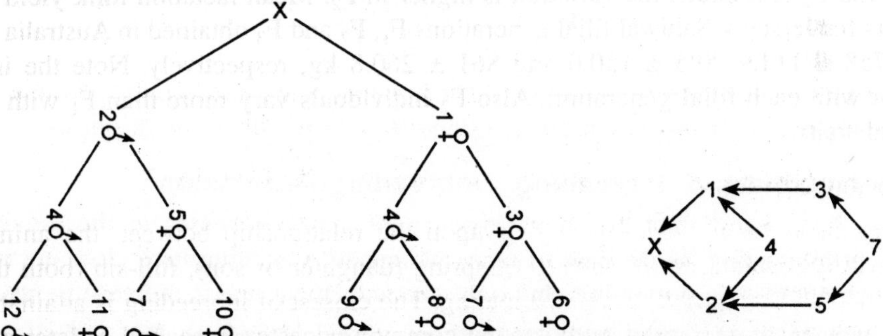

Fig. 18.5. Pedegree chart of the Simmental cow Gvozdika (X) and the same represented as an arrow diagram. Bulls Nos. 4 and 7 are the common ancestors of the cow.

Given in Figure 18.5 is the pedegree of the Simmental cow Gvozdika (X) of Terizino Stud Farm in Soviet Ukraine (Kravtchenko, 1973). She has two common ancestors—4 (grandparent) and 7 (great-grandparent). Animal 7 is itself inbred with an inbreeding coefficient of 0.125. The mode of calculating the inbreeding coefficient of the individual 'X' is shown in table given below. Drawing an arrow diagram is very helpful in this regard :

Common ancestor	*n*	*n + 1*	$(½)^{n+1}$	F_a	F_x
No. 4	2	3	0.1250	0	0.1250
No. 7	4	5	0.0313	0.1250	0.0352
				Total	0.1602
			or	16.02 per cent	

18.2.2. Coefficient of Relationship

Another common estimated parameter of inbreeding is the coefficient of relationship. Wright suggested the following formula for estimating this parameter :

$$R_{xy} = \frac{\Sigma\ 0.5^n(1 + F_A)}{\sqrt{(1 + F_x)(1 + F_y)}}$$

Kravtchenko N.A. (1973). *loc. cit.*

where, R_{xy} = coefficient of relationship between individuals X and Y.

n = the number of arrows connecting individual X (sire) to Y (dam) through the common ancestor A via each pathway. In tracing the connection between 'X' and 'Y' via A, no intermediary animal should be passed through more than once in the same path.

F_x = inbreeding coefficient of x

F_y = inbreeding coefficient of y

F_A = inbreeding coefficient of A.

For calculating genetic relationship between an individual and its ancestor such as meritorious bull, Kravtchenko (1973) suggested the following modification of the Wright's formula.

$$R_{xA} = \Sigma(½)^n \sqrt{\frac{1.F_A}{1.F_x}}$$

Explanations of the various notations are the same as for the Wright's formula, except n which is the number of arrows connecting individual x to the ancestor in question (A).

18.2.3. Consequences of Inbreeding : Inbreeding Depression

It has already been mentioned that homozygosity of genes increases in the population as a consequence of inbreeding. In the case of recessive genes inbreeding itself does not increase the number of such alleles in a population, but only uncovers them due to increase in homozygosity.

Consequences of increased homozygosity of genes depend on the type of gene action involved. Since there is no dominance or recessiveness in additive genes, increase in homozygosity of such genes as a consequence of inbreeding may not actually affect quantitative traits, if selection does not operate. However, changes in expression of such traits may occur under influence of other types of gene action.

Whereas preponderance of recessive genes in homozygous condition results in overall decline in the performance of animals. Such a condition is known as inbreeding depression.

TABLE 18.1. Effect of inbreeding on growth and reproductive efficiency of Holstein Heifers

Coefficient of inbreeding	*Body weight at birth (kg)*	*Body weight at 24 months of age (kg)*	*Services per conception*
0.00	39	505	2.18
0.26	35	446	2.38
0.33	34	438	2.78
0.44	32	432	2.81
0.45	30	433	3.34
0.60	26	402	4.57

Note. There was a 15 per cent mortality among inbreds. Age at maturity in inbred animals was 412 days as against 383 days in control animals.

Source : Kravtcbenko (1973). *loc. cit.*

By causing increased homozygosity, inbreeding fixes characters—favourable or unfavourable—in the population. It is generally recognised that dominant genes are favourable and recessive ones unfavourable. It has now been clearly established that most of the genes that

---

Kravtchenko N.A. (1973). *loc. cit.*

have bad or lethal effects on animals are recessive in nature. Normally they remain "shadowed" by their dominant allele. But, due to inbreeding, homozygous recessive individuals for various lethal and sub-lethal genes may result.

Lethal genes cause death of the individual before or after birth. There are other sub-lethal genes that cause various deformities in dairy cattle or affect functioning of some vital system resulting in loss of vigour and production of animals. Some common genetic defects in dairy cattle include albinism, hairlessness, congenital dropsy and anatomical defects of genitalia, blindness, bull-dog syndrome, certain types of hernias and skeletal defects, closed anus, dwarfism, etc. All of them are caused by recessive genes.

The known features of inbreeding depression in dairy animals are—(*i*) birth of dead, deformed and under-developed calves; (*ii*) fall in fertility (Table 18.1); (*iii*) fall in growth rate (Table 18.1); (*iv*) reduction in body resistance; and (*v*) fall in production (O.A. Ivanova from Soviet Union reported a reduction of 270 kg in lactation yield and 0.01 per cent in fat content in inbred Kastromsky cows as compared to outbred cows).

18.2.4. Uses of Inbreeding

In spite of the risks involved, inbreeding can be, and has been an important tool in the hands of skilled breeder. Some of the uses to which inbreeding can be put are as follows :

1. As a means of retaining in progeny inheritance of valuable individuals, especially for such characters that are new, complex and rarely met;
2. As a means of increasing uniformity and stability of genotypes within a group of animals;
3. As a means of identifying and getting rid of undesirable recessive genes;
4. As a means of forming within populations, groups of animals (lines, families) homogeneous for particular traits, with a future aim of crossing between groups later. This practice is widely used in poultry and swine breeding.
5. As a means of increasing prepotency of animal, especially of bulls; prepotency being the ability of an animals to transmit its characteristics truely to the offspring to the exclusion of those from the other parent;
6. As a means of testing the real breeding worth (genotype) of a sire by crossing it to 20-30 of his own daughters. Progeny of such crosses gives some idea about the desirable and undesirable genes of the sire in question;
7. As a means of evolving new specialised breeds.

Inbreeding was widely used in evolving major British breeds of cattle with their typical characteristics. After analysing the breeding programmes associated with breed evolution in several countries over the last century, the Russian professor P.N. Kuleshov had come to the conclusion that practically there is no breed in whose development inbreeding has not been used at certain stage or other.

18.3. OUTBREEDING

Outbreeding or outcrossing involves mating of unrelated individuals within a breed. Selection in some form or other is practised simultaneously to outcrossing, as a result, such a practice is often termed as 'selective or purebreeding'. Progress in the development of the present day developed dairy breeds of the world has been mostly achieved by this method.

The basic objective of such a system of breeding is to stabilize breed characteristics or to improve certain traits in individuals. The latter means to breed animals for a specialised purpose, like milk production; the basic characteristics of breeds not being disturbed.

The important steps involved in such a breeding are : selection of individuals with well developed trait(s) in question; specific pairing of mates excelling in desirable traits of the breed; rational maintenance of animals in condition that are ideal for expression of the characteristics for which the breed is reared. Thus the old axiom, "best to best" is actually practised in this method.

At this stage, the reader's attention is drawn to the various methods of estimating the breeding value of animals and their selection described in Chapter 17.

Genetic progress through selective breeding is very slow, especially in dairy cattle due to long generation interval (interval between the birth of an individual and birth of its offspring). Assuming an optimistic figure of 5 per cent genetic gain per generation, it will take about 15 to 16 generations to double the present level of 1000 kg per lactation of some Indian Zebu breeds. This will be in the neighbourhood of 90 to 100 years.

Further, there is a sort of upper limit to the progress that can be achieved by outbreeding. This is so because, by this method of breeding attempts are made to concentration within individuals of desirable genes that are present within the breed. Since theoretically the desirable number of genes within a group (breed) are limited, so is the scope of improvement through selective pure breeding.

At the National Dairy Research Institute of India, selective breeding has been carried out in Tharparkar and Sahiwal breeds for several decades. These herds have achieved their best level of production beyond which further progress has been improbable. (Sundaresan, 1976).

18.3.1. Consequences of Outbreeding—Heterosis

Inbreeding depression and heterosis or hybrid vigour are opposite phenomena with apparently the same genetic basis. Thus the genetic consequences of outbreeding (and crossbreeding) is just the opposite of inbreeding, namely, increase in the number of heterozygotes in the population for the various alleles. Hence the chances of outbreds and crossbreds breeding true are less.

In contrast to inbreeding, the outbred progeny is generally superior to either of the parents in vigour or hardiness, viability, growth and reproduction. Heterotic effect on milk production is not always apparent; milk yield of *Bos taurus* × *Bos indicus* breeds is generally midway between the two parent breeds.

Heterotic effect could arise because of increased heterozygosis in the first generation outbreds or crossbreeds. The parents might be homozygous recessive for different sets of alleles. In their progeny these different alleles tend to become heterozygous as shown below.

Parent A	×	Parent B
PPQQ rrss		ppqq RRSS

$$F_1 = PpQqRrSs$$

Sundaresan, D. (1976), cited by Katpatal (1977).

This results in masking of the recessive (undesirable) by its dominant allele (desirable) in the progeny. Heterozygosis in a large number of alleles thus makes the progeny superior to parents. Further, super-dominance (heterozygote superior to either of the homozygotes) and epistatic effect may also cause heterosis. Additive genes are not likely to be responsible for heterosis; as mentioned, milk yield of *Bos taurus* × *Bos indicus* breeds is generally midway between the parents. Greater the diversity between animals being mated (like members of two very different breeds); greater would be the heterotic effect. Such traits that are mostly depressed due to inbreeding show the greatest heterotic effect, *i.e.,* birth weight, survival rate, early growth and vigour. Traits that are lowly heritable like reproduction, are most affected, the degree of heterosis among highly heritable traits being less.

The heterotic effect will be most pronounced in the F_1 generation obtained by crossing different breeds; while further filial generations (*inter se* mating of F_1 and F_2) causes reduction in heterosis. In Sri Lanka while local Zebu × European cross F_1 cows yielded 1482 kg in the first lactation, the F_2 cows yielded 981 kg and the F_3 cows 957 kg per lactation (Buvanendran and Mahadevan, 1977). However, it should be remembered that other factors are also responsible for fall in production of late filial generations, apart from loss of heterotic effect. Taneja and Chawla (1978) estimated heterosis of various traits in Sahiwal × Brown Swiss crossbred grades which were as follows :

Age at first conception	— 29.2 per cent
Age at first calving	— 29.1 per cent
First gestation length	— 0.9 per cent
First lactation total yield	— 25.6 per cent
First lactation 305 day or less yield	— 35.0 per cent
First lactation length	— 15.3 per cent
First dry period	— 0.01 per cent
First calving interval	— 22.1 per cent

Note that magnitude of traits with negative signs is lowered and of those with positive signs is increased.

18.4. CROSSBREEDING

Mating of animals of two or more different breeds is known as crossbreeding. It is mainly used for production of commercial livestock, meat animals and birds. Its use in countries where highly developed purebreeds are available is to maintain heterosis which cannot be fixed by inbreeding techniques. But in India and in many other developing countries, crossbreeding has another very important use.

18.4.1. Why Crossbreeding

The native breeds of dairy cattle have low genetic potential for milk production due to various reasons. But they have such qualities as adaptability to hot climatic conditions, resistance to many diseases prevalent in the tropical regions and general thriftiness under inferior feeding and management conditions.

Taneja, V.K. and Chawla, D.S. (1978). *Indian J. Dairy Sci.* **31** : 3.

High producing specialised dairy breeds from temperate region when introduced into tropical and sub-tropical zones generally degenerated due to lack of adaptability to hot conditions and resistance to tropical diseases. Under such circumstances, crossbreeding of local breeds with improved exotic breeds has brought about considerable improvement in production with sufficient capacity to withstand the direct and indirect adverse effects of hot climate.

Such crossbreds are more or less a compromise between the ability to produce and the ability to adapt to hot conditions. The term "crossbreeding" is used here rather loosely; the various forms of this system are described earlier in this section (18.1). The breeder has the option to follow one or the other of the variants of crossbreeding (*see* also Figure 18.4) or a combination of 2 or more variants to suit his requirements.

Rotational crossing (2, 3 or more breed rotations) are mostly used for evolving new breeds of cattle. Half-breds may be used for production of commercial stock, though such a possibility is difficult in dairy cattle, especially in countries like India. Grading up the local non-descript cattle is a high potential and practical method of dairy animal improvement in developing countries. For grading up local cattle, well developed pure Zebu bulls (Sahiwal or Red Sindhi), or half-bred bulls (F_1-*Bos indicus* × *Bos taurus*) or even purebred *Bos taurus* bulls (where optimal conditions of dairying exist) can be used, depending upon the goals to be achieved in each region.

18.4.2. Experiences of Crossbreeding in India*

Crossbreeding indigenous cattle breeds with superior exotic dairy breeds for milk production has been adopted as a national policy in India. The genetic variability in most of the characters of economic importance in the indigenous breeds is larger than in the European breeds and selection for milk within the indigenous breeds has been fairly effective in bringing genetic improvement (Acharya, 1979). However, such a change will be extremely slow in bringing a substantial increase. Therefore, where feed resources and physical environmental conditions permit, introduction of superior European dairy inheritance to raise the initial level of production followed by inter-breeding among crossbreds and selection would be a rapid alternative. The results of the various crossbreeding experiments and the broad conclusions and decision that can be drawn out of them are discussed in this section. Details of various performance traits of different crossbreds have been presented in Chapter 16.

Choice of exotic breed. Jersey has been considered most suitable for Indian conditions because of its comparatively smaller size, early sexual maturity and higher fat content of milk. Further, as some workers suggest, it appears that Jersey had originated from *Bos indicus* and its ancestors migrated from the Indus Valley to Europe through Africa. It has also been indicated that Jersey has certain genes associated with the immunologically and biochemically polimorphic traits, usually missing in other European dairy breeds and prevalent in Zebu breeds, such as blood group allele Z and hemoglobin B type. This would make Jersey genetically more closer to indigenous breeds than other European breeds. (Acharya, 1979).

---

*The author quotes in this section material from (*i*) the article entitled *Strategy for Breeding Dairy Cattle in India for Enhanced Milk Production* by Dr. R.M. Acharya in Indian Dairyman (Sept. 1979) and (*ii*) the bulletin, *Breeding and Feeding for Milk Production in Operation Flood II*, published by the National Dairy Development Board, Anand, India. 1980. These sources may be consulted for the references quoted in this section.

TABLE 18.2. Relative Efficiency of Brown Swiss and Holstein Crosses with Sahiwal Expressed as per cent of Performance of Sahiwals

Trait	*BS × SW*	*HF × SW*
Birth weight	21.3	11.5
Weight at 6 months	10.9	22.2
Weight at 12 months	7.3	21.2
Weight at 18 months	5.3	16.5
Weight at 1st calving	4.8	5.7
Age at 1st calving	–21.3	–5.9
1st lactation yield	62.7	23.4
1st lactation length	4.5	–3.9
1st dry period	–44.8	–7.3
1st calving interval	–10.0	–4.9

BS × SW = Brown Swiss × Sahiwal.
HF × SW= Holstein × Sahiwal.
Source : Chawla and Mishra (1978). *loc. cit.*

Earliest attempts in India have been to compare different grades arising from one exotic breed, mainly based on data from Military Dairy Farms and other institutional herds. Most of such studies revealed that Holstein Friesian seem to perform best followed by Ayershire and Jersey. Recent studies of Nagarcenkar *et al.,* (1978), however have shown the Brown Swiss to be the best followed by Jersey and Friesian. Brown Swiss was put above Friesian by Chawla and Mishra (1978) based on the performance of the Karnal herd of the National Dairy Research Institute (Table 18.2).

TABLE 18.3. Comparative Performance of Different Grades of Sahiwal × Friesian Crossbred Cows on Military Dairy Farms

Characters	*Different grades of cows showing proportion of Holstein inheritance*								
	Sahiwal	*1/4*	*3/8*	*1/2*	*5/8*	*3/4*	*7/8*	*15/16*	*31/32*
First lactation yield (kg)	1772	1550	2114	2538	2415	2393	2268	2111	1837
Number of animals	118	24	96	117	72	307	218	95	10
First lactation length (days)	296	284	297	308	292	306	299	294	263
First calving interval (days)	421	417	445	432	433	468	462	442	463
Age at first calving (months)	38.3	37.5	38.5	36.2	35.9	36.3	36.0	37.0	36.8
Percentage culled till first calving	28	33	41	4	3	17	32	37	47
Mortality till first calving (%)	15	19	28	4	16	10	18	16	32
Adult mortality rate per year	9	9	11	2	7	11	11	11	12
Index life time production (relative to Sahiwal)	1	0.5	0.9	2.6	1.6	1.4	1.0	0.9	0.6

Source : Amble and Jain (1966). *loc. cit.*

Level of exotic inheritance. Data from Military Dairy Farms and other institutional herds have been extensively studied and reported by several workers (Acharya, 1970; Bhat, 1974; Sundaresan, 1976; Katpatal, 1978 and Bhat, 1978). It has been generally concluded that exotic inheritance between 50 to 62.5 per cent gives the best overall performance (Table 18.3). In majority of cases the halfbreds have performed the best and there is little gain beyond 50 per cent exotic inheritance. The 3/4 breeds arising from the same exotic breed involved at halfbred level or another exotic breed have shown no superiority in performance (Acharya, 1979).

TABLE 18.4. Ranking of Various *Bos Taurus* × *Bos Indicus* Crosses (in Descending Order) for Efficiency of Milk Production

Net energetic efficiency	*Dairy merit*
A. Southern Region	
1/2 F	1/2 F
1/2 RD	1/2 RD
1/2 J	1/2 J
5/8 F	5/8 F
3/4 F	3/4 F
7/8 F	7/8 F
B. Northern Region	
1/2 BS	1/2 BS
3/4 BS	3/4 BS
1/2 J	1/2 J
1/2 F	1/2 F
3/4 F	3/4 F
1/2 BS	1/2 BS
5/8 F	5/8 F
7/8 F	7/8 F
3/8 F	3/8 F
1/8 F	1/8 F
1/4 F	1/4 F

F = Friesian; RD = Red Dane; J = Jersey; BS = Brown Swiss.

Source : Rao, G.N. Murty, A.S.R. and Nagarcenkar (1977). *Military Farms J.* 9 : 24.

Analysis of the earlier data (already studied by others) by Nagarcenkar *et al.* (1978) has shown that for fat corrected milk/kg body weight, Brown Swiss × Sahiwal ½s were better than 3/4s. Friesian × Sahiwal 3/4s were similar to ½s, though the 5/8s were inferior. As regards net energetic efficiency and dairy merit, halfbred Friesian and Red Dane crosses ranked first in the Southern region while Brown Swiss halfbreds ranked first in the northern region (Nagarcenkar *et al.*, 1978). The overall rankings of the different grades and breeds of crosses, as per these authors are reproduced in Table 18.4.

In discussing milk production Acharya (1970) has opined that results of various studies can neither be pooled nor compared for universal application because of effects of breeds, herds, years, seasons, age, lactation number and genetic differences of sires. Nevertheless, most results do indicate clearly that milk yield and efficiency of milk production in relation to body weight has usually been higher in crossbred with 5/8 to 3/4 exotic inheritance (Singh and Desai, 1964 and 1966; Khanna, 1968; Katpatal, 1970; Narain and Garg, 1979).

Chawla and Mishra (1978) have shown (Table 18.2) that among Holstein × Sahiwal crosses, halfbreds were the best in age at first calving, first calving lactation milk yield, first dry period and first calving interval and there were no gains beyond 50 per cent Holstein inheritance. In

Brown Swiss × Sahiwal crosses the performance of 3/8s, though based on small numbers, was the best but halfbreds were decidedly superior to 3/4s, especially under field conditions (Table 18.5).

TABLE 18.5. Growth, Production and Reproduction Traits of Animals having 0%, 50%, 75% and 100% Brown Swiss Inheritance in Madupatty Station in Kerala under the Indo-Swiss Project

Trait	*0% BS (local cattle)*	*50% BS*	*75% BS*	*100% BS*
Birth weight (kg)		24.7	29.1	37.7
Body weight at 6 months of age (kg)		84.6	102.9	116.5
Body weight at I yr. of age (kg)		140.8	185.1	169.4
Body weight at 2 yr. of age (kg)		245.5	295.1	289.6
Body weight at 3 yr. of age (kg)		326.6	372.5	434.6
Postnatal mortality (%)	11.1	12.8	13.6	26.4
Age at first insemination (months)	28.2	23.9		24.2
Age at first calving (months)	38.5	34.5		34.7
AI Index (inseminations per calf)	2.2	1.8		2.1
Service period (days)	178	147		192
Intercalving period (days)	458	433		475
Lactation length (days)	155	292		300
Lactation yield (kg)	723	2499		2660

Source : Nair, P.N.R. (1973). *Evolutionary crossbreeding as a basis for cattle development in Kerala State (India).* Ph.D. thesis. University of Zurich. Juris-Druck + Verlag Zurich

The aforementioned facts suggest that it is desirable to maintain crossbred population with around 50 per cent exotic inheritance. This can by far be attained through the difficult path of *inter se* mating of selected males and females, especially the male.

Crossbreds in different climates. It appears that in general the crossbreds perform well under mild climatic conditions. The animals in arid and semi-arid conditions were slightly inferior to those under mild climatic conditions but were superior to those under hot-humid conditions. Under mild and arid/semi-arid conditions the crossbreds arising out of larger breeds (Holstein Friesian and Brown Swiss) had higher averages for lactation yield and milking average than Jersey crosses (Table 18.6).

But under hot humid conditions differences among them were small or non-existent and in some cases Jersey crosses were even better than the Holstein and Brown Swiss crosses. If corrections are made in milk yield according to fat content, Jersey crosses may prove to be superior to the crosses of Holstein and Brown Swiss.

Record performances. When crossing of different breeds or strains is involved, of the various crosses only one or two may "nick" well; progeny of such "nicked" ones turning out to be the best performers. Thus certain strains or genetic groups or breeds might exhibit specific combining ability. Nothing has been concluded based on Indian experiences with respect to the specific combining ability of European with Indian breeds. An exception is Acharya's (1970) attempt that indicated that Hariana seems to combine better with Ayershire and Holstein Friesian when compared to the other Indian breeds.

TABLE 18.6. First Lactation 300 days Milk Yield of Native and Different Halfbred Cows under Different Climatic Conditions

Location	*Mean daily temper-ature*	*Mean daily relative humidity*	*Average THSI**	*No. of months with THI above 75*	*First lactation 300 day milk yield (kg)*			
					*Native***	*FN*	*BN*	*JN*
Hissar (Hot-arid)	25.3	51.8	70.6	5	722±13	2618±56	2246±72	1954±78
Izatnagar (Hot semi-arid)	24.6	61.5	71.1	6		1835±55	1435±42	1485±37
Haringhata (Hot-humid)	26.3	73.5	75.1	8	455±24	2194±38	2148±35	2030±56
Rahuri (mild)	24.5	59.4	66.7	2	1064±61	3329±88		2536±68
Jabalpur (Hot-semi-arid)	24.5	55.0	71.8	6	676±38	2411±91		1972±61
Guntur (Hot-humid)	27.7	74.9	76.4	8	298±58	2390±87	1885±223	2011±285

$$\text{*Average THSI} = \frac{(S \times 0.75 \times \text{THI at 3 P.M.}) + (24 - 0.75S) \times \text{THI at 9 A.M.}}{24}$$

S = hours of bright sunshine; THI = 0.72 (dbt °C + wbt °C) + 40.6

**Native breed Hariana at Hissar, Izatnagar, and Haringhata; Gir at Rahuri and Jabalpur; and Ongole at Guntur. FN = Friesian × Native; BN = Brown Swiss × Native; JN = Jersey × Native.

Source : Thomas. C.K. and Acharya. R.M. (1981) *Indian J. Anim Sci.* **51** : 351.

However, there are a few instances where particular crosses between *Bos taurus* and *Bos indicus* resulted in production of record performers; their performances were so unbelievably good that they have been termed "freaks" by some. Mention may be made here of four individual cows KS65 belonging to the National Dairy Research Institute (Fig. 18.6), cows RS 110 and RS 10 (Fig. 18.7) belonging to the Punjab Agricultural University; and cow $8F_2$ 68 belonging to the Haryana Agricultural University. Their individual performances (Table 18.7) were much higher than the toppers of the herds to which they belonged.

Nutrition and Feeding. Careful management and adequate feeding of cows are essential components of any breeding programme. For crossbred cattle, attempts have been made to assess nutritional requirements. Ranjhan *et al.* (1977) have proven that for Holsteins versus cross-bred respectively the requirements were—metabolisable energy : 175.1 Mcal versus 153.0 Mcal: digestible crude protein per kg metabolic body weight 4.21 gm versus 3.73 gm; metabolisable

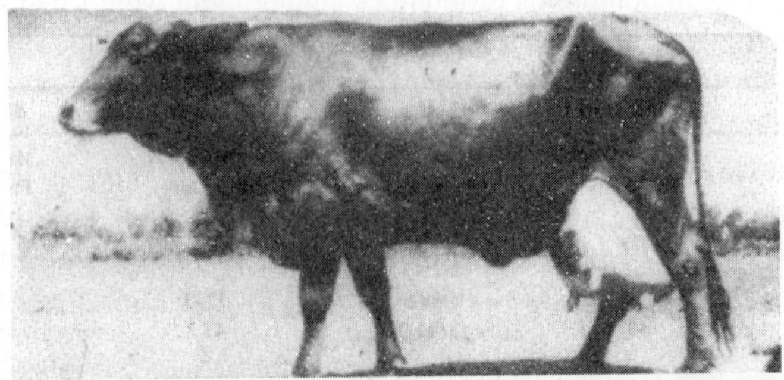

Fig. 18.6. Karanswiss cow (½ Brown Swiss-½ Sahiwal) belonging to NDRI. Karnal. Her best peak, yield was 43 kg and best lactation yield-5731 kg (*Photo Courtesy* : NDRI, Karnal).

Fig. 18.7. Half breed Red Dane × Sahiwal cow No. Rs. 110 belonging to Punjab Agril. University, Ludhiana. Her best peak yield was 50.9 kg and first lactation yield was 5205 kg (*Photo courtesy* : Dept. Animal Science, PAU, Ludhiana).

energy per litre of fat corrected milk (4 per cent) : 0.579 Mcal versus 1.165 Mcal. Nitrogen requirement per 100 gm milk nitrogen secreted was 143.3 g in crossbreds and 156.6 g in Holsteins.

TABLE 18.7. Performance of Some Outstanding Crossbred Cows

Particulars	*Cow Number*			
	RS 110	***Karan Swiss 65***	***RS 10***	***$8F_2$ 68***
Breed of dam	Sahiwal	Sahiwal	Sahiwal	Hariana
Breed of sire	Red Dane	Brown Swiss	Red Dane	Friesian
Age at first calving (months)	28	28	29	37
First lactation 305 days yield (kg)	5205	3820	3243	—
Highest peak yield (kg)	50.9	43.0	43.3	31.1
Best lactation	2	5	4	6
Best lactation 305 days yield	—	5731	6150	4264
Owners	Punjab Agric. University, Ludhiana.	NDRI. Karnal	Punjab Agric. University, Ludhiana.	Haryana Agric. University, Hisar.

It was proved that crossbreds required less energy for maintenance but more for milk production and that protein requirements were lower in crossbreds for maintenance as well as for milk production. Krishna Mohan *et al.* (1977) observed that there were no significant differences between various feeding standards studied for crossbreds. Their study indicated that to attain proper growth rate, crossbred heifers could be fed according to the standards prescribed either by National Research Council (1971) or by Sen and Ray (1964).

Disease problems. As early as in 1937, Macquckin observed that halfbred Sahiwal × Friesian compared well with their Sahiwal contemporaries in survival rate. Amble and Jain (1966) found the lowest rate of mortality in halfbreds before and after first calving and during their life time. The 5/8 crosses also evidenced low mortality rates before first calving. Abortion, still births and premature births were also reported to be lowest in halfbreds, followed by purebred

Sahiwals. The incidence of abnormal births increased as exotic inheritance increased beyond 50 per cent.

A comprehensive study (IVRI, 1975) has shown that incidence of infectious diseases, protozoan infections, helminthic infections, pneumonia, actinobacillosis, coccidiosis, mastitis and cow pox was more frequent among crossbreds than in purebred zebu. This study has also showed a higher incidence of reproductive disorders (conditions which lead to difficult breeding, including retention of placenta) among Indian cattle than among the crossbreds, while both groups were similar in frequency of abortions and still births.

18.4.3. Crossbreeding in Other Tropical Countries

Successful attempts have been made in other tropical countries in various time zones to use crossbreeding of *Bos indicus* with *Bos taurus* cattle for evolving new dairy breeds that are more suitable to the local conditions of climate and husbandry practices. It is worthwhile to study their experiences.

The Australian Milking Zebu (AMZ) has been developed using Sahiwal (males) and Jersey (females) as the foundation breeds (Hayman, 1977); Red Sindhi, originally used as a foundation breed was later discarded. The first filial generations obtained in the above fashion were mated *inter se* upto F_3. All the cows completing first lactation successfully formed the female mates in each generation while sons of only top producing females of the previous generation were used as sires.

Then the meritorious F_3 bulls from meritorious dams were progeny tested on a Field Station herd as well as on selected farmer's herds, after subjecting them to screening-tests for resistance to artificial hot climate stress as well as tick infestation. The sires going into the ultimate breeding programme would have been proven for resistance to heat stress and tick infestation, fertility and milk production. Performance of AMZ heifers was found to be superior to contemporary Jersey heifers under tropical New South Wales conditions (Table 18.8).

TABLE 18.8. First Lactation Performance of Jersey and Australian Milking Zebu (Amz) Heifers in New South Wales, Australia

Description	*Number of animals*	*Milk yield (kg)*	*Fat yield (kg)*	*Age at calving (months)*
Jerseys born in Lismore (Farmers' herds) reared and milked at Badgery's Creek (Field station)	31	1805±82	87±9.7	28±0.66
AMZ born, reared and milked at Badgery's creek	35	1917±209	88±9.6	28±0.39
AMZ born at Lismore, reared and milked at Badgery's creek	19	2056±103	103±12.2	34±0.95*

*Higher value was considered to be due to other factors.

Source : Hayman (1977). *loc. cit.*

The Jamaica Hope breed developed in Jamaica, West Indies is another example of a new dairy breed for tropics. After trying various European dairy breeds at the Hope Farm, near Kingston, Jamaica, crossbreeding Jersey with Sahiwals was settled for. Simultaneously, improvement of feeding and management and progeny testing of sires was undertaken (Wellington and Mahadevan, 1977).

Wellington, K.E. and Mahadevan, P. (1977). Development of the Jamaica Hope breed of dairy cattle. In Animal Breeding. FAO Animal Prod. & Health Paper 1, FAO, Rome.

TABLE 18.9. Production Parameters of Jamaica Hope Cattle at Bodles (Farm) and on Six Farmers' Herds

Trait	*Bodles (1950-64)*		*Farmer's herds (1960-66)*	
	Number of records	*Mean*	*Number of records*	*Mean*
Age at first calving (months)	903	34.2	368	35.9
Length of lactation (days)	2153	322	1634	324
Length of calving interval (days)	1751	439	1394	405
305-day milk yield (kg)	2153	3218	1665	2755
Milk yield per day of lactation (kg)	2131	10.9	388	10.0
Milk yield per day of calving interval (kg)	1654	8.2	308	8.2
Butter fat content (%)	879	4.88	—	—

Source : Wellington, K.E., Mahadevan, P. and Roache, K.L. (1970). *J. Agric. Sci.* **74** : 643. Hayman (1977) *loc. cit.*

The basic population of the breed at one time comprised of different grades of Jersey with Sahiwal (1/8 to 3/4 Jersey, 5/8 to 7/8 Jersey being the best) and purebred Jerseys to the exclusion of even most of Holstein derivatives. Because of the Jersey grades, low age at first calving, shorter calving interval and smaller body size they were considered economical and ideal under Jamaican conditions. Sire selection was based on dam and progeny performance in an organised farm and farmer's herds. Ultimately, the breed stabilised at a level of about 80 per cent Jersey, 15 per cent Sahiwal and 5 per cent Holstein. It has impressive production records in organised as well as farmer's herds (Table 18.9).

On the Indian Ocean Coast of Kenya (hot humid), continuous 2-breed rotational crossbreeding involving Sahiwal and Ayrshire resulted in the formation of either 67 per cent Sahiwal—33 per cent Ayrshire or 33 per cent Sahiwal--67 per cent Ayrshire groups. The former group was found to be superior only in calf survival, birth weight and calving interval, while the latter group was found superior in age at first calving, lactation yield and weight gains (Gregory and Trail, 1981).

Attempts have been made in Colombia and other tropical South American countries with variable success to cross local breeds like Criollo (descendents of cattle originally imported by Spaniards and Portuguese into Latin America) with mostly Holstein Friesians (de Vaccaro, 1977). In Kenya, based on the experience gained over the past few decades a crossbreeding programme was formulated for different ecological zones of the country (Meyn and Wilkins, 1977) involving local, Sahiwal, Jersey and Friesian cattle. Jersey × Sahiwal crosses performed even better than purebred Jerseys (1446 kg against 1044 kg first lactation yield), while Sahiwal × Friesian crosses were almost equal in yield but better in reproductive efficiency to Friesians.

While the general trend of results of Zebu × European crossings in Sri Lanka is similar to those in other countries, Buvanendran and Mahadevan (1977) recommended use of different *Bos*

Gregory, K.E. and Trail, J.C.M. (1981). *J. Dairy Sci.* **64** : 1178.
deVaccaro, L.P. (1977). Dairy cattle breeding in tropical South America. In *Animal Breeding*. FAO Animal Prod. & Health Paper 1., FAO, Rome.
Meyn, K. and Wilkings. J.V. (1977). Breeding for milk in Kenya. In *Animal Breeding*. FAO Animal Prod. & Health Paper 1, FAO, Rome.
Buvanendran and Mahadevan (1977). *loc. cit.*

TABLE 18.10. Recommendations of System of Breeding for Cattle under Different Ecosystems

Ecosystem		*Proposed breeding system*		
Feed resources	*Physical environment*	*Bull breed*	*Level of Eurpean dairy inheritance*	*Further breeding programme*
1. (*i*) Extensive grazing on poor natural vegetation with almost no supplementation with cultivated fodder and/or grain.				
(*ii*) Crop residues being the major source of nutrition and a little supplementation with cultivated fodder/grain only to lactating animals.	All	Superior Indian dairy breeds/ Halfbred Jersey	0.25 per cent	Grading/ Interbreeding with selection among crossbreds.
2. Inadequate cultivated fodder availability, feeding on crop residues with supplementary grain/cultivated fodder, essentially stall feeding.	All environmental conditions especially hot-humid.	Jersery	50 per cent	Interbreeding with selection among crossbreds.
3. Adequate cultivated fodder availability and supplementation with grain, stall feeding and adequate housing.	Mild hot-arid/ semi-arid	Holstein Friesian	50 per cent	-do-

Source : Acharya (1979). *loc. cit.*

tourus breeds for different agro-climatic zones of the country depending on local husbandry status and potential for milk production. In high dairy-potential medium-rainfall Coconut Triangle in the South-West, Friesians were said to be more suitable notwithstanding the tropical climate. In the vast central dry zone, the Jersey seems to have a distinct advantage because of not too ideal environmental and husbandry conditions. The Sri Lanka experience has further shown that performance of back-crosses of F_1 to parent European breed varies with climate and husbandry status.

18.4.4. Ultimate Crossbreeding Options

The general Conclusions of the experience gained in crossbreeding in India and elsewhere and the logical venturable options can be summarised as follows.

When upgrading of local non-descript cattle is planned, crossing them first to an improved Zebu breed like Sahiwal and using the resulting grades for crossing with European breeds is desirable in the long run. Such a programme ensures uniformly higher performance of F_1 and subsequent generations. This fact is proved by experience in Sri Lanka and Kenya. This is advantageous from provision of sound husbandry point of view too.

The overall livestock husbandry status is much below the desired level in most of the Third World countries. If developed Zebu breeds are used initially, the farmer, hither to having experience with only local poor yielders, would initially have to deal with a better (but still Zebu) specimens that do not demand too high a standard of feeding and management and finances, he would be better prepared for taking care in the future of more better yielding grades—Zebu × European crosses successfully.

Another alternative is to use as a first step in upgrading non-descript cattle, selected sire (based on performance of dams progeny) that are *Bos indicus* × *Bos taurus* halfbreds themselves. Organised government and other institutional farms can act as suppliers of such bulls.

Experience of crossbreeding work in Banaskantha and Kutch districts (backward, tribal areas) of Gujarat has demonstrated the wisdom in introducing exotic inheritance into local cattle only in a small way. Vaishnav (1978) opined that in such areas only cows with 25 per cent exotic inheritance may be developed by using halfbred bulls on non-descript cattle due to poor productivity of land, non-availability of inputs and finances to support crossbreds and inadequacy of technical skills in cattle husbandry at farmers' level. Similar was the experience in backward areas of West Bengal (Sharma, 1978).

There is nothing *sacrosanct* about sticking to about 50 per cent level of *Bos tourus* inheritance either. The level of inheritance of European breeds that should be incorporated in crossbreeding programmes of countries has to be decided depending on level of management and on the general climatic conditions in the locality. The proposals suggested by Acharya (1979) in the Indian context (Table 18.10) can be good guidelines in this respect.

Generally, in high dairy potential areas with moderate rainfall and good animal husbandry, Friesians can be advantageous. But in vast under-developed tropical areas (steppes, hilly and high humid tracts) with meagre feed resources and low animal husbandry status, Jersey is to be preferred.

Crossbreeding has to go hand in hand with selection of bulls (which themselves should be sons of high yielding dams) by progeny testing. Progeny testing on one or two government owned herds does not suffice. There is need to ensure participation of a large number of milk recorded herds of farmers and their full involvement in sire testing.

Thus the essentials for success in the use of crossbreeding include—(*i*) performance recording at farmers level; (*ii*) efficient artificial breeding service; (*iii*) good veterinary cover for the rural areas; and (*iv*) incentives to farmers for production of more milk. Indeed, these are essential pre-requisites for implementing any breeding programme on farmers' herds.

18.5. BUFFALO BREEDING PLANS

Asia is the natural habitat of most of the riverine buffaloes, the largest concentration being in India. Swamp buffaloes are abundant in South and South East Asia.

18.5.1. Importance of the Buffalo

Buffaloes contribute over 50 per cent to the total milk production in India, whereas they constitute around 1/3 of the total dairy bovine population. In recent times buffalo gained importance as an international object of research and development because of its triple uses—(*i*) as an efficient milk producer under not too favourable conditions of feeding and management obtaining in tropical Asia; (*ii*) as a steady worker in the rice belt of Asia; and (*iii*) as an efficient convertor of low grade roughages and by-products into meat.

Vaishnav, T.N. (1978). Crossbreeding programme—its scope and limitations. *Ibid*, p. 108.
Sharma, S.K. (1978). Influence of socio-economic factors on the feeding of crossbred cows in West Bengal area. *Proc. National Seminar on Crossbreeding*, NDDB, Anand, p. 208.
Acharya (1979). *loc. cit.*

Indian buffaloes, mostly of the Murrah breed, are exported to countries ranging from Italy, through South East Europe, to as far east as Japan. They are crossed with local stock for improving milk or meat production or both. In Europe, the improvement in milk production of buffaloes does also facilitate economical rearing of market stock. For this purpose, what is known as a "group-suckling of foster dam" is practiced with advantage in the Transcarpathian Republics of Soviet Union (Agabeili, 1978).

At this stage the reader may consult chapter 16 for knowing growth, production and reproduction characteristics of different breeds of buffaloes.

18.5.2. Selective Breeding

The best dairy breeds of buffaloes are native to India and Pakistan. Hence, the only possible option is selective breeding—breeding best males to best females. Reputed riverine buffalo breeds like Murrah from North Western India can be used and are being used for upgrading local buffaloes elsewhere in India as well as in other buffalo rearing countries.

Although milk yield is an important economic trait, dairy animal's real profit function is the result of interplay of several traits like growth rate, size, reproductive efficiency, production level and length of productive life. Though, the buffalo is a relatively superior yielder to several zebu breeds, its productive life is very short. This is mainly due to its delayed age at first calving (due to slow growth rate) and long intercalving periods (due to lower reproductive efficiency).

Age at first calving, days open, days dry and herd life are some important factors that are mainly dependent upon managemental decisions. At a given genetic potential for production, it is pertinent to determine the optimum levels of those traits at which given managemental practices help to obtain efficient exploitation of genetic potential.

Most of the reproductive traits have little genetic variability (*see* Chapter 16). Improvement of fertility and regular breeding can be attained by efficient management, improved nutrition and systematic health control.

The heritability estimates of milk production traits in buffaloes, although variable, are more or less akin (Table 15.1) to those in cattle to indicate that the same breeding programmes would be appropriate. Improvement in milk yield by selection without progeny testing within a closed herd will be slow. Progeny testing as a basis of selection of males is therefore essential. Use of sons of real high yielding dams could be a plausible starting point.

In view of quite high repeatability of milk yield (Table 15.5), selection could be based on first lactation. Since the relationship between milk yield for 120 days and total lactation yield is positive and high, selection can be made earlier with reliable results (Nagarcenkar, 1978). This can also be useful in early progeny testing of males.

In selection of dams of breeding sires (for production of future bulls) it is desirable to wait until the completion of three or four lactations to ascertain the good herd life of the female, *i.e.*, her health, freedom from udder problems, regularity of breeding etc.

Since growth rate and body weight are characters of high heritability and can be measured in both the sexes, progeny testing can be dispensed with for these traits (NDDB, 1980). Body

---

Agabeili, A.A. (1978). Genetic basis of selection of Caucasian buffaloes. *Proc. Indo-Soviet Symp. Buffalo Breed.* NDDB. Anand, p. 150.
Nagarcenkar (1978). *loc. cit.*
NDDB (1980). *loc. cit.*

weight has also high correlation with physical maturity, age at first calving, as well as the first lactation yield. Hence body weight should be considered important when selecting future bulls for progeny testing. This will also help in improvement for meat production, though, it may be essential to conduct progeny testing for traits related to carcass quality and dressed weight yields.

As stated earlier, maximum genetic progress can be attained by identifying young bulls, first from outstanding dams (both in milk production and reproduction) and then further screening them by progeny testing. Since only a small proportion of the population can be bred by these bulls, a more practical proposition would be to use bulls produced by mating progeny tested sires to females identified on the basis of their own performance and who are the daughters of proven bulls.

TABLE 18.11. Increase in Milk Production at Different Rates of Genetic Gains in Buffalo Herds at Various Levels of Production

Base production	*One bull selected out of*	*Rate of genetic gain*					
		1%		*1.4%*		*1.8%*	
		Uncorrected	*4% FCM*	*Uncorrected*	*4% FCM*	*Uncorrected*	*4% FCM*
Large breeds with 6.5 per cent Butter Fat (Murrah, Nili Ravi etc.)							
1200 kg	5	1261	1734	1286	1368	1311	1803
	10	1325	1822	1338	1988	1433	1970
	15	1392	1914	1477	2031	1566	2153
	20	1464	2013	1583	2177	1704	2343
1600 kg	5	1681	2311	1714	2356	1748	2404
	10	1766	2428	1837	2526	1918	2626
	15	1857	2533	1968	2706	2087	2870
	20	1952	2684	2111	2903	2272	3124
2000 kg	5	2101	2889	2143	2947	2185	3004
	10	2208	3036	2296	3157	2388	3284
	15	2320	3150	2460	3382	2608	3587
	20	2440	3355	2639	3629	2839	3904
Medium breeds with 7.5 per cent Butter Fat (Surti Mehsana etc.)							
900 kg	5	946	1443	964	1470	934	1501
	10	994	1516	1034	1577	1075	1640
	15	1045	1594	1108	1690	1175	1792
	20	1098	1674	1187	1810	1284	1985
1100 kg	5	1156	1763	1179	1798	1202	1833
	10	1215	1853	1264	1928	1314	2004
	15	1277	1947	1354	2065	1434	2187
	20	1342	2046	1452	2214	1569	2393
1300 kg	5	1366	2083	1393	2124	1421	2167
	10	1436	2190	1493	2277	1553	2368
	15	1509	2301	1601	2442	1697	2588
	20	1586	2419	1715	2615	1854	2827

Correction factors : 6.5 per cent = 1.375
7.5 per cent = 1.525

Source : Nagarcenkar, R. and Basu, S.B. (1974). *Indian Dairyman* **26** : 237.

Thus the complete selection plan would involve carrying out progeny testing programme with sufficiently good number of bulls along with screening a large buffalo population (at least by periodical test recording of yields) to select elite dams for producing future young sires. Since the scope of carrying out progeny testing on institutional herds is limited these programmes must be extended to farmers's buffaloes. As regards the size of the progeny one may aim at obtaining complete records on a minimum of 40 daughters per sire.

An idea about the improvement in milk production that can be attained by selecting one out of 5, 10, 15 and 20 buffalo bulls of different breeds for different durations and at different rates of annual genetic gain can be had from the estimates of Nagarcenkar and Basu (1974) given in Table 18.11. Clearly, there are excellent potentialities for improvement among buffaloes, if adequate inputs are provided.

18.5.3. Cross Breeding

Where the local population of buffaloes is non-descript or low yielding (Southern and Eastern parts of India and swamp buffaloes of South East Asia), crossbreeding—grading with developed riverine breeds of India may be considered. Such attempts have generally been encouraging (Table 18.12). Heavier breeds like Murrah and Nili-Ravi can be used in areas having good dairy potential and husbandry status, while medium sized breeds like Mehsana and Surti can be tried in low husbandry status areas.

In India, there are two outstanding examples of such a breeding programme. The Mehsana breed in Gujarat is said to be a cross between the Murrah and Surti breeds, evolved over a century or so by the local traditional milk producers. This breed is considered to be having the higher milk yield potential of the Murrah and the regular breeding tendency of the Surti and is somewhat intermediate in size and appearance to the parent breeds. Similarly, commercial milk producers in the coastal Godavari delta region in Andhra Pradesh have, by continuous mating of Murrahs with the local stock over several decades, evolved a better milk producing buffalo breed called Godavari.

TABLE 18.12. Production and Reproduction Performance of Swamp and Riverine Buffaloes and Their Crosses in South East Asian Countries

Traits	*Swamp*	*Riverine*	*Crossbreds*
Lactation yield (kg)	210-828	566-2459	492-1684
Lactation period (days)	121-333	217-523	216-309
Butter fat %	8.3-10.0	7.2-10.0	7.1-8.6
Age at first calving (years)	4.0-6.0	3.0-6.0	3.2-4.2
Gestation period (days)	312-332	305-315	303-308
Calving interval (days)	414-912	450-580	344

Source : From Tumawasorn (1981).

One out of every 20, 60, 100 and 1500 buffaloes in Burma, Malaysia, Philippines and Indonesia, respectively were Murrahs being used for crossbreeding (Chantalakhana, 1979). In

---

Nagarcenkar and Basu (1974). *loc. cit.*

Chantalakhana, C. (1979). Performance of swamp, riverine and crossbred buffaloes in SE Asia. In : *Buffalo reproduction and AI.* pp. 143-151. FAO, Rome.

recent times Murrah bulls, cows and heifers are being imported into Vietnam and Compuchea for similar purpose. Riverine × swamp crossbred buffaloes are fertile and considering their reproductive performance, milk production and draft capacity, such a crossbreeding practice is recommendable, though more detailed information on performance of such crossbreds under various conditions is desirable before advocating any large scale crossbreeding plans.

It was observed in most countries that the performance level of crossbreds is intermediate between the two parent breeds (Table 18.12). Several reports indicate that crossbred buffaloes are superior to swamp buffalo in growth, meat production, working capacity and ability to thrive on low grade roughages (Sastry, 1984). Improvement plans for the Carabao in the Philippines· constitute grading up with Murrah and Nili-Ravi as well as triple-crossing involving Philippinese, Murrah and Thai buffaloes aimed at ultimate evolution of an animal with 28.125 per cent Philippinese, 56.250 per cent Thai and 15.625 per cent Murrah inheritances (Eusebio, 1981). The grades are considered to be helpful in commercial production of milk in sub-urban areas, while the triple crosses would be the improved triple-purpose animals for the rural areas.

18.5.4. Cow Versus Buffalo

In conclusion, the need for paying equal attention to improvement of buffalo and cattle breeding in the Indian context may be emphasised again. It is generally presumed that the cross-bred cattle will eventually surpass the buffalo as the main milk producing species.

Crossbreeding is essentially an exercise in improving very large numbers of stock with very poor genetic potential for milk production. It is bound to be not all that fast a method because of the involvement of large numbers spread over large territories. Selective breeding of buffaloes, on the other hand, is an attempt to improve relatively lesser number of individuals with much better genetic potential for milk production. Both options could well be equally time consuming.

The average Indian farmer still considers the buffalo as the producer of milk and the cow as the mother of the bullock. With the present services and manpower at our command, replacing the buffalo in the farmer's household with an efficient and ideal crossbred cow can easily take several decades. Further, such a step is fraught with such obstacles as low husbandry status at the farmer's level, meagre feed resources, inadequate health cover and social and individual preferences.

Thus the buffalo is going to remain as the important milk producer in the country for long long time to come. Hence the need to pay equally great attention to buffalo breeding, as the efforts at crossbreeding of cattle.

FURTHER READING

Section V

Acharya, R.M. 1979. Strategy for breeding dairy cattle in India for enhanced milk production. *Indian Dairyman* 31 : 607.

Bhat, P.N. 1978. Genetic parameters of milk production and scope of increasing milk production in buffaloes *vis-a-vis* cattle. In: *FAO Animal Production and Health Paper 13*, FAO, Rome.

Balaine, D.S. (Ed) 1979. *Dairy cattle breeding in the tropics*. Haryana Agricultural University, Hissar.

FAO, 1977. *Animal Breeding*. FAO Animal Production and Health Paper I, FAO Rome.

Johansson, I. 1962. *Genetic aspects of dairy cattle breeding*. Oliver and Boyd, London.

Johansson. I. and Rendel, J. 1968. *Genetics and animal breeding*. W.H. Freeman & Co., San Francisco.

Katpatal, B.G. 1978. Dairy cattle crossbreeding in India. *Indian Dairyman* **30** : 189.

Eusebio. A.N. (1981). Programme of Philippine Carabo research and development centre. In: *Recent advances in buffalo research and development.* F.F.T.C., Taipei.

Sastry, N.S.R. (1984). Milk production from buffalo. In : *Proc. International Conference on milk production in developing countries.* 2-5 April (1984). CTVM. Edinburgh. p. 497.

Tumawasorn, S. (1981). Crossbreeding of Swamp buffalo for milk and draft. In : *Recent-advances in buffalo research and development.* F.F.T.C., Taipei.

Lush, J.L. 1945. *Animal Breeding Plans.* Iowa State College Press, Ames, Iowa.

Nagarcenkar, R. 1978. Riverine buffaloes of India and possibilities of genetic improvement *vis-a-vis* cattle. In: FAO *Animal Production and Health Paper 13*. FAO, Rome.

NDDB, 1978. *Proceedings of the National Seminar on Crossbreeding.* NDDB, Anand.

NDDB, 1978. *Proceedings of the Indo-Soviet Symposium on Buffalo Breeding*, NDDB, Anand.

NDDB. 1980. *Breeding and Feeding for Milk Production in Operation Flood II.* NDDB, Anand.

Pirchner, F. 1969. *Population genetics in animal breeding.* W.H. Freeman & Co. San Francisco.

Rice, V.A., Andrews, F.N., Warwick, E.J. and Lepres, J.E. 1957. *Breeding and improvement of farm animals.* McGraw Hill Book Co. Kogakusha Co. Ltd. Tokyo.

Sinnut, E.W., Dunn, L.C. and Dobzansky, T. 1960. *Principles of genetics.* McGraw Hill Book Co., New York.

Snedecor, G.W. and Cochran, W.G. 1976. *Statistical methods.* Oxford and IBH Publishing Co., New Delhi.

Sundaresan, D. 1976. *Livestock breeding in India.* Vikas Publishing House, New Delhi.

Section VI : HOUSING, HYGIENE AND HEALTH

CHAPTER 19

Dairy Bovine Housing and Hygiene

The principal functions of any system of housing for dairy cattle are : (*i*) Provision of health sustaining and comfortable environment to the cows and calves; (*ii*) Provision of desirable working conditions for labour and supervisory staff on the farm; (*iii*) Security from thieves and protection from extremes of weather conditions; and (*iv*) Integration of housing with feeding, watering, milking and manure handling systems. In other words, a system of housing is more or less synonymous to the system of management of dairy cattle. The present view is that, the houses for cattle and buffaloes are necessary evils and they should be as little 'house-like' as possible, but at the same time serve right the animals and man.

In India, as not much of attention is paid so far to this important aspect, we find different types of animal houses constructed without necessary care in planning and designing. Farm buildings should be properly located, constructed, spaced out and grouped. Economy in their initial cost and maintenance and the health and comfort of animals should receive special attention. An efficient housing system should lead to an efficient management so as to reduce labour requirements to the minimum.

19.1. DESIGN CONSIDERATIONS FOR ANIMAL HOUSES

Research work has been sparse and sporadic in the tropics to assess the relative suitability of one type of housing or the other to dairy animals. Though, isolated reports on the effect of housing on physiological well-being of cattle and buffaloes exist, no long-term studies on housing systems *vis-a-vis* production of animals and economy of livestock production have been conducted. Whatever little evidence is available indicated that in warm climates shelter of some type or other leads to more comfort to animals, as judged by their physiological reactions. Based on experiments conducted in some central and southern parts of United States McDowell (1972) states that

McDowell. R.E. (1972). *Improvement of livestock production in warm regions.* W.H. Freeman & Co., San Francisco, pp. 51-53.

shelters prove to be beneficial to cattle in hot and drier regions. However, their value is questionable in more humid areas. In India, long term studies on animal housing in different seasons in the semi-arid north-western region were conducted at Haryana Agricultural University, Hisar (Sastry and others) and other institutions.

The basic justification for animal shelter is that it should alter or modify the environment for the benefit of animals enclosed in it. The animal shelter should normally buffer the extremes of climatic conditions to reduce peak 'stress' on the animals housed. The main climatic factors from which protection is to be provided are high and low ambient temperature, environmental humidity, solar radiation, wind and rain. Further, a good shelter must permit a dry and comfortable surface for the animals to rest, that does not permit breeding of internal and external parasites and other pathogens within the shed.

From physiological point of view the dairy manager's interest is directed more towards creating a micro-environment inside animal sheds, which may provide protection to animals against stressful environment. Under tropical conditions prevailing in India and other warmer regions the basic principle that should guide the dairy farmer or manager embarking on building a farm is to reduce heat gain and promote heat loss from the structures of the animal house by radiation and conduction during summer. During winter the structures, especially those parts that come in constant physical contact with animals, like floors and walls, should not get too cold and should give protection from cold winds.

As far as possible the reduction in heat liberation inside the building should be achieved by keeping the density of animals in pens low. To reduce the heat gain and to promote heat loss by radiation and conduction a number of measures are recommended. These are—external shade (by planting trees), reduced ground reflection (by coverage of landscape around sheds with grass, shrub, bushes or even colouring of reflecting surfaces), attached shade (by projection of roof), cooling of exterior surfaces of buildings (effective only to a limited extent), minimising solar projection (by suitable materials for roof, walls etc.), aiding convection just over and underneath surfaces of structures exposed to solar radiation (by creating wind movement near about the roof with suitable ventilators), provision of adequate ceiling height and insulation, and even resorting to evaporative or refrigerant cooling.

For guidance one may refer to Table 19.1 in which reflectivity of solar radiation by different surfaces is given.

The relative effectiveness of one or the other measure individually or in combination has to be judged prior to making recommendation depending on the agro-climatic situation of the place and socio-economic structure of the people.

Animal growth, production and reproduction are the ultimate concern of the farmer. Animal production and reproduction are not generally influenced by slight and normal seasonal variations in environmental factors. But when animals are exposed to extreme conditions for longer periods, the performance of animals is impaired; resulting in death in extreme cases.

Thus, animal houses should be so designed that they are cheaper and afford protection from extreme weather conditions. Expensive fittings and designs on a permanent basis should not be resorted to, to face only temporary brief periods of climatic extremes. Often it will suffice to make some temporary additions and alterations (like provision of Hessian cloth curtains as protection against bitter cold droughts on a very cold day) to protect animals against short periods of inclement weather. By using suitable materials for construction of buildings and landscaping around them effect of solar radiation and high ambient temperature on animals can be minimised.

TABLE 19.1. Percentage of Incident Solar Radiation Diffusely Reflected by Various Surfaces*

Nature of Surface	*Reflected Radiation (%)*	*Nature of Surface*	*Reflected radiation (%)*
Dry bare ground	10-25	Desert	24-28
Wet bare ground	8-9	Salt flats	42
Dry sand	18-30	Brick	25-50
Wet sand	10-18	Asphalt	15
Dry black earth	14	City area	10
Wet black earth	8	Aluminium	85
Rock	12-15	Paint—white	75
Green fields	3-15	Paint—green	50
Green leaves	25-30	Paint—aluminium	45
Dark forest	5	Paint—grey	25
Dry grass	32	Pine wood	40
Green grass	5-11		

* Data compiled from: 1. *Smithsonian Meteorological Tables,* Smithsonian Institutions, 1951, pp. 442-3; 2. *Hand Book of Chemistry and Physics,* Chemical Rubber Publishing Co. Clevland, 1944, pp. 2294-6; and 3. Lee, D.H.K., *Manual of Field studies Heat tolerance of domestic animals.* F.A.O. Development Paper No. 38,1953, F.A.O., Rome.

19.2. TYPES OF DAIRY CATTLE HOUSING

It is impossible to design an ideal set of buildings which would be equally suitable for livestock farming all over the country owing to the different methods of animal husbandry and/or farming priorities, which in turn are governed by climatic, geographical, economical and other conditions. All the same, there are broadly speaking two types of houses used for dairy cattle, namely, conventional barns and loose houses.

Conventional barns or Tie-barns. These are also called stanchion barns or tie barns. These refer to housing in which the cows are confined in a row on a platform and secured at neck by stanchions or neck-chains. The cows are fed as well as milked in this barn. The barns are completely roofed and the walls are also complete with windows and/or ventilators located at suitable places.

Animals and men caring for cattle in barns are less exposed to harsh weather conditions. There is also a feeling that the animals can be kept cleaner and diseases controlled better in barns. Barns may be constructed in temperate Himalayan region where the winters are prolonged and severe. In warmer parts, the air in barns tends to be humid and barn floors damp during autumn and rainy seasons; and hence, are not recommended.

Loose housing. This comprises of keeping animals loose in an open paddock or pasture throughout the day and night except at milking time. The open paddock is provided with shelter along one side under which the animals can retire when it is very hot or cold or during rains. A common watering tank is provided and fodder is fed in common mangers. Concentrates are fed at the time of milking, which is done in a separate milking barn. Under loose housing system a milking barn or parlour is always constructed, in which the cows are secured at milking time and are milked. It is ideal to store dry fodder in the sheltered area where animals are allowed a limited access through movable fences; this is known as self-feeding. The open paddock is enclosed by means of half walls or wooden or plain-wire fences of convenient height (Fig. 19.1 and 19.2).

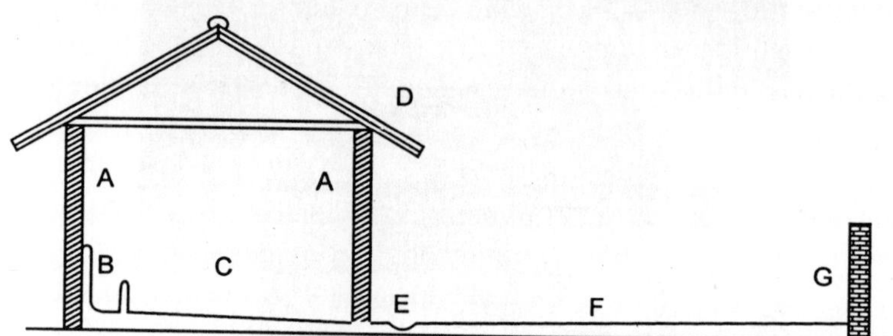

A. Roof supporting structures, **B.** Manger, **C.** Covered area, **D.** Roof, **E.** Gutter, **F.** Open area and **G.** Compound wall.

Fig. 19.1. Section of a loose house showing internal details.

Kooner *et al.* (1969) have recommended loose housing for dairy cattle under Indian conditions. This type of housing is suitable to most parts of the country except perhaps to temperate Himalayan region and in heavy rainfall areas. However, slight modifications of structure or use of extra fittings is often necessary to protect animals during extreme cold or hot seasons. Loose houses are cheaper to construct, easier to expand and flexible in their utility. Feeding and management of stock is easier in these because of common feeding and watering arrangements; and animals are comfortable as they can move about freely. Experience at the Haryana Agricultural University

Fig. 19.2. Features of a loose house for cattle—the covered and the open area showing position of manger and water tank.

Kooner. D.S. Sundaresan, D. and Murthy, V.V.N. (1969). *Dairy cattle housing.* Bull. No. 23, PAU. Ludhiana.

Fig. 19.3. Purebred Jersey and Friesian cows in a loose house at the Government Cattle Breeding Farm, Hisar, India, on a hot summer day. Note tree barrier in the back ground.

farm indicated that such a situation seems to allow buffaloes and crossbred cows to overcome heat stress better by voluntary movement under the shelter or *vice versa* as the situation demands. This helps in overcoming the problem of summer sub-fertility in buffaloes to a certain extent.

Fig. 19.4. Karanswiss cows in a loose house at cattle yard, National Dairy Research Institute, Karnal. Note trees in the back ground.

Fig. 19.5. Crossbred heifers in a loose house in which the shelter is located in the middle of the paddock. Note the extra shade provided by trees in the paddock. (NDRI Cattle yard).

At least 10 to 15 per cent more stock than the standard can be accommodated in each house for shorter periods without unduly affecting their performance. In loose houses it is an essential pre-requisite to have shady trees both within the paddock area and immediately around the building. In a tropical country like India green coverage increases the efficiency of a loose house in providing protection to dairy cattle against summer heat. Excellent dairy animals, including purebred *Bos taurus* cattle are being maintained efficiently in such houses on well known farms in this country (Figs. 19.3, 19.4 and 19.5).

19.3. LOOSE HOUSES *VERSUS* TIE-BARNS

Before investing his money on construction of buildings for his dairy cattle, a farmer would definitely wish to know which type of housing — loose houses or tie-barns; he should go for and for what reasons. Greater attention should be paid to the suitability or otherwise of loose houses or barns during extreme climatic conditions, like peak of summer and winter. *Responses of buffaloes to type of housing in semi-arid regions of India* has been reviewed by Satry and Georgie (1985) as follows.

19.3.1. Housing Standards

Details of construction of barn-type house for cattle are given by Vasavada (1964). Based on deliberations among experienced animal husbandry specialists, rather than on empirical findings, the Indian Standards Institution has brought out standards on space requirements of animals (ISI, 1968a, b; 1969; 1970a, b). These are adapted to a large extent without discrimination by state organizations. A bulletin on dairy cattle housing brought out by the Punjab Agricultural University (Kooner *et al.,* 1969) gives details of loose housing, which the authors recommend for the semi-arid regions. The loose housing system, in principle, allows animals to be housed loose and comprises of an open paddock with a partial shelter. This type of housing is in wide use in hot and arid southern USA. Loose house system is also recommended by Sastry *et al.* (1982a), who gave guidelines on the layout, design and space requirement in

Sastry, N.S.R. and Georgie, G.C, 1985. *Indian J. Anim, Sci.* **55** (7): 553-556.

loose house. In all the major organized farms in the area, the loose house is largely in vogue. However, information regarding how much superior it is in comparison to more closed type or completely open type house for providing comfort and sustain production in bovines has not been codified.

19.3.2. Microclimate in Animal Houses

The conditions inside animal houses are affected to a considerable extent by the meteorological factors of the external environment, the construction itself (Padmanabhamurty, 1983) and the animals housed in them. The macroclimate-microclimate relationships are different in different seasons (Singh *et al.,* 1984a). Meteorological data from several studies on the microclimate in the open, in the loose house with brick pavement and partial asbestos-roofed shelter, and in the brick-walled asbestos-roofed barns and sheds, as well as hollow brick-roofed barns are presented in Table 19.2.

While the maximum temperature inside the barns was similar to that in the open in all seasons, it was invariably higher in the loose house by about 2°C in all seasons (Fig. 19.6). Humidity, though higher in the loose house than in the open, was highest in the barn house. In another set of experiments (Karki, 1981; Mehla, 1982; Singh, 1982) from which data comparing loose house and a large asbestos-roofed shed with cross-ventilation (through ground to roof-height wire-net covered window spaces) was available, again the maximum temperature was higher in the loose house than in this shed and in the open. The atmospheric humidity was also higher in the loose house during the hot seasons than in the shed, though in cold seasons the humidity was similar. The minimum temperature tended to be, in comparison to that in the open, consistently higher by 1° to 2.5°C in the loose house, by 2° to 4°C in the barn and 2° to 5°C in the shed, with the difference between inside and outside values being far bigger in cooler season (Table 19.2).

The failure of direct sunshine to reach inside closed type houses prevents the rise of ambient temperature much beyond that in the open, whereas the reflected radiation in addition to direct radiation caused higher ambient temperature conditions in loose house. The minimum temperature remains higher in closed housing conditions because of inadequacies in the loss of absorbed heat. This is usually because of the absence of thermal radiation during the time beyond mid-night, due to presence of walls in closed houses.

Poor ventilation adds to this effect. Similar considerations also affect the humidity conditions; any type of building elevates humidity levels inside. Bempong (1983), comparing tree shade with loose house, found that the temperature-humidity index (THI) in the open (tree shade) and loose house is similar (Table 19.2) giving an indication that thermal load on the animal under these two conditions may not be very different in humid summer. However, Singh *et al.,* (1977) at the Indian Veterinary Research Institute (IVRI), Izatnagar, reported lower humidity level in the loose house, and Razdan (1965) and Thomas (1969) at the National Dairy Rsearch Institute, Karnal, found lower humidity inside the barn than in the open during dry and mild hot climates. This is some what at variance with reports from the Haryana Agricultural University (HAU), Hisar (Table 19.2). These seem to be the results of design differences of the houses, especially the loose house at IVRI was not so open as that at HAU. Satya Pal *et al.* (1973) found that asbestos-roofed loose house produced a more moderate microclimate than the open, and Sastry *et al.* (1973) showed that the microclimate in the loose house can be further improved upon by an additional shelter of hessian in the open area of the loose house (Table 19.3).

TABLE 19.2. Microclimate in Different Types of Animal Housing

Season/climate classification	*Housing conditions*						*Reference*
	Open	*Tree shade*	*Loose house*	*Asbestos roofed shed*	*Asbestos roofed barn*	*Hollow brick-roofed barn*	
1	*2*	*3*	*4*	*5*	*6*	*7*	*8*
Average temperature (°C)							
Dry	19.9	—	—	—	—	36.6	Razdan (1965)
Humid	32.2	—	—	—	—	27.9	Razdan (1965)
Cold	15.8	—	—	—	—	19.3	Razdan (1965)
Mild hot	30.6	—	—	—	—	29.0	Razdan (1965)
Maximum temperature (°C)							
Summer	40.7	—	42.1	—	40.6	—	Yadav(1982)
Winter	20.7	—	22.4	—	20.3	—	Thomas *et al.* (1978)
Rains	36.1	—	37.7	—	36.1	—	Yadav (1981)
Summer	—	38.5	40.0	—	—	—	Bempong (1983)
Hot dry	41.2	—	42.6	39.4	—	—	Karki (1981), Singh (1982)
Hot humid	36.4	—	38.3	36.7	—	—	
Mild cold	27.1	—	26.7	24.9	—	—	Mehla (1982), Singh (1982)
Severe cold	20.0	—	19.7	18.7	—	—	
Minimum temperature (°C)							
Summer	25.4	—	27.6	—	29.1	—	Yadav (1982)
Winter	4.9	—	5.3	—	8.6	—	Thomas *et al.* (1978)
Rains	25.3	—	26.3	—	26.9	—	Yadav(1981)
Hot dry	23.2	—	26.4	28.1	—	—	Karki (1981), Singh (1982)
Hot humid	26.9	—	28.3	29.1	—	—	
Mild cold	10.6	—	15.0	15.2	—	—	Mehla (1982), Singh (1982)
Severe cold	5.6	—	9.3	10.2	—	—	
Relative humidity (%)							
Summer	40.1	—	48.2	—	51.7	—	Yadav (1982)
Summer (*a*) Morning	—	66.7	68.7	—	—	—	Bempong (1983)
(*b*) Evening	—	45.4	42.6	—	—	—	
Winter	51.3	—	54.6	—	60.7	—	Thomas *et al.* (1978)
Rain	54.4	—	52.4	—	57.3	—	Yadav(1981)
Vapour presure (mmHg)							
Hot dry	11.7	—	18.3	15.5	—	—	Karki (1981), Singh (1982)
Hot humid	22.5	—	26.1	24.4	—	—	
Mild cold	8.2	—	10.5	10.3	—	—	Singh (1982), Mehla (1982)
Severe cold	7.3	—	8.5	8.2	—	—	
Dry	19.9	—	—	—	—	18.7	
Humid	19.5	—	—	—	—	19.8	Razdan (1965)
Cold	7.9	—	—	—	—	8.1	
Mild hot	17.3	—	—	—	—	15.8	
Temperature humidity index (%)							
Summer (*a*) Morning	—	80.6	79.8	—	—	—	Bempong (1983)
(*b*) Evening	—	87.0	87.7	—	—	—	

All recordings made with animals present.

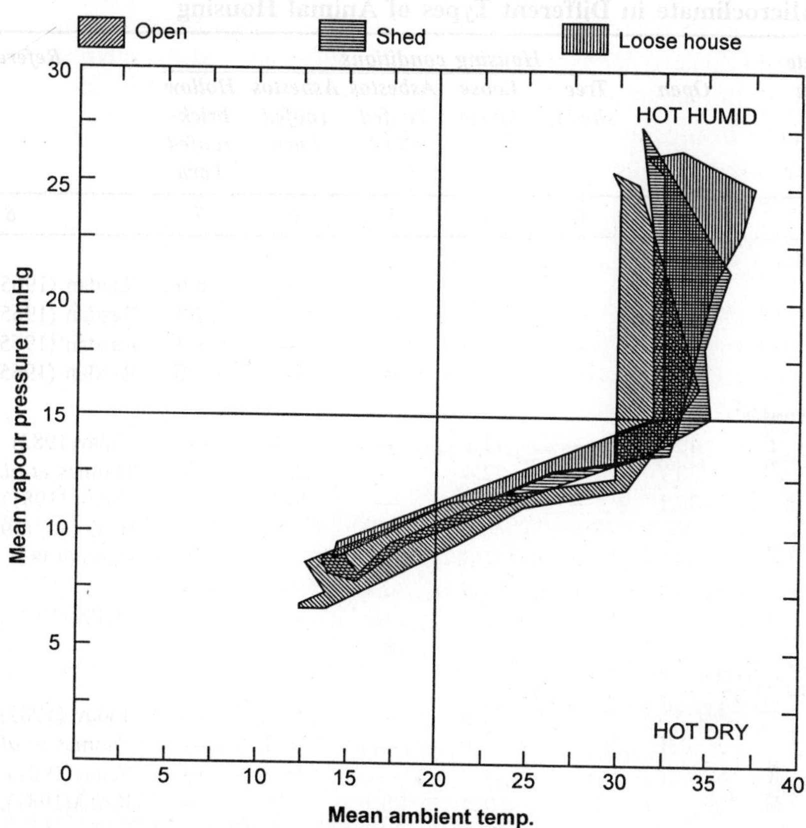

Fig. 19.6. Climograph showing weekly mean ambient temperature and vapour pressure values in the open in closed type barns and loose houses.

TABLE 19.3. Effect of Alterations of Sheds and Loose Houses on Microclimate during Summer

	Asbestos-roofed half-walled sheds				*Asbestos-roofed loose house*	
	Unaltered		*Altered with wet KHAS Panels above half walls*		*Unaltered*	*Altered with double-layered hession shelter over the open + water sprinkling on animal*
	Misra et al. (1963)	*Misra and Sengupta (1965)*	*Misra et al. (1963)*	*Misra and Sengupta (1965)*	*Sastry et al. (1973)*	
Maximum temp. (°C)	42.3	—	32.9	—	36.3	35.8
Minimum temp. (°C)	30.0	—	39.6	—	27.5	28.1
Ambient temp. (°C)						
07.00 hr	30.8	31.0	31.5	33.0	—	—
16.00 hr	38.4	42.5	32.7	36.5	—	—
Relative humidity (%)						
07.00 hr	46.0	42.0	61.5	48.0	—	—
16.00 hr	29.7	20.0	52.1	45.0	—	—
Vapour pressure (mmHg)	—	—	—	—	18.5	19.5

Some earlier reports from Mathura (Misra *et al.*, 1963; Misra and Sengupta, 1965) described the microclimate in some other types of houses for buffaloes and the attempts to improve the animal comfort by providing a cooler atmospheric condition in these houses (Table 19.3).

The provision in summer of wet *khas tatti* panels hung from roof level to the level of half-walls of the sheds reduced the ambient temperature of sheds in the morning but increased it in the afternoon; the humidity levels were always higher in such sheds. On the other hand, in a similar house with no such additional wet panels, though the reduction in ambient temperature was only marginal, humidity was not unduly increased (Table 19.3).

However, taking into consideration the need to protect the animals from intense thermal radiations, as well as the necessity to have adequate ventilation in any protective shelter provided to the animals, the loose house is theoretically the most suitable and with the least involvement of capital investment.

19.3.3. Animal Respones to Type of Housing

Physiological reactions. Animal responses like rectal temperature, pulse rate and respiration rate under different housing conditions were well within physiological norms in growing calves as well as lactating buffalo-cows, and changes, if any, under different housing regimes were minimal (Thomas *et al.*, 1978; Gangwar *et al.*, 1980; Yadav, 1981; Karki, 1981; Yadav, 1982; Mehla, 1982). However, the direction of change is of interest (Table 19.4).

TABLE 19.4. Differences in Physiological Reactions Under Several Housing Types with Reference to Loose House

Housing systems compared	*Seasons*	*Rectal temp. (°C)*		*Resp. rate counts min*		*Pulse rate counts min*	
		Morning	*Evening*	*Morning*	*Evening*	*Morning*	*Evening*
Barn-Loose house	Summer	–0.2	+0.2	–2.1	+2.5	–0.9	–0.6
(Lactating buffalo-cows)	Rainy	–0.5	0.0	–2.1	–0.6	–3.3	–2.6
	Winter	–0.4	+0.3	+1.0	+5.3	–3.2	–0.5
Shed-Loose house	Hot dry	0.0	–0.1	+5.2	+14.4	–0.4	+1.8
(Growing buffalo heifers)	Hot humid	–0.3	–0.1	+1.7	–0.1	–1.0	+1.9
	Mild cold	+0.7	+0.2	+1.5	+6.5	–0.9	+3.0
	Severe cold	+0.2	+0.3	+1.2	+6.1	+3.0	+3.0

(+) Increase or (–) Decrease in the loose house compared to the first mentioned housing in the first column. Compiled on the basis of data from Thomas *et al.* (1978), Yadav (1981, 1982), Karki (1981) and Mehla (1982). Morning : 07.00 hr, evening between 14.00 hr and 16.00 far.

In general, during hot or humid conditions the morning rectal temperature was lower, and in cold conditions higher. This indicated that the loose house environment was better than the barn or the shed during the night (Table 19.4). The evening rectal temperature which reflects the day time heat load on the animals was also lower in the growing animals kept in the loose house than in those kept in the shed during hot weather. But the barn-type house caused a lower heat load in hotter climates than the loose house for the lactating cows. The direction of rectal temperature response was generally in favour of the loose house in cold weather. The extent of

Thomas, C.K., Razdan, M.N. and George, G.C. (1969). Effect of shelter and different environmental conditions on nutrient utilization by Sahiwal × Brown Swiss and Purebred Sahiwal cattle. Indian J. Dairy Sci. 22 : 173.

changes in respiration and pulse rate were of little physiological consequence when you consider the normal range of changes one encounters with these parameters.

Systems of housing as well as alterations in management practices in addition to the housing provided to the animals have also received the attention of several workers (Roy *et al*, 1962: Misra *et al.*, 1963; Srivastava *et al.*, 1978; Gangwar *et al.*, 1980; Radadia *et al.*, 1980c; Karki, 1981; Mehla, 1982; Singh, 1982). Comparison between these studies has been made on the basis of extent of change in physiological reactions (Table 19.5).

TABLE 19.5. Differences in Physiological Reactions Due to Some Additional Managemental Practices under Different Housing Types

Housing type and management practice	*Season or month*		*Difference in*		
			Rectal temperature	*Resp. rate (counts/min)*	*Pulse rate (count/min)*
Half-walled shed[1]	Summer				
Wet *khas tatti* over half-walls		M	—	+1.5	+0.4
		E	—	–0.8	–2.8
Asbestos-roofed shed[2]					
Inside during day	Hot dry	M	0.0	+7.5	–0.3
In the open in		E	+0.9	–0.1	+1.2
the night during hot months	Hot humid	M	0.0	+2.7	+1.2
		E	0.3	+7.2	–0.5
Vice versa in cold months	Mild cold	M	+0.5	+1.5	–2.9
		E	–0.4	+0.2	+1.8
	Severe cold	M	+0.5	+2.8	+3.2
		E	+0.1	+2.4	±4.9
Loose house[3]					
Wallowing	Hot dry	M	–0.7	–3.2	—
		E	–1.0	–3.3	—
	Hot humid	M	–0.6	–2.3	—
		E	–0.6	–2.4	—
Spray cooling	Hot dry	M	–0.5	–3.4	—
		E	–0.8	–2.3	—
	Hot humid	M	–0.4	–2.4	—
		E	–0.4	–2.4	—
Loose house[4]	May-June				
Showering			–1.1	–34.6	–24.6
Cool drinking water			–0.8	–29.8	–19.0
showering + cool drinking water			–1.3	–36.6	–25.6

On the basis of data compiled from: [1]Roy *et al.* (1962), Misra *et el.* (1963), [2]Karki (1981), Mehla (1982), [3]Srivastava *et al.* (1978), Gangwar *et al.* (1980), and [4]Radadia *et al.* (1980c). M, Morning; E, evening; (+), increase or (–) decrease due to the alteration described in the first column.

A consistent trend of lower rectal temperatures in animals in all loose house systems is evident, whereas alternating between the open and the shed resulted in higher evening rectal temperatures in all seasons except mild cold regime. The extent of drop in rectal temperature varied between 0.4° and 1.3°C Though Radadia *et al.* (1980c) recorded considerable drops in

Tripathi, V.N., Thomas, C.K., Sastry, N.S.R., Pal, R.N. and Gupta, L.R. (1972). Effect of shelter and water sprinkling on buffaloes; Growth rate. Indian J. Dairy Sci. 42 : 743.

both respiration and pulse rates by showering or providing cool drinking water. Other reports showed that the effect of such measures on these parameters are marginal.

This extent of drop may not be altogether desirable because of the considerable reduction in metabolic rate indicated by this reduction. Also, in the loose house without any additional management practice to ameliorate heat load, the pulse rate and respiration rate were not unduly high (76.4 ± 0.48 and 55.4 ± 2.45 counts/min) in the lactating buffaloes studied by Radadia *et al.* (1980c). It can thus be said that loose house system can very considerably be improved in the direction of better thermal comfort to the animals by additional management adjuncts like wallowing or spray cooling as practiced by Srivastava *et al.*, (1978), and Gangwar *et al.* (1980) or by water showers and provision of cool (20.5°C) drinking water as described by Radadia *et al.* (1980c), though the last may not be so easily practicable under all farming conditions.

Feed and water economy. Under a high ambient temperature environment, suppression of appetite leading to reduced feed intake is well recorded. Also an enhanced water loss occurs from surface through evaporation resulting in increased water intake. These criteria were used in comparing housing management systems by several groups of workers (Table 19.6), though no specific conclusion can be drawn as to how much of the increased feeding or even a further reduction or an increase in water intake, not to talk about quantities involved. These, however, seemed to

TABLE 19.6. Alterations in Water and Feed Consumption on Account of Housing Management

Description	*% change in**		*Data source from which values are computed*
	Water intake	*Feed/dry-matter intake*	
Housing types			
Open vs. barn	– 2.7	—	Satya Pal *et al.* (1973)
Barn vs. loose house	+11.9 S	+11.0	Yadav (1982)
	–8.3 R	–24.4	Yadav (1981)
Shed vs. loose house	+26.0 S (dry)	+14.7	Karki (1981)
	+21.2 S (humid)	+ 6.2	Singh *et al.* (1984)
	–1.1 W (mild)	+ 0.8	Mehla (1982)
	–2.8 W (severe)	+ 7.9	Singh *et al.* (1984)
Managemental adjuncts to housing			
Day-night alteration between open and shed depending on season	+20.5 S (dry)	+10.5	Karki (1981)
	+22.9 S (humid)	+ 0.1	Singh *et al.* (1984)
	+14.0 W (mild)	+ 0.5	Mehla (1982)
	–16.1 W (severe)	+ 2.7	Singh *et al.* (1984)
Half-walled shed covered by wet *khas tatti*	–12.0	+44.1	Roy *et al.* (1962)
Cool drinking water	–22.7	+16.4	*Rifodin et al.* (1980a)
Showers in loose house	–15.7	+16.4	-do-
Showers + cool drinking	–14.0	+24.2	-do

Increase (+) or decrease (–) over the first-mentioned housing type or control condition. S = Summer; R = Rains; W = Winter. Other unspecified data are for hot season.

Thomas, C.K., Tripathi, V.N. and Sastry, N.S.R. (1975). Effect of thermal protection and concentrate supplementation on the growth rate of Murrah buffalo heifers. Indian J. Dairy Sci. 45 : 828.

Thiagarajan, M. and Thomas, C.K. (1991). Housing effects on crossbred cows in hot-humid climate. Indian J. Anim. Sci. 61 : 1222.

be optimal for maintenance of productive efficiency. Part of the evaporative heat loss is achieved without much energy cost and provision of *ad lib* water is of least importance to the management cost apart from the fact that this condition is usually implied in any good management system.

Thus the fact that in loose house demand for water is considerably higher than in shed in summer (varying between 20 and 26% increase in water intake) may not be taken as a point against loose house system. On the other hand, that loose house system has almost invariably (exception in rainy season with high humidity; Yadav, 1981) increased appetite and feed intake is a much weighty point in its favour. The extent of increase in feed/dry-matter intake is greater when further management assistance to provide thermal comfort is provided, considering the data (Table 19.6). However, it must not be forgotten that much of the differences are due to differences in types of animals, their production levels, period effects, etc.

The per cent time spent on feeding activity is also better under loose house (Table 19.7; Singh *et al.*, 1984b). The crude fibre digestibility during hot dry period in growing buffalo heifers (Singh, 1982) was 54% in loose house compared to that (43%) in shed and in the open during night and in the shed during day (47%) though the corresponding figures for hot humid period were 54, 58 and 62% (Singh *et al.*, 1984b).

Growth. Young buffalo-calves of 3 to 4 months of age showed better growth response when individual pens were provided to them within a large shed with high roofs. The pens were provided with a dried dung + slaked lime bedding in the winter, but no special bedding in summer (Bose and Thomas, 1979).

TABLE 19.7. Feeding Patterns of Growing Buffalow Heifers Housed Differently

		Shed	*Season-dependent day-night alteration between shed and open*	*Loose house*
Time spent on feeding (min/day)	S	196	213	261
	W	196	209	304
Time spent on rumination (min/day)	S	421	443	431
	W	549	540	799
Percentage time spent on feeding activity per day	S	42.85	45.56	48.06
	W	51.74	52.01	76.60

S, Summer; W, Winter.
Source : Singh *et al.* (1984).

However, much of the work on growth response of buffalo-calves were made with older animals housed in groups (Tripathi *et al.*, 1972; Mehla, 1982; Karki *et al.*, 1983; Singh *et al.*, 1984a, b; Tripathi *et al.*, (1972) obtained significant differences in various growth response measures under loose house and when the open area in the loose house was covered by tarpaulins in the summer, with calves (6 months-1 year old) receiving two sprinklings of water at 10.00 and 14.00 hr for 5 min. Almost the same trend was obtained by Thomas *et al.* (1975) even when the plane of nutrition was revised by 10% above the requirement. The weekly weight gain they reported for animals under tarpaulin (hessian) covered loose house was 3.22 kg as against 2.79 kg for simple loose house. Use of paved floors as against dirt floors practiced in villages for growing buffaloes was also studied during summer (Sastry *et al.*, 1981) as well as winter (Grewal *et al.*, 1982). Calves reared on brick-lined floors fared better (grew by 7% and 14% faster) than calves reared on dirt floors. The latter practice was also less hygienic, especially during berseem (high moisture containing leume) feeding period, *i.e.*, winter.

A comparison of growth responses under shed, loose house and alternating between shed and open, in much older buffalo heifer-calves are available from reports of Mehla (1982), Singh (1982) and Karki *et al.* (1983). These reports showed that relative growth rate as well as daily weight gain was higher for calves housed in the loose house during winter and summer. The growth efficiency was also studied by Singh (1982) in terms of dry matter, digestible crude protein and total digestible nutrient intakes per unit weight gain (Table 19.8), which also indicated that economic efficiency of growth was better under loose house.

TABLE 19.8. Growth and Efficiency of Feed Utilization of Growing Buffaloes Kept under Different Housing Conditions (From Singh, 1982)

Parameters	*Type of housing*	*Seasons*				
		Hot dry	*Hot humid*	*Mild cold*	*Severe cold*	*Overall*
Daily weight gain (kg)	S	0.452	0.459	0.429	0.327	0.417[a]
	S+O	0.446	0.524	0.464	0.355	0.452[ab]
	LH	0.617	0.483	0.438	0.535	0.518[a]
DM intake/kg gain (kg)	S	6.53	8.06	8.71	13.43	9.28
	S+O	6.32	7.06	8.10	11.97	8.36
	LH	5.66	8.04	8.60	8.35	7.66
DCP intake/kg gain (kg)	S	0.67	0.75	0.69	1.01	0.78
	S+O	0.70	0.62	0.67	0.97	0.74
	LH	0.58	0.70	0.75	0.67	0.68
TDN intake/kg gain (kg)	S	4.30	5.40	4.61	8.11	5.62[a]
	S+O	3.65	4.70	4.45	7.06	4.97[ab]
	LH	3.26	4.04	4.67	5.12	4.52[a]
Cost/kg gain (Rs)	S	—	—	—	—	9.21[a]
	S+O	—	—	—	—	8.77[ab]
	LH	—	—	—	—	6.90[b]

Note. S = Inside shed; S+O = inside shed during day and in the open at night in summer and *vice-versa* in winter; LH = inside loose house; overall average bearing same or no superscripts for each parameter do not differ significantly.

Lactation performance. Lactating buffaloes did not differ significantly in milk yield and fat and solids-not-fat per cent when housed in barn or loose house in the winter (Thomas *et al.*, 1978). This is not surprising because winter ambient conditions are not sufficiently severe to affect milk production in the semi-arid regions of India, even though some amount of deviation of energy consumed to maintain body temperature balance is expected, which usually is provided through the increased feed intake levels in the winter. But, it is the feed intake suppression in summer that may cause reduction in milk production level. Provision of wet *khas tatti* in half-walled sheds improved milk production from 4.5 to 5.8 kg/day (Roy *et al.*, 1962). Several aspects of lactation responses of buffaloes in loose house were studied at Ludhiana (Srivastava *et al.*, 1978; Gangwar *et al.*, 1980) and at Hisar (Radadia *et al.*, 1980b) in summer (Table 19.9).

The milk yield responded with 10 to 25% increase with all management practices designed to ameliorate thermal stress except in the case of spray cooling. Though milk let-down time increased with thermal ameliorative measures and consequent to increased yields, milking time also increased, the milk flow rate was favourably affected. This was particularly so in the case of wallowing group. Thermal ameliorative measure as adjunct to housing in loose house also favourably affected milk composition. All these, indicated that optimising lactational performance within a loose house system is linked to adoption of some additional environmental cooling process, which has to be selected through further studies. However, even simple loose houses do seem to be superior (by 8 to 21%) to housing inside barns, in summer (Yadav, 1982), rainy season (Yadav, 1981) as well as in winter (Thomas *et al.,* 1978) as regards influencing milk yield, fat and solids-not-fat content in milk of buffaloes (Table 19.10). The village practice (day-in and night-out in summer and vice versa in winter) also appear to be beneficial.

TABLE 19.9. Lactational Performance of Buffaloes in Loose House in Summer

Parameters	*Management system*	*Authors*	
		Srivastava et al. (1978), Gangwar et al. (1980)	*Radadia et al. (1980b)*
Milk yield (kg/day)	LH	6.90	4.37
	LH + spray cooling	7.01 (1.6)	—
	LH + showers	—	5.46 (34.9)
	LH + wallowing	8.19 (19.7)	—
	LH + cool drinking water	—	4.89 (11.9)
	LH + showers + cool drinking water	—	5.48 (25.4)
Butter-fat %	LH	—	7.46
	LH + showers	—	8.12
	LH + cool drinking water	—	8.60
	LH + showers + cool drinking water	—	8.55
Solids-not-fat %	LH	—	9.40
	LH + showers	—	9.57
	LH + cool drinking water	—	9.61
	LH + showers + cool drinking water	—	9.81
Milk let-down time (min)	LH	3.110	—
	LH + spray cooling	3.60	—
	LH + wallowing	3.95	—
Milking time (min)	LH	5.45	—

LH, Loose house; values in parentheses indicate percentage increase over loose house conditions.

TABLE 19.10. Differences in Milk Production of Lactating Murrah Buffaloes Housed in a Conventional Barn and a Loose House During Different Seasons

Parameters	*Type of House*	*Seasons*		
		Summer	*Rains*	*Winter*
Daily milk yield (kg)	LH	8.7	8.2	6.5
	B	7.6	7.9	6.3
Milk fat %	LH	7.4	6.6	6.5
	B	7.2	6.5	6.3
Solids-riot-fat %	LH	10.1	9.6	10.1
	B	9.8	9.3	9.9

LH, Loose house; B, barn.
Based on : Yadav (1981, 1982), Thomas *et al.* (1978).

Reproductive performance. The least studied aspect of animal response to housing in buffaloes is their reproductive performance. This is unfortunate because of well-recognized drop in reproductive efficiency during summer in these animals. Housing and management practices should have thus received attention on a priority basis, aimed at alleviating this production bottleneck. Pioneers in this field, Roy *et al.* (1962) observed that protection from solar radiation clearly improved the seminal attributes of buffalo-bulls as well as the incidence of oestrous activity in the buffalo-cows during summer. The reproductive performance of lactating buffaloes in a loose house system with thermal ameliorative measures like spray cooling or wallowing in summer was best when wallowing was allowed (Srivastava *et al.*, 1978; Mehta *et al.*, 1979, 1981).

Field experience too indicated that loose housing system may be congenial for better reproduction management of buffaloes. The data from herds at the Government Livestock Farm, Hisar and Haryana Agricultural University, Hisar (Gupta *et al.*, 1981), Punjab Agricultural University, Ludhiana (Mehar Singh and Chauhan, 1980) and NDRI, Karnal (1982), showed that the loose house system reduced the magnitude of inter-season differences in conception rates in comparison to tethered-in-barn (with grazing allowed for some hours) system (Sastry, 1984). This could be the consequence of more successful heat detection possible in the loose house due to easiness in bull parading and observation of cow behaviour.

19.3.4. Conclusions

If the above discussion seems to be leading to the conclusion that a loose house is the system of choice for buffalo rearing in the semi-arid regions of India, it is with a reason—physical comfort and productive performance have been found to be better in such a type of housing. And possibilities for further optimising productive efficiency through the aid of temporary measures to ameliorate heat load on the animals, wherever explored, have been found adoptable in this housing system. Combined with this, cost of construction of such a housing system can be made very competitive (Sastry, 1980). However, it has to be admitted that comparative studies with other systems of housing and other possible environmental control systems have not been fully exhausted. Other aspects of shelter engineering like orientation of housing, construction material to be used, standards of ventilation and shelter requirements, optimal space requirements, health and hygiene, and aspects of environmental pollution have not at all received the attention of any research worker or the institutes of standardization.

19.4. SELECTION OF SITE FOR FARM BUILDINGS

When planning an entirely new farm, the choice of location for the buildings is the first consideration. Then come the mode of arrangement of different buildings on the farm site. The following factors should be considered in deciding about the location of farmstead.

Nearby market. Whatever is produced on the dairy farm has ultimately to reach the market. Hence, nearer the market, lesser will be the transportation charges and lesser will be loss due to spoilage of highly perishable milk and milk products during transportation and storage. Thus farms should be located nearer to towns, but not in towns itself as this will offset some other advantages considered below. However, with the spreading of milk collection networks in the country's milk shed areas under the various State Dairy Development Corporations, it may be possible to have dairy farms even in distant villages, where good conditions exist for dairy farming.

Water supply. Water should be available cheaply and in plenty. With the availability of electric power for lifting water, many farmers can locate their farms farther from conventional water sources if it is advantageous otherwise.

Topography. The topography should be high level with no abrupt slopes. A relatively level area requires less site preparation, thus lowering building costs.

Drainage. The soil should be porous and the slope gentle so that drainage is efficient and the farm premises remain dry. Animal health is more easily maintained when the animal sheds are well drained.

Size and shape of area. The area for the farms depends on the number of cows to be kept on the proposed farm. Roughly speaking, one hectare of land would be sufficient to produce sufficient fodder for 10 adults and another 0.2 hectare would be required for housing and other buildings. Where irrigation facilities are not available perhaps one hectare of land can support less than 10 adult. Preferably the area of the farm should be square in shape. One should avoid having farmstead on both sides of a road or rail track.

Sun exposure and wind protection. The farmstead should be so located to obtain the maximum sun exposure in the north and the minimum sun exposure in the south and protection from the strong prevailing winds. A site with many trees around, is ideal as trees act as wind-breakers and also provide natural shade. A more detailed discussion on shades and shelters follows a little later.

Miscellaneous points to be considered. The farmstead should be located preferably nearer to an all-weather road with the fields arrayed around it. There should be round-the-clock supply of electricity to the farm. Small scale village units can be run, however, without much need for electricity. Other facilities like availability of telephone, nearby school for children of farm workers, post office, shopping places, entertainment places and other places that make life pleasing should also be sought. However, such things are still a luxury in most of the developing countries for the present.

19.5. ARRANGEMENT OF FARM BUILDINGS

In planning a new farm or altering an old one, buildings, fence, lots, gates, trees etc. should be added according to a well-laid master plan; for once constructed, buildings are difficult and expensive to remove. In general, the animal sheds and other commonly used buildings should be located around a central court and should be so arranged that most of them can be seen from the house of the farmer or farm manager. In arriving at the best pattern of arrangement of individual buildings within the site some of the popular patterns like E, U, L or C may be followed; but preferably one should decide about his model after considering the following factors.

Location of the farmer's house. The farmer's or the farm manager's dwelling as well as the office is often called as homestead and its location should be considered first. The homestead should be located — (*a*) on a high area which is well drained and commands a view of other buildings, (*b*) where it is easily accessible, and (*c*) nearer to road.

Orientation. Though the homestead should be oriented facing a road, other buildings may be turned at an angle to the road or reversed to take advantage of the prevailing winds and sunlight. In general, animal sheds are located with the long axis east to west with the open

Kelly, C.F., Bond, T.E. and Ittner. KR. (1950). *Agric. Engr.* **30** : 601.
Kelly, C.F., Bond. T.E. and Ittner, N.R. (1950). *Agric. Engr.* **38** : 726.

paddock side of the building facing towards north. Environment underneath shades oriented in east-west direction tend to be cooler than that underneath a shelter oriented in a north-south direction (Kelly *et al.*, 1950). But in a later report, the same group of authors (Kelly *et al.*, 1957) as well as McDowell (1972) felt that a shelter oriented in a north-south direction with open area on either side of it (which facilitates free movement of animals) will be better for cattle in hot climates because—(*a*) this facilitates drying of ground under shade more easily thus aiding in better sanitation, and (*b*) animals can easily move towards the shades side of the shelter (which is west side before noon and east side after noon) where they can give off some radiation to the cooler air existing there.

However, our experience at Hisar (arid climate in north-western India) has shown that during summer air temperature even in the shady side of the shelter is almost as high as right underneath the shelter (asbestos, 3.5 metres above ground). Hot surface winds that generally blow in these parts during summer may be partially responsible for this. However, north-south orientation (in fact any other orientation) with plenty of good shady trees around, seem to give good results. There are a few shelters of this type at the National Dairy Research Institute, Karnal, where crossbred cows and buffaloes are housed.

Direction of prevailing winds. The homestead should be located on the windward side of farm buildings considering the direction of summer and winter winds. Buildings should preferably, be constructed towards one side of hills whenever possible so that the hill acts as wind-break. When this is not possible it is desirable to arrange suitable tree planting for this purpose. Usually, a tree wind-break is located 20 to 40 meters from the buildings with three to seven rows of trees six to ten meters apart. Eucalyputs, casuarina, acacia, *Ashoka*, *Gulmohur*, yellow *champa* and other similar quick and tall growing trees that retain green foliage during the hot season are suitable for this purpose. Pods and leaves of *acacia* and *sesbania* species are greedily eaten by cattle and buffaloes.

Saving labour and time. The buildings should be located so as to require minimum time and walking when doing farm chores. The buildings in which most of the time is spent — such as the milking barn, workshop etc should be closest to the homestead. The buildings should be near enough to one another to permit efficiency of labour without making a fire hazard. Similarly, orientation and spacing of animal sheds should be so arranged that they are nearer to and easily accessible from grain, hay and straw stores, fodder chaffing yards and pastures.

Fire protection. Farm buildings should be enough far apart so that fire will not spread easily from one building to another. Large buildings should be about 10 to 15 meters apart. In acquiring added fire protection through spacing buildings farther apart, one should avoid extreme distances that may reduce operational efficiency. Fortunately, under loose housing system, fire hazards are less as shelters are fewer under this system. Also animals can easily move away from fire into the open paddocks as they are kept loose and not tethered.

19.6. PLANNING AND PHASED CONSTRUCTION

It is sound axiom to *"grow into"* rather than *"go into"* dairy farming business. A person new to dairying should start on a small scale and increase the size of his enterprise over the years. The advantage of such a practice is that when the farmer acquires valuable experience in herd management, it will give him confidence in operating on a larger scale.

McDowell, R.E. (1972). *loc. cit.*

However, the farmer should be able to visualise how his herd will grow over years. On the basis of this, a master plan should first be formulated for the complex of buildings. Construction should be undertaken in a phased manner; starting with the most required buildings and adding more in later years as the necessity for them arises.

Details of how the herd strength increases over the years are shown in Table 19.11. In this table, ten new cows each are added to the herd in the fifth and eighth years. The farmer may like to add more animals if the going is good, he having acquired sufficient expertise by this time. An extra pair of bullocks each is also added in the fifth and eighth years to take care of areas under fodder cultivation, in case tractor is not available. If artificial insemination facilities are not available the farmer has to procure a breeding bull also by this time. Some unwanted cows are also culled.

TABLE 19.11. Herd Strength in a Dairy Farm at the end of Each Year Starting with 20 Cows in the First Year

Type of animal	*Years*									
	1st	*2nd*	*3rd*	*4th*	*5th*	*6th*	*7th*	*8th*	*9th*	*10th*
Newly purchased cows	20	—	—	—	10	—	—	10	—	—
Reared cows	—	—	—	5	4	4	5	10	10	12
Old cows	—	19	18	22	34	36	34	37	54	60
Total cows	20	19	18	22	34	36	39	57	64	72
Bulls	—	—	—	—	1	1	1	2	2	2
Heifers > 2 years of age	—	—	6	5	5	6	11	12	13	19
Heifers 1-2 years old	—	7	6	6	7	12	13	14	21	24

Some explanation about the basis of calculating these figures is perhaps warranted. The proportion of male and female calves born each year would be 50 : 50. It is presumed that all male calves are sold at the time of birth or within one year after birth. All the farm born female calves will be reared, at least until decision is made to retain them in the herd or cull them. There will be culling for poor performance as well as mortality due to various causes, which under Indian conditions is likely to be as high as 15 per cent during the first year of life, 10 per cent in second year and about 5 per cent among the older stock. These factors have been taken into consideration while carrying forward the herd strength from one year to another.

One of the standard practice of converting herd strength into adult units is followed by considering (*i*) animals above 3 years of age — a cow, bullock or bull as one unit; (*ii*) heifers above 2 years of age as 0.75 unit, (*iii*) heifers of 1 to 2 years age as 0.50 unit and (*iv*) calves below one year of age as 0.35 unit. An adult buffalo may be taken as 1.1 unit because of higher body weight.

The number of calving taking place in each year is likely to be less than the number of adult cows in the herd. This shortfall is to the extent of 20 to 30 per cent depending on the reproductive efficiency of the herd, which in turn depends on the breed and genetic merit of the stock. Crossbred cows are efficient breeders and 75 to 80 per cent of them are likely to calve each year. On the other extreme, among buffaloes, the per cent calving in each year can be as low as 65 to 75 per cent. This fact has to be taken into consideration while calculating herd strength. This was, however, not done in Table 19.11 so as to make it less complicated.

One can take help of Table 19.11 to know the strength of his herd in a given year and phase the construction programme of farm accordingly. Phased construction has an economic aspect too. Housing is an expensive item. It should be taken up as profits accumulate and dairying as a profession gets stabilised. For this reason, buildings at the beginning may be constructed using cheap and locally available materials and changes towards better should be made by using costlier but better materials after one is well established.

The roof can be made of thatch or *Kadbi.* Wooden poles and coarse timber can be used for the roof supporting structure. Second hand plumber tubing or angle irons cost less than new fittings. Only floor in the covered area can be paved. Floor in the open area may be left unpaved *(Kutcha).* Wooden or wire fencing can be erected instead of walls around the open area. Even raising earthen ridges for walls of paddock will do. All these are likely to reduce the initial investment on buildings. But such structures need frequent renovations.

19.7. SPACE REQUIREMENT

Standards for space requirements for different classes of dairy cattle that are currently in vogue are based on experience and convention rather than on actual experimentation. One study in India (Verma and Tripathi 1980) revealed that in loose houses, provision of 1.76 m^2 paddock area per two year old buffalo heifer as against 3.86 m^2 does in no way hamper their growth and health.

The Indian Standards Institution (ISI) has brought out certain standards of space requirement for farm animals. These standards are basically for loose houses, though dimensions of barn type houses were also given. There is a sort of insistence that animal houses in this country should be constructed according to these standards. Hence, discussions on different aspects of housing for dairy cattle in this chapter are mostly based on the ISI standards. The ISI recommendations generally agree with the figures suggested by several other workers (as discussed by McDowell, 1972) for cattle in warmer regions. The study of Verma and Tripathi (1980) mentioned above also suggests that ISI specifications are quite handy. Cattle and buffaloes in warm regions mostly lie down during night time (Hafez, 1962; Verma and Tripathi, 1980; Sastry *et al.*, 1981; Barua *et al.*, 1982 and Grewal *et al.*, 1982). Hence the need to provide adequately large open or paddock area per animal in loose houses, as it is in the open area rather than under the shelter summer nights are more bearable. For this very reason feed mangers must be provided in the paddock area also so that animals can feed at night.

Floor space requirements. The requirement of floor space per individual animal is given in Table 19.12. In the last column of this table is given the maximum number of animals that can be housed in each pen. Under emergencies ten per cent more stock can be put in each pen for short periods; otherwise more pens should be constructed to accommodate the extra animals, keeping in view the number of animals that can be housed per pen. Constructing a larger pen with an idea to house larger number of animals is not advisable as feeding and management of animals will not be efficient; bossy animals may keep the docile ones away from feed and water. And higher yielders are generally more docile.

1970; IS : 2733., 1964 and Murthy *et al.* (1969).

Barua, K.C., Pal, R.N. and Sastry, N.S.R. (1982). *Haryana Agric. Univ. J. Res.* **11** : 58.

Grewal, S.S., Sastry, N.S.R. and Yadav, R.S. (1982). *Indian J. Anim Sci.* **52** : 58-60.

Hafez, E.S.E. (1962). *The behaviour of domestic animals.* Billiere, Tindall and Cox Ltd., London, p. 619.

McDowell, R.E. (1972). *loc. cit.*

Sastry N.S.R., Yadav, R.S., Sharma, V.B. and Juneja, I.S. (1981). *Haryana Agric. Univ. J. Res.*, **11** : 452-457.

Varma, A.K. and Tripathi, V.N. (1980). *Haryana Auric. Univ. J. Res.* **10** : 594-599.

In certain buildings like calving pens, milking barns etc the space — more so its shape and size to be provided per individual has to be slightly different than the standards described above, due to obvious reasons. Details of such special space provisions are shown in different layouts given in later sections.

Feeding and watering space requirements. Apart from providing sufficient floor space in animal houses, care should be taken to see that the dimensions of mangers and water troughs are such that all the animals in the pen have a free access to these.

TABLE 19.12. Floor Space Requirement for Dairy Cattle and Buffaloes

Type of animal	*Floor space requirement* m^2		*Maximum No. of animals/pen*	*Height of shed at leaves cm*
	Covered area	*Open paddock*		
Bulls	12.0	24.0	1	
Cows	3.5	7.0	50	175 cm in medium and heavy rainfall and 220 in dry areas
Buffaloes	4.0	8.0	50	
Down-calvers	12.0	12.0	1	
Young-calves	1.0	2.0	30	
Old heifers	2.0	4.0	30	

Note : For further details see ISI Bulletins — IS : 4466 (Parts I, II & III), 1968; IS : 5605 (Parts I & II).

It is generally believed that each animal in a loose house must have manger space (or feed bunk space) of sufficient width. However, for two year old buffalo heifers provision of only 53 cm manger space as against 73 cm does not seem to interfere with the feed consumption and growth performance of animals nor cause any aggressive behaviour amongst them (Verma and Tripathi 1980). But fodder has to be put more frequently (4 times a day) when smaller manger space is provided.

Earlier Kooner *et al* (1969) suggested that, if cattle have access to roughages at all times, 30 to 40 cm manger space per cow will be adequate. On the other hand, if roughages are fed in batches, as is generally done in India, each cow needs from 60 to 75 cm of manger space so that all the animals will have access to eat at the same time. Such a proposition is also evident from the works of Friend *et al.* (1977) and Keys *et al.* (1978). Keeping in view of the above findings the feeding and watering space requirements per animal are suggested (Table 19.13).

Dimensions of water troughs shall be the same as that of feed mangers and water should be made available in them round the clock. Generally, all the animals in a pen may feed at the same time, but not all animals drink water simultaneously. Thus, while the length of the mangers

ISI (1968). *IS : 4466(Part I)-1967.* Indian Standards Institution, New Delhi.
ISI (1968). *IS : 4466 (Part II)-1968. Ibid.*
ISI (1969). *IS : 4466 (Part III)-1968. Ibid.*
ISI (1970). *IS : 5605 (Part I)-1970. Ibid.*
ISI (1970). *IS : 5605 (Part II)-1970. Ibid.*
Murthy, V.V.N., Kooner, D.S. and Sundaresan, D. (1969). *Indian Dairyman,* March, 1969.
Varma, A.K. and Tripathi, V.N. (1980). *loc. cit.*
Friend, T.H., Polan, C.E. and McGilliard, M.L. (1977). *J. Dairy Sci.* **60** : 108.
Keys, J.E., Pearson, R.E. and Thomson, P.D. (1978). *J. Dairy Sci.* **61** : 448.
Kooner, D.S., Murthy, V.V.N. and Sundaresan, D. (1969). *loc. cit.*

should be sufficient to provide feeding space for all the animals in the pen, it will suffice if the length of water troughs is sufficient for about ten per cent of animals in that pen.

TABLE 19.13. Feeding/Watering Space Requirements of Dairy Cattle and Buffaloes

Type of animal	*Space per animal (cm)*	*Total manger length in a pen for 50 animals*	*Total water tank length in a pen for 50 animals*	*Dimensions of manger/tank (cm)*		
				Widh	*Depth*	*Height of innrer wall*
Adult cattle and buffaloes	60-75	3000-3750	300-375	60	40	50
Calves	40-50	2000-2500	200-250	40	15	20

Note. Height of inner wall and outer wall of mangers as well as depth or width of mangers in this table are mainly based on the suggestions of the Indian Standards Institution (ISI).

In loose houses, great care must be taken to see that feed and water is always available in the mangers and water troughs, respectively. Otherwise, docile animals in the pen may get starved.

19.8. BUILDINGS REQUIRED FOR DAIRY FARMS

On dairy farms buildings are required for housing animals of different ages and physiological conditions. Also required are some ancillary buildings for stores etc. The location and arrangement of individual buildings within the chosen site of the farm as well as their functional and structural details are described in this section.

19.8.1. Buildings Required for Animals

Milking barns. This is a fully covered barn in which the milch stock are milked. There is a lot of activity in the miling barn and hence it should be located at a central place with all other buildings arranged around it. This is all the more important on larger farms. There shall be individual standings or stalls for stanchions in one (Figure 19.7) or two (Figure 19.8) rows. Double rows of stanchions can be arranged either tail-to-tail (Figure 19.8) or face-to-face (Figure 19.9); the former being better.

Herringbone-type or round milking parlours can also be used for two to four cows or buffaloes when hand milking is practiced. The number of standings required on a farm are thus equal to half to quarter of the number of milch stock.

The length and width of the standing shall be decided according to the size of the animal and may vary from 1.5 to 1.7 meters in length and 1.05 to 1.20 meters in width. Buffaloes of

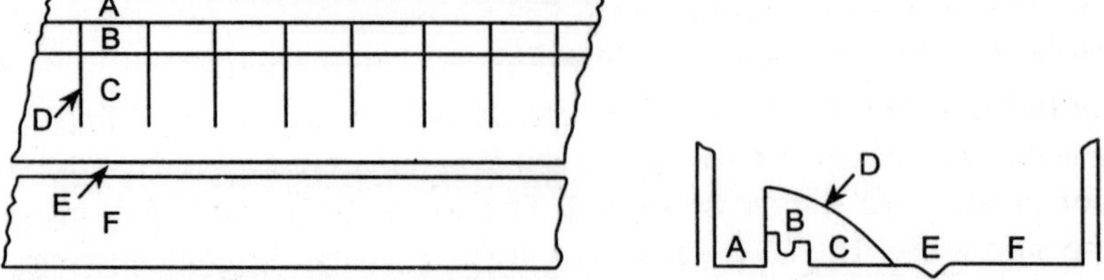

Fig. 19.7. Plan (I) and section (II) of a single row milking barn. A. Feed alley, B. Manger, C. Standing D. Standing partition, E. Gutter, and F. Passage.

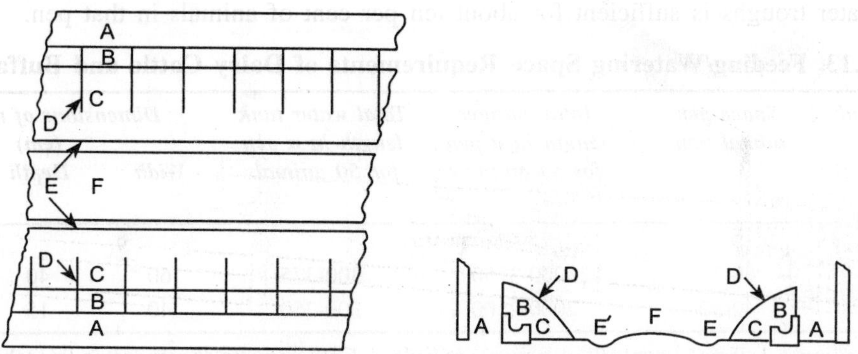

Fig. 19.8. Plan (I) and section (II) of a double row tail to tail type milking barn. Legend same as in Fig. 19.7.

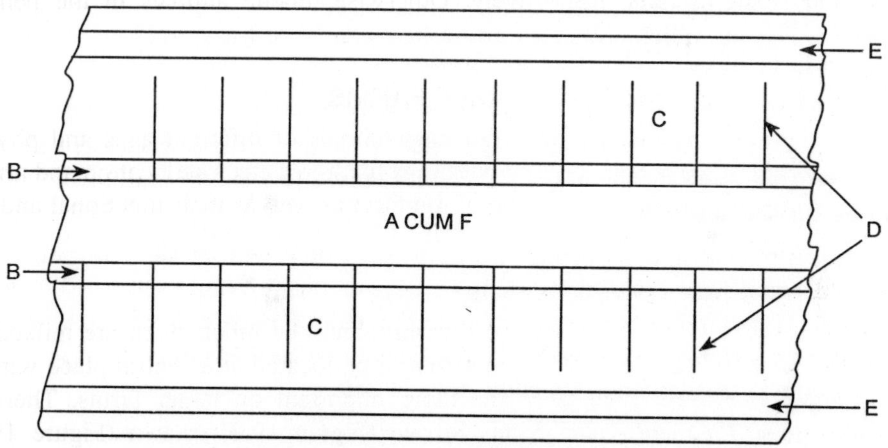

Fig. 19.9. Plan of double row face-to-face type milking barn. Legend same as in Fig. 19.7.

heavier breeds like Murrah, Nili and Jaffarabadi as well as crossbred cows (Brown Swiss and Friesian) require standings that are longer and wider than those for smaller breeds of animals. The width of the standing is about 80 per cent of the length. The width of the central passage shall be 1.5 to 1.8 meters. The central passage shall have a gentle slope from the centre outwards — towards the drain or gutter. There shall be two continuous mangers, one on each side along the head side of the standing rows and a 0.75 meter wide feeding alley beyond each manger. There shall be a shallow 'U' shaped drain 30 cm wide, one on either side of the central passage; in case of single row barns one drain towards the standings side of the passage is sufficient.

The floor of the standing should be paved and should have a slope of 1 in 40 towards drains. There should be walls with large windows along the length and at each end of the barn to support the roof. The roof of the shed should be gabled. The eaves of the roof shall project out at least 50 cm beyond the side walls. The side walls need not necessarily be complete; large open spaces may be left in the side walls at suitable intervals. When walls are complete up to the roof, as is necessary in cooler and/or heavy rainfall areas, there shall be windows and ventilators at suitable places in the walls.

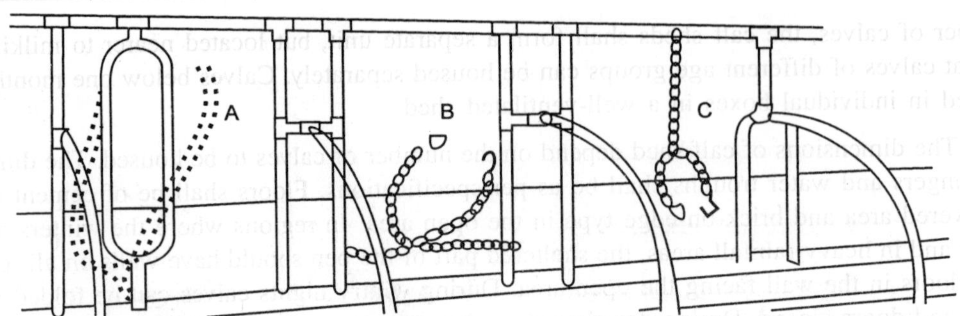

Fig. 19.10. Different types of devices used for securing individual cows by neck in the standings. A. Metal yoke stanchion. B. Double chain, and C. Single chain.

Some device should be arranged to secure individual animals by neck in each standing (Figure 19.10). The milking barn may, if necessary, be extended to accommodate down-calvers as well as calves. Further, milch cows that are in heat can be left after milking in the standings and inseminated right there.

Shed for milch/dry cows. These are sheds for housing milch cows and dry cows separately. On small farms a partition can be raised in one shed for housing dry cows separately instead of constructing a separate shed for them. On farms with only a few animals, all the cows—milking or dry—can be housed together.

These are simple sheds comprising of a covered portion and an adjoining open paddock. The manger and water tank dimensions shall be as specified in Table 19.13. The floor in the covered area should preferably be of cement concrete. Brick-on-edge or stone-slab flooring can also be used; at least a *Moorum* or *Kankar* flooring should be laid. *Moorum* flooring is quite good for the open paddock, though brick-on-edge floor is better.

Maternity pens. Pregnant cows are to be transferred into maternity pens or calving pens or calving boxes two to three weeks before the expected date of calving. The number of calving pens required is about 10 per cent of the number of breedable female stock on the farm. These should be located nearer to living quarters of farmer and/or milking barn so that the down-calvers are constantly observed. The pens can be constructed either in a row or in groups of two or four.

The dimensions of each calving pen shall be about 3 m × 4 m for covered area and another 3 m × 4 m for the open paddock. The covered area shall have a 1.25 – 1.5 m high wall all around, barring a 1.5 m wide gate opening into the open lot. A manger and a water trough of proper size should be constructed in each pen. The floor shall be of *Moorum,* brick-on-edge or cement concrete paved with 1 in 60 slope towards the drain.

Sick animal shed. Sheds, more or less similar to maternity pens in structure, shall be located well away from the other sheds so that these sheds are inaccessible to other animals. Sick animal sheds should be cleaned towards the end, *i.e.*, after cleaning all other sheds so that no infective material may be carried by workers accidentally from the sick animal sheds to the normal ones.

Calf sheds. The calf sheds can be located either at the end of the milking barn or somewhere near it. This facilitates taking calves to their dams quickly at milking time (if weaning is not practised) and hand feeding of milk to calves (if weaning is practised). If there are a large number of calves, the calf sheds shall form a separate unit, but located nearer to milking barn, so that calves of different age groups can be housed separately. Calves below one month can be housed in individual boxes in a well-ventilated shed.

The dimensions of calf shed depend on the number of calves to be housed. The dimensions of mangers and water troughs shall be as per specifications. Floors shall be of cement concrete in covered area and brick-on-edge type in the open area. In regions where the winters are quite chilly and in heavy rainfall areas, the sheltered part of the pen should have walls on all four sides with doors in the wall facing the open area. During wintry nights calves can be folded into this room and doors closed. During day time the calves can move through the opened doors into the open area to have the benefit of sunshine. Provision of a thin bedding of wheat straw *(bhusa)* or any other suitable material in the calf pen keeps calves warm and cozy.

National Dairy Research Institute (Arora, 1978) suggested that considerable reduction in mortality and higher rates of gain can be achieved if calves (especially buffalo calves) are housed individually in cages with raised floors made of expanded metal (Figure 19.11) during wet season. Bose and Thomas (1979) from Haryana Agricultural University too found the calves to fare better when housed in individual cages during winter. Housing calves individually during summer or giving thatched roof and slotted floor (Saseendranath *et al.,* 1983), however, does not seem to give any added advantage. Older calves housed on paved floors grew marginally better and utilised feed better as compared to those reared on dirt floor both in summer (Sastry *et al.,* 1981) and winter (Grewal, *et al.,* 1982).

Fig. 19.11. Individual cages for calves with raised expanded metal floor (*Courtesy* : NDRI Karnal).

Arora, S.P. (1978). *Calf raising.* Publ. No. 135, NDRI. Karnal.
Bose, A.M. and Thomas, C.K. (1979). *Indian J. Anim Sci.* **49** : 419.
Grewal, S.S., Sastry, N.S.R.. Yadav, R.S. (1982). *loc. cit.*
Sastry, N.S.R., Yadav, R.S., Sharma, V.B. and Juneja, I.J. (1981). *loc. cit.*
Saseendranath, M.R., C.K. Thomas & P.I. Geevarghese, 1983. *Indian J. Anim. Sci.* **53** : 591.

Young or growing stock shed. Heifer calves from about six months of age to breeding age are to be housed separately from the suckling calves. Generally, no male calves are kept on farms beyond six months of age. When large number of young stock is there, they should be divided into different age groups and each group housed separately. When their number is small, the young stock shed may be an extension of the dry animal shed or it may be a separate unit nearer to it.

In the latter case, the sheds shall be constructed in a single row of sheds facing each other. The constructional details of this shed are the same as of cow sheds, except for the difference in dimensions of mangers and water troughs.

Bull sheds. The bull sheds should be constructed towards one end of the farm. There shall be one shed for each bull; the number of bulls required being one for about every 50 breedable females on the farm, if natural breeding is practised. When artificial insemination service facilities are available nearby, there may be no necessity to keep bulls on the farm. The bull shed shall have covered area of 3 m × 4 m dimensions, leading into a paddock of 24 square metres.

19.8.2. Ancillary Structures on a Dairy Farm

Stores. Stores for concentrate feeds, hay, straw and equipment are a must on livestock farms. The size and type of buildings for stores can be decided on the basis of the quantity of feeds and fodders that are to be stored at a time on the farm. Approximate storage space required for different type of feeds is given in Table 19.14. These figures may be helpful in determining the sizes of feed stores, though a sound judgment of the situation is all that is needed. Sometimes the size of the concentrate store required is determined on the presumption that 0.2 m^3 storage space is required per adult unit.

TABLE 19.14. Approximate Storage Space (m^3) Required per Quintal of Certain Cattle Feeds

Hay (loose)	1.60
Hay (baled)	0.70
Hay Chopped	0.60
Straw (loose)	3.00
Straw (chopped)	0.70
Bran	0.50
Grains & Oil cakes	0.17
Stover (Maize, Bajra, Jowar etc)*	3.85

Source : A. Ensiminger, *Animal Science,* Interestate, U.S.
* Authors' observation at Haryana Ag. University;

There shall be one main concentrate store-cum-feed mixing room at a distant place; but it is advisable to have a small ration room nearer to milking barn for storing prepared concentrate mixture temporarily for a day or two. On smaller farms perhaps one feed store near the milking barn will be sufficient. Floors in the feed store should be made damp and rodent-proof.

Silos. Silage is an important item of feed for dairy animals. Silage is obtained by ensiling green fodder in silos under anaerobic'conditions. Silage is fed at the rate of 10-15 kg and 5-8 kg, for adult and young stock, respectively, per day.

Tower, pit and trench silos are in use; but under Indian conditions trench silos will be more practicable and convenient. The capacity of different sizes of trench silos is given in Table 19.15. The figures given in the first three columns are the cross-sectional dimensions of the trench.

TABLE 19.15. Capacity of Trench Silos

Bottom width (cm)	*Top width (m)*	*Depth (m)*	*Capacity (Qtls of silage) per linear (m)*
1.5	2.6	1.2	15.4
1.8	2.4	1.8	23.9
1.8	3.0	2.4	34.0
2.1	3.3	1.8	27.1
2.1	3.7	2.4	40.9
2.4	3.9	1.8	33.4
2.4	4.4	2.4	48.3

Source : Kooner, D.S., V.V.N. Murthv and D. Sundaresan. *Dairy Cattle Housing., Farm Bulletin No. 23. Punjab Ag. University, Lundhiana, 1969.*

Milk house. Milk house is another important item of dairy farm buildings. This is the place where the milk is collected, weighed, and stored in cans for brief periods—before delivering to milk plants. Dimensions of milk house depend upon the quantity of milk handled daily (Table 19.16).

This space can accommodate the milk recording equipment, milk cooling device in the form of bulk cooler, can racks, milking pail rack, sinks, washing outfit and furniture of the milk recorder. On very large farms the different components of the milk house, *i.e.*, recording-cum-

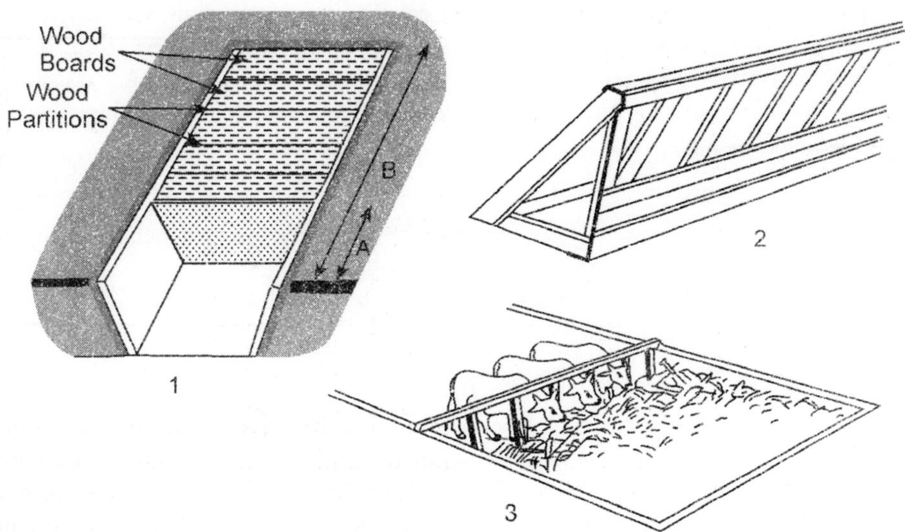

Fig. 19.12. Details of a trench silo. (1) with a movable fence, (2) its movable fence portion, (3) cows feeding through a movable fence.

milk cooling room, milk utensils and equipment room and washing room, may be constructed as separate units (see layouts 19.15 to 19.18)

TABLE 19.16. Floor Space Requirements for Milk House

Daily milk output	*Floor space* (m^2)	*Remarks*
Under 100 litres	3.7 × 3.0	Additional 0.37 m^2 for every 40 litres output, over and above 700 litres
100-200 litres	3.7 × 3.7	
200-450 litres	3.7 × 4.4	
450-700 litres	3.7 × 5.1	

Source : Kooner D.S. *et at.* (1969). *loc. cit.*

The doors and windows of the milk house shall be made fly-proof. The flooring of the room should be of cement concrete, impervious and reinforced with iron strips located at suitable distance to make it hard wearing. The walls should be lined with white glazed tiles up to a height of 1.5 m. A suitable platform or concrete slab (like that of a laboratory bench) shall be provided for keeping the milk testing apparatus.

Hay/straw sheds. Since the common dry fodders — hay or crop by-products — are available at the harvesting seasons only, one has to store these items for use throughout the year. Space requirement shall depend upon the manner in which the dry fodder is stored (see Table 19.14). Adult animals consume about 5-6 kg of dry fodder per day while young animals consume about 1 kg between 6 to 12 months of age, about 2 kg between 1 to 2 years, and 3 – 3.5 kg between 2-3 years of age per day. Annual requirement of hay needed can be worked out on the basis of the number of days in a year when hay is used. The size of the shed needed to store this hay or straw can be calculated by referring to Table 19.14.

Sheds, preferably with walls on three sides, are better for storing straw or hay. Storage of hay or straw in stacks or heaps in the open results in lot of wastage due to spoilage. Sometimes a simple framed shed with a gabled roof is used. The hay shed shall be away from animal sheds because of possible fire hazards. On larger farms the hay sheds can be made into a sort of self-feeding hay bunks by placing movable partitions on one side edge of the hay shed. The animals can eat hay at will from only the portions accessible through this wooden partition and so do not spoil the hay excessively.

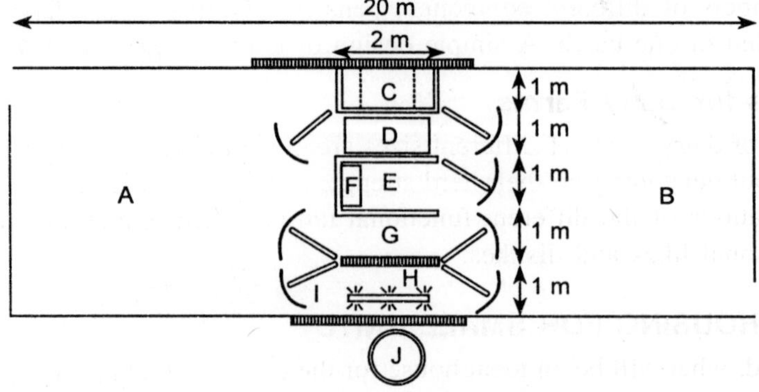

Fig. 19.13. Handling yard for a medium sized dairy farm. A. Assembly pen, B. Collecting pen, C. Weighing Machine, D. Weighing platform, E. Operators lounge, F. Working table, G. Trevis or crush for cattle, H. Spraying chamber, I. Overhead sprayer, J. Spraying equipment.

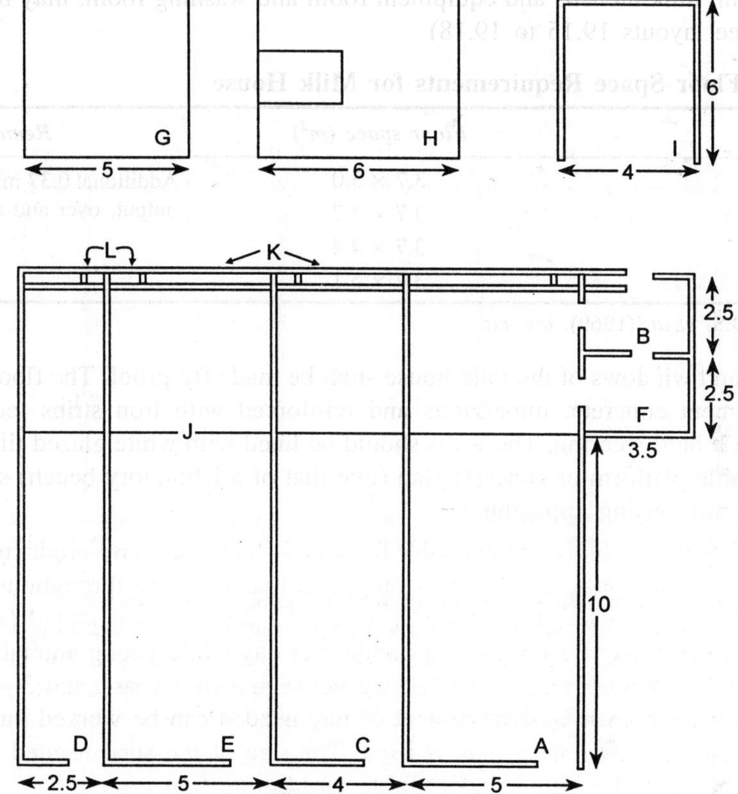

Fig. 19.14. Layout showing the functional details of building for a 10-cow dairy farm. A. Milch cow pen cow-milking barn, B. Milk house-cum-record room, C. Calf-pen, D. Maternity pen, E. Dry cow pen, F. Store, G. Hay/Straw/Cart shed, H. Chaffing shed, I. Silo, J. Gutter, K. Manger, L. Water tank, All dimensions in metres.

Handling yards. On larger farms it is advantageous and economical to construct one flexible handling yard that can be used for the three processes of animal handling—namely collecting, fixing and control of individual animals for weighing, vaccination, marking etc. The size and the numbers of different component pens of handling yards, depend on the heads of stock to be handled in one batch. A simple design of handling yard is shown in Figures 19.13.

19.8.3. Layouts for Dairy Farms

Suitable layouts for dairy farms of different sizes are given in Figs., 19.14 to 19.17. Farmers may consider any layout depending on their herd strength. The layouts are intended to show the sizes, numbers and locations of the different functional units. A farmer may modify these to suit his location and personal likes and dislikes.

19.9. CATTLE HOUSING FOR SMALL UNITS

Often it is debated, what will be an ideal house for the one or two animals kept by a small farmer in India, for whom dairying is only a secondary occupation. Considering a small farmer's economic position, perhaps, the place where he presently keeps his cow or pair or bullocks, may be the best; only it should be kept clean, dry and airy.

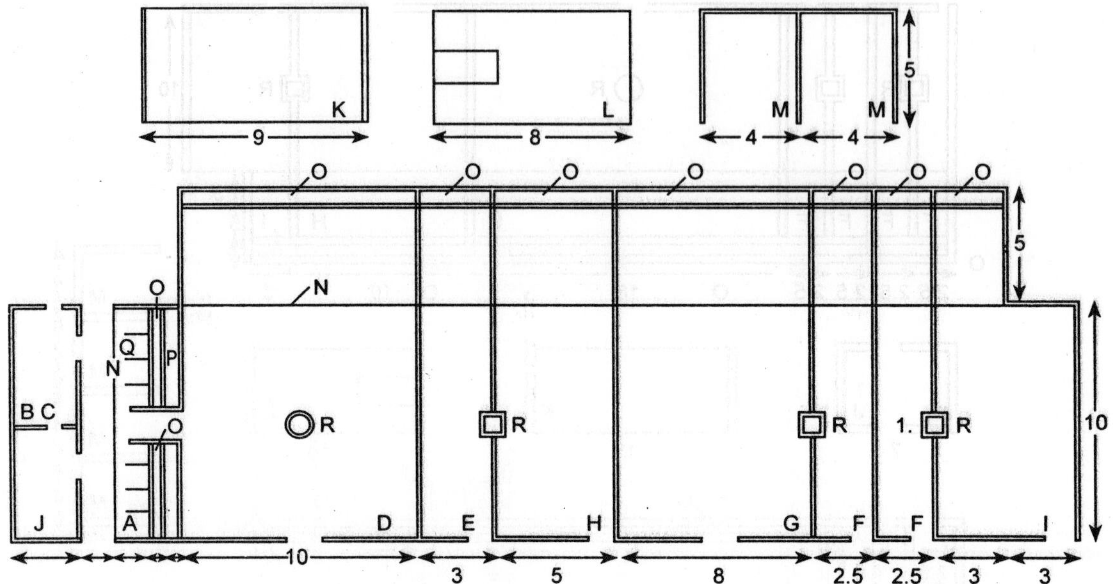

Fig. 19.15. Layout showing the functional details of buildings required for a 25 cow dairy farm, A. Milking barn, B/C. Milk house-cum-record room, D. Milch cow pen, E. Young calf pen, F, Maternity pens, G. Dry cow/bullock pen, H. Older calf pen, I. Bull pen, J. Store, K, Hay/ straw/ cart/tractor shed, L. Chaffing shed, M. Silopits, N. Shallow gutter, O, Feed manger, P. Feed passage, Q. Stanchion, R. Water tanks. All dimensions in metres.

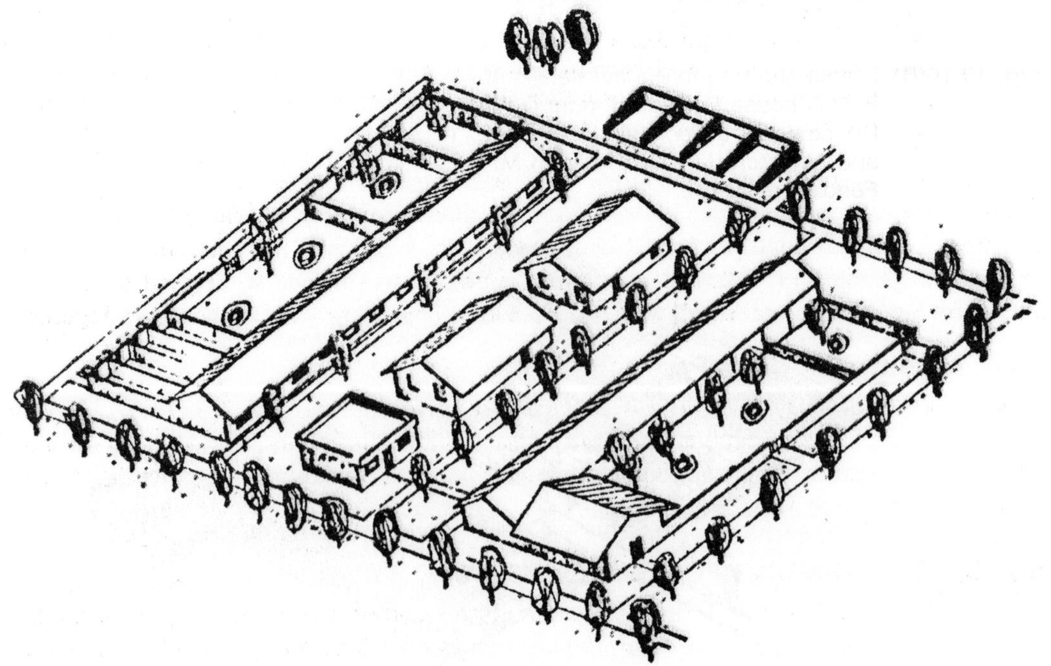

Fig. 19.16(A). Bird'seye view of the farm constructed according to the layout given in Fig. 19.16(B).

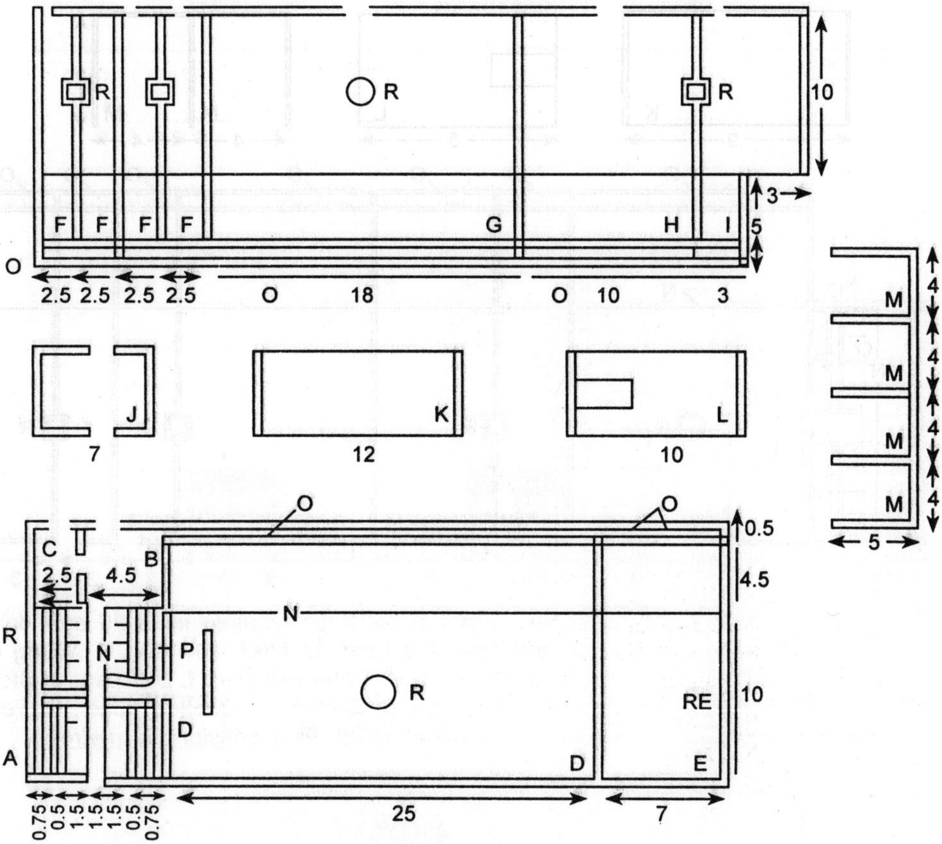

Fig. 19.16(B). Layout showing functional details of buildings for 50 cow dairy farm. A, Q. Milking barp B. Milk house C. Record room D. Milch cow pen E. Young calf pen F. Maternity pens G. Dry cow cum bullock pen H. Older calf pen I. Bull pen J. Office/laboratory K. Concentrate and hay store L. Chaffing shed M. Trench silos N. Gutter or drain" O. Feed manger P. Feeding passage.

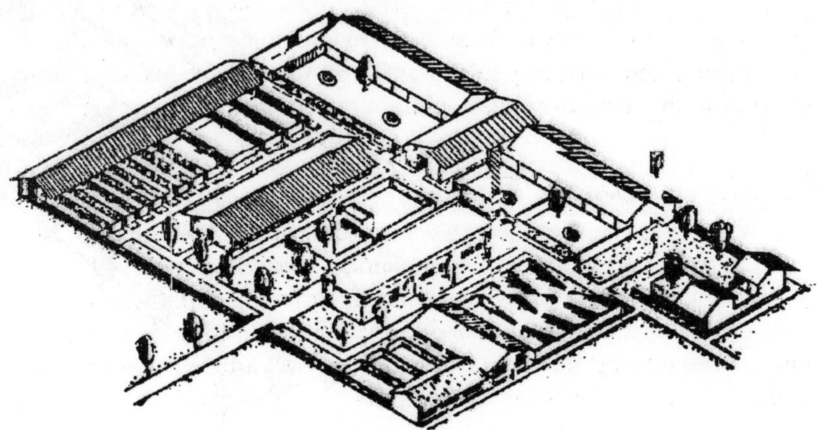

Fig. 19.17 (A). Bird'seye view of the farm constructed according to the layout given in Fig. 19.17(B).

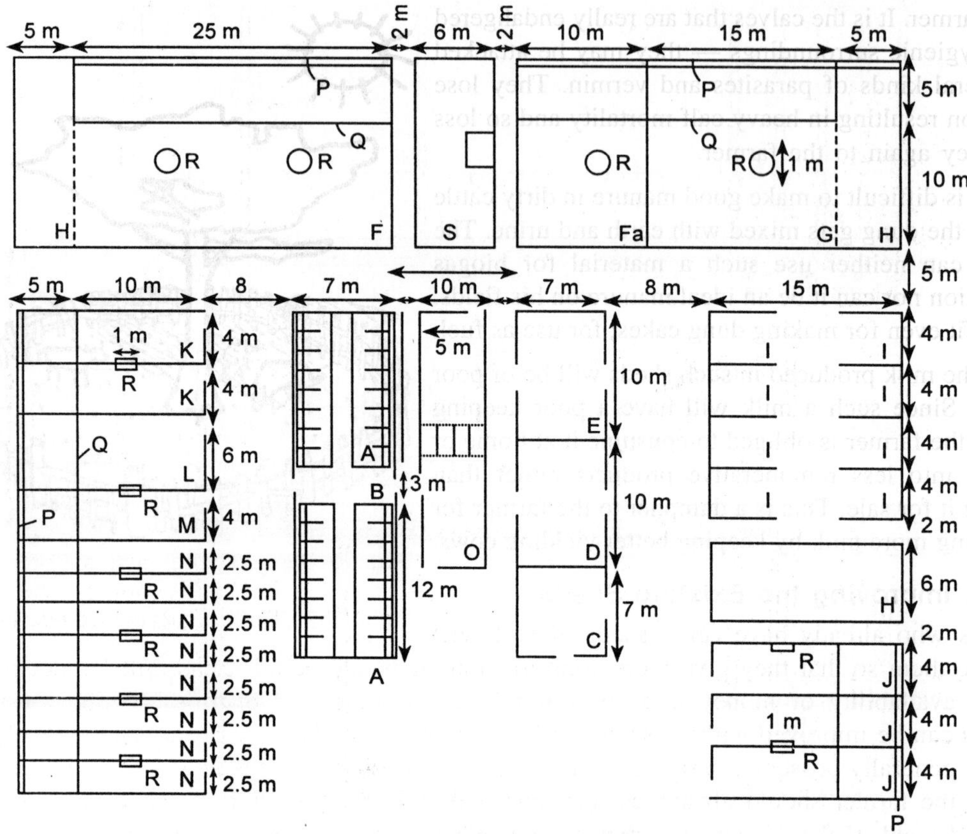

Fig. 19.17(B). Layout showing the functional details of buildings for a 100 cow dairyfarm. A. Milking barn B. Milk house C. Office/record room D. AI laboratory E. Stores F. Milch cow shed G. Dry cow shed H. Additional hay/straw store I. Trench silos J. Bull pens K. Young calf pens L. Older calf pen M. Pen for older heifers. N. Maternity pens O. Handling yard P. Feed manger Q. Gutter or drain R. Water tank S. Hay/straw/cart/tractor shed.

However, there is considerable scope for introducing modern dairy cattle management techniques in villages. As has already been discussed at the beginning of this Chapter, the system of dairy animal management is closely related to the nature of cattle housing, or more appropriately, the cow shed. Hence, the need to suggest some improved and/or improvised cow sheds for small farmers.

19.9.1. The Traditional Cattle 'Shed'

The farmers in India keep their cattle under great diverse conditions, such as: tethering under a tree, tethering them very close to their living quarters, enclosing them within a partially sheltered wood-fenced paddock (Figure 19.18) or even making small sheds specially for cattle near the living quarters. Under such systems of housing the animals have to face two problems, *viz.* (*i*) over-crowding, (*ii*) remaining standing on a mixture of earth, excrement urine and water as a result of which they cannot lie down comfortably.

When animals are dirty, not only their wounds (especially those on limbs) heal slowly but the animals may fall sick easily. And sick animals mean loss of production and so loss of money to the farmer. It is the calves that are really endangered by unhygienic surroundings — they may be attacked by several kinds of parasites and vermin. They lose condition resulting in heavy calf mortality and so loss of money again to the farmer.

It is difficult to make good manure in dirty cattle shed as the dung gets mixed with earth and urine. The farmer can neither use such a material for biogas production nor can it be an ideal manure on his fields. It is unfit even for making dung cakes; for use as fuel.

The milk produced in such sheds will be of poor quality. Since such a milk will have a poor keeping quality the farmer is obliged to consume it at home or convert into less remunerative products rather than sending it for sale. This is a dampner to the farmer for producing more milk by keeping better yielding cows.

Fig. 19.18. Traditional enclosure for cattle in Africa. (**Source.** F.A.O. Better Farming Series, 197b).

19.9.2. Improving the Existing Sheds

Farmers who already have cattle sheds of sorts can improve them so that they give more comfort to animals and permit clean milk production. Of course, availability of money is a constraint here. There are ways by which the housing of animals can be improved without spending lot of money by using wood, mud and straw that the farmers generally possess at home. If one has to spend money on the alterations, before spending money, the farmer should always calculate what the expenditure will be and what will be the expected yield. If the farmer feels that the expenditure does not yield enough in return it is not worthwhile to incur it, as he may become poorer and will feel discouraged.

Reproduced in Figures 19.19 and 19.22 are designs recommended for rural milk producers in tropics by specialists of various institutions like the Food and Agricultural Organisation of United Nations, State Agricultural Universities, National Insitutions etc.

The farmer should make the changes gradually, that too by using cheaper locally available materials. Once the farmer starts producing more milk and earning more money, he will be able to afford to spend money to make modern cattle sheds. Some of the suggested alterations for improving the farmers' existing cattle sheds are as follows.

1. The surface of the floor in the shed should be made even by leveling off ridges and filling up of pits. The surface can be paved with bricks or at least made hard by good ramming.
2. The farmers cattle sheds may be having a roof made of straw, or mud or GI sheets or asbestos sheets. Straw and mud roofs may be leaky and harbour insects and pests. In rainy season they may leak. Money permitting, it is better to replace such roofs with better materials.
3. Farmers' traditional cattle sheds are made very close to or in between living quarters and the roof is made too low. This makes the shed dark inside and the air movement is poor. So as to make it better the roof should be raised to proper height, which is 6 feet at eaves of the shed.

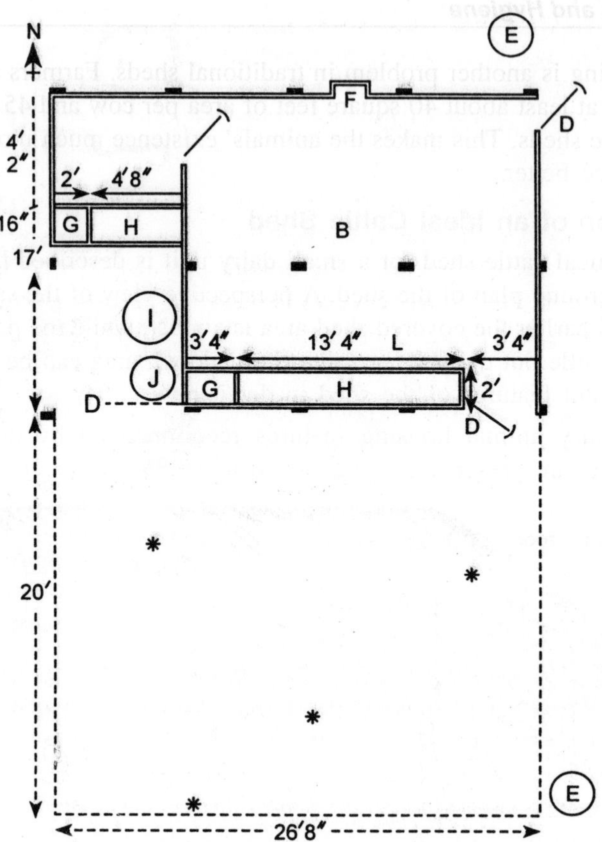

Fig. 19.19(A). Plan of a proposed cattle shed for small dairy units (six buffaloes or seven cows with followers) by Haryana Agricultural University. A. Calf shed, B. Cow shed, C. Open paddock, D. Gates, E. Site of gas plant, F. Wall cupboard for keeping records and medicines, G. Water trough, H. Feed trough (manger). I. Site for hay or straw stack, J. Water reservoir (a drum or a small tank), K. Site for mannure pit, L. Rings fixed to the manger for tethering.

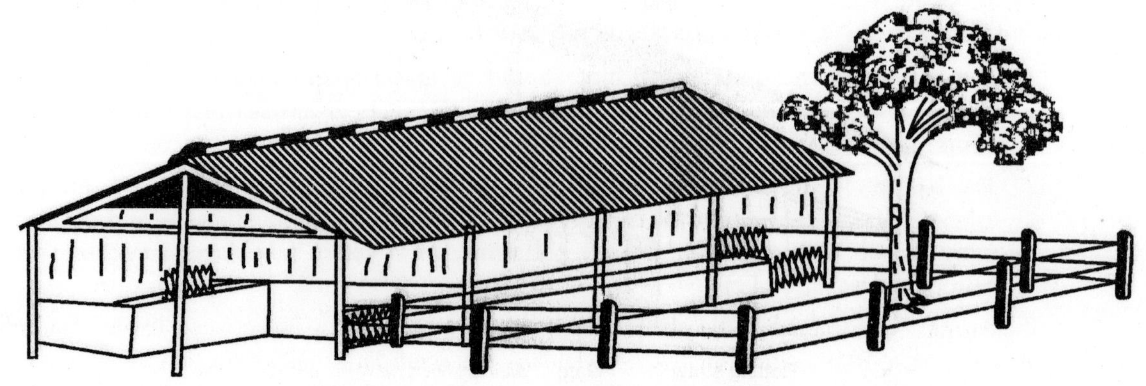

Fig. 19.19 (B). A view of the proposed cattle shed (Fig. 19.20 A) along with the paddock.

4. Over-crowding is another problem in traditional sheds. Farmers should take care to see that there is at least about 40 square feet of area per cow and 45 square feet per buffalo in their cattle sheds. This makes the animals' existence much more comfortable making them produce better.

19.9.3. Construction of an Ideal Cattle Shed

A simple and economical cattle shed for a small dairy unit is described here under. Depicted in figure 19.19A is the ground plan of the shed. A perspective view of the same is shown in Figure 19.19B. As mentioned earlier the covered shed area is a must, while the paddock area is optional. If farmers take their cattle out daily for grazing the paddock area can be dispensed with. Let us consider some important features of the shed in detail now.

Small farmer dairy animal housing features recommended by different other research/ extension organisations are presented in Fig. 19.20 to 19.25.

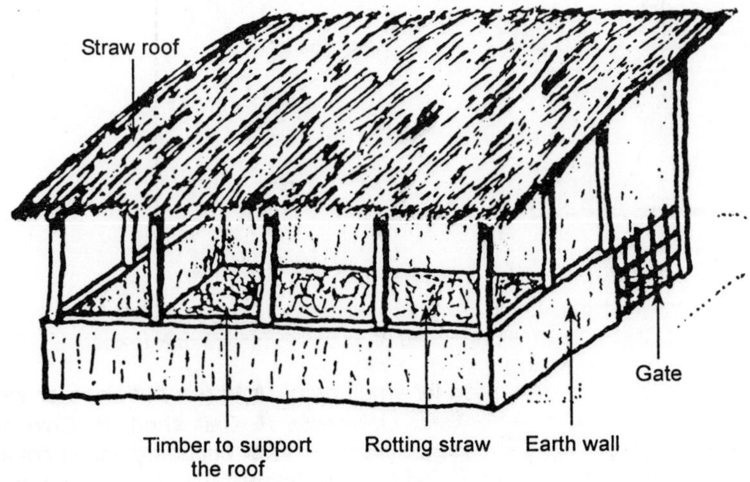

Cowshed seen from the front

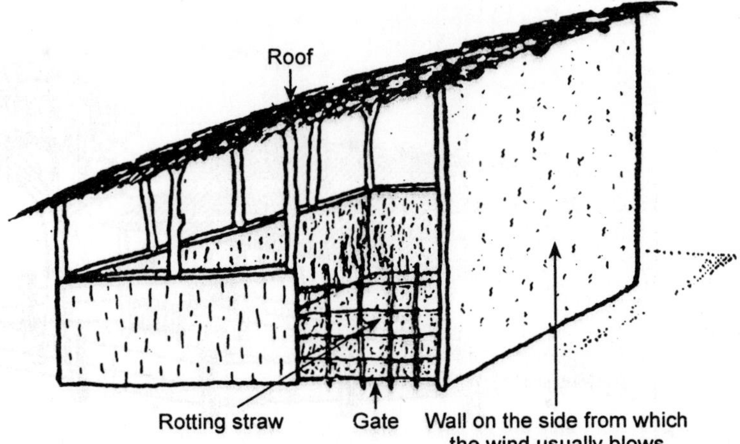

Cowshed seen from the side

Fig. 19.20. A simple house for a couple of cows to be housed under tropical conditions. (**Source :** FAO Better Farming Series, 1970)

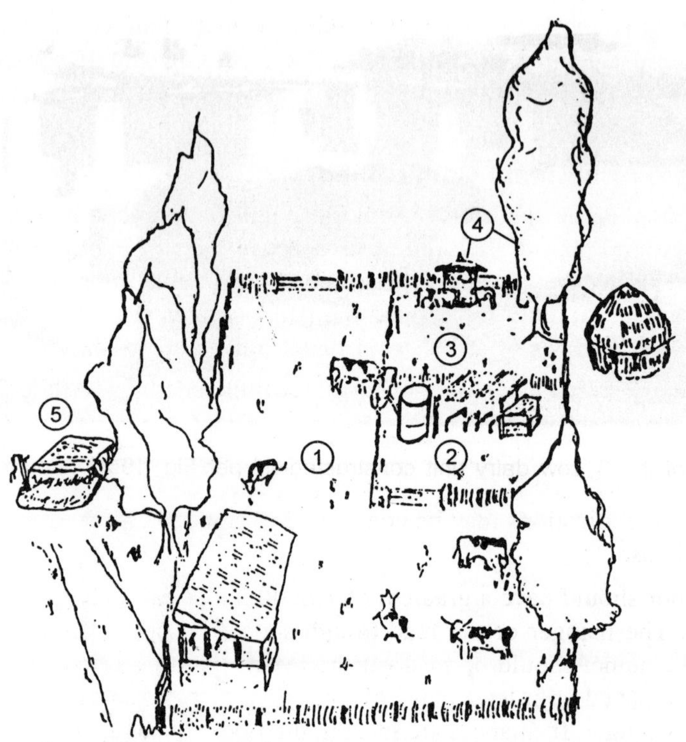

Fig. 19.21. A perspective view of the whole farm of a peasant milk producer 1. Coral-Paddock with a simple shelter, 2. Site of milking, 3. Calf area, 4. Hay stack-feeder, 5. Manure pit. (**Source** : FAO Better Farming Series, 1980)

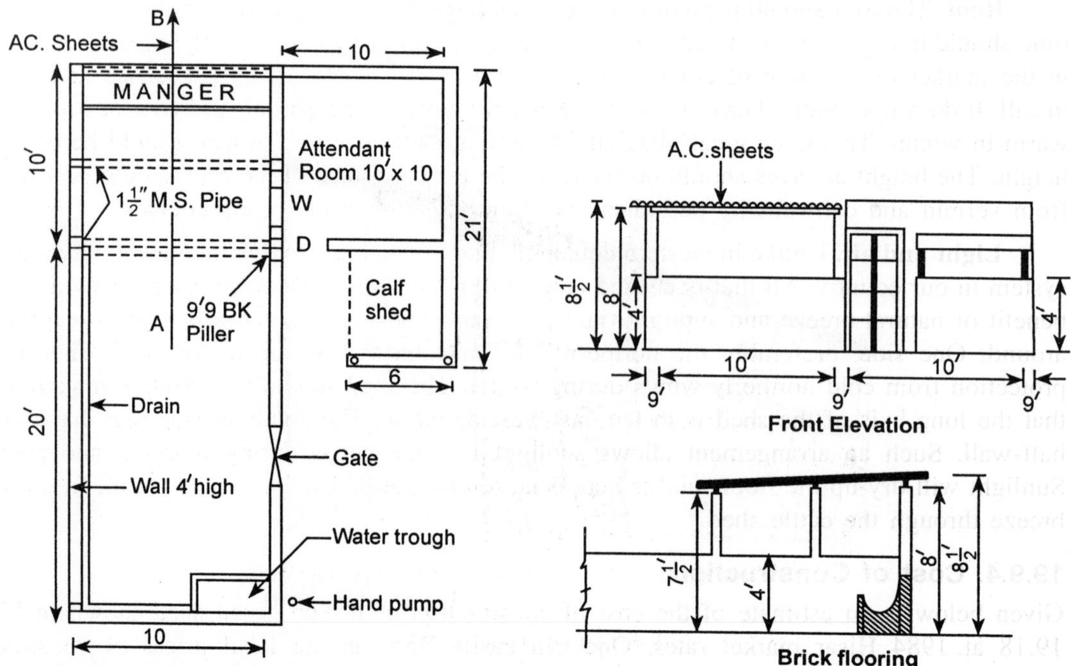

Fig. 19.22. A Design of 3_cow dairy unit (*Courtesy :* NDRI, Karnal).

Fig. 19.23. A view of the 3_cow dairy unit constructed as per Fig. 19.22 (*Courtesy* : NDRI, Karnal).

The following specifications may be considered in constructing the various structures inside the dairy animal sheds.

Floor. The floor should have a gradient of 1 in 40, *i.e.,* a fall in height of one unit for every 40 units of length. The manger or the feed trough should be put at the top end of the floor. A narrow 'U' shaped channel should be made at the lower end of the floor for drainage. The floor should preferably be paved or at least made of bricks-on-edge. A paved floor can be kept cleaner, and drier and it lasts longer. If finances are meagre the floor should at least be made hard by good ramming using *Kankar* or *Moorum* as the material. The surface of the floor should be even and all attempt be made to keep it dry.

Roof. The roof should prevent rain, cold and heat from getting into the shed and at the same time should not interfere with ventilation. Nowadays a new asphalted roofing material is available in the market in the form of corrugated as well as plain sheets. It is light, cheap and easy to install. It does not require heavy roof supporting structures. It keeps the shed cool in summer and warm in winter. This is an ideal material for roofing cattle sheds. The roof should be of proper height. The height at eaves should be six feet. The under surface of the roof should be kept free from vermin and cobwebs by periodical cleaning and spraying with insecticides.

Light and air. Unlike in western countries, there is no need for having elaborate ventilation system in our country. All that is needed is to design the walls and roof in such a way as to take benefit of natural breeze and sunlight. The shed can at best have half-walls (3 to 4 feet tall) all around. One side, preferably the northern side, may have walls up to the roof. This affords protection from cold northerly winds during winter. The shed should be oriented in such a way that the long axis of the shed is in the east-west direction. The southern side may have only a half-wall. Such an arrangement allows sunlight into the sheds during morning and evening. Sunlight will dry up the floors and is beneficial to cows also. The half-walls permit free flow of breeze through the cattle shed.

19.9.4. Cost of Construction

Given below is an estimate of the cost of construction of the proposed shed shown in Figure 19.18 at 1984 Hisar market rates. One can easily find out the local prices of construction materials and make similar calculations for any place on any given date.

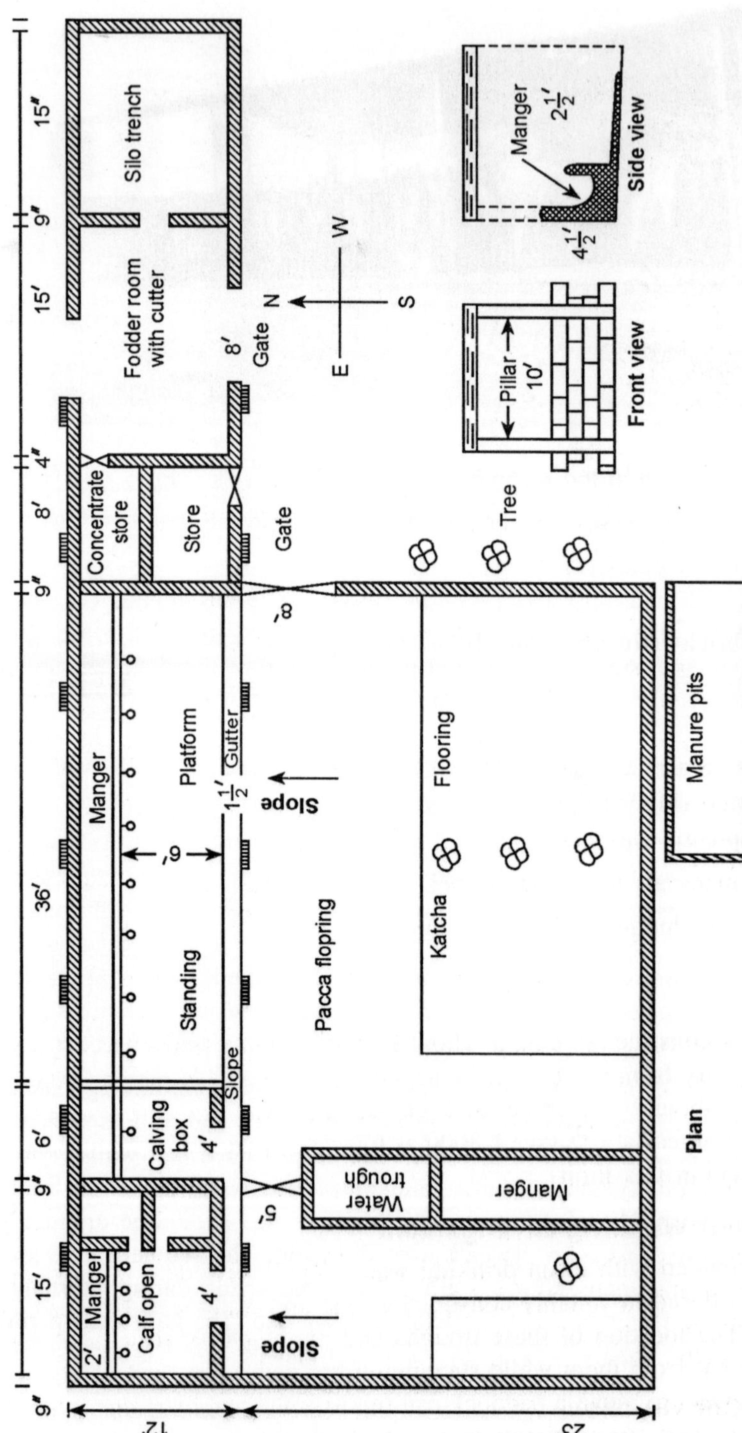

Fig. 19.24. Shed for 10 cow/buffalo dairy unit recommended by Punjab Agricultural University (*Courtesy* : Dr. S.S. Gill).

Fig. 19.25. Cows feeding at the manger in a 12 cow dairy unit (*Courtesy* : NDRI, Karnal).

		₹
1.	Cost of 40 wooden poles (to be used as roof supporting structures and rafters) @ ₹ 20 each	800.00
2.	Cost of 9000 bricks (for walls and floor) @ ₹ 300 per 1000 bricks	2700.00
3.	Cost of 40 bags of cement @ ₹ 30 per bag	1200.00
4.	Cost of (*a*) 50 corrugated asphaltic roofing sheets (4′ × 2′6″) @ ₹ 10.50 per sheet (*b*) 12 corrugated ridges (2′6″ long) @ ₹ 6.50 each, and (*c*) 2 kg of asphaltic washers	650.00
5.	Cost of other materials like sand, gravel, iron brackets etc.	500.00
6.	Labour charges including that of a mason	1000.00
	TOTAL	6850.00

This comes to a construction cost of about ₹ 15 per square foot area or ₹ 600 per cow. Of course, the cost will vary from place to place depending on the local market prices. If the farmer can put in his family labour and procure sand, gravel etc. locally the cost can be reduced further. Most of the financing organisations give ₹ 8000 as loan for cattle sheds. Thus, the proposed cattle shed can be made within this limit.

19.9.5. Feeding and Watering Arrangement

Animals must be provided with clean drinking water and fresh fodder throughout the day and night. These can be offered in suitably constructed feed and water troughs. Feeding troughs are knows as mangers. The location of these troughs and the height of their walls are so designed that the animals can eat from them while standing either inside the shed or out in the paddock. Farmers who do not want to make a paddock can increase the height of the outer (paddock side) wall of the manger so that stray animals may not use the manger.

Feed and water trough can be made with one-brick thick wall bound by cement. The inside and outside of the water trough should be plastered with cement. These can also be made with stone slabs. Empty petrol drums cut longitudinally and mounted on a wooden frame can also be used, so as to reduce the cost.

Water and feed should be made available always. However, some farmers may be producing lesser quantities of fodder. Such farmers can feed the animals individually by tethering them to the rings fixed at the base of the manger. If animals are fed in groups when fodder is in short supply, bossy cows will drive away docile ones, with the result the latter will starve, resulting in lowered milk production. The insides of the feed and water troughs should always be kept clean and they should be white-washed at fortnightly intervals.

Urban and sub-urban dairy units are also an important aspect of dairying in this country. Interesting is the design for house-cum-cow shed proposed for the milk colony at Dhanas on the outskirts of Chandigrah in north western India. The cattle shed on the ground floor comprises of the shed for cows/buffaloes and a small room for calves, a fodder store and a central platform for chaffing fodder and milking rations. The courtyard in the rear has a water trough and an open manger. The first floor comprises of a two bedroom apartment for the owner. This house-cum-cattle shed design goes well with tradition in India where the owner lives close to his animals. However, more attention has to be paid to the ventilation in such designs.

A study of the designs shown above will reveal to the reader how simple a cattle shed need be and what type of innovations can be made in constructing them. Either the farmer can choose one of the above designs or can form his own design based on the principles and designs discussed above.

19.10. WATER SUPPLY TO DAIRY FARMS

Water will be required on a dairy farm (*i*) for the animals to drink, (*ii*) for washing of animals, (*iii*) for washing sheds and barns, and (*iv*) for irrigating fodder crops. While constructing the animal houses due attention has to be paid, however, to the first three aspects.

Drinking water for dairy cattle should be free from contamination by pathogenic microorganisms or eggs and larvae of animal parasites. Great care should be taken to see that mineral poisons such as lead or arsenic should not find their way into drinking water provided to animals. The water should also be devoid of any undesirable smell or taste, which fact can be judged by the willingness with which animals drink it. Turbid water or water with lot of suspended matter should be avoided as such water might have been exposed to contamination by animal excreta or other organic wastes.

Wells, springs, rivers and streams are common sources of water. It is better to provide piped water supply to dairy cattle, as far as possible.

19.10.1. Water Requirement of Dairy Cattle

Daily water requirement of dairy cattle for drinking purposes basically depend on their body size, physiological condition, season and nature of feed consumed by them. Lactating cows require higher quantities of water in direct proportion to their milk production. The voluntary intake of animals increases as ambient temperature rises. Winchester and Morris (Cited by McDowell, 1972) reported that a lactating cow weighing 500 kg consumed 22.7, 26.5, 31.0 and 32.0 litres of water respectively at 10°, 15°, 20° and 30° temperature. Further, water consumption during summer varies with mode of housing animals. Murrah buffaloes housed in a loose house consumed on an average 10 litres more water in a day than those housed in a tie-barn. Another group of buffaloes housed inside a shed during day, but out in the open at night consumed the same quantities as those housed in the barn. This could be due to higher air temperature attained inside loose house at mid-day in summer. Buffaloes, by virtue of their proneness to heat stress during summer require far more water than cows. Further, more water is required for cleaning buffaloes because they are washed daily.

According to R.G. Linton (1965), dairy cows need 1.8 to 2.3 litres (4 to 5 lbs) of water for each 0.9 litre (1 lb) of milk produced both for maintenance and production. A cow weighing about 454 kg (1000 lb) would require a total quantity of 33.3 to 42.8 litres (8 to 9 gallons) of water out of which nearly 6.7 litres (1 gallon) are required for milk production.

Water requirement of dairy buffaloes, worked out at the Haryana Agricultural University by dual experimentation are presented in Table 19.17. Considering the above mentioned figures, it is safe to presume that cows and buffaloes require roughly 45 and 65 litres of water respectively, per head for drinking purposes alone. For washing purposes, cows require much lesser amounts of water than buffaloes (Table 19.17), as cows are not washed daily. Only the udders and perhaps soiled hind quarters of cows need be washed daily in preparation for milking.

It is ideal, however, to provide clean drinking water *ad libitum* to dairy animals throughout the day and night, in all seasons, by special water troughs or tanks constructed in loose houses. In such places where water scarcity exists, adequate quantities of drinking water, distributed over three or four times in a day, should be provided manually or mechanically. The quantity of water required for washing purposes depends on the system of husbandry and manure collection followed. However, rough guidelines in this regard can be obtained from table 19.17.

TABLE 19.17. Water Requirements for Drinking and Washing of Buffaloes Under Loose Housing System of Management

Type of animal	*Season*	*Water intake during 24 hours per buffalo (litres)*	*Water requirement for washing of each buffalo (litres)*	*Total water requirement in a day per buffalo (litres)*
Calf	Winter	11.8	15.2	27.0
	Summer	28.4	23.0	42.0
Heifer	Winter	27.5	28.3	47.0
	Summer	55.3	45.8	102.0
Adult	Winter-Dry	45.1	23.3	68.4
	Lactating	58.9		82.1
	Sumar-Dry	55.5	36.4	91.9
	Lactating	63.9		101.5

Data pertaining to young stock, after Kidan (1976) and that pertaining to adult stock, after Pal *et al.* (1973) and Radadia *et al.* (1980).

Kidan, N.B. (1976). *Water requirements of young buffalo calves and heifers and other operations at the farm.* M.Sc Thesis. Haryana Agric. Univ., Hissar.

Pal, S., Pal. R.N. and Thomas, C.K. (1973). *Indian J. Dairy Sci.* 26 : 253.

Radadia. N.S.. Pal. R.N. Juneja, I.J. and Sastry, N.S.R. (1980). *Haryana Agric. Univ. J. Res.* 10 : 144.

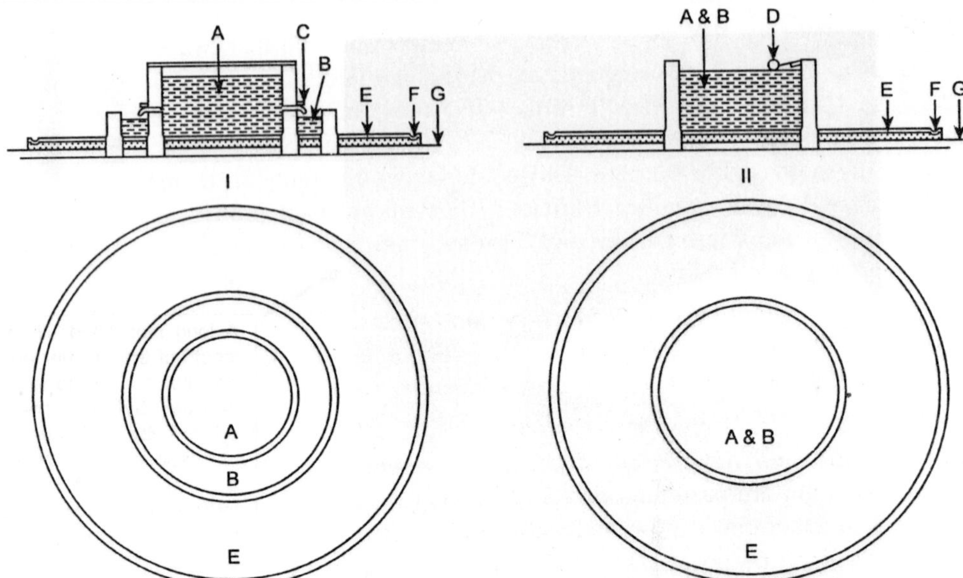

Fig. 19.26. Section and plan of round water tanks of storage tank type (I) and continuous filling type (II) A. Storage water tank, B. Drinking water tank, (A & B-same tank serves as storage as well as drinking water tank), C. Tap, D. Tap with float valve, E. Platform, F. Gutter, G. Ground level. Height and capacity of tank depend on type of livestock for which it is used.

Water tanks. Water troughs or tanks should be constructed in each shed for the convenience of animals. The water troughs should be built with reinforced concrete, bricks with cemented junctions, stone slabs with cemented joints or plain thick galvanized iron sheets. The water tanks may be round (generally in larger pens) or rectangular and shall be located at a convenient place in the shed as shown in different layouts given elsewhere in this chapter. Its dimensions shall be more or less the same as that for mangers for different classes of animals (*see* Table 19.13).

A two metre wide paved/platform shall be provided adjacent to the water troughs to withstand heavy treading by animals and permit easy washing and cleaning. Details of round types of water tanks are shown in Figure 19.26. In smaller pens a tank can be made by raising an impervious partition towards one end of the continuous manger (Fig. 19.27).

Each water tank should have one hydrant, the outflow from which may be controlled by a float valve. The hydrant can also be left without a tap and water let into the troughs at regular intervals so that the water trough may get filled as the animals empty them. Hydrants with spouts to attach a hose pipe should also be fixed in each shed at convenient places. The pens can be washed by flushing water through a hose.

19.11. HYGIENE AND SANITATION

Disposal of excreta and other wastes from dairy animal habitations is a bothersome yet important activity as this material not only has a good manurial value but can also provide ideal breeding place for pathogens and disease carrying insects if left in site.

Manure should promptly and completely be removed from the sheds so that it may not cause any disease, should be well conserved so that it may not lose plant nutrients, and then it should conveniently be applied to lands without any costly involvement of labour and transportation.

A long pipe fitted all along the length of the manger above and parallel to the inner wall of manger byfixing in position either by erecting them independently or by means of a series of angle iron brackets jutting out from the roof supporting posts. This pipe prevents the animals from getting into the manger or jumping out over the manger.

Fig. 19.27. In smaller pens a water tank can be made by raising an impervious partition towards one end of the continuous manger as shown in this figure.

19.11.1. Manure

Cattle dung has 77.5 per cent water, 20.3 per cent organic matter, 0.34 per cent Nitrogen, 0.16 per cent phosphoric acid, 0.40 per cent potash and 0.31 per cent lime. Manure can be solid (dung, feed wastes, soiled bedding etc) and liquid (urine, wash water and other liquid wastes). Under ideal management conditions solid manure is usually collected and removed from sheds twice daily. Provisions are to be made to carry off and store liquid manure as and when the same appears.

In loose house, one of the two alternatives can be used for collection of manure : (*a*) collecting solid and liquid manures separately, or (*b*) flushing out both manures together with plenty of water by means of hose pipes. In the first method using shovels, dung and other solid wastes can be lifted into wheel barrows and then carried out of the shed. Bullock carts or even a trolley can be used in big sheds. The liquid manure and wash water run out of the shed by a shallow 'U' shaped channel located longitudinally to the shed at the junction of open and closed areas. The drains shall be constructed out of the same materials as for the floor. The width of the drains may vary between 30 and 40 cm. A slope of 1 in 40 shall be provided to the drains. Shallow 'U' shaped drains with a depth of 6 to 8 cm are preferable over drains with cut sides (rectangular or triangular cross-section). The drains in a row of pens should be made continuous by routing it through holes in the intervening partition walls. Outside the shed, liquid manure from each shed can be connected by means of drains (preferably closed or sub-surface drains) to a main farm drain. This main drain leads the liquid waste ultimately to a liquid manure storage tank through an inspection chamber and settling chamber (Fig. 19.28).

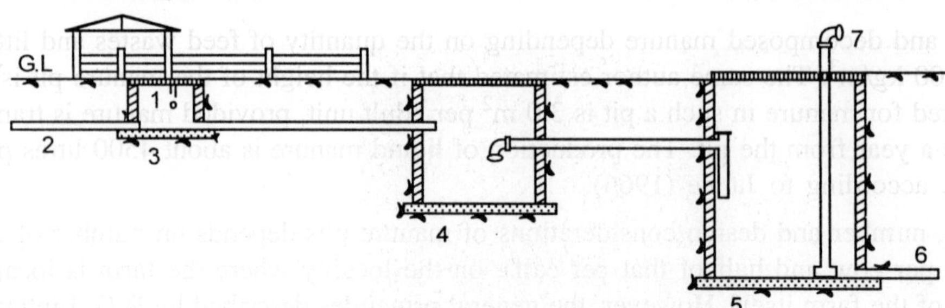

Fig. 19.28. Drainage system for clearing liquid manure from animal farms. 1. Drain from a building opening into the inspection chamber, 2. Drain from another individual building, 3. Inspection chamber with a man hole, 4. Settling chamber, 5. Liquid manure storage tank for the whole farm, 6. Outlet towards fields (if the tank is located at a higher level), 7. Outlet with pumping device. (If the tank is located at the same or at a lower level).

Where facilities are available for supply of plenty of water under pressure both liquid and solid manures can be flushed off the floors of a loose house. However, for this purpose elaborate construction of an underground network of sufficiently wide drains is a prerequisite. The mix wash water can be directly led into fields of fodder grasses or can be fed as a slurry into bio-gas plants.

The latter method may be more suited for buffalo farms as buffaloes void dung with greater water content, more so during the season in which berseem and other succulent fodders are available. In heavy rainfall areas solid manure may get mixed with rain water and hence it may be difficult to collect separately. At the Arey Milk Colony, Mumbai, fodder grasses are being cultivated economically by irrigating them with wash water from buffalo sheds.

Elaborate system of traps and other devices used in western countries for separating liquid excreta from wash water and conserving urine separately may not be feasible under Indian conditions as it has been observed that during summer the quantity of urine voided by animals is comparatively less and the same gets evaporated within a short time after voiding due to high ambient temperature.

When solid manure is collected separately, it should be stored properly in manure pits, so that the same gets well decomposed and there is no fly menace.

19.11.2. Manure Pits

Manure pits should be located as far off as possible from animal habitations after duly considering the labour required in transporting manure from the sheds. This is necessary as a safeguard against foul odours existing near milk parlours and as a measure against fly menace. Fresh manure lying near animal sheds is an ideal breeding ground for a variety of insects and flies, especially under tropical conditions. They not only do not permit clean milk production but also pester dairy cows physically as well as carry infectious diseases.

For these very reasons of hygiene, manure pits should be at a minimum distance of 10 m from wells, rivers and tanks. Further, they must be impermeable to water. The size and number of manure pits required depends on the production of manure on the farm, which is on an average 30 to 40 kg per day per adult unit. Manure weighs about 4.5 quintals per cubic metre. The volume-density of manure is as follows (J. Jahne, 1966) — (*a*) Fresh, loose manure—700 to 1000 kg/m^3; Stored and decomposed manure depending on the quantity of feed wastes and litter in it — 1000 to 1300 kg/m^3. The same author estimated that if the height of the manure pit is 2.5 m, the area required for manure in such a pit is 3.0 m^2 per adult unit, provided manure is transferred to fields twice a year from the pit. The production of liquid manure is about 4500 litres per year per adult unit, according to Jahne (1966).

The size, number and design considerations of manure pits depends on number of animals (2.2 m^3 space per cow and half of that per calf), on the locality where the farm is located and on conditions of the farm itself. However, the general principles described by R.G. Linton in this regard hold good for all localities and structures.

Broadly speaking, it is better to have rather two moderate sized pits than one single large one. For reasons of ease in dumping in and lifting out manure, the height of the retaining walls of the pit should not be more than 1.5 to 2.0 m. The walls should preferably be made of bricks or concrete. The bottom of the pit should invariably be made of impervious material like cement concrete and there should be a fall to one end (so that the liquid portion can collect at this end) and from this end a drain should lead to the liquid manure chamber. A simple roof should be put over the pit to prevent leaching of valuable soluble plant nutrients from the manure.

The roof should at the same time permit free circulation of air over the manure; lest anaerobic conditions may be created within the manure. This leads to development of high temperature within and consequent deterioration of the manure. The manure should be well turned periodically to ensure uniform decomposition of fresh manure being added to the pit periodically and consequent destruction of eggs and larvae of parasites that are normally present in dung of animals.

Allnutt's manure pit as described by R.G. Linton with a permanent or temporary roof (Fig. 19.29) perhaps is ideally suited to our conditions. In villages, where means are not generally available for construction of such structures, a simple improvisation can easily be made. A 0.5 m deep and 4 m × 3 m area pit dug out on an elevated land should suffice (Fig. 19.31). Only a roof has to be made over it using locally available materials.

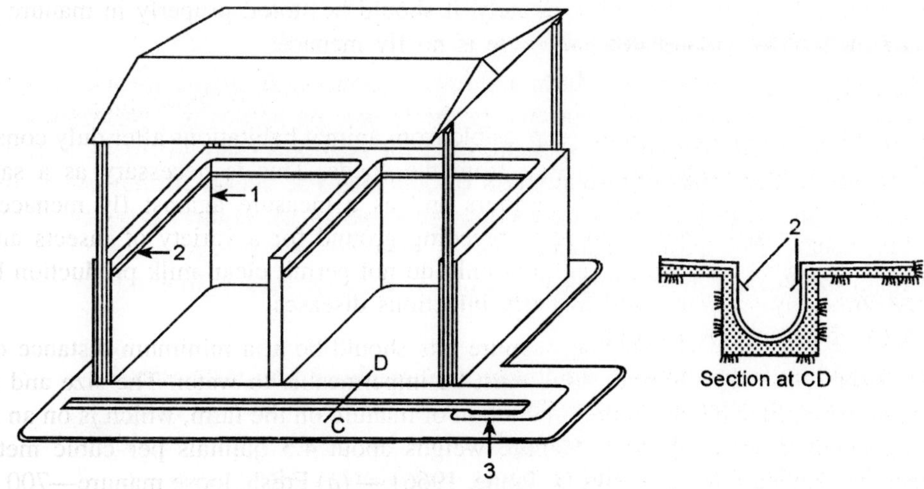

Fig. 19.29. Allnutt's manure pit.

Reproduced below is a description of the Allnutt's manure pit (Fig 19.29), from 'Linton's Veterinary Hygiene'. "Allnutt's Device is designed to overcome the difficulty of turning the manure. He advocates a two stalled receptacle, walled in on three sides and set on a cement concrete or other strong impervious platform and having a gutter running across the open front; a media partition wall of the same height as the outer wall divides the receptacles into two equal compartments. The surrounding walls and the partition are provided on their inner surface a few inches from the top, with a baffle or ledge projecting inwards in order to prevent migrating larvae which have crawled up the walls from escaping over the edge. It is essential for the success of the system that at no time the manure stored in the compartments reaches the height of the ledge.

The gutter is constructed in a manner similar to that in the Barber trap. It extends at both ends slightly beyond the retaining walls and is closed at one end; at the other end it drains into a sump (pool), or exit-drain, at the entrance of which is a two-inch ledge to maintain a continuous level of fluid in the gutter. The gutter can thus be kept always half-filled with a solution of cresol to kill migrating larvae and to prevent mosquitoes from ovipositing, which would occur if water alone was used. A vertically sliding shutter fitted to the front of each compartment hinders debris falling into the gutter. If the walls of the manure pit are likely to be exposed too much rain and thus kept moist, a condition which will aid larvae climbing the projecting ledge, the gutter should be extended all around the walls so that it will trap any of the maggots which may have surmounted them and would otherwise escape.

The idea underlying the provision of two compartments in the Allnutt's pits is that while one is being filled and tightly packed, fermentation and decomposition can proceed in the other which is already full. The system allows the front and top layers of manure to be periodically removed and deposited in the centre of the stack with the least difficulty".

19.12. CONSTRUCTION DETAILS OF DIFFERENT STRUCTURES

The construction of buildings starts from the foundation and rises gradually from plinth to walls and roof. A plan should be made before hand regarding the location of different internal structures like mangers, water tanks etc. Care has also to be taken regarding choice of material and construction of floors, roofs and walls.

19.12.1. Construction Materials

The materials used for construction of farm buildings depend on their cost, local availability, type of buildings to be erected and other reasons. In the tropics properties of construction materials should also be considered while selecting them, since the materials used for floors, walls etc. have an important role in minimising thermal stress on animals. For details of thermal properties of common building materials used for construction see Table 19.18. Figures given in this table are average values only. The thermal conductivity of materials varies with density and moisture content in them which varies from sample to sample. This should be kept in mind while using them.

19.12.2. Construction of Floors

The floors should be impervious to water and easy to clean. The floor may be of—(*i*) cement concrete or paved with cement concrete flooring tiles; (*ii*) brick-on-edge flooring, (*iii*) stone-slab flooring; or (*iv*) *moorum* or *kankar* flooring. The best but costliest floors are of cement concrete which may be made only in milking barns and stores, where strength, cleanliness and imperviousness of floors are important.

TABLE 19.18. Density and Thermal Conductivity of Certain Construction Materials

Construction material	*Density of the raw material (kg/m^3 on dry basis)*	*Thermal conductivity (kcal/through metre/°C)*
Hollow bricks	1200	0.46
Solid construction bricks	1300	0.50
Laid brick layer	1600	0.61
Stone natural	1300	0.45
Light (porous and other types) concrete	800-1400	0.25-0.53
Heavy reinforced concrete	1800-2400	1.00-1.40
Asbestos sheets	1800	0.40
Cement plaster	2000	1.00
Lime-cement plaster	1800	0.82
Lime mortar	1700	0.75
Asphalt	2000	0.60
Bitumen	1200	0.15
Concrete under-floor fillings	2200	1.30
Ceramic facing tiles	2000	0.90
Sand	1700	0.30
Gravel	1800	0.30
Wet ground	1500-2000	0.90-1.50
Clay	2000	2.2
Aluminium	2700	175.00
Construction steel	7800	50.00
Window glass	2400	0.70
Plastic	15-150	0.04-0.05
Rubber crumb	400	0.07
Cork sheets	750	0.17
Straw fillings	150	0.05
Light woods like oak, fir	600-800	0.15-0.20
Heavy woods like Sisso	800-1000	0.25-0.30

Source. E. Motes, *Microclimate in Animal Houses* (Russian), Kolos, Moscow, 1976, p. 60.

In general, in sheds for cattle and buffaloes, floors may be made of stone-slabs or brick-on-edge linings, *i.e.*, stone or bricks laid and set in cement or lime mortar. *Moorum* or *kankar* floors are cheapest but are messy and require constant maintenance. Certain bituminous floorings were tried in western countries and were found to provide a warmer surface. Wooden floors or thick rubber mats are also warm and can be tried with advantage in temperate Himalayan regions.

Floors should be laid over a firm plinth, the height of which depends upon the nature of the ground, the location and importance of the building, rainfall, dampness, drainage and possibility of nuisance from insects, scorpions, snakes etc. Its height is normally 45 to 90 cm; can be higher in damper areas. It may be constructed with bricks or stone masonry.

The floors shall have a gradient of 1 in 40 to 1 in 60 towards the drains so that wash water can run into drains easily. Special care should be taken to make the surface of floors rough and non-slippery in milking barns and passages leading to milking barn, calf pens, feed bunks etc, where cement concrete floors are laid. Brick-on-edge floors are sufficiently rough and may not require any special efforts in this direction. The surface of cement floors can be roughened by imprinting the impression of a piece of expanded metal or suitable wire mesh on the surface while the concrete is still moist.

Elevated slatted floors made of bamboo splits or timbers are sometimes used in portable houses for young calves. Studies at the National Dairy Research Institute showed that elevated slatted floors are better suited for rearing buffalo calves during wet season (Arora, 1978). Studies at Haryana Agricultural University revealed that growing buffalo calves fare better on paved floors in comparison to unpaved floors, both during summer (Sastry *et al.,* 1981) and winter (Grewal *et al.,* 1982).

19.12.3. Walls

Under loose housing system walls are few and not so extensive. The covered area has walls only at the ends. The open area is enclosed by one and a half metres high half walls. Walls should be erected over foundations of sufficient strength.

The walls may be constructed of stone, brick, mud or bamboo or any other material suitable for the locality and climate. Stone or brick walls are costly but durable and hygienic. Bamboo and mud walls are economical and useful but are temporary and are difficult to keep hygienic. Concrete walls 10 to 12 cm thick and reinforced with steel bars along their length and height are strongest and best but are most expensive. Walls supporting the roof and wall portions with which animals come in direct contact must be robust and materials such as brick, stone or cement concrete may be best at least for the lower parts of such walls.

For ordinary walls, thickness should not exceed 35 cm, the side walls being fully thick and the lengthwise ones slightly less thick. Partition walls and walls lining the open area should be 22.5 cm (two-brick) thick. Height of walls shall be 2 to 2.5 metres for houses with sloping roofs.

Walls and partitions can also be made of galvanised corrugated iron or asbestos sheets by fixing to posts 2.5 to 3.0 metres apart. Usually the inner face of walls is plastered and the outer one left un-plastered in low rainfall areas.

19.12.4. Roof

The roof should be light, strong, durable, weather-proof, good looking and a bad conductor of heat (*see* Table 19.18) and free from tendency to condense moisture inside. There are two types of roofs — sloping and flat. Flat roofs are preferred in low rainfall, dry areas while sloping roofs are desirable in medium to heavy rainfall areas.

There are several materials available for roofing farm buildings including tiles, slates, GCI sheets, corrugated asbestos and aluminum sheets, wood, thatch, bamboo etc. GCI sheets, asbestos and aluminum sheets have many advantages, the chief one being fire resistance, easy fixing, reasonably long life, hygiene and cheapness in the long run. These being light materials they do not require heavy roof supporting structures. Tiles and slates are cumbersome to fix and require heavy supporting structures. Wood makes the most comfortable roof, being a good insulator; but is liable to fire risks and is quite costly.

Thatch and bamboo are most readily available and cheap materials; these are good insulators and can be put over rough and cheap trusses. However, these are unhygienic especially during

---

Arora, S.P. (1980) *loc. cit.*
Grewal *et al.* (1982) *loc. cit.*
Sastry *et al.* (1981) *loc. cit.*

monsoon and harbour insects, flies, cobwebs and vermin and are highly prone to fire hazards. Though, initial investment on thatch roofs is small, they are costly in the long run due to high cost of maintenance and frequent replacements.

Some new asphaltic roofing sheets are coming into the market now a days. They are claimed to be cheaper than even thatch, light but strong, do not leak and are fire and insect proof. By virtue of their structure these are bad conductors of heat and hence can be used with advantage for roofing cattle sheds. The slope of a roof is expressed as its pitch — angle or slope with the horizontal. The pitch should be 35° for thatch roof, 25° to 30° for tiled roof and 12° to 18° for a sheet roof. The slope is generally kept steeper in heavy rainfall areas. The pitch should not exceed 45° at any rate.

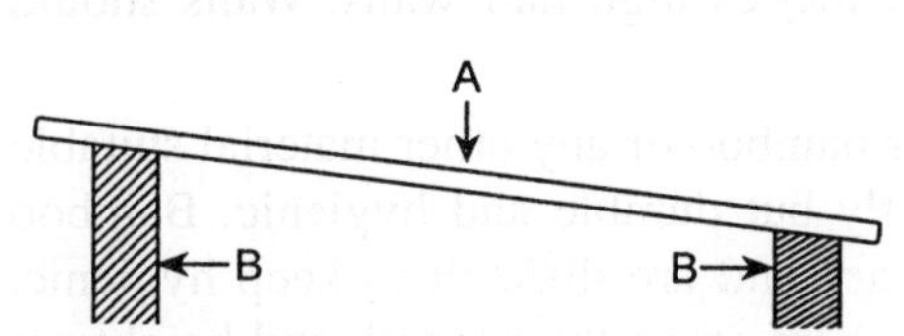

Fig. 19.30. Single-slope roof. A. Asbestos sheet, B. Roof supporting pillars.

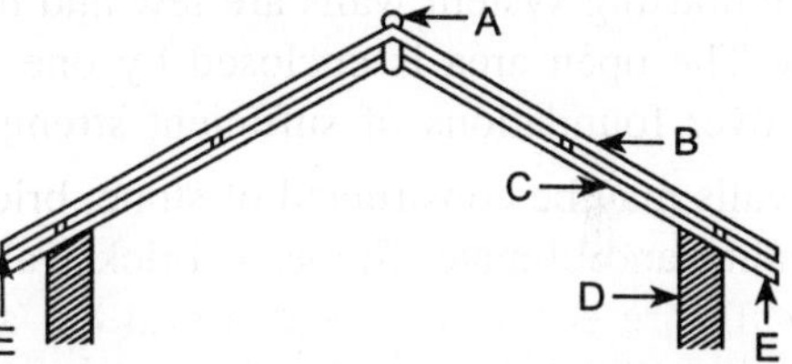

Fig. 19.31. Couple-closed roof. A. Ridge, B. Roof material. C. Truss, D. Roof supporting pillar. E. Eaves.

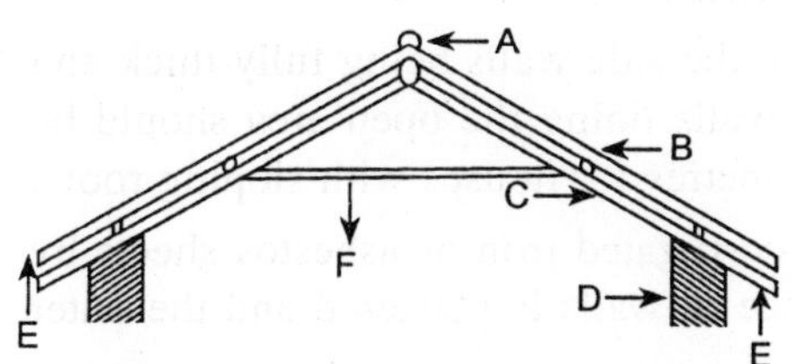

Fig. 19.32. Collar beam truss. F. Collar beam

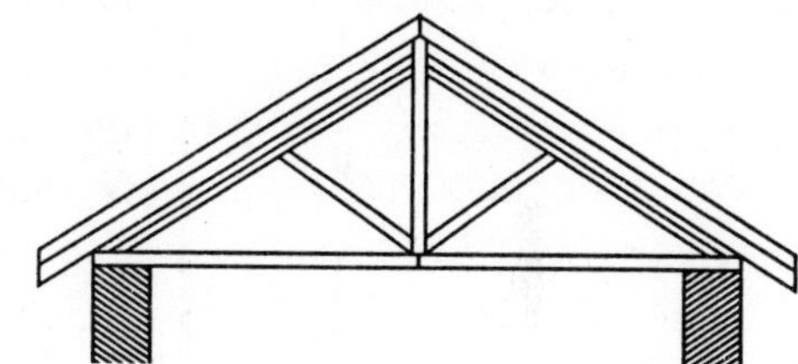

Fig. 19.33. King-post truss.

The roof can be single-sloped or double-sloped. Single-sloped roof can be used for huts and hay sheds upto a span (distance between side walls) of 3 metres. Double-sloped roofs are used for more important buildings with wider spans. For single-sloped roofs the supporting walls or posts are higher on one side than the other and rafters are fixed directly on opposite walls or posts (Fig. 19.30). For double-sloped roofs it is customary to fix roof-trusses or roof frames. A couple-closed roof (Fig. 19.31) or a collar beam roof (Fig. 19.32) is suitable for spans of four to six metres. A king-post (Fig. 19.33) has to be used for still wider spans.

In a loose house, the roof has to be mostly supported on pillars. Pillars may be built either of stones, columns of bricks laid in cement mortar, cast iron pipes or hard wooden posts. Each of them shall be placed at intervals of two to three metres depending on the span and type of roof. The approximate width or diameter of pillars shall be as given below.

1. Brick	45 × 35 cm (or 2 lengths and 3 widths of 9″ × 4.5″ bricks).
	40 × 30 cm (or 2 lengths and 3 widths of 20 × 10 modular bricks).
2. Timber	10 × 10 cm rectangular posts or 10 cm diameter round poles.
3. Stone	10 × 10 cm or 8 × 15 cm
4. Iron pipes	10 cm diametre.

In hot regions a ceiling of stout country cloth (*dungaree*), old gunny bags, tarpaulin, compressed or loose straw or wooden planks should be fixed to the underside of the roof for heat insulation. For similar reasons, the upper surface of roof may be painted white (which reflects back radiation) while the under surface painted with darker colours.

It is preferable to fit all the roofs at their eaves with a 15 cm half-galvanised sheet gutter to convey and discharge rain water at a suitable spot for easy drainage. In dry regions, the water can be diverted and stored in tanks for future use.

19.12.5. Mangers

Mangers are very important structures in loose houses as the whole feeding plan may flop because of mangers of improper size. Mangers can be made either in the covered area or in the open lot. Manger dimensions should be as per standards given in Table 19.13. The manger shall be of continuous type.

Mangers can be made of stone slabs, wooden planks, bricks lined with cement mortar or of cement concrete. The flooring material of the manger may be the same as that for floor of shed but the surface shall be rounded off and finished smooth. The tops of manger walls shall be arched rather than flat (Fig. 19.34). Mild iron plates may be fixed over the curb (inner wall) of the manger or recessed into curb wall in case the curb is made of reinforced cement concrete.

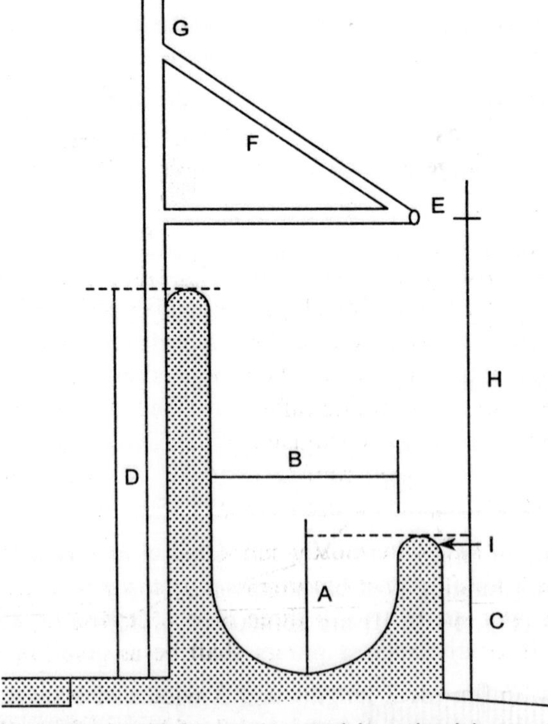

Fig. 19.34. Features of feed manger to be constructed in animal houses. A. Depth of manager, B. Width of manager, C. Height of inner wall, D. Height of outerwall (1.0 to 1.2 m), E. Horizontal pipe, F. Angle iron brackets, G. Roof supporting structure. H. Space between horizontal pipe and top of inner wall (Animals feed through this space), 1. Mould iron plate cover top of the inner wall. This prevents excessive wear of the inner wall of the manger.

Note that the floor of the manger is at a higher level than the floor of the paddock. Also note how the metal tubing is fixed above the inner wall of the manager. This prevents animals from getting into the manger.

A long pipe of suitable diameter shall be fitted all along the length of the manger just above and parallel to the inner wall of manger (Fig. 19.27). This pipe can be fixed in position either by erecting them along side and parallel to the inner wall of the manager or by means of a series of angle iron brackets jutting out from the roof supporting posts. This pipe prevents the animals from getting into the manger or jumping out over the manger and is very necessary in sheds for young animals.

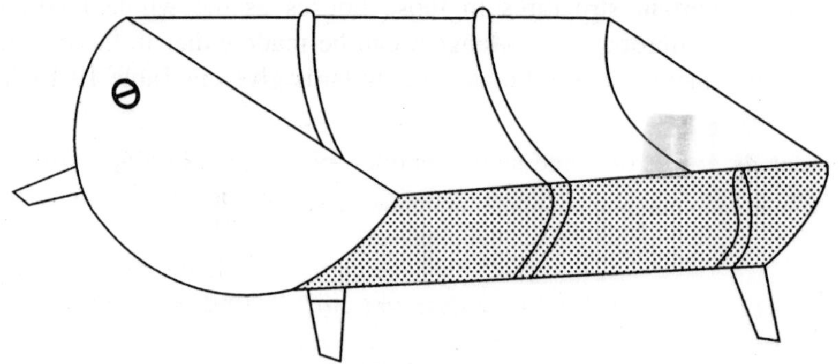

Fig. 19.35. A split oil drum mounted on legs can also be used as manger in small shed. Only there should be no sharp cut edges that may injure the animals.

Gates and Passages. There are two types of gates on a farm, the ones leading into individual pens and the gates on farm fences and roads. The latter should be wide enough to allow passage of trucks, tractors and carts. Suggested widths for different types of gates and passages on dairy farms are given in Table 19.19.

TABLE 19.19. Recommended Minimum Widths for Service Passages and Gates

Kind or passage or Gate	*Use*	*Minimum width*	*Type of passage or gate*
Feed passage along the manger	For feed cart	1.5 m	Metalled
Passages leading to individuals sheds	Movement of animals	1.2 m	Gravel
Farm roads	Trucks, tractors etc.	2.5 m	Metalled
Shed gates	To individual pens	1.2 m	Hinged-single
Farm gates	Drive through	2.5 m	Hinged-double

Gates shall be made of iron or strong wood. Hinged gate (Fig.19.36-1) is better on dairy farms. It is suspended to a hinging post by hinges and closed to a shutting post by a hasp and staple. Double bar gates (Fig. 19.36-II) are sometimes preferred for large animals.

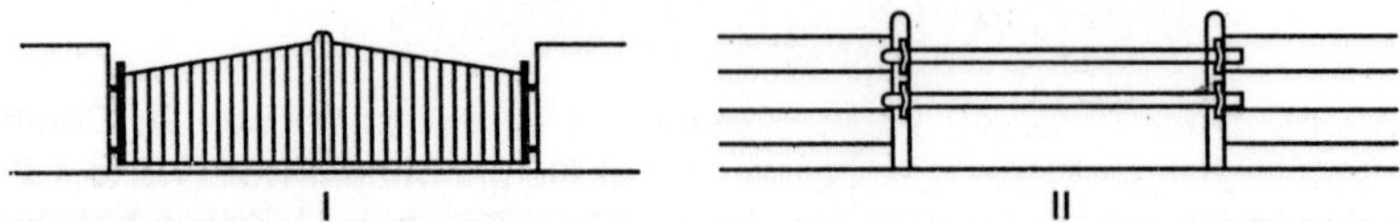

Fig. 19.36. Common types of farm gates—I. Double hinged gate; II. Double bar gate.

19.12.6. Foot Bath and Spray Race

These are important in protecting cattle from contagious diseases and pests. A foot bath (Fig. 19.37) is a tank 6 × 3 m wide at the bottom, 12 × 4 m on top and 0.3 m deep and constructed near the entrance into the farm. This tank is filled with a germicidal solution. Animals and carts entering and leaving the farm have to wade through this solution and in the process, the animals' feet or the wheels of vehicles get disinfected. Thus, no disease producing germs will be brought into the farm through incoming animals and vehicles.

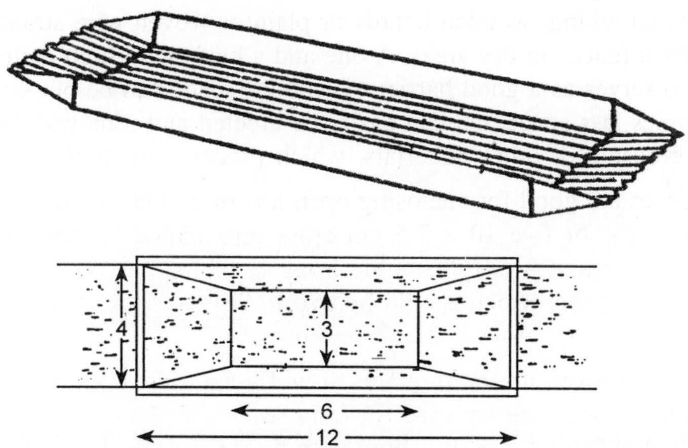

Fig. 19.37. Foot bath to be constructed at the entrance of the farm.

Spray races are for spraying insecticide solutions on animals for controlling insect pests. A spray race is a 2.5 × 1.8 metre passage with a slope of 1 in 40 towards one end. Showers or sprayer nozzles are fixed overhead at a hight of 2 metres above the platform and along the sides. The surface of the platform is kept rough to prevent the animals from slipping. Animals are taken on to the platform one by one along an enclosed passage and sprayed with insecticides or bactericides to protect them against flies, insects and bacteria. The excess liquid spray draining out of the platform can be collected in a tank and can be reused. A spray race should preferably be built as a component of a handling yard (Fig. 19.38) so that utility of the latter is increased.

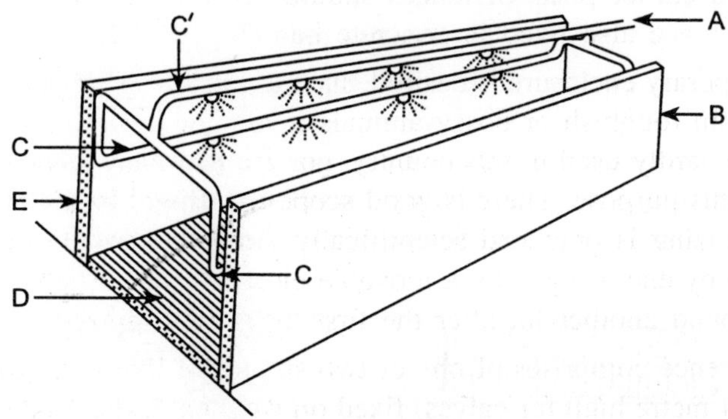

Fig. 19.38. Spray race A. From source of spraying solution, B. Entrance for animal, C. Lines of sprayers, D. Roughened (by grooves) floor of spray race, and E. Exit for animals.

9.12.7. Farm Fences

Farm fence locates the boundary, protects the crops and property from intruders, keeps the animals inside, sometimes serves as a shelter and improves the attractiveness and value of the farm. The fence must be light, strong and durable. The nature and size of holding, the crops raised, and the materials available in the locality determine the type of fence. On farms fences are required for enclosing the farm premises, fields and pastures. Also, fences can be used instead of walls to enclose the open lots of cattle sheds. Barbed wire fences are necessary for enclosing farm area. For enclosing open lots of animal sheds, fences can be made of iron or wooden rails, wooden poles, metal tubing, wooden boards or plain or woven wire strands. Live thorny hedges can be used as farm fences in dry areas. A one and a half metre high earthen embankment along the boundary also serves as a good barrier. A wooden fence can be put up in areas where timber is cheap. It occupies less space and can easily be erected and removed. Wood should preferably be treated with creosote or other materials, which process will prolong the life of the fence.

Wooden fences are good for enclosing open lots of animal sheds. The fence can be of post-and-rail type with four or five 10 × 7.5 cm cross rails nailed to fence posts 22 cm apart or a palisade type. Second-hand iron rails, angle irons or tubing can also be used for fences. When iron is used the fence posts which should be two to three metres apart should be fixed in a firm bed of concrete.

Wire fences, however, are more practical and longer lasting if properly done. Plain woven wire, wire mesh or barbed wire are used for farm fences. Barbed wire should be used for fencing farm premises. It should be 1.5 metres high with 4 or 5 strands 0.4 to 0.3 metres apart. The line posts of fence should be three to four metres apart depending on the type of soil and material of posts. Details of wire fences for adult and young stock pens are given in Table 19.20.

TABLE 19.20. Details of Plain Woven Wire Fences for Animal Pens

Kind of stock	*Height (cm)*	*Wire Gauge*	*Space between strands (cm)*	*Remarks*
Calves/Sheep/Goats	125	10	18	Top most strand can be barbed
Cattle/Buffaloes	150	9	36	Top most strand can be barbed

In case of fences for calf pens, two strands fixed diagonally will give extra strength to the fence. The end and corner posts of fences should be well fixed in a concrete bed and further strengthened by a brace also fixed in concrete bed (Fig. 19.39).

Where a temporary enclosure is desired, such as enclosing a pasture or when existing fences need bolstering from roughish or bossy animals it may be desirable to install an electric fence. Electric fences are hardly used in this country, nor are there any standard materials available yet in the market for this purpose. There is good scope for using electric fences where good pasture is available and grazing is practiced scientifically. Animals can be confined to a lot on pasture for specific period by enclosing it by a movable electric fence. The same fence can be removed and fixed later around another lot after the first lot is fully grazed.

The electric fence comprises of one or two strands of live wire (one at 0.8 metre height for adult cattle and 0.4 metre high for calves) fixed on wooden posts. As the animal comes in contact with the live wire the power gets earthed through the animal and it gets a mild shock in the process. This discourages the animals from crossing the wire.

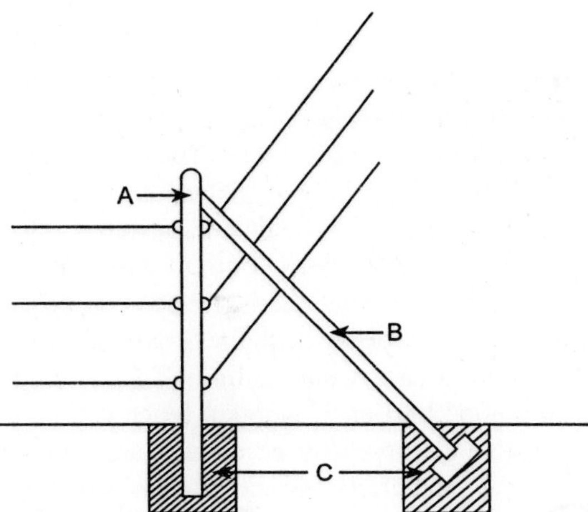

Fig. 19.39. Method of fixing and strengthening corner post of a fence. A Corner post, B. Brace for strengthening. C. Concrete bed.

Electric fences can be dangerous to men and animals. Fence controllers, wires, switches, etc. should be purchased from standard firms and got installed by experts. Men working on farms and people around should be educated about the wire fence and necessary safety precautions against accidents to both men and animals taken.

CHAPTER

20

Bovine Herd Health

"Pure milk from healthy animals is a luxury of the rich, whereas it ought to be the common food of the poor"

—MAHATMA GANDHI

From profit point of view and from public health point of view a herd health programme among bovines is important. Without keeping diseases under control, progress in dairy cattle breeding is seriously impeded and the effects of good management and feeding cannot manifest themselves. Poorly managed health can change profit into considerable loss.

20.1. SIGNS OF GOOD AND ILL HEALTH

The animal, which is not in good health, will manifest several behavioural changes and other symptoms, which can be observed by any experienced person. The managers, farm employees, especially the milkers, owners, all can contribute towards watching for signs of ill health and taking necessary preventive steps at the earliest. The behavioural changes due to diseases have been dealt in detail in the chapter on behaviour. Some of the symptoms of ill health in bovines are described below.

1. The General posture of the animal, its movement and behaviour will change in case of illness. Animals separating from the general herd, showing weariness, lack of alertness and slow in responding to external stimuli and keeping head down are likely to be sick.
2. Eating with relish and rumination are two important signs of good health. Stoppage of rumination and going off-feed are the earliest signs of ill health in bovines. Early morning when feed is offered, cattle and buffaloes rush to the manger and consume greedily. If any animal lags behind showing no interest, it should be further examined.

The authors are grateful to Dr. M.R. Saseendranath, Professor and Head of Dept. of Veterinary Epidemiology and Preventive Medicine, and Dr. Usha Narayana Pillai, Associate Professor of the Dept. of Veterinary Clinical Medicine, College of Veterinary and Animal Sciences, Kerala Agricultural University, Mannuthy, Thrissur, for their contributions in revising this Chapter.

3. Muzzle and nostrils will be moist and devoid of any discharges in healthy animals. In febrile conditions muzzle will be dry. There will be nasal discharge in many diseases.
4. The 'Core' body temperature as indicated by 'rectal' temperature is generally stable. When this is altered it can indicate diseases. While interpreting it, possible deviations due to exercise, environmental temperature, effect of feeding, physiological state, age etc. should be taken into account. The average normal temperature of cattle and buffaloes are given below :

   Cattle : 101.5°F (38.6°C)
   Buffalo : 98.3°F (36.8°C) in winter to 103°F (39.4°C) in summer
5. Variation in nature and rate of pulse can be indicative of ailments. Here also effects of factors like environmental temperature; excitement, feeding, physiological state, activity etc. have to be discounted before arriving at a conclusion.

   Cattle = 50-60 per minute
   Buffalo = 40-50 per minute

   Pulse is taken by placing a finger lightly over the coccygial artery at the base of the tail. Using a stethoscope the heart sounds can also be heard and counted from the chest wall.
6. The rate of respiration and manner of breathing deviates in febrile conditions and ailments of respiratory system. The factors affecting pulse rate can also affect respiration rate. The normal respiration rates are :

   Cattle : 20-25 per minute
   Buffalo : 15-20 per minute

   Respiration rate is counted by watching the rise and fall of flanks with each respiration, by counting the hot gushes of expired air blowing against back of palm kept near the nostril or by means of a stethoscope placed on the chest wall or over the trachea on the neck.
7. Incidence of coughing, whistling, crackling, grunting with pain with breathing and abdominal type of respiration should be specially noted.
8. Eyes in healthy animals are bright and alert with a pink colour of the eye membranes. Sunken eyes with fixed staring look often accompany the onset of fever. Redness in the eye, lachrymation and glued eyes should be noted specially. Paleness of eye membranes is indicative of anaemia and yellow colouration, jaundice. Trouble in one eye is indicative of a local condition whereas bilateral cases can be a sign of systemic ailment.
9. Dung in healthy bovines is semi-solid in consistency with a dark-green colour. The consistency can change with the feed. For example in early monsoon when immature grass is fed, there is a more loose consistency. Constipation, diarrhoea, presence of mucous, blood clots etc. may be watched. Where parasitic infestations are doubted, a sample of the dung may be subjected to microscopic examination.
10. Urine of healthy animals is clear and straw coloured. If variation to deep yellow, bloody or coffee colour is noticed it can be a sign of disease. However, certain drugs also produce changes in urine colour. Peculiar odour, sediments, pain at the time of micturition and absence or difficulty in passing urine should be noted and properly examined.

11. Purulent or creamy discharges from the reproductive tract sticking to the tail and vulva should be noted. Abortions are also indicative of reproductory diseases.
12. The skin of animals should be soft, elastic and pliable. Skin texture can be felt by grasping a fold of skin over the side of neck between thumb and forefinger. A thick, coarse and non-elastic skin is usually associated with parasitic and chronic diseases.
13. In a healthy animal the hair coat is glossy, lustrous and well set. In febrile conditions the hairs are raised. In unhealthy animals suffering from chronic diseases, the hairs are devoid of their shine. They become brittle and fall off. Patches usually indicate rubbing to relieve irritation caused by parasites. The conditions of the coat will vary with housing condition and grooming, but when bovines are infested with worms, or having wasting diseases, their coat loses bloom.
14. Change in quality and quantity of milk produced is an early indicator of disease. Milk yield in dairy cows fall when they get diseases apart from other reasons of management and feeding. So if there is a drop in the yield of a cow, it should be investigated.

20.2. DISEASES OF CATTLE AND BUFFALOES

Any deviation from the normal physiological functioning of any or all the tissues and organs of the animal body is disease. In dairy bovines diseases have big economic implications as morbid animals produce much less than normal or not at all, which means direct economic loss to the farmers. If there is a mortality that is even a bigger loss, as the farmers have to get replacement stock as well. Add to this the cost of medicines and veterinarian's fees.

20.2.1. Kinds of Animal Diseases

One way of classifying diseases is according to the systems they affect like nervous, respiratory, circulatory, alimentary, genital, urinary, muscular, skeletal and cutaneous. Another system of classification is based on the cause of the ailment as described below.

1. **Infectious diseases**. They are caused by specific pathogenic organisms — bacteria, virus or protozoa. They need not always be transmitted directly between animals. Wooden tongue, tetanus and lumpy jaw are examples. All the infectious diseases are not contagious, but all contagious diseases are infectious.
2. **Contagious diseases** are those transmitted from animal to animal by direct or indirect contact.
3. **Parasitic diseases** are due to the infestation of different kinds of parasites. They are either internal parasites (*e.g.*, Ascariasis) or external parasites (*e.g.*, Mange).
4. **Metabolic diseases** are caused by some impairment in metabolism that affects important body processes. Examples are ketosis and milk fever.
5. **Deficiency diseases** are caused by prolonged deficiencies of certain nutrients like vitamins and minerals in the diet. Examples are rickets and night blindness leading to xerophthalmia.
6. **Poisoning** occurs due to entry of poisons into the system either by ingestion or through other routes. These substances destroy tissues or interfere with certain vital functions of the body. Examples are cyanide poisoning and organo-phosphate poisoning.
7. **Allergic disorders** are caused by an exaggerated response of the immunological system of the animal to certain external antigenic materials. Examples are serum shock, photo-sensitization etc.

8. **Injuries** are physical injuries done to the body by cuts, blows, falls and other accidents and results in broken skin surface, wounds traversing through one or more tissue layers, fractures, dislocations, burns, haematomas etc.

9. **Congenital defects** are hereditary defects due to some defective genes or they may be due to improper development of the foetus. They are seen at birth. Examples are *atrecia ani,* white heifer disease and freemartins.

10. **Miscellaneous troubles.** Certain ailments that cannot be classified in any of the previously mentioned groups come under this heading. Tumours, prolapse of the uterus, bloat etc. are some of the examples.

20.2.2. Portals of Entry of Pathogenic Organisms

The body of the animal has natural defense mechanism against entry of pathogenic organisms. The organisms will have to overcome these natural barriers to gain entry and cause diseases. The chief portals through which the organisms gain entry are :

1. **Alimentary canal.** A large number of non-pathogenic and pathogenic organisms enter through the mouth. Once they are swallowed, they are exposed to the acidity of the gastric juice in the stomach and are destroyed. But some of the organisms stick to the mucous lining of the mouth and throat. Unless the organisms are endowed with special characters, they are destroyed by the lysozyme activity in the mouth. The phagocytic action of the tonsils and the flushing action of saliva assist in removing the organisms out of the body. In the intestines, the mucous has a cleansing and protective action and is believed to possess lysozyme activity. Bile is also known to have destructive action on many organisms. In spite of all these barriers, the commonest venue of infection is through mouth along with food and water.

2. **Respiratory tract.** The peculiar anatomy of the nostrils and nasal cavities, the backward lashing action of the ciliated epithelium lining the nasal mucous membrane, the mucous secretion and the high lisozyme activity of the nasal secretions comprises the natural barriers to infection entering the respiratory tract. In spite of all these 'security' measures, many organisms still reach the lungs through the inhaled air. Droplets of saliva or nasal discharge or dust particles in the air containing pathogenic organisms when inhaled, can lead to respiratory infection. When diseased animals sneeze or coughs a spray of small droplets of saliva or nasal discharge containing pathogenic organisms is expelled into the air. These droplets may be inhaled by neighbouring animals. Inhalation of dust laden with pathogens can cause the disease.

3. **Skin** acts as a mechanical barrier by virtue of its hair coat. It has been shown that skin rids itself rapidly of many bacteria, which are not its normal inhabitants. The auto-sterilising action of skin is probably due to the chemical properties of sweat and acidic reaction of the skin. This power is considerably reduced when the skin is dirty.

4. **Uro-genital tract** forms another important portal of entry for infection. Natural defense consists of mucous secretions with bactericidal action and the flushing action of the urine with its slight acidic reaction. Some local glandular secretions in the uro-genital tract have known bactericidal activity. Most of the venereal diseases like trichomoniasis are transmitted at the time of coitus. The urinary (bladder and kidney) infections may occur by the ascent of organisms through the urethra.

5. **Conjunctiva** is protected by the mechanical flushing action of tears and its removal through the lachrymal ducts and a high degree of lysozyme activity. Despite this, some diseases like brucellosis and viral conjunctivitis (pink eye) can arise from conjuctival infection.
6. **Mammary gland.** Several forms of mastitis caused by different micro-organisms occur due to introduction of these into the udder through the teat-canal.
7. **Umbilicus.** Post-natal infections through the umbilicus commonly result in navel-ill and joint-ill in calves.
8. **Placenta.** The foetus may get infected through placenta as in the case of congenital tuberculosis and brucellosis.

Thus, despite barriers, the avenues of infection are many. The chances of animal succumbing to disease are more when the vitality is down. Therefore, the first step in disease control is good feeding and management aimed at keeping the animals in a vigorous condition. Coupled with that, sound hygiene and sanitation reduces the chances of infection through eliminating potential sources.

Some of the most important diseases of cattle and buffaloes are dealt here.

20.3. METABOLIC DISEASES (PRODUCTION DISEASES)

Production diseases mean those disease that affect the production of animals like milk, meat, etc. Metabolic diseases achieve their greatest importance in dairy cows and pregnant cows. Increased demand of lactation precipitates these diseases, so it is also considered as lactation diseases. E.g. Milk fever, ketosis, hypomagnesaemia tetany, low milk fat syndrome, etc.

20.3.1. Parturient Paresis (Milk Fever)

Milk fever occurs most commonly in high-producing adult lactating dairy cattle during the first 48 hours of calving, but also occurs several weeks before or after calving. Most marked in cows at their 3rd to 7th parturition. Jersey cows are most susceptible. Complete milking in the first 48 hour after calving as opposed to normal suckling by a calf appears to be a precipitating factor. A depression of the level of ionized calcium in tissue fluid is the basic biochemical defect in milk fever. It occurs at 3 main stages of the lactation cycle.

- Prepartum – last few days of pregnancy.
- During parturition – Results in uterine inertia.
- Within first 48 hour after calving and danger period may extend up to 10th day post partum.

Some of the risk factors associated with hypocalcaemia are :

- Increasing age of cow.
- High producing dairy cattle.
- Dry period nutrition – feeding >100 g of calcium daily during dry period is associated with increased incidence.
- Housing.

Three stages of milk fever in cattle are commonly recognized. First stage, *i.e.*, stage of excitement is characterized by brief stage of excitement and tetany with hypersensitivity and muscle tremor of head and limbs. There may be shaking of head, protrusion of the tongue and grinding of teeth. Animal is usually ataxic and falls down easily while walking.

Second stage is prolonged sternal recumbency. This stage is characterized by depression of consciousness, drowsy appearance in sternal recumbency, with lateral kink of neck or head turned towards flank region. Muzzle is dry, skin and extremities cool, and the rectal temperature is usually subnormal. Heart rate will be increased with decreased intensity. Pupilary light reflex (PLR) is incomplete, size of the pupil may be normal to maximum dilatation. Ruminal stasis with secondary bloat and constipation are characteristics.

Third stage is stage of lateral recumbency. The animal is usually in comatose stage. The heart sounds are inaudible and the rate may increase up to 120 bpm. There may be complete flaccidity of limb.

Some animals exhibit concurrent hypomagnesaemia and hypophosphataemia. Persistence of tetany beyond the first stage is suggestive of concurrent hypomagnesaemia. In cases of concurrent hypophosphataemia, the animal usually respond to calcium therapy in all respect but unable to stand after treatment. The condition can be easily diagnosed by history of calving, typical clinical signs, serum calcium estimation (< 8 mg/dl, sometime it may be reduced to 5 mg/dl) and response to calcium therapy.

Treatment should be employed at the earliest to avoid complications. Calcium borogluconate 100-200 g is the treatment of choice. For cattle, IV administration of 400-800 ml of 25% solution of calcium borogluconate is the usual dose. Under dosing increases the chances of incomplete response, with inability to rise or relapse. As calcium is cardiotoxic, heart rate should be monitored during therapy. If severe arrhythmia develops stop the injection and give atropine IV.

Response to calcium therapy include :

- Belching.
- Muscle tremor especially on the flank and extending to the whole body.
- Sweating of muzzle.
- Defecation.
- Urination.
- Increased intensity of heart sound.

The care of the cow and calf following milk fever is important :

- Cow that remains recumbent should be provided with plenty of clean dry bedding.
- Animal should be rolled from side to side at least four times daily.
- Protect the animal from bad weather.
- Lifting of animal using appropriate cow lifters is useful.
- Calf should be removed from the cow.
- Only sufficient milk drawn for the maintenance of calf.

Various methods for the control of milk fever in ruminants, especially dairy cows are available :

- Dietary management during prepartum period. Diet providing less than 20 g calcium per day is highly effective. Calcium phosphorus ratio of 1 : 3 is recommended.
- Diet high in anions (chloride and sulfur) reduces the incidence of milk fever.
- Avoid over fattening.
- Avoid stress.

- Avoid stress.
- Calcium gel, which contains 49% calcium chloride before calving, at calving and at intervals within 1 day after calving, is effective.
- A single dose of 10 millions units of Vit. D3 IM given 2-8 days before parturition has been considered optimal.
- Oral dosing of Vit. D2 20 million unit/day for 5 days immediately prior to calving is also effective.
- 25 hydroxycholecalciterol – 8 mg IM 3-10 days before calving repeated at weekly intervals.
- 1-alpha hydroxy Vit. D3 – 350 μg between 24-100 hours before calving.

20.3.2. Downer Cow Syndrome

Downer cow is a complication of milk fever. Most common in dairy cows, which have had milk fever and characterized by prolonged recumbency even after two successful treatment with calcium. Cows, which have had at least 3 calving or 4-8 years of age group, are more prone to this condition. It is more prevalent in winter and early spring season.

Cattle may also become persistently recumbent for many reasons other than complication of milk fever such as peracute coliform mastitis and carbohydrate engorgement.

Various risk factors associated with incidence of downer cow syndrome includes :

- Long delay in the treatment of milk fever.
- Prolonged recumbency before treatment of milk fever (> 4-6 hour) results in ischemic neurosis of muscle due to obstruction of the blood supply.
- Traumatic injury to pelvis and pelvic limb are present in 25% or more of downer cow. The sciatic and obturator nerves are vulnerable to injury by pressure from the moving of calf through pelvic canal.
- Serum electrolyte imbalances may be associated with prolonged recumbency.
- Incorrect dosage of calcium for the treatment of milk fever may result in incomplete response and failure of cow to stand.
- Persistent hypophosphataemia has been regarded as a cause of downer cow syndrome associated with milk fever.
- Long term low level hypomagnesaemia has been associated with downer cow.
- Hypokalemia associated with milk fever (increase cell membrane permeability) cause myotonia results in downer cow syndrome.

In typical cases, the affected cows either make no effort or are unable to stand following treatment for parturient paresis. Downers are usually bright and alert. Appetite is reduced, but the cows eat and drink moderately well. The temperature may be normal and the heart rate may be normal or elevated to 80-100 bpm. Respirations are usually unaffected. Defecation and urination are normal but urine may contain protein. Animal usually crawl or 'creep' along the ground. But non alert downers do not usually eat or drink and thought to have brain damage.

The course of the disease is variable and depend on the nature of lesions and quality of the care provided. About 50% of downer cows will stand within 4-5 days. The prognosis is poor for those, which are still recumbent after 7 days of treatment.

Before confirming as downer cow, all other possible causes for recumbency must have been eliminated. Repeated careful clinical examination is essential to find out the etiology.

- The most important aspect of treatment is to provide most comfortable bedding.
- Roll the cow from side to side to minimize the extent of ischemic necrosis.
- Parentral solutions containing potassium, calcium, magnesium and phosphorus have been recommended.
- Cow lifting devices have been used to assist downer cow to stand (hip lifters, body slings etc.)
- Water floatation tank can be used for management of cows.
- Parentral administration of analgesics is useful to reduce pain associated with prolonged recumbency.
- The early detection and treatment of milk fever will reduce the incidence and severity of downer cow syndrome.

20.3.3. Ketosis in Ruminants (Acetonemia in cattle)

Ketosis is a metabolic disease of dairy cows characterized by hypoglycemia, ketonemia and ketonurea. The current concept is that it is a multifactorial disorder of energy metabolism in the bovine with result of negative energy balance.

The ruminants absorb very little dietary carbohydrate as hexose sugar because dietary carbohydrates are fermented in the rumen to short chain fatty acids *viz.* acetate, propionate and butyrate. So glucose needs of ruminants must largely be met by gluconeogenesis from propionate and amino acids.

In the high producing dairy cows, there is often a negative energy balance in the first few weeks of lactation. The highest dry matter intake does not occur until 8-10 weeks after calving but peak milk production occur at 4-6 weeks and energy intake may not keep up with demand. In response to negative energy balance and low serum concentration of glucose, cows will utilize/mobilize body fat resulting in increased concentration of non estrified free fatty acids and β-hydroxy butyric acid (BHBA). Free fatty acids are transported to the liver and oxidized to acetyl-CoA. Acetyl Co-A is oxidized via TCA cycle to acetoacetyl Co-A in presence of adequate supply of oxaloacetate from the precursor propionate. If propionate and consequently oxaloacetate is deficient, acetyl Co-A is metabolized to acetoacetyl CoA and then to acetoacetate and BHBA, which accumulate as ketone bodies resulting in ketosis.

Different types of bovine ketosis are :

1. Primary ketosis (production ketosis).
2. Secondary ketosis (other diseases which result in decreased food intake).
3. Alimentary ketosis–due to excessive amount of butyrate in silage.
4. Starvation ketosis–poor quality feed stuffs.
5. Ketosis due to specific nutritional deficiency : Cobalt and phosphorus.

Ketosis may be either clinical or subclinical. Subclinical ketosis occurs more commonly than clinical ketosis. Two major forms of bovine ketosis are described–Wasting and nervous forms.

Wasting form is the most common form characterized by gradual but moderate decrease in appetite and milk yield. Cow first refuses to eat grain, then ensilage but continue to eat hay. Rapid

loss of body weight resulted in 'Woody' appearance. The faeces are firm, and dry. The temperature, pulse and respiratory rates are normal but rumen motility may be depressed. A characteristic odour of ketones is detectable on the breath and milk. Spontaneous recovery occurs in some animals.

Nervous forms develop quite suddenly. The syndrome is suggestive of delirium rather than frenzy and characteristic sign include :

- Circling.
- Straddling or crossing of the legs.
- Head pushing.
- Apparent blindness.
- Aimless movements and wandering.
- Licking of skin and inanimate object.
- Chewing movements, salivation.

Nervous signs are mainly due to production of isopropyl alcohol, a breakdown product of acetoacetic acid (Table 20.1).

TABLE 20.1. Hypoglycaemia, Ketonemia and Ketonuria are Characteristic of the Disease

Parameters	*Normal*	*Ketotic animal*
Glucose —Primary	50 mg/dl	20-40 mg/dl
—Secondary		>40 mg/dl
Blood ketone—Primary	10 mg/dl	10-100 mg/dl
—Secondary		<50 mg/dl
Urine ketones	70 mg/dl	80-1300 mg/dl
Milk ketone	3 mg/dl	40 mg/dl
Rothera's test (Cowside test)	—	+ve

Standard recommendation for treatment consists of :

Replacement therapy

- IV injection of 500 ml of 50% solution of glucose.
- 25% Dextrose 2-3 doses found effective.
- Propylene glycol and glycerine—225 g twice daily for 2 days followed by 110 g daily for 2 days.

Hormonal therapy. Glucocorticoids therapy has advantages over other drugs because of ease of administration and rapid effect. Dexamethasone 10 ml (40 mg) IM produced response within a week but there is reduction in milk yield especially in high producers. Insulin—facilitates cellular uptake of glucose suppresses fatty acid metabolism and stimulate hepatic gluconeogenesis. Protamine zinc insulin 200-300 IU/animal s/c or combination of glucose and insulin are also effective. Anabolic steroids—Trenbolone acetate 60 mg and 120 mg are effective but the drug is banned for use in food animals.

Miscellaneous Treatments. Vitamin B_{12} and cobalt, cysteamine (biological precursor of co-enzyme A) sodium fumerate and Niacin, an antilipolytic which increases blood glucose and insulin. It should be supplemented from 2 weeks prior to parturition to 12 weeks post partum.

It is difficult to make general recommendations for the control of the disease because of multifactorial etiology.

- Cows should neither have been starved nor be over-fat at calving.
- Adequate calorie intake especially in early lactation.
- Provide adequate exercise.
- Ration should contain adequate amount of cobalt, phosphorus and iodine.
- Provide sodium propionate 100 g daily for 6 weeks commencing at calving.
- Propylene glycol 350 ml daily for 10 days after calving.
- Monensin @ 25 mg/day.
- Use of somatotropin produces some beneficial results.

20.3.4. Low Milk Fat Syndrome (LMF)

The goal of the dairy producer is to generate the maximum amount of milk at the lowest possible cost and then sell it for the highest possible price. So the production system have became more intensive and this resulted in a significant increase in stress on dairy cow and make them more vulnerable to various production diseases/metabolic diseases. Among this, LMF is found to be apparently associated with increased production under varying dietary and management conditions. The levels of milk fat percentage in many crossbred cows were lower than the standards prescribed by Prevention of Food Adulteration Act, so the milk is usually rejected by co-operative milk societies. LMF can be described as the secretion of normal volume of milk but its milk fat reduced often to less than 50 per cent of normal. It occurred after feeding a high ratio of concentrates over roughage to dairy cow. It was also noted when normal rations were fed, supplemented with protected polyunsaturated fatty acids (PUFA).

Triglycerides then packed in fat globules and secreted into the milk. The ratio of short to long chain fatty acid is very important in the synthesis of triglycerides because the different chain length were required to occupy the various positions on the glycerol–3-phosphate acceptor molecule. So the deficiency of short chain fatty acids synthesis within the mammary gland or long chain fatty acid supply to the mammary gland, milk fat synthesis might be depressed resulting in low milk fat.

Etiology

1. Feeding of low quantity roughage or high grain ration resulted in LMF associated with the changes in the proportion of VFA in rumen.
2. Finely ground forage or concentrates.
3. Deficiency of Vit B_{12} $\rightarrow$ accumulation of methylmalonate.
4. Subclinical hepatic dysfunctions.
5. Endocrinological changes associated with high propionate ($\uparrow$ insulin)
6. Complication of lactic acidosis, liver abscess $\rightarrow$ reduced fiber digestion-decreased milk fat.
7. Heat stress *i.e.*, heat modulated suppression of lipoprotein metabolism.

Factors affecting milk fat

1. Season — lowest in summer.
2. Age — As age increases the % of SNF and fat decreases.
3. Stage of lactation — lowest in 3rd month for HF lowest in 2nd month for Jersey.
4. Time of milking — evening milk more fat %.
5. Stage of milking — milk fat level tended to be highest in the final stage, so incomplete milking will give milk with lower fat level.

Prevention and control

1. For maximum production of milk fat, diet should contain optimum quantity of neutral detergent fibre (NDF) and acid detergent fibre (ADF). The recommended level of NDF and ADF in the dairy ration is 27% and 18% respectively.
2. Forage should not be chopped too fine
3. Ration should contain proper ratio of concentrate and roughage.
4. Feeding management *i.e.* order of feeding. Hay and other forages should be fed before grains—to achieve better buffering of rumen. Grain ration should be divided into 4 to 6 feeding a day, grain should be coarsely ground to decrease the surface area and thus slow the rate of fermentation.
5. High fibre feeds should be incorporated into grain mixture like ground corn, whole cotton seeds, beet pulp, etc.
6. Ruminal buffers—to maintain ruminal pH
7. Sodium bicarbonate or sodium sesqui carbonate at 0.75% of DM or Magnesium oxide at 0.25% of DM. This can be added to the grain mix or the total mixed ration.
8. The level of unprotected fat should not exceed 5% of DM.
9. Increase the frequency of feeding
10. Oral supplementation of 2.5–5% acetic acid @ 30 ml daily.

20.3.5. Hypomagnesemic Tetany (Lactation tetany, Grass tetany, wheat pasture poisoning)

Hypomagnesemia tetany is commonly reported in ruminants and it is characterized by hypo-magnesaemia and hypocalcaemia. Clinically it is characterized by tonic-clonic convulsions and muscular spasms and death due to respiratory failure. Various factors associated with hypomagnesaemia include :

- Dietary deficiency of magnesium.
- Increased body loss of magnesium.
- Short period of starvation.
- Feeding lush green pasture.
- Use of potassium and nitrogen rich fertilizers reduces the availability of soil magnesium.
- Presence of large amount of ammonia in the rumen.
- No effective homeostasis mechanism to maintain its concentration.
- Presence of chelating agents in the diet.

- Increased sodium, which facilitate absorption of magnesium. Na : K ratio is very important. Absorption is reduced when Na : K ratio less than 3 : 1.

Disease usually develops when serum magnesium falls below 1 mg/dl (1.7-3 mg/dl normal). It is mainly concerned with impulse transmission of the neuromuscular system, release of acetylcholine, sensitivity of the motor end plate threshold of the muscle membrane and activation of cholinesterase system. In acute lactation tetany, animal may suddenly stop the grazing, and show unusual alertness, twitching of muscles and ears. Severe hyperesthesia and slight disturbances precipitate attacks of continuous bellowing and frenzied galloping, followed by staggery gait, convulsion, opisthotonos, nystagmus, champing of the jaws, frothing of mouth, pricking of ears and retraction of the eyelids.

Subacute lactation tetany is characterized by wildness of facial expression and exaggerated limb movements, spasmodic urination and defecation, muscle tremor and mild tetany of hind limb. Animal with this form of disease may recover spontaneously within a few days.

In chronic hypomagnesaemia, animal usually does not show clinical signs or shows vague syndrome including dullness, unthriftness, and reduced appetite. In lactating cows this may lead to paresis and milk fever like syndrome that is poorly responsive to calcium treatment.

Determination of serum and CSF magnesium level is useful for early diagnosis of this condition.

Normal serum magnesium level in cattle is 1.7-3 mg% and that will be reduced to 1-2 mg%. Tetany develops when serum magnesium falls below 1.2 mg%. CSF magnesium level also falls below 1.25 mg% as against normal value of 2 mg%.

Combined calcium-magnesium preparation is more effective as treatment of choice. 500 ml of solution containing 25% calcium borogluconate and 5% magnesium hypophosphite IV.

- 200-300 ml of 20% magnesium sulfate IV.
- 200 ml of 50% magnesium sulfate s/c.
- 200-400 ml 15% magnesium gluconate I/V.

As control measures :

- The high risk animal should be moved to low risk pastures.
- Oral feeding of magnesium oxide @ 60 g/day.
- Oral feeding of magnesium phosphate @ 53 g/day is also a safe and effective way.
- Top dressing of pasture with magnesium rich fertilizers.
- Provide sufficient shelter in winter.

20.4. INFECTIOUS DISEASES

20.4.1. Foot and Mouth Disease (Syn. Aphtous Fever)

Foot and mouth disease (FMD) is an acute febrile, highly contagious disease of cattle and buffaloes. Exotic and crossbred cattle are highly susceptible than indigenous breeds of cattle. This disease is prevalent in almost all countries where intensive cattle rearing is going on, though some of the countries like Australia and New Zeland had never experienced this disease. In India an analysis of the reported outbreak of the disease during the last two decades has shown that on an average about 6000 outbreaks occur each year involving as many as three lakhs of cattle and buffaloes. This disease is not fatal in adult cattle, but its after-effects cause great economic losses. According to a study conducted by a British team in the year 1976 the losses on account of reduction of milk yield, mortality and lost bullock power due to FMD in India amounts to rupees 400 crores (4 billions) per annum (Adalakha, 1981). While this disease has been under control in most of the developed countries, it continues to cause havoc in many developing countries. While exact figures pertaining to economic losses due to this is not available, it can only be presumed to be enormous.

FMD is caused by seven different types of FMD virus, *viz.*, *'O'*, *'A'*, *'C'*, *'Asia-I'*, *'SAT-I'*, *'SAT-II'*, and *'SAT-III'*, and their sub-types. In India *O, A, C, and Asia-I are* responsible for this disease. Among this, *'O'* has been most preponderant and widespread followed by *"A "* and *'Asia-I'*. Type *'C'* shows low profile. Each strain is immunologically different. Recovery from attack by any one strain confers an immunity against that particular strain but not by others. The virus is present in all excretions and secretions.

The disease starts with high fever (104° F-106° F) followed by anorexia, depression, and resulting fall in milk yield. As the temperature falls in one or two days time vesicles appear in mouth and in the inter-digital space, which in later stages rupture resulting in ulcers. At this stage there will be profuse salivation, ropes of stringy saliva hangs from mouth. At the cleft of foot the vesicles rupture resulting in red sores, and animal tends to have a recumbent position. Vesicles may also appear on skin of teat and udder, which may lead to mastitis. Abortion, metritis, infertility etc., may be the after-effects of FMD.

Since it is a viral disease cleaning and dressing the inflamed area with mild antiseptic lotion and dressing with antibiotics for prevention of secondary complications are the only adaptive treatments. In advanced debilitated cases supportive treatments with dextrose, vitamins etc., may be given.

This disease can be controlled either by slaughter of all infected animals and burial or burning immediately. But in Indian condition this is not practicable. Next alternative strategy is only to increase the immune status of the animal. Vaccination of all animals with killed polyvalent vaccine depending on the prevalence of types in that particular area may be undertaken. In India polyvalent vaccine containing 'O, *'A'*, 'C, and *'Asia-1* types are being used. Yearly two vaccinations at six months interval are recommended. Maintenance of sanitary conditions and disinfections of contaminated premises with 1 to 2 per cent sodium hydroxide or formaline or 4 per cent sodium carbonate may be done.

20.4.2. Rinderpest (Cattle Plague)

Rinderpest is an acute, contagious viral disease of cattle and buffaloes, characterised by high fever and focal, erosive lesions confined largely to the mucosa of the alimentary tract. This disease causes great economic losses throughout the world, but in India due to Rinderpest eradication scheme, the incidence has come down to a great extent.

Purebred *Bos taurus* cattle, crossbred cattle and buffaloes are highly susceptible than indigenous breeds of cattle. Morbidity and mortality rates are high in susceptible population. This disease is caused by a myxovirus, which occur in more than one strain, which are immunologically

---

Adalakha. S.C. (1981). Epidemiological Status of Foot & Mouth Disease in India. Summer Institute on "Epidemiological surveilance of techno-economies of animal disease control".

identical and are very fragile. The main portals of entry are nasal and oral mucosa. Infected animals shed the virus in all excretions and secretions for a period of up to 10 days.

Incubation period in this disease ranges from 3-9 days. There is high fever at fourth to fifth day. which reaches a peak in 24 to 48 hours and last for 2 to 7 days without localized lesions. Anorexia, fall in milk yield, lachrymation etc. also accompany the fever. Inflammation of buccal and nasal mucosa and conjunctiva follows the temperature. The lachrymation and nasal discharge become purulent, saliva becomes stringy. Discrete necrotic foci develop in the buccal mucosa. inside lip on the gum and in the gastro-intestinal tract. They in a later stage coalesce and form ulcers.

After a period of 3-5 days, the temperature falls to subnormal level. Dyspnoea, cough, dehydration and further fall in body temperature leads the animal to death.

Post-mortem reveals lesions in the gastro-intestinal tract. The gastro intestinal mucosa is congested, folds of abomasum thickened and dark in colour. Posterior part of intestine shows intense congestion giving "zebra marking" appearance.

Diagnosis mainly depends on the symptoms, the nature of spread, and on characteristic lesions like erosion and inflammation of the alimentary tract mucosa by diarrhoea, salivation and nasal discharge. Confirmation is possible by viral demonstration, animal transmission experiment, and by detecting antibodies. Use of electron microscopy, fluorescent antibody technique, enzyme linked antibody test, immuno diffusion test, and Measel's haemaglutination test etc. are used for viral demonstration.

Control of this disease mainly depends on maintenance of sanitary conditions, and increasing the antibody titre of individual animals by vaccination procedure. As a precautionary measure in epidemic areas all the susceptible animals should be vaccinated. There are different types of vaccines available at present.

1. Goat-adaped virus vaccine, advisable for the use in zebu cattle.
2. Rabbit-adapted virus vaccines. It is not so good as goat-adapted vaccine, because of its more attenuation.
3. Cell culture vaccine. This vaccine is being widely used in field conditions. It is suitable for exotic and crossbred cattle and cheap. It does not cause much reaction. Calves can be vaccinated when they are 6-8 months and repeated within 12 months. Adult cattle are vaccinated once in three years, but it is better to carry out vaccination once in two years in endemic areas. Dose is 1 ml subcutaneously.

Implementation of strict quarantine measures, immediate slaughter of all in-contact and infected animals and proper disposal of carcasses are necessary for control. Movement of susceptible animals should be prohibited.

20.4.3. Ephemeral Fever (Three Day Sickness)

Ephemeral fever is a viral disease of cattle and buffaloes characterised by muscular pain. shivering, stiffness, lameness and enlargement of peripheral lymph nodes.

This disease is acute in nature, with sudden onset of fever (105-106° F), anorexia, fall in milk-yield, cessation of rumination and nasal and occular discharges. Second day onwards, muscular pain becomes more evident with lameness, but temperature rise remains. Third day onwards febrile reactions subside though lameness persists for 2-3-days more, shifting of lameness from limb to limb being characteristic.

Diagnosis can be done based on symptoms, its transient nature and its rapidity of spread. Animal inoculation test is used for laboratory confirmation.

Symptomatic treatment for pyrexia, muscular pain and anorexia helps in easy recovery, though recovery occurs without any treatment. Fly control helps in reduction of spread of the disease.

20.4.4. Pox Disease

Pox disease is a viral disease of domesticated animals and human beings. It is characterised by fever and vesicular eruptions on body surface, especially on skin of teat and udder in cattle and buffaloes. Pox is transmitted by cutaneous inoculation or by contamination of infected material on abraded skin or on mucous membrane. Milkers transmit the disease by milking both infected and healthy animals.

Pox lesions start to appear after 2-5 days of infection. Initially there will be rise in body temperature and red spots develop on the skin. These spots later become pastules and vesicles. The serous fluids in the vesicles become purulent, hence called pustular stage.

In about 10-14 days time these pustules become dry and scabs are formed. These crusts and scabs are high source of virus. In males the lesions are often unnoticed, but on scrotum and inside thigh region the same type of lesions develop.

Treatment consists of cleaning the lesion with antiseptic lotions and dressing. Potassium permanganate (1 : 1100) lotion is ideal for washing the affected area. Boric acid ointment (10 per cent) can be used for dressing. The affected animal should be isolated and milked by separate milkers. The milk may be consumed after proper boiling.

20.4.5. Mastitis

Mastitis is the inflammation of the mammary gland caused by physical, chemical and biological agents. Mainly it is caused by biological agents such as different species of bacteria, fungus, virus, yeast etc. It is characterised by physical and chemical changes in milk and pathological changes in glandular tissue. Important changes in milk are discolouration, presence of clots, and presence of large number of leucocytes.

This disease is of high economic importance because of its effect on milk production by dairy animals including the impairment of capacity for milk production in subsequent lactations. It has also much public health importance as various pathogenic organisms are being excreted in the milk of cows suffering from mastitis.

The most common of bacteria associated with mastitis are *streptococci* species and *staphylococci* species. Other important causative agents are *Corynybacterium pyogenes, C. bovis, Mycobacterium species and Escherichia coli.*

The main route of entry of causative agent is through teat canal, though hematogenous spread also occurs as in tuberculous mastitis.

Infection occurs through three phases :

1. Invasion, when organism gets into mammary gland through teat canal.
2. Infection, when organism multiplies in mammary tissue and establishes, and
3. Inflammation, stage of exhibition of symptoms.

Depending on severity and on-set of symptoms mastitis are classified into per-acute, acute, sub-acute and chronic mastitis. In per-acute and acute cases, onset of symptom will be quick. The gland will be reddened, swollen, high temperature, painful on touching with discolouration of secretion. In per-acute cases there will be systemic reaction like fever, depression, anorexia, etc., in addition to udder infection. Persistent abnormalities with mild inflammation form in sub-acute, but recurrent attacks of inflammation with little changes in milk constitute in chronic cases.

This disease can be diagnosed from the clinical manifestation in acute, per-acute, sub-acute and chronic cases. But sub-clinical cases of mastitis can be diagnosed by using laboratory tests.

Milk sample should be collected hygienically. Examination of milk samples for cells, bacteria and chemical changes helps in identification of the causative agent. Cell counts being an indication of inflammation helps in diagnosis of mastitis.

Strip-cup Test. Used to detect early cases of mastitis. Strip-cup consists of a black coloured enamel plate on which the milk is stripped to detect the clots present in it.

California Mastitis Test is a widely used one in field condition. This is an indirect test to show the number of leukocytes, which are present in inflamed condition.

Treatment of mastitis mainly depends on the type of organism causing the disease. Antibiotic sensitivity test helps in this regard. Based on the test, select the antibiotic. Completely milk out the affected quarter, then infuse the quarter with that particular drug. In case of acute and per-acute cases, better to give the antibiotic both locally as well as parenterally with supporting treatment.

Control of bovine mastitis mainly depends on the maintenance of hygienic conditions of surroundings where the dairy animals are kept. Keep the animals clean and healthy. Early detection and prompt treatment is the factor, which reduces the loss due to mastitis. Treatment of all cows during drying off, though a costly affair, is indicated in herds where incidence is more. Culling of chronic cases helps in reducing the chance of spreading the disease. Dipping of teat after each milking helps to reduce the incidence of mastitis. Vaccination has so far proved to be of limited value in control of mastitis as it is caused by multiple agents.

20.4.6. Haemorrhagic Septicaemia (Pasteurellosis)

Pasteurellosis is a bacterial disease of dairy cattle and buffaloes caused by *Pasteurella maltocida.*

Pasteurellosis occur when the animal is subjected to stress, as the organism is present in the system, especially in the upper respiratory tract. Disease spreads from these carriers to healthy animals through feed and water. The organism does not survive outside for more than 24 hours.

The disease runs an acute form in cattle and buffaloes and the animal may die in 24 hours. Sudden rise in body temperature to 106-108° F, profuse salivation, severe depression and the death occurs. Again this disease may manifest in different forms depending on the organs affected like the edematous form, pneumonic form, and alimentary form. In edematous form, along with pyrexia, there will be a hot painful swelling at head, jowl region or brisket region. In pneumonic from, severe dyspnoea is the pronounced symptom. But in intestinal form, animal exhibits severe colic symptoms with or without diarrhoea along with pyrexia.

Generalised haemorrhage in visceral organs is the important lesion on post-mortem examination. Examination of blood smear or oedema fluid for the bipolar organism is the first and foremost diagnostic method. Cultural examination and biological testing method also help in diagnosis. The best clinical material will be heart blood and bone-marrow as the organisms do not survive more than 24 hours in other materials.

Sulphadimidine along with antibiotics provide better treatment.

Vaccination of all animals one month prior to monsoon with oil-adjuvant or broth vaccine helps in controlling the disease. Two to four ml of oil-adjuvant vaccine deep intra-muscularly, depending on size of the animals, provides immunity for five to six months. In endemic areas, yearly vaccination has to be done.

20.4.7. Anthrax

Anthrax is a bacterial disease caused by *Bacillus anthracis.* Cattle, sheep and buffaloes are highly susceptible to this disease. It is characterised by high fever, haemorrhage of subcutaneous and subserous tissues, and enlarged spleen. The organism is voided through all excretions and secretions of infected animals, which survive outside for years together, by forming spores. The infection is rarely transmitted directly from one animal to another. Blood sucking flies sometimes transmit the disease mechanically. Ingestion of infected pastures, inhalation of spores while grazing and infection through skin wounds are other routes of infection.

The disease may manifest in per-acute form, acute form or skin form. In per-acute form, the onset of disease may be sudden resulting in rise in temperature and running a rapid fatal course.

Death may occur in 1 or 2 hours in per-actue cases but in acute cases death occurs in 24 hours of infection. There is bleeding from natural orifices.

In case of skin infection, localised form of infection is formed. This type is more common in human beings than in animals. Cutaneous anthrax forms swelling of an oedematous nature, which will be hot and painful in the beginning but later, becomes cold and painless.

This disease can be diagnosed by demonstration of the organism in peripheral blood, body excretions and secretions, or a smear from visceral organs. Animal inoculation of clinical material helps in diagnosis. Sub-cutaneous injection of fresh fluid or tissue portions causes death in guinea pig within 24 to 48 hours. *Ascolis precipitation* test is another reliable method. For this, to act as antigen, precipitating serum of high titre and tissue extract is needed, which contain anthrax protein.

Antibiotics such as penicillin, oxytetracyclin, and tetracyclin are all effective in treating anthrax.

In endemic areas vaccination is advisable. Anthrax spore vaccine is widely being used. Dosage is 1 ml subcutaneously. which provide immunity for one year. Proper disposal of carcasses of the animals died of anthrax and maintenance of hygiene are essential for control of anthrax.

Disposal of carcasses of animals died of anthrax. The disposal of carcasses by safe, hygienic and efficient means is very important in the case of anthrax. All dead animals should be burned preferably where they lie if the site is suitable or a nearby site may be chosen for burning or burrial. Under no circumstances a carcass of an animal suspected to have died due to anthrax be opened or autopsied. This is because opening the carcases will lead to large number of the anthrax bacteria which are present in the vegetative form within the body of the animal, to sporulate on coming in contact with air. The site of burial or burning should be selected where there will be no danger of pollution of streams, rivers, canals or other water supplies. Burial should be at least at a depth of two metres. Everything that came into contact with the infected animal should be carefully disinfected before it is removed. Old ropes, bedding and other objects used for handling the dead animal may be burned along with the carcases. The surface of the soil around the edge of the pit for burial where the animal was laid just before burrial may be scraped off and thrown into the pit so that any blood or discharge that might have fallen is rendered harmless.

All attendants handling the anthrax carcass should wash their hands and arms and dip their feet in disinfectants before they leave the place of disposal.

While the vegetative forms of the bacilli are killed easily by a temperature of 60°C for 30 minutes, the spores are highly resistant and will survive steaming or boiling at 100°C for five minutes. Strong disinfectants like 5% Lysol requires to be in contact with spores for two days to destroy them. Strong solutions (5 to 10%) of formalin or sodium hydroxide are most effective. Three per pent solution of peracetic acid is an effective sporicide and if applied to soil in appropriate amounts (8 liters/m^2) is an effective sterilizing agent. Infected clothing should be sterilized by soaking in ten per cent formaldehyde.

20.4.8. Tetanus

Tetanus is a highly fatal disease, it is not frequently seen in cattle and buffaloes. It is caused by the exotoxin produced by *Ciostridium tetani* and characterised by hyperaesthesia, tetany and convulsions. The tetanus bacilli mainly get inside body through wounds, where in favourable conditions, they multiply and produce toxin, which is absorbed into the system resulting in the disease. The neurotoxin produced at the site of infection reaches central nervous system through peripheral nerves. There will be potentiation of normal sensory stimuli and a state of constant muscular spasticity is produced.

A general increase in muscle stiffness and muscle tremor is observed after an incubation period of 1-3 weeks. There is trismus with restricted jaw movement, saliva may drool out and hence the name "locked jaw", prolapse of third eyelid, unsteady and staggering gait. As the disease progresses, muscular tetany increases. Death occurs due to asphyxia because of the fixation of muscles of respiration.

The symptoms and history of injury or surgery are characteristic evidence for suspecting the case of tetanus. This disease may be confused with strychnine poisoning, lactation tetany, milk fever, but all these cases respond to concerned treatment.

The main aim of treatment is to eliminate bacteria and neutralise the toxin, relax muscle and release asphyxia. Parenteral administration of penicillin in heavy doses, treatment of local wound with antiseptic lotion and local application of penicillin may facilitate elimination of bacteria. Local injection of tetanus antitoxin also helps in some cases. Two to three lakh units of tetanus antitoxin before onset of disease is rewarding. Relaxation of muscle by oral chloral hydrate, or subcutaneous administration of magnesium sulphate solution is good, but has only short duration effect. Chloropromazine at the rate 0.4 mg/kg. B.W. twice daily for 8-10 days is useful. Supportive treatment is also advisable as the animal may not be able to eat and drink.

20.4.9. Black Quarter (Black Leg)

This is an infectious disease causeed by *Ciostridium chauvoei,* characterised by inflammation of muscles, toxemia and death. Cattle and sheep are highly susceptible to this disease.

The exact nature of pathogenesis is still not completely known. It is reported that in cattle this disease occurs by ingestion of spores of the organism from feed or soil. Also wound infection occurs in cattle. The age group of 6 months to 2 years is highly susceptible to this infection.

This disease usually runs a short course. Affected animals die in one or two days. Though it is amenable to treatment in the early stages of the disease, the cases will be usually presented at later stages. There is high fever, depression, anorexia, with hot painful swelling in the affected leg.

Other parts of the body like cardiac muscle, diaphragm, brisket and udder will also be affected. The swelling soon becomes cold and painless. There is crepitating sound on pressing the affected part of muscle. This is due to gas production by this anaerobic bacteria. This is also a pathognomonic symptom of the disease. Death usually occurs due to toxemia.

Skin of the affected part becomes dark coloured. Swelling when cut open, foul smelling sero-sanguineous fluid oozes out. Regional lymph nodes are swollen, oedematous and may show haemorrhage.

Typical black quarter cases can be diagnosed from symptom itself. The animal will show sudden rise in temperature, anorexia, local swelling followed by death. This may be confused with other clostridial infections, anthrax, lightning stroke etc. But identification of the organism, local incidence of the disease, season of the year, age group affected, and presence of characteristic lesion will help to reach a conclusion. For confirmation, a piece of affected muscle tissue should be collected, dried and sent to nearest laboratory for confirmation by animal inoculation test and other cultural studies. Smears from the affected tissue will help in demonstration of the organism.

If the case is presented in the beginning, treatment is beneficial. Penicillin at a higher level, 10,0001.U./kg. B. W. should be administered, followed by long acting penicillin like Benzathin penicillin. Local administration of antibiotic is also helpful in easy recovery. In enzootic areas where the incidence of the disease is more, annual vaccination of all cattle of that particular age group, *i.e.,6* months to 2 years controls the disease. Formalin killed alum-precipitated vaccine is more satisfactory. Carcasses died of black leg should be destroyed by burning or deep burial to limit soil contamination.

20.4.10. Tuberculosis

Tuberculosis is a chronic contagious disease of cows and buffaloes caused by *Mycobacterium bovis,* characterised by weakness, emaciation, development of tubercles and abscess formation with resulting caseation and calcification. This disease is world-wide in distribution, though some countries have limited the spread of this disease through strict measures. Incidence of disease in buffaloes is more than that in cattle. Again incidence is higher in adult animals, but all age groups are susceptible to infection.

Infected animal is the main source of infection. The organisms are excreted in exhaled air, sputum, faeces, milk, urine, vaginal and uterine discharge. Inhalation and ingestion are the main routes of infection. The morbidity rate is more in overcrowded herds; in unhygienic conditions with poor housing.

At the point of entry or at the local lymph node the organism produces primary foci of infection from where it spreads to other parts of body, called post primary dissemination. Mainly thoracic lymph nodes, lung and pleura get affected. Because of the hematogenous spread, any visceral organ can get affected. In the early stage of the disease, there may not be any cardinal signs except general weakness and depression. In later stages, depending on the organ involved there may be symptoms.

When lung is involved there is dry cough, dyspnoea emaciation and intermittent fever. Persistent diarrihoea in case of digestive tract involvement. Hard, painless enlargement of the gland when udder gets affected. All superficial lymph nodes will be enlarged. Since it is a chronic disease, as the disease progresses animal becomes emaciated or debilitated and death follows.

Bovine tuberculosis is suspected primarily when the animals go down in condition in spite of feeding well and good management. Clinical signs, post-mortem findings and other laboratory findings like tuberculin reaction, demonstration of the organism in the body excretions and secretions, animal inoculation test etc., aid in confirmation of the diagnosis.

Tuberculin Test. Tuberculin is a refined biological product containing specific protein synthesised by growing tubercle bacilli. This test is highly specific for diagnosing the case in very early stages, but may not give positive reaction in advanced cases. Tuberculin tests are of different types.

Single intra-dermal test. This is done by injecting 0.1 ml of tuberculin into anal fold. The reaction is read between 72 and 96 hours after inoculation. A positive reaction constitutes a diffuse swelling at site of injection.

Double intra-dermal test. This test is more commonly used for diagnosis of tuberculosis. In this case 0.1 ml of tuberculin is injected intra-dermally at the middle of neck. It is repeated at the same site after 48 hrs of first injection. The resulting swelling is read 24 hrs after second injection. In positive cases considerable thickness, with warm painful oedamatous swelling occurs at 72 hrs of injection.

Stormont test. This test is used for detection to poorly sensitized animals. This test is performed similarly to that of single intra-dermal test in the neck with a further injection at the same site 7 days later. An increase of skin thickness of 5 mm or more 24 hours after this second injection is a positive reaction.

Subcutaneous test. This test is carried out by subcutaneous injection of 3 ml of tuberculin. One degree centigrade rise in body temperature in 9 to 12 hours of injection indicates positive cases.

All these tests, if positive, indicate presence of primary focus in body but not the advanced cases. Post-mortem findings include presence of characteristic nodular appearance and contain cheesy mass. In doubtful cases smears from cheesy mass should be made and examined for the presence of the organism. The lesions can also be examined histo-pathologically.

Treatment with streptomycin, penicillin, para amino salicylic acid and iso-niazid have given encouraging results in human beings. But due to long duration and cost of treatment it is not advisable in animal treatment.

Control of tuberculosis can be achieved mainly through "test and slaughter" programme. In herds, all animals should be tested once in three months. Slaughter all positive reactors. In two subsequent tests if animals in a herd are found negative that herd will be free of tuberculosis. All calves above three months should be tested. Prevention of spread of infection if any, is another important step in control. This can be achieved by providing good sanitary and hygienic conditions. Testing of animals before adding to a new healthy herd is another important point to do.

B.C.G. vaccination can be done in cattle to limit the extent of tuberculosis. But test and elimination is the only reliable method to control tuberculosis.

20.4.11. Johne's Disease

Johne's disease is a bacterial disease caused by *Mycobacterium paratuberculosis,* characterised by chronic diarrhoea and progressive emaciation. Though mortality is low, loss due to ill health and decreased productivity causes severe economic losses.

Infection mainly occurs through ingestion of contaminated food and water. As the incubation period is very long, symptoms may develop after several months. Generally incubation period is 6 months to 2 years. However, affected animals will excrete organisms in faeces 15 to 18 months before expression of symptoms. In addition to this, the organisms can survive in dung and water for months together.

The disease will start with progressive emaciation and intermittent diarrhoea which usually do not respond to treatments. Appetite of the animal will be normal, but milk production will go down. Faeces will be soft, thin and without blood, epithelial debris and mucus. Course of the disease varies from weeks to months. Ultimately death occurs due to severe dehydration in advanced cases.

At early stage of the disease it is very difficult to diagnose the case of Johne's disease because of the lengthy incubation period and as the symptoms are not so specific. It has to be differentiated from cases like parasitic diarrhoea, Rinderpest etc. But, the case of parasitic diarrhoea could be differentiated by detecting the parastic ova in dung. The nasal discharge and conjuctivitis, which are present in rinderpest cases are not occurring in Johne's disease. Allergic test using "Johnin" is being applied. Double intra-dermal test is used. But this test is not reliable because of false positive reaction. This test is performed by injecting 0.2 ml of "Johnin" intra-dermally in the cervical area. This test should be read at 48 hours as most significant reaction occurs at that time. Increase of skin thickness by 3 mm or more is taken as positive. Cultural examination of faeces for isolation and identification of bacteria is another method. Serological tests like complement fixation test and fluorescent antibody techniques are also helpful to a limited extent only, because these tests do not give positive reaction in early cases and non-clinical carriers. Histopathological studies of terminal ilial or ileo-caecal lymph nodes help in diagnosis.

The intestinal mucosa will be thickened three to four times and corrugated. No ulceration or discontinuation of mucosal surface occurs. The mesentric or ileo-caecel lymph nodes will be oedematous and enlarged. This organism is not amenable to treatment, though streptomycin has activity against the organism. It causes only temporary relief.

Control of this disease can be done by eradication of infected animals by slaughtering and disposal. Maintenance of hygienic condition on farm for prevention of further spread, if the disease is present, vaccination of healthy population to increase the resistance etc. are other control measures. Vaccination can be done by sub-cutaneous injection of 5-10 mg of a live culture of Johne's bacilli suspended in suitable base.

20.4.12. Actinobacillosis (Wooden Tongue)

This disease is caused by an organism called *Actinobacillus lignieresis,* which is an obligatory parasite of upper respiratory tract and alimentary tract mucosa. It affects mainly soft tissues rather than hard tissues like bone as in actinomycosis. Lymphnodes, lungs and sub-cutaneous tissues are mainly affected.

This disease occurs through ingestion of contaminated feed, water, and pasture with discharge from infected animals. Injury on mucosa helps in easy entrance of the organism into the system. Tongue is commonly affected. Gum, palates, and neighbouring lymph glands are also affected. Spread of disease occurs through lymphatic ducts.

The lingual actinobacillosis is usually acute. The animal is unable to eat. Excessive salivation and gentle chewing of tongue are seen. On examination, the tongue is swollen, painful and hard at base. Nodules start developing on dorsum and side of tongue, which later becomes ulcer. The same type of lesions are also noted on the walls of fore-stomach, skin, liver and lungs.

Examination of smears from discharge reveals the presence of organisms. This helps in diagnosis along with symptoms. Histological examination and agglutination tests are also used in confirmation of the diagnosis.

Surgical removal of pus from the nodules and providing facility for drainage helps in curing. Similar to actinomycosis, administration of iodides or antibiotics like penicillin, streptomycin, tetracyclin or chloramphenicol are effective in treatment of actinobacillosis.

20.4.13. Actinomycosis (Lumpy Jaw)

This disease is caused by organism called *Actinomyces bovis* and the infection usually invades bones of the head in bovines. Usually mandible and less frequently bones of maxillary sinuses are affected. Occasionally organs like liver, udder and reproductive organs of both males and females are affected.

This organism is usually present in the digestive tract. Whenever they get a chance through wound they gain entrance to the system and produce the disease. The exact mode of infection is not known. Pus and saliva of affected animals are infective.

This infection is sub-acute or chronic and is slowly progressive producing marked inflammatory reaction and results in rarefying osteomyelitis. As the destruction of bone occurs, the teeth become loose and sensitive leading to anorexia. Animal loses condition. As the infection progresses, the extensive periostitis stimulates an inflammatory reaction in the surrounding tissues producing marked swelling at the region. The infection may also gain entrance to the maxillary sinus and also involve the tissues of hard palate and maxilla. The sinus may become filled with granulation tissue.

Clinical signs depend on the part of body that is affected. In the early stage of involvement of mandibles, there may be occasional loss of bolus of food from mouth, accompanied by salivation. These are indicative of sensitive teeth. Palpation of mandible reveals thickening and in some cases separation of the gum from teeth. As the disease progresses there is gradual thickening or enlargement of the mandible. Bulging at maxillary region is an indication of affection of maxillary sinus. Here, the bone, may be softer or may be absent due to rarefaction. In case of udder involvement, there is a uniform enlargement and hardening of the gland.

Diagnosis can be made from typical symptoms of bone involvement. Demonstration of organism in discharge helps in confirmation.

Iodides are the common drug of choice in actinomycosis, which reduce the severity of fibrous tissue reaction. Potassium iodide 6 to 10 gm daily orally as drench for 7 to 10 days may be given. Treatment should be discontinued if iodism develops. Lachrymation, loss of appetite, anorexia etc. are the common symptoms of iodism. Penicillin and streptomycin have been found to be effective in early stage of disease.

20.4.14. Ringworm

Ringworm is a chronic, localised superficial mycotic dermatitis caused by *Trichophyton verrucosum* mainly and *T. mentagraphytes* to a lesser extent. It is characterised by scaliness and partial loss of hair. Incidence of disease is higher in animals of two years or younger than in other age groups.

The spores of *Trichophyton verrucosum* locate in skin lesions of cattle. The affected animals may rub its body against any inanimate object. When healthy animals come in contact with these objects they get this infection. Direct contact with affected animals also forms a main reason for transmission of disease.

Localised infection may develop mainly on face, neck, or any other part of the body. The mycelia grow rapidly among the host cells of the *stratum corneum.* Infundibuli of hair follicle also get affected by the organism. This weakens the hair and causes it to break resulting in alopecia. Due to chronic inflammation epithelial cells proliferate, keratinize and form scale and scabs.

Due to the irritation animal will be restless, and will lead to anorexia resulting in loss of condition. In crowded lots morbidity is 50-70 per cent while mortality is less.

The lesions in bovine ringworm are typical, distinctive and characteristic for accurate diagnosis. Single or multiple circular, dry, scaly, gray, depilated crust is the significant clinical sign. For diagnosis crust or hair from periphery of lesions are scraped and heated gently on slide with 10 per cent sodium or potassium hydroxide for a few seconds. After 15-20 minutes, slide is examined under microscope for presence of fungus. Histopathological sections are made from periphery of lesions and stained. The spores and mycelia can be seen on culturing of the specimen in Sabouraud's Dextrose Agar. It has to be differentiated mainly from mange.

This disease cures spontaneously. However, curative treatment is effective. Clean the affected area, remove hair and crusts. Apply Tr. Iodine or ointments containing salicylic acid 3 per cent, chrysarobin 2 per cent or iodine 2 per cent. Oral administration of Greseofulvin for 30 days is effective in wide-spread infections. Segregation of the affected animal is a must for prevention of spread of infection.

20.4.15. Brucellosis (Contagious Bovine Abortion; Bang's Disease)

This disease is caused by *Brucella abortus* and is characterised by abortion in late pregnancy and a subsequent high rate of infertility. This disease is worldwide in occurrence. It has got both economic and public health importance as it affects human beings. Death of calves, abortion, reduction in milk yield and temporary or permanent infertility leads to great economic loss. In human beings, it causes undulant fever. In India the overall infection rate among cattle is 14 per cent and 7.7 per cent in buffaloes.

Infection usually occurs through ingestion of feed and water contaminated with uterine discharge of a recently aborted cow. The organism can also penetrate intact mucous membrane and skin. Bulls with lesions in testicle, epididymis or seminal vesicle infect the cow at natural service or at artificial insemination. Infection occurs in cattle of all age groups but persists in sexually mature animals only. In the initial stage there will not be any body reaction. As disease progresses organisms get localised at various parts of the body. Uterus, udder and supra-mammary lymph nodes are common sites of localisation in females and in males genital organs. Abortion usually occurs after fifth to eighth month of gestation. Often retention of placenta also occurs in aborted cows.

Diagnosis is arrived at by cultural isolation of the organism from uterine discharges, fresh placenta, or aborted foetus. Serum agglutination test is more reliable for establishing a diagnosis in a case. Semen plasma agglutination test has been used in bulls. In milking animals *Milk Ring test* can be applied.

Prevention and control of disease depends on periodical screening of all animals in the particular herd and isolation or elimination of infected animals, simultaneously immunising healthy stock.

All animals in main herd should be subjected to serum agglutination test in every three to six months. If reactors are found more, they should be isolated, or if their number is less they should be destroyed and properly disposed off.

Vaccination is done with *Brucella abortus* strain-19 live vaccine. It protects healthy animals living in a contaminated environment. The better age group of vaccination is 4-8 months of age. But in zebu cattle, it can be done at the age of 6 months to 1 year as they mature very late. As this vaccine strain is infective to human beings it should be handled with great care. Dose is 5 ml subcutaneously.

20.4.16. Listeriosis

Listeriosis is an infectious, contagious disease caused by *Listeria monocytogenes,* affecting men and animals. It is primarily pathogenic for central nervous system, but sporadic abortion takes place between 4th and 7th month of gestation.

The disease is more observed in stall-fed animals. Silage feeding has also got some role in incidence of the disease, as the pH of spoiled silage is suitable medium for the growth of the organism. Cattle of 6 months to 2 years are commonly affected.

After incubation period of 2 to 3 weeks, affected animal exhibits distress and mild fever. The head is inclined laterally and shows unilateral paralysis. Circling of the animal is common in this disease. In 10 to 14 days of infection animal becomes prostrated, comatosed and finally dies. Pregnant cows may abort. The morbidity rate varies from 5-10 per cent, but mortality is nearly 100 per cent.

At necropsy, gross changes are not discernible. Cut surface of the medulla and pons are hyperaemic. The brain stem exhibits multiple foci of infection.

Diagnosis can be done based on symptoms, necropsy findings, and isolation of organism from brain stem (Ground brain tissue has to be stored at 4°C for weeks together before culturing).

Affected animals may be treated in early stages with sulphonamides or antibiotics. Supportive treatment has to be provided until self-feeding is resumed. All infected animals should be isolated. Suspected feeds, silage etc., have to be discarded. No effective vaccine is available for immunizing cattle against Listeriosis.

20.4.17. Leptospirosis

In cattle and buffaloes this disease is caused by different serotypes of *Leptorpira grippotyphosa, L. pomona, L. hardjo, L. icterohemorrhagine.* Subsequent to infection with Leptospira during convalescence 10 to 25 per cent of the animals abort in last three months of gestation. In cattle it is a fatal disease causing haemorrhage, haemoglobinuria and icterus. Organism may be shed in urine during convalescent period.

Direct microscopical examination of tissue preparation or body fluid helps in detection of organism under dark field illumination. Cultural examination, animal inoculation, serological test, fluorescent antibody technique etc., are helpful in diagnosis.

Control of this disease is very difficult as the organisms are being shed in urine and several agents are acting as carriers. Streptomycin, penicillin and tetracycline are useful for effective treatment. Vaccine pepared from locally prevailing serotypes are useful for inducing immunity, but again it will be of short duration.

20.4.18. Vibriosis

Vibriosis is caused by organism *Vibrio foetus.* It is transmitted mainly through coitus. Abortion is the characteristic symptom, which occurs at 5 to 7 months of gestation. After one abortion the animal becomes resistant to infection.

There is temporary infertility and results in increase in number of natural services or AI per pregnancy in infected herd. The bull harbours the infection on prepuce or penis and deposits the organism in vagina during coitus.

Genital vibriosis is suspected when a cow/heifer returns to service especially when this coincides with introduction of a new bull. Organisms can be isolated from stomach contents, and lungs of foetus. Agglutination test of vaginal mucus is positive after 40 days of infection.

The animal acquires immunity after one infection. However, hygienic measures have to be observed for prevention of spread of infection.

20.5. PARASITIC INFESTATIONS

The more common parasitic infestations of cattle and buffaloes are ascariasis, trichostrongylosis, fascoliasis, amphistomiasis and schistosomiasis, though to a small extent cestode infection also occurs but is not much pathogenic.

20.5.1. Ascariasis

Ascariasis caused by *Ascaris vitulorum* is a common disease of young animals. This disease is characterized by prenatal infection as well as colostral transmission. Sometimes it is responsible for heavy mortality, especially where hygienic conditions are not maintained. Diarrhoea, emaciation, loss of appetite, pot-bellied appearance, stunted growth and death are common symptoms. During the life cycle, this parasite gets a migratory phase and causes much damage to intestine, liver, and lung. Adult animals harbour infection and act as carriers.

20.5.2. Parasitic Gastro-Enteritis

Several species of **trichostrongylidae** cause parasitic gastro-enteritis. They are found in abomasum and small intestine. Some important species found in abomasum are *Mecistocirrus digitatus, Haemonchus placet, Ostertagia* sps., and *Trichostrongylus axel.* The species found in small intestine belong to the genera *Cooperia, Trichostrongylus* and *Nematodrius.* Mixed infection with *Bunostomum phlebotomum, Oesophagostomum radiatum* and *Trichurs* species, are also common in cattle and buffaloes. Symptoms of gastro-intestinal disturbances in the form of constipation or diarrhoea, anaemia, emaciation and general weakness are seen. Edema in dependent parts is also usually seen. This infection mostly occurs after the onset of monsoon, when there is a favourable condition for the parasitic development.

20.5.3. Fascioliasis

Fascioliasis is another common disease of cattle and buffaloes. *Fasciola gigantiai* affects cattle, buffaloes, sheep and goats. This parasite requires intermediate hosts like fresh water snails *(Lymnea* species). The infected snails release cercaria into the water. The cercaria encyst on grass blades and form metacercaria, the final host getting infection on ingestion of metacercaria. When large number of metacercaria are ingested within a short time acute facioliasis is produced, which may be fatal. The migration of large number of immature flukes through liver causes traumatic hepatitis. In mild cases the flukes reach bile duct, mature and cause chronic fascioliasis. Symptoms of chronic cases are loss of appetite, loss of condition, anaemia, edema, constipation and diarrhoea in the extreme stages.

20.5.4. Amphistomiasis

Amphistomiasis is caused by several species of ruminal flukes. They belong to the genera *Paramphistomum, Cotylophoron, Calicophoron, Ceylonocotyle, Gastrothylax, Hschoedc rius* etc. The immature flukes are the really harmful ones compared to the adults. The immature form burrow into the intestinal mucosa, and draw a piece of mucosa into the suckers and pinch them off causing necrosis and haemorrhage. Affected animals show loss of appetite, anorexia, thirst, diarrhoea with foetid faeces and edema in dependent parts.

20.5.5. Schistosomiasis

Schistosomiasis is another disease which is prevalent among cattle and buffaloes in India. The disease mainly occurs in two forms, hepatointestinal form and nasal forms. Hepatointestinal form is caused by *Schistosoma indicum,* and *S. spindale* in cattle and buffaloes. Intermediate hosts are *Indoplanorbis* species of snails. The parasites are found in portal and mesenteric vessels. The symptoms include anaemia, loss of condition, alternative diarrhoea and constipation and gradual wasting. Faeces contain mucus blood and eggs of the parasites.

Nasal schistosomiasis is caused by *Schistosoma nasale.* This disease is also known as snoring disease. The parasites inhabit veins of nasal mucosa. The eggs are deposited in mucosa and are passed out in nasal discharge. Symptoms include profuse nasal discharge, respiratory distress and cauliflower-like growth.

Diagnosis. Practically all animals harbour some worms. Some may be pathogenic and some may be non-pathogenic. Moreover, healthy animals can tolerate mild infection without seriously affecting their health. Diagnosis cannot be based on mere presence of worm or egg. Quantitative assessment of worm or egg, condition of animal, history of animal, etc., is required to reach a final conclusion. Clinical signs supported by a knowledge of history and the epidemiology of parasites present in the area may be sufficient for diagnosis.

Treatment. As a routine all calves should be treated with piperazine salts at the age of 3rd week and repeated at 6th week. Thereafter broad spectrum anthelmintics like Phenovis, Nilworm. Fenbendazole, Helatac etc. should be used if there is infestation in the herd.

Maintenance of hygienic condition in and around animal shed, proper disposal of dung, provision of a well nourished diet to the animal so that animal's resistance increases etc. are other steps for control. Treat all animals whether infected or not.

20.5.6. Theileriasis

Theileriasis is one of the important diseases of bovines, especially of crossbred and exotic breeds. This is caused by *Theileria annulata* and it is transmitted by *Hyalomma anatolicum, Rhiphicephalus haemaphysaloides* and *R. appendiculatus* species of ticks.

Disease occurs in all age groups; cases have been recorded in calves as young as 7 days or less in age. Disease mainly occurs in summer and rainy seasons. In fact, this is mainly associated with the activity of ticks. In cattle mortality varies from 10 to 90 per cent depending on the strain of parasite.

This disease is seen in acute, sub-acute and chronic forms. Acute disease has been seen in all breeds of cattle and buffaloes. Indigenous cattle and buffaloes recover more readily, though they remain as carriers. Incubation period varies from 9 to 25 days and the disease may last for 3 to 4 days in acute form, or may prolong upto 20 days. There is a marked rise in body temperature 104°-107°F followed by depression, lachrymation, nasal discharge and swelling of the superficial lymphnodes. Animals become emaciated and there may be haemoglobinuria.

Blood smear examination helps in diagnosis. Both thick and thin smears are examined. Clinical signs and presence of ticks on the body of the animals may suggest the presence of infection. Smears from lymph node punctures can reveal the parasites especially in the lymphocytes as Koch's blue bodies. The following serological tests are also employed in Theileriasis—

(*i*) Complement fixation test, (*ii*) Capillary agglutination test, (*iii*) Indirect haemagglutination test., (*iv*) Agar gel diffusion test and (*v*) Allergic test.

No satisfactory treatment has been worked out so far. Acriflavin intravenous injection has some effect. Oxytetracycline, chlortetracycline, 'Berenil, 'Nivaquin', and 'Pamaquin' are effective and give clinical recovery. Blood transfusion, tonics, and haematinics also prove useful in assisting recovery. Control of this disease mainly depends on tick control.

20.5.7. Babesiosis (Texas Fever, Tick Fever, Red Water Fever)

Bovine babesiosis is another important disease of cattle and buffaloes. It is caused by *Babesia bigemina* in cattle and buffaloes and *B. bovis* in buffaloes. Young animals are naturally resistant upto 9-12 months, while old animals are susceptible. Indigenous cattle are resistant to this disease.

Babesiosis is transmitted by ticks. Its development and transmission is either by transovarian or stage to stage transmission. *Boophilus*, *Rhiphicephalus*, and *Haemophysalis* species of ticks are responsible for transmission of Babesiosis.

Incubation period is 1-2 weeks. Rise in body temperature from 106° to 108°F is common. Fever lasts for 2-7 or more days. Haemoglobinuria and profound anaemia develops. Initially there will be profuse diarrhoea followed by constipation. About 75 per cent of erythrocyles arc destroyed at the height of temperature and in such cases animal dies in 4-8 days time after onset of temperature. Recovering animals develop chronic disease and there may be intermittent fever and emaciation. But only haemoglobinuria is noticed. In the course of several weeks recovery takes place.

After a single *Babesia bovis* infection during calfhood, cattle remain immune for atleast 4 years.

Examination of blood smear at the height of temperature is the easy method to demonstrate the organisms within erythrocytes. The following serological tests are also used—

(*a*) Indirect fluorescent antibody test;

(*b*) Indirect haemagglutination test; and

(*c*) Capillary agglutination test.

Out of these, Indirect fluorescent antibody test is more quick and most sensitive.

The following drugs are used in the treatment of *Bovine babesiosis* :

(*a*) Trypan blue

(*b*) Berenil

(*c*) Pirevan (Babesan, Acaprin)

(*d*) Diampron

(*e*) Imidocarb dipropinate.

In *Babesia bigeminae* infection, inverse age resistance of cattle has been exploited for control by immunization. Young animals are inoculated by a mild strain of *Babesia bigemina* and the subsequent infection is controlled by chemotherapy. It has been shown that the piroplasms from carrier animals may show loss of virulence and injection of such materials in calves may produce a lower level of immunity than expected. Attempts to use killed vaccine using disintegrated parasites with adjuvant are also successful. Tick control is the important item in Babesiosis control.

20.5.8. Anaplasmosis

Anaplasmosis is a rickettesial disease, caused by *Anaplasma marginale* and *A centrale,* in cattle and buffaloes. *Anaplasma marginale* is worldwide in distribution and more pathogenic than *A. centrale.* This disease is transmitted by ticks, flies (tabanid flies, stable, hippoboscid flies) and mosquitoes. In addition to this, mechanical transmission during major and minor operations, vaccination etc., may also take place.

This disease is mainly seen in adult animals and infection does not occur until an animal is about 18 months of age. Incubation period is 15-36 days and the clinical signs resemble more to Babesiosis and Theileriasis. There is fever, which may terminate fatally, organism appearing in the red blood cell several days before onset of fever. Thirty to fifty per cent of the RBC may be parasitized as the fever progresses. Anorexia and anaemia are noted at the time or after fever. Mortality is high in exotic and crossbred cattle. Examination of blood smear helps to diagnose the anaplasmosis case. The following serological tests are also useful in diagnosis—(*a*) Capillary agglutination test, (*b*) Fluorescent antibody test, and (*c*) Bentonite flocculation test. Tetracycline compounds are very effective.

Killed *Anaplasma marginale* usually in adjuvant vehicle is used as vaccine in present days. However, it gives poor immune response.

20.5.9. Tick Control

Ticks themselves are important parasites. Tick control not only helps to reduce the damage by ticks but also the indirect transmission of other diseases. Since ticks are seen in all parts of the body, treatment has to be applied to whole body and usually carried out by dipping of animals in the tanks containing dip or by spraying the dip solution over animal. In one host tick it is very easy to control but two and three host ticks are very difficult to control.

The common acaricides used are—(*a*) Arsenic; (*b*) Asuntol (Co-Ral, Coumafos); (*c*) B.H.C.; (*d*) D.D.T.; (*e*) Diazinon; (*f*) Dieldrin: (*g*) Toxaphene; (*h*) Carbaryl and (*i*) Synthetic pyrethroids. In addition to this, the following measures can be undertaken—(*a*) Spraying of acaricides on pastures and in tracks and crevices, (*b*) Burning of pastures, (*c*) Cultivation of land.

There are some natural enemies for ticks such as ants and birds. *Hymenopterus* flies, *Ixodiphagus* and *Hunterellus hookenri* are the natural parasites of ticks. *Bos indicus* breeds of cattle are better equipped to resist ticks and tick-borne diseases. In the tick infested southern United States and South America many animals from these breeds were introduced for their tick resisting quality. They eventually evolved a breed known as *Brahman.* They were also used for crossbreeding with *Bos taurus* breeds. The new breeds thus evolved like *Santa gertrudis, Beef master, Braford* and dairy breed developed in Austrila on similar lines *Australian Milking Zebu* (AMZ), are resistant to ticks.

20.5.10. Coccidiosis

Coccidiosis is a protozoan disease occurring in young calves, especially under conditions of insanitation and overcrowding. This disease is characterised by mucoid, haemorrhagic dysentry. *Eimeria zurni* and *Eimeria bovis are* highly pathogenic species of coccidia in bovines, though a dozen species have been detected in India alone. It is a disease of economic importance because of loss of body weight, and condition, inefficient utilization of feed, mortality and expensive treatment. This disease is present wherever intensive cattle rearing are going on. It occurs in bovines of age group 1 week to 2 years, but more commonly at the age group of 6 months to 9 months.

The faeces of the infected animal is the source of infection, which contains oocysts of Coccidia. This oocyst under favourable conditions of atmospheric humidity, temperature and oxygen get sporulated. The new host gets infection on ingestion of these sporulated coccidal oocysts. After 16-30 days of ingestion, the parasite completes its life cycle in the epithelial cells of intestinal mucosa. The first sign of illness is the sudden onset of severe diarrhoea with foul smelling faeces which may contain mucous and blood. Severe straining is characteristic. Anaemia, as a result of severe haemorrhage in the intestine, weakness, staggering, dyspnoea, dehydration and emaciation are associated symptoms. The disease takes a course of 5 to 10 days.

Post mortem examination reveals severe congestion of intestinal mucosa with enteritis and thickening of the mucosa of the caecum, colon, ileum and rectum. Small spots formed by the schizonts may be visible in the villi of terminal ileum. Ulceration and sloughing of mucosa is seen in severe cases.

Coccidiosis is diagnosed from symptoms and from examination of faecal samples for the presence of oocysts. Histopathological examination of section of intestinal mucosa disclose presence of schizonts and sexual forms of parasites.

Sulphonamides have been used in the treatment of coccidiosis at the rate of 1 g per 15 lb body weight for 4 days. Nitrofurazone can be given in the feed as well as in drinking water. It may be mixed in the feed, at the rate of 0.04 per cent or at a dose level of 5 mg per pound body weight for 7 days. In drinking water it is given at the rate of 0.008 to 0.013 per cent for 7 days. Amprolium at the rate of 25 mg per kg body weight daily has been found effective in calves. In outbreaks of coccidiosis, all clinically affected animals should be isolated and should be treated. Supportive treatments have to be given in severe cases.

Coccidiosis can be controlled to a great extent by observing hygienic conditions by avoiding overcrowding. Animals should be grouped based on age. Calf pens should be kept clean and dry. Bedding should be removed and disposed of daily; this helps in prevention of sporulation of oocyst and becoming infective, which may lead to infestation. Feed and water troughs should be high enough to avoid faecal contamination. Carrier animals should be avoided from a herd.

Immunization against coccidiosis has been tried. Due to the multiplicity of strains it is found impracticable in field conditions. Immunity is specific to each species of Eimeria. In United States of America, immunization against poultry coccidiosis is being practiced using "COCCIVAC", which is nothing but sporulated oocysts of *Eimeria* species of poultry prevalent in USA. Research work is going on in this field intensively. So we can hope for a vaccine against coccidiosis in all domestic animals in the near future.

20.5.11. Trichomoniasis

This is a venereal infection caused by *Trichomonas foetus* transmitted during coitus. The infected bull does not show any symptoms, except balanitis for the first two weeks. Afterwards, it remains a permanent source of infection. Cows affected usually abort at 2 to 4 months of gestation. This leads to retention of fetal membrane and dead foetus, which get macerated. Endometritis and pyometra are common sequel. There is sterility, marked anoestrus, and persistent uterine discharge.

Diagnosis is made by demonstration of the organism microscopically or culturally in foetall stomach content, or in uterine discharges. Prepucial washing of the bull also reveals the presence of organism. Cervical mucus agglutination test is used for detection of the infected animals.

20.6. DEFICIENCY DISEASES

20.6.1. Hypovitaminosis B1 or Polioencephalomalacia or Cerebrocortical necrosis

Polioencephalomalacia occurs sporadically in goats, and other ruminants, and is characterized clinically by sudden onset of blindness, head-pressing, opisthotonus and convulsions, a rapid response following thiamin therapy in the early stages and pathologically by acute cerebral edema and laminar necrosis of the cerebral cortex.

Etiology. Thiamin inadequacy and dietary imbalances attributable to high sulfate concentration have been associated with the occurrence of the disease. The disease may affect goats from 2 months to 3 years of age and is commonly associated with milk-replacer diets in kids or concentrate feeding in older goats.

Thiamin inadequacy in ruminants could occur in situations as inadequate net microbial synthesis of thiamin in the rumen which may occur in concentrate-fed animals receiving inadequate roughage. Impaired absorption or phosphorylation of thiamin, the presence of a thiamin inhibitor in the tissues of the host, lack of sufficient or appropriate apoenzyme or coenzyme, apoenzyme binding for thiamin dependent systems, increased metabolic demands for thiamin in the absence of increased supply, and increased rate of excretion of thiamin resulting in its net loss from the body are other causes.

Animals with polioencephalomalacia were found to have elevated ruminal and fecal thiaminase activities, low erythrocyte transketolase activity, elevated thiamin pyrophosphate effect, low liver and brain thiamin levels and elevated plasma glucose levels compared with normal animals.

Animals changed very rapidly to high concentrate rations develop increased ruminal thiaminase levels.

Thiamin is an essential component of several enzymes involved in intermediary metabolism and a state of deficiency results in increased blood concentration of pyruvate, a reduction in the lactate to pyruvate ratio and depression of erythrocyte transketolase. These abnormalities affect carbohydrate metabolism and leads to edema of brain due to a reduction in ATP production.

In acute cases there is a sudden onset of blindness, muscle tremor, particularly of the head, champing of the jaws and frothy salivation, and head-pressing. Initially, the irritation signs may occur in episodes, and convulsions may occur, but within several hours the irritation signs may be continuous. The animal usually then becomes recumbent, there is marked opisthotonus, nystagmus, clonic-tonic convulsions and tetany of the forelimbs is common. The temperature is usually normal but elevated if there has been excessive muscular activity.

The frequency of the rumen movements remains normal for a few days, which is an important differential feature from lead poisoning in which the rumen is invariably static.

The menace reflex is always absent in the acute stage and its slow return to normal following treatment is a valuable aid in prognosis. The palpebral eye-preservation reflex is usually normal. The pupils are usually of normal size and responsive to light. In severe cases the pupils may be constricted. Dorsal strabismus due to stretching of the trochlear nerve is seen commonly.

Nystagmus is common and may be vertical or horizontal. Optic disc edema, is present in some cases but is not a constant finding.

The thiamin pyrophosphate (TPP) effect is increased upto 70-80% in animals with polio-encephalomalacia and the erythrocyte transketolase activity is decreased.

The thiamin concentrations of liver, heart and brain is also decreased. The levels of blood pyruvate and lactate are increased and thiamin pyrophosphate-dependent enzymes such as pyruvate kinase are decreased. The thiaminase activity of the faeces is increased.

The diagnosis is based on clinical findings as sudden onset of blindness, head-pressing, nystagmus and opisthotonus.

The treatment of choice is thiamin hydrochloride at 10 mg/kg body weight given intravenously initially and followed by similar doses every 3 hours for a total of five treatments. When treatment is given within a few hours of the onset of signs, a beneficial response within 1-6 hours is common and complete clinical recovery can occur in 24 hours. Transplants of rumen juice from roughage-fed cattle are often indicated to improve appetite and rumen function in those which are responding slowly.

Treatment is relatively ineffective in advanced cases, but unless an accurate history is available on the length of the illness, it is usually difficult to predict the outcome until 6-12 hours following treatment. Therefore, it is usual practice to treat most affected cases with thiamin at least twice and monitor the response.

The diet should contain at least 50% roughage and supplementation of the diet with thiamin are the measures to control polio-encephalomalacia.

20.6.2. Hypovitaminosis A (Xeropthalmia; Infertility etc.)

Vitamin A deficiency occurs either as a primary disease due to an absolute deficiency of Vitamin A or its precursor carotene in the diet or secondary disease in which the dietary supply of vitamin or its precursor is adequate but their digestion, absorption or metabolism is affected.

Vitamin A occurs in 2 forms, *i.e.*, carotene which occurs in green feed that does not pass the placental barrier hence it does not increases the hepatic stores of Vitamin A in new born calves, but ester form of Vitamin A (fish oils) will pass the placental barrier in cows. So feeding of these oils or parenteral injection of Vitamin A, before parturition will increase the stores of Vitamin A in the foetal liver. But antepartum feeding of carotene or its alcohol form will increase the Vitamin A content of colostrum.

Various factors take easy oxidation of Vitamin A or carotene in the presence of USFA, prolonged storage of feed, pelleting, heat, light, humidity, and presence of certain other substances affect the availability of Vitamin A content of feed.

Secondary Vitamin A deficiency occurs in chronic diseases of the liver or intestine because conversion of carotene to Vitamin A takes place in the intestinal epithelium and liver is the main storage site of Vitamin A. High concentration of phosphorus and nitrite in diet and excess ingestion of mineral oil (treatment of bloat) also affect availability of Vitamin A.

Vitamin A is essential for regeneration of visual purple, which is necessary for dim-light vision, normal bone growth, and maintenance of normal epithelial tissue. So most of the clinical signs are attributable to the disturbances of these functions.

Inability to see in dim light is the earliest sign of Vitamin A deficiency followed by thickening and clouding of cornea, especially in calves. Excessive keratinization and heavy deposition of bran like scales on the skin, are the major dermatological changes.

Loss of reproductive function, infertility, placental degeneration leading to abortion, still birth, or birth of dead or weak calves are often seen. Neurological signs like paralysis and convulsion may occur in some animals. Both pupils are widely dilated, fixed and will not respond to light and these signs are more pronounced in yearling cattle (12-18 month old). Congenital defects are also documented in calves and pigs (congenital blindness). Estimation of plasma Vitamin A (normal 25-60 µg/dl), retinal carotene, hepatic Vitamin A and measurement of CSF pressure (normal <100 mm of saline) will help in the early diagnosis of Vitamin A.

Early treatment with Vitamin A @ 440 IU/kg body weight gives some beneficial result. Ocular form of Vitamin A deficiency of cattle will not respond to treatment. As a preventive measure Vitamin A can be given at 40 IU/kg body weight.

20.6.3. Rickets (Deficiency of calcium, phosphorus and Vitamin D)

Dietary deficiency of calcium, phosphorus or Vitamin D or imbalance in Ca : P ratio is the major cause of osteodystrophies in cattle.

Calcium deficiency may be primary (absolute dietary deficiency) or secondary (conditioned by other factors like excess phosphorus in the diet).

Phosphorus deficiency { Primary → absolute dietary deficiency; Secondary → excess calcium in the diet }

Vitamin D deficiency { Primary → absolute deficiency; Secondary → excess carotene }

Major diseases includes Rickets, Osteomalacia and Osteodystrophic fibrosa.

Rickets is a disease of young growing animals characterized by defective calcification of growing bone. Clinically it is characterized by stiff gait, lameness, enlargement of ends of long bones, curvature of long bone, enlargement of costochondral junction (rachitic rosary) and delayed dentition. Rachitic bones have characteristic lack of density, and ends of bones have a wooly or moth eaten appearance on X-ray. Diet should be provided with adequate amount of calcium, phosphorus and Vitamin D. In severe cases, Vitamin D injection can be given.

Osteomalacia is a disease of mature animals in which endochondral ossification has been completed. The etiology and occurrence of osteomalacia are the same as for rickets but predisposing cause is the drain of lactation and pregnancy. Reduced productivity, licking and chewing inanimate objects, stiff gait, and lameness of shifting nature, crackling sound while walking and inward rotation of hock are the major clinical signs. The names milk leg and milk lameness are commonly applied to this condition when it occurs in heavily milking animals. Decreased density of long bone is characteristic finding on X-ray examination. Diet should be provided with adequate amount of Ca, P and Vitamin D. Deformity of bones is likely to be permanent. Osteodystrophia fibrosa is mainly due to secondary calcium deficiency associated with excess phosphorus feeding. It is rarely reported in cattle, but it occurs in horses, goats and swine.

20.7. COMMON POISONING CONDITIONS

Suspicion of poisoning is aroused when illness occurs in a number of previously healthy animals, all are affect at the same time and showing same signs and necropsy findings to the same degree of severity. Specimen for toxicological analysis should include a sample of suspected source of material, alimentary tract contents and sample of tissue to prove that absorption of the poison has occurred.

List of material to be collected in suspected cases of poisoning.

- Arsenic—kidney, skin and hair.
- Lead—Kidney, bones and blood.
- Phosphorus—Kidney and muscle.
- Mercury—Kidney.
- Copper—liver, kidney and blood.
- Fluorine—Bone, teeth and urine.
- Hydrocyanic acid—Stomach content, blood and muscle.
- Nitrate and Nitrite—Ingesta and blood.
- Strychnine—Blood, kidney and urine.

Principles of treatment in case of poisoning. The two main principles are :

- Removal of the residual poison from the alimentary tract or skin.
- Provision of physiological and chemical antidote to the poison that has been absorbed to the system.

Gastric leakage and emetics are of little or no practical value and removal of toxins from alimentary tract depends largely upon the use of adsorbants (activated charcoal @ 1-3 g/kg body weight) and purgatives (Magnesium sulphate @ 1 g/kg body weight.

20.7.1. Poisoning due to Chemicals

Organophosphorus compounds and carbamates (Organophosphate) poisoning

Organophosphorus compounds (OP) and carbomates act in essentially the same way therapeutically and toxicologically but bonding of the compound to the esterase enzyme is irreversible in the OP compounds and easily degradable with carbamates making them less dangerous. The introduction of these compounds into animal therapeutics as treatment for various disease conditions has increased their importance as possible causes of poisoning. Young animals are usually much more susceptible than adults.

Organophosphates are highly toxic and readily absorbed by ingestion, inhalation or by percutaneous absorption. The inactivation of cholinesterase by these OP compounds causes an increase in acetylcholine in tissue and nerve endings producing muscarinic and nicotinic responses of acetylcholine. The toxic effects of OP are largely functional and no residual defects in recovered animals are seen. Another form of toxicity is OP induced delayed neurotoxicity.

Salivation, dyspnoea, diarrhea and muscle stiffness with staggering gait are the major clinical signs in mild cases. In the acute cases, the additional signs include protrusion of tongue, constriction of pupils, impairment of vision, muscle tremor, bloat, collapse and death. The condition can be diagnosed by careful examination of history and symptoms. Atropine (about double the normal) is the rational and approved treatment. Atropine @ 0.25 mg/kg body weight (1/3 by I/V and 2/3 by I/M in cattle and 1 mg/kg body weight in sheep. Oximes like 2-PAM (2-pyridine aldoxime methiodide) and diacetylmonoxime (DAM) are also recommended as treatment of choice. Pralidoxime @ 50 mg/kg body weight may be given.

Organochlorine Poisoning

This group includes DDT, BHC, aldrin, dieldrin, chlordane, methoxychlor etc. Some of these compounds are readily absorbed through the skin. Very young animals of any species are more susceptible than adults and lactating and emaciated animals also show increased susceptibility. Toxic signs include increased excitability and irritability followed by muscle tremor, weakness, paralysis and terminal convulsions. Salivation and grinding of teeth are also observed in some animals.

Most of the substances accumulate in the fat depot, may be excreted in the milk or concentrated in the butter and cream (source of poisoning for others).

Treatment comprises sedation with Phenobarbital sodium, intravenous injection of glucose and calcium. Administration of a non-oily purgative is also effective.

Urea Poisoning

Urea is used as a feed additive for ruminants to provide a cheap protein substitute and also used as fertilizer on crop and pasture fields. Mixture of urea and molasses was used as feed supplements for cattle in earlier time. Accidental ingestion of large quantities of urea when they are unaccustomed to it is one of the major causes for poisoning. A dose of 0.44 g/kg body weight can produce toxic symptoms and dose of 1-1.5 g/kg body weight cause death. Animal usually develops tolerance.

Rapid production of large quantities of ammonia and its rapid absorption from the rumen result in onset of signs within 10-30 min. Clinical signs included abdominal pain, frothing at the mouth, hyper sensitivity, muscle tremor, weakness, dyspnoea, bloat and death usually occurs within 4 hours. Oral administration of vinegar (0.5-1 l in sheep and 4 l to a cow) or 5% acetic acid is useful. Rumen evacuation is also effective.

20.7.2. Poisoning by fungi (Mycotoxicosis, Aflatoxicosis)

Aspergillus spp. produce a series of hepatotoxins, of which Aflatoxin B_1 is the most important. The major source of this toxins is groundnut infestated with *A. flavus* but cotton seed meal, sorghum grain and corn can also be a source. The fungus grows on stored food, when temperature and humidity are high and not destroyed by milling.

Aflatoxins include AFB_1, B_2, G_1, G_2 and second generation metabolite M_1 and M_2

AFB_1 >100 µg/kg of feed is toxic for cattle. All animal species are susceptible but outbreaks mostly occur in pigs, sheep and cattle. As the toxin is excreted in cow's milk the disease has public health importance, as it is considered an important cause for human hepatocellular carcinoma.

Hepatosis and hepatic insufficiency are the major effects, but mutagenic and teratogenic effect also occurs. Clinical signs include blindness, walking in circles, frequent falling or ear twitching, grinding of teeth, diarrhoea, photosensitive dermatitis, severe tenesmus and anal prolapse. Abortion is one of the common features in affected animals.

The condition can be diagnosed by estimation of AFs in feed, blood, urine etc. Analysis of liver enzyme is also useful. Symptomatic therapy for hepatic insufficiency can be tried.

20.7.3. Poisoning by Plants

Hydrocyanic acid HCN (Cyanide) poisoning. Most case of HCN poisoning is caused by ingestion of plants containing cyanogenic glycosides. Glycosides as such are nontoxic but HCN may be liberated by the action of β-glycosidase and lyase present in the plant tissue or by the action of ruminal microorganism. The following are some of the plants, which contain enough glycosides.

Acacia	*Adenia*
Eucalyptus spp.	*Glyceria* spp.
Heavea brazillensis (Rubber)	*Manihot esculenta* (cassava, tapioca)
Mimosa spp.	Sugarcane, etc.

Ingestion of rubber and tapioca leaves is the major cause of HCN poisoning in Kerala.

Young growing leaves have maximum toxicity and it also vary between seasons and vary between different parts of the plant. Drying, hay making, chilling and freezing may reduce the toxicity through destruction of enzyme β-glycosidase.

The minimum lethal dose (LD_{50}) of HCN is 2 mg/kg body weight for cattle. Plants containing more than 20 mg HCN/100 g is likely to cause toxic effect.

The rate of formation of HCN from glycosides in the rumen also affects the toxicity of feed.

Acute intoxication is mainly by inhibiting cytochrome oxidase and thereby preventing cellular aerobic respiration. Chronic poisoning is characterized by goiter and myelomalacia. HCN is normally detoxified by its conversion to thiocyanate in the liver through the action of rhodanase.

Affected animals rarely survive for more than 1-2 hours. Clinical signs develop within 10-15 minutes of eating toxic material and die within a short time.

Clinical signs included dyspnoea, restlessness, anxiety, tremor (head and neck), moaning, recumbency, opisthotonus and clonic convulsions. Mucosa is bright red in colour. Small, weak and irregular pulse, dilatation of pupil and nystagmus occurred in later stages. Bright red colour to the blood and bitter almond smell to the breath are also characteristic findings. The course of disease may be as long as 1-2 hours. The condition can be diagnosed by history of feeding of suspected plant material, clinical signs, and Henrici test (picric acid test).

- Treatment is mainly by IV injection of mixture of sodium nitrite (5 g) and sodium thiosulphate (15 g) in 200 ml water for cattle.
- Sodium thiosulphate heavy dose (660 mg/kg body weight IV).
- Sodium thiosulphate heavy dose plus cobaltous chloride (10.6 mg/kg body weight).
- Sodium thiosulphate 30 g orally to all animals exposed but showing no signs.

Abrus precatorius. It is a toxic plant which contains a toxic principle Abrin – phytotoxic substance similar to ricin of castor oil seed. In certain parts of country, spikes coated with crushed seeds have been used for malicious poisoning/killing of cattle. Whole seeds given by mouth have no effect in cattle, but powdered seed is toxic to all species. Clinical signs included salivation, stiffness, non-coordination, muscular spasm, and convulsions with extensive painful swelling at the site of implant. Symptomatic treatment to reduce pain and inflammation may be given.

Lantana poisoning. It is mainly due to ingestion of plant *L. camara* that contain Lantadine A – a phytotoxin present in the plant. It has got both hepatotoxic and nephrotoxic effect. Acute cases are characterized by reduced feed intake, sluggish movement and loss of weight. Faeces become soft with blood and mucus coating. In some cases paralysis of limb will develop.

Chronic cases are characterized by constipation, yellowish discoloration of visible mucous membranes and hard, painful swelling of non-pigmented areas of skin. Later the skin may peel off leaving raw area (hepatogenous photosensitization). Edema of head and face is also seen.

Diagnosis is mainly by history of feeding toxic plants and symptoms. No specific antidote for this condition. Supportive therapy with glucose, B complex vitamins and liver extract favours recovery. Local treatment for skin lesion with emollients may be given. Exposure to sunlight should be avoided.

Mimosa poisoning (Anathottavadi). Toxic principle is Mimosin (hepatotoxic and nephrotoxic). Accidental ingestion of this plant may induce perineal, jowl, neck and sternal edema, anuria and constipation followed by diarrhoea. Gel like kidneys at post mortem is diagnostic. Symptomatic treatment with diuretics, antibiotics, and small doses of calcium give satisfactory result.

20.7.4. Zoo toxins (Animal bite and stings)

Disease caused by zoo toxins is limited to bites of snake, bee stings, tick bite and ingestion of preformed toxins like canthardin.

Snake bite. The common venomous snakes include Cobra, King cobra, Krait, Russell's viper and Saw-scaled viper.

Elapidae snake – Cobra and Krait etc.
Viperidae – Russel's viper, Saw scaled viper.
Toxic fractions – Necrotizing, anti coagulant fractions, Neurotoxic, cardiotoxic, haemolytic fractions.

Normally there is insufficient toxins to cause death in large animals but secondary bacterial infection may result in local swelling and death of animals. Morbidity rate is not high but mortality rate can go up to 20% in farm animals. Most of the snake bites cases occur during summer months and animal usually tend to be resistant because of large size (large doses of toxins are required to produce toxicity). The degree of envenomation depends on size and species of snake, size of the bitten animal, thickness of hair coat, location of bite and quantity of s/c fat.

Bite by adder-type causes painful local swelling, which may be sufficient to produce excitement and anxiety. Swelling of the head may be sufficient to produce dyspnoea. In some animals excitement may be followed by dilatation of pupil, salivation, tetany, depression, recumbency and terminal paralysis.

Excitement with convulsions, and death due to asphyxia are the major signs of bites of cobra-type snakes. Treatment includes the application of tourniquet above the bite to restrict the circulation and tourniquet should be released for a few minutes at 20 min intervals. Systemic treatment should include antivenin (effective against Cobra, Krait, Russel's viper and Saw scaled viper), antibiotics, antitoxin, anti inflammatory and analgesics. A portion of antivenin should be injected at the site of bite. Administration of fluids and sedatives are essential to counteract shock and pain.

Bee Stings (Apis mellifera). Toxic principle includes Apamine, melletin, and histamine. Multiple stings by bees may cause local swelling similar to angioedema especially in lips, eyelids, tongue and vulva. Swelling of head region may be sufficient to cause dyspnoea. Treatment includes local application of a weak solution of ammonia or sodium bicarbonate, nervous system stimulant if prostration is severe and tracheotomy if asphyxia threatens.

Tick Paralysis. The paralysis characteristic of the disease is caused by a toxins secreted by the salivary glands of female ticks. The severity of paralysis is independent of the number of ticks.

The toxins of *Dermacentor andersoni* affect liberation or synthesis of acetyl-choline at the motor end plates of muscle fibers, resulting in paralysis of peripheral neurons. The affected animals exhibit ascending, flaccid paralysis commencing with incoordination of the hind limb, followed by paralysis of the forelimb and chest muscles. The respiration is grossly affected and death due to respiratory failure. Removal of ticks during the early stage of the diseases may result in rapid recovery. Use of suitable acaricides is an effective prevention.

20.8. COMMON DIGESTIVE DISORDERS

The primary functions of the alimentary tract are the prehension, mastication, digestion and absorption of food and water and maintenance of the internal environment. Correspondingly there are four major modes of alimentary dysfunction *viz.* abnormality of motility, secretion, digestion and absorption. Major clinical signs of digestive dysfunction are inanition, salivation, vomiting, regurgitation, diarrhoea, constipation, alimentary tract haemorrhage, abdominal pain, tenesmus, abdominal distension, shock and dehydration. More than 60-70% of the cases are abdominal disorders. Many of these diseases, adversely affect the production of animals and hence the economy of farmers.

Major principles of treatment includes :

- Relief of abdominal pain.
- Relief of distension.
- Correction of abnormal motility.
- Correction of fluids and electrolytes.
- Reconstitution of rumen flora.
- Correction of acidity and alkalinity.

Major diseases includes stomatitis, oesophageal obstruction, acute lactic acidosis, bloat, vagus indigestion, etc.

20.8.1. Choke or Oesophageal Obstruction

Oesophageal obstruction may be acute or chronic and is characterized by inability to swallow, regurgitation of feed and water, drooling of saliva and bloat.

Obstruction may be intra luminal by swallowed material (potatoes, jack fruit waste) or extra luminal due to pressure on the oesophagus by surrounding organs, TB nodule, cervical or mediastinal abscess, thymoma etc. Oesophageal paralysis also results in oesophageal obstruction.

Obstruction is usually in the cervical oesophagus just above the larynx or at the thoracic inlet or at the base of the heart or the cardia.

Animal suddenly stops eating and shows anxiety and restlessness. Animal shows forceful attempt to swallow, regurgitate, salivation, coughing and continuous chewing movements. If obstruction is complete, severe bloating occurs.

- If animal is severely excited, the sedatives or analgesics should be administered before proceeding with specific treatment.
- Pass a stomach tube and allow the object to move into stomach/rumen.
- Manual removal through oral cavity is possible if the obstruction is in the upper oesophagus.
- Surgical removal (oesophagotomy).

20.8.2. Indigestion

It is one of the most common diseases encountered in dairy cattle and stall-fed beef cattle due to dietary abnormalities of minor degree like feeding indigestible roughage, moldy, over heated and frosted feeds and moderate excess of grain and concentrate intake. Sudden change of feed may also results in indigestion. Dietary changes may change the pH of ruminal content which markedly affect the motility of rumen (atony of rumen). High protein diet also depress the motility because of sudden increase in alkalinity of ruminal content. Simple accumulation of indigestible feed, physically interfere with ruminal motility.

Clinical findings include reduction in appetite, slight drop in milk production, mild depression, dullness and absence of rumination. There may be moderate tympany, faeces are usually reduced in quantity initially and there will be constipation followed by diarrhoea within 24-48 hours. There is no systemic involvement, temperature, pulse and respirations are usually within normal range.

Most cases recover spontaneously or with simple treatments like use of rumenotonics to stimulate reticuloruminal motility and appetite. Parasympathomimetic drugs like neostigmine @ 2.5 mg/45 kg body weight also have some beneficial effect. Metoclopromide is also used to increase ruminal contractions. Epsom salt is reasonably effective. Depending on the pH of r uminal content acidifying agents (Acetic acid) or alkalinizing agents (Mag. hydroxide can be used. Reconstitution of the flora by the use of cud transplantation from normal cow is highly effective.

20.8.3. Ruminal Lactic Acidosis (Rumen overload, acute carbohydrate engorgement of ruminants)

Ingestion of large amount of highly fermentable carbohydrate rich-feed such as grain, jack fruit, mangoes etc. without proper adaptation is the most common cause of this disease. Subacute ruminal acidosis is considered as an important problem in dairy herds. Morbidity will vary from 10-50% but the fatality rate may be upto 90% in untreated cases, while in treated case it may be upto 30-40%.

Ingestion of highly fermentable carbohydrates by ruminants is followed by marked change in microbial population in the rumen. There is an increase in the number of *S. bovis* which utilize the carbohydrate to produce large quantity of lactic acid which decrease rumen pH to 5 or less result in destruction of cellulolytic bacteria and protozoa. This low pH allows *lactobacilli* to utilize the carbohydrate in the rumen to produce lactic acid resulting in ruminal lactic acidosis. Lactic acid increases the osmolarity of ruminal content and water is drawn from systemic circulation causing haemoconcentration and dehydration. High concentration of lactic acid in the rumen cause chemical ruminitis and further it predisposes to mycotic ruminitis. Chemical ruminitis favours the growth of Fusobacterium and *Areanobacter pyogens* and directly enter into ruminal vessels and spread to liver where it produce hepatic abscesses. Other toxic substances like histamine, endotoxin, ethanol and methanol have been found in the rumen of experimentally engorged cattle.

Clinical signs include anorexia, depression, dehydration, ruminal stasis, profuse diarrhoea with sweet sour odour, weakness and ataxia, leading to recumbency. Rumen may or may not be full but atonic and fluid splashing sound audible on ballottment. Temperature usually subnormal (98-101°F) and heart rate may be increased. The prognosis is better in those cases with heart rates below 100/min. The respirations are usually shallow and increased (60-90/min). pH of rumen fluid is below 5. Severely affected animals have staggery drunken gait and their eye sight is impaired. Refluxes are usually sluggish. A rapid onset of recumbency suggests an unfavourable prognosis and death may occur within 24-72 hours. Laminitis and mycotic ruminitis are major complications in recovered animals.

Ruminal fluid pH below 5 is diagnostic confirmation. This condition should be differentiated from simple indigestion, parturient hypocalcaemia, peracute coliform mastitis and acute diffuse peritonitis.

The principles of treatment includes :

1. Correct the ruminal and systemic acidosis and prevent further production of lactic acid.
2. Restore fluid and electrolyte balance and maintain circulating blood volume.
3. Restore forestomach and intestinal motility.

When cattle are found engorging, the following procedures are recommended.

1. Prevent further access to feed.
2. Do not provide water for 12-24 hours.
3. Provide good quality palatable hay.
4. Provide adequate exercise to encourage movement of ingesta through the digestive tract.

In severe cases with ruminal distension, a heart rate of 110-130/min and a rumen pH of 5 or below 5, rumenotomy is the best course of action. Systemic acidosis can be treated with intravenous administration of a solution of 5% sodium bicarbonate @ 5 l/450 kg body weight, followed by isotonic sodium bicarbonate (1.3%) at 150 ml/kg body weight over the next 6-12 hours.

- Rumen lavage with warm water.
- Intra-ruminal administration of alkalinizing agents – magnesium hydroxide @ 500 g/450 kg body weight. This is useful in moderately affected cows.
- Ancillary treatment include antihistaminics, NSAID (shock), thiamine to promote metabolism of lactic acid and parasympathomimetic to stimulate gut motility. Control is mainly by prevention of accidental access to grain gradual introduction of high-level grain diets and careful feeding and management of dairy cattle during late pregnancy and early lactation. Use of ionophores (salinomycin, monensin, lasalociad) in feed alter rumen metabolism and control ruminal acidosis.

20.8.4. Ruminal Tympany (Bloat)

Ruminal tympany is abnormal distension of the rumen and reticulum caused by excessive retention of the gases of fermentation either in the form of a persistent foam mixed with rumen contents or free gas separated from the ingesta. Primary ruminal tympany (frothy bloat) is a major problem in cattle pastured on bloating forages (legumes) and feedlot cattle fed on high grain rations. Bloating forages are most dangerous in prebloom stage. Feedlot bloat is common when feed contains 80% or more grain and ground fine.

Secondary ruminal tympany (free gas bloat) occurs usually in animals due to interference with eructation because of physical obstruction of oesophagus or eructation mechanism as in reticular adhesions.

In frothy bloat, the gas bubbles remain dispersed throughout the rumen contents increasing the volume of ruminoreticular contents and inhibit eructation while in free-gas bloat gas bubbles coalesce and separate from rumen fluid but the animal cannot eructate the free gas because of abnormalities of reticulorumen or oesophagus.

Bloat is a common cause of sudden death (found dead) in cattle. In primary pasture bloat, obvious distension of the rumen occurs quickly, and more on the upper left paralumbar fossa, but in some cases the entire abdomen is enlarged. Frequent lying down and getting up and kicking at the belly is also observed. Dyspnoea is marked and accompanied by mouth breathing, protrusion of the tongue, salivation, and extension of the head. The skin over the left flank is very tense and cannot be grasped. Passage of stomach tube in frothy bloat reveal froth and failure to release significant amount of gas while in secondary free gas bloat, large quantities of gas released with ease. In severe cases animal may die in a few hours. Excessive quantity of froth or free gas in the rumen is diagnostic confirmation. The approach to treatment depends on the circumstances in which bloat occurs, whether the bloat is frothy or free gas and whether it is life threatening.

- Animal should be removed immediately from source of bloating pasture.
- Emergency rumenotomy may be done in severe cases.
- In less severe cases passage of stomach tube or trocar and canula to release rumen gas.
- The owners may be advised to tie a stick in the mouth like a bit on horse bridle to promote salivation.
- Antifoaming agents—A non toxic mineral oil or synthetic surfactants—polaxalene 25-50 g may be given orally.

Control is mainly by :

- Management strategies to reduce rate of fermentation.
- Use of grass-legume mixture at the rate of 50:50.
- Provide forage supplements prior to grazing.

20.8.5. Rumen Alkalosis (Alkaline indigestion)

Alkaline indigestion is mainly due to increase alkalinization of ruminal content associated with high protein feeding, urea poisoning and ingestion of plants containing high nitrate content. Clinically the condition is characterized by depressed appetite, decreased rumination, recurrent tympany, diarrhoea and staggering gait. The pH of rumen fluid is usually between 7 and 8.5 and have distinct smell of ammonia. Treatment includes oral administration of kitchen vinegar (2.5% acetic acid) @ 1-2 lit/animal or lactic acid 50-70 ml diluted with 8-10 lit water and change the diet to readily fermentable carbohydrate.

20.8.6. Indigestion in Calves Fed Milk Replacers (Ruminal drinkers)

It is a form of indigestion occuring in veal calves and is characterized clinically by recurrent ruminal tympany, inappetence, unthriftiness and production of clay-like faeces. Insufficient closure of reticular groove while drinking milk is the major cause of this condition. As a result milk directly enter into rumen, instead of flowing directly into abomasum. The pH of rumen decreases and lactate concentration increases rapidly.

On clinical examination, temperature, heart rate and respiratory rates are within normal range. Increased size of ventral half of abdomen, Ballottement of the left abdominal wall reveals fluid splashing sounds. Large volumes of foul or acid smelling, grayish white fluid can be siphoned off from the rumen. Presence of casein clot in the rumen content is diagnostic.

Clinical recovery occurs within several days after returning to normal feeding practices or weaning of calf to hay and concentrates return the calf to normal very quickly.

20.9. HERD HEALTH PROGRAMMES

The age-old dictum *"An ounce of prevention is worth a pound of cure"*, is the watchword for herd health management.

20.9.1. Objective

The objective of herd health programme is to minimise non-genetic culling and mortality and economic losses due to diseases. This should be achieved through a herd health programme centred on systematic prevention of diseases and other health problems rather than the sporadic treatment of various conditions. This needs the co-operation of an efficient dairy manager and a veterinarian. The manager should make daily observations of his herd, keep up-to-date health records feed and manage well and provide a sanitary and healthy environment in the cowshed. Appropriate vaccination programmes, prompt and correct diagnosis and treatment, pregnancy test etc. are the responsibilities of the veterinarian. Effectiveness of a programme should be evaluated in measurable terms and necessary alterations should be made if deficiencies exist. Some targets for achievement in herds in developing countries are given below.

1. Calf mortality—below 8 per cent.
2. Adult mortality—below 3 per cent.
3. Non-genetic culling—below 15 per cent.

20.9.2. Health-Care Programmes

Depending on the circumstances, health-care programmes vary from farm to farm. However, the basic principles remain the same. They comprise of the following practices.

Disease prevention

(*a*) Preventing the animals from getting exposed to infectious organisms. Good sanitation and cleanliness, quarantine of incoming animals and eradication of certain diseases etc. are important steps in this direction.

(*b*) Maintaining a high degree of resistance.

(*c*) Vaccination against diseases for which effective vaccines are available.

(*d*) Providing adequate nutrition, and

(*e*) Ensuring a comfortable environment.

Reducing the spread of diseases that do occur by

(*a*) Quick accurate diagnosis and prompt treatment.

(*b*) Isolation of sick and incontact animals.

(*c*) Elimination of carriers.

(*d*) Proper disposal of carcasses, and

(*e*) Control of vectors.

Maintaining and using an accurate health record system. Good records are useful in the diagnosis of problems, help alert herdsmen to potential problems (prolapse of the uterus, milk fever etc. have high probability of repeating) and ensure that important details like vaccinations are not overlooked. Some of these aspects are dealt in more detail below.

20.9.3. Quarantine

Quarantine means the segregation of apparently healthy animals (especially animals being introduced into the herd or a country for the first time), which have been expose to the risk of infection from those animals, which are healthy and unexposed to such risk.

The idea is to allow sufficient time, for any contagious disease that the quarantined animals may have in the incubation stage, to become active and obvious. Hence the quarantine period depends on the incubation period of the disease. In practice a span of 30 days has been generally accepted as a reasonable period. But in the case of certain diseases this period is more. For example, for rabies this is 6 months.

During quarantine, animals should be screened for internal parasites by faecal examination and deworming carried out on the 23rd/24th day, if need be. They should be dipped/sprayed on the 25th/26th day to remove ectoparasites.

20.9.4. Isolation

Isolation means segregation of affected and in-contact animals in the event of outbreak of a contagious disease from the apparently healthy ones. Preferably such segregated animals should be housed in a separate isolation ward situated far away from the normal animal houses. If a separate shed is not available, the animals for isolation should be tied at one end of the shed as far away from the apparently healthy stock as possible.

Attendants and equipments for sick animals should be, ideally, separate. If due to practical reasons this is not possible, the sick animals should be attended to daily only after the healthy stock has been tended to. The equipments should be thoroughly disinfected after use in the isolation group. The attendant should wash his hands, feet and gumboots in antiseptic lotion and change his clothes. The isolated animals are brought back to the healthy herd only after they are fully recovered and the chances of passing on infection removed.

20.9.5. Vaccination

Vaccination is a procedure for artificially inducing active immunity in animals against specific infectious diseases by introducing biological agents called vaccines into their systems. The term vaccine denotes an antigenic substance from the particular micro-organism which when introduced into the animal system produces an immunological reaction producing antibodies which combat an attack from that organism at a later date.

This may consist of live-attenuated or killed virus or bacteria or their metabolic products.

After vaccination it takes some time for the antibodies to appear in full strength. During the interval, there is often increased susceptibility to the disease. The duration for which immunity

TABLE 20.2. Chart Showing Programme for Vaccination in a Bovine Dairy Farm and Nature of Immunity

Name of the disease	*Type of vaccine*	*Time of vaccination*	*Duration of immunity imparted*
1. Foot and mouth disease	Polyvalent tissue culture vaccine.	At about 6 months of age with booster dose 4 months later, thereafter yearly once (September-October).	Oneseason
2. Rinderpest	Tissue culture vaccine or lapinised avianised vaccine for exotic and crossbred cattle, caprinised vaccine for zebu.	Once in 3 years.	3-4 years
3. Anthrax	Spore vaccine.	Once in an year (Pre-monsoon period).	One season
4. Haemorrhagic septicaemia	Oil adjuvant vaccine.	Once in an year (Pre-monsoon period).	One season
5. Black quarter	Polyvalent vaccine.	Once in an year (Pre-monsoon).	One season
6. Tuberculosis	BCG vaccine.	At about 6 months of age; to be repeated every 2-3 years.	One to 2 years

is imparted varies from vaccine to vaccine. Against certain diseases, one or more booster dosi are required for achieving immunity strong enough to combat the disease fully.

A vaccine is not used as a rule in a locality where an acute infection has already broken out because of the increased susceptibility and delay before protection is established. Vaccination is used to develop an immune zone all around an area of actual infection.

This prevents the spread of the disease. Vaccination is also carried out routinely on animal farms as an insurance against outbreaks in the herd. A chart showing a programme of vaccination against important contagious diseases of cattle and buffaloes is given in Table 20.2.

Immune sera against certain bacterial and viral diseases are also available. They are prepared by repeated inoculation into suitable animals of increasing doses of live, attenuated or dead microorganism, which leads to the production of a high titre of antibodies against it. Serum of such animals is harvested and used to protect valuable animals in the face of an actual infection.

20.9.6. Elimination of Carriers

Some animals carry disease-producing organisms without any apparent symptoms and shed them in body secretions and excretions. They are known as 'carriers'. Generally convalescent animals are carriers for a short period. In some diseases, however, the carrier state may remain for years and the animals become potential danger to susceptible ones. Sometimes the animals may not become affected, the organism not assuming a pathogenic role. Yet, it remains a potential source of disease both to itself and to others. Such animals are referred to as 'healthy carriers'.

Common diseases for which carriers have been observed in bovines are brucellosis and tuberculosis. Carriers of disease in the herd should be detected and eliminated so that the herd may be completely free from the disease. Certain **diagnostic screening tests** can be used in the herd to detect carriers. The tests should be periodically conducted, carriers detected and removed from the herd.

Tuberculin test. Tuberculin consists of the toxic products of *Mycobacterium tuberculosis* grown in artificial culture. On intra-dermal injection of the material, allergic symptoms are set up and these constitute a 'reaction.' In healthy animals, tuberculin even in large doses gives no reaction. This is a reliable test for diagnosing non-clinical cases of tuberculosis in cattle and buffaloes.

Tuberculin test may be carried out half-yearly in the initial stages and later on, annually. January is the ideal month for conducting the test in India.

A small area of skin on the side of the neck is clipped and cleaned with spirit. A quantity of 0.1 ml of concentrated tuberculin is injected into the deeper layers of the skin with an ordinary hypodermic needle. If correctly done, a bead-like swelling is detectable by the finger. The first injection is known as a sensitising dose. The animal is examined 2 days later, and if there is a diffuse swelling that persists, it is regarded as a 'reactor' and no further injection is necessary. If there is no reaction after 48 hrs, a further dose of 0.1 ml is given near the first site in the same manner. Final examination is made on the third day and if at this time a diffuse swelling with evidence of tenderness is noticed, the animal is noted as a 'reactor'. If the results are inconclusive, the animal is marked as 'doubtful' and the test is repeated on a later date. All positive cases should be destroyed.

Agglutination Test for Brucellosis. This is a serological test based on the principle of antigen (dead bacteria) antibody (agglutinins present in the serum or milk whey) reaction, resulting in the agglutination of the bacterial cells in positive cases. When the agglutinins present in the serum and other body fluids of animals suffering with brucellosis or carriers of the same is added to a suspension of *Brucella abortus* organisms (dead) the latter will cluster together; the reaction being known as agglutination.

Healthy animals do not show such agglutination. By serially diluting the serum or milk whey and noting the maximum dilution at which agglutination takes place the strength of the antibody titre in the fluid can be assessed. A well established diagnostic laboratory is needed for carrying out the test. The farmer can only collect blood serum periodically and take it to the lab for the test.

Johnin test. Johnin is an extract of *Mycobacterium paratuberculosis* (Jhone's bacterium) prepared on the same lines as tuberculin. It is used to test bovines for Johne's disease.

Mastitis Control Programme

1. Ensure correct technique of milking.
2. Follow a proper milking routine that encourages maximum let-down with minimum teat and udder damage.
3. Routine use of screening tests—

   Strip-cup test comprises of letting the first few streams of milk from each quarter on to the black disc of strip cup. This will show up any clots which occur only in the fore-milk in mild cases.

   California mastitis test (CMT). *Addition* of an anionic detergent like alkyl sulphates or sulphonates and 'Teepol' to mastitis milk causes formation of typical gel streaks or clumps according to the degree of abnormality of milk. A plastic paddle with four separate wells for the four quarters is used for carrying out the test. By tilting the paddle to an almost vertical position, surplus milk is allowed to run over leaving about 2 ml each. To this is added, approximately about same quantity of the test fluid (like Mastaid solution of Glaxo) from a plastic container avoiding foam or bubbles. Rotating the paddle in a horizontal plane while reactions are noted immediately after the milk and fluid mix. Formation of typical streaks and clumps indicate mastitis. After cleaning with water the paddle can be immediately used for the next cow or buffalo without drying.

   Either strip-cup test or CMT should be used to screen all the cows in a herd at least once in a month and preferably more frequently.
4. Teat dipping after milking may be practised.
5. Drying up therapy with suitable antibiotics may be done.
6. When mastitis occurs, identify causative organisms and treat promptly.
7. Reduce udder edema by exercise and/or use of medicines.

20.9.7. Disposal of Carcasses

Prompt disposal of carcasses of animals died/destroyed by contagious diseases is of great importance in controlling them. There are some rules, which should be strictly adhered to—

1. Never throw a carcass into a stream or river or on the railway track;

2. Do not allow the carcass to remain for long in the shed as biting insects and rodents can reach it and spread the disease.
3. Unless approved by a veterinarian it is not safe to open carcasses of animals died of a contagious disease.
4. Disinfect the shed properly after the carcass is removed.
5. Dispose of the materials like dung and bedding of the animal, 5 cm soil from where the place the animal was lying (if not cemented) and left over feed along with the carcass.
6. The carcass should be carried in a trolly and never by dragging along the ground.
7. Burying and burning are the two efficient methods of carcass disposal.

Burial of carcasses. The most common method, this is reasonably safe if done deep enough and in soil from which there is no drainage to neighbouring places. Deep burial is necessary to prevent worms carrying bacterial spores to the surface as well as jackals and foxes from digging up and spreading the carcass. The topmost point of the carcass should be a minimum of 1.5 m below the level of the soil surface.

Drainage of water out of the burial place can be considered negligible if the general water level is at least 2.5 m below the ground. After the carcass is in the grave, its skin should be slashed at several places to discourage gravediggers to exhume the carcass for its skin.

Burning. The most sanitary method of disposal of carcasses is to incinerate them. Burning can be done by means of wood and kerosene oil. Site for burning having been decided upon, a cross-shaped trench about 0.5 m deep comprising in length and breadth of the carcass is dug. It is first filled with wood, some iron bars placed across it and the carcass placed over it. Some kerosene oil is poured over and the wood is put on fire. The carcass will be completely consumed and with it all the infective material. In towns and cities the so called carcass utilisation or rendering plants are available. After removal of the skins they are usually disinfected by immersion in a disinfecting solution. The remainder of the carcass is 'fired out' for its fat, which is used in soap industry. Farmers can inform these plants so that these utilization plants can collect the carcass.

20.9.8. Sanitation

It is an integral part of any successful health programme. The presence of voluminous dung, urine, spilt milk, discarded foremilk, uternine and nasal discharges, feed residues etc. cause sanitation problems. Proper cleaning removes most of the germs and parasites along with the dirt. The remaining germs are so few in number and weakened so that they are rendered practically harmless. Abundant supply of water under pressure is the easiest and most effective means of cleaning animal sheds.

Sunshine for disinfection. Direct rays from the sun especially the ultraviolet spectrum are one of the most effective germ killers while being entirely safe and inexpensive. The farm buildings should be so oriented as to take full advantage of this natural resource. Under ordinary conditions a north south orientation of the long axis of the building with open front and back can allow the entry of direct sunlight into the sheds in the mornings and evenings. The dairy utensils should also be kept in the sun for drying and disinfection after cleaning.

TABLE 20.2. Common Disinfectants and Their Usage

Name of Disinfectant	*Nature and Utility*	*Concentration required*	*Mode of application*	*Remarks*
Bleaching powder (Chloride of lime, calx chlorinate)	Although, a powerful disinfectant the value of bleaching powder and chlorine containing disinfectants is limited as they get rapidly inactivated in the presence of organic matter. It can be used for disinfection of animal houses when a contagious disease has occurred. It should not be used in milking barns as its strong odour may taint milk.	It should contain not less than 30 per cent available chlorine.	Dusting	Bleaching Powder; must be stored in air-tight bins as damp surroundings and exposure to light and air cause it to decompose rapidly.
Boric Acid	It is a weak antiseptic and is likely to harm nervous system if absorbed into body in large quantities. It is used as wash for eyes and other sensitive parts of body. Can be used as an udder wash.	6 per cent solution.	Splashing	Now-a-days antibiotic solutions are replacing boric acid as eyewash solution.
Caustic soda (Lye, Sodium hydroxide)	For general use in farm buildings and animal houses caustic soda is a very effective disinfectant as it is an excellent cleansing agent as well as a powerful germicide. It is highly destructive to virus of foot and mouth disease, hog cholera etc. It is not effective against tuberculosis and Johne's disease organisms.	2 per cent solution for general use and 5 per cent solution against spores of anthrax and black quarter.	Splashing	Rubber gloves, goggles and protective clothing should always be worn when caustic soda solution is being used as it burns skin and damages fabrics.
Cresols	The cresols are only slightly soluble in water and are therefore generally emulsified with soap. Effective against a wide range of organisms including acid-fast tuberculosis and Johne's disease bacteria, but not effective against viruses and spores. Good for disinfecting floors, walls, equipment etc. but not in milking barns because of its phenolic odour.	2-3 per cent	Splashing	Use correct strength and preferably hot solutions, whenever practicable. Use only soft water for preparing solutions as hard water precipitates soap. Lysol is a solution of cresol with soap.
Heat (Burning, boiling water or steam)	Heat, especially moist heat, of sufficient intensity kills several types of germs though some bacterial spores are resistant even to boiling water. Steam under pressure is most effective. Burning is used to destroy rubbish, infected useless articles, carcasses of infected animals etc. Boiling water, generally with some chemicals added to it, is used to disinfect barns and equipment. Steam is used to sterilize milk cans, milking pails and other dairy equipment.	Boiling water, steam at 15 lb pressure	Burning filth use of blow-lamps, splashing of hot water.	Sometimes use of heat may be impracticable.

Lime (Calcium oxide, burnt lime, quick lime)	It is a deodorant as well as a disinfectant. It can be used for sprinkling on manure and animal discharges, on floors or as a white wash of milk of lime (also known as slacked lime).		Sprinkling, scrubbing or sometime dusting	Always use freshly prepared lime only.
Phenol (Carbolic acid)	Phenol is effective against several types of bacteria but not so effective on spores and viruses. Its activity is not reduced by the presence of organic matter but oil or alcohol reduces its disinfectant power. It is very toxic, corrosive, and irritant at germicidal concentration. Phenol is not popular now a days as better compounds have become available.	1-2 per cent	Splashing	Great care should be taken in using phenol to protect eyes, skin and clothing.
Quaternary ammonium compounds (QAC) *Dairy* sanitizers	These are cationic detergents. They have no effect on spores and viruses. They can be used to disinfect dairy utensils, udders, milker's hands and towels for wiping udders. Cetrimide, a white powder is an example of quaternary ammonium compounds.	0.1 per cent solution (0.5 per cent cream, for applying on teats and hands to prevent mastitis)	Wiping of udder with clothes wetted in 0.1 per cent solution, washing of milkers' hands.	Utensils should be scrubbed with boiling water before rinsing with QAC.
Soap	Soap is an anionic detergent. It is a very weak germicide. But its great usefulness is in cleaning various surfaces including skin. It can be used preparatory to the application of a disinfectant.		Scrubbing	It should preferably be used only as a surface-sanitizing agent.
Sodium hypo-chlorite	It is a chlorine compound. It is an excellent disinfectant but is not effective against TB bacteria and its effectiveness is reduced by presence of organic matter. Its uses are the same as that of quatenary ammonium compounds.	200 parts per million of available chlorine. About 300 ml sod. hypo chlorite and about 200 g of washing soda in 100 litres of hot water for washing utensils etc. For udder wash-about 60 ml in 10 litres of clean water.	Rinsing of utensils, wiping of udder. Scrubbing	It should be stored in airtight containers as hypochlorites deteriorate rapidly when exposed to air.
Washing soda (Soda ash, sodium carbonate)	It is good for disinfection of barn premises upon which an outbreak of virus diseases like foot-and-mouth disease has occurred. It is a good detergent too.	4 per cent solution.		Lye is better against foot-and-mouth disease virus than soda ash. It should be used as a hot solution.

Heat. Application of heat by steam, boiling water, by burning etc. are effective methods of disinfection, though, sometimes it is not practicable to use heat.

Chemical disinfectants. The nature, strength, utility and specificity of some of the commonly used disinfectants are given in Table 20.3. It must be remembered that not all pathogenic organisms are destroyed by the same disinfectant. Besides, effectiveness of chemical disinfectants is greatly reduced in the presence of organic matter. Therefore, a thorough cleaning of the area should be done before applying the disinfectant.

Disinfection of cattle and buffalo sheds. Under normal circumstances, daily scrubbing and washing of the floors, water troughs and mangers and the action of sunlight falling in the houses are sufficient enough to keep them moderately germ free. This should be supplemented with periodic whitewashing of the walls, mangers and water troughs.

Emergency Steps. When there is a disease outbreak, a more elaborate procedure of disinfection should be followed. All floors and walls up to the height of 1.5 m, interior of water troughs and mangers and all other fittings and equipments coming in contact with animals should be disinfected. The first step is to remove all the filth and feed residue.

All dung, litter, feed residue should be removed and stacked so that the heat generated inside kills the germs. The floors, walls up to 1.5 m height, interior of mangers and water troughs etc., are thoroughly scrubbed and washed with water. In case of outbreak of anthrax, the dung, litter etc. should first be disinfected *in situ* by thorough sprinkling of suitable disinfectant. If the floor is of earth, the top 10 cm earth should be removed and disposed of along with litter.

After removal of filth, the place should be scrubbed and washed with 4 per cent hot washing soda solution (*i.e.,* 4 kg washing soda in 100 1 of boiling water). The approved disinfectant solution should be liberally sprinkled or sprayed and left so as to act for 24 hours. After this, the shed is again washed with clean water and left to dry with wind and sunlight. The interiors of water troughs, mangers etc. should be whitewashed. The house is then fit for moving in the animals.

20.9.9. Disinfection of pastures

Use of chemical disinfectants on pastures is impracticable. What can be done is to remove all infective material like carcass, aborted foetus dung etc. from over the pasture and preventing animals from grazing an infected pasture for a period of 3 to 4 months. Pastures should be well drained and free from slushy patches. Internal parasites flourish in pastures that are damp the year round. Any dangerous area, whether from infective materials, toxic soil or poisonous plants should be fenced off immediately and corrective step taken urgently. In severe cases of contamination with disease germs, pasture should be ploughed up and left fallow for about 6 months during which period the pathogens would be destroyed by sun and wind.

FURTHER READING

Anon. 1981. *Handbook of Animal Husbandry,* 2nd Ed., I.C.A.R., New Delhi.

Blood, DC, Henderson. J.A. Radostitis. DM., Arundelo. J.H. and Gay, C.C. 1983. *Veterinary Medicine.,* 6th Ed. E.L.B.S. & Bailliere Tindall Cox.

Bose, A.M. and Thomas, C.K. 1979. Effect of individual calf pens and antibiotic feed additives on the performance of buffalo calves. Indian J. Anim. Sci. 49 : 419.

Clark, J.A. (Ed.) 1980. *Environmental aspects of housing for animal production.* Butterworths. London.

Esmay, M.L. 1969. *Principles of animal environment.* The AVI Publishing Co., Connecticut.

Hahn, G. Le Roy. 1981. Housing and management to reduce climatic impacts on livestock. *J. Anim, Sci.,* **52** : 175.

Hall, H.T.B. 1977. *Diseases and parasites of livestock in the tropics.* Longman group. London.

Jensen, R. and Mackey, D.R. 1971. *Diseases of feedlot cattle. 2nd* Ed., Lea & Febiger, Philadelphia.

Kooner, D.S., Murthy. V.V.N. and Sundaresan, D. 1969. *Dairy cattle housing.,* Bull. No. 21. P.A.U., Ludhiana.

Lee, D.H.K. 1953. *Physiological objectives in hot weather housing.* U.S. Govt. printing office, Washington. D.C.

Parker, W.W. 1978. *Health and disease in farm animals.* 2nd Ed., Pergmon Press, New York.

Ravindran Nair (Ed.) 2002. Current Indian Veterinary Index 3 : 1. cyberi@communications.aluva.

Ristic. M. and Mclntyre, I. 1981. *Diseases of cattle in the tropics.* Marins Nijhoff Publishers, The Hague.

Sainsbury, D. 1967. *Animal health and housing.* Williams and Wilkins Co., Baltimore.

Sastry, N.S.R. and Georgie, G.C. 1985. A review of studies on housing buffaloes in semi-arid region of India. *Indian J. Anim. Sci.* **55** (7) : 553.

Scorgie, N.J. and Wills. G.A. 1965. *Linton's veterinary hygiene,* 4th Ed., Scientific Book Agency.

Winkler, J.K. 1982. *Farm animal health and disease control.* 2nd Ed., Lea & Febiger, Philadelphia.

Section VII : HERD MANAGEMENT

CHAPTER 21

Behaviour of Bovines

The sum total of the overt and composite functioning of animals individually and in groups is termed *animal behaviour.* The science dealing with animal behaviour is termed as *animal ethology*.

The farm animal scientists of today must have a working knowledge of behaviour of the species he is dealing with. It has become increasingly clear that there are many ways in which ethology can help with the welfare of animals and thereby ensuring more profit from more sensible husbandry. Ethological studies can bring out the importance in the lives of domestic animals, of such factors as social companionship, dominance by other animals, regular feeding and resting regimes and the rhythms of other activities such as sleeping and breeding.

21.1. MOTIVATION OF BEHAVIOUR

Behaviour is dictated by the motivation most of which are derived from internal factors. Some of these factors are best appreciated as drives.

Behaviour is basically inherited through genes and modified through the environment. Genes express through the nervous and endocrinological systems to manifest basic behaviour patterns of the species. Conditioned reflexes and learning and other modifying effects of the environment react with the basic genetic behavioural 'core' to manifest different patterns in each individual animal. There are circumstances in which an animal has its behaviour dictated by conflicting factors. Most often, a group of factors are responsible for the behaviour of an animal at a time.

21.2. THEORETICAL CONCEPTS

Many concepts of a hypothetical nature have been postulated to explain the basis of behaviour. It is very difficult to have one theory to explain the entire process of behaviour. At the same time, it is possible to see a common basis in many units of behavioural patterns. The obvious Inter-relationship between these different facts of behavioural expression encourage the postulation of the various concepts which together become the theoretical basis of ethology. These theories help us in making a systematic scientific study of animal behaviour.

The concept of instinct deals with that element of behaviour which has been inherited. Instinctive behaviour is to help animals adapt to an environment. With this, the animal can produce a large repertoire of adaptive responses without prior learning or experience. Although amenable to some degree of environmental modification during the life time of the animal, in the main, instinctive behaviour manifests itself as a cascade of pre-set stereotyped behavioural patterns. These are set in motion by proper environmental stimuli (releaser) leading to adaptation of the individual to the environment concerned.

Another set of behavioural activities are termed *displacement activities*. When an animal is prevented to express normally an instinctive behaviour which has been already stimulated, it expresses itself in some other form. When equilibrium is maintained by two conflicting drives sufficiently long enough, the displacement activity will be organised. Some of these activities like weaving in horses are irrelevant behavioural activities serving no specific purpose than spending the energy liberated by stimulation of an instinctive behaviour which has been prevented from expressing itself due to factors like domestication and confinement.

21.2.1. Drives

Certain specific motivations which are the result of many innate forces like heredity, endocrine function, experience etc. and external stimuli are termed drives. Motivations have specific directions and goals. The term drive exists only as a concept and it has no substance. The well recognised drives are sexual, feeding and maternal ones. In addition, certain minor drives like play and exploration are also known.

The abstract idea of 'drives' can be explained more precisely through the neuroendocrinological mechanisms controlling major behavioural main and side 'currents'.

21.2.2. Imprinting and Learning

A neonatal and early infancy animal has a low threshold for learning and therefore 'impressionable'. When such an animal is subjected to a meaningful stimulus, an indelible impression is left in the brain of the animal known as 'imprinting'. It involves the recognition of its mother and consequent behavioural attachment to it. It is similar to learning, but differs from it in that it is very rapid and very stable and can remain powerful even without repeated reinforcements. The infant animal learns the major behavioural characteristics of its species as well as those of its mother through imprinting.

Animals do learn through experience. As they advance in age, learned responses come to dominate the earlier unlearned ones. Early learning is through a process of exploration and trial and error. Animals also learn by watching others and imitating them. Feeding behaviour, display of estrus etc. are learned and improved by mimesis (imitation). Repetition is the core to learning as it reinforces the impression retained in the brain.

Conditioned reflexes are a type of learning process. In this the natural stimulus which elicits a reflex action is replaced by one or more conditioned stimuli. An example is the stimutus of massaging the udder (conditioned stimulus) that can replace the natural (unconditioned) stimulus of suckling in producing let-down of milk.

In certain circumstances an animal may get conditioned not to make a response known as *inhibition*. Any faulty technique of A.I., if persisted with, can produce such an inhibition in breeding bulls.

21.2.3. Stress and Strain

Behavioural change is one of the major identifiable evidence of strain in farm animals. When an animal fails to adapt to an environment, abnormalities in behaviour manifest. Whenever an animal is subjected to an external *stress* impinging on the complex adaptive apparatus of the animal, the animal system as a whole experience *strain*. Many environmental factors like unfavourable ambient temperature, husbandry conditions, food supply, stock density, space allocation and confinement are stressors that can easily defy adaptation and create *strain* in an animal. These strains can produce distortions in existing behavioural manifestations or result in generation of new behaviour patterns. Thus study of farm animal behaviour in relation to stress is important.

21.3. PHYSIOLOGICAL BASIS OF BEHAVIOUR

The senses—seeing, hearing, touching, feeling, tasting and smelling—are the windows of the animal organism to the world. The respective sensory organs feed the brain—the phenomenal biological computer—various types of information about the environment. The brain integrates, interprets and mobilizes through its motor and endocrine systems, various responses to adapt, adjust and combat each situation according to its merit and the innate wisdom of the organism. The expressions of the sensory and endocrine systems result in behavioural manifestations. Behaviour, as has been defined earlier, should be seen as the overt and composite reaction of the whole animal to its environment. Thus the neuro-endocrine system consisting of brain, spinal cord, peripheral nerves and the endocrine glands, is primarily responsible for the production and control of behaviour.

The extro-receptors and intro-receptors convey a received stimulus to the brain through peripheral nerves by producing excitation. In many cases the neural paths lead to the cerebral cortex. Cerebral cortex has four main sensory areas which receive and interpret nerve impulses. These are :

1. The visual area.
2. The auditory area.
3. The olfactory area, and
4. The somasthetic or body sense area.

These specialised areas of cerebral cortex are primary sites for sensory reception.

The hypothalamus is a major site of behavioural control. It is a cross-road where there is convergence of many stimuli. Various parts of the fore-brain convey neural activities to the hypothalamus, the cerebral cortex interrelating closely with it. Stimuli also arrive through the spinal nerves. Information brought to it through circulation also influence the hypothalamus.

It is at the level of hypothalamus that nerve impulses related to adaptive mechanisms get integrated and controlled. The neural links involving the sub cortex and the hypothalamus and the other parts of the brain form the important complex that controls behavioural patterns. In this, hypothalamus plays a central role.

Once the *nuclei* (functional unit in hypothalamus) receives impulses and chemical signals through various pathways, it processes the information and issues appropriate nervous or endocrine signals to more specialised parts of the body which are geared to function under the control of hypothalamus.

Thus, the behaviour of the animal is the result of its interaction with the environment—the attempt of the animal to adjust to its environment.

21.4. MAINTENANCE BEHAVIOUR

Self preservation is the most primary and innate behaviour exhibited by all living organisms. In large animals behaviour aimed at self-preservation or self-maintenance is mostly expressed instinctively.

21.4.1. Feeding Behaviour

Feeding is one of the basic behavioural expressions needed for self preservation. Animals exhibit varying behaviour with respect to eating and drinking, grazing, selectivity of food, diurnal timing of ingestion etc. Bovines are ruminants and depend on ingestion of bulky feeds like grass for their nutrition. The nutritional needs of the body forms the motive force behind the ingestive drive. Considerable proportion of the behaviour associated with this drive appears to be set out in the genetic code of the species. In other words, heredity determines a large part of the ingestive behaviour. The species specific feeding behaviour controlled by continuous specific neural signals is innately programmed. Repeated experiences of similar sort results in learning, and the superimposed effect of learning results in formation of feeding habits. Management systems play an important role in this.

21.4.1.1. *Feed Intake*

Feed intake is basically controlled by hunger and thirst and are controlled by the respective centres located in the hypothalamus. But when sight of the feed, its flavour, memory of previous experience, texture of the feed and similar qualitative aspects of the feed and their effect on appetite is taken into account, we can realise an existence of a comprehensive feeding drive organisation with cortical contribution.

Feed intake is used by the bovines in maintaining homeostasis. In order to maintain ingestive energy equilibrium, the input of caloric-fuel should equal internal heat production, external work production (behaviour), production of energy reserves (fat deposition) and production of growth and products like milk. Appetite is the major controlling form which helps the organism to maintain a stable or favourable energy balance.

Food seeking behaviour like grazing, feed selection and associated ingestive behaviour are controlled by the efferent impulses from a feeding centre located in the lateral hypothalamus. Another set of neurons in the ventromedial hypothalamus forms the satiety centre. Depending on external stimuli, the satiety centre inhibits the lateral centres for promoting feed intake. Food intake by cattle is basically a maintenance behaviour which will continue uninhibited unless turned off on the command from satiety centre. The feeding centre is stimulated by afferent impulses originating from the animals' bodily needs for nutrients as well as from the sensory stimulation provided by availability of food. There are certain special receptors in the central nervous system which detect blood-borne 'clues' of obtaining sufficient nutrients like glucose, lipids, metabolic heat etc. and trigger the satiety centre. Signals originating from gastrointestinal tract or a depletion-detecting system do not seem to play any important role in satiation. Animals eat when the satiety centre is not active.

Thus, the feed ingestion in farm animals appears to be controlled by a feed-back mechanism in which the level of certain metabolites in the blood circulating in the brain gives the significant information to the brain centres controlling appetite.

Productivity of an animal influences feed-intake. A high producing cow has greater drive to eat than a low producer. This productivity-linked feeding-drive-hypertrophy is superimposed over the need for homeostasis. Calves with good appetite become good growers and adult producers.

Similarly certain social factors also influence quantity of feed consumed. Calves reared in individual pens or hutches consume less than those reared in group.

Physiological drives alone do not control all about feeding behaviour. It is strongly influenced by environmental and social associations which cause positive or negative reinforcements in feeding behaviour. Palatability of feed is another factor influencing ingestion.

21.4.2. Appetite for Specific Nutrients

Experiments have been conducted to see whether the animals are able to correct specific nutrient deficiencies if they are given free access to it. It has been seen that cattle do not possess much ability to correct a specific mineral deficiency even if given free choice access to the minerals. Organic nutrients are also not consumed according to the need of the body. The major exceptions are water and salt. When these nutrients are deficient in the system, the animal experiences thirst and salt-appetite.

Thirst is controlled by a hypothalamic centre. Besides this, the regulation of water and salt is linked in urine production and is humorally controlled by the posterior pituitary hormone anti-diurectic hormone (ADH) and aldosterone produced in the adrenal cortex.

Dryness of mouth and throat and a kidney-produced hormone angiotensin control thirst. All mammals relish to eat salt probably liking the taste. Apart from this, there is a specific appetite for salt regulated by need.

21.4.3. Grazing

While grazing, cattle cover on an average 4 km per day. The distance travelled increases if the weather is hot or rainy or if they are troubled by flies. In hot arid regions, during summer, cattle shift from a predominantly day-time grazing to a night-time grazing. Grazing is seldom uniform. Ingestive activity has been found to have peaks during (*i*) early dawn, (*ii*) mid-morning (*iii*) early afternoon and (*iv*) near dusk. During summer, early dawn and near dusk are the major periods of ingestive activity.

In a day, cattle spend on an average 4 to 9 hours grazing (Figure 21.1). The cow is not equipped to graze very close to the ground. Generally, grass is snapped at a height more than one cm from the ground.

Cattle show a natural tendency to select certain plants and grasses and even these are eaten if they are not over-mature. This selectivity will result in irregular growth in pastures. Herbage with hairy and coarse appearance are generally avoided by bovines. They also avoid foliage contaminated by urine and dung. Taste is the major factor in deciding which plant to eat, but definitely, smell, sight and touch also play a role. Calves show the capacity to consume fairly large quantity of sour feed-stuffs which have acidity. But adult cattle prefer sweet to acidic and salty to sour feedstuffs.

Fig. 21.1. Grazing behaviour of cattle.

Dairy cows eat a large quantity of feed shortly after milking. If a choice between hay and silage is given, cows spend roughly 2/3 of their eating time with silage and 1/3 with hay.

21.4.4. Rumination

Cattle roughly spend 4 to 9 hours on rumination in between periods of ingestion. During the day, cattle ruminate 15 to 20 times. Bovines regurgitate, masticate well and reswallow food which they have previously ingested into the reticulo-rumen. Thus, the animals can quickly consume large volumes of forage from pastures where they are exposed to predators and inclement weather. Later on, they can move to sheltered areas, relax and ruminate leisurely. Bovines prefer to lie down and chew the cud. But, if the floor is wet or if exposed to rain, they may ruminate standing or walking slowly.

The period of rumination varies from a few minutes to even more than one hour. The peak rumination activity is immediately after night-fall.

When the fodder consumed is soft and of good quality, rumination time is shortened and animals spend more time ruminating when the fodder is rough and tough.

There are many factors which affect rumination like estrus, pain, maternal anxiety, hunger, labour pain or illness. There is a belief that the longer rumination is interrupted, the more difficult it will be for the animal to resume it.

21.4.5. Drinking Behaviour

Cattle usually drink from one to four times a day in temperate climate. The frequency is increased in hot weather. Daily water consumption depends on the following factors.

1. Body size of the animal.
2. Physiological condition—lactating animals require more water depending on the milk yield.
3. Season—during hot season cows and buffaloes consume more (20-25 per cent). water.
4. Nature of feed—animals receiving predominantly dry feeds require more water than those getting succulent fodders. Cows on an average require 45-50 litres and buffaloes 65-70 litres per day for drinking.

Cattle drink using their muzzles keeping the nostrils above the surface of water.

Higher intake of protein induces animals to drink more water. *Bos taurus* cattle are known to drink more water than *Bos indicus*.

21.4.6. Behaviour of Reactivity

This includes behaviours in reflex form. These responses are typically prompt and brief and form a first line of defence in the individual animal. These responses are quick but predictable. Therefore, a knowledge of these will be useful to herdsmen.

One of the important reactive behaviours is agonistic behaviour which embodies activity of fight and flight. It includes all forms of behaviour by animals which is in conflict with other animals. These include aggressive and passive behaviour.

21.4.7. Agonistic Behaviour

This is that type of behaviour encompassing all behavioural aspects of conflict including submission and evasion. Fighting between cows can cause physical injury and reduce milk yield because the submissive and meek cows do not get sufficient opportunities to feed properly when group-fed. Certain managemental failures can cause psychological disturbances to cows causing a reduction of upto 5 per cent in total yield.

Aggression in bovines usually has a set pattern. Most encounters have a sequence : *approach*, *threat* and *physical contact* (or fighting).

Approach. There are two types of approaches, passive and active. A passive approach should be more appropriately called a chance meeting. An active approach involves seeking out a particular opponent and engaging it.

Threat. When one bovine happens to have a chance meeting of a herd-mate, the latter poses a mild threat which often discourages the approaching animal from engaging in physical contact. The threat/submission encounter ends when one animal lowers its head and turns away. Often this is so subtle that it goes unnoticed.

Physical contact. If the animal threatened is slow to submit or if it fails to see, it may be butted usually at the rump or side. If the attacking animal has horns, the upswinging butt (or bunt) may cause serious injury to the recipient. A subordinate animal quickly retreats when butted or when butt is imminent.

In active approach, the animal seeks out its opponent and makes a deliberate threat. If the animal being threatened accepts the challenge by assuming a similar aggressive posture indicated by lowering of head, with horns pointed on the other and eyes crossed, fighting will take place. The fight posture of cows is similar to the 'fight or flight' posture of bulls. Bull in addition paws the ground, rubs its head and neck on the ground and ploughs up earth with its horns. The animals may stand at one or two metres displaying an aggressive posture for some time presumably playing a 'war-of-nerves' on each other.

In actual physical contact, (fighting) different variations described as below by Dickson *et al.* (1967) can be observed :

(*a*) ***Forceful.*** "Characterised by a threatening swing of the head in the direction of the subordinate animal, followed by submission or avoidance on the part of the recipient".

Dickson, D.P., Barr, G.R. and Wieckert, D.A. (1967). Social relationships of dairy cows in feed lot. *Behaviour*, **29** : 195.

(*b*) ***Bunting.*** "When a bovine uses its head to attack or physically displace another one" it is called bunting.

(*c*) ***Pushing.*** The dominant animal physically displacing another one by some part of the body other than the head.

(*d*) ***Non-forceful.*** When an animal was observed purposely avoiding another animal.

(*e*) ***Contact.*** Head to head combat between two bovines with no clear-cut winner.

21.4.8. Social Hierarchy

Based on the outcomes of agonistic encounters, *hierarchial system* or *dominance order* develops in a herd. In bovines the dominance order is relatively stable. In a study by Sambraus and Osterkorn (1973) mean reversal of ranking of 14 per cent per year had been observed. Thus while there is a basic stability of social hierarchy in a herd, there are also social dynamics leading to a gradual change.

The way in which this order is established and the criteria often varies. It has been observed that age, body weight, seniority within the herd, chest girth, height at the withers etc. have a consistent relationship with dominance.

It has been observed that animals close to each other in the dominance order fight more than those further apart probably because of the continual competition between two cows of similar aggressive ability and the social tension generated from it.

In bovines, various types of social dominance systems can be seen. In small closed herds, a type of *linear hierarchy* is seen where the hierarchy is continuous and linear; where each animal is subordinate to its superior and dominant over its inferior. In large herds, *complex hierarchies* can be seen.

21.4.8.1. *Dominance development*

In closed herds of cows, the major social dynamics is the upward mobility of the younger cows in social order and the social tensions resulting from it. Reinhardt and Reinhardt (1975) observed on inverted U-shaped relationship between dominance and age, cows gaining in dominance up to the age of about 9 years and declining thereafter. Challenging of established hierarchies came from younger herd members. In an open herd, where often new individuals meet for the first time, the social hierarchy is determined based on strength, weight and agility and a fighting spirit. In large herds, there is formation of sub-groups of dominance hierarchy. This may or may not indicate the extent of the animals' capacity for individual recognition.

Individual recognition is a must for maintenance of social hierarchies. If there is a limit to the capacity of cows for individual recognition, then herd sizes beyond that may lead to perpetuation of social tension. There are always possibilities of cows not recognising each other coming across and may have to decide their dominance through actual physical combat. Had it been a smaller herd, the animals would have recognised each other and the subordinate would have submitted to her without actual fighting. In fact, decrease in production per cow in herds of 350 or more, despite ample feed and water has been observed in New Zealand.

It has been observed that individual recognition is possible in non-visual conditions through

---

Sambraus, H.H. and Osterkorn, (1973). The social stability of a herd of cattle. *Z. Tierpsychol.* **35** : 418.

Reinhardt, V. and Reinhardt, A. (1975). Dynamics of social hierarchy in a dairy herd. *Z. Tierpsychol.* **38** : 315.

olfactory (smell) means. But, in competing for feed, the subordinate cows could not be held away from feed through smell or sight in the absence of physical contact.

21.4.9. Physical Factors in Dominance Behaviour

Certain physical attributes like age and indirectly body weight have been found to be correlated with social rank. Bouissou (1972) investigated the effect of body weight and presence of horns during hierarchy formation. The results are presented in Table 21.1.

TABLE 21.1. Influence of Body Weight and Horns in Heifers on Hierarchy Formation in the Herd

A. Dominance in pairs of heifers having weight or horn differences

Nature of difference	*Difference in the pairs*	*No. of pairs*	*No. of pairs in which the dominant is*	*Per cent*
Horns	With horns opposed to dehorned (Weight difference less than 25 kg)	40	The heifer with horns 34	85
Weight	Weight difference more than 25 kg Both horned.	20	The heavier 18	90
	Weight difference more than 25 kg Both dehorned.	21	14	62

B. Dominance in pairs of heifers having horns and weight difference

Difference in the pair	*No. of pairs*	*No. of pairs in which the dominant is*	*Per cent*
Weight difference more than 25 kg the heavier has horns, the lighter is dehorned.	40	The heavier with horns 39	97.9
Weight difference more than 25 kg the lighter has horns, the heavier is dehorned.	20	The lighter with horns 15	75

The above result goes on to show that both presence of horns and live-weight play an important role in earning social position in the agonistic hierarchy in bovines. Out of these, possession of horns is a better asset than body weight for winning encounters.

Aggressiveness and Temperament. Collis (1976) observed a significant correlation of 0.61 between aggressiveness (number of aggressive encounteres initiated) and social rank. In his study, body weight had only a correlation coefficient of 0.05 with social rank. This might be due to different stages of pregnancy among the cows in the study as pregnancy increases body weight, but makes the animal avoid agonistic encounters.

In an attempt to relate temperament with assumption of social rank, milkers' temperament score was studied, but no clear association between these could be established (Dickson *et al.* 1970).

Bouissou, M.F. (1972). Influence of body weight and presence of horns on social rank in domestic cattle. *Anim. Behav.* **29** : 474.

Collis, K.A. (1976). An investigation of factors related to the dominance order of a herd of dairy cows of similar age and breed. *Appl. Anim. Ethol*, **2** : 167-174.

Dickson, D.P. Barr, G.R. Johnson, L.P. and Wieckert, D.A. (1970). Social dominance and temperament of Holstein cows. *J. Dairy Sci.* **53** : 904.

21.4.9.1. *Hetero-sexual herds*

In free-range hetero-sexual herds of cattle, many hierarchies can be seen : among adults, among juveniles, among males and among females separately. All adult males dominate all adult females who in turn dominate the juveniles. But the juvenile males starting from the age of 18 months challenge the adult females and by 30 months completely dominate over them.

Gradually development of unisex encounters become the order. The least will be encounters by females on males followed by males on females. The maximum will be between members of the same sex.

21.4.10. Effect of the Social Hierarchy

When animals are allowed unrestricted physical contact, there is a high correlation between social rank and performance in food competition, utilisation of space and sexual priority. In ideal management systems, care must be taken to create settings which avoid unhealthy competition. For example, in the loose housing system, construction of simple barriers which discourage direct physical contact at the feed manger will be very useful.

Studies so far carried out have not shown any strong correlation between dominance rank and milk production especially when conditions are socially stable. Many of the dominant cows are heavy-weights which deposit most of the nutrients in their body and secrete less in milk. They spend more time near the food source, but do not have the physiological urge to consume more or the physiological set-up to produce more. On the other hand, a high producing cow may be lighter due to outgo of enormous quantities of nutrients through milk. This disadvantages her in social order and allows only limited time near the feed source. This disadvantage, the cow tries to overcome by brisk eating during the time available. The extent to which it is affected by the social order varies depending on the nature and quantity of feed available.

To say social order has no effect on animal production is contrary to facts. A high degree of correlation was not observed because there are many forces like production capacity of the cow, dominance order, time spent at feed manger, actual feed consumption etc. acting in different directions. It is common observation that timid high producers under conventional stanchion system go down in their producing ability when brought under the loose housing system. It is also observed that dividing the herd to sub-groups on the basis of body weight improves production when feeding is restricted

Learning. It has been observed that calves raised in individual pens fare badly when introduced into a loose housing group in comparison to group-raised ones.

In comparison to the socially experienced heifers, socially naive animals fought more, were less submissive at the first encounter and took more time to establish the social hierarchy. Thus, there is considerable evidence indicating that social skills necessary for dominance can be learned.

21.5. SEXUAL BEHAVIOUR

The sex drives in males and females are manifest in significantly different arrays of behavioural features. Essentially, the behavioural manifestation of sex depends upon the action of sex hormones on the central nervous system. There should be a level of the respective sex hormones to produce a sexual behavioural display. This is favoured by the presence of appropriate environmental stimulation.

21.5.1. In Bulls

Sex-drive in male farm animals is commonly known as *libido*. Testosterone, the male sex hormone produced by the testes is primarily responsible for libido. It develops at puberty and persists at a fairly constant level until senility diminishes it gradually.

Variations in libido among bulls has been observed. Breed-type, hereditary factors, age, disabilities, pain and experience gratifying or adverse, are factors known to influence libido. It is clearly established that low libido trait can be genetically transmitted. In selecting bulls, those with low libido have to be eliminated. Impaired or inferior libido can also result from obesity or certain skeletal defects like arthritis or hoof trouble especially of the hind limb. Bulls with 'apprehensive' type of temperament are found to have inferior libido.

Levels of libido can be quantified by measuring reaction times and the number of ejaculations per unit time when stimulus pressures are maintained at an adequate constant level. Time required for recovery of libido after stimulus satiation provides critically important information to the level of sex drive in the bull.

Young inexperienced bulls approach the cows hesitantly and spend a long time on 'beating around the bush'. Timidity and unfamiliarity in strange surroundings are likely reasons for reduced levels of sexual expression in novice young bulls.

21.5.1.1. *Sexual pattern in the bull*

When a bull is consorting a cow in oestrus, a series of male behavioural patterns are visible.

21.5.1.2. *Courtship*

The bull detects a cow in pro-oestrus two days before oestrus and will try to tag on to her by remaining in the general vicinity. This is termed as *guarding* of the female by the bull. This behaviour can be expressed well only on open range system. A condensed version of this behaviour is evidenced by cattle in confinement.

Onset of oestrus proper in the cow will give intense excitement to the bull which will follow the cow closely, licking and smelling her external genitalia. One of the characteristic male sexual behaviour is the olfactory reflex of *flehmen* (l'amour) which is characterised by curling of the upper lip exposing the gum region while the head is held horizontal, neck extended and nostrils widened. The bull may sometimes butt the cow's hind-quarters. Bulls often pump their tail-head up and down during the courtship passing at times small quantities of faeces which become spread-around their hindquarters.

Bulls also show, like males of other farm animals certain behavioural displays like (*i*) threat displays and challenges, (*ii*) territorial activities and (*iii*) female tending.

The threat displays are manifestations of masculinity in which the animal bends its shoulders and neck with protrusion of the eyeballs and erection of hair along the dorsum. This is in fact the physiological state of fight or flight.

Challenging is part of agonistic behaviour among males. Under free-range system, the male at the top of the peck-order may perform all the breeding with the available females. But this is not usually the case under domestication. Challenging in bulls is displayed by bellowing, pawing and horning the ground throwing dirt over the back and shoulders and rubbing the side of his face and horns vigorously against the ground.

Territorial activities are less visible under domestication. Under free-range system, pawing and horning behaviour of bulls bares patches of earth at different locations in his 'territory' which form a claim of possesion of all the females in that area.

Female tending behaviour is manifested when the bull maintains close association and bodily contact with the cow while grazing with her. While consorting with her, the bull occasionally butts the hind quarters nudging her to move forward. The female in 'standing heat' responds to this by standing firm putting some weight backward. This positive response by the female provides reciprocal stimulation for the bull.

Once mating is completed, the bull continues to tend the cow for some more time.

Chin-resting. This often appears before mounting with the bull standing behind the cow placing his chin and neck on the rump of the cow applying mild forward pressure. This is again another nudging activity. Non-receptive cows respond by moving away.

21.5.2. In cows

Oestrus or in layman's language *heat* is the normal mating state of the female animal. Oestrus is a physiological state involving hormonal and other changes, but its main external manifestation

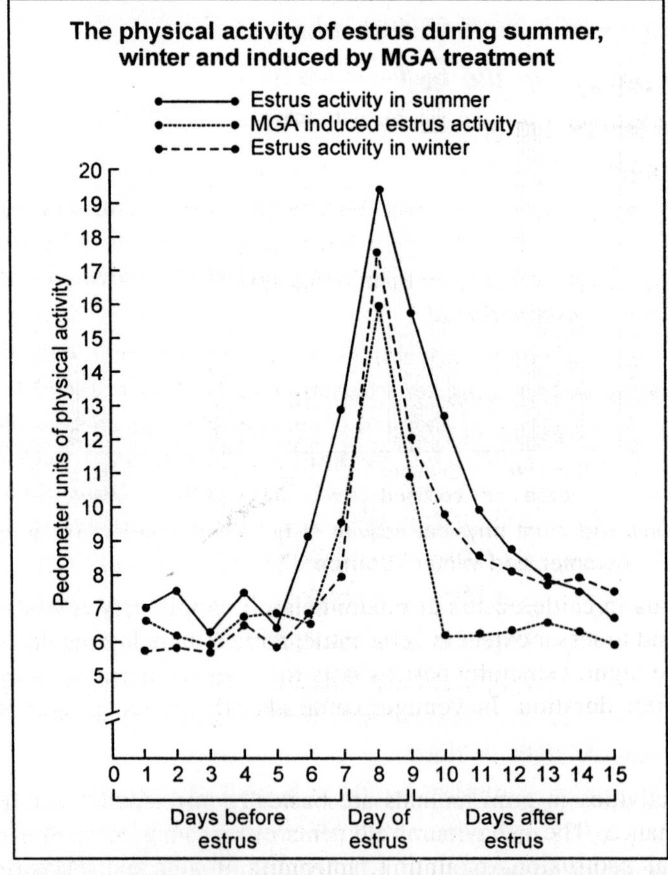

Fig. 21.2. Physical activity during eastrus in buffalo cows as shown by pedometer readings. (**Source** : Barua, K.K.; Pal, R.N. and Sastry, N.S.R. 1981).

is behavioural. When the cow is in oestrus, the normal behaviour and routines in many respects get disturbed. The cow may become more excitable and its normal feeding, ruminating and resting patterns get affected. Relative social hierarchical positions are temporarily ignored as the cow will approach both dominant and subordinate animals indiscriminately. This results in increased agonistic interactions.

The essential characteristic of oestrus behaviour is acceptance of the male in mating. But this is associated with signs like bellowing, grooming activities like licking other animals etc. The oestrus cows indulge in great deal of mounting of other cattle, in which case the non-oestrus cows refuse to stand for mounting. On the other hand, when the oestrus cow is mounted by other cows, she makes no effort to escape; instead she stands quietly to be mounted or even leans backward on the mount. When several cows in a group have been prompted to mounting activity through the first such act of an oestrus cow, it may become difficult to identify the cow in heat. However, when one animal in particular is being mounted on by others, it can be identified as the cow in heat. General physical activity of buffalo cows went up during oestrus as shown by pedometer readings (Fig. 21.2) as against normal physical activity (Fig. 21.3).

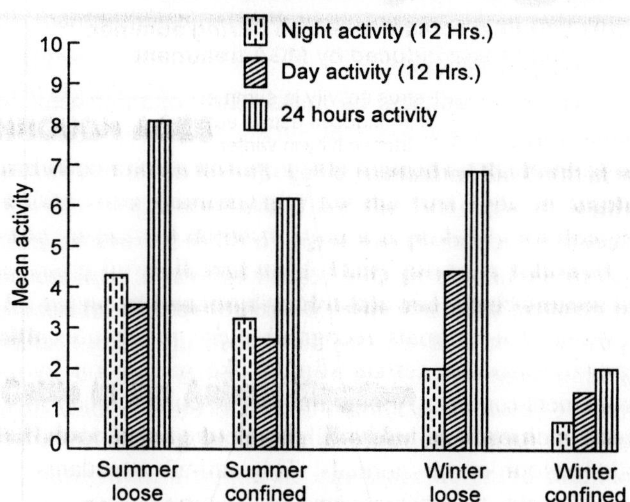

Fig. 21.3. Day and night physical activity of buffaloes housed lose and confined in summer and winter (**Source** : Same as for Fig. 21.2).

Onset of oestrus in cattle occurs at random starting apparently at any time of day or night. In water buffaloes and to lesser extent in zebu cattle, there is a greater incidence or commencement of oestrus during the night. Generally oestrus lasts for a period of 12-24 hours. In tropical regions this may be of shorter duration. In younger cattle also the period is shortened.

21.5.2.1. Mating

Although mating activities in farm animals are basically instinctive, training has been observed to improve performance. The behavioural elements culminating in mating consist of courtship, erection and partial protrusion, mounting, intromission, ejaculatory thrust, ejaculation and dismounting. Copulation in cattle is brief. The bull mounts orienting himself so that the female is appropriately covered for intromission to be effected. The bull claps the cow tightly with its forelegs abducted into her flanks, the clasping increasing in intensity at the moment of thrust and ejaculation. Ejaculation is followed by relaxation of muscular tension and the bull dismounts slowly.

Increasing the amount of pre-coital stimulation increases semen volume. Sudden changes in environment (change of place, attendant, locality of collection etc.) makes some bulls apprehensive. Fear and apprehension act against proper expression of sexual behaviour. This effect is usually transient and will disappear on getting accustomed to the new surroundings. Inhibition may also develop as a result of repeated frustration, faulty management, distraction by dogs, crows, wrong techniques of semen collection etc. If a bull has been conditioned to a certain stimulus, absence of the same can prevent expression of the sexual behaviour. The best example is that of an A.I. bull which used to be given a toffee before mounting by its attendant. When the attendant was changed, the bull refused to give collection with any amount of coaxing. Only when the 'trick' was told by the previous man and the toffee given, the bull agreed to oblige.

21.6. MANAGEMENT OF BEHAVIOUR

Under domestication it is often necessary to control and manage behaviour of farm animals. Restrictions on these animals can be by several means as indicated below :

(*i*) Giving training to regulate behaviour.

(*ii*) Control by manoeuvring the animals by anticipation of their behaviour.

(*iii*) The use of physical force.

(*iv*) By use of mechanical disadvantages.

(*v*) By the use of drugs.

21.6.1. Training

Animals can be trained to control behaviour and follow a routine procedure like in the case of feeding, milking, weighing etc. The animals recognise these procedures by repetition of the routines. Animals also learn to associate certain stimuli like putting on lights, sounds, human voice etc. with certain management routines. A large number of conditioned reflexes get established soon. Skilful exploitation of conditioned reflexes will enable the husbandsman to exercise a great deal of control over the behaviour of his animals. The animal attendants can use sounds of feeding, buckets, whistles, calls etc. to assemble and lead animals.

An animal unfamiliar to the routines learns quickly if allowed to mix with the group which is trained to it.

21.6.2. Anticipation of Behaviour

A knowledge of an animals anticipated behaviour is very useful in handling loose animals. The handling of loose cattle which have not been subjected to much manipulation previously presents a very real problem to the handlers. Loose cattle can be best controlled when they are gathered into tight groups. A grown calf unaccustomed to being handled exhibits a behavioural pattern when approached. The calf first looks keenly with pricked ears towards the approaching 'danger' with apprehension. Closer approach induces a state of alarm or fear. In a group, this alarm is transmitted by signals and they herd together tightly, heads held up and moving rapidly they bellow. This state of fear is described as *adrenalized*. The calves offer greater resistance to handling when adrenalized. Trying to catch the calves or any other factor which increase the stimulation of the calves can result in development of a state of panic in which calculated behaviour is abandoned and 'resort to flight' takes place. This flight due to panic can be immediate or after remaining apprehensive for some time.

To control the behaviour of panicked calves, the operator must cautiously move towards the calf quietly and evenly paced. The calf will try to move away trying to be at the centre of the tightly packed group. On approaching it there, it will try to run away backward or forward. The handler should accordingly modify his approach to cut-off the escape route. After one or two such attempts, the calf will bolt blindly to a comer. The expert handler makes use of this understanding of behaviour to time his attempt of catching the animal when it is confined to the corner. The best method of restraint is to catch the lower jaw and to raise it and pull it sharply to one side. Cattle apparently consider objects held in the hand as part of the body. Thus, by holding long sticks in either hand, one handler can extend his presence considerably to either side and thereby cut-off a very large escape route for an individual or group of cattle kept loose.

21.6.3. Control by Force

Rope. One common method used for controlling cattle is by means of ropes. Cotton ropes, nylon ropes, coir ropes, jute ropes etc. are commonly used. Ropes are used as lassos, halters, milk-man's rope casting rope and as means of tethering cattle.

21.6.4. Trevis, Crush or Stock

Trevises of various type and materials are available for controlling cattle. The best type is the one which allow an animal to be funneled down a narrow passage into the trevis from a holding area. At the exit also there should be a collecting yard which should be visible to the animal entering and passing through the narrow chute. This encourages them to move forward to join the group. Trevises which follow the above principle allow large number of cattle to be examined individually and closely in a short space of time with the least danger to the animals and the handlers. Such operations in large herds are required for vaccination, drenching, blood sampling, ear-tagging, tattooing, A.I., pregnancy diagnosis, branding, spraying, diagnostic test and dehorning. Many of these operations can excite the animals and make them behave in a manner injurious to themselves. They may try to push forward, kick, jump or at times slip and fall down. The trevises should be

Fig. 21.4. A race for controlling a number of cattle at a time for examination and handling.

constructed with solid sides so that it is impossible for an animal to put its hind leg between spars; otherwise it can lead to injury. Similarly, the neck when tied in front or secured in collar should be quickly releasable in case the animal slips and falls. Trevises in which a series of animals can be restrained behind one another are known as races. This form of crush is very suitable for a close inspection or manipulation of large number of animals (Fig. 21.4).

21.6.5. Use of Mechanical Disadvantages

There are many age old methods of mechanical control in which an animal's behaviour is controlled by applying some mild force to a part of its body in such a way as to cause a mechanical disadvantage. A knowledge of the physical mechanics of movement is beneficial in applying the minimum force to achieve the desired effect. For example, holding the head raised frequently restricts the forward movement (Fig. 21.5).

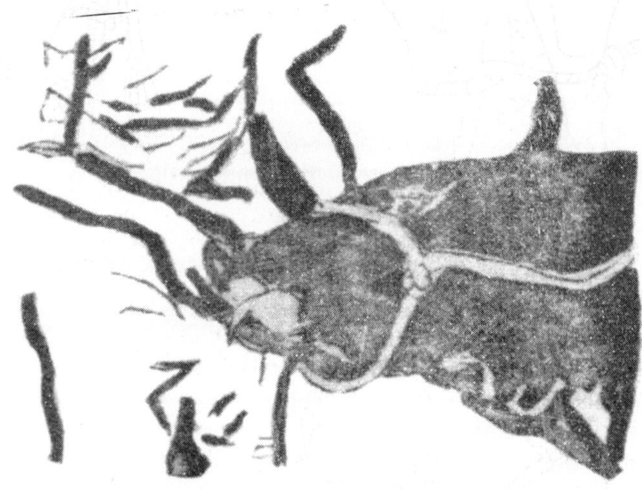

Fig. 21.5. Restarting cows by head leverage method.

Many times it may be necessary to control kicking. The most common method adopted is the milk-man's rope (Fig 21.6). Another method is to pass the tail between the hind legs round the hock on the same side as the operator. Pull the tail whenever the animal lifts its leg to kick, as such pulling brings the weight of the animal on to the leg being lifted and thus checks the kicking. Udder kinch (Fig. 21.7) which applies pressure above the abdomen can fairly control the kicking with the hind limb. Use of anti-cow-kicking devise can also control kicking. Raising the tail head vertically can also immobilize the hind limbs of cattle for short periods. (Fig. 21.8).

Mechanical devices like cradles to prevent self grooming, anti-self sucking device to prevent the vice of self-sucking, mouth gags to keep the mouth open for examination, bull-leader to lead bulls etc. are mechanical devices used in routine management of farm bovines.

21.7. USE OF DRUGS

Drugs are at times used to control animal behaviour. There are an array of chemical agents the pharmachological action of which are varied in nature and intensity. They can alter the behaviour of animals by various degrees of depressant action on the nervous system resulting in immobilization. Now-a-days tranquilizers are commonly used when individual animals have to

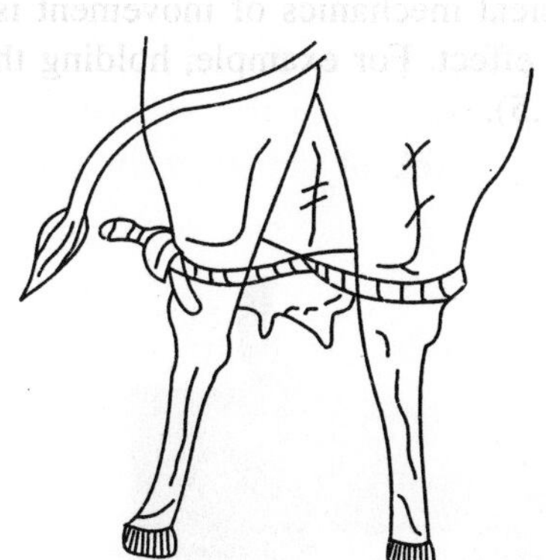

Fig. 21.6. A milk-man's rope in position.

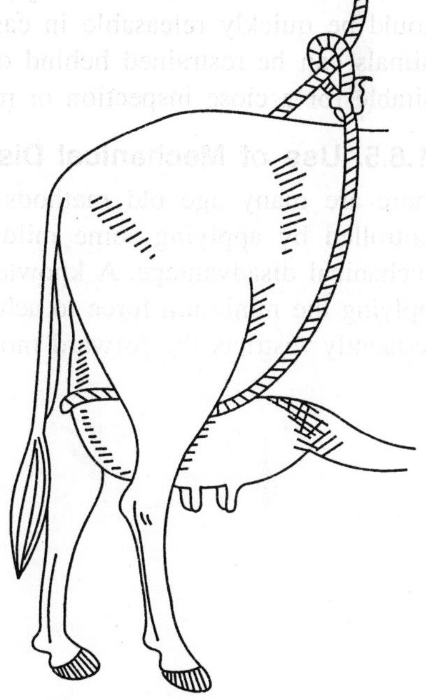

Fig. 21.7. Udder kinch.

Fig. 21.8. Restraining the movement of the hind limbs by pulling up the tail.

be subjected to forms of handling or restraint for a period to which they have no training. Otherwise, such a treatment would result in a state of panic with struggling and may cause injury and damage to the animal and sometimes the operator. Tranquillization during the post-operative period, after fractures have been treated with plaster-cast, after reduction of prolapsed uterus etc. are widely used to prevent animals from struggling and rubbing resulting in damage. Similarly, there is much to commend the administration of sedatives before transportation. The use of a drug filled projectile by means or a 'Capchur gun' is useful in the case of violently vicious bulls free on the ranges.

21.7.1. Electronic Devices

An electronic equipment known as '*Stockstill'* makes handling procedures more quick and painless. Manufacturers claim that the *Freemix Stockstill* can hold animals motionless with the application of minute electrical pulses allowing stock handling operations such as branding, dehorning and castration to be carried out quickly, easily and with reduced pain to the animal. The pulse causes the animal's skeletal muscles to contract-preventing movement—and also produce pain-blocking-effect which greatly reduces the amount of pain and stress felt by the immobilized animal. The low voltage used is very safe. The 'stockstill' is operated by placing two electrodes—a needle in the skin under the base of the tail and a needle in the cheek or clamp in the mouth.

CHAPTER

22

Routine Farm Operations

A typical dairy herd comprises of cows in different states of physiological condition (lactating and non-pregnant, lactating and pregnant, dry and pregnant, dry and non-pregnant), bulls and young stock at various stages of growth (suckling, pre-weaning, post-weaning, older heifers). Besides, a typical dairy unit in developing countries will invariably have a few work animals.

Herd management simply means taking care of all such stocks. This care includes common practices like feeding, health cover, grooming, etc. and special management practices for individual classes of animals, like milch cows, dry cows. etc. The reader might already be aware of the feeding, reproduction and other aspects discussed in other chapters of this book.

There are some activities that are routinely carried out on dairy farms. Proper execution of these operations has a direct or indirect bearing on the overall success of the enterprise. Described in this section are details of such operations.

22.1. IDENTIFICATION

Proper identification or marking dairy stock is the first pre-requisite for keeping their production and re-production records (Section 15.2). It is desirable that an owner should not only be able to identify each animal he owns but that he should be able to prove identification to the satisfaction of others. This is all the more important for making claims on insured animals.

The methods commonly used for identifying cattle are tattooing; branding; tagging ear, neck, chain or tail tags. Of these the most reliable is tattooing for calves to be followed by branding when the calves are 10-12 months old. Tags of any type are likely to fall off or even tear off the ear lobe.

22.1.1. Branding

It involves searing numbers, letters, designs or a combination of these on the skin with a hot iron or with chemicals. Branding is most suited for marking cattle and buffaloes. Calves should be branded by the time they are about one year old, using smaller sized brands, lest the marks may get distorted as the animal grows. A sunny day in the non-fly season should be chosen for branding animals.

Sets of branding irons are available in the market. These comprise of figures 1 to 9 and 0, and letters A to Z. If the part to be heated brand is of copper it will be better as it retains heat better and hair does not stick to it as to iron. Same branding irons can be used for hot and chemical branding, though, for the latter, branding irons with a shallow groove on the contact surface of letter or figure are better. Irons can be heated in a portable forge kept burning by coal or wood. For branding chemically, ready-made branding inks are available in the market. Both methods are equally effective if carried out properly, though chemical branding is less permanent and bad if carried out by an inexperienced person. For branding, animals should be well controlled, preferably by casting and tying their legs.

If hot iron branding is practised the desired iron is heated to bright red but not white heat, and is applied to the skin on sides of thighs with light pressure for not more than three seconds. A little mustard oil in which zinc oxide is mixed may be smeared over the burns to encourage healing. For chemical branding, the iron should be dipped in branding ink, extra ink drained off the iron and the wet iron applied to skin.

The figures should be big enough to be read at some distance and should be separated by at least 2.5 cm from each other to prevent sloughing of skin. Generally, the serial number of the animal is branded on the milking side, *i.e.*, on left thigh and the quarter number (month, and year of birth) and herd symbol are put on the right thigh. A type of code is used for putting quarter number; the first or first two digits of brand indicate month of birth while the last digit indicates the last figure of the year of birth (*see* Figs. 22.1 and 22.2). Sometimes instead of month of birth only the quarter of birth and year of birth are branded. Hence the name—quarter number. The four quarters of the year are 1 (Jan., Feb., March), 2 (April, May, June), 3 (July, August, September), and 4 (October, November, December).

Fig. 22.1. Quarter number brand showing the month of birth as eight and year as one ending with a 2 (*e.g.* 1982).

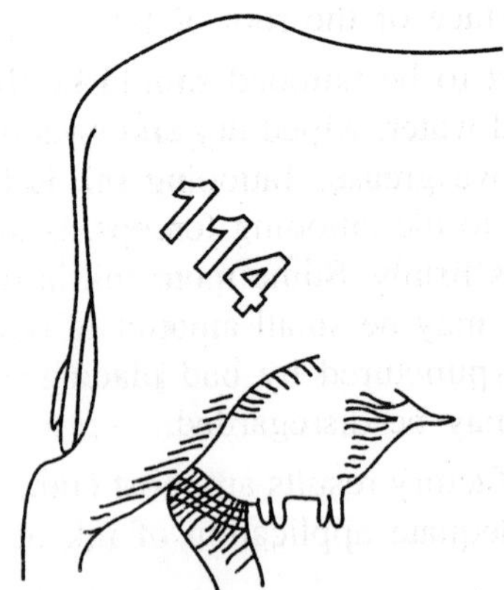

Fig. 22.2. Quarter number brand showing the month of birth as eleven and year as one ending with a 4 (*e.g.* 1984).

Cryobranding or freeze branding is now-a-days considered more effective especially for animals with black coat as this procedure leaves white brand marks. Copper brands with dry ice in 95 per cent alcohol or liquid nitrogen as coolant are used in this method. Liquid nitrogen is now fairly easily available in India thanks to the spread of use of deep-frozen semen in artificial insemination. Bath *et al.,* 1978 described cryobranding procedure using two coolants on clipped and unclipped skin.

For obvious reasons, on unclipped surface, contact time between the cool brand and skin is 1½ to 2 times more than that on clipped surface. The contact time recommended by these authors on unclipped surface using liquid nitrogen varies from 5 to 20 seconds from new born to adult cattle. Prior to exposure to the coolant, the brand site should be cleaned by thorough brushing and the site soaked with alcohol before applying each digit.

22.1.2. Tattooing

It consists of piercing outlines of desired numbers or letters on the skin inside ears and then incorporating a black pigment into these punctures. The various letters and figures are outlined by steel points each of which carries a small amount of coloured paste into the subcutaneous tissues and cartilage of ear. These pastes contain insoluble carbon (black) or green pigments that are inert to tissues.

Tattoo figures cannot be read without handling and close inspection of the animal and the marks are not easily legible when imprinted on pigmented skin. When the tattooing is properly done, the figures are clear and somewhat permanent, though often marks fade out in course of time. For this reason, tattooing is most suited for marking new born calves and pigs.

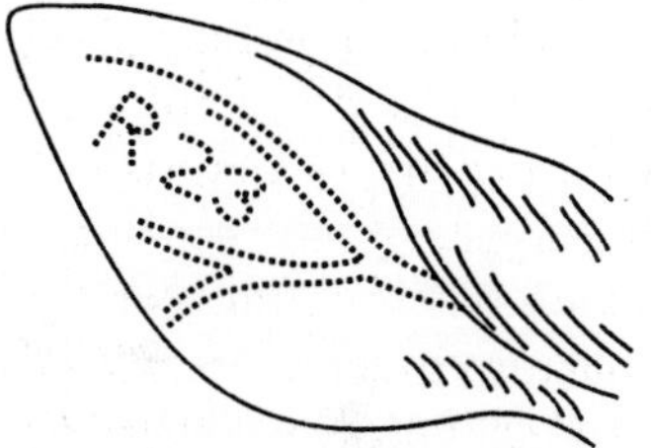

Fig. 22.3. Ear tattoo. Note that where the tattoo is made, there are no blood vessels.

Tattooing sets comprising a tattooing forceps, tattooing ink and series of letters and figures are available in market. Using the equipment, the imprints are made conveniently on the inner side of the ear avoiding ear veins (Fig. 22.3); in buffaloes and cattle with pigmented skins, it may be done on the under surface of the root of tail.

The part to be tattooed should be thoroughly scrubbed with soap and water, wiped dry and swabbed with methylated spirit to remove grease. Tattooing ink is liberally applied over the part. The desired number of figures, fixed to the tattooing forceps, is then firmly imprinted over the part by pressing the jaws of the forceps firmly. Some more ink is rubbed into the tattooed punctures with the bulb of the thumb. There may be small amount of oozing of blood from the skin in some cases. If the vein has not been punctured by bad placing, there may be a little haemorrhage, which will not be serious and may be disregarded.

Unsatisfactory results are most commonly caused by dirty equipment, dirty and greasy skin, bad ink, inadequate application of ink or a mis-application of the tattoo.

22.1.3. Tagging

Tags or labels are made of light metal or strong plastic with the numbers stamped on them. Tags are fixed generally to the ear with a special tagging forceps.

There are two types of tags—self-piercing type, and non-piercing type. The former variety has sharp ends and can be directly fixed to ear with a forceps while for the latter a hole must

Bath, D.L., Dickinson, F.N., Tucker, H.A. and Appleman, R.D. 1978. *Dairy Cattle : Principle, Practices, Problems and Profits.* 2nd Ed. Lea and Febiger, Philadelphia, 1978, pp. 491-494.

first be made with a tag punch or pen knife. Tagging is mostly used for marking sheep and goats, though young calves can also be tagged.

A hole is first punched in the upper edge of the ear as close to the head as possible. This should be far enough from edge so that the tag, when in place, will neither be tight on the ear or swinging free; yet leaving sufficient space for growth of the ear. Modern tag punches have a gauge to regulate the distance of the hold from the edge of the ear.

One end of the tag is then placed through the hold and clinched or locked in position with the pincers. The numbered side of the tag should be on the back (top) of ear (Fig. 22.4). Some antiseptic should be applied to the wound to encourage healing and prevent infection. With the self-piercing tags, the same procedure is followed except for prior punching of hole.

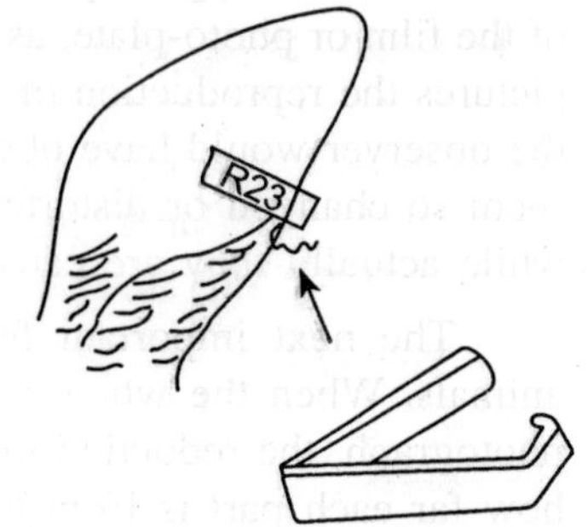

Fig. 22.4. Ear tag in position. The arrow shows the distance that allows ear to grow. A- an ear tag.

22.1.4. Other Methods of Marking

Cattle can be marked by branding on horns with small hot iron brands, by neck-chain and metal/ plastic tag or leather neck-strap and metal/plastic tag. These are not permanent but are useful for shorter duration, such as for market stock or experimental animals.

Cattle can be marked for temporary purposes by 'Keeling' with a pigment having a greasy basis (often with red or blue paint); these marks last long enough to identify animals going to show, sale or market, but are not permanent. Animals can also be identified by keeping a record of their natural marks—spots, patches, body colours, chestnuts, acquired marks—scars of wounds etc, or photographs.

22.1.5. Photographing Cattle

Generally, the animals are photographed to get the following views—side view or profile, and rear and hind views. Sometimes a so called 3-view showing the animal from front at 45° angle is also taken. According to Russian Professor N.A. Revin, whatever may be the view of the animal that is being taken, the animals should be photographed in such a way that the angle of 'vision' of the camera should not exceed the angle of 'vision' of the human eye. This angle is about 19° for the maximum dimension of the animal, that is, its length including its head. In practice, while taking the side view of the animal the distance between the objective of the

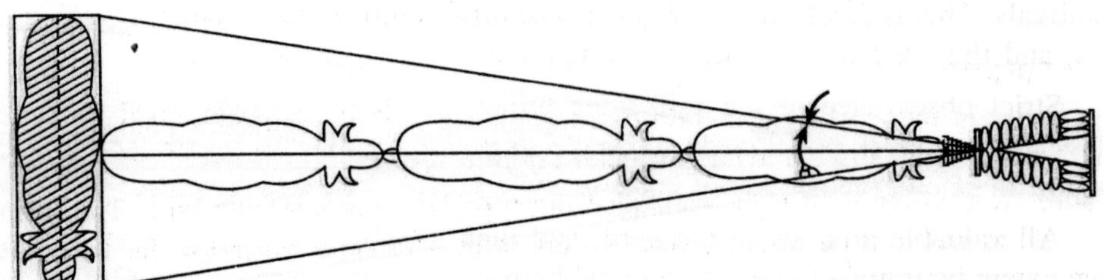

Fig. 22.5. The optimum distance to be maintained between the animal and the objective of the camera while taking the side view of the animal.

camera and the object (animal) should be 3 times latter's maximum dimension, namely, its length including its head (Fig. 22.5). This arrangement gives the desired angle of vision. The photographer should avoid taking the picture from a closer position aimed at maximum utilisation of the format of the film or photo-plate, as this will result in obtaining a picture with too wide an angle. In such pictures the reproduction of the animal does not correspond with that view of the animal, which the observer would have obtained if he had seen the actual animal. The form of the animal may seem so changed or distorted, that an observer is likely to attribute these defects to the animal, while actually they are caused by the improperly taken photograph.

The next important feature is the position of the camera while actually photographing animals. When the whole image of an animal is exhibited from a given point, as is done in a photograph, the reduction in dimension of various body parts from their natural size depends on how far each part is from the central point of view. Hence, the objective of the camera should lie opposite to the centre of the total length of the animal including its head, so that head, neck, fore and hind limb, and rump of the animals get a uniform perspective reduction in size (Fig. 22.6).

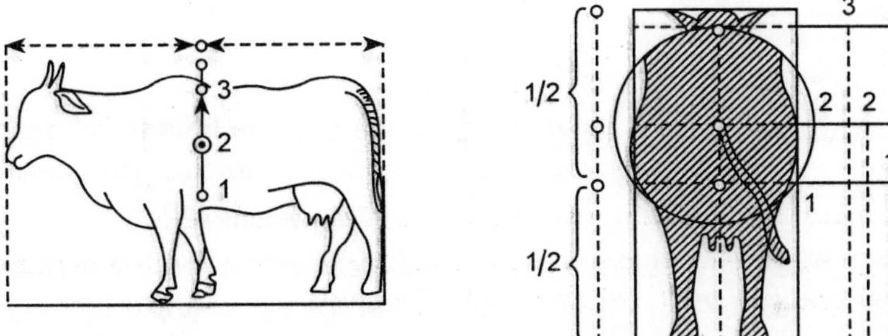

Fig. 22.6. The correct (2) and wrong (1 and 3) points opposite which the objective of the camera has to be positioned while taking the side view and hind view of the animal.

The centre of the total length is also the point on which the height at which the objective of the camera has to be positioned. It may be mentioned here that the form and proportions of the barrel or trunk of an animal are of greater importance than its body extremities in determining the peculiarities of its body construction as a whole. Hence, the objective of the camera should be placed not opposite to the central point of the total height of the animal, but opposite to the central point of the depth (vertical thickness) of its barrel. Taking from a higher point results in an enlarged view of the upper parts and reduced dimensions of the lower part of the picture. The reverse will be true if the picture is taken from a lower point.

By far, it is better to follow the same three principles for taking the front and rear views of animals. This is important so that the various body parts are reproduced in the same reduced sizes, and thus, the three views are comparable.

Strict observance of the said three principles alone, however, does not give best results unless the animal is positioned and made to pose properly. Excellent animals sometimes look haggard in photographs, simply because their photos are taken in wrong postures.

All valuable animals of the herd, and their number will comparatively be less, should to some extent be trained to be led and held before strangers and in strange places. This goes a long way in teaching the animal to behave at such times. Animals that are very temperamental can be quietened by giving suitable doses of tranquilizers or other depressants in consultation with a veterinarian.

At the time of taking photograph the animal must be made to stand squarely. Marks made with chalk on the floor can be used as bearing for this purpose as indicated in figure 22.7. The head of the animal should be held up at its natural sweep in such a way that the longitudinal axis of head and neck, as well as that of body are in a line.

A word about orientation of the camera while photographing animals. After positioning the camera at the appropriate place as described above, great care should be taken to see that it does not tilt in any direction. The camera should be so held that the plane of the film or photo-plate contained in it must absolutely be vertical and parallel to the plane of the general symmetry of the animal (broad side of the animal).

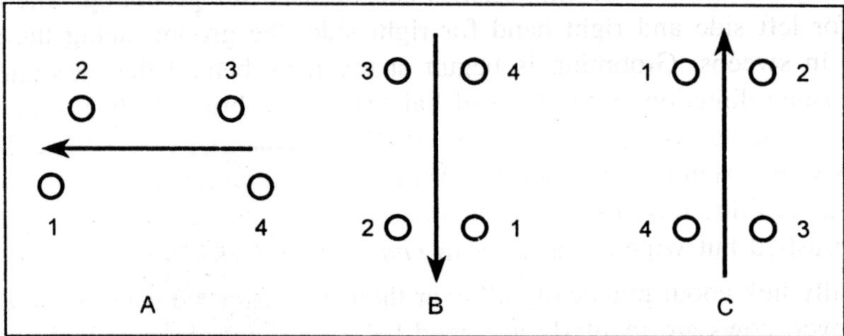

Fig. 22.7. The correct position—1. left fore, 2. right fore, 3. right hind and 4. left hind feet of the animal while taking side (*A*), front, (*B*) and hind (*C*) views.

22.2. Exercising, Grooming and Bedding

22.2.1. Exercising

Exercise keeps animals thrifty and active. This helps normal body metabolic processes and maintenance of good health. Too much standing at one place weakens the leg muscles, causes the toes to overgrow and puts abnormal strain on legs and feet. Exercise is most necessary for male breeding stock.

Just how much exercise is important or necessary for dairy cattle and buffaloes is not known, though moderate exercise is considered to increase milk production, "Motion" instead of 'exercise' does seem to be a more correct expression of the physical activity needs of dairy cows. The Russians use such a term. In many instances cows have been kept confined for long periods with little or no apparent harmful effect. However, it is reasonable to assume that some activity is beneficial and certainly not harmful to dairy animals. Outdoor exercise with exposure to sunlight is important in providing supplies of vitamin-D.

Exercise for the herd is not a problem in loose housing situations. Where cows are confined to closed quarters, it is highly recommended that they be turned into an open lot or on to pasture at least once and preferably twice daily. Cleaning of the barns becomes easier with the cows away. The utility of the exercise that the animal gets while grazing in tropics is doubtful because of the low intake of energy in relation to the energy spent.

Breeding bulls must be regularly exercised to keep them in good condition. This can be done by providing loafing area of about 120 square meters for each bull. The bull can also be paraded for an hour each day. On large farms and semen banks where large number of bulls are kept, the bull exerciser can be used to give physical activity to a number of bulls simultaneously. Bulls not exercised regularly put on fat and become sluggish breeders.

22.2.2. Grooming

Grooming comprises brushing the hair coat of animals. It is performed for cleanliness and appearance, to massage and stimulate the cutaneous blood and lymph circulation, to remove waste products like skin secretion, scurf and loose hair and to remove lice and other skin parasites from hair coat. Vigorous brushing also keeps skin in loose and pliable condition and brings out the natural oil in the hair.

Unlike horses, farm animals are groomed rarely and less elaborately. However, grooming is essential for all show animals. Dandy brush, made of stiff whisk fibre, whisk made of paddy straw or hay or body brush made of stiff bristles can be used for grooming. The brush is held in the left hand for left side and right hand for right side; the groom facing the animal's tail. Brushing is done in sweeps. Grooming is begun at the neck behind the ears and brushing is carried out in the same direction as the flow of hair. However, brushing has to be done against the fall of hair for removing dried mud, dung etc. sticking to hair. One has to lean hard upon the brush during the sweeps, standing well away from animals. At the end of each sweep the wrist is given a turn so as to bring out of coat the dust and debris which the brush has collected. Face should never be brushed but wiped with a clean *khadi* or *flannel* cloth.

Cows generally lick about practically all over their body and remove a good deal of dirt in the process. However, cows are regularly groomed before milking so that no dust falls in milk. Grooming cows includes brushing the rump, sides of thighs and buttocks, washing of their udders with warm water containing an appropriate antiseptic (usually one of the chlorine compounds) and final wiping of udder with boiled damp cloth. Wiping is done to remove the superfluous water which would otherwise drop into the milk pail. It is advisable, however, to have the surface of the udder damp (but not wet) as this prevents dry dust etc. from falling. These operations should be carried out immediately before milking. The actual brushing of sides of thigh etc. should be finished in the loose houses and then cows taken into the clean and dust-free milking sheds.

Work bullocks too need grooming before and after performing work. This keeps them active and the massaging will permit them to recover from tiredness and fatigue quicker. Generally only the legs, including shoulder and hips of bullocks are brushed using paddy straw or hay whisps.

22.2.3. Bedding

In tropical countries like India provision of bedding to farm animals is not that important as in colder temperate countries. However, winters in northern parts of the country are generally cooler, often accompanied by cold winds. Cold is not that dangerous, but cold drafts certainly are; hence the necessity to protect animals against them. Young calves, especially buffalo calves, succumb if unprotected against cold winds.

Bedding or litter is used primarily for the purpose of keeping animals clean and comfortable. Bedding also soaks up urine, makes manure easier to handle and absorbs plant nutrients, fixing both ammonia and potash in relatively insoluble form; thus protecting them against losses by leaching. Now-a-days rubber mat bedding is available which is convenient to use and comfortable for the animals.

During winter, suitable bedding should be provided in young calf pens. Wheat *Bhoosa*, sugarcane bagasse, chopped grass or crop residues are ideal for animal bedding. The bedding should be about 10-15 cm thick.

22.3. DEHORNING, CASTRATION AND "EXTRA" TEAT REMOVAL

22.3.1. Dehorning

Dehorning means removing horns of animals. Disbudding means the arrest of horn growth at an early age when the horn root is in the form of a 'bud' or button. Horns serve no useful purpose on dairy cattle; they can be a nuisance and cause many body and udder injuries. Horned cattle require more space. Hornless cattle are no longer discriminated against in show rings and sales. Even dehorning buffaloes is becoming a common practice on organised farms in India.

Dehorn calves as early as possible. Dehorning is done properly when horn buttons are very small; a neat and clean job can be done with little discomfort to the animal. Plan to dehorn when the calf is four to ten days old, or as the horn buttons can be easily detected—the earlier, the better. In older cattle the horn is hard and brittle and rather difficult to cut.

There are several ways of dehorning—with chemicals, saws, clippers, hot irons, dehorning spoons and tubes. Regardless of the method used, practise cleanliness and disinfect instruments (except hot irons) between animals in order to lessen infection. Less insect trouble is encountered when dehorning is done in the spring or post-monsoon season. If the operation is performed in fly season, apply a fly-repellent to the wound.

Chemical method. Use of chemicals should be limited to small herds kept under close supervision. Caustic potash (KOH) or caustic soda (NaOH) are the common chemicals used. These are available in the form of sticks, paste or liquor.

Clip the hair from around the horn buds of calves and surround the area with a ring of heavy grease or vaseline to protect the eyes against the chemical. Rub the chemical (held in a caustic holder) over the button until blood appears; protecting the hand while doing so (Fig. 22.8).

Hot-iron method. This comprises applying a specially designed hot iron to the horn buds of young calves (Fig. 22.8). The dehorning irons are heated in a portable forge. Where electricity is available the electric dehorning iron should be used. It keeps an even temperature. The hot iron method is bloodless and may be used in any season of the year, but it can be used on young calves only.

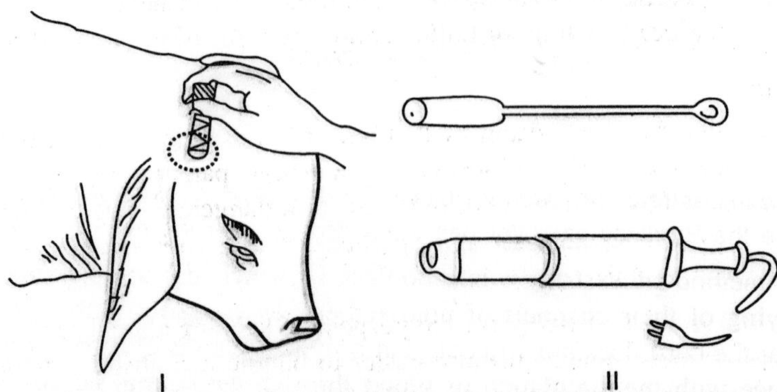

Fig. 22.8. Method of debudding calves I. Chemical method. Note how the caustic pencil is held and applied over the prepared site. II. Hot iron method, fire heated (above) and electric dehorner. The heated part is applied over the bud.

Use of dehorning saw and clippers. Clippers are satisfactory for removing the horns of younger cattle but the hard and brittle horns of mature cattle can best be removed with a saw. Any handy wood-cutting saw can be used for dehorning. Special dehorning clippers or shears are available, but these are heavy and rather cumbersome to use.

Confine the animal to be dehorned in a suitable cattle trevis or throw it down and tie its legs. Calves can be handled by throwing or by snubbing them to a post. Whatever instrument is used (saw or clippers), remove the horn with about 1 cm of the skin around its base.

Use of elastrator. The elestrator is an instrument for use in stretching a specially made rubber ring which may be used in dehorning cattle. A great deal of pain is associated with this method, which is used on cattle with horns 5 to 10 cm long. Stretch the rubber ring with the elastrator instrument and place over the horn well down into the hair. Smaller horns drop off in three to six weeks; larger horns may take two months.

22.3.2. Castration

Castration is removal of testicles, the glands that produce male germ cells. In India and other South-East Asian countries bullocks are extensively used for agricultural operations and haulage. Buffalo is particularly valued as a draught animal in rice growing tracts. Male cattle and buffaloes meant for work should be castrated so that they can be rendered docile and amenable. Castration also results in fineness of neck especially in buffaloes, which facilitate the fitting of yoke.

Male calves are castrated at the age of one year. The older the calf is, the more it suffers at the time of castration. Castration shou'ld be performed during cold season, strictly avoiding rainy season for fear of fly menace. Castrated animals should be rested for a few days in clean and comfortable pens. Operation wounds should be watched and dressed if necessary once or twice.

Under Indian conditions the best method for castrating cattle and buffaloes is with a pair of Burdizzo pincers (Fig. 22.9). If surgical splitting of scortum and removal of testicles is desired, this is better to be done by a veterinarian. Elastrator rings can also be used—these are very painful to the animals.

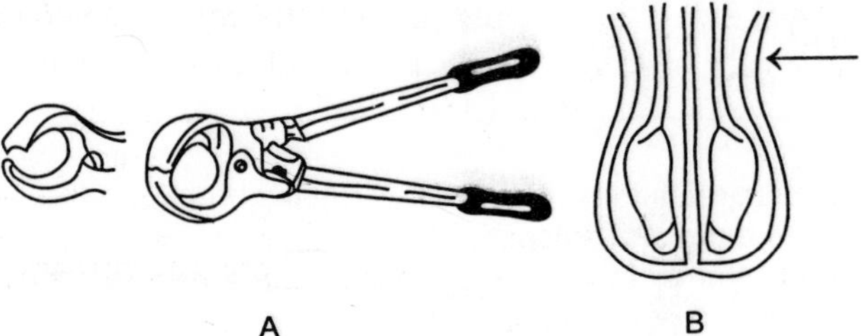

Fig. 22.9. Burdizzo castrator (A) used for bloodless castration of cattle (B) Arrow showing the place where the Burdizzo castrator has to be applied over the neck of scrotum.

Burdizzo method of castration is bloodless in which the testicles are made functionless through destroying of their channels of nourishment. In using Burdizzo pincers care should be taken to see that the cord does not slip out, that only one cord is clamped at a time and that there is no interference with the circulation of blood through the central portion of the scortum.

Throw the animal down and secure its feet together. Work the cord to the side of the scrotum, and then clamp the Burdizzo at about three to five centimetres above the testicle (Fig. 22.9) where it is held for a few seconds. Then repeat this operation on the same cord at a location about one centimetre below the first one. Repeat the same procedure on the other testicle. Paint a little of antiseptic on the clamp impressions.

Vasectomisation of bulls involves surgically obliterating just the vas deferans, leaving blood and nerve supply to the testis intact. Because of this, vasectomised bulls (also known as teaser bulls) show normal sexual activity, though their semen does not contain sperms. Hence they cannot impregnate cows and, thus, are used as teasers and for detection of cows in heat. Vasectomisation is best got done by a veterinary surgeon. Such bulls can be kept at the rate of about one for every 50 breedable cows. For villages, the village Panchayat can maintain one teaser bull for heat detection among the cows. Such a practice was found to be a great aid in better breeding of buffaloes in the Anand Milk Union Limited area villages.

22.3.3. Removal of "Extra" Teats

A normal udder should have four squarely placed teats of near Uniform size. But animals with one or even two extra teats are also seen. Such extra teats may be blind (a discrimination) or leaky (*serious disqualification*). They are also unsightly and hence should be removed even when the heifer calf is below six months old. At this age, it is easier to control the animal and job can be done cleanly.

After controlling the calf well, the region is thoroughly cleaned, suitable disinfectant like iodine applied, the previously marked extra teat drawn with one hand, the teat is snipped off at the level of its junction with the body with a sharp pair of scissors and some more disinfectant applied (Fig. 22.10). In older heifers putting a stitch or two and use of anti-haemorrage technique may be necessary.

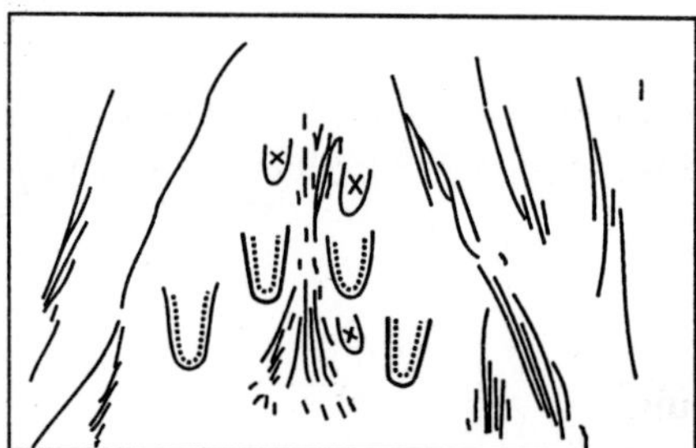

I

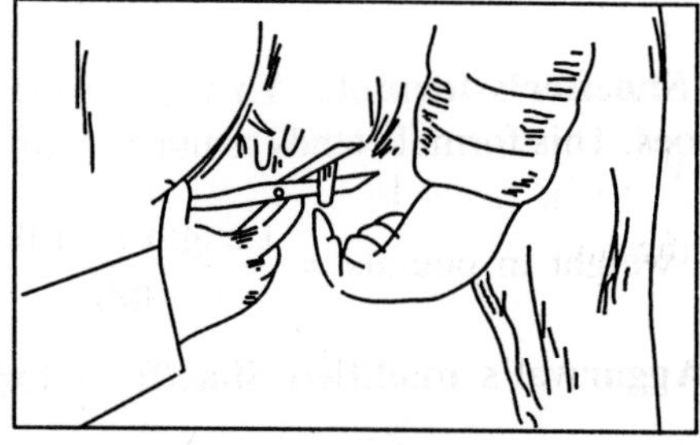

II

Fig. 22.10. I. Marking the extra teats to be removed. II. Holding the teat to be removed with thumb and fore finger of one hand and snipping off the marked extra teat with a pair of scissors. Note that the operator stands in a bent position towards one side facing the rear side of the animal.

22.4. WEIGHING CATTLE AND BUFFALOES

Scientific feeding of dairy animals is to be based on their body weight, and hence the necessity to weigh them periodically. Periodic weighings of stock gives an idea about their condition and well being. General loss of weight of animals in the herd is an indication of the existence of some chronic trouble like worm-infestation, nutritional deficiencies, infectious diseases like Tuberculosis, Johne's disease, etc.

Special platform scales are required for weighing large animals. On large farms a proper weighing yard should be made with collecting pens, cutting gates, passages and all such fittings which permit quick and efficient weighing of large animals.

All the animals to be weighed should preferably be starved for at least twenty four hours and kept away from water for six to eight hours before actual weighing. Otherwise, heavily fed and/or watered animals may weigh more than what they actually are; the extra weight being that of feed and water. A practical way of doing this will be to feed the animal for the last time in the morning, drain off watering sources on the same evening and weigh them the next morning. Some people weigh animals on three consecutive days and take the average as the correct weight.

When weighing on platform scales, let the animals on to the platform, allow the needle to come to a halt and only then record the weight. Slings can be used for hanging calves to spring balances. A simple sling can be made of a good cotton rope or a leather belt or one cycle tyre. Pass the sling under the belly of the animal. Adjust one strand of the sling just behind the elbows and the other just before the stiffles. Then lift the animal off the ground by holding the ends; the latter should be entangled to the hook of the balance.

22.4.1. Use of weight formulas

There is some relationship between the weight of an animal and its length, girth etc. After taking a number of observations on different animals, certain formulas have been developed to predict the weight of an animal from its length and girth. These formulas can be used for determining weight of animals when no balance is available. It is advisable to use formulas for the animal for which they are recommended. However, a formula is not a substitute for actual weight, the former gives rather a variable rough estimate of the weight. The girth of the animal is known by measuring the circumference of chest just behind elbow by means of a measuring tape. The distance between the point of shoulder and point of pin bones is taken as length, a measuring rod is needed for measuring length.

Formulas :

(1) **Shaeffer's formula.** This is the most common formula which can be used for cattle and buffaloes. This formula tends to under-estimate weights of very young and very heavy animals.

$$\text{Live weight in pounds} = \frac{\text{Length} \times \text{Girth}^2}{300}$$

(2) **Aggarwal's modified Shaeffer's formula**

$$\text{Live weight in Seers} = \frac{(\text{Girth})(\text{Length})}{Y}$$

where Y is equal to 9.0 if girth is less than 65″, 8.5 if girth is between 65-80″ and 8.0 if girth is over 80″ and *Seer* is equal to 0.93 kg.

(3) **Mullick's formula** for Buffaloes

$$X = 25.256\ (Y'') - 960.232$$

X = Estimate of body weight in lbs.

Y = Heart girth in inches.

(4) **Formulas of Bhandari and Team**

For dry and non-pregnant buffaloes

$$\text{BW in lbs} = -\,2123.73 + 26.356\ (\text{Girth in inches}) + 21.50\ (\text{Length in inches})$$

For dry and pregnant buffaloes

BW in lbs = – 1934.48 + 20.52 (Girth in inches) + 25.90 (Length in inches).

For lactating buffaloes

BW in lbs = – 2387.60 + 27.12 (Girth in inches) + 24.55 (Length in inches).

(5) **Formula for Hariana Cattle**

Body weight (kg) = 3.3 (Chest girth in cm) + (Posterior girth in cm) + 0.7 (Length in cm) – 490.

22.5. FARM OPERATIONS CALENDAR

The various practices to be carried out on the dairy farms routinely (daily or periodically) should be well planned and carried out punctually, so to say. Though minor changes in time of feeding, milking etc. may not be catastrophic but greater and longer lasting changes are; the dairy cows being a creature of habit. Any changes contemplated in timing of any practice should be brought about gradually.

It is ideal to draw up a schedule of daily farm operations for each farm so that the personnel working on farms can carry out various operations at regular timings. A model of such schedule is given in Table 22.1 This can be modified slightly by farmers/managers in different agro-climatic regions of the country to suit their local conditions.

TABLE 22.1. Schedule of Day to Day Operations on Dairy Farms (about 100 cows)

Approximate time (hours)	*Farm operations*
0300-0330	1. Cleaning/brushing of milch animals.
0330-0500	1. Feeding half of the daily concentrate ration just before milking. 2. Milking cows.
0500-0530	1. Delivery of raw milk (in cans) to the milk pick-up van of dairy plants and receiving previous day's empty cans. 2. Washing and disinfection of milking barns.
0530-0800	1. Cleaning of milch cow sheds. 2. Feeding of dry/green fodder to milch stock. 3. Cleaning farm premises. 4. Isolation of sick animals. 5. Isolation of *in-heat* cows for artificial insemination.
Note. Use milkers at the rate one for every 12-14 cows, for all the above operations. Milkers go off duty by 9.00 a.m. and farm labour come on duty.	
0800-1200	1. Cleaning calf, maternity, dry-stock, bullock and bull sheds. 2. Feeding half of the daily concentrate ration to calves, pregnant cows and bulls. 3. Exercising and grooming of bulls. 4. Treating sick animals. 5. Breeding cows that are *in-heat.* 6. Harvesting, chaffing and feeding of green fodders to all the stock. Mangers in all sheds should be filled with green fodder.
Note. Animals should be taken for grazing (if practised) between 9.00 a.m. and 2 p.m. in winter, and between 6 a.m. and 10 a.m. and again between 5 p.m. and 7 p.m. in summer.	
1200-1300	1. Lunch-cum-rest period for labourers.

1300-1500	1. Miscellaneous jobs of dairy farm : stock identification : periodical vaccination; preparation of concentrate mixture; repair of farm fences, fitting and repair of equipments; rope and halter making; weekly scrubbing and white washing of drinking water tanks; manure disposal/conservation; hay and silage making, periodical spraying of animal houses with suitable pesticides; periodical deworming of stock; clipping hair from sides and hind-quarters of cows; grooming; toe trimmings; dehorning of calves; attending to sale and purchase of livestock and their transportation; fitting and training of cows for show.

Note. 1. The dairy manager should plan the jobs well in advance in such a way that they are evenly distributed over the week. Some jobs may require longer time and the labour have to work extra time on such occasions.

2. Milkers come on duty by 1430 hours and remain upto 1830 hours, whereas general farm labour go off duty by 1700 hours.

1430-1500	1. Washing/brushing of milch cows by milkers.
1500-1630	1. Feeding the other half of daily concentrate ration to milch cows just before milking.
	2. Milking.
	3. Cleaning calf, maternity, dry-stock and bull sheds and feeding the other half of concentrate ration to calves, pregnant cows and bulls.
1630-1700	1. Delivery of milk (in cans) to milk pick-up vans of milk plants and collection of morning's empty cans.
	2. Washing and disinfection of milking barns.
	3. Feeding dry and green fodder to calves, dry stock and bulls.
1700-1830	1. Cleaning of milch cow shed.
	2. Feeding green/dry fodder to milch stock.
	3. Cleaning farm premises.
1830-0300	1. Night watchman comes in duty.

CHAPTER

23

Record Keeping

23.1. RECORD KEEPING AND HERD MANAGEMENT

The complicated job of management of a dairy herd requires sound planning for synchronisation of all the inputs and all the factors that are chain linked to each other (Guha, 1978). As for example, *"if we do not select good bulls we do not get good progenies, if we do not detect heat we cannot get the cows served or inseminated, if we do not get semen of good quality we do not get desired conception rate, if we do not get the cow to conceive we do not get the calf, if calf is not born we do not get milk, and at the end of the chain, if we do not get milk from cows efficiently we cannot get the desired volume of production."* Besides, if we do not know what is the performance of individual cows, we can not get rid of uneconomic cows and if we do not know which are the best cows we cannot do selection. If we do not supply required nutrients to cows we cannot get desired volume of production and level of reproductive efficiency. If we cannot grow fodder the cost of milk production goes high. If we cannot control disease and mortality, the proposition becomes uneconomic; it again affects both production and reproduction. If we cannot give sanitary and comfortable housing, again we end up with uneconomic performance. As such, it requires a day to day evaluation, planning, co-ordination and execution keeping the ultimate object in view.

Such an evaluation, planning, coordination and execution is possible only when the manager has all the facts pertaining to the stock before him. Such facts can be obtained from production and reproduction records. In other words, such records are to the dairy farmer/manager and breeder, what a good map of the locality to the army general at the war front. Great strides in development of dairy industry in UK, USA, the Netherlands and other countries have been taken only after initiation of scientific keeping and exploitation of performance records.

Guha, H. (1978). Efficient herd management and evaluation terminologies and information that help. *Indian Dairyman*, **30** : 613-617.

Thus, record keeping-management axis can aid decision making on the following lines (Choudhary, 1978).

23.1.1. On-herd Uses of Records

(*i*) Each animal in the herd is identified with respect to their production performances.

(*ii*) Close management and appropriate feeding levels can be provided on the basis of production level.

(*iii*) Efficiency of culling and selection for breeding purposes is increased, which in turn will considerably affect the performance level—increase profit rates.

(*iv*) Relative influences of feeding, management and breeding can be assessed on production performances.

(*v*) Livestock marketing can be promoted on the basis of performance records.

23.1.2. Off-herd Uses of Records

(*i*) Comparison of herd performances between and within breeds is possible on the basis of accurate recording.

(*ii*) Superior stock can be identified for extensive use in breeding programmes.

(*iii*) Herd and breed registration programmes can be implemented more effectively.

(*iv*) Realistic research and development planning and plan implementation becomes possible.

23.2. PERFORMANCE RECORDING ON FARMER'S HERDS

Recording practices in different countries have been developed depending upon the purpose and ultimate use of records. Milk recording system based on weighings and tests carried out by a recorder at intervals of 3-4 weeks throughout the year covering all milking animals in the herd is practised in most of the countries. Variations in the above system in both directions are also practised. In some countries where cattle population is concentrated, recording is done at fortnightly intervals and in other countries at 6 to 8 weeks intervals. The standard of service is the result of compromise between demand for accuracy on the one hand and the demand for economy on the other. Considerable economy in recording cost can be achieved under our conditions by doubling the length of visits of the recorder (*i.e.* 48 hours instead of 24 hours) and also the intervals between visits (Choudhury, 1978).

The twin objectives of performance recording in rural areas are : (*a*) it should help the dairy manager/farmer to produce greater quantities of milk at lower costs; (*b*) it should provide data to the administrators (for planning), research organisations (for processing and analysing the enterprise scientifically), breeders (for formulating selection and breeding programmes) and extension personnel (for giving feedback information to the farmers enabling them to manage cows better).

Thus the starting point and the ultimate beneficiary of herd recording should be the farmer. From experience in Kenya, Lindstorm (1977) emphasises that the farmers must still be convinced

---

Choudhary, A.I. 1978. Record keeping, herd books and cattle insurance : Programmes, problems and their solutions. *Management of exotic and crossbred cattles*. Proc. of a Subject Matter Seminar, 11-12 March, 1978. HAU. Hissar, p. 4.

Lindstorm, U.B. 1977. Milk recording in developing countries. *FAO Animal Production and Health Paper*, FAO. Rome, 1977, pp. 31-39.

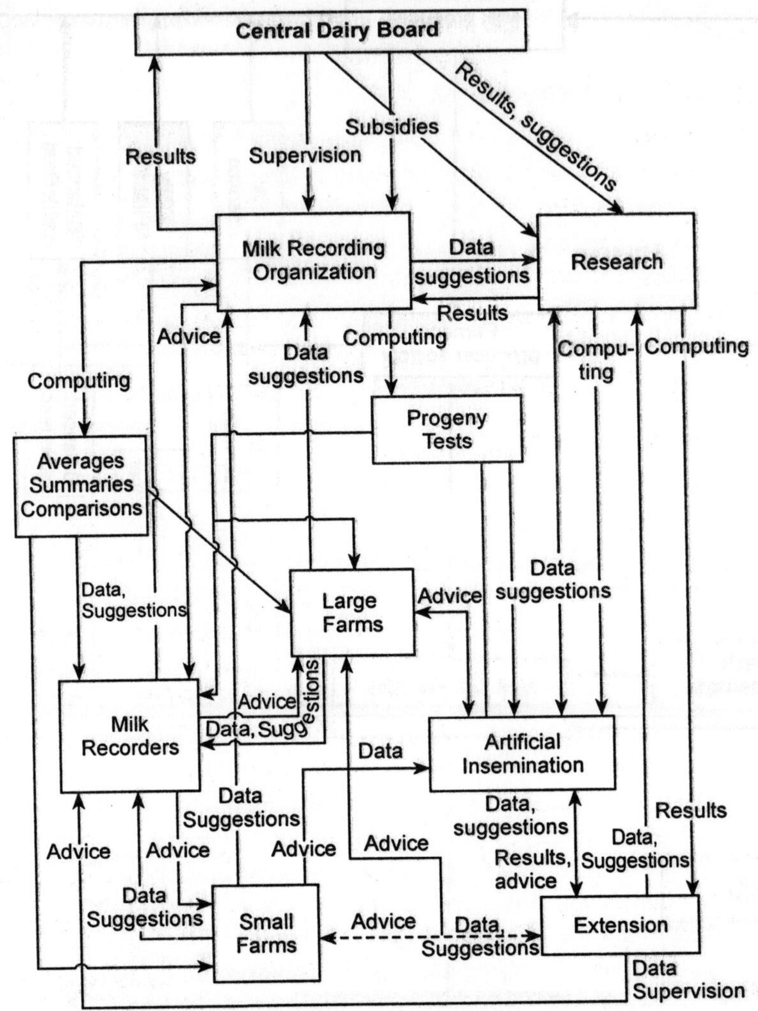

Fig. 23.1. Schematic outline of desired flow of information in milk recording and associated activities. (**Source** : FAO Animal Prod. & Health Paper No. 1)

that they have something, to gain from keeping records, even as a government recognises that milk recording should complement breeding programmes. The outline of desired flow of information according to this author is depicted in Figure 23.1. The National Dairy Development Board at Anand had developed an ambitious data collection, processing and utilisation of similar information in their Operation Flood areas (Roy Choudhury, 1978). Use of computers is envisaged. The goals to be achieved from this project can be best understood from Figure 23.2.

In India, as in most of the other developing countries, barring on government farms, practically no performance recording exists at the farmers level. Illiteracy, poverty, ignorance, lack of incentives etc. are the common causes for this situation. Based on the Kenyan experience Lindstorm (1977)

Roy Choudhury, P.N. 1978. Recording of Farm and field data. FAD/SIDA Seminar on Reproduction and Artificial Insemination. *FAO Production and Health Paper* 13. FAO. Rome, p. 355.

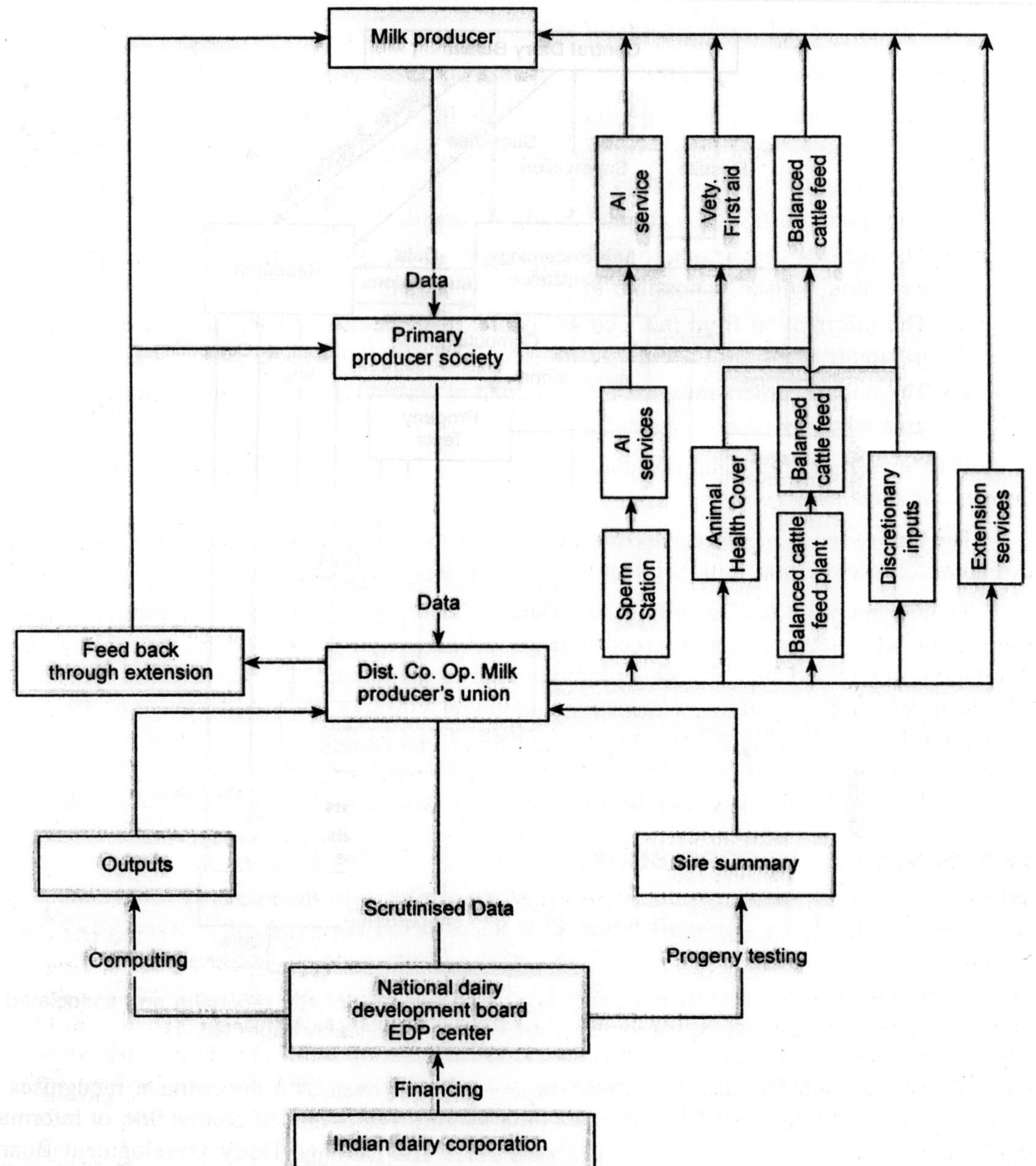

Fig. 23.2. Schematic outline of flow of information and associated activities in field recording system (Roy Choudhry, 1978).

suggests creation of the following incentives to farmers as means of developing field recording systems in rural areas of the Third World Countries :

(*i*) Paying for the fat and protein content of milk.

(*ii*) Giving free advice on feeding and management to farmers supplying records.

(*iii*) Providing farmers participating in recording with AI of good bulls free of charge or at low cost.

(*iv*) Paying for daughter records of selected young bulls.

Other major practical requirements of performance recording in developing countries are :

(*i*) The scheme must be simple *i.e.* involve little "paper work" for the farmer.

(*ii*) The scheme must appeal to the average farmer; he/she must feel that it benefits in managing the animals better.

(*iii*) The direct costs to the farmer should be small.

(*iv*) The records kept should enable the farmer to identify his best and poorest producers, and should make it possible to progeny test bulls.

(*v*) The information from the records should be made available to the farmer quickly and in simple form, and should be utilised for extension and research purposes.

(*vi*) The milk recorders must visit each farmer at least once a month to maintain contact and give advice.

Thus emphasis should be on quick and efficient flow of the information collected, to the farmers for their own good.

The type of recording sheets used on farmer's farms in Kenya (Lindstorm, 1977) is reproduced in Figure 23.3. Note that milk production recorded therein is based on test day production.

Though monthly test recordings were found to be fairly accurate in North America and Europe for selection and monthly recording for progeny testing (McDaniel, 1969), in Kenya, a developing tropical country, it seems preferable not to go beyond a 14-day interval for recording if a relatively high accuracy in selecting individual cows is desired (Lindstorm, 1976). The correlation between actual yield and estimate made from 14-day test recordings was 0.991 with 95 per cent confidence limits for estimated individual yields of – 9.1 to + 12.2 per cent (– 251 to + 337 kg). Such practices may be valid in many other countries with similar conditions.

23.3. INDIVIDUAL RECORD BOOKS

While there is need for using concise single sheet records in farmer's herds for reducing paper work and for clarity, on organised farms all types of detailed records of different stock have to be kept. They cover milch, breeding, calf and other registers. The essence of the basic information needed on individual cows comprises of the list given in Table 23.1 (Ellis, 1979). In view of the ambitious programme of recording of farm and field data in India taken up by the National Dairy Development Board, Anand (Roy Choudhury, 1978) it is advisable that existing and venturing dairy farmers consult the above organisation so as to follow a uniform pattern of performance recording in the country. The full complement of proformas for various records are given in Table 23.2 to 23.14.

On any farm it is useful to keep a sort of general *Farm Log Book* handy. Day to day and even hour to hour happenings of all sorts (calving, service, vaccination and so on) may be jotted

McDaniel, B.T. 1969. Accuracy of sampling procedures for estimating lactation yields : a review, *J. Dairy Sci.*, **52** : 1742-1761.

Lindstorm, U.B. 1976. Studies on milk records from Kenya. I. Accuracy of milk recording with various intervals. (Cited from Lindstorm, 1977).

Ellis, P.R. 1979. Management, health control and recording systems for improved dairy herds. Paper presented at the *Second FAO Expert Consultation on Breed Evaluation and Crossbreeding—Dairy Cattle Breeding in Humid Tropics*. Hissar, India. Feb., 1979.

A.—Permanent information (one sheet per cow)

Cow's ear (tattoo) No. ....
Cow's birth date ....

Cow's breed ....
Sire of cow ....
Dam of cow ....

Lact. No.	Service dates	Bull used	Calving date			Calf		Lactation length (days)	305 day production (kg)	Total production (kg)	Percentage fat or protein	Days dry
			Day	Month	Year	Sex	Ear No.					
1	....											
2	....											
3	....											

B.—Test Day Production

Test day	Cow No.	Milk yield kg (AM or PM)	% protein (% fat)	Remarks on calving, services, drying off, illnesses

Fig. 23.3. Simple recording sheets used in performa recording of farmer's cows in Kenya (Reproduction from FAO Animal Production on Health Paper No. 1, FAO, Rome, 1977, p. 36)

down immediately in this Log Book. Later, the events can be reported leisurely in the proper record book, *e.g.*, a calving in the Calving Register, a vaccination in the Health Record etc. A likely Performa for this Farm Log Book can be seen in Table 23.2. The author observed a similar book being effectively used on dairy farms in USSR. Such a practice will allow keeping records in duplicate (once in the Log Book and once in the specific register) as an insurance against loss or destruction. Sheets of the proposed Log Book can be hung in the manager's office and each sheet filed away the next day after the transfer postings are made.

Essentially two main methods of milk recordings are adopted (with many modifications for specific purposes)—*the cow-day method* and *the lactation method. The cow-day method* is particularly applicable to commercial milk producers and the end result is obtained by relating the total production of milk in a herd for a 12 months period to the number of cows, which is adjusted to feeding cost. This is an estimate of average herd performance. *The lactation method* meets the requirements of pedigree breeders where emphasis is given to lactation yields as distinct from annual yield. The breeder is more concerned with specific performance of individuals within his herd and the comparisons between them so as to assess the results of his breeding plan.

TABLE 23.1. A Basic Cow Record (Ellis, 1979)

Key data :	Identity number (unique)	
	Status (Pedigree, non-pedigree or grading up)	
	Line Number (farm cow number)	
	Cow	Name
		Breed
	Sire	Name
		Number
		Breed
	Dam	Name
		Identity Number (unique)
		Breed
	Date of Birth	
	Age at first calving	
	Weight at first calving	
For each Lactation :	Lactation number	
	Date of calving	
	Milk yield	
	Days in milk	
	Fat per cent	
	Protein per cent	
	Test of quality	
	Date of actual drying-off	
	Date of conception	
	No. of services	
	No. of mastitis cases	
	No. of infertility treatments	
	No. of lameness codes	
	No. of other health codes	
	Date sold/died and codes	

Modification of both the methods are practised. Varying standards of lactation, *viz.*, natural and curtailed are used to facilitate comparisons, exclusion of abnormal records, grouping by age, use of past records either due to fresh introduction of animals in the herd or disposal during the year, standardisations to mature equivalent, fat correction, lactation period corrections etc. are common variations. These details are discussed in chapter 17, on selection. There is little doubt that in work connected with herd improvement, the lactation performance is preferred to facilitate breed studies by herd book societies and research organisations. The lactation yield and butter fat records on each cow supplemented by age at calving (number of lactation), lactation length, season/month of calving, reasons for termination of lactation etc. may be useful informations for comparing performance on lactation as well as life time basis.

TABLE 23.3. A Likely Specimen of a Farm Log Book

FARM LOG BOOK

(A Hour-to-Hour Record of Happenings on Farm)

Dated : 11-6-1989

06.00-08.00 Cow R 361 calved, calf tattooed 846; Calf 346 died sent for PM	**18.00-22.00** Cow 393 calved, calf not tattooed.
08.00-10.00 Cow 301 bred, Bull PT 64 Cow 296 bred, Bull PT 64 Cow 103 bred, Bull PT 83 Semen collected from Bull PT 64 and 83	**22.00-02.00** Dystokia-cow B65—veterinarian called
10.00-12.00 Fodder harvested 38 Qts Maize All calves attaining 6 months vaccinated FMD	**02.00-06.00**
12.00-14.00 Cow 382 treated for Mastitis	**NOTES** Arranged transport of milk to Delhi by private van of Jain and Sons.
14.00-16.00 Sold 3 calves (Nos. 38, 42 and 60) and old cows Nos. 72 and 61.	
16.00-18.00 Vaginal prolapse p.p. cow R 361, veterinarian called	

TABLE 23.3. Cattle History and Pedigree Sheet

Identification of the animal	*Date of birth or date of purchase and age at purchase*	*Sire No.*	*Dam No.*	*1 2 3 4 5*	*Paternal grand dam No.*	*Maternal grand size*	*Maternal grand dam*
			Dam's milk yields Dam's lactation length Dam's days dry		1st L.Y. 1st L.L.		1st L.Y. 1st L.L.

GROWTH

1 2 3 4 5 6 7 8 9 10 11 12 13 14 15 16 17 18 19 20 21 22 23 24 25 26

27 28 29 30 31 32 33 34 35 36 37 38 39 40 41 42 43 44 45 46 47 48 49 50 51 52 (Weeks)

13 14 15 16 17 18 19 20 21 22 23 24 (Months)

REPRODUCTION AND PRODUCTION

SERVICE			*CALVING*			*MILK YIELD*								*REPRODUCTION*	
Sr. No.	*Bull No.*	*Date*	*Date*	*Sex of Calf*	*Weight dam at calving*	*Total*	*300 days*	*Lactation length*	*Date dried*	*Days dry*	*Peak yield*	*Date of peak yield*	*Reason for incomplete lactation*	*Service period*	*Calving interval*
1.															
2.															
3.															
4.															
5.															
6.															
7.															
8.															
9.															
10.															

IMMUNISATION AND HEALTH TEST RECORD

Date	*Immunisation*		*Health Test*		
	Against which disease	*Remarks*	*For which diseases*	*Result*	*Remarks*

GENERAL INFORMATION ABOUT THE COW

Record all information relevant to the history of cow like veterinary aid (excluding immunisation), breeding pecularities, udder condition, lactation features, etc.

Date	*Remarks*

TABLE 23.4. A Simplified Cattle History Sheets for Use in Farmers' Herds or even on Organised Farms

Name of the owner farm :					*Tattoo No. :*		*Breed :*			
Address :					Date of birth : Purcahse date : Value : Rs.					
Pedigree Particulars	*Dam :*		*Breed :*		*T. No. :*		*Milk yield :*			
	Sire :		*Breed :*		*T. No. :*		*Sire dam's yield :*			
Calving Order	*I*		*II*		*III*		*IV*		*V*	
Service	*Date*	*Bull No.*	*Date*	*Bull No.*	*Date*	*Bull No.*	*Date*	*Bull No.*	*Date*	*Bull No.*
1st service 2nd service 3rd service										
Calving Date										
Sex/Tatto No. of calf										
Lactation Order	I		II		III		IV		V	
Days in milk										
Milk yield (lit)										
Days dry										
305 days yield (lit)										

TABLE 23.5. Cattle Breeding Register

Sr. No.	*Cow No.*	*Date of calving*	*First service*			*Second service*			*Third service*			*First pregnancy diagnosis (Date/Result)*	*Second pregnancy diagnosis (Date/Result)*	*Expected date of calving*	*Date on which to be dried off*	*Actual date of calving*	*Weight of calf*	*Sex of calf*	*No. of calf*	*Remarks*
			Date	*Time*	*Bull No.*	*Date*	*Time*	*Bull No.*	*Date*	*Time*	*Bull No.*									

TABLE 23.6. Calving Register

Cow No.	*Due date*	*Date of calving*	*Sire*	*Calf No.*	*Calf sex*	*Birth weight*	*Breed*	*Remarks*

TABLE 23.7. Calf Register

Cow No.	*Date of birth*	*Date of numbering*	*Calf Number (ear)*	*Sire*	*Dam*	*Sex of calf*	*Birth weight*	*Disposal*		*Remark*
								How	*Date*	

TABLE 23.8. Growth Record of Young Stock

Totto brand no.	*Date of birth*	*Weight at birth*	*Weight at weeks*												*Weight at months*												*Weight at 1st service*	*Weight at 1st calving*	*Remarks*
			1	*2*	*3*	*4*	*5*	*6*	*7*	*8*	*9*	*10*	*11*	*12*	*4*	*5*	*6*	*8*	*10*	*12*	*14*	*16*	*18*	*20*	*22*	*24*			

TABLE 23.9. Daily Feeding Register for the month ……… 20…

Date	*Number of animal*	*Concentrate (kg)*			*Green fodder (Qt)*			*Dry fodder (Qt)*			*Other feeds*		
		Received	*Issued*	*Balance*	*Received*	*Issued*	*Balance*	*Received*	*Issued*	*Balance*	*Received*	*Issued*	*Balance*

TABLE 23.10. Fodder Crop Register

Plot No.	Name of crop	Human labour (Rs.)	Labour cost (Rs.)	Bullock labour (Rs.)	Tractor charges (Rs.)	Material seed etc. (Rs.)	Electric charges (Rs.)	Canal charges (Rs.)	Land revenue (Rs.)	Total (col. 4 to 10) (Rs.)
1	2	3	4	5	6	7	8	9	10	11

Interest on cash (Rs.)	Total cash costs (Rs.)	Land rent (Rs.)	Depreciation (Rs.)	Interest on fixed costs	Total FC 14+15+16 (Rs.)	Total Expenditure (13 + 17)	Area of the crop (hec)	Cost of yield per hectare (Rs.)	Crop yield	Cost per qunital (Rs.)	Remarks
12	13	14	15	16	17	18	19	20	21	22	23

TABLE 23.11. Herd Health Register

Date	Animal No.	When trouble started history	Symptoms	Diagnosis	Treatement	Result cured dead/sold	Name of the veterinarian	Cost of treatment	Date of vaccination	Remarks

TABLE 23.12. Insemination Slip

Date………… **Morning**

Cow No.	Shed No.	Date last calved	Total heats	Total insemination	Last insemination	Bull allotted	Remarks

Date………… **Evening**

Cow No.	Shed No.	Date last calved	Total heats	Total insemination	Last insemination	Bull allotted	Remarks

TABLE 23.13. Daily Milk Record for the Month of ....20...

Sr. No.	*Cow No. or Name*	*Date of calving*	*1st*		*2nd*		*3rd*		*4th etc.*		*30th*		*31st*		*Total*
			A.M.	*P.M.*	*A.M.*	*P.M.*	*A.M.*	*P.M.*	*A.M.*	*P.M.*	*A.M.*	*P.M.*	*A.M.*	*P.M.*	

TABLE 23.14. Lactation Record

Cow No.	*Milk Yield (Average fat per cent in brackets)*												*Lactation yield (Average lactation fat per cent)*	*Lactation period*	*Dry period*	*Remarks*
	Jan.	*Feb.*	*March*	*April*	*May*	*June*	*July*	*Aug.*	*Sept.*	*Oct.*	*Nov.*	*Dec.*				

TABLE 23.15. Daily Livestock Register

Date	*Cows*		*Bulls*	*Calves*		*Heifers bullocks*	*Total live-stock*	*Milk yield*			*Addition during day*		*Deduction during day*		*Remarks*
	Milch	*Dry*		*Males*	*Females*			*A.M.*	*P.M.*	*Total*	*No.*	*From where*	*No.*	*How*	

CHAPTER

24

Management of Producing Cows

Etgen and Reaves (1978) described the milch cow (or buffalo) as a sort of machine. "To obtain most economically produced and high quality product (*milk*) out of this machine, several things have to be taken care of First, the machine (*cow*) should be located (*housed*) in a comfortable building. The machine (*cow*) should be fed with low cost but good quality raw products (*feeds, fodders, water,* etc.). The machine (*cow*) has to be used very carefully and well maintained (*handling, washing, grooming, health programme*, etc.). By constant endeavours (*selection* and *breeding*) the design and performance (*anatomy* and *physiology*) and so the efficiency of the machine (*cow*) has to be ever improved. Obsolete (*old* and *low yielders*) machines are to be dispensed with (*culled*) and newer ones (*herd replacements*) to be procured. Finally, the end product (*milk*) should be harvested (*milked*) in such a way that its quality (*bacterial status*) and quantity (*yield*) should be of highest order".

The above gives an idea about the broad components of managemental practices involved in the upkeep of the producing herd. Let us observe some of these in greater detail.

24.1. COMPONENTS OF MILCH-COW MANAGEMENT

The aim of the farmer should be to see that there is as high a proportion of cows in the herd that are in milk at any given time, since more cows in milk means more total production of the herd. The regularity of reproduction is one of the most important factors that determines the proportion of cows in milk to total cows. The higher the calving interval (or service period) the lesser will be the proportion of milch cows in the herd (Table 24.1).

Etgen, W.M. and Reaves, P.M. 1978. Dairy Cattle Feeding and Management, 6th ed. John Wiley & Sons, New York.

TABLE 24.1. Production Status of Herds Depending on the Length of Calving Interval (Guha, 1978)

Calving Interval (months)	*Days in Milk*	*Days dry*	*Per cent cows of herd in milk*	*Per cent cows dry in herd*	*Fertility index**
12	305	60	83	17	1.00
13	305	90	77	23	0.92
14	305	120	71	29	0.87
15	305	150	66	34	0.80
16	305	180	62	38	
17	305	210	58	42	
18	305	240	55	45	0.67
19	305	270	52	48	
20	305	300	50	50	

*Fertility index = Calves produced per cow per year, as per the Australian practice. (Lamond and Campbell, 1970), which figure can be higher than one if intercalving periods are less than 12 months.

Guha, H. 1978. Efficient herd management and evaluation—Terminologies and information that help. *Indian Dairyman*, 30 (9) : 613-17.

The components of different managemental practices that help in greater number of cows remaining in high state of milk production for longer periods include the following :

(1) Feeding balanced rations as per nutrient requirements, especially making available good quality green fodders round the year (*see* Chapter 5 & 6). High yielders require special attention. The producing herd may be divided into classes—high yielders, medium yielders, low yielders, and housed in separate sheds. This helps in feeding milch cows more efficiently, especially in large herds.

(2) Maintenance of high standard of health in the herd (*see* Chapter 20), including hygiene and sanitation and housing in clean comfortable houses (*see* Chapter 19).

(3) Prevention of exposure to all possible managemental stresses on producing cows by : (*i*) careful and gentle handling and movement of stock in and out of the farm; using suitable tranquilisers if necessary, (*ii*) avoiding slippery floors and sharp bends and protruding objects in paths along which cows are moved; (*iii*) avoidance, as far as possible, of noise near about cows due to chaffing machines, vehicles, barking dogs, boisterous labourers etc. avoiding over-stocking in the sheds and provision of ample feeding and watering space (*see* Chapter 19); (*v*) grouping of cows according to size (age) and/or level of production; (*vi*) retaining in the herd cows with docile temperament only and culling out of the herd nervous and aggressive cows. Such cows can disturb good cows also. The temperament of dairy cows has a considerable influence on behaviour of cows in yard, consumption of rations, ease of milking as well as milk yield and composition (Fig. 24.1) of Red Sindhi, crossbred and buffalo cows (Nayak and Mishra, 1982). Observations on a large number of crossbred cows at Karnal (Gupta and Mishra, 1979), too revealed that

Lamond, D.R. and Campbell, E.A. 1970. Dairy Cattle Husbandry. Augus and Robertson, Sydeny, p. 100.

Nayak, S. and Mishra, M. 1982. Personal communication.

Gupta, S.C. and Mishra, R.R. 1979. Effect of dairy temperament of Karan Swiss cows. *Indian J. Dairy Sci.* **32** : 32-36.

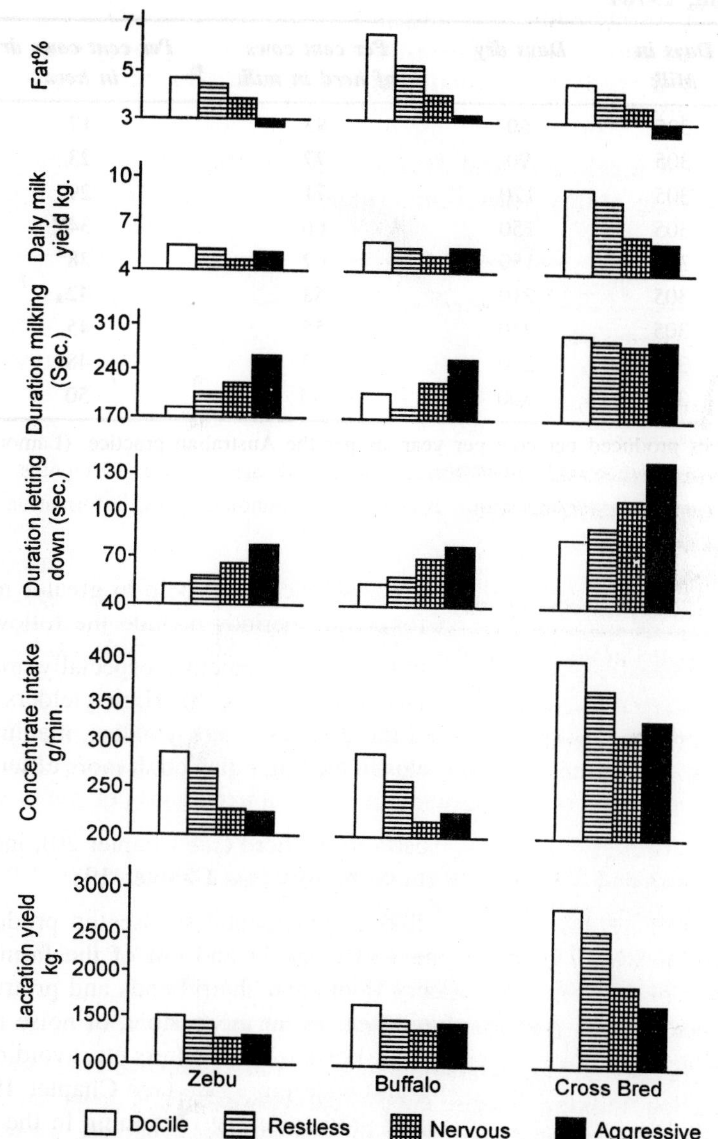

Fig. 24.1. Milking and feeding behaviour of dairy animals in relation to their temperament score docile, restless, nervous and aggressive (from data of S. Nayak and M. Mishra, 1982, a personal communication).

temperament score of cows is negatively associated with almost all milking behaviour traits (Table 24.2). It should be noted, however, that good feeding and watering arrangements and adequate space per animal can reduce the aggressive behaviour of cows considerably. Contrariwise, poor management can even make docile cows nervous or even aggressive with the resultant drawbacks.

(4) Stick to a regular schedule of different managemental practices on a dairy farm, otherwise cows become uneasy and restless. A model of daily dairy farm operations calendar is given in Table 22.1. Farmers and managers can slightly alter the same to suit their local conditions and

requirements. Any changes in routine should be affected gradually spread over a period of a week or so, since cattle are creatures that settle down to routines, any sudden change is liable to cause primary disturbance in the behaviour pattern and this can, in turn, result in accident or loss of production, partially or totally.

TABLE 24.2. Simple Correlation and Regression Coefficients of Various Components of Milking ability with Temperament at Test of Cows (Gupta and Mishra, 1979)

Particulars	*r*	*b*
1. Temperament × Initial milking rate	0.679**	0.218
2. Temperament × First mid-milking rate	0.698**	0.384
3. Temperament × Second mid-milking rate	0.735**	0.484
4. Temperament × Peak flow rate	0.689**	0.214
5. Temperament × Average rate of milk flow/min	0.765**	0.159
6. Temperament × Average milk yield/milking	0.293*	0.275
7. Temperament × Let down time	0.789**	9.980
8. Temperament × Milking time	0.438**	0.457
9. Temperament × Concentrate feeding rate/min	0.280*	8.618

Note. Temperament adjudged in the following sequence : Docile-1, Slightly restless-2, Restless-3, Nervous-4, Aggressive-5.

Apart from effecting changes in routines gradually and persuasively, farmers can take advantage of the dry period, and to some extent the early part of lactation, during which time cows seem to accept changes in routine without much apparent loss of production (Russel and Williams, 1978).

(5) Gentle but brisk and deft carrying out of pre-milking, milking and post-milking operations inside and near the milking parlour (see Chapter 10).

(6) Protection of cows and buffaloes, especially high yielders, against thermal stress conditions of summer. This is a key factor for success of dairying in tropics.

(7) Maintenance of high reproductive efficiency in the herd.

The reader might have already acquainted himself from other chapters of this book, principles of feeding, reproduction, milking, disease control etc. However, it is proposed here to discuss in greater detail two practical aspects of milch cow managements, namely, protection against hot weather conditions and augmentation of herd reproductive efficiency.

24.2. SUMMER MANAGEMENT OF PRODUCING COWS

Exposure to hot climate for varying periods affects animals directly (in contrast to indirect effect of climate via fodders and pests) evoking short and long term reactions in dairy animals (see Chapter 4). These have an adverse effect on their production and reproduction efficiency (Fig. 24.2). The summer depressing effect may have the cumulative effect of heat stress and shortage of fodders and their poor quality. From Fig. 24.2, it could be seen that the intensity of summer-effects on production of cows was in high grade crossbreds, low grade crossbreds and

Russel, K. and Williams. S. 1978. *The Herdsman's Book.* Farming Press Ltd. Ipswich. p. 147.

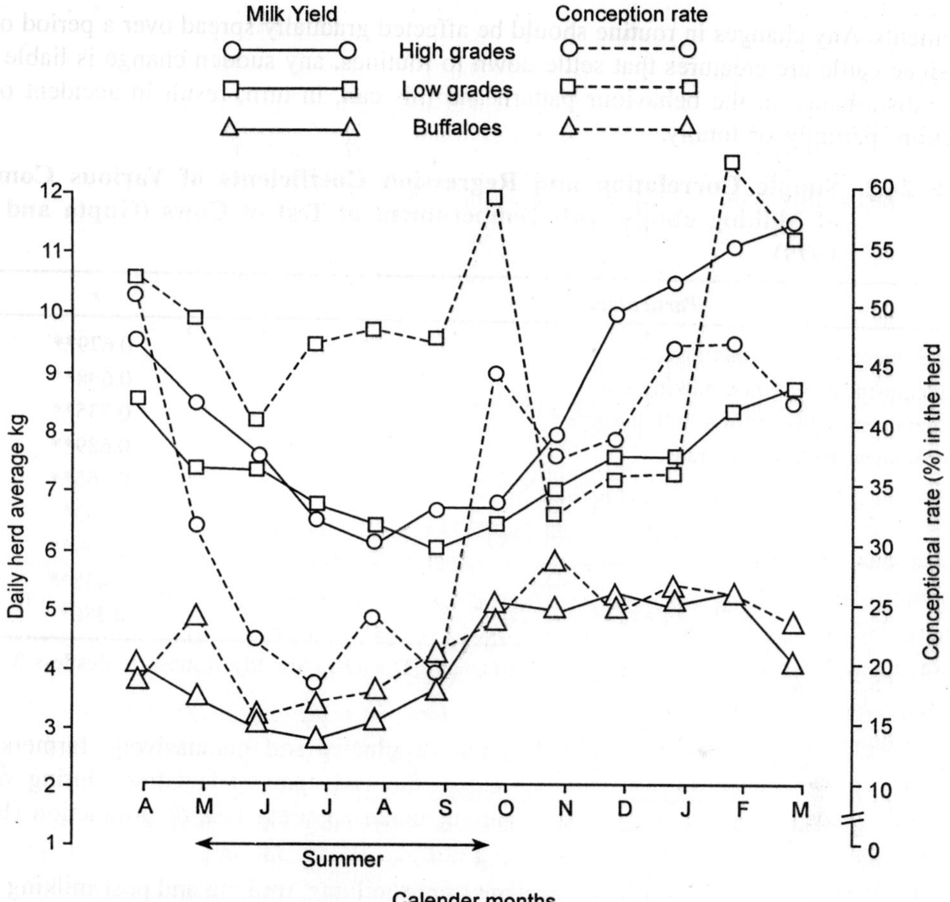

Fig. 24.2. Daily milking average and monthly conception rates in a herd of high grade crossbred (1/2 to 7/8), low grade crossbred (1/8 to 3/8) and buffaloes during different calender months at the Military Dairy Farm, Ambala, in the semi-arid north west. Daily average ambient temperature and relative humidity respectively in this locality are 27.4°C and 75.5 per cent in AMJ, 28.4°C and 93 per cent in JASO; 15.8°C and 82 per cent in NDJ; and 26.0°C and 72 per cent in FM (Drawing from data of Singh, 1981). The letters stand for different calendar months.

buffaloes in that order. (There is another but often ignored influence on animals of climate × husbandsman interaction effect). As a homeothermic measure the animal will try to get rid of excess heat, or cut down heat production in the body by reducing feed intake and utilization, or it may bring into operation both these means. Such a strain on the part of the animals is against the interests of the farmer, whose aim should be to see that :

(*i*) Cows do not get exposed to heat in the first place;

(*ii*) provide means that aid in heat loss by animals; and

(*iii*) take steps to see that the animals consume feeds adequately.

Hence the need for special effort on behalf of the farmer/manager to see that his animals are protected against unfavourable effects of the summer season. Such farmer's efforts perhaps can be put down as follows :

Objective	*Management effort*
A. Protection against heat stress	1. Green cover—landscaping around sheds
	2. Proper housing/shady trees/shades.
	3. Washing/sprinkling/watering.
	4. Provision of cooled drinking water.
	5. Cooling devices : Fans, *Khaskhas* panels with fans, desert coolers, air conditioners.
B. Improving feed consumption.	1. Feeding *green* roughages, silage, hay.
	2. Provision for night feeding.
	3. Grazing only if green grass is available—early in the morning and late in the afternoon.
	4. Use of wholesome concentrate mixtures.

The items listed above are mostly self-explanatory. Most of the management practices have established scientific basis and are experimentally proved. While it is out of the scope of this chapter to go into details, some aspects may be studied briefly.

Green cover (lawns, bush, creepers trees etc.) around buildings reduce reflection of incident radiation inside the sheds. Trees that can be planted for this purpose are discussed in Chapter 19.

In India and abroad, on *Bos taurus* cattle, crossbreds, zebu as well as on buffaloes, the efficacy of washing/showering/wallowing during hot part of the day in summer in reducing their rectal temperature by 1-1.5°C has been clearly established by several workers. Showering in the shed itself seems to be as efficacious as wallowing, without the attendant problems of hygiene and taking the buffaloes to the wallow in open sun, as is the common practice in villages (Sastry and Pal, 1973). Showering thrice between 12 noon and 2.30 P.M. or providing drinking water cooled in earthern pitchers (21°C), or a combination of both the measures during summer, sustained lactating buffaloes; physiological reactions within norms and allowed them to utilize dry matter better and produce moderately higher yields of milk of better quality (Radadia *et al.*, 1980a and b). In fact in the 40-head buffalo farm of the Haryana Agricultural University due to practicing the above mentioned thermal protection measures great seasonal fluctuations in monthly herd milk production have been smoothened out considerably. Mean monthly milk production of this herd in each season during 1979-80 was as follows (Sastry, 1980) :

Sastry, N.S.R. and Pal, R.N. 1973. You can make the buffaloes breed round the year. *Indian Farming*, **23** : 35-38.

Radadia, N.S., Pal, R.N., Juneja, I.J. and Sastry, N.S.R. 1980a. Effect of certain summer managemental practices on lactating Murrah buffaloes. I. Feed and Water Intake, *H.A.U.J. Res.* **10** : 144-147.

Radadia, N.S., Pal. R.N., Sastry, N.S.R. and Juneja. I.J. 1980b. Effect of certain summer managemental practices on lactating Murrah buffaloes. II. Milk yield and composition. *H.A.U.J. Res.* **10** : 152-156.

Sastry, N.S.R. 1980. Problems of seasonal 'Peaks' and 'Troughs' in buffalo milk production. Proc. Summer Inst. *Buffalo Management Systems*, 20-30. June, 1980, HAU, Hissar, p. 128.

Season	*Mean monthly milk production (lit.)*	*Mean service period (days) of buffaloes calving in these seasons*
Cold (NDJ)	3805.9	265
Spring (FM)	3354.1	141
Hot-dry (AMJ)	3472.2	162
Hot-humid (JA)	3675.5	133
Mild (SO)	3976.1	169

Maintenance of uniform monthly milk production of the herd was possible also due to overcoming the problem of summer sub-fertility in this species by the said thermal protection measures. The latter helped in a more even distribution of calvings in different months. Note that the ranges of service period (last column) were much smaller to generally reported values. Similar thermal protection measures were found adequate in ameliorating hot climate and sustaining good performance of European cattle in the Near East (Ansel, 1977).

24.3. MAINTENANCE OF HERD REPRODUCTION EFFICIENCY

Reproductive and Productive Life Relationships. A study of the different aspects of reproduction in cattle (Chapter 7) and variability, heritability and repeatability of reproduction traits of dairy bovines (Chapter 16), would make it clear to the reader that high levels of reproduction efficiency in the herd can by far be achieved by proper management and feeding of stock. Renowned bovine reproduction specialists like Salisbury and Van Demark (Salisbury *et al.*, 1978) maintain that, besides directly affecting reproductive efficiency of a herd of cows, managemental practices can greatly enhance or hinder various measures of prevention and control of reproductive disorders. It is necessary to pay stricter attention to management practices than to any other factors that affect fertility.

Efficient reproduction is very important to the overall economics of the farm. Reproduction in dairy bovines produces herd replacements as well as initiates milk flow. Thus, for the greatest overall returns per cow a long life, regular reproduction and a minimum of non-reproductive (infertile) periods are essential.

Considering a 9 year herd life (3 years upto first calving and 6 years for completion of 4 lactations) to be optimal for a dairy cow, the cow will actually be producing milk for only about 3½ years. This is the useful life of the cow as against her total herd life, which is different for different types of bovines (Table 24.3). It is the useful life which actually interests the farmer. The proportion of useful or productive life of the cow in its total herd life is dependent directly on her reproductive performance. With longer age at first calving and longer dry periods (contributing to longer calving interval) zebu and buffalo cows have smaller useful lives in the herd (Table 24.3). The more the useful life can be extended by good husbandry practices, the more profitable the cow will be.

Ansel, R.H. 1977. Maintaining European dairy cattle in the Near East. F.A.O. *Animal Production and Health Paper I,* Animal Breeding, F.A.O. Rome, pp. 73-79.

Salisbury, G.W., Van Demark, N.L. and Lodge, J.R. 1978. *Physiology of Reproduction and Artificial Insemination of Cattle,* 2nd ed. W.H. Freeman and Company, San Francisco.

And the useful life of the dairy bovines can be extended by reducing age at first calving, service period and dry period through good husbandry. The solution of this problem is largely upto the individual farmer or farm manager. The various means of sustaining high reproduction in dairy herds are discussed in Chapter 7.

TABLE 24.3. Total Herd Life and Actual Productive Life of Different Dairy Bovines Obtaining in Tropics as Influenced by their Breeding Performance

Type of bovine	*Age at first calving (days)*	*Days in milk upto end of fourth lactation (days)*	*Days dry upto end of fourth lactation (days)*	*Total herd life upto end of 4th lactation (days)*	*Productive life as per cent of total herd life*	*Yield per day of life at actual levels of production (lit.)*	*Yield per day of life (uniform rate of 2000 lit. lactation yield) (lit)*	*Time the cow should spend in the herd to produce 10,000 litres of milk at performance level given in the previous column*
Zebu	1282	1392	360	3034	42.5	3.0	2.6	10 yrs, 10 months
Crossbred	1011	1224	284	2519	48.6	5.6	3.1	9 yrs.
Buffalo	1390	1172	556	3118	37.5	1.9	2.5	11 yrs, 2 months

Note. Calculations are based on actual performance of animals on organised farms in India and East Africa as reported in literature.

CHAPTER 25

Management of Dry Cows

From the day the cow stops giving milk to the day it calves next, can be considered as dry period. The dry cow does not give milk but it is pregnant. In abnormal cases (infertility) the dry cow may be non-pregnant.

25.1. WHY DRY PERIOD

The mammary gland functions incessantly during the entire lactation period, which results in considerable wear and tear in the secretory tissue of the udder (see Chapter 10). Cows, especially high yielders, lose condition (body weight) during lactation. This is so because their own body tissue may be broken down to maintain the high milk flow; great quantities of nutrients, especially calcium and phosphorous, being drained through the milk.

In milch buffaloes of the Murrah breed yielding above 1400, 1100 to 1400 and below 1100 litres per lactation (with 6 to 7 per cent fat), respectively, body weight loss during the first five months of lactation was found to be 5.8, 5.5 and 2.0 per cent of the body weight at the time of calving (Bhalaru *et at.*, 1980). Similar phenomenon in purebred *Bos taurus* cows too is well established.

The importance of the dry period is to give rest to the cow's udder and an opportunity for the cow itself to recuperate its condition lost during the lactation. Repair and regeneration of the secretory cells of the udder takes place too. Allowing dairy cows a dry period has been shown to result in significantly higher production during the succeeding lactation; there being a moderate positive correlation between the two characters (see Chapter 16). On the other hand, uninterrupted lactation between calvings is known to impair succeeding lactation production (Etgen and Reaves,

---

Bhalaru, S.S., Dhillon, J.S. and Tiwana, M.S. 1980. Body weight changes during early lactation and milk production in buffaloes. Progress Rep. V. Workshop AICRP Buffalo, Udaipur, 13-14 Oct, 1980, pp. 16-17.

Etgen, W.M. and Reaves, P.M. 1980. *Dairy Cattle Feeding and Management.* John Wiley and Sons, New York, 1978, p. 289.

1980). Thus dairy cows need be dried off before the ensuing calving and should be, what is known in Western parlance, "steamed up" for the next lactation.

An ideal dry period is some thing around 50 days, though in Zebu and buffalo cows it is mostly much longer than this. Also, unlike in case of dairy breeds of the West, no special efforts are needed on the part of the farmer to dry cows; the latter have normally shorter (though variable) lactations which end on their own accord, resulting in too long a dry period (Williamson and Payne, 1978). The *Bos taurus* × *Bos indicus* crosses fall somewhat midway between the two parent types with respect to this character, the variation among the crosses reflecting the character of the ingoing *Bos taurus* and *Bos indicus* breeds.

25.2. DRYING OFF

Principally, there are three methods of drying cows, *viz.*, abrupt cessation of milking, intermittent milking, and incomplete milking. Australians, Lamond and Campbell (1970) feel that from the viewpoint of stopping lactation with no damage to the udder and avoiding udder infection, sudden cessation method is the better one, especially for not so high yielders.

Sudden cessation of milking causes build up of pressure in the udder which has no immediate permanent effect on rate of secretion and so on drying off for a few days. After 4 or 5 days, involution or regression of the secretory cells commences suddenly which is completed in another few days. The fluid trapped within the udder is ultimately resorbed. This method alone may not be successful in cows yielding 10 liters or more even at the drying off stage.

Intermittent milking. Milking once every second or third day or even less frequently, is not so effective a method and may even result in mastitis. But the method is preferred for cows already having mastitis; the udder is emptied and infused with antibiotic cream at each milking. Incomplete milking—emptying gradually smaller and smaller quantities of milk from the udder spread over a week or so, can at best be preferred for very high yielding cows, in which sudden cessation of milk may cause swelling and pain in the udder.

A more rational approach to drying high yielding cows is to reduce ration, especially protein, a couple of weeks prior to the proposed date of drying. In all cases, it is a beneficial precautionary step to wipe the teats with alcohol immediately after the last milking and then apply an antiseptic collodion seal (available in market). This will go a long way in preventing mastitis, especially in cows with weak sphincters.

25.3. CARE OF THE DRY-PREGNANT COW

Feeding the dry-pregnant cow should be aimed at making up the condition of the cow lost during the lactation. (The general principles of feeding and nutrition are discussed in Chapter 16). Special attention should be paid to the mineral content of the ration, especially calcium and phosphorous, as the lactating cow uses quite extensively the calcium and phosphorous present in her bones during the peak part of her lactation. However, reports suggest that the additional protein and energy requirements of the developing foetus and enlarging uterus during pregnancy are not great (Gomes, 1978). Further, the cow uses her feed more efficiently during lactation, or

---

Williamson, G. and Payne, W.J.A. 1978. *An Introduction to Animal Husbandry in Tropics*. The ELBS and Longmans, London. p. 285.

Lamond, D.R. and Campbell. E.A. 1970. *loc. cit.*

that maintenance requirements are less during gestation than during the previous six months. Cows that have been properly conditioned (fed) during the dry period produce upto 25 per cent more milk and fat than cows not conditioned (Moore and Gildow, 1953). An addition of 20 kg of body weight seems to cause an increase of 300 to 500 kg of milk in the subsequent lactation in case of well known *Bos taurus* dairy breeds. Among better known dairy animals in India, a 0.5 kg gain per day in body weight during dry period seems to be optimal for satisfactory milk production in the ensuing lactation (Khirwar, 1980).

It should also be remembered that dry cow incorporates in itself feeding the fast growing (in the cow's womb) unborn calf. The unborn calf draws upon the body of its dam all the nutrients including minerals and vitamins. Further, nutrients are required for rebuilding the secretory tissue of the udder. Besides, nutrition during the dry period seems to influence the quality and quantity of colostrum so badly needed for the new born calf (Moore and Gildow, 1953) and has a role in the reproductive health of the cow (Bath *et al.*, 1978). Inadequate nutrition may cause milk fever and retention of placenta, with all the consequences of intra-uterine infections. Hence, the need to closely watch the condition of the dry cow and feed it adequately (see Chapter 16).

The dry cows should be housed in a comfortable paddock at a lesser stocking rate than for the milch or growing stock, so that rushing at feeding time with resultant accidents is avoided. They should be least disturbed. In villages, disturbance and chasing by dogs and children has to be checked. About a fortnight before the expected date of calving the pregnant cow should be transferred to the maternity pen and watched closely. By this time the cow looks decidedly heavy, her pelvic ligaments (sacro-sciatic region) remain slackened, her teats remain turgid and her vulva enlarged and very much folded (Fig. 7.6). The Calving may take place about ten days prior to or later than the expected date of calving.

As the time of parturition approaches, the cow becomes restless and gets up and down frequently. A period of 10-12 hours may elapse from the commencement of restlessness until the calf is born. If the restlessness starts in the morning the cow should calve by late afternoon and if in the late afternoon by next morning. If it does not happen so, the farmer should investigate whether anything is wrong (Even otherwise down-calvers should be inspected routinely every three hours or so).

If some trouble is suspected it is better for the unskilled farmer to seek veterinary assistance. Farmers with some training can attempt a brief and simple examination after washing hands thoroughly with a detergent and lubricating them with a suitable obstetric cream kept handy on the farm.

The arm should be gently inserted into the vagina. If the cervix (neck of the womb) is either still closed or only partially open the cow should be left alone for some more time until further strengthening of the labour pains. The cow should be left alone also if the cervix is open and if the presentation of the Calf is normal—fore-legs first with the chin resting on them. This is the limit to which the farmer can go. If abnormal presentation is suspected or if prolonged labour proves futile, immediate veterinary assistance is a must.

Gomes, W.R. 1978. Gestation. In : *Physiology of Reproduction and Artificial Insemination of Cattle.* Salisbury, G.W. Van Demark, N.L. and Lodge, J.R.; W.H. Freeman and Company, San Francisco, p. 160; 1978.

Moore, M. and Gildow, E.M. 1953. *Developing a profitable dairy herd.* Windsor Press, p. 105.

Khirwar, S.S. 1980. *Feeding lactating buffalo.* Proc. Summer Inst. Buffalo Management Systems, HAU, Hissar, 2-30 June, 1980, p. 54.

Bath, D.L., Dickinson, F.N., Tucker, H.A. and Appleman, R.D. 1978. *Dairy Cattle : Principles, Practices, Problems, Profits.* 2nd ed. Lea and Febiger, Philadelphia, p. 411.

CHAPTER 26

Dairy Bull Management

The adage, *Bull is half the herd*, is literally true in the sense that the dairy bull contributes to the offspring half of the latter's genotype. Not only the bulls should be of genetically superior quality, but they also have to be kept in prime breeding condition by proper feeding and management.

Aspects of feeding, housing, disease control etc., of bulls have already been discussed in the appropriate chapters. Attention is paid in this section to certain specific aspects of herd sire management. Though artificial insemination is spreading gradually in most of the developing countries, still there is a long way to go in covering at least majority of breedable cow population by this method. Hence the need for keeping bulls on individual farms and in villages. Though generally bulls of questionable quality are still maintained in villages, some progress has been made in India by stationing better quality bulls at the village *Panchyats* for use in the village.

26.1. CARE OF THE YOUNG BULL

As soon as the tentative selection of bull calves is made based on their pedigree, they should be reared intensively by providing good growing rations, rations that are better than the average ration for a heifer calf. Special attention should be paid to an adequate supply of vitamins and proteins especially when green fodders available are of poor quality. The aim should be to rear the bull calf as quickly as possible, so that he can be tested early for his fertility and ability to beget desired progeny.

The bull calf should be separated from breedable cows and heifers by the time he attains puberty, which can be anything from 1½ to 2½ years in Zebu and buffalo breeds and lesser in crossbreds, the age at puberty is greatly influenced by the intensity of rearing. The bull calf should be dehorned within a few days of birth by debudding with chemical or hot iron. Though dehorning may adversely tell upon the appearance and value of the bull in the show ring, the practice is considered to make the bull less dangerous. However; there is evidence to the contrary too, polled bulls were found to be extremely dangerous in open yards and near fences in Australia (Lamond and Campbell, 1970).

The bull should be ringed by the time he is about one year old, by which time he begins to show his strength. A smaller ring may be put at this age, to be replaced by a bigger ring when he matures. If bigger ring is put on yearling bulls the ring may cut through his nasal septum or may become entangled in fences, posts etc., and also interfere with grazing. Nose rings are made in two semi-circular pieces hinged together and are of aluminium, copper or some alloy which does not rust. The free ends of the two parts either dovetail into one another or are in a form of point-and-socket, secured either by a flush spring or by a screw with counter sunk head, so that the joint is smooth (Fig. 26.1).

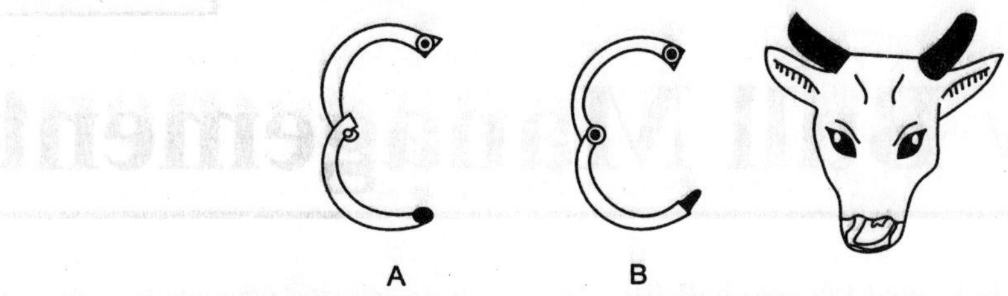

Fig. 26.1. Bullnose ring—(A) Ordinary type, (B) Self-piercing type, and (C) A ring in position.

Since the nose is extremely sensitive to touch, ring in the nose enables the attendant to keep the neck extended and the head raised while restraining or parading. Nose ring is an essential item in control of bulls. Bull leading poles (Fig. 26.2) can be conveniently hitched to the nose ring and this is mostly felt necessary also. The bull can be effectively controlled by means of a chain or rope around the horns threaded through the nose ring. By this arrangement the bull cannot get his head down and yet there is no great weight on the ring to damage the nasal cartilage.

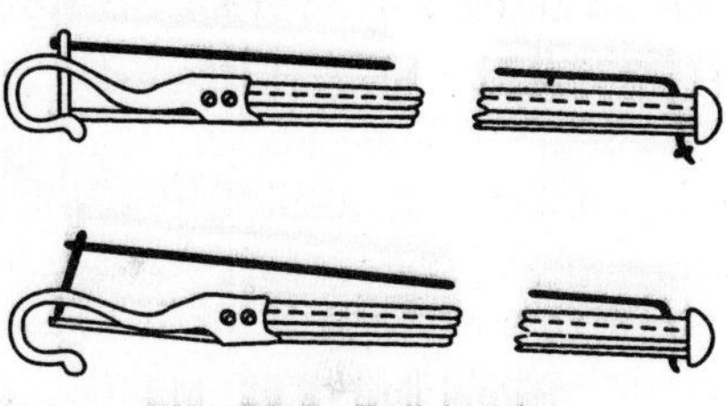

Fig. 26.2. Bull leader.

The young bull should be trained for handling and leading. It is much easier to maintain control on a mature bull if he was properly trained when young. Even when the bull is 4-6 months old a simple halter may be put over his face and he be accustomed to handling. After the nose ring is put he should be led either by a chain or pole as described in the previous paragraph.

26.2. FEEDING MATURE BULL

A good rule to feed mature bulls is to feed daily about 1 kg hay and 1/2 kg concentrate per 100 kg b.w. Thus a 400 kg bull should get 4 kg hay and 2 kg concentrate. These amounts should be adjusted according to the body condition of various bulls because there is individual variation in response. Excess fatness in mature bulls should be avoided at all costs as it reduces libido and may cause severe stress and strain on their feet and legs.

Excess calcium in bull rations can cause problems particularly in older bulls. When legume roughage is fed the concentrate mixture should not contain a calcium supplement. Herd concentrate mixes usually contain added calcium to fulfil needs of lactating cows which are losing calcium from their body in milk. However, bulls do not lose calcium, and in time excess calcium may

---

Lamond, D.R. and Campbell, E.A. 1970. *Dairy Cattle Husbandry*. Angus and Robertson, Sydney, 1970, p. 126.

cause vertebra and other bones to fuse together. Therefore, bulls may need a different concentrate mixture than that of milch cows.

The ideal condition for a herd bull is good forward condition—fully muscled with a thin fat layer. Feeding regime may be adjusted to maintain this condition.

26.3. CARE OF MATURE BULLS

While leading, the attendant should never walk in front of the bull, but must lead from the side holding the nose always higher than natural level. If the nose is allowed to drop, the bull may get inclination to butt. While handling and leading, all bulls should be considered as potentially dangerous and no complacency should be shown at any time even in case of old and well acquainted bulls.

Growing as well as mature bulls (Fig. 26.3) should regularly be exercised (Fig. 26.4) so that they do not put on fat and thus remain in thrifty condition. This will also help in keeping their toes well worn. Over-grown toes may hinder walking as well as mounting behaviour of bulls. Such a situation may necessitate trimming of hooves with a sharp knife and/or a rasp.

Fig. 26.3. Breeding bull in optimum condition.

Fig. 26.4. Several bulls are given exercise simultaneously using a bull exerciser.

The breeding bull should never be allowed to run with the herd. They should be housed in separate paddocks, individually. This helps in controlling the number of services by the bull for recording the breeding data, in keeping reliable breeding records and in maintaining the vigour and libido of the bull, especially if he is young. Further, such a practice will help in checking the spread of veneral diseases and improving conception rates by breeding at the appropriate time of oestrous cycle.

Sometimes the hair around the prepuce grows long and gets matted which may interfere with the ejaculation and contribute undesirable foreign matter to the semen. Hence, the hair should be periodically snipped off with a pair of scissors. The hairs should not be cut too close to the skin as the cut ends may cause itching and irritation to the prepuce; about 1 cm length may be ideal.

26.4. MAINTENANCE OF SEXUAL LIBIDO

Showing of consistently good sexual libido and production of high quality semen by bulls is possible only under ideal conditions of feeding and management. There are several factors which can reduce libido in bulls—youth or old age, inexperience, tiring exercise or too frequent usage, semen collection at unusual places in unfavourable conditions and using unsuitable fittings, faulty feeding, obesity or run down condition, inherent defects, temporary injuries or chornic defects of legs, back and penis, etc. All such problems should be rectified as soon as noticed.

Certain bulls are sensitive to A.I. whereas others seem able to withstand considerable rough handling. Intelligent management at service time keep them keen. The well known reflexes of mounting the cow, projecting the penis thrusting and ejaculation can easily be retarded or even inhibited in a bull by unnatural methods of handling him. The basis of his natural behaviours at time of semen collection or natural breeding is the gratification of the physical sense. The majority of bulls will serve if they are permitted to serve in familiar surroundings and are handled by the same attendant provided these are associated with previous satisfactory experience.

"The sexual reflexes are inhibited if their performance is accompanied not by pleasurable sensations but by sensations which are painful, uncomfortable or even distracting. In every sensitive bull, inhibitions may develop quickly, even when collections are taken carefully. Repeated careless collections will "put off" the most hardy bulls. Inhibition will be reinforced the more often the unfavourable conditions are repeated, and the shorter the interval between the collections. If inhibition has started, it is a great mistake to persist with unsuccessful attempts at collection. The animal should be given a rest from collections for as long as possible. Inhibitions are also reinforced if repeated in the same surroundings. The sight of A.V. or the operator and the colour of the teaser can be factors. Slight inhibitions can be overcome by changing these surroundings. (Hammond *et al.*, 1947).

Overwork is common in young bulls (which are initially very keen) allowed free access to cows and heifers. The number of services and not the number of cows served is the important consideration. No bull should be allowed to serve each cow more than twice in a heat period.

A young bull may be placed with 2 or 3 cows per week. A mature bull may ejaculate many times per week without effect on libido or semen quality. In fact, it is very difficult to reduce sperm quality of bulls by frequent ejaculation. Usually, they lose libido before they exhaust supplies of semen.

It is important at the first sign of reduced libido to tease the bull as much as possible by delaying service.

Bulls become bored in their surroundings, particularly if in small paddocks and may lose interest. The presence of another bull, injection of male sex hormone or thyroid hormone or a change in surroundings should overcome this.

Under Indian conditions overusage and underfeeding of the village bulls is the most common cause of reduced sexual activity. This is due to leaving the bull free with the grazing cows which results in the bull tiring itself by frequent mountings. This affects the semen quality adversely. The problem can be overcome by keeping them separately and allowing 2-3 services or collections per week (Bartlett, 1965). Younger bulls may be allowed just one service or collection per week.

Feeding has a significant-effect on rate of sexual development and libido of males. When energy intake is restricted, the rate of growth is decreased, testes growth is retarded, age at puberty is increased and sperm output is decreased (Hafez, 1980).

Testis size is too important since there is a high correlation between testis size and sperm producing potential (Amann, 1970). Testis size increases rapidly as puberty approaches and by one year of age (later in zebu) bulls would have reached approximately 50 per cent of their mature potential. Testicular measurements were found to be suitable criteria for predicting the reproductive performance of buffalo bulls even at an early age (Bedi *et al.*, 1982). Since testis size is also a highly heritable trait (Coulter *et al.,* 1976) and it can be easily monitored by measuring scrotal circumference, it should be evaluated in all males considered for breeding.

For bulls used for artificial insemination service the method of handling and controlling them at the time of semen collection is equally important though different practices are followed in different countries. Bartlett (1965) describes three suitable ways of managing the bulls while collecting the semen.

1. To place the cow or dummy in a breeding rack and leave the bull out of his pen by a lead strap or bull pole.
2. Making ceiling track to which the lead chain of the bull is attached on swiveled pulley and the bull is thus led to the cow or dummy; he is less dangerous as he cannot get his head down.
3. To lead the bull along a fence, the leader walking on an opposite side or along a raised walk so that the animal has no chance of attacking him.

Bartlett, J.W. 1965. Feeding and Management of Sires. In : *The Artificial Insemination of Farm Animals.* E.J. Perry (Ed.) Oxford and IBH Publishing Co., Delhi, p. 406.

Hafez, E.S.E. 1980. Artificial Insemination. In : *Reproduction in Farm Animals.* IV. Edn. Lea and Febiger, Philadelphia, p. 523.

Amann, R.P. 1970. Sperm Production Rates. In : *The Testis* : A.D. Johonson, W.R. Gomes and N.L. Van Demark. (eds.) New York, Academic Press, Volume I. pp. 443-482.

Bedi, K.S., Singh, G.B. Sharma, R.D. and Dugwekar, Y.G. 1982. Semen Production in relation to sexual and physical maturity in buffalo bulls. Buff. Seminar on Reproduction and meat production. Abstr. Tanuku (AP), 15-17, Jan. 1982, p. 29.

Coulter, G.H., Rouns Ville, T.R. and Foote, R.H. 1976. *J. Anim. Sci.* **43**, 9.

26.5. PROTECTION AGAINST HOT-CLIMATIC CONDITIONS

It is well established that libido and quality of semen produced by bulls vary considerably from season to season. Under tropical conditions, the low of sexual libido and poor semen quality are associated with summer season especially in purebred exotic and crossbred bulls (Bhosrekar, 1982) and buffalo bulls (Pandey and Raizada, 1979; Mehar Singh and Chauhan, 1980). Sexual libido of bulls is also reported to be sluggish during winter where the winters are severe.

To overcome such problems during summer, bulls should be housed in cool, well ventilated, dry sheds. Showering or splashing cold water on bulls 2 or 3 times during hot part of the day and protection against direct and reflected radiation were found to be very useful. For the latter purpose Pandey and Raizada (1979) suggested housing animals in half-walled sheds from 8.00 to 18.00 hours from April to June, *i.e.,* before the rains-set in and humidity increases. During day time the openings between the pillars of the shed should be closed by hanging jute screens or mats made out of maize or sorghum stalks or palm leaves or bamboos, etc. Between dusk and dawn the hangings should be raised up so that the animals will get the benefit of cool breeze. Alternatively, they can be tethered in the open at night, which practice was found to reduce heat increment in growing buffaloes considerably (Karki and Sastry, 1981). After the rains set in, *i.e.,* during July to September, the weather remains hot and humid in the whole Indian sub-continent. During this period a shade overhead is all that is necessary. A group of tall shady trees can suit the purpose admirably. The bulls during summer should be taken for exercise or grazing between 5.00 and 7.00 hours and/or between 17.00 and 19.00 hours.

The above measures for protection against thermal stress should be provided not only for the breeding bulls but also for the teaser bulls, as good libido in the latter is also necessary for locating animals in heat. For especially precious animals, use of fans, desert coolers or even air conditioners can be thought of.

Bhosrekar, M.R. 1982. Personal Communication.

Pandey, M.D. and Raizada, R.C. 1979. Overcoming summer sterility in buffalo bulls and cows. F.A.O. Animal Production and Health Paper No. 13, F.A.O., Rome, pp. 235-246.

Mehar Singh and Chauhan, F.S. 1980. Problems and Practices of Artificial Insemination in buffaloes with special reference to village buffaloes, Proc. Summer Inst. Buffalo Management Systems, 2-30 June, 1980. Haryana Agri. Univ. pp. 182-195.

Karki, D.B. and Sastry, N.S.R. 1981. Project on Modes or Summer Housing for buffaloes (unpublished data).

CHAPTER

27

Management of Work Bullocks

Management of work animals is in many ways like taking care of a motor vehicle—put adequate fuel (feeds and water), do not overburden the engine (physiological mechanism of the animal) and stick to rigid maintenance (care and management). Only then the beast can give out its best.

According to the estimates of the F.A.O. (1978), there are presently 110 million working bullocks and 16 million buffalo bullocks in the country. Bullocks are still a major source of motive power in agricultural operations. They are used for such varied operations as ploughing, harrowing, threshing, harvesting, lifting water and transporting people and agricultural produce. Nearly 80 per cent of cultivators in India are dependent on bullocks for motive power. As high as 43 per cent cattle in the rural area are working animals.

27.1. WORKING CAPACITY OF BULLOCKS

The average speed of a pair of bullocks pulling a cart is around 4-5 km per hour. If not overdriven, they can travel 25 to 30 km or even 40 km daily under favourable conditions. The pace and distance covered by bullocks depends not only on the weather (hot weather being unfavourable) but also on their condition, type of harness and equipment used, and to a considerable extent, on the drivers. Bullocks can be induced to work well by those who thoroughly understand their peculiarities and to whom they are accustomed. At any rate bullocks should not be used for more than 5 hours a day for tiring works like ploughing and heavy transport along uneven surfaces. For less tiring jobs like light carting on metalled roads and threshing they can be made to work a little longer.

Thomas, C.K. and Pearson, R.A. 1985. Heat tolerance of draught cattle. Draught Animal News, C.T.V.M. No. 5 : 3-4.

Farmers generally consider a pair of good bullocks to be sufficient to carry out all the cultivation and transport job during one year on a 5 acre holding.

Besides, good bullocks should be of good size, height and length; the last mentioned character indicates good speed. The animal must have an easy natural gait and good pace. A thick and muscular neck indicates heavy draught capacity while a slender and sinewy neck denotes speed. A tight skin is generally considered to denote great sensitivity and fiery temper as in some south Indian draft breeds like—*Amrit Mahal, Kangayam* and *Hallikar.*

27.2. FROM WHAT AGE CAN BULLOCKS WORK ?

Indigenous bullocks of about 4 years of age are the best. Animals of lighter breeds and crossbreds may be ready for work even earlier. At this age bullocks are strong enough to work well. Also they have several years of working life to come. Too young animals will not be strong enough. They would not have finished growing and their bones will not be strong. Initiation to work at an early age will not permit the animal grow any longer. The tendency of yoke injuries and yoke galls were found to be greater if too young animals are put to work. On the contrary, too old animals will be difficult to train. Even if trained they will be in working condition only for a few years than bullocks trained when young.

Bull calves intended to be converted into bullocks should be castrated when they are between 1 to 1½ years old. In no case the castration should be delayed beyond this age. There is no scientific basis for the common belief that early castration prevents further growth resulting in bullocks that are less hardy and smaller in size. On the contrary, bull calves castrated at proper age turn out to be better bullocks, their docility and robustness being markedly improved.

27.3. CHOOSING WORK BULLOCKS

The bigger a bullock the stronger it will be for work. Too small an animal cannot do much work. A good bullock should not have any quarter or limb less developed or ill-formed. The hump should be firm and plump, the back level, legs strong, the bones thick and the whole body muscular. Malformation of hooves and excessive turning out of toes are undesirable. The hooves should be hard, black and waxy. The two halves of the hoof should be even, while the cleft should be narrow. Lameness from sprains and bony enlargements of the knee and hock are to be specially looked for. The neck must be examined for swellings or yoke galls, which can often be so severe as to incapacitate the animal altogether. An ideal draught bullock should have good carriage and disposition, exhibiting health and vigour. Animals that keep their head erect when standing and their tail high while at work are considered to be more alert and brisk.

The two members of a pair of bullocks should be of the same size and of similar temperament. If one is bigger than the other, the yoke cannot be properly fitted. In that case the yoke bothers both the animals but the bigger one may have to bear most of the strain. The pair cannot work well, which means that not all the strength of the stronger bullock is fully used.

27.4. TRAINING OF YOUNG BULLOCKS

Time spent on well trained bullocks is never a waste. A month or two before commencement of training a punch has to be made through the nasal septum with the help of a punch or trocar. This is necessary for passing nose rope through it at a later date.

The first step in training is to get the bullocks used to being led by man. The trainer should be a calm and gentle person who should pat and urge the animals while training; periodically offering them some fodder or grain first to make them feel at ease. Abuses, beating and rough treatment given to young bullocks at this stage make them shy away at the sight of man. The bullock is going to become the friend of the man for a long time, so man has to be friendly to his animal friend.

The second step in training is to teach bullocks to carry a yoke. Always train the same pair together, putting always each bullock on the same side, *i.e.,* the bullock on the left side must always be on the left, and the one on the right must always be on the right. When the pair get accustomed to carry a simple yoke on their necks they can be got used to pulling a piece of log weighing around 75 kg by tying it to the middle of the yoke with a rope or chain. After the bullocks are accustomed to pulling a weight tied to the yoke they can be taught to obey commands to stop, to start, to run, to turn right and left and to walk straight. For this, one has to make them do the same manoeuvre a number of times, simultaneously uttering the command and obliging the animals to obey the command.

27.5. CARE AND FEEDING OF BULLOCKS

Bullocks are used heavily during some seasons like sowing and harvesting and lightly during other times. The plane of feeding bullocks too had to be varied accordingly. The general principles of feeding are described in Chapter 6. During idle season, the aim of feeding should be to keep bullocks in good condition. During busy season they have to be fed with extra concentrates. Grooming, especially the legs and quarters of bullocks with wisps made of straw before and after heavy work is desirable.

Bullocks should not be made to work in direct sun during the hot part of the day. During summer working hours may be restricted to 5.00 to 10.00 hours in the morning and 16.00 to 19.00 hours in the evening. A brief rest after working for an hour or so is advisable, when the animals should be provided with shade, water and a few mouthfuls of feed. Bullocks should not be overworked, lest they may lose weight. Continuous overworking may ultimately ruin the animals.

The bullocks should be properly shoed before using them for work on hard ground, as otherwise their feet may get injured. The hoof should first be prepared by filing and then the shoe fitted to the hoof by driving small nails. The hoof should never be cut to fit the shoe. The shoe should be made to fit the natural shape of the hoof. Animals working in the fields require renewal of shoes once in 2 or 3 months, while those on road work may require change of shoes every month. Animals having softer hooves require shoeing more frequently than those with hard hooves.

FURTHER READING

Section VII

Anil, K.S. and Thomas, C.K. 1996. Comparative draught performance of cattle and buffaloes. I. Physiological reactions. Indian J. Anim. Sci. 66 : 398-401.

Batch, D.L. Dickenson, F.N., Tucker, H.A. and Appleman, R.B. *Dairy Cattle : Principles, Problems and Profits.* 2nd ed. Lea & Febiger, Philadelphia.

Carthy, J.D. 1966. *The Study of Behaviour.* The Institute of Biology's Studies in Biology, No. 3, Edward Arnold, London.

CCG. 1971. *Cattle keeping in India.* Central Council of Gosamvardhana. New Delhi.

Davis, R.F. 1965. *Modern dairy cattle management*. Prentice-Hall India; New Delhi.

Dethier, V.G. and Stiller, E. 1961. *Animal behaviour*. Prentice Hall, London.

Etgen, W.M. and Reaves, P.M. 1978. *Dairy cattle feeding and management*, 6th ed. John Wiley & Sons, New York.

Fraser A.F. 1974. *Farm Animal Behaviour*. The English Language Book Society and Bailliere Tindall, London.

Hafez, E.S.E. and Schein, M.V. 1969. The behaviour of cattle. In : *The Behaviour of Domestic Animals*. Ed. E.S.E. Hafez. 2nd ed. Chao 9. Bailliere Tindall and Cox, London.

ICAR 1978. *Handbook of animal husbandry*. Indian Council of Agric. Research, New Deihi.

Kumaravelu, N., Thomas, C.K. and Sreekumar, D. 1997. Farm work performance of Kangayam bullocks in two locations. Indian Vet. J. 74 : 951-954.

Kumaravelu, N., Sreekumar, D. and Thomas C.K. 1998. Some facts of farming system involving Kangayam cattle. Indian J. Anim. Prod. Mgmt. 14 : 87-90.

McDowell, R.E. 1972. *Improvement of livestock production in warm areas*. Freeman and Company, San Francisco.

Sreekumar, D. and Thomas, C.K. 1990. Draught efficiency and thermal strain of Kangayam and Jersey-Red Sindhi crossbred bullocks in hot-humid tropics. Indian J. Anim. Sci. 60 : 582.

Sreekumar, D. and Thomas, C.K. 1990. Haematological changes due to work in Kangayam and crossbred bullocks in hot-humid tropics. Indian J. Anim. Sci. 60 : 587.

Thomas, C.K. 1995. Methodology of study draught. Proc. IVth Scientists' Meet. Project Directorate of Cattle (ICAR), Mannuthy.

Thomas, C.K. and Pearson, R.A. 1986. Effect of ambient temperature and head cooling on energy expenditure, feed intake and heat tolerance of Brahman and Brahman × Friesian Cattle working on tread-mills. Animal Prod. 43 : 83-90.

Williamson, G. and Payne, W.S.A. 1978. *An introduction to animal husbandry in the tropics*. 3rd ed. Longman Group Ltd., London.

Index

A

B

C

D

E

F

G

H

I

L

M

N

O

P

R

S

T

U

V

W

Z